Microbial Enzyme Technology in Food Applications

Books Published in *Food Biology* series

Food Biology Series

Microbial Enzyme Technology in Food Applications

Editors

Ramesh C. Ray
ICAR - Regional Centre of Central Tuber Crops Research Institute
Bhubaneswar, India

and

Cristina M. Rosell
Food Science Department
Institute of Agrochemistry and Food Technology
Avda Agustin Escardino Paterna
Valencia, Spain

CRC Press
Taylor & Francis Group
Boca Raton London New York

CRC Press is an imprint of the
Taylor & Francis Group, an **informa** business
A SCIENCE PUBLISHERS BOOK

CRC Press
Taylor & Francis Group
6000 Broken Sound Parkway NW, Suite 300
Boca Raton, FL 33487-2742

First issued in paperback 2020

ISBN-13: 978-1-4987-4983-1 (hbk)
ISBN-13: 978-0-367-78256-6 (pbk)

This book contains information obtained from authentic and highly regarded sources. Reasonable efforts have been made to publish reliable data and information, but the author and publisher cannot assume responsibility for the validity of all materials or the consequences of their use. The authors and publishers have attempted to trace the copyright holders of all material reproduced in this publication and apologize to copyright holders if permission to publish in this form has not been obtained. If any copyright material has not been acknowledged please write and let us know so we may rectify in any future reprint.

Library of Congress Cataloging-in-Publication Data

Names: Ray, Ramesh C., editor. | Rosell, Cristina M., editor.
Title: Microbial enzyme technology in food applications / editors Ramesh C. Ray, ICAR--Regional Centre of Central Tuber Crops Research Institute, Bhubaneswar, India, and Cristina M. Rosell, Food Science Department, Institute of Agrochemistry and Food Technology, Avda Agustin Escardino Paterna, Valencia, Spain.
Description: Boca Raton, FL : CRC Press, [2016] | Series: Food biology series | "A science publishers book." | Includes bibliographical references and index.
Identifiers: LCCN 2016036651| ISBN 9781498749831 (hardback : alk. paper) | ISBN 9781498749848 (e-book : alk. paper)
Subjects: LCSH: Food--Biotechnology. | Food--Microbiology. | Food--Preservation. | Microbial enzymes.
Classification: LCC TP248.65.F66 M526 2016 | DDC 664/.024--dc23
LC record available at https://lccn.loc.gov/2016036651

Visit the Taylor & Francis Web site at
http://www.taylorandfrancis.com

and the CRC Press Web site at
http://www.crcpress.com

Preface to the Series

Food is the essential source of nutrients (such as carbohydrates, proteins, fats, vitamins, and minerals) for all living organisms to sustain life. A large part of daily human efforts is concentrated on food production, processing, packaging and marketing, product development, preservation, storage, and ensuring food safety and quality. It is obvious therefore, our food supply chain can contain microorganisms that interact with the food, thereby interfering in the ecology of food substrates. The microbe-food interaction can be mostly beneficial (as in the case of many fermented foods such as cheese, butter, sausage, etc.) or in some cases, it is detrimental (spoilage of food, mycotoxin, etc.). The *Food Biology* series aims at bringing all these aspects of microbe-food interactions in form of topical volumes, covering food microbiology, food mycology, biochemistry, microbial ecology, food biotechnology and bio-processing, new food product developments with microbial interventions, food nutrification with nutraceuticals, food authenticity, food origin traceability, and food science and technology. Special emphasis is laid on new molecular techniques relevant to food biology research or to monitoring and assessing food safety and quality, multiple hurdle food preservation techniques, as well as new interventions in biotechnological applications in food processing and development.

The series is broadly broken up into food fermentation, food safety and hygiene, food authenticity and traceability, microbial interventions in food bio-processing and food additive development, sensory science, molecular diagnostic methods in detecting food borne pathogens and food policy, etc. Leading international authorities with background in academia, research, industry and government have been drawn into the series either as authors or as editors. The series will be a useful reference resource base in food microbiology, biochemistry, biotechnology, food science and technology for researchers, teachers, students and food science and technology practitioners.

Ramesh C. Ray
Series Editor

Preface

The aim of food processing is to produce palatable and sensory attractive food products, increase the shelf life of food and increasing food varieties, while maintaining the nutritional and healthcare needs. The use of food grade microbial enzymes or microbes (being the natural biocatalysts) is desirable due to their ambient processing conditions, and because they are safe for human cnsunpiom (GRAS). This extensively discuses the use of enzymes in conventional and non-conventional food and beverage processing as well as in dairy processing, brewing, bakery and wine making. The book is divided into four sections;

  I. History of microbial enzymes;
 II. Microbial enzymes for food processing: Characterization, production and applications;
III. Microbial enzymes in food fermentations;
 VI. Advancements in enzyme technology.

Enzymes from microorganisms have been first reported in the year 1878. In time, our knowledge on the usage of microbial enzymes is increased enormously, giving rise to the production of cheese, yogurt, vinegar, etc. and other foods. In the past few decades, the developments in bioprocessing tools and techniques have significantly expanded the potential for bulk enzyme applications, e.g., production of bioactive peptides, oligosaccharides and lipids, flavor and colorants, besides the conventional industrial applications in food and beverage processing. The chronological development of enzyme technology and applications of such technologies in food processing have been reviewed by Mishra et al. in Chapter 1.

There are seven chapters in Part 2 related to enzyme applications for processing of starch and/or related carbohydrate derivatives. In Chapter 2, Martínez and Gómez have described the amylase family and the constituent enzyme uses in starch processing to develop products such as glucose or maltose syrup, fructose syrup, starch/maltodextrin derivatives and others. The Chapter 3 by the same authors concentrates on starch-active de-branching and α-glucanotransferase enzymes and their uses in the production of resistant starch, cyclodextrin, cycloamyloses, cluster dextrins and others. In the Chapter 4, Desai et al. have discussed the structure, characterization, mechanism of action and fermentative production of glucose isomerase. Likewise, the enzymes involved in sucrose transformation: hydrolysis, isomerization, transfructosylation and transglycosylation, and above all, the applications of these enzymes in production of fructo-oligosaccharides and fructans have been elucidated in Chapter 5 and 6 by Harish and Uppuluri. The adverse effects associated with lactose intake by lactose-sensitive consumers can be overcome by the addition of lactase (β-galacosidases) to milk products. The Chapter 7 contributed by Plou et al. focuses on the mechanism of action of β-galactosidases on lactose removal from the milk/milk products as well as in the synthesis of galacto-oligosaccharides. Vohra and Gupta, in chapter 8, have elegantly described the biotechnological applications of various pectinolytic enzymes in food and beverage processing.

The remaining six chapters in this section broadly discuss on other aspects related to food processing. For example, the Chapter 9 (by Tavano) focuses on classification of proteases, source and applications. Transglutaminase is an enzyme that catalyzes the formation of isopeptide bonds between proteins. Kieliszek and Błażejak in Chapter 10 have discussed applications of transglutaminase cross-linking in the production of cheeses, dairy, meat products, food films and bread, and in improving properties such as firmness, viscosity, flexibility and water binding in these foods. Likewise, Celligoi et al. in Chapter 12 have critically reviewed the recent research related to properties, functions and food applications of lipase, and Sooch et al. in Chapter 13 discussed the types, structure, applications of microbial catalase and future perspectives. Rodríguez-Couto, in Chapter 14, has focused on catalytic mechanism, properties, fermentative production of laccase and its applications such as in biosensors with immobilized laccases in determining total phenolic content (i.e., tannins) in wines and beer, juice clarifications, baking, removal of aflatoxins in foodstuffs and others. Moschopoulou has elaborated the characteristics, production and use of microbial milk coagulants (proteinases), especially those from fungi that are used in the cheese industry since 1960s in the Chapter 11.

The Part 3 (Microbial Enzymes in Food Fermentation) includes five chapters: Brewing (Chapter 15 by Serna-Saldivar and Rubio-Flores), baking (Chapter 16 by Dura and Rosell), wine making (Chapter 17 by Rodriguez-Nogales et al.), dairy processing (Chapter 19 by Mohanty and Behare) and cassava fermentation (Chapter 18 by Behera and Ray). In these chapters the authors have discussed the implications of various intrinsic and supplemented enzymes in improving the processability, quality, functionality and shelf life of the final products. A thorough revision of the current uses of microbial enzymes in these specific sectors is included, explaining the scientific basis of the enzyme functionality.

The Part 4 comprises seven chapters covering recent developments in enzyme technologies. Some of these developments include extended use of the biocatalysts (as immobilized/encapsulated enzymes) and microbes (both natural and genetically modified) as sources for bulk enzymes, solid state fermentation technology for enzyme production and extremophiles as the source of food enzymes. The Chapters 20 and 21 by Jianping Xu's group have focused on 'recombinant enzymes', their regulations and applications in food processing, especially in meat- and fruit and vegetable juice industries. Extremophiles are naturally adapted to survive and grow in extreme environments, therefore, known as extremozymes. Sharma and Satyanarayana have provided an updated and comprehensive overview of food processing enzymes that are produced from extremophilic microbes in general (Chapter 23). Marine environment is such an environment that involves high salinity, high pressure, low temperature, and special lighting conditions, which make the enzymes generated by marine microorganisms significantly different from the homologous enzymes generated by terrestrial microorganisms. Arunachalam Chinnathambi and Chandrasekaran Muthusamy have reviewed the progress made in the last decade on microbial enzymes from marine sources and their applications (Chapter 22). Two chapters (24 and 25) in this book have exclusively dealt with 'solid state fermentation' for production of microbial enzymes. Carboué et al. discussed briefly a comparison of solid state and submerged fermentations, bioreactors and techno-economic feasibility of enzyme production in this system. In the subsequent chapter, Desobgo et al. have described the scaling up and modeling approaches of solid state fermentation.

In the last chapter (Chapter 26), Levic et al. have comprehensively reviewed the research on 'enzyme encapsulation technology' and how food industry has been economically benefitted by adopting such technologies. The various methods such as encapsulation using inorganic carriers (glass, magnetic particles, zeolites), synthetic carriers (polymers), hydrogels, nano materials, spray drying, have been described.

The editors are immensely thankful to all the authors for prompt response in accepting our invitations and timely delivering the quality manuscripts.

Ramesh C. Ray
Cristina M. Rosell

Contents

PART 1

HISTORY

1

Microbial Enzymes in Food Applications

History of Progress

Swati S. Mishra,[1],* Ramesh C. Ray,[2] Cristina M. Rosell[3] and Debabrata Panda[1]

1. Introduction

It seems now clear that a belief in the functional importance of all enzymes found in bacteria is possible only to those richly endowed with faith.

—Marjory Stephenson (Biochemist)

Enzymes are very important for sustainability of life in all life forms. They act as catalysts in chemical reactions. Microbial enzymes are of great importance in the development of industrial bioprocesses as they play a crucial role as metabolic catalysts. Enzymes have been applied in food preservation for millennia, and today they are enabling various food industries to provide the quality and stability of their products, with increased production efficiency. Microbial enzymes in food applications have not only diversified the food industry but also produced economic assets. The increasing demand for sustainable food has given an increasing drive to the use of microbial enzymes, knowingly or unknowingly since ages. Microorganisms have always been the largest and useful sources of many enzymes (Demain, 2008). They

[1] Department of Biodiversity and Conservation of Natural Resources, Central University of Orissa, Koraput 764020, India; E-mail: swatisakambarimishra@gmail.com; dpanda80@gmail.com
[2] ICAR-Regional Centre of Central Tuber Crops Research Institute, Bhubaneswar 751019, India. E-mail: rc_rayctcri@rediffmail.com
[3] Food Science Department, Institute of Agrochemistry and Food Technology, Avda Agustín Escardino Paterna, Valencia, Spain; E-mail: crosell@iata.csic.es
* Corresponding author

also provide environmental-friendly products to consumers, reducing consumption in energy, water and raw materials and generating less waste. Enzymes contribute to industrial processes by reducing energy consumption and maximizing its efficiency while contributing to its sustainability profile.

Although not in isolated form, enzymes have been used traditionally in dairy, baking, brewing and winemaking for centuries (Kirk et al., 2002). Their applications keep the bread soft and fresh for long, increase the dough volume and give a crispy crust (Rosell and Dura, 2016). Since time immemorial, enzymes are used in beer and wine to lower the calorie and alcoholic contents and also for more clarity and enhancement of flavour. Though used for centuries unknowingly, the revolution in food industry has been established by the use of enzymes or a whole microbial cell as the biocatalyst. The microbial enzyme has a high industrial and commercial application (Adrio and Demain, 2005). Microbes have proven to be the most useful and largest source of enzymes (Demain and Adrio, 2008). The current article covers the major developments in essential microbial enzyme production and applications in food industry.

2. Enzymes

An enzyme in purified form is a protein which is synthesized as an intra- and extra-cellular compound and may or may not possess non-protein prosthetic group (Vallery and Devonshire, 2003). Enzymes enhance the reaction rate with high specificity as they catalyzes biochemical reactions. All enzymes known (except ribozymes) are proteins which are high molecular-weight compounds made from chains of amino acids linked by peptide bonds. Enzymes are classified by the type of reaction they catalyse and the substance (called substrate) they act upon. It is customary to attach the suffix 'ase' to the name of the principal substrate upon which the enzyme acts (Bennett and Frieden, 1969). For example, lactose is acted upon by lactase, proteins by proteases and lipids by lipases. Also enzymes have common names, such as papain, from papaya.

2.1 History of enzymes

Enzymes in history were known as 'biocatalysts', which helped to accelerate the biological or biochemical reaction. The term 'enzyme' was first used in 1877, by Wilhelm Friedrich Kuhne, Professor of Physiology at University of Heidelberg, in his paper to the HeidelbergerNatur-Historischen und Medizinischen Verein, suggesting that such non-organized ferments should be called enzymes (Kuhne, 1876). It was derived from a Greek term 'ενζυμον' meaning 'in leaven' or 'in yeast' (Kuhne, 1877). Though enzymes have been used by mankind since centuries, they were technically termed as 'enzymes' only in the 18th century.

Before the nature and function of enzymes were understood, the practical applications were established as there were many ancient uses of enzymes, like barley malt for conversion of starch in brewing or calf stomach as a catalyst in the manufacture of cheese. Later on, many scientists reported on enzymes in different forms, for example, Spallanzani in about 1783, showed that the gastric juice secreted by cells could digest meat *in vitro* and whose active substance was named as pepsin by

Schwann in 1836 (Perham, 1976). The first enzyme to be discovered was 'diastase' by a French scientist, Payen in the year 1833, when he found it catalyzes the breakdown of starch into glucose in malt. James B. Sumner of Cornell University obtained the first enzyme in pure form, called 'urease', in 1926. He received the Nobel Prize in 1947 for isolating and crystallizing the enzyme urease from jack bean. The discovery of a complex procedure for isolating pepsin by John H. Northrop and Wendell M. Stanley of the Rockefeller Institute for Medical Research earned them the 1947 Nobel Prize as well. This precipitation technique has been used to crystallize several enzymes (Pfeiffer, 1954). Table 1 shows the periodic development of microbial enzymes over the centuries.

Table 1. Periodic development in microbial enzyme (for foods and beverages) utilization over centuries.*

Period	Events
2000 BC	Fermentation was developed mainly for use in brewing, bread baking and cheese-making by the Sumerians and Egyptians.
800 BC	The enzyme chymosin and calves' stomach were used for cheese-making.
1836 AD	Schwann discovered pepsin.
1856 AD	Berthelot showed enzymes require cofactor or co-enzyme for activity.
1860 AD	Berthelot demonstrated hydrolytic enzymes, including invertase (β- fructofuranosidase) obtained from *Saccharomyces cerevisiae.*
1878	Term 'enzyme' was derived from the Greek term 'ενζυμον' meaning 'in yeast'; also the components of yeast cells was identified which cause fermentation.
1894	Takamine first time patented a method (*koji* process; SSF) for preparation of diastatic enzymes (mostly α-amylase) from the mould that was marketed under the name 'Takadiastase'.
1913	Patents were awarded to French scientist A. Boidin and Belgian scientist Jean Effront for production of bacterial (*Bacillus subtilis* and *B. mesenterieus*) amylases and diastases as still culture (surface film).
1926	Enzymes were initially shown to be proteins.
1946	Commercially amylases were produced using *Aspergillus oryzae* strain by Mould Bran Co., Iowa, USA in SSF process.
1950	Amylase production in SmF using Tank Bioreactor at Northern Regional Laboratory, USDA, Illinois, USA; shift over from SSF to SmF.
1959	Bio 40-protease from *B. subtilis* was introduced in the market.
1950–1980	Spectacular increase in industrial enzyme production, particularly amylases and proteases (to a larger extent) and pectinases, lactase, invertase, lipase and cellulases (to lesser extent).
1965	International Union of Biochemistry set up 'Enzyme Commission' to publish enzyme classification.
1980s	Animal feed with improved nutrient availability and digestibility were developed through enzyme preparations.
1982	A product of gene technology, alpha amylase, was developed for application in food for the first time.
1988	An early approval of a product of gene technology, recombinant chymosin for food use was approved and introduced in Switzerland.
1990	Gene technology was used by developing two food processing aids—an enzyme for use in cheese-making in the US and a yeast used in baking in the UK.
2000-onwards	Re-designing microbial enzymes by tailoring their protein sequence.

*Source: Rose (1980), Behera and Ray (2015), Joshi and Satyanarayana (2015), Panda and Ray (2015), Panda et al. (2016)

2.2 Microbial enzymes

Like all living cells, microbes also produce enzymes which are hydrolyzing, oxidizing, reducing or metabolic in nature, but the amount of enzyme produced differs in various species and strains. Hence for commercial production of specific enzymes, a particular strain is to be selected that has the maximum enzyme activity. Enzymes from microbial sources are more advantageous than their equivalents from plant and animal sources because of lower production cost as compared to others, production on a large-scale, better scope for genetic manipulation, rapid culture development, less material use, being environment friendly and due to a wide range of physical and chemical characteristics; hence, they are preferred in various industrial applications (Hasan et al., 2006). Important progress in the food industry is mainly attributed to the use of microbial enzymes. Nowadays, enzymes are increasingly used in food processing with widespread interest in clean label foods and due to environmental concerns.

2.3 Microbial enzymes and their uses since centuries

Rudimentary use of microbial enzymes for food applications actually dates back to at least 6000 B.C. when neolithic people cultured fermented grapes to make wine and Babylonians used microbial yeast to make beer. Over time, mankind's knowledge on use of microbial enzymes increased, enabling the production of cheese, yogurt, vinegar, etc. and other foods. Rennet is an example of a natural enzyme mixture from the stomach of calves or other domestic animals that was successfully used in cheese-making for centuries. According to historical records, Rennet was first discovered by the Egyptians some 4000 to 5000 thousand years ago. They were using the dried intestines of animals, particularly stomachs, as containers for storing liquids. The rennet enzyme released by these stomachs caused milk to curdle, thus making it possible to be preserved. The art of cheese-making has developed over the centuries and natural rennet has always been inseparably linked to cheese. Rennet contains a protease enzyme that coagulates milk, causing it to separate into solid (curds) and liquid (whey).

The wonder drug penicillin was extracted by Sir Alexander Fleming in 1928 from mould and then around 1940, large-scale production of penicillin was started. Thereafter, the era of microbiology blossomed as a science and understanding of fermentation and its application increased (Caplice and Fitzgerald, 1999). Hence, mid 19th century saw the proper understanding and utilization of these microbial enzymes in food application. One of the important reasons was the industrial revolution in Europe that resulted in large-scale migration of people to cities which caused food scarcity and prompted discovery of methods for bulk food preparation and commercialization. Not until after World War II, however, did the biotechnology revolution begin, giving rise to modern industrial biotechnology and a fillip to microbial enzyme technology in food application.

2.4 Microbial enzymes for food industries

Many enzymes, like pectinase, lipases, lactase, cellulases, amylases, proteases, glucose oxidase, glucose isomerase, invertase, etc. are used variously in the food industry

(Table 2). The use of microbial enzymes was revolutionized by the action of enzyme to hydrolyze starch (amylazes) and (isomerise) glucose into fructose.

Table 2. Microbial enzymes in food application in industries.

Food Industry	Enzyme	Purpose/Function
Dairy	Rennet (protease)/chymosin	Coagulant in cheese production
	Lactase	Hydrolysis of lactose to give lactose-free milk products
	Protease	Hydrolysis of whey protein
	Catalases	Removal of hydrogen peroxide; mayonnaise
Brewing	Cellulases, β-glucanases, α-amylases, proteases, maltogenic amylases, α-acetolactate decarboxylase	For liquefaction, clarification and to supplement malt enzymes
Alcohol Production	Amyloglucosidase	Conversion of starch to sugar
Baking	α-amylases	Breakdown of starch, maltose production
	Amyloglycosidases	Saccharification
	Maltogen amylase	Delays process by which bread becomes stale
	Protease	Breakdown of proteins
	Pentosanase	Breakdown of pentosan, leading to reduced gluten production
	Glucose oxidase	Stability of dough; egg matonnaise
Wine and fruit Juice	Pectinase	Increase in yield and juice clarification
	Glucose oxidase	Oxygen removal
Meat	Protease	Meat tenderising
	Papain	
Protein	Proteases, trypsin, aminopeptidases	Breakdown of various components
Starch	α-amylase, phytase, glucoamylases, pullulanase,	Modification and conversion (i.e., dextrose or high fructose syrups)
	hemicellulases, xylanase maltogenic amylases, glucose isomerases, α-gluconotransferase	Dextranases, betaglucanases
Inulin	Inulinases	Production of fructose syrups
Beverages	Tannase	Tea processing; hydrolyzes esters of phenolic acids, including the gallated polyphenols found in tea
Cassava fermented foods and beverages	Linamarase	Breakdown of linamarin and lotaustralin

The advent of twentieth century marked the isolation of enzymes from microbes and subsequent large-scale commercial production for application in food industry. Microorganisms were isolated from the environment and their enzyme activities were optimized and often enhanced by manufacturers through genetic modification and/or process optimization. Many improved equipments and instruments, such as bioreactors with advanced techniques like immobilization, encapsulation, use of recombinant microorganisms and application of enzyme biosensors in the production of food-processing enzymes facilitated large-scale production of enzymes in food applications.

3. Enzyme Production in Bioreactor

Generally a bioreactor is defined as an apparatus in which biological reaction or process is carried out on an industrial/commercial scale. In other words, bioreactors are equipments designed to produce one or more useful products by using substrate, nutrients and microbial cells in a defined concentration and under controlled conditions (Mitchell et al., 2006; Behera and Ray, 2015). The continued success of biotechnology depends significantly on the development of bioreactors, which represent the focal point of interaction between the life scientists and the process engineers (Cooney, 1983). Hence, bioreactors are irreplaceable equipments for important biotechnological processes, such as commercial-scale enzyme production and other bio-products of demand.

3.1 Bioreactor and ancient world

Bioreactors have naturally occurred since time immemorial in the form of ponds, calf stomach and termite gut (Cooney, 1983; Brune, 1998). Around 600 A.D., the Mayans used to produce fermented beverage from *cacao* (chocolate), the Babylonians and Sumerians were known to produce beer before 5000 B.C. and even wine was made from historic times. The Egyptians used to make bread and cheese before 6000 B.C. The importance of bioreactors was long established by mankind as an art form rather than as a scientific device. With the advent of urbanization, human beings started to migrate to cities. This called for sanitization and waste water management from the health perspective; hence, the biological treatment of waste in bioreactor was the first engineered major achievement in bioreactor designing. Later, the bioreactor was developed to produce various bio-products, such as enzymes, organic acids and fermented foods through introduction of microorganisms. The practice of the microbial enzyme production through solid state fermentation (SSF) in industrial processes is adopted from our traditional knowledge to prepare fermented foods (Hesseltine, 1977). The traditional knowledge of making *tofu*, pickles, sausages, *miso*, *koji*, *idli*, *tempeh*, blue cheese, etc. constitutes the basis for the recent development of SSF for industrial production of commercial bio-products (Panda et al., 2016). Hence, the utilization of SSF technique has continued for decades to produce several important biochemicals and enzymes in bulk with expenditure of less energy in an environment-friendly process (Pandey, 1992). The significant development of bioreactors over time is described in Table 3 giving the approximate year of its occurrence.

The production of biochemicals and alcohols was discovered in literature only in the second half of the 19th century. In 1880, lactic acid production was accomplished in the US and it is known to be the first optically active compound to be produced industrially through fermentation (Sheldon, 1993). The first process in production of

Table 3. Periodic development of bioreactors and modern technologies for microbial enzyme production.

Events	Year
Vinegar production	2000 year B.C.
Prototype bioreactor with immobilized bacteria	17th century
Oldest bioreactor to use immobilized living organisms and so called generator	1832
Conversion of alcohol into vinegar	1862
Use of resting cells for transformation	1897
Chemical nature of DNA and RNA and the function of gene in synthesis of particular enzyme was known	1950
Evolutionary phase for biochemical engineering	1960–1980
Recombinant DNA technology	1970
First step change (growth phase in biochemical engineering)	1980–2000
In vitro engineering of microbial enzymes	2000-onwards

microbial enzymes was developed by Jokichi Takamine in 1894. He used the mould *Aspergillus oryzae* which was then used in the *koji* production (SSF) system used for production of takadiastase (a mixture of amylaze and proteases) (Takamine, 1914). The design and selection of bioreactor is always unique as per the requirements and the target bioproduct, but the basic principles, like adequate oxygen transfer, low shear stress, adequate mixing, etc. are common (Wang and Zhong, 2006). Enzyme production using microorganisms can be broken down to two major steps: (1) multiplication of microorganisms and production of desired enzymes—the optimization of fermentation parameters, and (2) the characterization and purification of the product from the broth—the downstream step. Even though these two steps involve different strategies, fermentation and downstream processing are strongly interlinked to each other, and therefore, process development has to be done in an integrated approach. In contrast to the 1970s, when surface cultures were used, nowadays submerged fermentation in large-scale bioreactors is employed because of higher productivity.

3.2 Bioreactor and recent advances

Important enzymes produced by SSF are α-amylase, β-galactosidases, cellulase, hemicellulase, pectinase, proteases, tannase, caffeinase, etc. (Aehle, 2004; Panda et al., 2016). Thermophilic bacteria and fungi are the potential microorganisms for enzyme production in SSF processes. The strains which are applicable are considered as GRAS (Generally Regarded As Safe) as per the regulatory aspect of food enzyme USFDA (United States Food and Drug Administration) (Olempska-Beer et al., 2006; Kappeler et al., 2006) and as Quality Presumption of Safety (QRS) in EU (European Union) (Barlow et al., 2007). However, recombinant DNA approach has proved successful for overproduction of several bioproducts, i.e., improvement of lactic acid bacterial strains for dairy application (Demain, 2007). Genetic improvement of microorganisms not only facilitates improved production rate but also economically beneficial products. Around 1973, foreign genes were inserted into eukaryotic cells for higher protein and nucleic acid expression which further required improved bioreactors and computerized process-controllers (Hochfeld, 2006). For the production of ethanol or organic acids without catalytic activity which are small sized bio-metabolites, bioreactors with ultra- or microfiltration recycling are widely used nowadays (Cheryan and Mehaia, 1983; Boyaval et al., 1987).

In course of time, it was discovered that microorganisms had the ability to modify/break down certain compounds by simple, well-defined biochemical reactions that were further catalyzed by enzymes. These processes are called biotransformations. The essential difference between fermentation and biotransformation is that there are several catalytic steps between the substrate and the product in fermentation, while there are only one or two steps in biotransformation (Turner, 1998). The distinction lies also in the fact that chemical structures of the substrate and the product resemble one another in a biotransformation, but not necessarily in fermentation. New methods and processes have brought a revolutionary change in the field of biotechnology and food industry (Papoutsakis, 2003). Understanding these features and adapting them for engineered processes enables better process control and lead to novel methods for producing new biochemicals using cell-derived catalysts (Chakrabarti et al., 2003). Technological advances in enzymatic biotransformations have led to the acceptance of enzymes as 'alternative catalysts' in the industry (Lilly, 1994), resulting in a better understanding of biochemical engineering.

3.3 Types of bioreactors used for food enzyme production

Bioreactors may be broadly classified as: (i) solid state bioreactor and (ii) submerged state bioreactor.

3.3.1 Solid state bioreactor

In a chemical laboratory, petri dishes, jars, wide mouth Erlenmeyer flasks, Roux bottles and roller bottles are used while for continuous agitation of the solid medium, rotating drum bioreactors, perforated drum bioreactors and horizontal paddle mixer bioreactors are used (Raj and Karanth, 2006). Raimbault (ORSTOM) reactor, which is a special type of column bioreactor, is also used in the laboratory scale SSF (Rodriguez-Leon et al., 2013). Industrial-scale solid-state bioreactors are designed according to the requirements of aeration and agitation or mixing. The important solid-scale bioreactor systems are: (1) SSF bioreactor without forced aeration, (2) SSF bioreactor with forced aeration and no mixing and (3) SSF bioreactor with continuous mixing and forced aeration. The SSF bioreactor without forced aeration is a tray bioreactor frequently used for the production of enzymes in the food industry and includes lipase and amylase (Ramos-Sanchez et al., 2015). The SSF bioreactor with forced aeration and no mixing is a special type of bioreactor where *in situ* manual agitation is carried out. However, the generation of metabolic heat during bioprocessing is its main limitation. The SSF bioreactor with continuous mixing and forced aeration is a modern type of bioreactor known for efficient heat transfer through convective and evaporative cooling. Important enzymes, such as amylase, cellulase, protease and lipase are known to be successfully produced in the rotating drum bioreactors (a type of bioreactor with continuous mixing and forced aeration) (Pandey et al., 1999; Kumar and Ray, 2014).

3.3.2 Submerged state bioreactor

Like solid state bioreactors, the submerged state bioreactors are grouped into three categories: (1) submerged bioreactor with no agitation and aeration (anerobic system), (2) submerged bioreactor with agitation and aeration (aerobic system) and (3) bioreactor with aeration and no agitation (Raj and Karanth, 2006; Kaur et al., 2013). However, among the three groups of bioreactors, the last type (bioreactor with aeration and no agitation) is the most convenient process for production of food enzymes. In addition, the other important bioreactors used nowadays are stirred tank bioreactor, airlift bioreactor, fluidised bed bioreactor, micro-carrier bioreactor, membrane bioreactor and photo bioreactor (Behera and Ray, 2015). Airlift bioreactors are frequently used for production of industrial food enzymes (Williams, 2002).

4. Cell/Enzyme Immobilization or Encapsulation

With the increase in market requirements for food packaging and preservation, the quest for optimum performance of enzymes has given enzyme engineering, particularly enzyme/cell immobilization or cell encapsulation, prime importance in production of biocatalyst with improved properties (Cao et al., 2003; Hamilton, 2009). Immobilization implies associating the enzymes or cells with an insoluble matrix so that it is retained for further economic use, i.e., giving the optimal immobilization yield and having the activity stability in long term (Miladi et al., 2012). Over the last few decades, intensive research in the area of enzyme technology has shown promise, i.e., the immobilization of enzymes (for extracellular enzymes) and cells (for intracellular enzymes). Immobilization enzymes and cells are widely used in the fermentation industry. Also biosensors are designed on the principle of immobilization of enzymes as it is convenient, economical and a time-efficient process of isolation and purification of intracellular enzymes.

Immobilization is used in the food industry for multifarious purposes, like baking, brewing, dairy, milling and in the beverage industry (Fernandes, 2010). Development of high fructose syrup through immobilized cell technology was a major achievement in the 1970s (Jensen and Rugh, 1987; Chen et al., 2012). Production of whey syrup using β-galactosidase (Harju et al., 2012) and production of L-amino acids by aminoacylase (Watanabe et al., 1979) were also important developments in the food industry. The first industrial use of an immobilized enzyme was in the application of amino acid acylase by the Tanabe Seiyaku Company of Japan, for resolution of racemic mixtures of chemically synthesized amino acids. Specific immobilization methods include adsorption, affinity immobilization, entrapment, encapsulation, covalent binding and cross-linking.

4.1 Immobilization techniques in food application

Immobilization is an accepted technology because of its capacity to act as catalyst in the efficient manufacture of novel products, such as wine (Divies et al., 1994; Silva et al., 2002), alcoholic and malolactic fermentation of apple juice (Nedovic et al., 2000), beer making (Willaert and Nedovic, 2006), meat processing and preservation

(Mc Loughlin and Champagne, 1994), flavour, non-reducing carbohydrates (Schiraldi et al., 2003) and sweetener enhancement (Kawaguti and Sato, 2007), dairy products like kefir (low alcoholic Russian fermented milk) and cheese (Witthuhn et al., 2005; Katechaki et al., 2009). Yeast immobilization for baking is used extensively (Plessas et al., 2007) as it makes the bread mould-free and improves the overall quality of bread.

Recent fusion proteins (Ushasree et al., 2012) and nanotechnology are used for immobilization of cells and enzymes in food processing because of their efficiency in increasing enzyme loading and diffusional properties and reducing mass transfer limitation. Nano particles are used as carrier material for enzyme immobilization (Kim et al., 2008). The cell/enzyme immobilization technique basically concentrates on economic, fast, non-destructive and food-grade purity which helps the food industry in obtaining improved quality, aroma and fine taste in the final product.

5. Extremophiles as a Potential Resource for Enzymes

Extremozymes have a wide application in agricultural, chemical and pharmaceutical industries with substantial economic potential. Enzymes from extremophiles have special properties that make them unique and valuable resources in biotechnology. Thermophilic extremophiles have attracted most attention. Hyperthermophiles (temperature > 80°C) and thermophiles (60–80°C) help in obtaining thermostable proteases, lipases and polymer-degrading enzymes, such as cellulases, chitinases and amylases that have found their way into industrial applications. Thermophilics are known to produce many enzymes, for example, thermophilic amylase and glycosidases are used in glucose and fructose production as sweeteners, starch processors, saccharifying enzymes, etc. (Di Lernia, 1998). Xylanases are used for paper bleaching (Ishida, 1997); lipase is applied for waste water treatment and detergent formulation (Becker, 1997) and proteases in food processing, amino acids production and detergent manufacturing (Hough and Danson, 1999). DNA polymerases from thermophilics are used in genetic engineering and molecular biology (Madigan and Maars, 1997) and dehydrogenases for oxidation reactions (Tao and Cornish, 2002).

Psychrophiles are extremophiles which survive at a temperature < 15°C. It is widely used in detergents as laundry applications can be performed at lower temperatures. Several food application industries also acquire benefits of these enzymes that are active at low temperature (Abe and Horikoshi, 2001). Halophiles can survive in hypersaline habitats as they have the ability to maintain osmotic balance (e.g., 2–5 M NaCl or sodium chloride). Compatible solutes and glycerol are used in pharmaceuticals and membrane in cosmetic industry and carotene in the food industry (Madern et al., 2000). Alkaliphilic pH > 9 cellulase and protease are widely used in detergent, amylase and lipases as food additive (Horikoshi, 1999). Acidophiles pH < 2–3 amylase and glucoamylases are used in starch processing, protease and cellulose as the feed component (Saeki, 2002).

There are two strategies for production of enzymes from extremophiles—first, by increasing the microbial biomass, the biocatalysts can be easily harvested and process optimized for enzyme production; alternatively, by genetic engineering, the gene encoding the biocatalyst can be cloned and expressed in a suitable host (Schiraldi and Rosa, 2002).

6. Genetic Engineering of Microorganisms for Enhanced Enzyme Production

Nowadays, genetic engineering has become a very useful tool to increase the production of enzymes as it is more cost-effective and also to obtain enzymes better adapted to the conditions used in modern food production methods. The former can also be reached by screening microorganisms from diverse environments, but that process is really time-consuming and success is not guaranteed. Alternatively, they can be expressed on filamentous fungi and yeasts. The genetic engineering technologies like recombinant DNA technology, protoplast fusion and mutation are commonly used for the enhanced production of enzymes.

6.1 Recombinant DNA technologies

In late 1970s, methodologies for isolation of discrete DNA segments for manipulation and insertion into living cells was possible with the discovery of restriction and ligase enzymes. This potential manipulation of industrial microorganisms brought a revolution in food enzymes and its industrial attributes. The first recombinant enzyme approved by the USFDA (United States Food and Drug Administration) for use in food was bovine chymosin expressed in *Escherichia coli* K-12 (Flamm, 1991). Lipolase was the first commercial recombinant lipase obtained by cloning the *Humicola lanuginose* lipase gene into the *A. oryzae* genome introduced by Novo Nordisk in 1994. This has its application in the food industry as an emulsifier, surfactants for detergents, contact lens and skin care products. The use of recombinant DNA technology allowed obtaining of novel enzymes adapted to specific food-processing conditions and increased the production levels of enzymes by transferring their genes from native species into industrial strains (Liu et al., 2013). Enzymes of known properties can be modified by modern methods of protein engineering or molecular evolution (Olempska-Beer et al., 2006), resulting in almost tailor-made enzymes. For instance, amylases and lipases with properties designed for specific food applications have been developed. The same technique has been applied to obtain microbe strains with an increased enzyme production ability by deleting native genes encoding extracellular proteases. Detailed information about the construction of recombinant production strains and methods of improving enzyme properties was reviewed by Olempska-Beer et al. (2006). Recombinant enzymes have been expressed in bacteria (*Escherichia coli*, bacillus and lactic acid bacteria) when they are not large proteins or proteins that require post-translational modifications.

6.2 Protoplast fusion

The discovery of parasexual cycle by Pontecorvo and Roper (Pontecorvo et al., 1953) has proved beneficial in biotechnology and genetic engineering for improvement of fungi of industrial interest. The application of this parasexual cycle in industrial production of fungi was highly significant as most of the fungi do not have a sexual cycle. This is widely used in genetic engineering for improved production of enzymes. It involves the fusion of two genetically originated protoplasts from different somatic cells so as to obtain parasexual hybrid protoplasts. Fusion of *Aspergillus flavipes* with *Aspergillus* sp. protoplasts results in diploids putatives with increased pectinase production (Solis et al., 1997). Three *Trichoderma* species—*T. reesei*, *T. harzianum* and *T. viride* strain were used for inter-generic protoplast fusion for citric acid production (El-Bondkly, 2006). *Aspergillus niger* recombinants by protoplast fusion presented glycoamylase activity 2.5 times higher than in the parental strain (Hoh et al., 1992). The protoplast fusion application helped in developing as a promising technique for obtaining strains with increased enzyme production.

6.3 Mutation

Mutation is one of the successful approaches in strain improvement and enhancement of enzyme properties for industrial application of microorganisms. Mutagenic agents have led to strain development either with physical or chemical mutagens. Lipase production from *Aspergillus japonicus* MTCC 1975 by mutation using ultra-violet irradiation, nitrous acid (HNO_2), N-methyl-N'-nitro-N-nitroso guanidine showed 127, 177 and 276 per cent higher lipase yield, respectively than their parent strain (Karanam and Medicherla, 2006). Cellulase was produced 2.2-fold higher than the wild strain by the mutants (*Aspergillus* sp. SU14-M15) when treating spores of *Aspergillus* sp. SU14 repeatedly with different mutagens, such as Co60 γ-rays, ultraviolet irradiation and N-methyl-N'-nitro-N-nitrosoguanidine (Vu et al., 2011). Lipase production from *Aspergillus niger* by nitrous acid induced mutation showing 2.53 times higher activity (Sandana Mala et al., 2001). Environmental adaptability and better bioproduct productivity have made mutation an important technique for enhanced enzyme production.

7. Examples of Enzymes from Recombinant Strains in Food Application

One example in the use of genetic engineering to obtain new sources of enzymes is the production of β-fructofuranosidases or invertases (EC 3.2.1.26) from *Pichia pastoris* yeast (Veana et al., 2014). Invertases allow production of fructose which is preferred in the food industry over sucrose owing to its sweeter taste and restricted crystallization. *Aspergillus niger* GH1 has been reported to be an invertase producer, but Veana et al. (2014) described the use of a synthetic gene to produce the invertase in the methylotrophic yeast *Pichia pastoris*, which has an optimum pH of 5.0 and optimum temperature of 60°C.

The other alternative is to combine microorganism screening and genetic engineering, which are applied to produce extracellular α-amylase (Ozturk et al., 2013). For instance, a high α-amylase-producing *Bacillus subtilis* isolate A28 was selected and its α-amylase gene was cloned and expressed in *E. coli* by a ligase-independent method (Ozturk et al., 2013). The α-amylase isolated and purified from the recombinant strain was highly active at pH range of 4.5–7.0, and Ca^{2+} ions did not stimulate its activity. The pH stability and thermostability of the recombinant amylase makes this enzyme most appropriate for starch processing, brewing and other food industries. Nevertheless, when selecting enzymes, it is necessary to know the specific country regulations regarding the use of genetically modified enzymes for specific food applications.

8. Future Prospects and Conclusion

The current article presents a holistic approach on the history and subsequent developments of enzyme production technology for application in food processing. Over the centuries different enzymes have served to satisfy our palates as well as to help maintain good health. Now with the growing population and increased demand for food, microbial enzymes and industrial production of food has acquired utmost importance. Maintaining the quality and economic benefits for the manufacturer and consumer simultaneously is the most important aspect to be considered in the present scenario. Genetic manipulation through recombinant DNA technology, protoplast fusion and mutation has been proved to be useful for improved production of food enzymes. Genome mining of essential enzyme-producing extremophiles can provide a brighter avenue towards overproduction of the enzymes. Functional genomics, proteomics and metabolomics are now being exploited for the discovery of novel enzymes for food applications. In this scenario, the powerful potential of bioinformatics cannot be ignored to gain insights into the structural/functional and phylogenetic relations of enzymes and also the recombinant production of tailor-made enzyme. Presently, there is a need for new, improved or/and more versatile enzymes in order to develop more novel, sustainable and economically competitive production processes. Hence, in a nutshell, the history, the present and future aspects of microbial enzymes in food application are a precious discovery that mankind has made for all times to come.

References

Abe, F. and Horikoshi, K. 2001. The biotechnological potential of piezophiles. Trends in Biotechnology, 19: 102–108.

Adrio, J.L. and Demain, A.L. 2005. Microbial cells and enzymes—a century of progress. pp. 1–27. *In*: J.L. Barredo (ed.). Microbial Enzymes and Biotransformations: Methods in Biotechnology. Humana Press: Totowa, NJ, USA. Vol. 17.

Aehle, W. 2004. Enzymes in Industry, Production and Applications. Wiley-VchVerlag, Weinheim.

Barlow, S., Chesson, A., Collins, J.D., Dybing, E., Fl Ynn, A., Fruijtier-Pölloth, C., Hardy, A., Knaap, A., Kuiper, H., Le Neindre, P., Schans, J., Schlatter, J., Silano, V., Skerfving, S. and Vannier, P. 2007. Introduction of a qualified presumption of safety (qps) approach for assessment of selected microorganisms referred to European food safety authority. European Food Safety Authority, 587: 1–16.

Becker, P., Abu Reesh, I., Markossian, S., Antranikian, G. and Markl, H. 1997. Determination of the kinetic parameters during continuous cultivation of the lipase producing thermophile *Bacillus* sp. IHI-91 on olive oil. Applied Microbiology and Biotechnology, 48: 184–190.

Behera, S.S. and Ray, R.C. 2015. Solid state fermentation for production of microbial cellulases: Recent advances and improvement strategies. International Journal of Biological Macromolecules. Accepted Manuscript, 86: 656–669.

Bennett, T.P. and Frieden, E. 1969. Modern Topics in Biochemistry, Macmillan: London, pp. 43–45.

Boyaval, P., Corre, C. and Terre, S. 1987. Continuous lactic acid fermentation with concentrated product recovery by ultrafiltration and electrodialysis. Biotechnology Letters, 9: 207–212.

Brune, A. 1998. Termite guts: the world's smallest bioreactors. Trends in Biotechnology, 16: 16–21.

Caplice, E. and Fitzgerald, G.F. 1999. Food fermentations: Role of microorganisms in food production and preservation. International Journal of Food Microbiology, 50(1-2): 131–149.

Cao, L., Langen, L.V. and Sheldon, R.A. 2003. Immobilised enzymes: Carrier-bound or carrierfree? Current Opinion in Biotechnology, 14: 387–394.

Chakrabarti, S., Bhattacharya, S. and Bhattacharya, S.K. 2003. Biochemical engineering: Cues from cells. Trends in Biotechnology, 21: 204–209.

Chen, C., Chi, Y.J., Zhao, M.Y. and Xu, W. 2012. Influence of degree of hydrolysis on functional properties, antioxidant and ACE inhibitory activities of egg white protein hydrolysate. Food Science and Biotechnology, 21: 27–34.

Cheryan, M. and Mehaia, M.A. 1983. A high performance membrane bioreactor for continuous fermentation of lactose to ethanol. Biotechnology Letters, 5: 519–524.

Cooney, C.L. 1983. Bioreactors: design and operation. Science, 19: 728–740.

Demain, A.L. and Adrio, J.L. 2008. Contributions of microorganisms to industrial biology. Molecular Biotechnology, 38: 41–45.

Demain, A.L. 2007. The business of biotechnology. Industrial Biotechnology, 3: 269–283.

Di Lernia, I., Morana, A., Ottombrino, A., Fusco, S., Rossi, M. and De Rosa, M. 1998. Enzymes from *Sulfolobus shibatae* for the production of trehalose and glucose from starch. Extremophiles, 2: 409–416.

Divies, C., Cachon, R., Cavin, J.F. and Prevost, H. 1994. Immobilized cell technology in wine production. Critical Reviews in Biotechnology, 14(2): 135–153.

El-Bondkly, A.M. 2006. Gene transfer between different *Trichoderma* species and *Aspergillus niger* through inter generic protoplast fusion to convert ground rice straw to citric acid and cellulases. Applied Biochemistry and Biotechnology, 135: 117–132.

Fernandes, P. 2010. Enzymes in food processing: a condensed overview on strategies for better biocatalysts. Enzyme Research, Article ID 862537, 19 pages, doi: http://dx.doi.org/10.4061/2010/ 862537.

Flamm, E.L. 1991. How FDA approved chymosin: A case history. Bio/Technology, 9: 349–351.

Hamilton, S. 2009. Introduction to a special issue on food and innovation. Business History Review, 83: 233–238.

Harju, M., Kallioinen, H. and Tossavainen, O. 2012. Lactose hydrolysis and other conversions in dairy products: technological aspects. International Dairy Journal, 22: 104–109.

Hasan, F., Shah, A.A. and Hameed, A. 2006. Industrial application of microbial lipase. Enzyme and Microbial Technology, 39(2): 235–251.

Hesseltine, C.W. 1977. Solid state fermentation, Part1. Process Biochemistry, July/August: 24–27.

Hochfeld, W.L. 2006. Producing Biomolecular Substances with Fermenters, Bioreactors and Biomolecular Synthesizers. CRC Press. Boca Raton. Florida, pp. 372.

Hoh, Y.K., Tan, T.K. and Yeoh, H. 1992. Protoplast fusion of β-glucosidase-producing *Aspergillus niger* strains. Applied Biochemistry and Biotechnology, 37: 81–88.

Horikoshi, K. 1999. Alkaliphiles: Some applications of their products for biotechnology. Microbiology and Molecular Biology Reviews, 63: 735–750.

Hough, D.W. and Danson, M.J. 1999. Extremozymes. Current Opinion in Chemical Biology, 3: 39–46.

Ishida, M., Yoshida, M. and Oshima, T. 1997. Highly efficient production of enzymes of an extreme thermophile, *Thermus thermophilus*: A practical method to overexpress GC-rich genes in *Escherichia coli*. Extremophiles, 1: 157–162.

Jensen, V.J. and Rugh, S. 1987. Industrial-scale production and application of immobilized glucose-isomerase. Methods in Enzymology, 136: 356–370.

Joshi, S. and Satyanarayana, T. 2015. *In vitro* engineering of microbial enzymes with multifarious applications: Prospects and perspectives. Bioresource Technology, 176: 273–283.

Kappeler, S.R., Van Den Brink, H.J., Rahbek-Nielsen, H., Farah, Z., Puhan, Z., Hansen, E.B. and Johansen, E. 2006. Characterization of recombinant camel chymosin reveals superior properties for the coagulation of bovine and camel milk. Biochemical and Biophysical Research Communications, 342(2): 647–654.

Katechaki, E., Panas, P., Kourkoutas, Y., Koliopoulos, D. and Koutinas, A.A. 2009. Thermally-dried free and immobilized kefir cells as starter culture in hard-type cheese production. Bioresource Technology, 100: 3618–3624.

Karanam, S.K. and Medicherla, N.R. 2008. Enhanced lipase production by mutation induced *Aspergillus japonicas*. African Journal of Biotechnology, 7(12): 2064–2067.

Kaur, P., Vohra, A. and Satyanarayana, T. 2013. Laboratory and industrial bioreactors for submerged fermentation. pp. 165–178. *In*: C.R. Soccel, A. Pandey and C. Larroche (eds.). Fermentation Processes Engineering in the Food Industry. USA: CRC Press.

Kawaguti, H.Y. and Sato, H.H. 2007. Palatinose production by free and Ca-alginate gel immobilized cells of *Erwinia* sp. Biochemical Engineering Journal, 36: 202–208.

Kim, J.B., Grate, J.W. and Wang, P. 2008. Nanobiocatalysis and its potential applications. Trends in Biotechnology, 26: 639–646.

Kirk, O., Borchert, T.V. and Fuglsang, C.C. 2002. Industrial enzyme applications. Current Opinion Biotechnology, 13: 345–351.

Kuhne, W. 1876. Über das Verhalten verschiedener organisirter und sog. UngeformterFermente. Über das Trypsin (Enzym des Pankreas), Verhandlungen des Heidelb. Naturhist. Med. Vereins. N.S.I3, Verlag von Carl Winter's, Universitäatsbuchandlung in Heidelberg.

Kuhne, W. 1877. Uber das Verhalten verschiedener organisirter und sog. Ungeformter Fermente, Verhandlungen des Heidelb.Naturhist.-Med. Vereins, Neue Folge, 1(3): 190–193.

Kumar, D.S. and Ray, S. 2014. Fungal lipase production by solid state fermentation: An overview. Journal of Analytical and Bioanalytical Techniques, 6: 230. doi: 10.4172/2155-9872.1000230.

Lilly, M.D. 1994. Advances in biotransformation processes. Chemical Engineering Science, 49: 151–159.

Liu, L., Yang, H.Q., Shin, H.D., Chen, R.R., Li, J.H., Du, G.C. and Chen, J. 2013. How to achieve high-level expression of microbial enzymes: Strategies and perspectives. Bioengineered, 4: 212–223.

Madern, D., Ebel, C. and Zaccai, G. 2000. Halophilic adaptation of enzymes. Extremophiles, 4: 91–98.

Madigan, M.T. and Marrs, B.L. 1997. Extremophiles. Scientific American, 4: 66–71.

Mc Loughlin, A. and Champagne, C.P. 1994. Immobilized cells in meat fermentation. Critical Reviews in Biotechnology, 14(2): 179–192.

Miladi, B., El Marjou, A., Boeuf, G., Bouallagui, H., Dufour, F., Di Martino, P. and Elm'selmi, A. 2012. Oriented immobilization of the tobacco etch virus protease for the cleavage of fusion proteins. Journal of Biotechnology, 158: 97–103.

Mitchell, D.A., Berovic, M. and Krieger, N. 2006. Solid-state Bioreactors. Germany: Springer-Verlag Berlin, Heidelberg.

Nedovic, V.A., Durieux, A., Van Nedervelde, L., Rosseels, P., Vandegans, J., Plaisant, A.M. and Simon, J. 2000. Continuous cider fermentation with co-immobilized yeast and leuconostoc oenos cells. Enzyme and Microbial Technology, 26: 834–839.

Olempska-Beer, Z.S., Merker, R.I., Ditto, M.D. and Dinovi, M.J. 2006. Food-processing enzymes from recombinant microorganisms—A review. Regulatory Toxicology and Pharmacology, 45: 144–158.

Ozturk, M.T., Akbulut, N., Ozturk, S.I. and Gumusel, F. 2013. Ligase-independent cloning of amylase gene from a local *Bacillus subtilis* isolate and biochemical characterization of the purified enzyme. Applied Biochemistry and Biotechnology, 171: 263–278.

Panda, S.K. and Ray, R.C. 2015. Microbial processing for valorization of horticultural wastes. pp. 203–221. *In*: L.B. Shukla, S.K. Panda and B. Mishra (eds.). Environmental Microbial Biotechnology. Springer-Verlag.

Panda, S.K., Mishra, S.S., Kayitesi, E. and Ray, R.C. 2016. Microbial-processing of fruit and vegetable wastes for production of vital enzymes and organic acids: Biotechnology and scopes. Environmental Research, 146: 161–172.

Pandey, A. 1992. Recent process developments in solid-state fermentation. Process Biochemistry, 27: 109–117.

Pandey, A., Benjamin, S., Soccol, C.R., Nigam, P., Krieger, N. and Soccol, V.T. 1999. The realm of microbial lipases in biotechnology. Biotechnology and Applied Biochemistry, 29: 119–131.

Papoutsakis, E.T. 2003. Murray moo-young: the gentleman of biochemical engineering. Biotechnology Advances, 21: 381–382.

Perham, R.N. 1976. The protein chemistry of enzymes. *In*: H. Gutfreund (ed.). Enzymes: One Hundred Years, FEBS Letters, 62 Suppl. E20–E28.

Pfeiffer, J. 1954. Enzymes, the Physics and Chemistry of Life. pp. 171–173.

Plessas, S., Bekatorou, A., Kanellaki, M., Athanasios, A., Koutinas, A.A., Marchant, R. and Banat, I. 2007. Use of immobilized cell biocatalysts in baking. Process Biochemistry, 42: 1244–1249.

Pontecorvo, G., Roper, J.A., Hemmons, L.M., MacDonald, K.D. and Bufton, A.W.J. 1953. The genetics of *Aspergillus nidulans*. Advances in Genetics, 5: 141–238.

Raj, E.A. and Karanth, G.N. 2006. Fermentation technology and bioreactor design. *In*: K. Shetty, G. Paliyath, A. Pornetto and R.E. Levin (eds.). Food Biotech., 2nd edition, USA: CRC Press.

Ramos-Sanchez, L.B., Cujilema-Quitio, M.C., Julian-Ricardo, M.C., Cordova, J. and Fickers, P. 2015. Fungal lipase production by solid-state fermentation. Journal of Bioprocessing and Biotechniques, doi: 10.4172/2155-9821.1000203.

Rodriguez-Leon, J.A., Rodriguez- Fernandez, D.E. and Soccol, C.R. 2013. Laboratory and industrial bioreactors for solid state fermentation. pp. 181–199. *In*: C.R. Soccol, A. Pandey and C. Larroche (eds.). Fermentation Processes Engineering in the Food Industry. CRC Press.

Rose, A.H. 1980. Microbial Enzymes and Bioconversions. Academic Press, London, pp. 655.

Rosell, C.M. and Dura, A. 2016. Enzymes in baking industries. pp. 171–204. *In*: M. Chandrasekaran (ed.). Enzymes in Food and Beverage Processing 2016. CRC Press, Taylor & Francis Group.

Saeki, K., Hitomi, J., Okuda, M., Hatada, Y., Kageyami, Y., Takaiwa, M., Kubota, H., Hagihara, H., Kobayashi, T., Kawai, S. and Ito, S. 2002. A novel species of alkaliphilic bacillus that produces an oxidatively stable alkaline serine protease. Extremophiles, 6: 65–72.

Sandana Mala, J.G., Kamini, N.R. and Puvanakrishnan, R. 2001. Strain improvement of *Aspergillus niger* for enhanced lipase production. Journal of General and Applied Microbiology, 47: 181–186.

Schiraldi, C., Di Lernia, I., Giuliano, M., Generoso, M., D'agostino, A. and De Rosa, M. 2003. Evaluation of a high temperature immobilised enzyme reactor for production of non-reducing oligosaccharides. Journal of Industrial Microbiology & Biotechnology, 30: 302–307.

Schiraldi, C. and De Rosa, M. 2002. Production of biocatalysts and biomolecules from extremophiles. Trends in Biotechnology, 20: 515–521.

Sheldon, R.A. 1993. Chirotechnology, Marcel Dekker, New York, pp. 105.

Silva, S., Ramon-Portugal, F., Silva, P., Texeira, M.F. and Strehaiano, P. 2002. Use of encapsulated yeast for the treatment of stuck and sluggish fermentations. Journal International des Sciences de la Vigne et du Vin, 36: 161–168.

Solis, S., Flores, M.E. and Huitron, C. 1997. Improvement of pectinase production by interspecific hybrids of *Aspergillus* strains. Letters in Applied Microbiology and Biotechnology, 24: 77–81.

Takamine, J. 1914. Enzymes of *Aspergillus oryzae* and the application of its amyloclastic enzyme to the fermentation industry. Industrial and Engineering Chemistry Research, 6: 824–828.

Tao, H. and Cornish, V.W. 2002. Milestones in directed enzyme evolution. Current Opinion in Chemical Biology, 6: 858–864.

Turner, M.K. 1998. Perspectives in Biotransformations, in Biotechnology (Vol. 8), (eds.) Rehm H.J., Ree, G., Biotransformations I, (ed.) Kelly, D.R., Wiley-VCH, Weinheim, pp. 9.

Ushasree, M., Gunasekaran, P. and Pandey, A. 2012. Single-step purification and immobilization of MBP–phytase fusion on starch agar beads: Application in dephytination of soy milk. Applied Biochemistry and Biotechnology, 167: 981–990.

Vallery, R. and Devonshire, R.L. 2003. Life of Pasteur. Kessinger Publishing, LCC.

Veana, F., Fuentes-Garibay, J.A., Aguilar, C.N., Rodriguez-Herrera, R., Guerrero-Olazaran, M. and Viader-Salvado, J.M. 2014. Geneenco ding a novel invertase from a xerophilic *Aspergillus niger* strain and production of the enzyme in *Pichia pastoris*. Enzyme and Microbial Technology, 63: 28–33.

Vu, V.H., Pham, T.A. and Kim, K. 2011. Improvement of fungal cellulose production by mutation and optimization of solid state fermentation. Mycobiology, 39(1): 20–25.

Wang, J.S. and Zhong, J.J. 2006. Bioprocessing for value added products from renewable resources. pp. 131–161. *In*: S.T. Yang (ed.). Amsterdam: Elsevier.

Watanabe, T., Mori, T., Tosa, T. and Chibata, I. 1979. Immobilization of aminoacylase by adsorption to tannin immobilized on aminohexyl cellulose. Biotechnology and Bioengineering, 21: 477–486.

Willaert, R. and Nedovic, V. 2006. Primary beer fermentation by immobilized yeast—A review on flavor formation and control strategies. Journal of Chemical Technology and Biotechnology, 81: 1353–1367.

Williams, J.A. 2002. Keys to bioreactor selection. *In*: Bioreactions. CEP, 34–41.

Witthuhn, R.C., Schoeman, T. and Britz, T.J. 2005. Characterization of the microbial population at different stages of Kefir production and Kefir grain mass cultivation. International Dairy Journal, 15: 383–389.

PART 2

CHARACTERIZATION, PRODUCTION AND APPLICATIONS OF MICROBIAL ENZYMES

Insight of the α-Amylase Family of Enzymes

Endo- and Exo-Acting α, 1-4 Hydrolases

Mario M. Martínez[1],* and *Manuel Gómez*[2],*

1. Introduction

Starch-containing crops are an important constituent of the human diet and constitute the raw material of a large proportion of the food consumed by the world's population. Besides the use of starch-containing plant parts directly as a food source, starch is obtained and used as such or else, physically, chemically or enzymatically processed into a variety of different products, such as starch hydrolysates, glucose or maltose syrups, fructose syrups, starch or maltodextrin derivatives, or cyclodextrins. Among the large number of plants that produce starch, maize, wheat and potatoes are the most processed raw materials with 7.76, 7.87 and 6.33 million tons processed in 2013, respectively. Specifically, 4.82, 3.91 and 1.26 millions tons of maize, wheat and potato starch, respectively, were obtained in 2013, according to the data collected by European Starch Industry Association in 2013.

2. Starch Structure and Properties

Starch is a polymer of glucoses linked through the C_1 oxygen, known as the glycosidic bond. This bond is stable at high pH but is hydrolysed at low pH. At the end of the

---

[1] Whistler Center for Carbohydrate Research, Department of Food Science, Purdue University, 745 Agriculture Mall Drive, West Lafayette, IN 47907, USA.
[2] Food Technology Area, College of Agricultural Engineering, University of Valladolid, 34004 Palencia, Spain.
* Corresponding authors: mario.martinez@iaf.uva.es (alt: mariomartinez175@gmail.com); pallares@iaf.uva.es

polymeric chain, a latent aldehyde group is present, known as the reducing end. Starch is formed by two types of glucose polymers, amylose and amylopectin. Amylose is a linear polymer consisting of up to 6000 glucose units with α,1-4 glycosidic bonds. The number of glucose residues varies with the origin. Amylose from, potato or tapioca starch has a degree of polymerization (DP) of 1000–6000, while amylose from maize or wheat has a DP varying between 200 and 1200. The average amylose content in starches can vary between almost 0 to 75 per cent, but a typical value is 20–25 per cent. Amylopectin consists of short α,1-4 linked linear chains of 10–60 glucose units and α,1-6 linked side chains with 15–45 glucose units. The average number of branching points (α,1-6 linkages) in amylopectin is 5 per cent, but varies with the botanical origin. The complete amylopectin molecule contains on an average about 2,000,000 glucose units, thus being one of the largest molecules in nature. The most commonly accepted model of the structure of amylopectin is the cluster model, in which the side chains are ordered in clusters on the longer backbone chains (Buleón et al., 1998; Myers et al., 2000).

Starch granules have a complex hierarchical structure, which can be described by at least four levels of organization (i.e., molecular, lamellae, growth ring and granular levels), ranging in length scale from nanometer to micrometer. Several detailed comprehensive reviews on the heterogeneous organized structures of granular starch have been published (Oates, 1997; Tester et al., 2004; Jane, 2006; Le Corre et al., 2010). In tuber and root starches, the crystalline regions are solely composed of amylopectin, while amylose is present in the amorphous regions. In cereal starches, the amylopectin is also the most important component of the crystalline regions, but certain amounts of amylose are also found. The amylose in cereal starches is complexed with lipids that form a weak crystalline structure. While amylopectin is soluble in water, amylose and the starch granule are insoluble in cold water. Thus, it is relatively easy to extract starch granules from their plant source.

When starch is heated in excess of water, the granules first swell until a point is reached at which the swelling is irreversible. This swelling process is termed gelatinization. During this process, amylose leaches out of the granule and causes an increase in the viscosity of the starch-slurry. Further increase in temperature leads to maximum swelling of the granules and increases viscosity. Finally, the granules break, causing a complete viscous colloidal dispersion, resulting in a decrease of viscosity. Subsequent cooling of concentrated colloidal starch dispersion results in the formation of an elastic gel; process known as retrogradation. This occurrence is primarily caused by the amylose, since amylopectin, due to its highly branched organization, is less prone to retrogradation (Ai and Jane, 2015).

3. Starch-active Enzymes (α-Amylase Family of Enzymes)

A variety of different enzymes are involved in the synthesis of starch. Sucrose is the starting point of starch synthesis. It is converted into nucleotide sugar ADP-glucose that forms the actual starter molecule for starch formation. Subsequently, enzymes, such as soluble starch synthase and branching enzyme synthesize the amylopectin and amylose molecules (Smith, 1999). In bacteria, an equivalent of amylopectin is found in the form of glycogen. This has the same structure as amylopectin. The major difference lies within the side chains: in glycogen, they are shorter and about twice as high in number (Geddes, 1986). A large variety of plants and bacteria employ extracellular or intracellular enzymes which convert starch or glycogen that can thus serve as energy and carbon sources. However, these enzymes will not be discussed in this chapter, since they are not extensively used for industrial applications.

A classification scheme for glycosidic hydrolases and transferases based on structure rather than specificity has been established while comparing their amino acid sequences (Henrissat and Bairoch, 1993, 1996). These enzymes have been grouped into more than 80 families, where the members of one family share a common three-dimensional structure and mechanism, and display from a few to many sequence similarities. According to this classification, families 13, 70 and 77 contain structurally- and functionally-related enzymes catalysing hydrolysis or transglycosylation of α-linked glucans (MacGregor et al., 2001). Many of these enzymes act on starch and in particular, α-amylase is one of the most important and widely-studied. In this way, this collection of enzymes, considered to consist of families 13, 70 and 77, is often known as the α-amylase family of enzymes, but is, in fact, composed of enzymes with almost 30 different specifities. In some cases, specificities of different enzymes in different families overlap and this, together with amino acid sequence similarities, has led to confusion in identification (MacGregor et al., 2001).

The α-amylase family is formed by a group of enzymes with a variety of different specificities, having in common the quality to act on one type of substrate, with the glucose residues linked through an α,1-1, α,1-4, or α,1-6 glycosidic bond. Members of this family share a number of common characteristics but at least 21 different specificities have been found within the family. These differences in specificities are based not only on fine differences within the active site of the enzyme but also on the differences within the overall architecture of the enzymes. The α-amylase family can roughly be divided into two subgroups—exo- and endo-acting hydrolases that hydrolyse the α,1-4 and/or α,1-6 glycosidic linkages consuming water; and glucanotransferases that break an α,1-4 glycosidic linkage and form a new α,1-4 or α,1-6 glycosidic one (van der Maarel and Leemhuis, 2013). Most of these enzymes have a retaining mechanism, that is, they maintain the anomeric configuration of the hydroxyl group at the C_4 atom on the non-reducing end of the newly formed dextrins and therefore generate oligosaccharides with α-configuration. On the other hand, some starch-active enzymes, such as β-amylase, are inverting enzymes, therefore they change the anomeric configuration of the hydroxyl group at the C_4 position from α to β, and therefore give rise to oligosaccharides with β-configuration (Fig. 1). The enzymes that accomplish these criteria and belong to the α-amylase family were listed by van der Maarel et al. (2002). A complete overview of all glycoside hydrolases known to date can be found on the carbohydrate-active enzymes database (CAZY, http://www.cazy.org) developed by the Glycogenomics Group at AFMB in Marseilles, France. In addition, detailed information and recommendations for enzyme nomenclature by the Nomenclature Committee of the International Union of Biochemistry and Molecular Biology (NC-IUBMB) can be found on a website with sophisticated search function for the desired purpose (http://www.enzyme-database.org).

It is noteworthy that due to the large amount of enzymes used industrially in the food industry belonging to the α-amylase family, they will be discussed in two chapters separately. Starch-active hydrolases with catalysing activity mainly on glucose residues linked through an α,1-4 will be approached in this chapter. On the other hand, starch-acting hydrolases with catalysing activity mainly on α,1-6 glycosidic bonds, also called debranching enzymes, together with α-glucanotransferases that form new α,1-4 or α,1-6 glycosidic bonds, will be discussed in a different chapter (Table 1). Nevertheless, some industrial processes resort to the combination of several enzymes and thus the enzymes will be noted indistinctly in each section but approached in depth in the corresponding chapter.

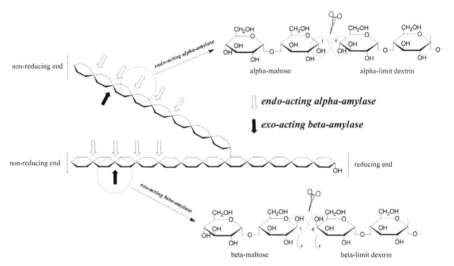

Figure 1. Catalysing effect of endo-acting α-amylase and exo-acting β-amylase on starch. (The effect of hydrolysis from both enzymes on the anomeric configuration of the hydroxyl group at the C_4 and C_1 atoms of dextrin and maltose generated respectively is highlighted.)

3.1 Sources of the starch-active enzymes

The α-amylase family of enzymes produced by plants, animals and microbes play a crucial role in their metabolism. Earlier, amylases from plants and microorganisms were employed traditionally as food additives/ingredients, for instances as barley amylases in the brewing industry and fungal amylases for the preparation of oriental foods (Sivaramakrishnan et al., 2006). Despite the extensive distribution of amylases, microbial sources are preferred for industrial production due to their advantages, such as cost effectiveness, consistency, vast availability, higher stability, less time and space required for production and ease of process modification (Burhan et al., 2003). Besides, the improvement in their production has facilitated the fulfillment of the industrial demands.

Table 1. Starch-active enzymes of the α-amylase family with their corresponding EC number. (Letters in bold indicate the hydrolases acting mainly on α,1-4 glycosidic bonds approached in this chapter.)

Enzyme	EC Number
Glucan branching enzyme	2.4.1.18
Cyclodextrin glycosyltransferase	2.4.1.19
4-α-glucanotransferase (Amylomaltase)	2.4.1.25
Maltopentaose-forming amylase	3.2.1.-
Pullulan hydrolase type III	3.2.1.-
α-Amylase	3.2.1.1
β-Amylase	3.2.1.2
Glucoamylase (amyloglucosidase)	3.2.1.3
α-Glucosidase	3.2.1.20
Amylo-1,6-glucosidase	3.2.1.33
Pullulanase	3.2.1.41
Cyclomaltodextrinase	3.2.1.54
Isopullulanase	3.2.1.57
Maltotetraose-forming amylase	3.2.1.60
Isoamylase	3.2.1.68
Maltohexaose-forming amylase	3.2.1.98
Maltotriose-forming amylase	3.2.1.116
Maltogenic amylase	3.2.1.133
Neopullulanase	3.2.1.135
Limit dextrinase (R-enzyme)	3.2.1.142

Microbial enzymes comprise fungal and bacterial amylases. The specific microbial source for each type of enzyme will be discussed in each section. Nevertheless, it is convenient to highlight a continuous demand to improve the thermal stability of all the described enzymes. In addition, acid-stable α-amylases active at elevated temperatures also have a high demand; however, commercial enzymes that meet these requirements are still scarce. Acid-stable α-amylases have been reported in fungi, bacteria and archaea (Sharma and Satyanarayana, 2013). The present search for these enzymes is forcing the enzyme producers to establish new industrial requirements by coping with different strategies. The first approach would be to screen for novel microbial strains from extreme environments, such as hydrothermal vents, salt and soda lakes, and brine pools (Rana et al., 2013). This is being done successfully by industry and academia and has resulted in the submission of several patent applications, such as the thermostable pullulanase from *Fervidobacterium pennavorans* (Bertoldo et al.,

1999) or α-amylase from *Pyrococcus woesei* (Antranikian et al., 1990). However, other factors, such as their activity with high concentrations of starch (more than 30 per cent of dry solids) or their protein yields during industrial fermentation do not fulfill the industrial criteria and has led to shying away from their commercialization at a large-scale (van der Maarel et al., 2002). The second approach would be to use the nucleotide or amino acid sequence of the conserved domains with a view to design PCR primers, which can be used to screen genomes for the presence of genes putatively encoding the enzyme of interest (Tsutsumi et al., 1999; van der Maarel et al., 2002). However, one of the most successful approaches is to engineer commercially available enzymes. A short overview of some of the results obtained by engineering the protein was given by van der Maarel et al. (2002). Added to this information, an extensive and detailed literature on different microbial sources for the production of α-amylases is also available (Vihinen and Mantasala, 1989; Pandey et al., 2000; Park et al., 2000; Kumar and Satyanarayana, 2009; Hii et al., 2012; Han et al., 2014).

Nowadays, a large number of microbial amylases are available commercially and they have nearly replaced chemical hydrolysis of starch in the starch-processing industry (Rana et al., 2013). Thereby, only starch-active enzymes from microorganisms will be discussed as, approached from the perspective of the food industry.

4. Endo-acting α-Hydrolases

The endo-acting hydrolases comprise a group of enzymes with two common features: retention of the anomeric configuration of the hydroxyl group at the C_4 and C_1 atoms at the non-reducing end of the newly formed dextrins and oligosaccharides respectively (Fig. 1) and an endo-mechanism of attacking the starch-derived molecules at glucose residues in the interior parts. The most well-known endo-acting hydrolase is the enzyme α-amylase (EC 3.2.1.1) which hydrolyses α,1-4 linkages in the interior part of starch polymer randomly, leading to the formation of linear and branched oligosaccharides, or α-limit dextrins. However, another endo-acting α-hydrolase with the ability to hydrolyse faster cyclic dextrins (also called cyclodextrins or cyclomaltodextrins) than starch called cyclomaltodextrinase (CDase, EC 3.2.1.54) has also been reported.

4.1 α-Amylase

α-Amylase may be derived from several bacteria, yeasts and fungi. However, the starch-processing industry, one of the biggest markets for α-amylase, generally prefers bacterial α-amylase over fungal amylase due to their different pH and temperature optima. Microbial α-amylase has been highly purified and crystallized from *Bacillus amyloliquefaciens*, *Bacillus lincheniformis*, *Bacillus subtilis* var. *saccharitikus*, *Bacillus coagulans*, *Pseudomonas saccharophila* and *Aspergillus oryzae* (Robyt, 2009). Their optimum temperature varies depending on the microorganisms, being in many cases around 30–50°C. Based on their optimum pH, α-amylases are classified as acidic, neutral or alkaline. Their optimum pH varies from 2 to 12, although most α-amylases are active in the neutral range (Sharma and Satyanarayana, 2013). The search for enzymes with higher thermal- and acid-stability to accomplish the desired kinetic properties for diverse applications is encouraged in terms of process economy and feasibility. One of the approaches for finding enzymes is through unusual forms of bacteria growing in unusual extreme environments and capable of producing thermostable and acid-stable enzymes. In this way, a large number of hyperthermophilic *Archaeabacteria*, such as *Pyrococcus furiosus*, *Thermococcus profundus*, *Thermococcus hydrothermalis*, *Sulfolobus solfataricus* and *Sulfolobus acidocaldarius* have been reported. Subsequently, many of them were cloned and sequenced. These thermostable α-amylases have the optimal enzyme activity at 90°C or higher and often only begin to show activity at 40°C or 50°C. In particular, *Pyrococcus furiosus* secretes an α-amylase with an optimum temperature of 100°C and a maximum temperature of 140°C. The optimum pH values vary between 5 and 9. Robyt (2009) compiled the organisms, their optima temperature and pH for several of these thermostable enzymes.

The endo-mechanism of α-amylases rapidly fragments starch chains into smaller chains (Fig. 2) and thereby rapidly decreases the starch-iodine blue color and the viscosity of starch pastes. This is the reason why endo-acting amylases are sometimes called liquefying amylases. α-Amylases from different origins have supported differences in their degree of multiple attack by comparing the drop of the blue

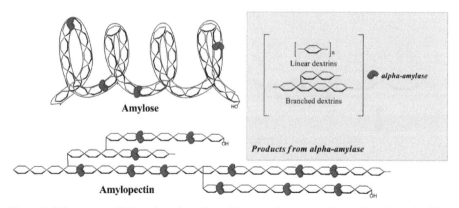

Figure 2. Substrate specificity and products formed by α-amylase on amylose and amylopectin. (The glycosidic bond hydrolysed in each case is marked with different symbols referred to the enzyme.)

starch-iodine color versus the increase in reducing value (reducing sugars) (Robyt and French, 1967). These authors showed that each of the α-amylases had different degrees of multiple attacks, being also different from that produced by acid-catalysed hydrolysis (Robyt and French, 1967). These particular features of α-amylase attack have produced an important misunderstanding; the final products of α-amylase are glucose and maltose. α-Amylases from different sources produce different products in different amounts, with glucose being invariably a very minor final product and almost never an early product (Robyt, 2009).

These different action mechanisms of the endo-acting α-amylases have to be understood by taking into account the model of the active sites, which are located in a cleft that accommodates the binding of inner glucose units of the polysaccharide chain. The sub-sites themselves are composed of side chains of amino acid residues situated on loops in the enzyme structure (Robyt, 2009). The first reported detailed action patterns of an α-amylase on α- and β-limit dextrins and glycogen was that from *Bacillus amyloliquefaciens* (earlier called *Bacillus subtillis*) α-amylase (Robyt and French, 1963). In order to explain the relatively high initial yields of maltotriose and maltohexose, these authors postulated a nine-unit binding subsite that formed products by a dual product specificity (Fig. 3). This model was later confirmed by Thoma et al. (1971). Suganuma et al. (1978) determined that *Aspergillus oryzae* α-amylase has seven glucosyl-unit-binding subsites, with the catalytic groups located between the third and fourth sub-sites from the reducing end. Products originated by α-amylases therefore depend on the number of glucosyl unit binding subsites, their relative affinities for binding glucosyl units and their location in relation to the position of the catalytic groups that produce the actual hydrolytic cleavage of the α,1-4 glycosidic bond. The molecules generated are also determined by whether the substrate is amylose, amylopectin, glycogen, β-amylase limit dextrin of amylopectin

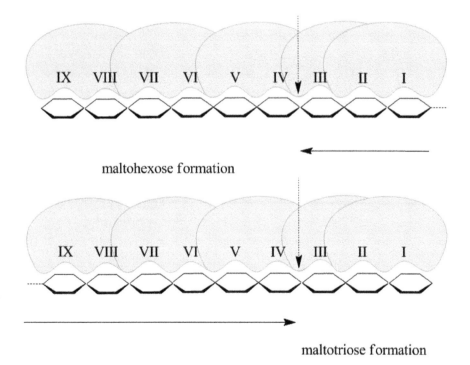

maltohexose formation

maltotriose formation

Figure 3. Glucose-binding subsites at the active-site of *Bacillus amyloliquefaciens* α-amylase after Robyt and French (1963). (Subsites for endoacting amylases are numbered with Roman numbers, beginning with the subsite that would bind the reducing-end glucosyl unit. The arrow represents the location of the catalytic center in relation to the glucose-binding sites, where the hydrolysis takes place.)

or glycogen, or specific individual linear or branched maltodextrins. Thus, different α-amylases produce different products, depending both on the nature of the enzymes' active sites and on the substrates (Robyt, 2009).

4.2 Cyclomaltodextrinase (CDase)

Cyclomaltodextrinase (CDase, EC 3.2.1.54) is an endo-acting α-hydrolase with the ability to hydrolyse much faster cyclodextrins than starch. This substrate preference is the distinctive mark from α-amylase (Park et al., 2000). CDase has a similar hydrolytic activity towards cyclodextrins than cyclodextrin glycosyltransferase (CGTase, EC 2.4.1.19). However, CGTase catalyses the cleavage of starch into cyclodextrins, unlike CDase, which does not produce cyclodextrins from starch.

CDase has been obtained from several bacteria, such as *Bacillus macerans*, *Bacillus coagulans*, *Thermoanaerobacter ethanolicus*, *Bacillus* sp., *Bacillus sphaericus*, *Bacillus* sp., *Flavobacterium* sp. and *Bacillus sphaericus* (Park et al., 2000). Most of these enzymes have an optimum temperature below 50°C, producing mainly maltose from CDs. CDases capable of hydrolysing pullulan have been also reported from *Flavobacterium* sp. and *Bacillus sphaericus*, whereas CDase from *Xanthomonas campestris* possesses the combined activities of α-amylase and cyclodextrin/pullulan-degrading enzymes, that is, having high activity on cyclodextrins, soluble starch, pullulan and amylose, with the cyclodextrins being the best substrates (Abe et al., 1994). The relative rates of hydrolysis of the various substrates differ among the enzyme sources, but are generally in the order of cyclodextrin/oligosaccharides>starch>pullulan (Park et al., 2000). Alongside the hydrolytic activity, some CDases, such as those from *Flavobacterium* and *Bacillus* sp., show remarkable transglycosylation activity (Bender, 1993; Kim et al., 2000).

CDases possess a broad specificity for cyclodextrins, pullulan and starch. In particular, the hydrolysis of pullulan and cyclodextrins leads mainly to panose and maltose, respectively, which together with the transglycosylation activity of some of CDases, can result in the formation of branched oligosaccharides. This occurrence was observed with *Flavobacterium* CDase on pullulan, which gave rise to the production of a remarkably high amount of a branched tetrasaccharide (Bender, 1994). However, to the best of our knowledge, the industrial production of branched oligosaccharides resorts to other enzymes of the α-amylase family; thus no commercial products through the use of CDase are available until date.

4.3 Brief insight of the industrial applications

Amylases are among the most important hydrolytic enzymes for all starch-based industries and the commercialisation of amylases is the oldest, having first been used in the 1980s. Several amylase preparations are available with various enzyme manufacturers for specific use in different industries. A large amount of commercial applications of endo-acting α-hydrolases has been reported in several reviews (van der Maarel et al., 2002; van der Maarel, 2010; Rana et al., 2013). However, the major market for endo-acting hydrolases lies in the conversion of starch hydrolysates (glucose syrups, crystalline glucose, high fructose corn syrups, maltose syrups or reduction of their viscosity), even though baking and brewing industries are also significant.

4.3.1 Application of endo-acting hydrolases in the conversion of starch hydrolysates

Because of the high sweetening property of starch hydrolysates, a large-scale starch processing industry emerged in the mid-1900s. These products are used in huge quantities in the beverage industry as sweeteners for soft drinks and to a lesser extent for solid foodstuffs. The traditional acid hydrolysis method for the production of glucose has been totally replaced by the enzymatic treatments at the end of the 20th century (Crabb and Mitchinson, 1997; Crabb and Shetty, 1999). A brief overview of the production of starch hydrolysates is depicted in Fig. 4. Most enzymatic conversion processes start with the liquefaction of gelatinized starch into soluble and short-chain dextrins. Liquefaction usually requires the use of a highly thermostable endo-acting amylase for optimizing costs. Initially, the α-amylase from *Bacillus amyloliquefaciens* was used during liquefaction, but this has been replaced by the α-amylase of *Bacillus stearothermophilus* or *B. licheniformis* (van der Maarel et al., 2002). These thermostable amylases will hydrolase both amylose and amylopectin of a 30–35 per cent dry solids starch slurry with an endo-action and therefore decrease its viscosity. The liquefaction proceeds until a dextrose equivalent (DE) of 8–12 is

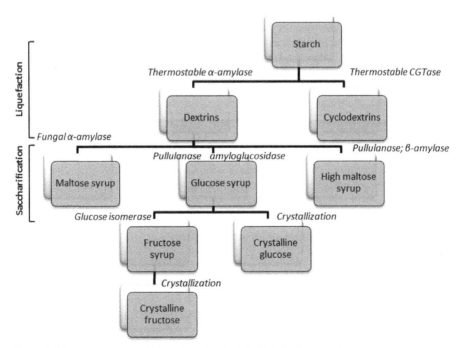

Figure 4. Schematic overview of the various steps in industrial starch processing.

reached. The drawback of the α-amylases used currently is that they are not active at a pH below 5.9 at the high temperatures used (95–100°C). Therefore, the pH has to be adjusted from the natural pH 4.5 of the starch slurry to pH 6 by adding NaOH (sodium hydroxide). Also Ca^{2+} needs to be added because of the Ca^{2+}-dependency of these enzymes (van der Maarel, 2010).

4.3.2 Application of endo-acting hydrolases in other industries

The baking industry is the other big market for α-amylases. Since enzymes are produced from natural sources, they are finding great acceptance among the consumers. Thus several enzymes have been suggested to act as dough and/or bread improvers. Therefore, another important application of endo-acting enzymes on a large scale is the production of bakery products, such as bread. α-Amylases can be used both in flour supplementation and added during dough preparation in order to enhance the rate of fermentation, reduce the viscosity of the dough (resulting in improvement in the volume and texture of the product) and generate additional fermentable sugars, which improve the taste, crust, color and toasting quality of the bread (Kumar and Satyanarayana, 2008). Moreover, when the bread is removed from the oven, a series of changes start that eventually lead to the deterioration of quality. These changes include increase of crumb firmness, loss of crispiness of the crust, decrease in moisture content of the crumb and loss of bread flavor. All undesirable changes that occur upon storage are called staling. Among the different factors which foster bread staling, retrogradation of the starch fraction, especially retrogradation of amylopectin, is considered crucial (Kulp and Ponte, 1981; Champenois et al., 1999). Besides generating fermentable compounds, α-amylases also have an anti-staling effect in bread baking, thus reducing the retrogradation of amylopectin (Dragsdorf and Varriano-Marston, 1980; Lin and Lineback, 1990; Rojas et al., 2001) and improving the softness retention of baked goods (De Stefanis and Turner, 1981; Cole, 1982; Sahlstrom and Brathen, 1997). Despite this possible anti-staling effect, the use of endo-acting α-amylases as anti-staling agent is not widespread because even a slight overdose of α-amylase can lead to sticky breads. It is noteworthy that maltogenic α-amylase is being used successfully for this purpose. However, due to the fact that this enzyme possesses both an endo- and exo-acting mechanism of action, it will be discussed in the next section.

Another important industry that exploits the catalytic mechanism of the endo-acting α-amylase is the brewing industry. One of the most important steps in beer-making is mashing. This process consists of mixing the crushed malt with hot water and letting the mixture stand while the enzymes degrade the proteins and starch to yield the soluble malt extract, also known as wort. When brewing adjuncts are prepared for use in a mash, defined as any carbohydrate source other than malted barley which provides sugars to the wort, α-amylase must be present to hydrolyse the starch when it gelatinises. This ensures that the viscosity of the adjunct mash does not become too high and prevents possible retrogradation of the starch as it is added to the colder malt mash (Ryder and Power, 2006). The α-amylase from malt is most used for this purpose. However, exogenous microbial α-amylases from *Bacillus amyloquefaciens* and *Bacillus lincheniformis* have also been applied to replace the malt enzymes or to strengthen the action of a low-malt mash enzyme (Ryder and Power, 2006). The use of these exogenous enzymes allows more adjunt to be used, improving the flavor and increasing the cooking capacity, since they are more heat stable than those from malt.

5. Exo-acting Hydrolases

Exo-acting hydrolases include a group of enzymes with an exo-mechanism acting on the non-reducing ends of starch polymer chains, starch polymer-derived chains and glycogen. Some of these exo-acting enzymes possess inversion capacity by changing the anomeric configuration of the hydroxyl group at the C_4 position from α to β and therefore giving rise to oligosaccharides with β-configuration (Fig. 1). According to their inversion capacity, these enzymes can be grouped in two groups. The first group comprises the exo-acting β-hydrolases [β-amylase (EC 3.2.1.2) and glucoamylase, also named amyloglucosidase (EC 3.2.1.3)], that release β-maltose and β-glucose respectively. Meanwhile, the second group comprises the exo-acting α-hydrolases, formed by α-glucosidase (EC 3.2.1.20) and maltogenic α-amylase (EC 3.2.1.133), that release glucose and maltose in their α-configuration respectively and amylases producing specific maltodextrin products. Within this last group are included maltotriose-forming enzyme (maltotriohydrolase, EC 3.2.1.116), maltotetraose-forming enzyme (maltotetrahydrolase, EC 3.2.1.60), maltopentose-forming enzyme (EC 3.2.1.-) and the maltohexaose-forming amylase (maltohexaohydrolase, EC 3.2.1.98) that retain the anomeric configuration and therefore release products with an α-configuration. In addition, cyclodextrin glycosyltransferase (CGTase, EC 2.4.1.19) also presents an exo-mechanism of action. However, this enzyme presents a strong transglycosylation activity. Thus, their catalytic mechanism and applications will be approached in detail in a separate chapter.

5.1 Exo-acting β-hydrolases

Exo-acting β-hydrolases, comprised by β-amylase and glucoamylase, catalyse starch, glycogen or related polysaccharides and oligosaccharides hydrolysis by a mechanism that produces the inversion of the configuration at the anomeric center. This mechanism is depicted in Fig. 1, where the action of exo-acting β-amylase gives rise to β-maltose and β-limit dextrins.

Unlike other members of the α-amylase family, only a few attempts have been made to study β-amylases from microbial origin, as β-amylases have generally been obtained from plant sources, such as sweet potatoes, soybeans, barley, and wheat (Balls et al., 1948; Robyt, 2009; van der Maarel, 2010), noticeably more cost-effective than those from microorganisms. However, β-amylases from microbial sources generally possess greater thermal resistance than plant β-amylases. In addition, β-amylases have higher optima pH than α-amylases and do not require Ca^{2+} for stabilization or enhancement of activity (Pandey et al., 2000). Bacterial strains belonging to the genus *Bacillus*, e.g., *B. polymyxa*, *B. megaterium*, *B. cereus*; *Pseudomonas* (aerobic) and *Clostridium* (anaerobic) sp.; actinomycete strains belonging to *Streptomyces* sp. and fungal strains belonging to *Rhizopus* sp., have been reported to synthesize β-amylase (Hyun and Zeikus, 1985; Fogarty and Kelly, 1990; Kwan et al., 1994; Swamy et al., 1994; Ray et al., 1997; Reddy et al., 1998). All these β-amylases produce β-maltose and high molecular weight β-limit dextrins (Figs. 1 and 5). The limit dextrins are formed when the enzyme reaches an α,1-6 branch bond, which cannot be hydrolysed. Robyt (2009) reported that approximately half of an amylopectin molecule is converted to β-maltose whereas the remaining half keeps as β-limit dextrin. In other words, the β-amylase removes all glucose units that are part of the outer chains of amylopectin, leaving the inner part intact. The relatively large amount of maltose that is formed tends to inhibit the enzyme, thus slowing the reaction.

Meanwhile, glucoamylase can be derived from plant, animal and microbial sources (Pandey, 1995). Nevertheless, all known commercial glucoamylases are elaborated from filamentous fungi and especially important are *Aspergillus niger, A. awamori, A. awamori* var. *kawachi, Rhizopus delemar* and *R. niveus* (Robyt, 2009). An extensive list of microorganisms used as the sources was already reported by Pandey (1995), Pandey et al. (2000) and Kumar and Satyanarayana (2009). Glucoamylase have optimum pH values of around 4.5–5.0. The optimum temperature for glucoamylase activity from most sources has generally been found to be between 40 and 60°C (Pandey et al., 2000). However, there are a few reports describing glucoamylases from a thermophilic fungus *Humicola lanuginosaw* with optima pH of 4.9–6.6 and stable up to 11.0 (Pandey et al., 2000). Besides, the glucoamylases from *Coniophora cerebella* (King, 1967) and *Corticum rolfsii* (Kaji et al., 1976) were stable up to pH 9.0, whereas *Mucor rouxianus* showed pH stability up to 8.0 (Pandey et al., 2000). Conversely to other amylases, glucoamylases can catalyse the hydrolysis of both α,1-4 and α,1-6 glycosidic bonds in starch (Fig. 5), although the α,1-4 bond is hydrolysed

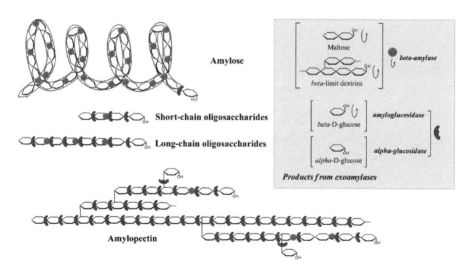

Figure 5. Substrate specificity and products formed by β-amylase, glucoamylase and α-glucosidase on amylose, amylopectin and starch-derived oligosaccharides. The glycosidic bond hydrolysed in each case is marked with different symbols referred to the enzymes. Circular arrows indicate which final products were hydrolyzed with the inversion of the anomeric center of the non-reducing end from α- to β-configuration.

approximately 500 times faster than the α,1-6 linkage (Hiromi et al., 1973; Sierks and Svensson, 1994; Davies et al., 1997). In this way, glucoamylase can completely hydrolyse starch to D-glucose, having a key role in the starch industry. Nevertheless, glucoamylases cannot hydrolyse trehalose and most β-linked substrates. This could be explained by the fact that most α-linked disaccharides dock easily into the active site of glucoamylase, whereas exclusively β-linked disaccharides, except for β-cellobioside and β-gentiobioside, do not (Coutinho et al., 1997a,b,c). In addition, when the concentration of maltose is sufficiently high, glucoamylase can also perform a transglycosylation reaction to form isomaltose, by coupling two glucose residues together via α,1-6 linkage (Ojha et al., 2015).

Both, β-amylases and glucoamylases act by binding the non-reducing ends of starch chains into a pocket in their structure rather than into a cleft, as do endo-acting hydrolases. However, like endo-acting hydrolases, exo-acting hydrolases also have a number of glucose unit-binding subsites (Robyt, 2009). In order to understand this mechanism and as an example of exposure, the glucose-binding subsites at the active-sites of a soybean β-amylase and a glucoamylase from *Aspergillus awamori* are depicted in Fig. 6. Soybean β-amylase has at least five subsites with the catalytic

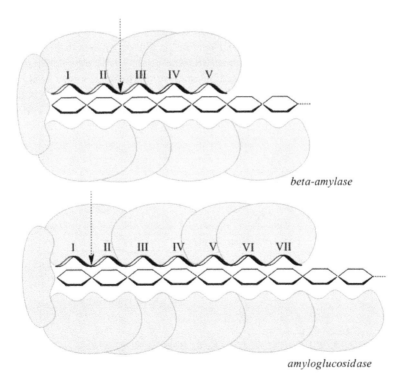

beta-amylase

amyloglucosidase

Figure 6. Glucose-binding subsites at the active-site of soybean exo-acting β-amylase (Kato et al., 1974) and glucoamylase from *Aspergillus awamori* (Savel'ev et al., 1982). Subsites for endoacting amylases are numbered with Roman numbers, beginning with the subsite that would bind the reducing-end glucosyl unit. The arrow represents the location of the catalytic centre in relation to the glucose-binding sites, where the hydrolysis takes place.

groups located between the second and third glucose-unit-binding subsites from the non-reducing end (Kato et al., 1974). Meanwhile, glucoamylase from *Aspergillus awamori* has five to seven glucose unit-binding subsites with the catalytic groups located between the first and second subsites from the non-reducing end (Savel'ev et al., 1982). The location of the catalytic groups of the exo-acting β-hydrolases depends on their origin, differing extensively with respect to their multiple attacks (Robyt, 2009).

5.2 Exo-acting α-Hydrolases

5.2.1 α-Glucosidase

α-Glucosidase (EC 3.2.1.20) is an exo-acting α-hydrolase, which releases D-glucose residues from the non-reducing end side of substrate (amylose, amylopectin and oligosaccharides including maltose), similarly to glucoamylase. However, both enzymes are essentially differentiated by the type of final product originated in such a way that α-glucosidase produces α-glucose instead β-glucose. Neither this enzyme nor glucoamylase should be confused with glucose oxidase, EC 1.1.3.4, which catalyses the oxidation of glucose to hydrogen peroxide. α-Glucosidase is widely distributed in microorganisms, plant and animal tissues. Among the microbial source, the catalytic sites of α-glucosidase have been studied from enzymes extracted from *Saccharomyces carlsbergensis, Bacillus cereus, Aspergillus niger, Aspergillus oryzae, Mucor javanicus, Candida tsukubaensis, Schwanniomyces occidentalis, Streptococcus mutans Escherichia coli, Saccharomyces cereviswe, Bacillus cereus* and *Microbacterium* sp. (Chiba, 1997; Ojha et al., 2015). Many α-glucosidases are capable of hydrolysing not only synthetic α-glucosides and oligosaccharides but also α-glucans, such as starch and glycogen. Glucoamylase and α-glucosidase also differ in their substrate preference: α-glucosidase acts best on short maltooligosaccharides whereas glucoamylase hydrolyses better long-chain polysaccharides. To the best of our knowledge, this enzyme still has scarce applications in the food industry and is not being commercialised on a large-scale, so far.

5.2.2 Maltogenic α-amylase

Maltogenic α-amylase (EC 3.2.1.133) hydrolyses α,1-4 glycosidic linkages (and α,1-6 bonds in a lesser extent) in polysaccharides, removing successive maltose residues from the non-reducing ends of the chains (Fig. 7). This enzyme exhibits both an endo- and an exo-action mechanism, with a high degree of multiple attack action and with endo-action increasing with temperature increase (Grewal et al., 2015). A maltogenic α-amylase that hydrolyses cyclodextrins (CD) and starch, mainly to maltose, and pullulan to panose, was isolated and cloned from *Bacillus licheniformis* (Kim et al., 1992, 1994). Several other maltogenic α-amylases have been cloned from Gram-positive bacteria, including *Bacillus stearothermophilus* and *B. subtilis* (Cha et al., 1998; Cho et al., 2000). Kim et al. (1999) reported the cloning and physicochemical properties of a maltogenic α-amylase from a *Thermus* strain of superior stability at high temperatures than other maltogenic α-amylases. Maltogenic α-amylase differs from typical α-amylases by

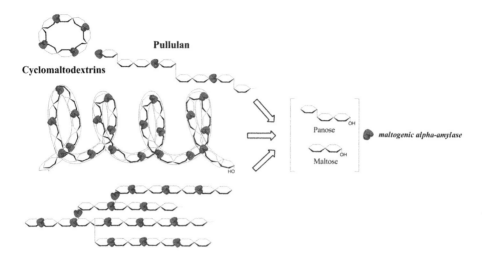

Amylopectin

Figure 7. Hydrolytic action of maltogenic α-amylase on different substrates as well as the main products formed after hydrolysis.

having substrate preference of cyclodextrins over starch and pullulan, thus possessing the relative hydrolytic activities cyclodextrins>pullulan>soluble starch. Added to that, maltogenic α-amylase exerts its high-degree multiple attack on starch granules during baking when they begin to gelatinize, leading to a shortening of the amylopectin side chains and therefore, preventing amylopectin recrystallization (retrogradation) and hindering water immobilization. This fact turns out to be very interesting for bread-manufactures in delaying bread staling (see also Section 5.3.2).

Maltogenic α-amylase not only exhibits hydrolytic activity but also, in certain conditions, possesses high transglycosylation capacity (Fig. 8). Thus, maltogenic α-amylase from *B. stearothermophilus* transfers glucose units and forms α,1-6 linkages. When this enzyme was incubated with 30 per cent liquefied corn starch, it mainly produced isomaltose, panose and branched tetraose, accounting for 66 per cent of the total sugar (Kang et al., 1997). Kim et al. (1994) also reported the isolation of a novel maltogenic α-amylase from *Bacillus licheniformis* which was able to produce branched oligosaccharides via formation of α,1-6 linkages in the

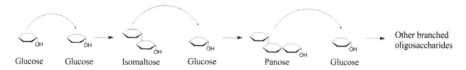

Figure 8. Repeated transglycosylation action of maltogenic α-amylase in presence of glucose for the production of branched oligosaccharides.

presence of an excessive amount of glucose. Furthermore, Lee et al. (2002) used a *B. stearothermophilus* maltogenic α-amylase in combination with *Thermotoga maritima* α-glucanotransferase for the production of branched oligosaccharides. Therefore, this enzyme appears to be highly suited for the production of branched oligosaccharides, also called isomaltooligosaccharides (IMO).

5.2.3 Amylases-producing specific maltodextrin

A wide range of amylases capable of producing specific maltodextrins have been reported to date. They will be commented upon according to the product released from the catalysis (from the smallest, maltotriose, to the highest, maltohexose).

An enzyme from *Streptomyces griseus* that produced primarily maltotriose from the non-reducing ends of starch chains was described by Wako et al. (1979). This enzyme was able to convert amylose to maltotriose, together with small amounts of glucose and maltose. Waxy maize starch, oyster glycogen and phytoglycogen gave 51 per cent, 40 per cent and 20 per cent maltotriose, respectively and corresponding amounts of limit dextrins, endorsing an exo-acting activity to this enzyme. Takasaki (1983) also reported an α-amylase from a *Bacillus subtilis* strain that produced maltotriose.

Robyt and Ackerman (1971) reported an enzyme from *Pseudomonas stutzeri* that formed maltotetraose, which was one of the first producer enzymes of a single specific product from starch, other than glucose and maltose. Like β-amylases, it is an exo-acting amylase, but produces maltotetraose with an α-anomeric configuration. It specifically hydrolyses the fourth α,1-4 glycosidic linkage from the non-reducing end of a starch chain, giving 42 per cent α-maltotetraose and 58 per cent limit dextrin (Robyt and Ackerman, 1971). However, this enzyme does not catalyse α,1-6 glycosidic linkages of amylopectin. Later, similar maltotetraose-producing amylases were reported from *Pseudomonas saccharophila* (Zhou et al., 1989) and *Pseudomonas* sp. (Fogarty et al., 1994).

An α-amylase elaborated by *Bacillus licheniformis* was reported to give relatively high yields (33 per cent) of maltopentaose and a range of other maltodextrins from DP1 to DP12, having an optimum temperature in the range of 70°C to 90°C (Morgan and Priest, 1981).

Kainuma et al. (1971) discovered an exo-acting amylase from *Aerobacter aerogenes*, producing maltohexaose. Later, Taniguchi et al. (1982) discovered another maltohexaose-forming amylase elaborated by a strain of *Bacillus circulans*. Added to that, Fogarty et al. (1991) described an α-amylase elaborated by *Bacillus caldovelox* that preferentially produces maltohexaose in yields of 40–44 per cent and no glucose.

An α-amylase from a hyperthermophilic archaebacterium, *Pyrococcus furiosus* was found to produce DP4, DP5 and DP6 maltodextrins as the primary products, whereas DP3 maltodextrin was also produced but at a much reduced rate, which is characteristic of α-amylase secondary reactions. This enzyme was purified and found to have optimal activity, with substantial thermal stability at 100°C (Laderman et al., 1993). Unlike other α-amylases, it is not dependent on calcium ions for activity or stability. Using molecular biology techniques, Conrad et al. (1995) produced 33 hybrids from the genes of *Bacillus amyloliquefaciens* and *Bacillus licheniformis*, encoding the α-amylases from these microorganisms. The hybrids comprised the entire α-amylase sequence with variable proportions from *Bacillus amyloliquefaciens* α-amylase and *Bacillus licheniformis* α-amylase. The hybrid enzymes fell into six groups that retained the extra-thermostability of the *Bacillus licheniformis* enzyme. In particular, a specific hybrid sequence was correlated with the enzymes' product specificity for forming and accumulating maltohexaose.

5.3 Brief insight of the industrial applications of exo-acting hydrolases

The catalytic action of microbial exo-acting hydrolases also turns out interesting for many applications in the food industry. However, as in the case of endo-acting hydrolases, their major market also resides in the starch conversion, baking and brewing industries.

5.3.1 Application of exo-acting hydrolases in the conversion of starch hydrolysates

Industrial starch processing is either stopped or followed by a saccharification step, depending on the type of product to be produced (Fig. 4). In this step, pullulanase, glucoamylase, β-amylase or α-amylase can be added to further degrade the liquefied starch into maltodextrins, maltose or glucose syrups used in food, beverages and fermentation industries.

The liquefied starch is pumped into a large stirrer vessel where glucoamylase is added. Most commonly used are glucoamylases from *Aspergillus niger* or a closely related species. As this enzyme has an optimum pH of 4.2, thus pH must be adjusted to induce some deactivation of the α-amylase used in the liquefaction process. Moreover, since glucoamylase is stable at 60°C, the temperature must also be adjusted. It is noteworthy that the temperature must be lowered quickly to avoid retrogradation of the liquefied starch, which would become resistant to enzyme-hydrolysis. Depending on the specifications of the final product, saccharification is usually performed for 12–96 hours at 60–62°C (Crabb and Mitchinson, 1997; Reilly, 1999). When the desired dextrose equivalent (DE) is reached, the solution is heated to 85°C for a few minutes to stop the reaction, by inactivating the enzymes (van der Maarel, 2010). To make maltose syrup, traditionally barley β-amylase was added. Nevertheless, due to its high cost, its lack of thermal stability and its inhibition by copper and other metal ions, barley β-amylase has been replaced by fungal acid-α-amylase. Another enzyme suggested to be used for the production of maltose syrups is thermostable cyclodextrin glycosyltransferase of the bacterium *Thermoanaerobacterium thermosulfirigenes* (van der Maarel, 2010), due to its exo-acting hydrolytic mechanism.

A practical problem during saccharification is that the glucoamylase is specific for cleaving α,1-4 glycosidic bonds and slowly hydrolyses α,1-6 glycosidic bonds present in maltodextrins turning out in the accumulation of isomaltose. A solution to this problem is to use a pullulanase that efficiently hydrolyses α,1-6 glycosidic bonds, having the same optimum pH and temperature as glucoamylase (Reilly, 2006). When glucoamylase and pullulanase are used for saccharification, high glucose syrups are produced. It is important to achieve a high dry solid content in order to make the production of high glucose syrups (> 95 per cent glucose) cost-effective. Nevertheless, this high dry solid content together with the transglycosylation activity of glucoamylase can easily form reversion products such as maltose and isomaltose at the expense of the amount of glucose. The current solution is to balance the amount of enzyme, the temperature and the time of incubation (Crabb and Mitchinson, 1997).

Glucose has about 75 per cent of the sweetness of sucrose, while its isomer fructose is twice sweeter than sucrose. Consequently, fructose is preferred in some foodstuffs and it can be metabolized without insulin (Synowiecki, 2007). Commercially, fructose is produced by the isomerization of glucose, using bacterial glucose/xylose isomerase (E.C. 5.3.1.5) at 50–60°C and pH 7–8. Glucose isomerase is the most expensive of all the enzymes involved in starch processing, and thus is reused until it loses most of its activity (Crabb and Mitchinson, 1997; Crabb and Shetty, 1999; Reilly, 1999). This enzyme does not belong to the α-amylase family.

During the production of glucose from liquefied starch, we mentioned how transglycosylation activity of glucoamylase can form negative reversion products. However, this feature can be beneficial in manufacturing IMOs from starch and which are being used as substitutes for sucrose due to their low energy (2 kcal/g) and low carcinogenic properties. In addition, they are known to improve the intestinal microflora by stimulating the growth of *Bifidobacteria* and suppressing the multiplication of harmful bacteria (Tomomatsu, 1994). When the concentration of maltose is sufficiently high, glucoamylase performs a transglycosylation reaction, forming isomaltose by coupling two glucose residues together via α,1-6 linkage. In a second step, the isomaltose is glycosylated to form isomaltotriose. Other IMOs, such as panose and isopanose, also are created. The glucoamylase used for their production is obtained from the fungus *Aspergillus niger*. Besides, glucoamylase and α-glucosidase purified from cells of *Microbacterium* sp. efficiently utilize maltose and convert it into IMOs up to DP6. While isomaltotriose and isomaltotetraose were the major products formed through conversion efficiency of 25 per cent, higher oligomers like pentaose and hexaose were also formed (Ojha et al., 2015). Another exo-acting enzyme with coupled transglycosylation and hydrolysis activities is maltogenic α-amylase obtained from *Bacillus stearothermophilus*, which has also been utilized industrially for efficient production of IMOs from liquefied starch (Kuriki et al., 1993; Kim et al., 1994; Lee et al., 1995; Yoo et al., 1995; Kang et al., 1997).

5.3.2 Application of exo-acting hydrolases in other industries

The baking industry has also resorted to the use of exo-acting hydrolases, alone or in combination with endo-acting hydrolases (α-amylase) for two main purposes. Firstly, the action of glucoamylase assists in saccharification of starch to glucose that can be easily fermented by yeast and improve the bread crust colour (Kumar and Satyanarayana, 2008). Secondly, both β-amylase and glucoamylase shorten the external side chains of amylopectin by cleaving maltose or glucose molecules, respectively. Therefore, these enzymes are suggested to delay bread staling by reducing the tendency of the amylopectin compound in bakery products to retrograde (Vidal and Gerrity, 1979;

Van Eijk, 1991; Wursch and Gumy, 1994). The synergetic use of α- and β-amylase is also claimed to increase the shelf-life of baked goods (Van Eijk, 1991). Nevertheless, α- and β-amylases and glucoamylases are inactivated at the beginning of baking. Thus they act only on small amounts of damage to starch, producing scarce positive effects on bread staling. On the other hand, α-amylases from bacteria possess high thermal stability, thus resisting the temperatures reached during baking. They act on gelatinised starch even after baking to produce sticky crumbs. In order to solve this problem, the use of other exo-acting hydrolases with intermediate thermal stability has been suggested. In this way, amylases producing specific maltodextrins, such as maltose (Diderichsen and Christiansen, 1988; Olesen, 1991), maltotriose (Tanaka et al., 1997) and maltotetraose (Shigeji et al., 1999a,b) are claimed to increase the shelf-life of bakery products. They delay retrogradation of the starch compound by producing linear oligosaccharides of 2–6 glucose residues. Among them, maltogenic α-amylase and maltotetraose-forming enzyme are mostly used in bread-making.

The use of enzymes, such as fungal α-amylase and glucoamylase, to assist in saccharification of starch to glucose has also been exploited by the brewing industry. However, glucoamylase from *Aspergillus niger* is mostly used. It leads to increase in the amount of fermentable sugars available in the wort with a view to producing special beers with more alcohol or lower calories if the alcohol level is reduced by dilution (Ryder and Power, 2006). However, brewers know the potential problems when active glucoamylase is present in beer. Thus it is important to use this enzyme during wort production; it is equally important to inactivate it during wort boil. This enzyme does not work well at mash pH (above 5.0) and the pH must be reduced to reach a compromise to shy away from favoring malt protease activity and reduce malt α-amylase activity. The initial temperature during saccharification should be just above 60°C. Then the temperature should be slowly raised to above 70°C to reach a compromise in fostering the availability of starch for the enzyme and its gradual inactivation (Ryder and Power, 2006). To get a better insight of the use of glucoamylase in the brewing industry see Chapter 19.

6. Conclusions and Further Trends

Starch-active hydrolases with catalyzing activity mainly on glucose residues linked through an α,1-4 have since long been employed as food additives/ingredients. Recently, the improvement in the production of these enzymes from microbial sources has led to the capability of fulfilling the demands of different processes in the food industry. In this way, microbial starch-active hydrolases are largely used in the starch converting, baking and brewing industries. Added to this, it is important to highlight the emerging food trends, which are making manufacturers to fulfill increasing demands for different products, e.g., low-calorie beers and breads with low staling. However, these industrial processes require enzymes with improved thermal and acid-stability. Therefore, the present search for these enzymes is forcing the enzyme producers to set the new industrial requirements by dealing with different strategies. However, commercial enzymes that accomplish these requirements are still scarce.

References

Abe, J., Onitsuka, N., Nakano, T., Shibata, Y., Hizukuri, S. and Entani, E. 1994. Purification and characterization of periplasmic α-amylase from *Xanthomonas campestris* K-11151. Journal of Bacteriology, 176: 3584–3588.

Ai, Y. and Jane, J.L. 2015. Gelatinization and rheological properties of starch. Starch/Stärke, 67: 213–224.

Antranikian, G., Koch, R. and Spreinat, A. 1990. Novel Hyperthermostable Alpha-amylase. Patent WO 90/11352.

Balls, A.K., Walden, M.K. and Thompson, R.R. 1948. A crystalline β-amylase from sweet potatoes. The Journal of Biochemical Chemistry, 173: 9–19.

Bender, H. 1993. Purification and characterization of a cyclodextrin-degrading enzyme from *Flavobacterium* sp. Applied Microbiology and Biotechnology, 39: 714–719.

Bender, H. 1994. Studies of the degradation of pullulan by the decycling maltodextrinase of *Flavobacterium* sp. Carbohydrate Research, 260: 119–130.

Bertoldo, C., Duffner, F., Jorgensen, P.L. and Antranikian, G. 1999. Pullulanase type I from *Fervidobacterium pennavorans Ven5*: Cloning, sequencing and expression of the gene and biochemical characterization of the recombinant enzyme. Applied and Environmental Microbiology, 65: 2084–2091.

Buleón, A., Colonna, P., Planchot, V. and Ball, S. 1998. Starch Granules: Structure and Biosynthesis. International Journal of Biological Macromolecules, 23: 85–112.

Burhan, A., Nisa, U., Gokhan, C., Omer, C., Ashabil, A. and Osman, G. 2003. Enzymatic properties of a novel thermostable, thermophilic, alkaline and chelator resistant amylase from an alkaliphilic *Bacillus* sp. isolate ANT-6. Process Biochemistry, 38: 1397–1403.

Cha, H.J., Yoon, H.G., Kim, Y.W., Lee, H.S., Kim, J.W., Kweon, K.S., Oh, B.H. and Park, K.H. 1998. Molecular and enzymatic characterization of a maltogenic amylase that hydrolyzes and transglycosylates acarbose. European Journal of Biochemistry, 253: 251–262.

Champenois, Y., Della, V.G., Planchot, V., Buleon, A. and Colonna, P. 1999. Influence of alpha-amylases on the bread staling and on retrogradation of wheat starch models. Sciences des Aliments, 19: 471–486.

Chiba, S. 1997. Molecular mechanism in (α-glucosidase and glucoamylase). Bioscience, Biotechnology and Biochemistry, 61: 1233–1239.

Cho, H.Y., Kim, Y.W., Kim, T.J., Kim, D.Y., Kim, J.W., Lee, Y.W. and Park, K.H. 2000. Molecular characterization of a dimeric intracellular maltogenic amylase of *Bacillus subtilis* SUH4-2. Biochimica et Biophysica Acta, 1478: 333–340.

Cole, M.S. 1982. Antistaling Baking Composition. Patent US4320151.

Conrad, B., Hoang, V., Polley, A. and Hofemeister, J. 1995. Hybrid *bacillus-amyloliquefaciens* x *Bacillus licheniformis* α-amylases: Construction, properties and sequence determinants. European Journal of Biochemistry, 230: 481–490.

Coutinho, P.M., Dowd, M.K. and Reilly, P.J. 1997a. Automated docking of glucosyl disaccharides in the glucoamylase active site. Proteins: Structure, Function, and Bioinformatics, 28: 162–173.

Coutinho, P.M., Dowd, M.K. and Reilly, P.J. 1997b. Automated docking of monosaccharide substrates and analogues and methyl α-acarviosinide in the glucoamylase active site. Proteins: Structure, Function, and Bioinformatics, 27: 235–248.

Coutinho, P.M., Dowd, M.K. and Reilly, P.J. 1997c. Automated docking of isomaltose analogues in the glucoamylase active site. Carbohydrate Research, 297: 309–324.

Crabb, W.D. and Shetty, J.K. 1999. Commodity scale production of sugars from starches. Current Opinion in Microbiology, 2: 252–256.

Crabb, W.D. and Mitchinson, C. 1997. Enzymes involved in the processing of starch to sugars. Trends in Biotechnology, 15: 349–352.

Davies, G.J., Wilson, K.S. and Henrissat, B. 1997. Nomenclature for sugar-binding subsites in glycosyl hydrolases. Biochemistry Journal, 321: 557–559.

De Stefanis, V.A. and Turner, E.W. 1981. Modified Enzyme System to Inhibit Bread Firming Method for Preparing Same and Use of Same in Bread and Other Bakery Products. Patent US4299848.

Diderichsen, B. and Christiansen, L. 1988. Cloning of a maltogenic alpha-amylase from *Bacillus stearothermophilus*. FEMS Microbiology Letters, 56: 53–60.

Dragsdorf, R.D. and Varriano-Marston, E. 1980. Bread staling: X-ray diffraction studies on bread supplemented with alpha-amylases from different sources. Cereal Chemistry, 57: 310–314.

Fogarty, W.M., BealinKelly, F., Kelly, C.I. and Doyle, E.M. 1991. A novel maltohexaose-forming alpha-amylase from *Bacillus caldovelox*: Patterns and mechanisms of action. Applied Microbiology and Biotechnology, 36: 184–189.

Fogarty, W.M., Bourke, A.C., Kelly, C.T. and Doyle, E.M. 1994. A constitutive maltotetraose producing amylase from *Pseudomonas* sp. IMD 353. Applied Microbiology and Biotechnology, 42: 198–203.

Fogarty, W.M. and Kelly, C.T. 1990. Recent advances in microbial amylases. pp. 71–132. *In*: W.M. Fogarty and C.T. Kelly (eds.). Microbial Enzymes and Biotechnology. Elsevier Science Publishers, London.

Geddes, R. 1986. Glycogen: A metabolic viewpoint. Bioscience Reports, 6: 415–428.

Grewal, N., Faubion, J., Feng, G., Kaufman, R.C., Wilson, J.D. and Shi, Y.C. 2015. Structure of waxy maize starch hydrolyzed by maltogenic α-amylase in relation to its retrogradation. Journal of Agricultural and Food Chemistry, 63: 4196–4201.

Han, R., Li, J., Shin, H.-D., Chend, R.R., Du, G., Liu, L. and Chen, J. 2014. Recent advances in discovery, heterologous expression, and molecular engineering of cyclodextrin glycosyltransferase for versatile applications. Biotechnology Advances, 32: 415–428.

Henrissat, B. and Bairoch, A. 1993. New families in the classification of glycosyl hydrolases based on amino acid sequence similarities. Biochemical Journal, 293: 781–788.

Henrissat, B. and Bairoch, A. 1996. Updating the sequence-based classification of glycosyl hydrolases. Biochemical Journal, 316: 695–696.

Hii, S.L., Tan, J.S., Ling, T. and Ariff, A.B. 2012. Pullulanase: Role in starch hydrolysis and potential industrial applications. Enzyme Research, 2012: 1–14.

Hiromi, K., Nitta, Y., Numata, C. and Ono, S. 1973. Subsite affinities of glucoamylase: examination of the validity of the subsite theory. Biochim et Biophysica Acta, 302: 362–375.

Hyun, H.H. and Zeikus, J.G. 1985. General biochemical characterization of thermostable extracellular β-amylase from *Clostridium thermosulfurogenes*. Applied and Environmental Microbiology, 49: 1162–1167.

Jane, J.-L. 2006. Current understanding on starch granule structures. Journal of Applied Glycoscience, 53: 205–213.

Kainuma, K., Ito, T., Suzuki, S. and Kobayashi, S. 1971. Isolation and action pattern of maltohexaose producing amylase from *Aerobacter-aerogenes*. FEBS Letters, 26: 281–285.

Kaji, A., Sato, M., Koyabashi, M. and Murao, T. 1976. Acid-stable glucoamylase produced in medium containing sucrose by *Corticium-rolfsii*. Journal of the Agricultural Chemical Society of Japan, 50: 509–511.

Kang, G.J., Kim, M.J., Kim, J.W. and Park, K.H. 1997. Immobilization of thermostable maltogenic amylase from *Bacillus stearothermophilus* for continuous production of branched oligosaccharides. Journal of Agricultural and Food Chemistry, 45: 4168–4172.

Kato, M., Hiromi, K. and Morita, Y. 1974. Purification and kinetic studies of wheat bran beta-amylase-evaluation of subsite affinities. Journal of Biochemistry, 75: 563–576.

Kim, I.C., Cha, J.H., Kim, J.R., Jang, S.Y., Seo, B.C., Cheong, T.K., Lee, D.S., Choi, Y.D. and Park, K.H. 1992. Catalytic properties of the cloned amylase from *Bacillus licheniformis*. Journal of Biological Chemistry, 267: 22108–22114.

Kim, I.C., Yoo, S.H., Lee, S.J., Oh, B.H., Kim, J.W. and Park, K.H. 1994. Synthesis of branched oligosaccharides from starch by two amylases cloned from *Bacillus licheniformis*. Bioscience, Biotechnology and Biochemistry, 58: 416–418.

Kim, M.J., Park, W.S., Lee, H.S., Kim, T.J., Shin, J.H., Yoo, S.H., Cheong, T.K., Ryu, S.R., Kim, J.C., Kim, J.W., Moon, T.W., Robyt, J.F. and Park, K.H. 2000. Kinetics and inhibition of cyclomaltodextrinase from alkalophilic *Bacillus* sp. I-5. Archives of Biochemistry and Biophysics, 373: 110–115.

Kim, T.J., Kim, M.J., Kim, B.C., Kim, J.C., Cheong, T.K., Kim, J.W. and Park, K.H. 1999. Modes of action of acarbose hydrolysis and transglycosylation catalyzed by a thermostable maltogenic amylase, the gene for which was cloned from a *Thermus* strain. Applied and Environmental Microbiology, 65: 1644–1651.

King, J. 1967. The glucoamylase of *Coniophora cerebella*. Biochemical Journal, 105: 577–583.

Kumar, P. and Satyanarayana, T. 2008. Potential applications of microbial enzymes in improving quality and shelf-life of bakery products. pp. 132–142. *In*: A. Koutinas, A. Pandey and C. Larroche (eds.). Current Topics on Bioprocesses in Food Industry. Asiatech Publishers, New Delhi, India.

Kumar, P. and Satyanarayana, T. 2009. Microbial glucoamylases: Characteristics and applications. Critical Reviews in Biotechnology, 29: 225–255.

Kuriki, T., Yanase, M., Takata, H., Takesada, Y., Imanaka, T. and Okada, S. 1993. A new way of producing isomaltooligosaccharide syrup by using the transglycosylation reaction of neopullulanase. Applied and Environmental Microbiology, 59: 953–959.

Kwan, H.S., So, K.H., Chan, K.Y. and Cheng, S.C. 1994. Purification and properties of β-amylase from *Bacillus circulans* S31. World Journal of Microbiology and Biotechnology, 10: 597–598.

Kulp, K. and Ponte, J.G. 1981. Staling white pan bread: Fundamental causes. Critical Reviews in Food Science and Nutrition, 15: 1–48.

Laderman, K.A., Davis, B.R., Krutzsch, H.C., Lewis, M.S., Griko, Y.V., Privalow, P.L. and Anfinsen, C.B. 1993. The purification and characterization of an extremely thermostable alpha-amylase from the hyperthermophilic archaebacterium *Pyrococcus furiosus*. Journal of Biological Chemistry, 286: 24394–24401.

Le Corre, D., Bras, J. and Dufresne, A. 2010. Starch nanoparticles: A review. Biomacromolecules, 11: 1139–1153.

Lee, H.S., Auh, J.H., Yoon, H.G., Kim, M.J., Park, J.H., Hong, S.S., Kang, M.H., Kim, T.J., Moon, T.W., Kim, J.W. and Park, K.M. 2002. Cooperative action of alpha-glucanotransferase and maltogenic amylase for an improved process of isomaltooligosaccharides (IMO) production. Journal of Agricultural and Food Chemistry, 50: 2812–2817.

Lee, S.J., Yoo, S.H., Kim, M.J., Kim, J.W., Seok, H.M. and Park, K.H. 1995. Production and characterization of branched oligosaccharides from liquefied starch by the action of *Bacillus licheniformis* amylase. Starch/Stärke, 47: 127–134.

Lin, W. and Lineback, D.R. 1990. Changes in carbohydrate fractions in enzyme-supplemented bread and the potential relationship to staling. Starch/Stärke, 42: 385–394.

MacGregor, E.A., Janecek, S. and Svensson, B. 2001. Relationship of sequence and structure to specifities in the α-amylase family of enzymes. Biochimica et Biophysica Acta, 1546: 1–20.

Morgan, F.J. and Priest, F.G. 1981. Characterization of a thermostable alpha-amylase from *Bacillus licheniformis* ncib-6346. Journal of Applied Bacteriology, 50: 107–114.

Myers, A.M., Morell, M.K., James, M.G. and Ball, S.G. 2000. Recent progress towards understanding biosynthesis of the amylopectin crystal. Plant Physiology, 122: 989–997.

Oates, C.G. 1997. Towards an understanding of starch granule structure and hydrolysis. Trends in Food Science & Technology, 8: 375–382.

Ojha, S., Mishra, S. and Chand, S. 2015. Production of isomalto-oligosaccharides by cell bound α-glucosidase of *Microbacterium* sp. LWT-Food Science and Technology, 60: 486–494.

Olesen, T. 1991. Antistaling Process and Agent. Patent WO9104669.

Pandey, A., Nigam, P., Soccol, C.R., Soccol, V.T., Singh, D. and Mohan, R. 2000. Advances in microbial amylases. Biotechnology and Applied Biochemistry, 31: 135–152.

Pandey, A. 1995. Glucoamylase research: An overview. Starch/Stärke, 42: 439–445.

Park, H.-H., Kim, T.-J., Cheong, T.-K., Kim, J.-W., Oh, B.-H. and Svensson, B. 2000. Structure, specificity and function of cyclomaltodextrinase, a multispecific enzyme of the α-amylase family. Biochimica et Biophysica Acta, 1478: 165–185.

Rana, N., Walia, A. and Gaur, A. 2013. α-Amylases from microbial sources and its potential applications in various industries. National Academy Science Letters, 36: 9–17.

Ray, R.R., Jana, S.C. and Nanda, G. 1997. Production of β-amylase from starchy wastes by a hyperamylolytic mutant of *Bacillus megaterium*. Indian Journal of Experimental Biology, 35: 285–288.

Reddy, P.R.M., Swamy, M.V. and Seenayya, G. 1998. Purification and characterization of thermostable β-amylase and pullulanase from high yielding *Clostridium thermosulfurogenes* SV2. World Journal of Microbiology Biotechnology, 14: 89–94.

Reilly, P.J. 1999. Protein engineering of glucoamylase to improve industrial performance: A review. Starch/Stärke, 51: 269–274.

Reilly, P.J. 2006. Glucoamylase. pp. 727–738. *In*: J.R. Whitaker, A.G.J. Voragen and D.W.S. Wong (eds.). Handbook of Enzymology. Marcel Dekker Inc., New York.

Robyt, J.F. 2009. Enzymes and their action on starch. pp. 237–292. *In*: J. BeMiller and R. Whistler (eds.). Starch, Chemistry and Technology. Academic Press, New York.

Robyt, J.F. and Ackerman, R.J. 1971. Isolation, purification and characterization of a maltotetraose producing amylase from *Pseudomonas stutzeri*. Archives of Biochemistry and Biophysics, 145: 105–114.

Robyt, J.F. and French, D. 1963. Action pattern and specificity of an amylase from *Bacillus subtillis*. Archives of Biochemistry and Biophysics, 100: 451–467.

Robyt, J.F. and French, D. 1967. Multiple attack hypothesis of α-amylase action: Action of porcine pancreatic, human salivary and *Aspergillus oryzae* α-amylases. Archives of Biochemistry and Biophysics, 122: 8–16.

Rojas, J.A., Rosell, C.M. and Benedito De Barber, C. 2001. Role of maltodextrins in the staling of starch gels. European Food Research and Technology, 212: 364–368.

Ryder, D.S. and Power, J. 2006. Miscellaneus ingredients in aid of the process. pp. 333–381. *In*: F.G. Priest and G.G. Stewart (eds.). Handbook of Brewing. CRC Press Taylor & Francis, Boca Raton.

Savel'ev, A.N. and Firsov, L.M. 1982. Carboxyl groups in the active center of glucoamylase from *Aspergillus awamori*. Biokhimiia, 47: 1618–1620.

Sahlstrom, S. and Brathen, E. 1997. Effects of enzyme preparations for baking, mixing time and resting time on bread quality and bread staling. Food Chemistry, 58: 75–80.

Sharma, A. and Satyanarayana, T. 2013. Microbial acid-stable α-amylases: Characteristics, genetic engineering and applications. Process Biochemistry, 48: 201–211.

Shigeji, M., Kimihiko, S. and Yoshiyuki, T. 1999a. Bread Quality Improving Composition and Bread Production Using the Same. Patent JP11266773A2.

Shigeji, M., Kimihiko, S. and Yoshiyuki, T. 1999b. Quality Improving Composition for Bread and Production Thereof Using the Same. Patent JP11178499A2.

Sierks, M.R. and Svensson, B. 1994. Protein engineering of the relative specificity of glucoamylase from *Aspergillus awamori* based on sequence similarities between starch degrading enzymes. Protein Engineering Design and Selection, 7: 1479–1484.

Sivaramakrishnan, S., Gangadharan, D., Madhavan, K., Nampoothiri, M.K., Soccol, C.R. and Pandey, A. 2006. α-Amylases from microbial sources: An overview on recent developments. Food Technology and Biotechnology, 44: 173–184.

Smith, A.M. 1999. Making starch. Current Opinion in Plant Biology, 2: 223–229.

Suganuma, T., Matsuno, R., Ohinishi, M. and Hiromi, K. 1978. A study of the mechanism of Taka-amylase A on linear oligosaccharides by product analysis and computer computer simulation. Journal of Biochemistry, 84: 293–316.

Swamy, M.V., Sai Ram, M. and Seenayya, G. 1994. β-Amylase from *Clostridium thermocellum* SS8: A thermophilic, anaerobic, cellulolytic bacterium. Letters in Applied Microbiology, 18: 301–304.

Synowiecki, J. 2007. The use of starch processing enzymes in the food industry. pp. 19–34. *In*: J. Polaina and A.P. MacCabe (eds.). Industrial Enzymes: Structure, Function and Applications. Springer, Dordrecht.

Takasaki, Y. 1983. An amylase producing maltotriose from *Bacillus subtillis*. Agricultural and Biological Chemistry, 49: 1091–1097.

Tanaka, N., Nakai, K., Takami, K. and Takasaki, Y. 1997. Bread Quality Improving Composition and Bread Producing Process Using the Same. Patent US5698245.

Taniguchi, H., Odashima, F., Igaashi, M., Maruyama, Y. and Nakamura, M. 1982. Characterization of a potato starch-digesting bacterium and its production of amylase. Agricultural and Biological Chemistry, 46: 2107–2115.

Tester, R.F., Karkalas, J. and Qi, X. 2004. Starch—composition, fine structure and architecture. Journal of Cereal Science, 39: 151–165.

Thoma, J.A., Rao, G.V.K., Brothers, C. and Spradlin, J. 1971. Subsite mapping of enzymes. Journal of Biological Chemistry, 246: 5621–5635.

Tomomatsu, H. 1994. Health effects of oligosaccharides. Food Technology, 48: 61–65.

Tsutsumi, N., Bisgard-Frabntzen, H. and Svendsen, A. 1999. Starch Conversion Process Using Thermostable Isoamylases from *Sulfolobus*. Patent WO 99/01545.

van der Maarel, M.J.E.C. 2010. Starch-processing enzymes. pp. 320–331. *In*: R.J. Whitehurst and M. van Oort (eds.). Enzymes in Food Technology. Blackwell Publishing Ltd., Ames, Iowa, USA.

van der Maarel, M.J.E.C., van der Veen, B., Uitdehaag, J.C.M., Leemhuis, H. and Dijkhuizen, L. 2002. Properties and applications of starch-converting enzymes of the α-amylase family. Journal of Biotechnology, 94: 137–155.

van der Maarel, M.J.E.C. and Leemhuis, H. 2013. Starch modification with microbial alpha-glucanotransferase enzymes. Carbohydrate Polymers, 93: 116–121.

Van Eijk, J.H. 1991. Retarding the Firming of Bread Crumb During Storage. Patent US5023094.

Vidal, F.D. and Gerrity, A.B. 1979. Antistaling Agent for Bakery Products. Patent US4160848.

Vihinen, M. and Mantasala, P. 1989. Microbial amylolytic enzymes. Critical Reviews in Biochemistry and Molecular Biology, 24: 329–418.

Wako, K., Hashimoto, S., Kubomura, S., Yolota, K., Aikawa, K. and Kanaeda, J. 1979. Purification and some properties of a maltotriose-producing amylase. Journal of the Japanese Society of Starch Science, 26: 175–181.

Wursch, P. and Gumy, D. 1994. Inhibition of amylopectin retrogradation by partial beta-amylolysis. Carbohydrate Research, 256: 129–137.

Yoo, S.H., Kweon, M.R., Kim, M.J., Auh, J.H., Jung, D.S., Kim, J.R., Yook, C., Kim, J.W. and Park, K.H. 1995. Branched oligosaccharides concentrated by yeast fermentation and effectiveness as a low sweetness humectant. Journal of Food Science, 60: 516–519.

3

Starch-active Debranching and α-Glucanotransferase Enzymes

Mario M. Martínez[1],* and *Manuel Gómez*[2],*

1. Introduction

Starch is an important constituent in the human diet and a raw material constituting a large proportion of the food consumed by the world's population. The reason for this prominence lies on its structure and physicochemical properties, which make the starch to be used as such or physically, chemically or enzymatically processed into a variety of different products. The glycosidic hydrolases and transferases responsible for such enzymatic modifications belong to the α-amylase family of enzymes (Table 1). As it was reported in the previous chapter, a complete overview of all these enzymes known to date, detailed information and the recommendations for their nomenclature can be found on the carbohydrate-active enzymes database (CAZY, http://www.cazy.org) and a website dedicated to enzyme nomenclature (http://www.enzyme-database.org), respectively. These enzymes are widely distributed in plants, animals and microorganisms. Nevertheless, the improvement in the production of microbial enzymes has facilitated fulfillment of the industrial demands and they are preferred to other sources (Burhan et al., 2003). The α-amylase family is formed by a group of enzymes with different specificities, having in common the ability to act on one type of substrate—glucose residues linked through an $\alpha,1$-1, $\alpha,1$-4, or $\alpha,1$-6 glycosidic bond. It can be divided into two subgroups: exo- and endo-acting hydrolases that hydrolyse the $\alpha,1$-4 and/or $\alpha,1$-6 glycosidic linkages consuming water; and glucanotransferases that break an $\alpha,1$-4 glycosidic linkage and form a new $\alpha,1$-4 or $\alpha,1$-6 glycosidic one (van der Maarel and Leemhuis, 2013).

[1] Whistler Center for Carbohydrate Research, Department of Food Science, Purdue University, 745 Agriculture Mall Drive, West Lafayette, IN 47907, USA.

[2] Food Technology Area, College of Agricultural Engineering, University of Valladolid, 34004 Palencia, Spain.

* Corresponding authors: mario.martinez@iaf.uva.es (alt: mariomartinez175@gmail.com); pallares@iaf.uva.es

Table 1. Starch-active enzymes of the α-amylase family with their corresponding EC number. (Letters in bold indicate the debranching and α-glucanotransferase enzymes approached in this chapter.)

Enzyme	EC Number
Glucan branching enzyme	2.4.1.18
Cyclodextrin glycosyltransferase	2.4.1.19
4-α-glucanotransferase (Amylomaltase)	2.4.1.25
Maltopentaose-forming amylase	3.2.1.-
Pullulan hydrolase type III	3.2.1.-
α-Amylase	3.2.1.1
β-Amylase	3.2.1.2
Glucoamylase (Amyloglucosidase)	3.2.1.3
α-Glucosidase	3.2.1.20
Amylo-1,6-glucosidase	3.2.1.33
Pullulanase	3.2.1.41
Cyclomaltodextrinase	3.2.1.54
Isopullulanase	3.2.1.57
Maltotetraose-forming amylase	3.2.1.60
Isoamylase	3.2.1.68
Maltohexaose-forming amylase	3.2.1.98
Maltotriose-forming amylase	3.2.1.116
Maltogenic amylase	3.2.1.133
Neopullulanase	3.2.1.135
Limit dextrinase (R-enzyme)	3.2.1.142

In the previous chapter, microbial exo- and endo-acting hydrolases that hydrolyse mainly the α,1-4 glycosidic bonds were discussed. Their catalytic action has been optimally used in the food industry, and, in particular, in the starch and baking industries. However, due to the industrial requirements and the demand for new products, their presence sometimes needs to be complemented with the catalytic action of starch-active hydrolases acting mainly on α,1-6 glycosidic bonds, also known as debranching enzymes, and enzymes with transglycosylation activity capable of forming new α,1-4 or α,1-6 glycosidic linkages, also named α-glucanotransferases. In this way, debranching and α-glucanotransferase enzymes, listed in bold letters in Table 1, will be approached from the perspective of the food industry.

2. Hydrolases Acting on α,1-6 Linkages (Debranching Enzymes)

The vast majority of starches, which are of industrial importance, contain around 80 per cent amylopectin. When starch is subjected to hydrolysis by α-amylase, the amylopectin fraction is only partly cleaved. The branch points containing α,1-6 glycosidic bonds are resistant to attack and their presence also offers a certain degree of resistance on adjacent α,1-6 linkages. Therefore, the prolonged action of α-amylase on amylopectin results in the formation of α-limit dextrins, which are not susceptible to further hydrolysis by α-amylase (Robyt, 2009). Likewise, when amylopectin is treated by β-amylase, hydrolysis stops when an α,1-6 glycosidic bond is reached, resulting in the formation of β-limit dextrins (Robyt, 2009). On the other hand, glucoamylases are able to hydrolyse branch points, but the reaction proceeds relatively slowly (Kumar and Satyanarayana, 2009). This fact can lead to a reduction in the yield of glucose and maltose. Therefore, the use of debranching enzymes would offer obvious advantages. Debranching enzymes catalyse the hydrolysis of α,1-6 glycosidic bonds in amylopectin and/or glycogen and related polymers. The affinity of debranching enzymes for the α,1-6 bond distinguishes these enzymes from other amylases which have primary affinity for α,1-4 glycosidic linkages.

There is an important confusion when grouping and naming these enzymes. Therefore, a schematic representation listing them is approached in Fig. 1. Firstly, these enzymes are classified in two major groups; indirect enzymes, which need the presence of other enzymes to act, such as amylo-1,6-glucosidase (EC 3.2.1.33), and direct enzymes, which can act without any further support. Direct enzymes can also be grouped according to their substrate affinity in two main groups, pullulanases and isoamylases (EC 3.2.1.68). At the same time, pullulanases can be classified according to the additional inability (Type I) or ability (Type II) to degrade the α,1-4 glycosidic bonds of other polysaccharides (Erra-Pujada et al., 1999). Type I pullulanases comprise

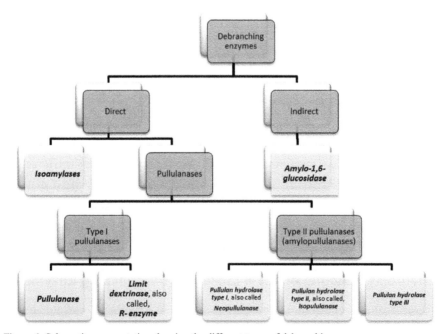

Figure 1. Schematic representation showing the different types of debranching enzymes.

pullulanase (EC 3.2.1.41) and limit dextrinase (also called R-enzyme, EC 3.2.1.142). These two enzymes differ in their action on amylopectin and glycogen and their rate of hydrolysis on limit-dextrins. Pullulanase is able to hydrolyse glycogen to produce complete hydrolysis of amylopectin, whereas limit dextrinase has no action on glycogen or its β-limit dextrins and produces an incomplete hydrolysis of amylopectin (Robyt, 2009). However, limit dextrinase is isolated from plants, and hence will not be discussed further. Meanwhile, Type II pullulanases, also referred to as α-amylase-pullulanases or amylopullulanases, are mainly comprised by pullulan hydrolase Type I, also called neopullulanase (EC 3.2.1.135), pullulan hydrolase Type II, also called isopullulanase (EC 3.2.1.57), and the recently discovered pullulan hydrolase Type III. More detailed information on debranching enzymes is presented as follows in Fig. 1.

2.1 Amylo-1,6-glucosidase

Amylo-1,6-glucosidase (EC 3.2.1.33) is not capable of acting without the presence of 4-α-D-glucanotransferase (also named dextrin glycosyltransferase, EC 2.4.1.25). This amylo-1,6-glucosidase/4-α-D-glucanotransferase complex exists in nature in a mammal's muscle (rabbit muscle) and yeast (*Saccharomyces cerevisiae*), which have been studied in detail and found to consist of two discrete enzymatic activities, both of which are essential for the debranching process (Brown and Illingworth, 1964; Lee and Carter, 1973; Brown et al., 1973; White and Nelson, 1975). Added to that, this complex acts together with phosphorylase to bring about total degradation of glycogen or amylopectin (Norman, 1981). Amylo-1,6-glucosidase only hydrolyses α,1-6 branch points if the side chain consists of a single glucose unit. For further degradation, the dextrin has to be modified by the action of 4-α-D-glucanotransferase enzyme. This glycosyltransferase preferentially moves maltotriose, and to a lesser extent, maltose residues from some donor oligosaccharides and from glycogen to other linear or branched acceptors of suitable structure (Brown and Illingworth, 1964), thus exposing the single α-1-6 linked glucose unit (Norman, 1981) (Fig. 2). Then, the amylo-1,6-

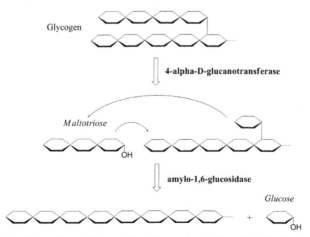

Figure 2. Hydrolytic and transglycosylation activity of the amylo-1,6-glucosidase/4-α-D-glucanotransferase complex on glycogen.

glucosidase hydrolyses the single α,1-6 linkage that becomes exposed at each outer branch point of glycogen or amylopectin, consisting of a single glucose unit (Hii et al., 2012). In both enzymes, the transferase portion of the debranching system is the rate-limiting step. These debranching enzymes are not of industrial importance in the food industry at present, so their properties and function will not be described further in this chapter.

2.2 Isoamylase

Isoamylase (EC 3.2.1.68) can split all the α,1-6 branching points in amylopectin, glycogen and β-limit dextrins, presenting an endo- and exo-acting mechanism of amylopectin and glycogen, respectively. However, the branching points of pullulan, a polysaccharide with a repeating unit of maltotriose, which is α,1-6 linked (Israilides et al., 1999), are cleaved very slowly. Therefore, a catalytic mechanism towards pullulan is usually not attributed to this enzyme. Isoamylase releases linear maltodextrins with retention of the α-configuration.

The name isoamylase was first given by Maruo and Kobayashi (1951) to an enzyme found in yeast cells and able to cleave α,1-6 bonds of intracellular glycogen. The enzymatic activity in these cells was too weak to allow detailed studies on the properties of the enzyme. Later, isoamylases were also isolated from *Cytophaga* sp. (Gunja-Smith et al., 1970; Marshall and Whelan, 1974; Manners and Matheson, 1981), *Flavobacterium* sp. (Sato and Park, 1980) and *Bacillus amyloliquefaciens* (Urlaub and Wober, 1975). However, the most selectively isolated and crystallized isoamylase is the one from bacteria *Pseudomonas amylodermosa* (Harada et al., 1968).

Pseudomonas isoamylase catalyses the hydrolysis of both the inner and outer branch linkages of amylopectin and glycogen, in opposition to pullulanase, which only hydrolyses the outer branch linkages. In addition, *Pseudomonas* isoamylase is able to hydrolyse the branching points of amylopectin and glycogen completely at rates 7 and 124 times higher than pullulanase. Therefore, *Pseudomonas* isoamylase is an effective and efficient debranching enzyme for starch and glycogen. Nevertheless, this enzyme is not capable of removing glucose units from branched dextrins as it acts very slowly on maltose units attached by α,1-6 bonds (Robyt, 2009). In particular, the hydrolysis of the α,1-6 branch bond of short chains occurs at a rate of around 10 per cent of the rate of hydrolysis of the branch linkages of longer chains (Robyt, 2009). This difference in the rate of hydrolysis is even appreciable between the α,1-6 linkages of maltose and maltotriose side chains, being the α,1-6 bonds of maltotriose side chains catalysed much faster than those of maltose (Harada et al., 1972; Kainuma et al., 1978).

2.3 Type I pullulanases (pullulanase)

Pullulanase (EC 3.2.1.41) attracts more interest because of its specific action on α-1,6 linkages in pullulan. Pullulanase is derived from a wide range of microorganisms, such as *Bacillus acidopullulyticus* (Jensen and Norman, 1984), *Klebsiella planticola* (Teague and Brumm, 1992), *Aerobacter aerogenes* (Robyt, 2009; Hii et al., 2012), *Bacillus deramificans* (Uhlig, 1998), thermophilic *Bacillus* sp. (Kunamneni and Singh, 2006), *Bacillus cereus* (Nair et al., 2007), *Geobacillus stearothermophilus* (Zareian et al., 2010), *Bacillus flavocaldarius* (Suzuki et al., 1991), *Bacillus thermoleovorans* (Messaoud et al., 2002), *Clostridium* sp. (Klingeberg et al., 1990), *Thermos caldophilus* (Kim et al., 1996) and *Fervidobacterium pennavorans* (Koch et al., 1997).

There are several differences in the modes of decomposition of pullulan, starch and glycogen by isoamylase and pullulanase, with both enzymes useful for structure elucidation and different applications in the food industry. The major difference is that pullulanase has the ability to hydrolyse the α,1-6 glycosidic bond in pullulan, because of which this enzyme was first used to clarify the structure of pullulan. It also hydrolyses the branch linkages of amylopectin, even though the relative rate of hydrolysis is much lower as compared to that of isoamylase. It also acts on the branch linkages of glycogen, but very slowly (Harada, 1984), since glycogen is randomly and more numerously branched with shorter branches as compared to amylopectin (Geddes, 1986).

2.4 Type II pullulanases (amylopullulanases)

In recent years, a considerable number of Type II pullulanases (also termed amylopullulanases) have been isolated from a wide variety of microorganisms, particularly from thermophilic ones, since scientific interest in this class of enzymes is being motivated by industrial applications (Erra-Pujada et al., 1999). *Aspergillus niger*, *Bacillus* sp., *Pyrococcus furiosus*, *Thermococcus litoralis*, *Pyrococcus woesei*, *Thermococcus celer*, *Thermococcus aggregans*, *Thermococcus guaymagensis* and *Thermococcus hydrothermalis* have been described as amylopullulanase producers (Erra-Pujada et al., 1999; Robyt, 2009; Hii et al., 2012). Thermophilic amylopullulanases have optimum activity between 80°C and 100°C and pH between 5.5 and 6.6 (Robyt, 2009), except for the enzyme produced from *Thermococcus litoralis* that has optimum temperature at 125°C. Moreover, several alkaline pullulanases Type II, which were active at pH ranging from 8.5 to 12.0, have been isolated and characterized from *Bacillus* spp. (Hii et al., 2012). The production of these enzymes appears to be inducible, with malto-oligosaccharides being the principal inducers.

As earlier commented, Type II pullulanases are comprised mainly by three kinds of enzymes, pullulan hydrolase Type I (neopullulanase, EC 3.2.1.135), pullulan hydrolase Type II (isopullulanase, EC 3.2.1.57) and pullulan hydrolase Type III. Neopullulanase and isopullulanase enzymes are only able to cleave α-1,4 glycosidic linkages in pullulan, releasing panose and isopanose, and have practically no action on starch (Aoki and Sakano, 1997). In particular, neopullulanases hydrolyse pullulan to produce panose as the main product with the final molar ratio of panose, maltose, and glucose of 3:1:1 (Imanaka and Kuriki, 1989). It can also perform transglycosylation with the formation of new α,1-4 or α,1-6 glycosidic bonds (Takata et al., 1992). This enzyme is also closely related to cyclomaltodextrinases (CDase, EC 3.2.1.54), based on similarity of their specific reaction towards pullulan (Hii et al., 2012). Added to that, neopullulanase and isopullulanase are also highly active on cyclodextrins (Duffner et al., 2000), like cyclomaltodextrinase. However, the latter degrades cyclodextrins much faster than linear dextrins (Hii et al., 2012). As these enzymes can precisely recognize the structural differences between α,1-4 and α,1-6 glycosidic linkages, they can be applied in structural analysis of oligo- and polysaccharides (Roy et al., 2003). Unlike all pullulan-hydrolysing enzymes as described above, pullulan hydrolase Type III, detected by Niehaus et al. (2000) from *Thermococcus aggregans,* has the ability to attack α,1-6 as well as α,1-4 glycosidic linkages in pullulan, leading to the formation of a mixture of maltotriose, panose, maltose and glucose. The enzyme is also able to degrade starch, amylose and amylopectin, forming maltotriose and maltose as the main products. This enzyme is active at above 100°C. However, hyperthermophilic Type II pullulanases have not yet been used in commercial production due to low yield and limited activity at high starch concentrations (> 30 per cent) (Crabb and Mitchinson, 1997).

2.5 Brief insight of the applications of debranching enzymes

The high affinity of debranching enzymes towards α,1-6 linkages in branched glycans makes them interesting for analytical purposes in research laboratories as well as in the starch-processing industry, for the production of high glucose and maltose syrups.

2.5.1 Application of debranching enzymes for analytical purposes

De Branching enzymes are particularly useful for elucidating the structures of glucans, such as glycogen and starch (Marshall and Whelan, 1974; Manners and Matheson, 1981; Robyt, 2009). Initially, β-amylases were used for this purpose, reaching a complete hydrolysis of amylose into maltose (Stetten and Stetten, 1640). However, the purification of sweet potato β-amylase led to an incomplete conversion of amylose into maltose, indicating the presence of a small amount of α,1-6 branch linkages (Hassid and McCready, 1943; Peat et al., 1949; Thomas et al., 1950). Later, Hizukuri et al. (1981) studied the structures of branched amyloses with pullulanase and with *Pseudomonas* isoamylase, finding that about 30 per cent of the branch linkages were hydrolysed and that the average degree of polymerization (DP) dropped. The incomplete conversion into maltose was attributed to the retrogradation of amylose during the reaction and to the presence of branch points containing only two glucose units (Hizukuri et al., 1981). Takeda et al. (1984) showed distinct characteristics of molecular size, inner-chain lengths and number of chains per molecule of amyloses from different botanical origin using *Pseudomonas* isoamylase. Currently, *Pseudomonas* isoamylase is being used for research in determination of the side chain composition of amylose and amylopectin throughout a debranching prior to measuring the chain length distributions of starch samples by mass spectrometry or anion exchange chromatography among others. However, it is important to highlight that the purity of the enzyme has to be high enough to have accuracy in the assays. Thus, the price of the enzymes for these purposes is an important factor to be considered. In contrast, enzymes for industrial applications do not need a perfect purity and prices are more affordable.

2.5.2 Application of debranching enzymes in the starch-processing industry

Depending on the type of product to be produced, the industrial starch processing is either stopped or followed by a saccharification step, where the liquefied starch is hydrolysed with glucoamylase or β-amylase for the production of glucose syrups or high maltose syrups, respectively. Alongside glucoamylase and β-amylase, the addition of isoamylase or pullulanase is common for improving the process (van der Maarel, 2010).

For the production of glucose syrups, the liquefied starch is combined with glucoamylase, with optimum pH and temperature of ≈ 4.2 and ≈ 60°C. However, due to the fact that this enzyme hydrolyses the α,1-6 glycosidic bonds present in maltodextrins very slowly, accumulation of isomaltose becomes a problem. A solution is to use a debranching enzyme that efficiently hydrolyses α,1-6 glycosidic bonds. Both *Pseudomonas* isoamylase and *Klebsiella* and *Bacillus* pullulanase have been studied for this purpose (Norman, 1982a,b). The use of debranching enzymes increases the level of glucose and a shorter incubation time with a lower amount of glucoamylase was needed. In addition, these authors reported that only *Bacillus* pullulanase could act at 60°C—the optimum temperature for glucoamylase—even though its optimum pH was 4.5–5.0, slightly higher than that of glucoamylase. In contrast, *Pseudomonas* isoamylase had a lower optimum pH, which was in accordance with the pH of glucoamylase. Nevertheless, pullulanase is considered to have greater temperature stability and a good glucoamylase compatible pH range (Reilly, 2006). Currently, no thermostable isoamylases have been described (van der Maarel, 2010). Moreover, the most established advantage of using pullulanase instead of isoamylase during starch saccharification process is that the time of addition of pullulanase is not as critical as the time for isoamylase addition. When isoamylase is used before glucoamylase, the amylopectin fraction is depolymerised too rapidly and therefore, is highly susceptible to retrogradation (Guzman-Maldonado and Paredes-Lopez, 1995), with the consequent creation of enzymatically-resistant structures and lower yield.

In the production of high maltose syrups (50–60 per cent yield) throughout β-amylase treatment, hydrolysis stops as a α,1-6 glycosidic bond is reached, resulting in the formation of β-limit dextrins. Adding a debranching enzyme, the branching point of amylopectin is cleaved and the yield of maltose greatly increases. However, among the debranching enzymes, several authors have postulated reasons for not using isoamylase for saccharification in high maltose syrup-making: (i) pullulanase has greater temperature stability and more β-amylase compatible pH range (Reeve, 1992); (ii) isoamylase is not capable to hydrolyse maltosyl and maltotriosyl side chains in α- and β-limit. Thus, simultaneous action of β-amylase and isoamylase cannot quantitatively convert amylopectin to maltose (Fogarty and Kelly, 1990); (iii) β-amylases are also found to be potential inhibitors in the isoamylase activity (Uhlig, 1998); (iv) the presence of maltotriose or maltotetraose competitively inhibits isoamylase action (Teague and Brumm, 1992). Nevertheless, the bacterial pullulanases from either *Klebsiella planticola* or *Bacillus acidopullulyticus* pullulanases are able to hydrolyse amylopectin and its β-limit dextrins and attack the partially degraded polymer.

Within the starch-processing industry, it is also noteworthy that pullulanase has also been used to prepare resistant starch (RS), which has nutritional benefits and a large market demand (Vorwerg et al., 2002). Resistant starch encompasses those forms of starch, which are not accessible to human digestive enzymes and are fermented in the colon, producing short chain fatty acids. Because of its unique physicochemical and functional properties, resistant starch finds use in a wide range of food products.

It has recently gained substantial importance due to its positive impact on health. The health benefits include improvement of colonic health and microflora, management of diabetes, lower glycemic index and blood cholesterol levels, reduced gall stone formation, increased mineral absorption and potential to modify fat oxidation (Ashwar et al., 2015). Type III resistant starch (RSIII, retrograded starch) can be prepared from high amylose starch by gelatinization followed by treatment of slurry with debranching enzymes, such as pullulanase. Then, the starch is cooled to produce the retrogradation of amylose chains (Ashwar et al., 2015).

3. α-Glucanotransferase Enzymes

Glucoamylase, α-glucosidase, maltogenic α-amylase and pullulan hydrolase Type I (neopullulanase) can perform a transglycosylation reaction forming new α,1-4 and α,1-6 glycosidic bonds. This feature is also exploited by the starch industry to produce branched oligosaccharides, also known as isomaltooligosaccharides (IMO) from maltose or glucose syrups. However, over the last 10 years, the demand for new products with health benefits, especially those related to a controlled release of glucose from the starch polymers to the bloodstream, has increased. This fact has made α-glucanotransferases gain prominence, yielding new commercial products, such as cycloamylose, cyclic cluster dextrin, cyclodextrins, thermoreversible starch, resistant starch and highly branched structures (van der Maarel and Leemhuis, 2013).

α-Glucanotransferases act on substrates, such as amylose, amylopectin, maltodextrins and glycogen. These enzymes cleave an α,1-4 glycosidic bond of the donor molecule and transfer part of the donor to a glycosidic acceptor with the formation of a new α,1-4 (amylomaltase and cyclodextrin glycosyltransferase) or α,1-6 (branching enzyme) glycosidic bond, retaining the α-configuration of the anomeric center. Since α-glucanotransferases use polymeric substrates, the hydroxyl acceptor can also be located downstream on the glucose-enzyme intermediate leading to an intra-molecular transglycosylation, giving to cyclic products (Robyt, 2009; van der Maarel, 2010; van der Maarel and Leemhuis, 2013). α-Glucanotransferases mainly comprise three types of enzymes; 4-α-glucanotransferase, also known as amylomaltase (EC 2.4.1.25), cyclodextrin glucanotransferase (CGTase, EC 2.4.1.19) and branching enzyme (BE, EC 2.4.1.18).

3.1 Cyclodextrin glucanotransferase (CGTase)

3.1.1 Catalytic action of cyclodextrin glucanotransferase (GGTase)

Cyclodextrin glucanotransferase, cyclomaltodextrin glucanotransferase or cyclomaltodextrin glycosyltransferase (CGTase, EC 2.4.1.19) is an extracellular enzyme that catalyses the cleavage of starch into cyclic, non-reducing, α,1-4 linked cyclomaltodextrins, also called cyclodextrins (Robyt, 2009). A number of reviews on the production, properties and use of cyclodextrins have been published over the last years (Qi and Zimmermann, 2005; van der Maarel and Leemhuis, 2013; Li et al., 2014; Lopez-Nicolas et al., 2014). CGTases are mainly obtained from bacteria. The first studied CGTase was that from *Bacillus macerans*, called initially *macerans* amylase (French, 1975), which later has been highly purified and crystallized (Kobayashi et al., 1978; Stavn and Granum, 1979; Robyt, 2009). This enzyme catalyses an intramolecular transfer of α,1-4 glycosidic bonds to form cyclic maltodextrins. Its action was ascribed to be exo and capable of binding six, seven or eight glucose units at the non-reducing end of a starch polymer chain, producing cyclic dextrins containing initially six (α-cyclodextrin), seven (β-cyclodextrin) and eight (γ-cyclodextrin) glucose units. The proposed catalytic mechanism has been completely elucidated by Nakamura et al. (1994). Firstly, the non-reducing end of a starch-derived chain enters into the active-site pocket. After binding the oligosaccharide chain, the enzyme cleaves the α,1-4 glycosidic bond and transfers the chain to the C_4 oxygen of the non-reducing end glucose unit to form a new α,1-4 glycosidic bond, and therefore producing the cyclic maltodextrin. Then, the cyclodextrin leaves the active site, allowing the enzyme to repeat the step again (Robyt, 2009) (Fig. 3). The formation of higher cyclodextrins from 9 to 13 glucose units, called δ-, ε-, ζ-, η- and θ-cyclodextrins, has also been reported (Pulley and French, 1961; French et al., 1965; Endo et al., 1997a,b; Terada et al., 1997). Among them, like α-, β-, and γ-cyclodextrins, δ-cyclodextrins turned out to be unbranched. However, the branching degree increases with the molecular size, reaching almost a 100 per cent of branching degree in the case of η- and θ-cyclodextrins. Robyt (1991) and Yoon and Robyt (2002) even reported the formation of larger cyclodextrins with a DP values of 6 to 60. However, as the reaction goes along, these cyclodextrins are converted into smaller cyclodextrins (α- and β-cyclodextrins).

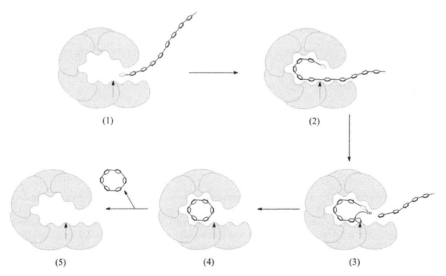

Figure 3. Mechanism for the formation of α-cyclodextrin from starch by cyclodextrin glycotransferase (CGTase) (Robyt, 2009). (1) The non-reducing end (grey hexagon) of a starch-derived chain enters into the active-site pocket; (2) the glucose unites bind to the active-sites and the α,1-4 glycosidic bond is cleaved; (3) the enzyme transfers the chain to the C_4 oxygen of the non-reducing end glucose unit to form a new α,1-4 glycosidic bond; (4) therefore the cyclic maltodextrin is produced; (5) then, the cyclodextrin leaves the active site, allowing the enzyme to repeat the step again.

Alongside the cyclization reaction, CGTases also catalyse acceptor or transfer reactions (glycosylation) between cyclodextrins and a carbohydrate acceptor, such as glucose, isomaltose or panose (Robyt, 2009). In the acceptor reaction (Fig. 4a), the enzyme opens the cyclic ring of the cyclodextrin and forms an enzyme-maltodextrin complex. Then, an acceptor reacts with enzyme-maltodextrin complex, displacing the enzyme and forming a α,1-4 glycosidic linkage while giving an acceptor-terminated maltodextrin. Later, the acceptor product can undergo further transfer reactions, also known as disproportion reactions (Fig. 4b). The linear maltodextrin is hydrolysed by the enzyme to give smaller linear maltodextrins, which form a new covalent maltodextrin-enzyme complex. Then, another maltodextrin partially enters the active-site pocket, forming a new α,1-4 glycosidic linkage and therefore producing a new maltodextrin with the acceptor exclusively located at the reducing end of the maltodextrin. Each of the products of the above disproportionation can undergo further reactions, giving a wide range of maltodextrins with different chain lengths, depending on the amounts and kind of maltodextrins and the molar ratios of the acceptor and the cyclodextrins (Yoon and Robyt, 2002).

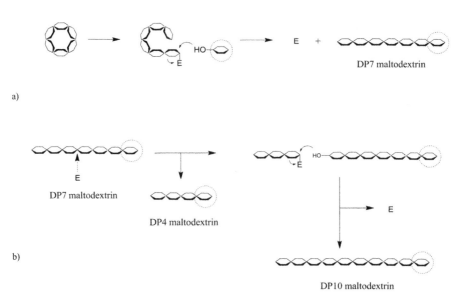

Figure 4. Mechanism for the acceptor and disproportionation reactions by cyclodextrin glycotransferase (CGTase) (Robyt, 2009). (a) Acceptor reaction; the enzyme opens the cyclic ring of the cyclodextrin and forms an enzyme-maltodextrin complex. Then an acceptor (glucose) reacts with enzyme-maltodextrin complex forming a α,1-4 glycosidic linkage. (b) Disproportionation reaction; a DP7 maltodextrin is hydrolysed by the enzyme to give smaller linear maltodextrins, which are susceptible to form a new covalent maltodextrin-enzyme complex. Then, another maltodextrin partially enters the active-site pocket forming a new α,1-4 glycosidic linkage and therefore giving rise to a new maltodextrin with the acceptor exclusively located at the reducing end of the maltodextrin. Acceptors are depicted in grey colour.

Bacillus macerans, Bacillus megaterium, Bacillus circulans, Bacillus stearothermophilus, Klebsiella pneumonia and *Brevibacterium* sp. have also been reported to produce CGTase (Kitahata et al., 1974; Bender, 1977; Biwer et al., 2002;

Kelly et al., 2009; Robyt, 2009). Some of these enzymes differ in the optimum pH and temperature of action as well as in the relative amounts of each cyclodextrins that are able to produce. Most of the CGTases have optimum pH around 5.0–5.5, except for that from *Brevibactenum* sp., with an alkaline optimum pH of 10.0. Regarding the temperature, most of the CGTases have an optimum temperature of 50–55°C. With regard to the cyclodextrins produced, *Bacillus macerans, Klebsiella pneumoniae* and *Bacillus stearothermophilus* generate α-cyclodextrin as the main product, with lesser amounts of β- and γ-cyclodextrins. However, β-cyclodextrin as the main product is produced by *Bacillus megaterium* and *Bacillus circulans,* whereas the enzyme produced by *Brevibacterium* sp. primarily forms γ-cyclodextrin as the main product. In addition, other thermostable CGTases have been found in *Bacillus stearothermophilus, Thermoanaerobacterium thermosulfurigenes, Thermoanaerobacter* sp. and *Thermococcus kodakaraensis*, which produce α- and β-cyclodextrins as main products (Li et al., 2014).

3.1.2 Applications of cyclodextrin glucanotransferase (CGTase)

For the industrial production of cyclodextrins, starch is first liquefied by a heat-stable α-amylase and then the cyclization and disproportionation occurs through the CGTase treatment (Leemhuis et al., 2010). A major drawback in the production of cyclodextrins on a large scale is that all the enzymes that are used today produce a mixture of cyclodextrins. Nevertheless, sometimes specific cyclodextrins, such as α-cyclodextrins, are required for complexation of guest molecules of specific sizes or for their particular nutritional properties. For example, α-cyclodextrins is marketed as a non-digestible, fully fermentable dietary fiber and is used in carbonated and non-carbonated transparent soft drinks, dairy products, baked goods and cereals (van der Maarel, 2010).

The glucose residues of cyclodextrin rings are arranged in such a way that the inside is hydrophobic, resulting in an apolar cavity, while the outside is hydrophilic. This enables cyclodextrins to form inclusion complexes with a variety of hydrophobic guest molecules. This has attracted considerable attention and led to the development of various applications of cyclodextrins. The formation of inclusion complexes leads to changes in the chemical and physical properties of the guest molecules, such as stabilization of light- or oxygen-sensitive compounds, stabilization of volatile compounds, improvement of solubility, extraction of cholesterol from foods, improvement of smell or taste, or modification of liquid compounds to powders (Szente and Szejtli, 2004). In addition, the capacity to catalyse glycosylation reactions is being exploited in the production of the intense sweetener called stevioside, which is isolated from the leaves of the plant *Stevia rebaudania*. CGTase treatment contributes to increase in the solubility and decrease in the bitterness of stevioside (Pedersen et al., 1995).

CGTase have also been implicated in prevention of staling of bakery products by cleaving maltose units from the non-reducing end of the amylopectin side-chains, thus slowing down their retrogradation (Beier et al., 2000). CGTases have also been reported to increase the volume of rice breads (Gujral et al., 2003) and cakes (Shim et al., 2007). In the starch industry, the combination of fungal α-amylase and the thermostable CGTase from the bacterium *Thermoanaerobacterium thermosulfirigenes* has been proposed for the production of maltose syrups (Wind et al., 1998).

3.2. 4-α-Glucanotransferase (amylomaltase)

3.2.1 Catalytic action of 4-α-glucanotransferase (amylomaltase)

4-α-Glucanotransferase (amylomaltase) transfer fragments of amylose or parts of amylopectin side chains on to the non-reducing ends of amylopectin with the formation of new α,1-4 glycosidic linkages by breaking an α-1,4 bond between two glucose units to subsequently make a novel α-1,4 bond (Fig. 5). Amylomaltase is very similar to cyclodextrin glycosyltransferase with respect to the enzymatic reaction. The major difference is that amylomaltase performs a transglycosylation reaction resulting in a linear product, while cyclodextrin glycosyltransferase gives a cyclic product. Amylomaltases have been found in the Eukarya as well as in the bacteria and Archaea, in which they are involved in the utilization of maltose or the degradation of glycogen (Takaha and Smith, 1999). In nature, amylomaltase participates in the intracellular metabolism of glycogen in bacteria and in the formation of amylopectin during starch biosynthesis in plants (Kaper et al., 2004). Nevertheless, amylomaltase is mainly isolated from the bacteria *Thermus thermophilus* (Euverink and Binnema, 1998; van der Maarel et al., 2005) and *Pyrobaculum aerophilum* (Kaper et al., 2005). In addition, the enzyme is also produced via heterologous overexpression in *Bacillus amyloliquefaciens*, but the enzyme remains inside the cell, making the downstream processing more complicated (van der Maarel, 2010).

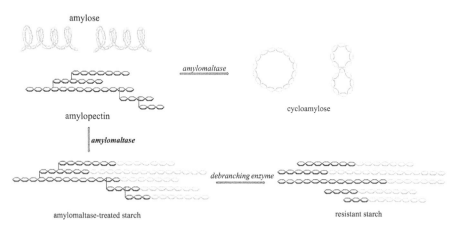

Figure 5. Transfer of fragments of amylose or amylopectin side chains onto the non-reducing ends of amylopectin by 4-α-glucanotransferase (amylomaltase) producing amylomaltase-treated starch and cycloamylose. Treatment of amylomaltase-treated starch with a debranching enzyme giving rise to a resistant starch like structure.

3.2.2 Applications of α-glucanotransferase (amylomaltase)

Amylomaltase is mainly used to convert starch into a thermoreversible gelling derivative. The starches that can be used for modification should contain amylose, therefore excluding waxy varieties. The most common starches employed are those from potato, maize, wheat, rice and tapioca. During the process, gelatinized starch is incubated with a relatively low amount of enzyme and during relatively long incubation periods. The average molecular weight, reducing power and branching percentage remain unchanged from the starting material. It is believed that a mutual rearrangement between the starch molecules occurs without an increase in oxidation-sensitive places or parts having reducing activity. The resulting product is free of amylose and possesses amylopectin with longer side chains (van der Maarel et al., 2005). Despite the existence of short side chains, the presence of long side chains, with a length of 35 glucose residues and longer, in combination with its relatively high molecular weight, make these starches form a white opaque network or gel. However, in contrast to regular starch, this gel is thermoreversible, that is, the amylomaltase-treated starch can undergo cycles of heating and cooling in which it dissolves upon heating and gels upon cooling, in opposition to normal starches, which cannot be dissolved in water after being retrograded. The gel strength and paste flow behavior of the amylomaltase-treated starches are related to the incubation conditions and enzyme to starch ratio (Hansen et al., 2008; Lee et al., 2008), whereas the melting temperature is independent of the reaction conditions and enzyme dosage. Amylomaltase-treated starch can be used in gum confectionary (Buwalda and Tomasoa, 2009) and as a fat replacer in mayonnaises (Mun et al., 2009) and low fat products, fostering a creamy texture (Buwalda and Sein, 2008; Alting et al., 2009; Sein et al., 2009). Thermoreversible properties of these starches are also found in gelatin, except for amylomaltase-treated starch which forms a turbid gel whereas gelatin gels are transparent (Euverink and Binnema, 1998; van der Maarel et al., 2005).

Once created, the amylomaltase-treated starch can be further treated with a debranching enzyme, such as pullulanase, leading to the formation of a resistant starch like product (Fig. 5) (Norman et al., 2007; Richmond et al., 2008). The debranched amylomaltase-treated starches consist of linear malto-oligosaccharides of such DP that they form crystallites by retrogradation. These crystallites are inaccessible to salivary and pancreatic α-amylases, lowering the glycaemic and insulin responses as compared to glucose and soluble fibers, as Kendall et al. (2008) reported in healthy volunteers.

Amylomaltase has also been used for the production of cycloamylose (Terada et al., 1999). Cycloamylose, showed in Fig. 5, is a large cyclic molecule consisting of 16 or more α,1-4 linked glucose residues with a hydrophobic interior in which hydrophobic molecules can be included, such as certain drugs (Tomono et al., 2002). Unlike cyclodextrins, cycloamylose can be formed by all the transglycosylating enzymes of the α-amylase family of enzymes (Terada et al., 1997, 1999; Takaha and Smith, 1999; Takata et al., 2003). The reason is that this reaction is not based on a novel catalytic mechanism but is a direct effect of the limited availability of acceptor molecules. Thus, to form cycloamylose, low concentrations of amylose are incubated with a relatively high amount of enzyme. Besides the potential use of cycloamyloses as a chaperone to prevent proteins from misfolding, no commercial application is known so far (van der Maarel, 2010).

3.3 Branching enzymes (BEs)

3.3.1 Catalytic action of branching enzymes (BEs)

Glycan branching enzymes (BE, EC 2.4.1.18) are involved in the biosynthesis of amylopectin in plants or glycogen (Zeeman et al., 2010). These enzymes initially break α,1-4 glycosidic bonds and create new α,1-6 branches within linear α,1-4 segments, resulting in products devoid of long α,1-4 chains. BEs are widely distributed in plant and animal tissues as well as in microorganisms. Glycogen branching enzyme (GBE) is responsible for the formation of the α,1-6 linkages in the glycogen molecule, whereas starch branching enzyme (SBE) is an analogue of glycogen branching enzyme in plants that introduces α-1,6-branches into amylose and amylopectin (Kim et al., 2008). However, there are major differences in the actions of GBE and SBE as manifested in the differences in the degree of branching between glycogen and amylopectin. The degree of branching is 8–9 per cent in glycogen and 3.5 per cent in amylopectin and the average chain-length of the branches is usually 10–12 glucose residues for glycogen and 20–23 glucose residues for amylopectin (Kim et al., 2008; van der Maarel, 2010). These differences are thought to be primarily due to different specificities between GBE and SBE in terms of the size of transferred chains. Added to that, some SBEs have been reported to act only on amylose (Palomo et al., 2011), whereas others act on both amylopectin and amylose (Palomo et al., 2009).

BE are obtained from microorganisms and in particular from bacteria. One of the most studied BEs is the one from *Bacillus stearothermophilus*, with an optimum temperature of 55°C (Takata et al., 1994). Three more enzymes with similar thermal stability have been studied from *Streptococcus mutans* (Kim et al., 2008), *Deinococcus geothermalis* and *Deinococcus radiodurans* (Palomo et al., 2009). However, food industries interested in these enzymes are demanding higher stability. In this way other microorganisms to produce these enzymes have also been studied. Shinohara et al. (2001) reported the first real BE from *Rhodothermus obamensis*, optimally active at 65°C and stable up to 80°C—a temperature at which the processing of gelatinized starch is feasible. Later, Gruyer et al. (2002) reported evidence for the presence of a branching enzyme active at 80°C in the hyperthermophilic archaeon *Thermococcus hydrothermalis*. More recently, Takata et al. (2003) overexpressed the *Aquifex aeolicus* glucan branching enzyme in *E. coli*, reporting that this enzyme is active at 80°C. This enzyme alongside the one from *Rhodothermus obamensis*, is the most commercially exploited BE so far.

3.3.2 Applications of branching enzymes (BEs)

BE are finding commercial applications mainly in the starch-processing industry as well as in the baking industry. The demand for new starchy products is prodding manufacturers to consider the use of BE in the production of cluster dextrin (CCD), particularly low viscosity dextrins which in turn do not retrograde and in the production of slowly digestible carbohydrates (Fig. 6).

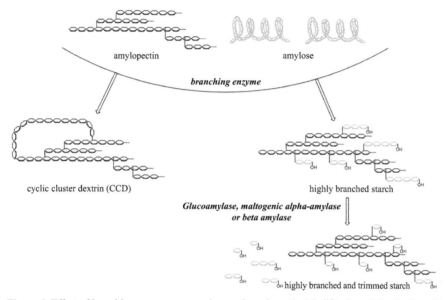

Figure 6. Effect of branching enzymes on amylose and amylopectin. Modification of slowly digestible starch and highly branched starch through glucoamylase, maltogenic α-amylase or β-amylase treatment in the production of highly branched and trimmed starch.

CCD is a branched cyclic dextrin used as a sport drink ingredient (van der Maarel, 2010). There are indications that fluids containing CCD influence the gastric emptying time positively in humans (Takii et al., 2005) and have helpful effects on the endurance of mice (Shinohara et al., 2001). Another application of BE is in the production of dextrins which at high concentrations have a low viscosity and do not retrograde. The BE treatment of liquefied starch leads to the transfer of parts of the amylose and long side chains of the amylopectin to other amylopectin molecules with the formation of new short side chains. This new branched structure leads to weakening of the interactions among amylopectin molecules, leading to solutions with relatively low viscosity (van der Maarel, 2010). Related to the creation of these branched structures, the action of BEs to convert regular starch, rapidly degraded by our digestive system into slowly digestible starch, is also being investigated, though it remains to be deeply explored. It is expected that by increasing the branching points, the pancreatic human α-amylase find it difficult to catalyse the starch hydrolysis. This led to a reduction in the amount of glucose produced and therefore a lower blood glucose level. In this way, van der Maarel et al. (2007) patented an application of BE from bacterium *Deinococcus radiodurans* in which amylopectin was converted into a branched maltodextrin with a side chain distribution dominated by small side chains. *In vitro* digestion analysis indicated that the product showed a slower glucose release. In the same framework, Fuertes et al. (2005), Dermaux et al. (2007) and Le et al. (2009) patented a slow digestible starch through BE modification followed by a β-amylase, glucoamylase or maltogenic α-amylase treatment, respectively. These enzymes have a limited hydrolytic capacity towards α,1-6 glycosidic linkages but trim the side-chains by cleaving the α,1-4 glycosidic linkages from the non-reducing ends. It was unanimously agreed that the shorter side chains make the product less prone to pancreatic α-amylase digestion (Fig. 6).

With regard to the baking industry, Okada et al. (1984) and Spendler and Jorgensen (1997) patented the potential use of branching enzyme in bread as an anti-staling agent, thus increasing the shelf-life and loaf volume of baked goods. These effects were achieved by modifying the starch material in the dough during baking. As already pointed out, retrogradation of the amylopectin fraction in bread limits its shelf-life (Goesaert et al., 2009). However, BEs create products that form a stable solution which cannot retrograde since the side chains are too short to have a strong interaction. Improved quality of baked products was obtained when the branching enzyme was used in combination with other enzymes, such as α-amylase, maltogenic amylase, cyclodextrin glycosyltransferase, β-amylase, cellulase, oxidase and/or lipase (Spendler and Jorgensen, 1997).

4. Conclusions and Future Trends

Improvement in the production and purification of microbial enzymes belonging to the α-amylase family of enzymes is being extensively exploited by the food industry. Among these enzymes, endo and exo-acting hydrolases capable of hydrolasing α,1-4 glycosidic bonds have been the most used and studied. Nevertheless, sometimes the continuous development for new products and processes needs to resort to the use of enzymes with different catalytic activity. Thus, microbial debranching enzymes, which catalyze mainly α,1-6 glycosidic bonds and microbial α-glucanotransferases, which possess transglycosylation activity forming new α,1-4 or α,1-6 glycosidic linkages, can turn out to be interesting. Moreover, the future prospects of these enzymes must be understood taking into account their catalytic action to accomplish the new demands requested by the industry (yield and process improvement) as well as by the consumers (functionality and health). From the industrial approach, hydrolysis of branching points by debranching enzymes is a key factor in the glucose and maltose production. Additionally, the action of α-glucanotransferases, such as CGTase, amylomaltase and branching enzyme, can be used to decrease the viscosity of starch dispersions, sometimes necessary during manufacturing. Meanwhile, consumers are demanding products with new functionalities. Therefore, debranching enzymes with different specificities, such as thermostables amilopullulanases, or α-glucanotransferases, with different transglycosilation activity, are gaining prominence. To conclude, both debranching enzymes and α-glucanotransferases have been studied with a view to producing healthier products. However, further studies should be carried out to better understand their action and the products obtained from their catalysis. In any case, the future use and production of these enzymes will be conditioned by (i) the feasibility of an economic production with low residual enzymatic activity and (ii) the suitability of their optimum values of pH and temperature to accomplish the emerging manufacturing criteria.

References

Alting, A.C., van de Velde, F., Kanning, M.W., Burgering, M., Mulleners, L., Sein, A. and Buwalda, P. 2009. Improved creaminess of low-fat yoghurt: The impact of amylomaltase-treated starch domains. Food Hydrocolloids, 23: 980–987.

Aoki, H. and Sakano, Y. 1997. A classification of dextran-hydrolysing enzymes based on amino-acid-sequence similarities. Biochemical Journal, 323: 859–861.

Ashwar, B.A., Gani, A., Shah, A., Wani, I.A. and Masoodi, A.F. 2015. Preparation, health benefits and applications of resistant starch—A review. Starch-Starke. DOI: 10.1002/star.201500064.

Beier, L., Svendsen, A., Andersen, C., Frandsen, T.P., Borchert, T.V. and Cherry, J.R. 2000. Conversion of the maltogenic α-amylase novamyl into a CGTase. Protein Engineering, 13: 509–513.

Bender, H. 1977. Cyclodextrin-glucanotransferase von *Klebsiella pneumonia*. Archives of Microbiology, 111: 271–282.

Biwer, A., Antranikian, G. and Heinzle, E. 2002. Enzymatic production of cyclodextrins. Applied Microbiology and Biotechnology, 59: 607–617.

Brown, D.H. and Illingworth, B. 1964. The role of oligo 1,4-1,4 glucanotransferase and amylo-1,6 glucosidase in the debranching of glycogen. pp. 139–150. *In*: W.J. Whelan and M.P. Cameron (eds.). Ciba Found Symposium Control of Glycogen Metabolism. J. and A. Churchill Ltd., London.

Brown, D.H., Gordon, R.B. and Brown, B.A. 1973. Studies on the structure and mechanism of action of the glycogen debranching enzymes of muscle and liver. Annals of the New York Academy of Science, 210: 238–253.

Burhan, A., Nisa, U., Gokhan, C., Omer, C., Ashabil, A. and Osman, G. 2003. Enzymatic properties of a novel thermostable, thermophilic, alkaline and chelator resistant amylase from an alkaliphilic *Bacillus* sp. isolate ANT-6. Process Biochemistry, 38: 1397–1403.

Buwalda, P.L. and Sein, A. 2008. Cream Substitute. Patent WO2008071744.

Buwalda, P.L. and Tomasoa, D.T.B. 2009. Gum Confections. Patent WO2009080838.

Crabb, W.D. and Mitchinson, C. 1997. Enzymes involved in the processing of starch to sugars. Trends in Biotechnology, 15: 349–352.

Dermaux, L., Peptitjean, C. and Wills, D. 2007. Soluble, Highly Branched Glucose Polymers for Enteral and Parenteral Nutrition and for Peritoneal Dialysis. Patent WO2007099212.

Duffner, F., Bertoldo, C., Andersen, J.T., Wagner, K. and Antranikian, G. 2000. A new thermoactive pullulanase from *Desulfurococcus mucosus*: Cloning, sequencing, purification, and characterization of the recombinant enzyme after expression in *Bacillus subtilis*. Journal of Bacteriology, 182: 6331–6338.

Endo, T., Nagase, H., Ueda, H., Shigihara, A., Kobayashi, S. and Nagai, T. 1997a. Isolation, purification and characterization of cyclomaltodecaose (epsilon-cyclodextrin), cyclomaltoundecaose (zetacyclodextrin) and cyclomaltroridecaose (theta-cyclodextrin). Chemical & Pharmaceutical Bulletin, 45: 532–536.

Endo, T., Nagase, H., Ueda, H., Shigihara, A., Kobayashi, S. and Nagai, T. 1997b. Isolation, purification, and characterization of cyclomaltotetradecaose (iota-cyclodextrin), cyclomaltopentadecaose (kappa-cyclodextrin), cyclomaltohexadecaose (lambda-cyclodextrin), and cyclomaltoheptadecaose (mu-cyclodextrin). Chemical & Pharmaceutical Bulletin, 45: 1856–1859.

Erra-Pujada, M., Debeire, P., Duchiron, F. and O'Donohue, M.J. 1999. The type II pullulanase of *Thermococcus hydrothermalis*: Molecular characterization of the gene and expression of the catalytic domain. Journal of Bacteriology, 181: 3284–3287.

Euverink, G.J.W. and Binnema, D.J. 1998. Use of Modified Starch as an Agent for Forming a Thermoreversible Gel. Patent WO9815347.

Fogarty, W.M. and Kelly, C.T. 1990. Recent advances in microbial amylases. pp. 71–132. *In*: W.M. Fogarty and C.T. Kelly (eds.). Microbial Enzymes and Biotechnology. Elsevier Science Publishers, London.

French, D. 1975. The Schardinger dextrins. Advances in Carbohydrate Chemistry, 12: 189–260.

French, D., Pulley, A.O., Effenberger, J.A., Rougvie, M.A. and Abdullah, M. 1965. Studies on the Schardinger dextrins: XII. The molecular size and structure of the δ-, ε-, ζ-, and η-dextrins. Archives of Biochemistry and Biophysics, 111: 153–160.

Fuertes, P., Roturier, J.M. and Petitjean, C. 2005. Highly Branched Glucose Polymers. Patent EP1548033.

Geddes, R. 1986. Glycogen: A metabolic viewpoint. Bioscience Reports, 6: 415–428.

Goesaert, H., Slade, L., Levine, H. and Delcour, J.A. 2009. Amylases and bread firming: An integrated view. Journal of Cereal Science, 50: 345–352.

Gruyer, S., Legin, E., Bliard, C., Ball, S. and Duchiron, F. 2002. The endopolysaccharide metabolism of the hyperthermophilic archaeon *Thermococcus hydrothermalis*: Polymer structure and biosynthesis. Current Microbiology, 44: 206–211.

Gujral, H.S., Guardiola, I., Carbonell, J.V. and Rosell, C.M. 2003. Effect of cyclodextrinase on dough rheology and bread quality from rice flour. Journal of Agricultural and Food Chemistry, 51: 3814–3818.

Gunja-Smitz, Z., Marshall, J.J., Smith, E.E. and Whelan, W.J. 1970. Glycogen-debranching enzyme from *Cytophaga*. FEBS Letters, 12: 96–100.

Guzman-Maldonado, H. and Paredes-Lopez, O. 1995. Amylolytic enzymes and products derived from starch: A review. Critical Reviews in Food Science and Nutrition, 35: 373–403.

Hansen, M.R., Blennow, A., Pedersen, S., Norgaard, L. and Engelsen, S.B. 2008. Gel texture and chain structure of amylomaltase-modified starches compared to gelatin. Food Hydrocolloids, 22: 1551–1566.

Harada, T., Misaki, A., Akai, H., Yokobayashi, K. and Sugimoto, K. 1972. Characterization of *Pseudomonas* isoamylase by its actions on amylopectin and glycogen: Comparison with *Aerobacter* pullulanase. Biochimica et Biophysica Acta, 268: 497–505.

Harada, T. 1984. Isoamylase and its industrial significance in the production of sugars from starch. Biotechnology and Genetic Engineering Reviews, 1: 39–63.

Harada, T., Yokobayashi, K. and Misaki, A. 1968. Formation of isoamylase by *Pseudomonas*. Applied Microbiology, 10: 1939–1944.

Hassid, W.Z. and McCready, R.M. 1943. The molecular constitution of amylose and amylopectin of potato starch. Journal of the American Chemical Society, 65: 1157–1165.

Hii, S.L., Tan, J.S., Ling, T. and Ariff, A.B. 2012. Pullulanase: Role in starch hydrolysis and potential industrial applications. Enzyme Research, 2012: 1–14.

Hizukuri, S., Takeda, Y., Yasuda, M. and Suzuki, A. 1981. Multi-branched nature of amylose and the action of debranching enzymes. Carbohydrate Research, 94: 205–213.

Imanaka, T. and Kuriki, T. 1989. Pattern of action of *Bacillus stearothermophilus* neopullulanase on pullulan. Journal of Bacteriology, 171: 369–374.

Israilides, C., Smith, A., Scanlon, B. and Barnett, C. 1999. Pullulan from agro-industrial wastes. Biotechnology and Genetic Engineering Reviews, 16: 309–324.

Jensen, B.D. and Norman, B.E. 1984. *Bacillus acidopullyticus* pullulanase: Applications and regulatory aspects for use in food industry. Process Biochemistry, 1: 397–400.

Kainuma, K., Kobayashi, S. and Harada, T. 1978. Action of *Pseudomonas* isoamylase on various branched oligosaccharides and polysaccharides. Carbohydrate Research, 61: 345–357.

Kaper, T., Talik, B., Ettema, T.J., Bos, H.T., van der Maarel, M.J.E.C. and Dijkhuizen, L. 2005. Amylomaltase of *Pyrobaculum aerophilum* IM2 produces thermoreversible starch gels. Applied and Environmental Microbiology, 71: 5098–5106.

Kaper, T., van der Maarel, M.J.E.C., Euverink, G.J. and Dijkhuizen, L. 2004. Exploring and exploiting starch-modifying amylomaltases from thermophiles. Biochemical Society Transactions, 32: 279–282.

Kelly, R.M., Dijkhuizen, L. and Leemhuis, H. 2009. The evolution of cyclodextrin glucanotransferase product specificity. Applied Microbiology and Biotechnology, 84: 119–133.

Kendall, C.W.C., Esfahani, A., Hoffman, A.J., Evans, A., Sanders, L.M., Josse, A.R., Vidgen, E. and Potter, S.M. 2008. Effect of novel maize-based dietary fibers on postprandial glycemia and insulinemia. Journal of the American College of Nutrition, 27: 711–718.

Kim, C.H., Nashiru, O. and Ko, J.H. 1996. Purification and biochemical characterization of pullulanase type I from *Thermus caldophilus* GK-24. FEMS Microbiology Letters, 138: 147–152.

Kim, E.J., Ryu, S.I., Bae, H.A., Huong, N.T. and Lee, S.B. 2008. Biochemical characterisation of a glycogen branching enzyme from *Streptococcus mutans*: Enzymatic modification of starch. Food Chemistry, 110: 979–984.

Kitahata, S., Tsuyama, N. and Okada, S. 1974. Purification and some properties of cyclodextrin glycosyltransferase from a strain of *Bacillus* species. Agricultural and Biological Chemistry, 38: 387–393.

Klingeberg, M., Hippe, H. and Antranikian, G. 1990. Production of the novel pullulanases at high concentrations by two newly isolated thermophilic clostridia. FEMS Microbiology Letters, 69: 145–152.

Kobayashi, S., Kainuma, K. and Suzuki, S. 1978. Purification and some properties of *Bacillus macerans* cycloamylose (cyclodextrin) glucanotransferase. Carbohydrate Research, 61: 229–238.

Koch, R., Canganella, F., Hippe, H., Jahnke, K.D. and Antranikian, G. 1997. Purification and properties of a thermostable pullulanase from a newly isolated thermophilic anaerobic bacterium, *Fervidobacterium pennavorans* Ven5. Applied and Environmental Microbiology, 63: 1088–1094.

Kumar, P. and Satyanarayana, T. 2009. Microbial glucoamylases: Characteristics and applications. Critical Reviews in Biotechnology, 29: 225–255.

Kunamneni, A. and Singh, S. 2006. Improved high thermal stability of pullulanase from a newly isolated thermophilic *Bacillus* sp. AN-7. Enzyme and Microbial Technology, 39: 1399–1404.

Le, Q.T., Lee, C.K., Kim, Y.W., Lee, S.J., Zhang, R., Withers, S.G., Kim, Y.R., Auh, J.H. and Park, K.W. 2009. Amylolytically-resistant tapioca starch modified by combined treatment of branching enzyme and maltogenic amylase. Carbohydrate Polymers, 75: 9–14.

Lee, E.Y.C. and Carter, J.H. 1973. Amylo-1,6-glucosidase/1,4-α-glucan: 1,4-α-glucan 4-α-glycosyltransferase: Specificity toward polysaccharide substrates. Archives of Biochemistry and Biophysics, 154: 636–641.

Lee, K.Y., Kim, Y.R., Park, K.H. and Lee, H.G. 2008. Rheological and gelation properties of rice starch modified with 4-alpha-glucanotransferase. International Journal of Biological Macromolecules, 42: 298–304.

Leemhuis, H., Kelly, R.M. and Dijkhuizen, L. 2010. Engineering of cyclodextrin glucanotransferases and the impact for biotechnological applications. Applied Microbiology and Biotechnology, 85: 823–835.

Li, Z., Chen, S., Gu, Z., Chen, J. and Wu, J. 2014. Alpha-cyclodextrin: Enzymatic production and food applications. Trends in Food Science & Technology, 35: 151–160.

Lopez-Nicolas, J.M., Rodriguez-Bonilla, P. and Garcia-Carmona, F. 2014. Cyclodextrins and antioxidants. Critical Reviews in Food Science and Nutrition, 54: 251–276.

Manners, D.J. and Matheson, N.K. 1981. The fine-structure of amylopectin. Carbohydrate Research, 90: 99–110.

Marshall, J.J. and Whelan, W.J. 1974. Multiple branching in glycogen and amylopectin. Archives of Biochemistry and Biophysics, 161: 234–238.

Maruo, B. and Kobayashi, T. 1951. Enzymic scission of the branch links in amylopectin. Nature, 167: 606–607.

Messaoud, E.B., Ben Ammar, Y., Mellouli, L. and Bejar, S. 2002. Thermostable pullulanase type I from new isolated *Bacillus thermoleovorans* US105: Cloning, sequencing and expression of the gene in *E. coli*. Enzyme and Microbial Technology, 31: 827–832.

Mun, S., Kim, Y.L., Kang, C.G., Park, K.H., Shim, J.Y. and Kim, Y.R. 2009. Development of reduced-fat mayonnaise using 4-α-GTase-modified rice starch and xanthan gum. International Journal of Biological Macromolecules, 44: 400–407.

Nair, S.U., Singhal, R.S. and Kamat, M.Y. 2007. Induction of pullulanase production in *Bacillus cereus* FDTA-13. Bioresource Technology, 98: 856–859.

Nakamura, A., Haga, K. and Yamane, K. 1994. The transglycosylation reaction of cyclodextrin glucanotransferase is operated by a ping-pong mechanism. FEBS Letters, 337: 66–70.

Niehaus, F., Peters, A., Groudieva, T. and Antranikian, G. 2000. Cloning, expression and biochemical characterisation of a unique thermostable pullulan-hydrolysing enzyme from the hyperthermophilic archaeon *Thermococcus aggregans*. FEMS Microbiology Letters, 190: 223–229.

Norman, B.E. 1981. New developments in starch syrup technology. pp. 15–51. *In*: G.G. Birch, N. Blakebrough and K.J. Parker (eds.). Enzymes and Food Processing. Springer, Netherlands.

Norman, B.E. 1982a. A novel debranching enzyme for application in the glucose syrup industry. Starch/Stärke, 34: 340–346.

Norman, B.E. 1982b. The use of debranching enzymes in dextrose syrup production. pp. 157–179. *In*: G.E. Inglett (ed.). Maize: Recent Progress in Chemistry and Technology. Academic Press, Inc., New York.

Norman, B., Pedersen, S., Stanley, K., Stanley, E. and Richmond, P. 2007. Production of Crystalline Short Chain Amylase. Patent US2007059432.

Okada, S., Kitahata, S., Yoshikawa, S., Sugimoto, T. and Sugimoto, K. 1984. Process for the Production of Branching Enzyme, and a Method for Improving the Qualities of Food Products Therewith. Patent US4454161.

Palomo, M., Kralj, S., van der Maarel, M.J.E.C. and Dijkhuizen, L. 2009. The unique branching patterns of *Deinococcus* glycogen branching enzymes are determined by their N-terminal domains. Applied and Environment Microbiology, 75: 1355–1362.

Palomo, M., Pijning, T., Booiman, T., Dobruchowska, J.M., van der Vlist, J., Kralj, S., Planas, A., Loos, K., Kamerling, J.P., Dijkstra, B.W., van der Maarel, M.J., Dijkhuizen, L. and Leemhuis, H. 2011. *Thermus thermophilus* glycoside hydrolase family 57 branching enzyme: Crystal structure, mechanism of action, and products formed. Journal of Biological Chemistry, 286: 3520–3530.

Peat, S., Whelan, W.J. and Pirt, S.J. 1949. The amylolytic enzymes of soya bean. Nature, 164: 499–500.

Pedersen, S., Jensen, B.F. and Jorgensen, S.T. 1995. Enzymes from genetically modified microorganisms. pp. 196–208. *In*: K.H. Engel, G.R. Takeoka and R. Teranishi (eds.). Genetically Modified Foods: Safety Issues. ACS Symposium Series 605.

Pulley, A.O. and French, D. 1961. Studies on the Schardinger dextrins. XI. The isolation of new Schardinger dextrins. Biochemical and Biophysical Research Communications, 5: 11–15.

Qi, Q. and Zimmermann, W. 2005. Cyclodextrin glucanotransferase: From gene to applications. Applied Microbiology and Biotechnology, 66: 475–485.

Reeve, A. 1992. Starch hydrolysis: processes and equipment. pp. 79–120. *In*: F.W. Schenck and R.E. Hebeda (eds.). Starch Hydrolysis Product: Worldwide Technology, Production and Applications. VCH, New York, USA.

Reilly, P.J. 2006. Glucoamylase. pp. 727–738. *In*: J.R. Whitaker, A.G.J. Voragen and D.W.S. Wong (eds.). Handbook of Enzymology. Marcel Dekker Inc., New York.

Richmond, P.A., Marion, E.A., Eilers, T., Evans, A., Han, X.Z.H.X., Ahmed, S. and Harris, D.W. 2008. Production of Resistant Starch Product. Patent WO109206.

Robyt, J.F. 2009. Enzymes and their action on starch. pp. 237–292. *In*: J. BeMiller and R. Whistler (eds.). Starch, Chemistry and Technology. Academic Press, New York.

Robyt, J.F. 1991. Strategies for the specific labeling of amylodextrins. pp. 98–110. *In*: R.B. Friedman (ed.). Biotechnology of Amylopectin Oligosaccharides. ACS Symposium series. Washington, DC.

Roy, A., Messaoud, E.B. and Bejar, S. 2003. Isolation and purification of an acidic pullulanase type II from newly isolated *Bacillus* sp. US149. Enzyme and Microbial Technology, 33: 720–724.

Sato, H.H. and Park, Y.K. 1980. Purification and characterization of extracellular isoamylase from *Flavobacterium* sp. Starch/Stärke, 32: 132–136.

Sein, A., Mastenbroek, J., Metselaar, R., Buwalda, P.L., van Rooijen, C. and Visser, K.A. 2009. Low Fat Water-in-Oil Emulsion. Patent WO2009101215.

Shim, J.H., Seo, N.S., Roh, S.A., Kim, J.W., Cha, H. and Park, K.H. 2007. Improved bread-baking process using *Saccharomyces cerevisiae* displayed with engineered cyclodextrin glucanotransferase. Journal of Agricultural and Food Chemistry, 55: 4735–4740.

Shinohara, M.L., Ihara, M., Abo, M., Hashida, M., Takagi, S. and Beck, T.C. 2001. A novel thermostable branching enzyme from an extremely thermophilic bacterial species, *Rhodothermus obamensis*. Applied Microbiology and Biotechnology, 57: 653–659.

Spendler, T. and Jorgensen, O. 1997. Use of a Branching Enzyme in Baking. Patent WO97/41736.

Stavn, A. and Granum, P.E. 1979. Purification and physicochemical properties of an extracellular cycloamylose (cyclodextrin) glucano-transferase from *Bacillus macerans*. Carbohydrate Research, 75: 243–250.

Stetten, D. and Stetten, M.R. 1640. Glycogen metabolism. Physiological Reviews, 40: 505–537.

Suzuki, Y., Hatagaki, K. and Oda, H. 1991. A hyperthermostable pullulanase produced by an extreme thermophile, *Bacillus flavocaldarius* KP 1228, and evidence for the proline theory of increasing protein thermostability. Applied Microbiology and Biotechnology, 34: 707–714.

Szente, L. and Szejtli, J. 2004. Cyclodextrins as food ingredients. Journal of Food Science and Technology, 15: 137–142.

Takaha, T. and Smith, S.M. 1999. The functions of 4-α-glucanotransferases and their use for the production of cyclic glucans. Biotechnology & Genetic Engineering Reviews, 16: 257–280.

Takata, H., Kuriki, T., Okada, S., Takesada, Y., Iizuka, M. and Imanaka, T. 1992. Action of neopullulanase. Neopullulanase catalyzes both hydrolysis and transglycosylation at alpha-(1-4) and alpha-(1-6) glucosidic linkages. Journal of Biological Chemistry, 267: 18447–18452.

Takata, H., Ohdan, K., Takaha, T., Kuriki, T. and Okada, S. 2003. Properties of branching enzyme from hyperthermophilic bacterium, *Aquifex aeolicus*, and its potential for production of highly-branched cyclic dextrin. Journal of Applied Glycoscience, 50: 15–20.

Takata, H., Takaha, T., Kuriki, T., Okada, S., Takagi, M. and Imanaka, T. 1994. Properties and active center of the thermostable branching enzyme from *Bacillus stearothermophilus*. Applied and Environmental Microbiology, 60: 3096–3104.

Takeda, Y., Shirasaka, K. and Hizukuri, S. 1984. Examination of the purity and structure of amylose by gel-permeation chromatography. Carbohydrate Research, 132: 83–92.

Takii, H., Takii Nagao, Y., Kometani, T., Nishumura, T., Nakae, T., Kuriki, T. and Fushiki, T. 2005. Fluids containing a highly branched cyclic dextrin influence the gastric empty rate. International Journal of Sports Medicine, 26: 314–319.

Teague, W.M. and Brumm, P.J. 1992. Commercial enzymes for starch hydrolysis products. pp. 45–79. *In*: F.W. Schenck and R.E. Hebeda (eds.). Starch Hydrolysis Products: Worldwide Technology, Production and Applications. VCH, New York, NY, USA.

Terada, Y., Yanase, M., Takata, H., Takaha, T. and Okada, S. 1997. Cyclodextrins are not the major cyclic α-1,4-glucans produced by the initial action of cyclodextrin glucanotransferase on amylose. Journal of Biological Chemistry, 272: 15729–15733.

Terada, Y., Fujii, K., Takaha, T. and Okada, S. 1999. *Thermus aquaticus* ATCC 33923 amylomaltase gene cloning and expression and enzyme characterization: Production of cycloamylose. Applied and Environment Microbiology, 65: 910–915.

Thomas, G.J., Whelan, W.J. and Peat, S. 1950. The role of Z-enzyme in beta-amylolysis. The Biochemical Journal, 47: 39–52.

Tomono, K., Mugishima, A., Suzuki, T., Goto, H., Ueda, H., Nagai, T. and Watanabe, J. 2002. Interaction between cycloamylose and various drugs. Journal of Inclusion Phenomena and Macrocyclic Chemistry, 44: 267–270.

Uhlig, H. 1998. Industrial Enzymes and Their Applications. Wiley-Interscience, New York, NY, USA.

Urlaub, H. and Wober, G. 1975. Identification of isoamylase, a glycogen debranching enzyme, from *Bacillus amyloquefaciens*. FEBS Letters, 57: 1–4.

van der Maarel, M.J.E.C. 2010. Starch-processing enzymes. pp. 320–331. *In*: R.J. Whitehurst and M. van Oort (eds.). Enzymes in Food Technology. Blackwell Publishing Ltd., Ames, Iowa, USA.

van der Maarel, M.J.E.C., Binnema, D.J., Semeijn, C. and Buwalda, P.L. 2007. Novel Slowly Digestible Storage Carbohydrate. Patent EP1943908.

van der Maarel, M.J.E.C., Capron, I., Euverink, G.J.W., Boss, H.T., Kaper, T., Binnema, D.J. and Steeneken, P.A.M. 2005. A novel thermoreversible gelling product made by enzymatic modification of starch. Starch/Stärke, 57: 465–472.

van der Maarel, M.J.E.C. and Leemhuis, H. 2013. Starch modification with microbial alpha-glucanotransferase enzymes. Carbohydrate Polymers, 93: 116–121.

Vorwerg, W., Radosta, S. and Leibnitz, E. 2002. Study of a preparative-scale process for the production of amylose. Carbohydrate Polymers, 47: 181–189.

White, R.C. and Nelson, T.E. 1975. Analytical gel chromatography of rabbit muscle amylo-1,6-glucosidase/4-α-glucanotransferase under denaturing and non-denaturing conditions. Biochimica et Biophysica Acta, 400: 154–161.

Wind, R.D., Uidehaag, J.C., Buitelaar, R.M., Dikstra, B.W. and Dikhuizen, L. 1998. Engineering of cyclodextrin product specificity and pH optima of the thermostable cyclodextrin glycosyltransferase from *Thermoanaerobacterium thermosulfirigenes* EM1. Journal of Biological Chemistry, 273: 5771–5779.

Yoon, S.H. and Robyt, J.F. 2002. *Bacillus macerans* cyclomaltodextrin glucanotransferase transglycosylation reactions with different molar ratios of D-glucose and cyclomaltohexaose. Carbohydrate Research, 337: 2245–2254.

Zareian, S., Khajeh, K., Ranjbar, B., Dabirmanesh, B., Ghollasi, M. and Mollania, N. 2010. Purification and characterization of a novel amylopullulanase that converts pullulan to glucose, maltose, and maltotriose and starch to glucose and maltose. Enzyme and Microbial Technology, 46: 57–63.

Zeeman, S.C., Kossmann, J. and Smith, A.M. 2010. Starch: Its metabolism, evolution, and biotechnological modification in plants. Annual Review of Plant Biology, 61: 209–234.

4

Glucose Isomerising Enzymes

Savitha S. Desai, Dhanashree B. Gachhi and
Basavaraj S. Hungund*

1. Introduction

Enzymes are biological macro-molecules which increase the rate of the reaction without themselves undergoing any change. There are vast varieties of enzymes that exist in Nature. Danish chemist Christian Hansen, in 1874, obtained the first specimen of rennet by extracting dried calves' stomach with saline solution. This was the first enzyme used for industrial purposes and which marked the beginning of history of enzyme technology.

The International Union of Biochemistry (1984) has classified the enzymes into six different classes based on the type of reaction that is catalysed. Isomerases are fifth class of enzyme in Enzyme Classification. Isomerases are a class of enzymes which catalyse geometric or structural changes within one molecule. According to the type of isomerism, they may be called *racemases, epimerases, cis-trans-isomerases, isomerases, tautomerases, mutases* or *cycloisomerases*.

One of the important isomerase enzymes in food industry is glucose isomerase (GI) which is in high demand because of its importance in commercial production of high-fructose corn syrup (HFCS). The enzyme is thermostable and does not require expensive cofactors, such as NAD or ATP for activity. Although the conversion of glucose to fructose can be carried out chemically, it leads to formation of non-metabolizable sugars and sometimes the reaction itself is non-specific. Chemically-synthesized fructose has limited commercial application as it has off-flavor and reduced sweetness. Where, as in the case of enzymatic conversion, the reaction is very specific, and despite utilization of ambient conditions of pH and temperature, side products

Department of Biotechnology, B.V.B. College of Engineering and Technology, Vidyanagar, Hubballi, 580031, Karnataka.
* Corresponding author: hungundb@gmail.com

are not formed. Because of these advantages of enzymatic conversion, it is a suitable method for isomerisation of glucose to fructose. At present, the processes involving glucose isomerase (GI) has undergone significant development in the industrial sector.

2. Glucose Isomerase

D-Glucose/xylose isomerase (D-xylose ketol isomerase; EC 5.3.1.5) commonly known as glucose isomerase (GI), is one of high-tonnage value enzyme. It catalyzes the reversible reaction of D-glucose to D-fructose and D-xylose to D-xylulose (Fig. 1). Production of xylulose from xylose meets the dietary requirement for saprophytic bacteria. Also, it is useful in bioethanol production from cellulose. Isomerization of glucose to fructose is of industrial significance in the production of high-fructose corn syrup. The enzymatic production of fructose from glucose is of high interest because the enzymatic process is highly efficient and very selective. Here the energy requirement is also less and the side products formed are less (Yu et al., 2013).

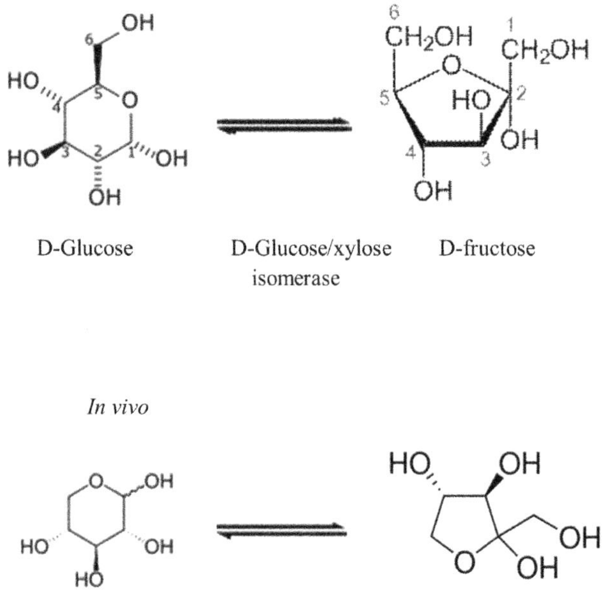

D-Glucose   D-Glucose/xylose   D-fructose
                        isomerase

In vivo

D-Xylose   D-Glucose/xylose   D-Xylulose

Figure 1. Reversible reactions of D-glucose and D-xylose (Wiseman, 1975).

2.1 Phosphoglucoisomerase

Phosphoglucoisomerase (PGI) (EC: 5.3.1.9) alternatively known as glucose-6-phosphate isomerase (GPI), or phosphohexose isomerase is an enzyme crucial for the inter-conversion of D-glucose 6-phosphate and D-fructose 6-phosphate (Fig. 2). It is an enzyme responsible for the second step of glycolysis which is isomerisation of glucose 6-phosphate to fructose 6-phosphate and is also involved in gluconeogenesis.

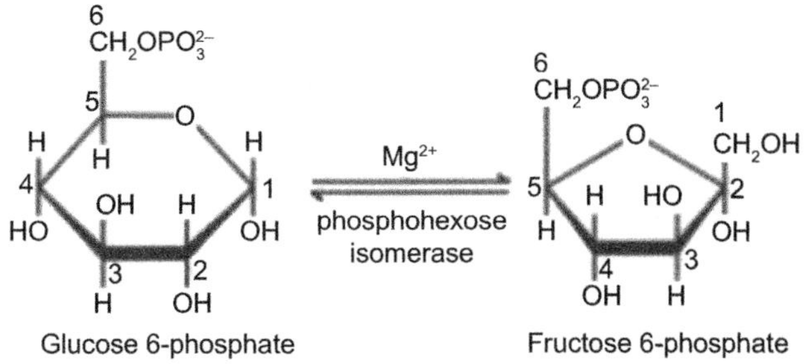

6
$CH_2OPO_3^{2-}$

5

O

H | H

4 | H

OH | H | 1

HO | OH

3 | 2

H | OH

Mg^{2+}
⇌
phosphohexose
isomerase

6
$CH_2OPO_3^{2-}$

1
O | CH_2OH

5 | H | HO | 2

H | OH

4 | 3

OH | H

Glucose 6-phosphate　　　　　　　　**Fructose 6-phosphate**

Figure 2. Reversible reaction of glucose 6-phosphate.

Phosphoglucoisomerase is involved in different pathways, like glycolysis (in higher organisms), gluconeogenesis (in mammals). Carbohydrate biosynthesis (in plants) also provides access for fructose to enter into the Entner-Doudouroff pathway (bacteria).

3. Structure of Glucose Isomerases

The primary and secondary structures of glucose isomerase reveal that it is a tetramer composed of four identical polypeptides of 43,000 Da or a dimer of related or identical subunits (dimer of dimmers) connected with non-covalent bonds and do not include any inter-chain disulfide bonds (Fig. 3). Two monomers form a tight dimer and two dimers form tight tetramer. It is seen that the interaction between mononers in a dimer are more extensive than that of dimmers in tetramer (Farber et al., 1987). Each monomer has two domains: a large N terminal domain (residues 1–320) and a small C-terminal loop (residues 321–387). Up to 90 per cent sequence homology is observed in the enzymes from *Streptomyces* sp. (Ramagopal et al., 2003). It was seen that the tertiary structure is responsible for the biological activity of GI rather than the secondary structure.

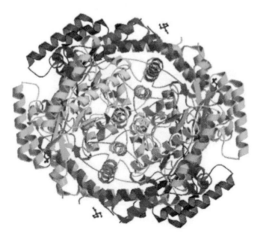

Figure 3. Structure of glucose isomerase from *Streptomyces rubiginosus* (Protein Data Bank open source, DOI: 10.2210/pdb1oad/pdb).

The enzyme also exists as trimer, as in the case of *Bacillus* sp. (extracellular GI). The sedimentation constants and molecular weights of GI vary from 7.55 to 11.45 and from 52,000 to 1,91,000 Da respectively (Chan et al., 1989). Isoenzymes of glucose isomerase from *Streptomyces phaechromogenes* have been reported by Basuki et al. (1992). These isoenzymes differ in their N-terminal amino acids and also in the peptide digest patterns as obtained after digestion with enzymes like trypsin, *Achromobacter* protease I, and chemicals like cyanogen bromide. It revealed that each of the isoenzymes was a tetramer of non-identical subunits.

Phosphoglucoisomerase is a dimer made up of two chemically identical polypeptide chains (555 amino acids). Secondary structure of each dimer consists of two separate β-sheets surrounded by α-helices. These alpha helices and β-sheets result in two asymmetrical domains within each subunit. A central five-stranded parallel β-sheet surrounded by loops and α-helices is present in small domain whereas the larger domain has four parallel β-strands and two anti-parallels near the N-terminus. Two helices and a loop exist near the carboxyl terminus of each subunit which protrudes into the adjacent subunit. The amino terminus contains 36 amino acid residues that exist in an alpha helix followed by a loop and a β-sheet (Fig. 4). A hook is possibly involved in extracellular function or attachment of the subunits protrude from the side of each subunit in a helix-turn-helix motif. Hence, one subunit evidently reveals the comparative positions of the carboxy terminus amino terminus and hook. It is thought that PGI has a unique secondary structure. Yamamoto et al. (2008) studied the crystal structure of glucose-6-phophate isomerase from *Thermus thermophilus* HB8, showing a snapshot of active dimeric state. The results suggest that the observed dimeric state of GPI from *Thermus thermophilus* HB8 in the crystal is biologically relevant and that this enzyme uses a common catalytic mechanism for the isomerase reaction. Ghatge

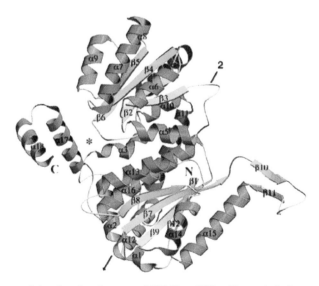

Figure 4. Structure of phospho-gluco isomerase (PGI) (Sun, 1999, with permission).

et al. (1994) studied the unfolding and refolding of the tetrameric glucose isomerase from *Streptomyces* sp. NCIM 2730. They have correlated the biological activity with protein transitions.

4. Reaction Mechanism and Analysis of Glucose Isomerase

4.1 Mechanism of action

The general reaction mechanism of PGI or GI for the reversible reaction of glucose-6-phosphate to fructose-6-phosphate/conversion of glucose to fructose involves acid/base catalysis. The basic mechanism involves the isomerisation of an aldose to a ketose. This includes opening of the ring, followed by isomerisation of the opened ring, before the ring closes (Fig. 3). A detailed step-by-step mechanism is seen as follows:

Step 1. The cyclic form of substrate binds to the active site of enzyme.

Step 2. The lysine residue acts as an acid donating a proton and resulting in the ring opening with C-2 becoming relatively acidic.

Step 3. Conserved His residues in catalytic diad (His/Glu) abstracts the acidic proton from C2, resulting in double-bond formation between C-1 and C-2, accompanied by forming a cis-enendiol intermediate (addition of proeon from Lys residue).

Step 4. Arg-272 causes glucose-6-phosphate to have a positive electrostatic potential and stabilizes the negative charge on the enediol transition state.

Step 5. Glu residue donates back the proton at the C-1 position.

Step 6. A pair of electrons is transferred from the oxygen at C-1 to C-2 and the resulting ketose receives a proton from the Lys residue, resulting in a ring closure, to give the product.

Researchers tried to understand the reaction mechanism of GI. It was earlier thought that GI follows enediol mechanism (Fig. 5) and its function is similar to sugar phosphate isomerases (Rose et al., 1969). Recent studies attribute the action of GI to a hydride shift mechanism. Different methods have been used to study the active site of GI and its reaction mechanism. These involve (i) chemical modification, (ii) X-ray crystallography, and (iii) isotope exchange. The key steps of the mechanism proposed for GI are ring opening of the substrate, isomerization *via* a hydride shift from C-2 to C-1, and ring closure of the product.

These include:

(i) *Chemical modification of glucose isomerase*: Certain chemicals directly act on the active site of the enzyme and modify them, helping the researcher to look into the active site of the enzyme. When DEPC (diethylpyrocarbonate) was used, it resulted in inactivation of GI, indicating the presence of histidine residue in the active site (Kume et al., 1983).

(ii) *X-ray crystallography*: It helps in understanding the three-dimensional structure of protein and allows real elucidation of the complex between the enzyme and its substrate

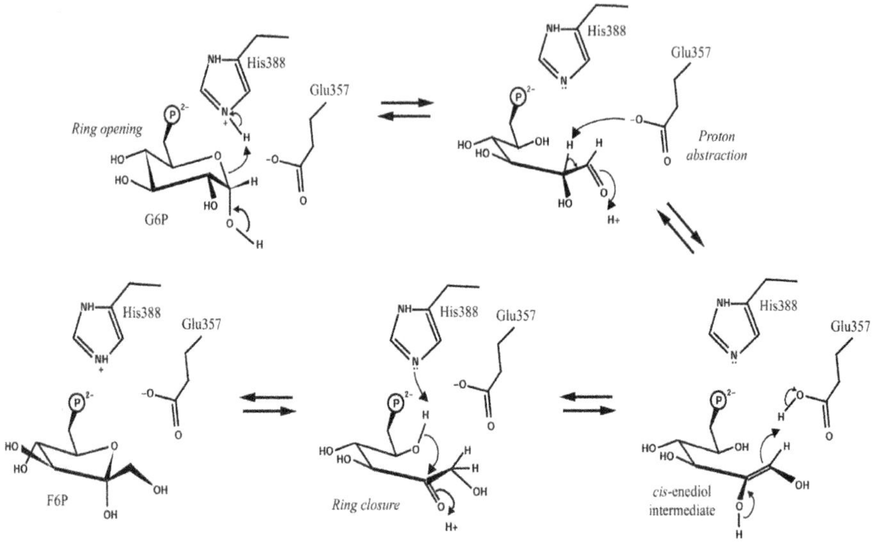

Figure 5. General mechanism of action of PGI or GI (Davies and Muirhead, 2003).

or inhibitor. GI from different bacterial species has been studied to understand and explain the reaction mechanism. The structures of GI from several *Streptomyces* spp. are very similar, especially at the active site. The structure of GI from *Streptomyces rubiginosus,* as determined at 4-Å (1 Å 5 0.1 nm) resolution (Carell et al., 1984) indicates that the enzyme contains eight b-strand–a-helix [(a/b) 8] units which is similar to those found in triose-phosphate isomerase. There are two domains, one smaller and one larger. The smaller domain forms a loop away from the larger domain but overlaps the larger domain of another subunit, so that a tightly-bound dimer is formed. The tetramer is thus considered to be a dimer of active dimers.

(iii) *Isotope exchange*: The available crystallographic data for GI propose a hydride shift mechanism. However, only structural data are not sufficient to reveal the reaction mechanism. There exists an ambiguity regarding the proton transfer mechanism in GI because of the absence of solvent exchange during investigations (Rose et al., 1969).

D-xylose isomerise enzyme from *Streptomyces rubiginosus* has been extensively studied till date at high resolution by the method of time-of-flight neutron diffraction, which helps to locate hydrogen atoms that define the ionization states of amino acids in the crystals of D-xylose isomerase. Katz et al. (2006) have determined the position and orientation of a metal ion-bound water molecule that is located in the active site of the enzyme; this water has been thought to be involved in isomerization. These studies also reveal the ionization states of the amino acids. High-resolution X-ray studies (at 0.94 A) reveal side-chain movement during catalysis and disorderliness in the side chain when a truncated substrate is bound. The combination of time-of-flight neutron diffraction and X-ray diffraction together help in understanding the reaction mechanism to a great extent.

5. Fermentative Production

5.1 Microbial strains used for glucose isomerase production

Several researchers reported the production of glucose isomerase from members of bacterial genera *Streptomyces, Bacillus, Clostridium, Anoxybacillus, Acidothermus, Thermus,* etc. Cloning and sequencing of thermostable glucose isomerase gene was reported from the thermophile *Thermus thermophilus* (Dekker et al., 1992) and from *Thermoanaero bacterium* (Liu et al., 1996). The presence of GI in a few yeasts, such as *Candida utilis* (Wang et al., 1980) and *Candida boidinii* (Vongsuvanlert and Tani, 1988) has been reported. *Aspergillus oryzae* is an important fungus which is reported to possess GI activity. The existence of GI in barley malt and wheat germ has also been reported (Pubols et al., 1963). Since GI finds several industrial applications which highlight commercial importance, most of the scientific advances in novel producer strains and process development exist in the form of patents. Mezghani et al. (2005) constructed a new stable strain from *Streptomyces* sp. for expressing the glucose isomerase. This construct was introduced *via* site specific recombination process into GI deficient *Streptomyces violaceoniger*. They used non-replicative vector PTS55 during the study. In the absence of selection pressure, the recombinant strain showed very good stability. This recombinant strain showed four- to nine folds greater activity as compared to *Streptomyces* sp. SK strain. Mu et al. (2011) studied thermostable GI from *Acidothermus cellulolyticus* and found that this enzyme could be a new source to obtain high-fructose corn syrup. Hartley et al. (2000) discussed the insights into protein engineering of GI for increased thermostability. They compared the thermostability of Class-I and Class-II enzymes. Though protein engineering was unsuccessful in developing thermostable GI enzyme, they suggested thermal inactivation pathway as the future guide. Karaoglu et al. (2013) reported production of new glucose isomerase from thermophile *Anoxybacillus gonensis*. They cloned the gene responsible for the production of this enzyme and engineered the heterologous protein expression in *Escherichia coli*.

Sarryar et al. (2004) reported expression and translocation of GI as a fusion protein in *Escherichia coli*. Fusion activity was done for GI from *Thermus thermophilus* with maltose-binding protein. Using the signal sequence of maltose-binding protein with slow folding property, the fusion protein was transported to periplasm of *E. coli*. Around 1.6 per cent of total GI activity was detected in the periplasm of *E. coli*. The effect of inducer concentration on translocation was insignificant. However, induction at 23°C increased periplasmic glucose isomerase fusion by 50 per cent. These studies showed that the properties of the system influenced the secretion efficiency of the fusion protein.

5.2 Parameters affecting enzyme production

Several parameters affect production of GI from various microorganisms. These include carbon and nitrogen sources, pH and temperature, presence of inducers, presence of metal ions, etc. Large-scale production of the enzyme is possible using the submerged aerobic fermentation process. There is need to develop an economically feasible medium and optimize the fermentation process. Previous research efforts were focused towards the use of inexpensive carbon and nitrogen sources, replacement of xylose by another cheaper inducer, optimization of process parameters and substitution of cobalt ions by other divalent metal ions. It was reported earlier that for maximal enzyme production from different microorganisms, a single medium with defined media composition could not be developed. Each microbial strain requires specific media components and production conditions.

5.2.1 Nitrogen source

The nitrogen source present in the medium greatly influences microbial growth and enzyme production. Both organic and inorganic sources of nitrogen are used. For optimum enzyme production, an optimization study needs to be conducted. Although complex sources of nitrogen are usually preferred by many organisms, the specific nitrogen requirement differs from organism to organism. Takasaki et al. (1966) reported the use of peptone, yeast extract and ammonium salts, while, urea and nitrate were not found suitable. Corn steep liquor was one of the most economical sources of nitrogen (Bhoslae et al., 1996) but it suffers from the limitation of seasonal availability and batch variation. There are efforts to evaluate suitable nitrogen sources which can substitute corn steep liquor. Soy flour can be one such alternative. Callens et al. (1988) found that addition of amino acids improves the yield of enzyme production through *Streptomyces violaceoruber.*

5.2.2 pH and temperature

It is a well-established fact that the type of nitrogen source affects the pH of the media. Consequently, this value of pH affects the yield of enzyme. Most studies on GI production set media pH between 7–8 without any control. Similarly, incubation temperature also plays an important role on the yield of the enzyme. *Streptomyces* sp., *Arthrobacter* sp., and *Actinoplanes missouriensis* were grown at 30°C for GI production. However, thermophilic *Bacillus* sp. was incubated at 50°C to 60°C for growth and enzyme production (Calik et al., 2009). The optimum duration of GI production varies from 6 h to 48 hours, depending on the type of organism used.

5.2.3 Inducer

During production of GI, most of the microorganisms have an obligate requirement for D-xylose which acts as an inducer, through its use is limited due to its high cost. Some studies report use of starch, glucose, sorbitol, glycerol, etc. as inducers. Takasaki and Tanabe (1966) showed that *Streptomyces* strain YT-5 was able to grow on xylan or xylan-containing material, such as corn cobs or wheat bran. This was the first unique attempt in selecting a microbial strain which could grow and produce GI in an economical medium. Many researchers reported isolation of microbial strains which were capable of producing GI, using glucose as an inducer instead of xylose. Such strains include *Actinoplanes* spp., mutant strains of *Bacillus coagulans*, *Streptomyces olivochromogenes*, etc. Another approach to eliminate the requirement of xylose as an inducer is to generate mutants which can produce GI constitutively.

Calik et al. (2009) reported production of GI in a xylan-based medium using *Bacillus thermoantarcticus*. The highest GI activity was obtained as 1630 Udm^{-3} in the medium containing birchwood xylan.

5.2.4 Metal ion requirement

Studies have shown the requirement for divalent cations in the fermentation medium for optimum production of GI. Generally, the enzyme-producing microbe may have a specific requirement for the presence of divalent cations. Presence of cobalt was essential for GI production by *Streptomyces* strain YT-5 whereas *Bacillus coagulans* required Mn^{+2} or Mg^{+2} for production of the enzyme (Takasaki and Tanabe, 1966). Some organisms, such as *Arthrobacter* spp. (Reynolds, 1974), as well as mutants of *Streptomyces olivochromogenes* do not require cobalt for optimal production. The combination of different ions has proved to increase the activity of enzyme production (Yassien et al., 2013). The combination of magnesium and cobalt ion also improved the GI activity. Zhang and associates (2013) studied the catalytic activity of recombinant glucose isomerase from *Thermobifida fusca* through addition of Co^{+2} which enhanced the activity and thermostability of the glucose isomerase.

6. Downstream Processing

Various studies have been conducted for the production of GI from different microorganisms, but the reports on enzyme purification are very few. For effective and economical use of GI, the technique of immobilization was adopted. The significance of immobilization process is that the enzyme does not require further purification and concentration. Studies on enzyme purification carry academic significance, since the fundamentals of enzyme structure, its chemical modification, structure-function relationship, etc. are involved. In a majority of cases, GI is produced as an intracellular enzyme. However, there are some reports on its extracellular production also. Extraction of intracellular enzyme from microbial cells involves mechanical disruption, sonication, grinding, or homogenization or cell lysis with lysozyme, cationic detergents, toluene, etc. Further purification steps involve ammonium sulfate precipitation, solvent precipitation, ion-exchange chromatography, gel filtration, etc. (Chen et al., 1980a, 1980b). Literature on the purification of GI by affinity chromatography methods is also available. In a study by Chan et al. (1989), the use of an affinity adsorbent xylitol-sepharose was reported to purify GI from *Streptomyces* sp. Use of other affinity matrices coupled with xylose or mannitol immobilized on silochrome-based adsorbents were also reported. Ghatge and co-workers (1994) applied the immuno-affinity chromatography technique for the purification of GI from *Streptomyces* sp. strain NCIM 2730.

7. Immobilization of Glucose Isomerase

The cost of production of GI can be reduced by effectively recovering it and using it many times. This can be done through the technique of immobilization wherein the enzymes are immobilized on support so that they can be effectively reused. There exist various methods by which enzymes are immobilized. It is essential to immobilize GI enzyme since it is intracellular and large quantities are required for glucose; if not immobilized, then it would be very expensive for utilization. In the market, the immobilized form of GI is in great demand.

There exist various techniques for immobilization which include adsorption to insoluble materials (Hanover and White, 1993), entrapment into polymeric gels (Seyhan and Alago, 2008) and encapsulation in membranes, cross-linking with bi-functional reagents (Hanefeld et al., 2009) or covalent linking on to insoluble carriers. GI obtained from different sources has been immobilized on different support materials, such as diethyl-aminoethyl cellulose (DEAE-C). polyacrylamide gel (Bhosale et al., 1996) and alginate beads.

Yu et al. (2011) studied the immobilization of glucose isomerase on to patented GAMM support prepared from inverse suspension polymerization with glycidyl methacrylate (GMA), alyl glycidyl ether (AGE), N, N-methylene bis acrylamide (MBAA) and acrylamide (MAA) used for isomerization of glucose to fructose and proved to be an excellent catalyst for isomerization. Chopda et al. (2014) carried out immobilization of GI on a glass surface and studied the effect of glass on enzyme activity. Several studies on immobilization of GI using biopolysaccharide/biopolymer microspheres have been reported (Zhao et al., 2016). They studied the enhancement of glucose isomerase activity by immobilization, using silica/chiotsan hybrid microspheres. They compared the activity of free GI and immobilized GI at pH range of 5.8–8.0 and the temperature range of 40–80°C. They found that the activity of immobilized GI was 90 per cent and hence concluded that the immobilized GI supported by silica/chitosan hybrid microspheres is an ideal candidate for biocatalysis. Palazzi et al. (2001) systematically studied the importance of intra-particle and inter-particle diffusional resistances in the process of glucose isomerization to fructose by immobilized GI enzyme.

7.1 Cell-free immobilization

Cell-free enzymes or soluble enzymes when immobilized on to a support, exhibit flow characteristics which are appropriate for continuous operations and offer substantial savings in terms of investment. DEAE cellulose was used as a matrix for immobilizing GIs from *Streptomyces phaeochromogenes* and *Lactobacillus brevis* (Bucke, 1981). The Clinton Corn Processing Company utilized *Streptomyces* GI immobilized on DEAE cellulose to produce HFCS. A half-life of 49 days was exhibited by GI from *Streptomyces* sp. when immobilized on porous alumina and was found fit for application in plug-flow reactors (PFR). When immobilized on controlled-pore alumina in the presence of Co^{2+}, the Co^{2+} could be easily removed. Single-step conversion of dextrin to fructose was done by co-immobilizing glucoamylase and glucose isomerase by the molecular deposition method (Ge et al., 1999). The system has the ability to function at pH 6.0 to produce fructose from starch and dextrin.

7.2 Whole-cell immobilization

For most of the commercially available immobilized GI, whole-cell immobilization is a better method since GI is an intracellular enzyme. The Clinton Corn Processing Company spray-dried whole cell containing GI to produce HFCS. Inclusion of inorganic salts in the fermentation broths of *Streptomyces* sp. or *Arthrobacter* sp. followed by filtration and drying provided a simple method to immobilize the cells (Reynolds, 1974). In a wider sense, current technology uses an immobilized GI in a system at a higher temperature (65°C) and higher pH without the addition of Co^{2+} ions. Novo Industries makes use of physical entrapment of whole cells in polymeric material whereas chemical entrapment of cells in a membrane followed by cross-linking with glutaraldehyde was used to obtain immobilized GI. An enzyme when immobilized on Indion 48-R from *Streptomyces* sp. strain NCIM 2730, resulted in enhancement of pH and temperature stability (Gaikwad et al., 1988).

8. Applications of Glucose Isomerase

Glucose isomerase has received increased attention from industries for its use in producing HFCS and for its potential application in the production of ethanol from hemicelluloses.

8.1 Production of high-fructose corn syrup

Many of the studies have shown that fructose possesses higher sweetening index than glucose and sucrose. HFCS contains nearly equal amounts of glucose and fructose and is 1.3 times sweeter than sucrose and 1.7 times sweeter than glucose. Glucose alone has not been accepted as a sweetening agent since it is the most preferred source of carbon and energy by the cells. Also, to prevent crystallization, the glucose syrup must be kept warm with appropriate precautions taken to prevent any microbial attack.

The production of HFCS from starch involves three stages: (i) gelatinization and liquefaction of starch by α-amylase, (ii) saccharification of starch by the combined action of amyloglucosidase and a debranching enzyme, and (iii) isomerization of glucose by GI. The final product obtained is a corn syrup containing a mixture of glucose and fructose with an increased sweetening capacity than that of sucrose. Some studies have shown that wheat, tapioca, rice, etc. can be used instead of starch.

Gelatinization is a process of conversion of starch granules to viscous suspension. Liquefaction is a process of starch hydrolysis and saccharification is the process of conversion of starch to glucose and maltose by hydrolysis.

Gelatinization can be brought about by heating starch with water. One the starch is gelatinized, it is further subjected to hydrolysis by acids or enzymes. Further, the hydrolysed products are saccharified by acidic or enzymatic hydrolysis. The granular starch is converted into slurry with cold water (30–40 per cent) at pH 6.0–6.5. To this slurry, 20–80 ppm of Ca^{2+} is added which stabilizes and activates the enzyme. Then, enzyme α-amylase is added with a dose of 0.5–0.6 kg tonne⁻¹ (about 1500 U kg⁻¹ dry matter) of starch. The contents are mixed as, gelatinization occurs rapidly through the combined effect of enzyme hydrolysis and mixing shear forces. This partially gelatinised starch is passed into a series of holding tubes maintained at 100–105°C and held for 5 min. This will complete the gelatinisation process. Hydrolysis of the required DE (Dextrose Equivalent) is completed in holding tanks at 90–100°C for 1 to 2 hours.

The liquefied starch is usually saccharified by debranching enzyme and glucoamylase to obtain the glucose and maltose syrup. At the end, the enzyme activity is destroyed by lowering the pH. Glucose produced is isomerized to fructose by glucose isomerase. When supplied with cobalt ions, α-D-glucopyranose is isomerized to α D-fructofuranose. Several genera of microbes like *Actinoplanes missouriensis*, *Bacillus coagulans* and various *Streptomyces* species can produce glucose isomerases. These enzymes offer advantages as they are resistant to thermal denaturation and can act at very high substrate concentrations, so that the enzyme is stabilized at higher operational temperatures. Batch process for Immobilized enzyme had disadvantages as it was very costly, resulted in many by-products and proved difficult in removal of added ions and the catalyst. Nowadays most isomerisation is performed in packed bed reactors (PBRs). Mg^{2+} and Co^{2+} act as cofactors for enzyme activity. Excess Mg^{2+} interferes with the purification as well as the isomerisation process. Further, Co^{2+} are fixed during the immobilisation step so that it is not added with the substrate.

For, the immobilised glucose isomerase to be used efficiently, the substrate solution must be purified so that it is free from impurities which might inactivate the enzyme. At equilibrium, at 60°C, about 51 per cent of the glucose in the reaction mixture is converted into fructose. However, since excessive time is required for equilibration, the flow rate is adjusted to produce 42–47 per cent fructose. After isomerisation, the pH of the syrup is lowered to 4–5 and it is purified by ion-exchange chromatography and treated with activated carbon. Then, it is normally allowed to get concentrated by evaporation. Though 42–47 per cent of fructose is most satisfactory, the coco-cola preparation uses 55 per cent of fructose concentrate. The market for HFCS is expanding and its production is achieved through immobilised-enzyme technology. The high-fructose syrups can be used to replace sucrose where sucrose is used in solution. However, HFCS is inadequate to replace the use of crystalline sucrose. According to the latest survey done by United State Department of Agriculture, the HFCS consumption in US was 1372 metric tons (dry weight) in the year 2013-14 and estimated to reach 1482 metric tons by 2015-16 (MacConnell, 2016).

8.2 Production of fructose syrup from whey permeate

Glucose isomerase can also be used for conversion of dairy waste into syrup which can be used as a sweetening agent. Andres Illanes (2011) reported hydrolysis and isomerization of the whey permeate with immobilized β-galactosidase and glucose isomerase into fructose syrup.

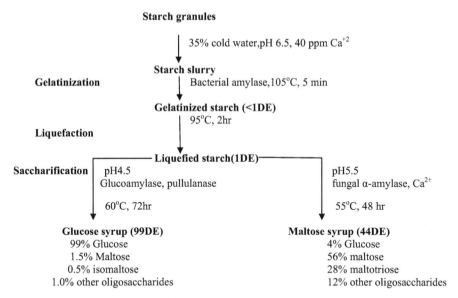

Figure 6. Schematic representation of conversion of starch into glucose syrup and maltose syrup.

8.3 Xylitol purification

Xylito ($C_5H_{12}O_5$) finds several applications in pharmaceutical, health-care and food industries. As a sugar-free sweetening agent, it can be used as a sugar substitute for diabetic patients. It is reported to possess anti-cancer and skin-softening properties. Presently, xylitol is used as a sweetener in food products such as chewing gum, candy, soft drinks and personal health products (mouthwash, tooth-paste, etc.). Studies show that xylitol is an inhibitor of GI and the enzyme specifically binds to xylitol. This property can be used to purify xylitol from the mixtures of xylitol, sorbitol and arabinitol. During industrial production of xylitol from hardwood pentosan arabinoxylan, a mixture of xylitol, sorbitol and arabinitol is obtained. Xylitol, sorbitol and arabinitol are difficult to separate from each other on the industrial scale and a specific separation technique would be of great value. Cross-linked enzymes have been used in many industrial applications. The first commercial cross-linked enzymes were prepared from glucose isomerase. These cross-linked enzymes were packed in a column and a mixture of arabinitol and xylitol were passed through it. Arabinitol did not bind to the enzyme. Xylitol could bind to the enzyme in the column and was eluted with $CaCl_2$ after washing the column with water. More than 95 per cent of both arabinitol and xylitol can be recovered in the effluent.

8.4 Production of ethanol

Glucose isomerase catalyzes the isomerization of both glucose and xylose. This ability of GI is used to isomerase xylose to xylulose. This D-xylulose is further fermented through conventional yeast to ethanol. The bioconversion of feedstock into usable carbon source and further to ethanol is very important. The feedstock consists of cellulose, hemicellulose and lignin. The economic feasibility of feedstock utilization depends on the hydrolysis of cellulose and hemicellulose to glucose and xylose (xylulose) and their subsequent fermentation to ethanol by yeasts (Wang et al., 1980). Low ethanol tolerance and catabolism of ethanol in the presence of oxygen limit the commercial application of yeasts (duPreez et al., 1986). GI has been used to produce xylulose from xylose, which otherwise represents a major metabolic block in the process of fermentation of xylose to ethanol by conventional yeasts, such as *Saccharomyces cerevisiae*, *Schizosaccharomyces pombe* and *Candida tropicalis* (Chaing et al., 1981; Chan et al., 1989; Ligthelm et al., 1988). A few yeasts, such as *Pachysolen tannophilus*, *Pichia stipitis*, *Candida utilis* and *Candida shehatae* are known to utilize pentoses through the oxidoreductive pathway, but the rates of fermentation are very low (duPreez et al., 1986). Li and his coworkers (2015) constructed xylose isomerase (XI) pathway with *Saccharomyces cerevisiae* strains using flocculating industrial strain (YC-8) as the host. Both strains showed better growth ability and fermentation capacity when xylose was used as the sole carbon source, thus showing a better performance in the fermentation of xylose to ethanol.

9. Conclusion

Glucose isomerase is in high demand because of its importance in commercial production of high-fructose corn syrup. The catalytic mechanism of GI is also reported by many researchers. Several microorganisms are reported to produce GI in various quantities. The enzyme is primarily produced by submerged fermentation. The parameters which affect its large-scale production include pH, temperature, type and concentration of substrate, presence of divalent metal ions and inducers, etc. Various techniques are used for downstream processing of the enzyme after fermentation. The enzyme finds several industrial applications and for repeated and economical usage, various techniques of enzyme immobilization are used.

Acknowledgement

The authors acknowledge the constant support and encouragement received from Dr. Ashok Shettar, Vice Chancellor, KLE Technological University, Hubballi (Formerly BVBCET, Hubballi).

References

Basuki, W., Iizuka, M., Ito, K., Furuichi, K. and Minamiura, N. 1992. Evidence for the existence of isozymes of glucose isomerase from *Streptomyces phaeochromogenes*. Bioscience, Biotechnology and Biochemistry, 56(2): 180–185.

Bhosale, S.H., Rao, M.B. and Deshpande, V.V. 1996. Molecular and industrial aspects of glucose isomerase. Microbiological Reviews, 60(2): 280–300.

Bucke, C. 1981. Industrial glucose isomerase. *In*: A. Wiseman (ed.). Topics in Enzyme and Fermentation Biotechnology, Industrial Glucose Isomerase. Ellis Horwood, Chichester, United Kingdom, 1: 147–171.

Callens, M., Kersters-Hilderson, H., Vangrysperre, W., Clement, K. and De Bruyne, A. 1988. D-xylose isomerase from *Streptomyces violaceoruber*: Structural and catalytic roles of bivalent metal ions. Enzyme and Microbial Technology, 10(11): 695–700.

Calik, P., Angardi, V., Haykir, N.I. and Boyaci, I.H. 2009. Glucose isomerase production on a xylan-based medium by *Bacillus thermoantarticus*. Biochemical Engineering Journal, 43(1): 8–15.

Carrel, H.L., Rubin, B.H., Hurley, T.J. and Glusker, J.P. 1984. X-ray crystal structure of D-xylose isomerase at 4 A resolution. Journal of Biological Chemistry, 259(5): 3230–3236.

Chaing, L.C., Hsiao, H.Y., Ueng, P.P. and Tsao, G.T. 1981. Ethanol production from xylose by enzymic isomerisation and yeast fermentation. Biotechnology and Bioengineering Journal, 11: 263–274.

Chan, E.C., Ueng, P.P. and Chen, L.F. 1989. Metabolism of D-xylose in *Schizosaccharomyces pombe* cloned with a xylose isomerase gene. Applied Microbiology and Biotechnology, 31(5): 524–528.

Chen, W.P. 1980a. Glucose isomerase (a review). Process Biochemistry, 15: 30–35.

Chen, W.P. 1980b. Glucose isomerase (a review). Process Biochemistry, 15: 36–41.

Chopda, V.R., Nagula, K.N., Bhand, D.V. and Pandit, A.B. 2014. Studying the effect of nature of glass surface on immobilization of glucose isomerase. Biocatalysis and Agricultural Biotechnology, 3: 86–89.

Dekker, K., Sugiura, A., Yamagata, H., Sakaguchi, K. and Udaka, S. 1992. Efficient production of thermostable *Thermus thermophilus* xylose isomerase in *Escherichia coli* and *Bacillus brevis*. Applied Microbiology and Biotechnology, 36(6): 727–732.

du Preez, J.C., Bosch, M. and Prior, B.A. 1986. Xylose fermentation by *Candida shehatae* and *Pichia stipitis*: Effects of pH, temperature and substrate concentration. Enzyme and Microbial Technology, 8(3): 360–364.

Farber, G.K., Petsoko, G.A. and Ringe, D. 1987. The 3.0 α crystal structure of xylose isomerase from *Streptomyces olivochromogenes*. Protein Engineering, 1(6): 459–466.

Gaikwad, S.M., More, M.W., Vartak, H.G. and Deshpande, V.V. 1988. Evidence for the essential histidine residue at the active site of glucose/xylose isomerase from *Streptomyces*. Biochemical and Biophysical Research Communications, 155(1): 270–277.

Ge, Y., Wang, Y., Zhou, H., Wang, S., Tong, Y. and Li, W. 1999. Co-immobilization of glucoamylase and glucoisomerase by molecular deposition technique for one-step conversion of dextrin to fructose. Journal of Biotechnology, 67: 33–40.

Ghatge, M.S., Phadatare, S.U., Bodhe, A.M. and Deshpande, V.V. 1994. Unfolding and refolding of glucose/xylose isomerase from *Streptomyces* sp. NCIM 2730. Enzyme and Microbial Technology, 16(4): 323–327.

Haitao, Y., Yanglong, G., Wu, D., Zhan, W. and Lu, G. 2011. Immobilization of glucose isomerase onto GAMM support for isomerisation of glucose to fructose. Journal of Molecular Catalysis B: Enzymatic, 72(1-2): 73–76.

Hanefeld, U., Gardossi, L. and Magner, E. 2009. Understanding enzyme immobilisation. Chemical Society Reviews, 38: 453–468.

Hanover, L.M. and J.S. White. 1993. Manufacturing, composition, and applications of fructose. American Journal of Clinical Nutrition, 58: 724–732.

Hartley, B.S., Hanlon, N., Jackson, R.J. and Rangarajan, M. 2000. Glucose isomerase: insights into protein engineering for increased thermostability. Biochimica et Biophysica Acta, 1543: 294–335.

Illanes, A. 2011. Whey upgrading by enzyme biocatalysis. Electronic Journal of Biotechnology, 14(6).

Karaoglu, H., Yanmis, D., Sal, F.A., Celik, A., Canakci, S. and Belduz, A.O. 2013. Biochemical characterization of a novel glucose isomerase from *Anoxybacillus gonensis* G2[T] that displays a high level of activity and thermal stability. Journal of Molecular Catalysis B: Enzymatic, 97: 215–224.

Katz, A.K., Li, X., Carrell, H.L., Hanson, B.L., Langan, P., Coates, L., Schoenborn, B.P., Glusker, J.P. and Bunick, G.J. 2006. Locating active-site hydrogen atoms in D-xylose isomerase: time-of-flight neutron diffraction. Proceedings of the National Academy of Sciences, 103(22): 8342–8347.

Kume, T. and Takahisa, M. 1983. Effect of radical ions on the inactivation of glucose isomerase. Agricultural and Biological Chemistry, 47: 359–363.

Li, Y.C., Li, G.Y., Gou, M., Xia, Z.Y., Tang, Y.Q. and Kida, K. 2015. Functional expression of xylose isomerase in flocculating industrial *Saccharomyces cerevisiae* strain for bioethanol production. Journal of Bioscience and Bioengineering, 20(20): 1–7.

Ligthelm, M.E., Prior, B.A. and du Preez, J.C. 1988. The oxygen requirements of yeasts for the fermentation of D-xylose and D-glucose to ethanol. Applied Microbiology and Biotechnology, 28: 63–68.

Liu, S., Wiegel, J. and Gherardini, F.C. 1996. Purification and cloning of a thermostable xylose (glucose) isomerase with an acidic pH optimum from *Thermoanaero bacterium* strain JW/SL-YS 489. Journal of Bacteriology, 178: 5938–5945.

MacConnell, M. 2016. Sugars and Sweetners Outlook. United States Department of Agriculture, Economic Research Service, USDA.

Mezghani, M., Borgi, M.A., Kammoun, R., Aouissaoui, H. and Bejar, S. 2005. Construction of new stable strain over expressing the glucose isomerase of the *Streptomyces* sp. SK strain. Enzyme and Microbial Technology, 37: 735–738.

Mu, W., Wang, X., Xue, Q., Jiang, B., Zhang, T. and Miao, M. 2011. Characterization of thermostable glucose isomerase with an acidic pH optimum from *Acidothermus cellulolyticus*. Food Research International, 47: 364–367.

Palazzi, E. and Converti, A. 2001. Evaluation of diffusional resistance in the process of glucose isomerization to fructose by immobilized glucose isomerase. Enzyme and Microbial Technology, 28: 246–252.

Pubols, M.H., Zahnley, J.C. and Axelrod, B. 1963. Partial purification and properties of xylose and ribose isomerase in higher plants. Plant Physiol., 38: 457–461.

Ramagopal, U.A., Dauter, M. and Dauter, Z. 2003. Phasing on anomalous signal of sulfurs: What is the limit? Acta Crystallographica D-Biological Crystallography, 59(6): 1020–1027.

Reynolds, J.H. 1974. The uses of precipitated nylon as an enzyme support: An α-galactosidase reactor. pp. 63–70. *In*: A.C. Olsen and C.L. Cooney (eds.). Immobilized Enzymes in Food and Microbial Processes. Plenum Press, New York.

Rose, I.A., O'Connell, E.L. and Mortlock, R.P. 1969. Stereochemical evidence for a *cis*-enediol intermediate in Mn-dependent aldose isomerases. Biochimica et Biophysica Acta-Enzymology, 178(2): 376–379.

Sarryar, B., Ozkan, P., Kirdar, B. and Hortacsu, A. 2004. Expression and translocation of glucose isomerase as a fusion protein in *E. coli*. Enzyme and Microbial Technology, 35(2-3): 105–112.

Seyhan, S. and Alagoz, D. 2008. Catalytic efficiency of immobilized glucose isomerase in isomerization of glucose to fructose. Food Chemistry, 111(3): 658–662.

Takasaki, Y. and Tanabe, O. 1966. Studies on sugar-isomerizing enzyme: Production and utilization of glucose isomerase from *Streptomyces* sp. Agricultural and Biological Chemistry, 30: 1247–1253.

Vongsuvanlert, V. and Tani, Y. 1988. Purification and characterization of xylose isomerase of a methanol yeast, Candida boidinii, which is involved in sorbitol production from glucose. Agricultural and Biological Chemistry, 52(7): 1817–1824.

Wang, P.Y., Johnson, B.F. and Scneider, H. 1980. Fermentation of D xylose by yeasts using glucose isomerase in the medium to convert D-xylose to D-xylulose. Biotechnology Letters, 2: 273–278.

Yamamoto, H., Miwa, H. and Kunishima, N. 2008. Crystal structure of Glucose-6-phosphate isomerase from *Thermus thermophilus* HB8 showing a snapshot of active dimeric state. Journal of Molecular Biology, 382: 747–762.

Yassien, M.A.M., Jiman-Fatani, A.M. and Asfour, H.Z. 2013. Purification, characterization and immobilization glucose isomerase from *Streptomyces albaduncus*. African Journal of Microbiology Research, 7(21): 2682–2688.

Yu, H., Gluo, Y., Wu, D., Zhan, W. and Lu, G. 2011. Immobilization of glucose isomerase onto GAMM support for isomerization of glucose to fructose. Journal of Molecular Catalysis B: Enzymatic, 72: 73–76.

Yu, Z., Lowndes, J. and Rippe, J. 2013. High-fructose corn syrup and sucrose have equivalent effects on energy-regulating hormones at normal human consumption levels. Nutrition Research, 33(12): 1043–1052.

Zhang, F., Duan, X., Chen, S., Wu, D., Chen, J. and Wu, J. 2013. The addition of Co^{+2} enhances the catalytic efficiency and thermostability of recombinant glucose isomerase from *Thermobifida fusca*. Process Biochemistry, 48: 1502–1508.

Zhao, H., Cui, Q., Shah, V., Xu, J. and Wang, T. 2016. Enhancement of glucose isomerase activity by immobilizing on silica/chitosan hybrid microspheres. Journal of Molecular Catalysis B: Enzymatic, 126: 18–23.

Sucrose Transforming Enzymes
Hydrolysis and Isomerization
Harish B.S. and *Kiran Babu Uppuluri**

1. Introduction

Sucrose, the most abundant disaccharide in Nature, contains glucose and fructose. In most of the plants sucrose accumulates in the leaves or gets stored in developing fruits, seeds, roots and tubers. Sucrose along with starch serves as a reserve carbohydrate in most of the plants (Nagem et al., 2004). Sucrose synthesis and cleavage in plants is significant in plant metabolism (Zheng et al., 2011; Figueroa et al., 2013). First it was believed that sucrose and its metabolism is a characteristic feature of plants only, but it was later found in oxygenic photosynthetic organisms. In cyanobacteria, sucrose serves as a carbon source and its metabolism is significant in nitrogen fixation. Recent studies show the presence of sucrose-related genes in non-photosynthetic organisms, such as proteobacteria, firmicutes and planctomycetes. These organisms would have acquired the genes of sucrose metabolism by horizontal gene transfer (Wu et al., 2015).

Sucrose, which is available in bulk quantities at relatively low costs, can be easily transformed into its monosaccharide units, or fructans, or dextrans, or sucrose isomers, or other high value glycosides, or oligosaccharides, by the action of various enzymes, collectively called as sucrases. Further, the conversion of sucrose into its sugar constituents and to other valuable products is one of the most important step in sustainable energy production as sugars are the key intermediates between biomass and biofuels (Pito et al., 2012).

Bioprospecting Laboratory, School of Chemical and Biotechnology, SASTRA University, Thanjavur, Tamil Nadu-613401, India.
  E-mail: harish.bs74@gmail.com
* Corresponding author: kinnubio@gmail.com

In this chapter, an overview of major enzymes involved in sucrose transformation (hydrolysis and isomerization), the microbial production of these enzymes and their applications in food-processing industries are discussed. For the sake of convenience, these enzymes can be segregated into four different groups based on the catalytic reaction and the resultant product.

- **Hydrolytic enzymes**—invertase, sucrose synthase and exo-inulinase
- **Isomerase**—sucrose isomerase
- **Transfructosylation/Fructan synthesizing enzymes**—inulosucrase, levansucrase and sucrose: sucrose fructosyl transferase
- **Transglycosylation enzymes**—sucrose phosphorylase and dextransucrase

The utilization of the energy stored in sucrose is initiated by cleaving the O-glycosidic bond catalyzed by two enzymes with different properties—invertase (EC 3.2.1.26) and sucrose synthase (2.4.1.13). Invertase is a hydrolase which cleaves sucrose to its monosaccharide constituents while sucrose synthase is a glycosyl transferase which, in presence of UDP, reversibly converts sucrose to UDP-glucose and fructose (Sturm and Tang, 1999; Qi et al., 2014). Besides these two enzymes, exo-inulinase (EC 3.2.1.80) from plants and microbes can also hydrolyze sucrose (Zittan, 1981; Catana et al., 2005). Sucrose can also be converted to its structural isomers (isomaltulose and trehalulose) by sucrose isomerase (EC 5.4.99.11). These isomers have advantages over sucrose in certain applications (Contesini et al., 2013).

Sucrose can be enzymatically converted to high-value carbohydrate molecules, such as fructooligosaccharides and fructo polymers by the action of various fructosyl transferases (inulosucrase, levansucrase and sucrose: sucrose fructosyl transferase) and sometimes by invertase at high sucrose concentration (Anwar et al., 2008).

Sucrose phosphorylase (EC 2.4.1.7) catalyzes the reversible phosphorolysis of sucrose into glucose-1-phosphate, releasing fructose, in which the product is key intermediate in the synthesis of nucleotide sugar and donor to the glycosyl transferases (Wildberger et al., 2015). Glycosyl units in sucrose can be utilized for the production of dextran by dextransucrases (EC 2.1.4.5) (Falconer et al., 2011).

Sucrose-converting enzymes belong either to glycoside hydrolase (GH) family or glycosyl transferase (GT) family. The catalytic mechanism of glycoside hydrolase family enzymes involved in sucrose conversion is believed to be double displacement/ ping-pong reaction with a covalent enzyme-glycosyl intermediate which is further hydrolyzed by suitable acceptors (Lammens et al., 2008). Three putative catalytic mechanisms for glycosyl transferase family enzymes have been suggested: (i) double displacement mechanism with a covalent enzyme-glycosyl intermediate; or (ii) $S_N i$ (substituition nucleophilic internal) mechanism involving a transition state with a transient oxocarbenium ion pair followed by concerted phospho-sugar bond breakage and glycosidic bond formation on the same face of the sugar; or (iii) $S_N i$-like mechanism with a transient oxocarbenium phosphate ion pair intermediate with a conformational shift as the incoming acceptor attacks (Vetting et al., 2008; Zheng et al., 2011).

2. Enzymes Involved in the Cleavage of Sucrose

Hydrolysis of sucrose is a common practice where the monosaccharide constituents are preferred. Fructose is sweeter than sucrose. The enzymatically hydrolyzed sugar mixture is colorless in comparison to the colored compounds formed by the acid hydrolysis. Invertase is the most important enzyme involved in the hydrolysis of sucrose to glucose and fructose (Fig. 1). Sucrose synthase and exo-inulinase are the other two enzymes involved in cleaving the O-glycosidic bond.

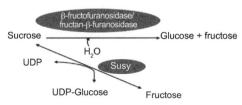

Figure 1. Enzymatic transformation of sucrose.

2.1 Invertase

Invertase/β-fructofuranosidase (EC 3.2.1.26) hydrolyzes the terminal non-reducing β-fructofuranoside residue in the β-fructofuranosides. Invertase is a hydrolase that cleaves sucrose to its monosaccharide constituents. Invertase belongs to glycoside hydrolase (GH) 32 families. The enzyme is a glycoprotein inhibited by cations, such as Hg^{2+}, Ag^{+}, Co^{2+} and Cu^{2+}. 2,5-anhydro D-mannitol, a furanose form of fructose can competitively inhibit the enzyme (Kulshrestha et al., 2013). The enzyme is named because of the inversion of the optical rotation after the hydrolysis of sucrose. Sucrose is dextrorotatory, but the resulting mixture of glucose and fructose is slightly levorotatory. Levorotatory fructose is having more molar rotation than dextrorotatory glucose (Kulshrestha et al., 2013).

2.1.1 Invertase sources

It is widely found in some animals, higher plants, filamentous fungi, yeast and bacteria (Plascencia-Espinosa et al., 2014). Plants like Japanese pear fruit (*Pyrus pyrifolia*), pea (*Pisum sativum*) and oat (*Avena sativa*) are extensively studied for the invertase extraction. Among all these, yeast invertases are the most extensively studied hydrolytic enzymes. *Saccharomyces cerevisiae* is a chief strain for commercial production of invertase. Invertases from non-conventional yeasts, other than *Saccharomyces*, such as *Candida utilis, Pichia anomala, Xanthophyllomyces dendrorhous, Rhodotorula glutinis, Schwanniomyces occidentalis* and *Candida guilliermondii* are also known to produce invertase (Kulshrestha et al., 2013).

2.1.2 Invertase types

Different isoforms of the enzyme are found in both plants and microbes with different biochemical properties and sub-cellular locations. In plants, it can be vacuolar (acid soluble), cytoplasmic (alkaline soluble) and cell wall bound invertase. In yeasts, it can be of two types: a heavily glycosylated secreted form and non-glycosylated intracellular form (Kulshrestha et al., 2013; Sainz-Polo et al., 2013).

2.1.3 Invertase catalytic mechanism

Yeast invertases are encoded by a family of SUC genes. The crystal study of *Saccharomyces* invertase shows that the enzyme folds to catalytic β-propeller and β-sandwich domains which are characteristic to the GH 32 enzymes. Yeast invertases in Nature have been found organized as monomers, dimers, tetramers, hexamers or octomers. Dimerization plays an important role in substrate specificity as it sets stearic constraints that limit the access of the active sites to the oligosaccharides of more than four units. Interaction between the dimmers make tetramers, hexamers or octomers (Esmon et al., 1987; Sainz-Polo et al., 2013).

The catalytic activity of invertase on sucrose is by a double-displacement mechanism, in which anomeric configuration of the substrate is retained (Table 1). The glycosidic oxygen of the sugar substrate is protonated by acid/base catalyst (Glu-204 in yeast invertase) followed by nucleophilic attack (Asp-23 in yeast invertase) on the anomeric carbon, forming a covalent fructose-enzyme intermediate. Subsequently this intermediate is hydrolyzed, releasing fructose and free invertase (Reddy and Maley, 1990; Lammens et al., 2008).

Table 1. Microbial production of invertase.

Producer	Substrate	Nature of Enzyme Usage	Activity	Vm	Km	References
Cladosporium cladosporioides	Pomegranate peel waste	Submerged fermentation	197.50 Units/mL	28.57 IU/mg	0.26 mg/mL	Uma et al. (2012)
Saccharomyces cerevisiae	Synthetic media	Immobilized on modified beidellite nanoclays	2200 U/g	-	37 mM	Andjelković et al. (2015)
Lactobacillus brevis	Synthetic media	Immobilized on alginate beads	1399 U/ml	-	-	Awad et al. (2013)
Candida guilliermondii	Sucrose	Submerged fermentation	82,027 U/mg of protein	10.9 µmol/min/mg of protein	0.104 mM	Plascencia-Espinosa et al. (2014)
Aspergillus niger	Agro-industrial wastes	Solid state fermentation	140 U/g	-	-	Ohara et al. (2015)

2.1.4 Invertase applications

Invertase has the following application in industries:

2.1.4.1 In food industry

Invertase is commonly used in the food industry where fructose is preferred, as it is sweeter and does not crystallize easily. The commercial applications of the enzyme include production of confectionery with liquid of soft centers, chocolates and in the fermentation of cane molasses into ethanol. The inversion of sucrose in food products increases the solubility of the total sugar, thus increasing the density of syrups containing sucrose. Moreover, the hygroscopic character of the inverted sucrose due to the presence of levulose retards the complete drying of products, such as fondant confections and icings. It also helps in controlling crystallization of sucrose and modifies the flavor (Paine et al., 1925). The enzyme is preferred over the customarily used acids and acid salts which have unfavorable effects—producing caramelized, off flavored and colored invert sugar. The determining factors in industrial usage of invertase include: pH, temperature, period for inversion and the commercial availability of the enzyme. Inversion of all the sucrose in the syrup is undesirable because of the danger of crystallization of dextrose. The influence of non-sugar components in the solubility equilibrium between the sucrose and the inverted sugar should also be considered (Uma et al., 2012).

2.1.4.2 In fermentation technology

The fermentation of sucrose-rich substrates to alcoholic beverages, lactic acid and glycerol requires the use of invertase (Kulshrestha et al., 2013). More than half of the world's ethanol production relies on sucrose-rich substrates, like sugar cane juice and molasses. Extracellular invertase of *Saccharomyces cerevisiae* can hydrolyze sucrose into glucose and fructose, and this can be used in the production of baker's yeast as well as in different alcoholic beverages. Glucose and fructose are taken into the cells by hexose transporters and is further metabolized through glycolysis (Badotti et al., 2008).

Industrially, *S. cerevisiae* is exposed to hyper osmotic conditions when it is grown to ferment high sucrose containing medium. This results in complex responses in the cells, including rapid reduction of internal cell volume due to water efflux. This may induce several genes governing high osmolarity glycerol response pathway that enhances production and intracellular retention of glycerol. There is a strong correlation between synthesis and retention of glycerol; and the fermentation ability of yeasts. At lower sucrose concentrations, there is no correlation between glycerol production and its fermentation ability, but as the concentration increases, there is a correlation between glycerol production and its fermentation ability. The ability to produce and retain glycerol increases when fermentable sugars are present in comparison to non-fermentable sugars with same osmotic pressure. The invertase activity and fermentation activity are linked inversely (Myers et al., 1997).

2.1.4.3 In pharmaceutical industry

In the pharmaceutical industry, it is used as a digestive aid, in milk powder for infants, in calf feed preparation, assimilation of alcohol in fortified wines and in manufactured inverted sugars as food for honeybees (Uma et al., 2012). For human consumption it is either used alone or used along within a multi enzyme formula. It also plays a key role in human disease prevention, physical rejuvenation and anti-ageing.

2.1.4.4 In analytical technology

Analyzing the concentration and purity of sucrose in food and beverage is one of the important quality assurance tests in the food and beverage industry. Immobilized invertase is widely used as a biosensor for detecting sucrose. Invertase is also an ideal enzyme for inhibition-based electrochemical sensors for the detection of heavy metal ions due to low cost, stability and high specific activity (Bagal-Kestwal et al., 2008).

2.1.4.5 Other applications

A glycoprotein secreted from hypo-pharyngeal glands of honeybees possess the invertase activity. It plays an important role in honey ripening as it converts sucrose to glucose and fructose which are physiologically usable by bees. Invertase is also used for the production of artificial honey which is a powerful antimicrobial and antioxidant that helps in preventing bacterial infestations and gut fermentation due to oxidation (Kulshrestha et al., 2013). Determination of invertase activity in the honey is used as one of the parameters to estimate its freshness (Vorlova and Pridal, 2002). Invertase also finds application in the cosmetic industry for the production of plasticizing agents.

2.2 Sucrose synthase

Sucrose synthase/NDP-glucose: D-fructose 2-alpha-D-glucosyltransferase (2.4.1.13) is a member of GT-4 glycosyl transferase sub-family within the metal-independent GT-B glycosyl transferase super family. In the presence of UDP, sucrose synthase reversibly converts sucrose to UDP-glucose and fructose. It is the only sucrose-cleaving enzyme which is capable of both synthesis and cleavage of sucrose. It is a versatile biocatalyst in the synthesis of nucleotide sugars and sucrose analogues which may be used as precursors for starch and cellulose biosynthesis. Sucrose synthase is one of the key plant enzymes involved in carbohydrate metabolism, controlling not only sucrose metabolism but also structural and storage pathways (Sauerzapfe et al., 2008).

In plants, sucrose synthase plays a vital role in pollen-tube growth, nitrogen fixation, biomass production and maturation of fruits and seeds during periods of abiotic stress (Zheng et al., 2011). Although it is capable of catalyzing both synthesis and cleavage of sucrose in a reversible and energy neutral manner, but *in vivo* the enzyme is believed to function predominantly in the cleavage of sucrose (Baier et al., 2007). This is a key step in providing substrates for rapidly-growing tissues and sink organs.

The breakdown of sucrose is also an important step in nitrogen fixation and a necessary prerequisite for nodule development and its function. The deficiency of sucrose synthase in root nodules renders nitrogen fixation inefficient (Röhrig et al., 2004). The presence of sucrose synthase at plasma membrane plays a vital role in directing the carbon flow for cell-wall biosynthesis (Baroja-Fernández et al., 2012). It also provides plant cell with activated sugar precursors for sucrose-starch transformations (Römer et al., 2001).

2.2.1 Sucrose synthase catalytic mechanism

The biochemical characterization of sucrose synthase enzyme has been done by recombinant expression of the gene responsible for the enzyme from plants and cyanobacteria in microorganisms, such as *Escherichia coli* and *S. cerevisiae* (Sauerzapfe et al., 2008; Figueroa et al., 2013). Sucrose synthase is encoded by *suc* genes. In higher plants, it is encoded by two genes: *sus1* and *sus2*, but in rice and in pea a third gene, *sus3*, is also present (Römer et al., 2004). The occurrence of sucrose synthase gene in cyanobacteria is neither ubiquitous nor consistent with the phylogeny of cyanobacteria (Kolman et al., 2012).

The catalytic mechanism of glycosyl transferases that utilize nucleoside diphosphate sugars remains unclear (Lee et al., 2011). The putative saccharide transfer mechanism by glucosyl transferases is done in three different ways:

- Double displacement mechanism, in which a covalent enzyme-glycosyl intermediate is involved
- S_Ni mechanism, that involves a transition state with a transient oxocarbenium ion pair that is not dissociated from donor and acceptor
- S_Ni-like mechanism, that combines a transient ion pair intermediate with a conformational shift, as the incoming acceptor attacks (Table 2).

Table 2. Microbial Production of sucrose synthase.

Producer	Nature of Production	Specific Activity	Reference
Anabaena sp.	Native	0.71338 U/mg protein	Porchia et al. (1999)
susy1 from *Solanum tuberosum* in *S. cerevisiae*	Recombinant	0.3 U/mg protein	Römer et al. (2004)
susy1 from *Solanum tuberosum* in *E. coli*	Recombinant	0.01 U/mg	Sauerzapfe et al. (2008)
susy1 from *Solanum tuberosum* in *S. cerevisiae*	Recombinant	0.28 U/mg	Sauerzapfe et al. (2008)
susy from *Thermosynechococcus elongatus* in *E. coli*	Recombinant	1.28 U/mg	Figueroa et al. (2013)

An S_Ni-like mechanism is proposed for enzymes in the GT-B enzyme family which includes sucrose synthase. As sucrose synthase binds to UDP-glucose and fructose or UDP and sucrose, the ternary enzyme substrate complex rapidly generates a stabilized oxocarbenium phosphate ion pair intermediate. The collapse of this intermediate towards sucrose cleavage or sucrose synthesis depends on the subtle changes in the position of carbenium ion with respect to UDP or fructose (Zheng et al., 2011; Wu et al., 2015).

The crystal structure of sucrose synthase from *Nitrosomonas europaea* shows a tetramer with four identical subunits, with all the subunits carrying four domains: Sucrose synthase N-terminal-1 (SSN-1); sucrose synthase N-terminal-2 (SSN-1); GT B (D) that typically binds to the nucleotide donor; and sugar-binding GT B (A). Catalytic triad of conserved homologous catalytic residues Arg 567, Lys 572 and Glu 663 implies the common origin with the GT-B family (Wu et al., 2015).

2.2.2 Sucrose synthase sources

Sucrose synthase is present in most of the plants and plays an important role in the plant carbohydrate metabolism. The occurrence of sucrose synthase is not ubiquitous in cyanobacteria as that in the plants. It is mostly found in filamentous heterocyst-forming strains, *Anabaena* sp., *Nostoc punctiforme* and *Nostoc commune* (Kolman et al., 2012). In unicellular cyanobacteria, sucrose cleavage plays an important role in several physiological functions, such as carbon metabolism, nitrogen fixation and stress tolerance. Sucrose synthase form cyanobacteria strains, such as *Microcystis aeruginosa, Gloeobacter violaceus* and *Thermosynechococcus elongates* and prefer ADP rather than UDP (Figueroa et al., 2013). Sucrose synthase from non-photosynthetic bacterium like *N. europaea* is also reported and characterized (Wu et al., 2015). To meet the large-scale industrial needs, efforts are on to express the enzyme responsible for the enzyme in microbes, using recombinant technology.

2.2.3 Sucrase synthase applications

Sucrose synthase is a versatile tool for carbohydrate engineering. It can be used for synthesis of nucleotide sugars, sucrose analogues and glycoconjugates (Römer et al., 2004). Sucrose synthase is a homotetrameric enzyme, catalyzing the reversible, UDP-dependent cleavage of sucrose to UDP-glucose and fructose and these products are precursors for plant-starch biosynthesis and bacterial glycogen synthesis.

2.2.3.1 For synthesizing nucleotide sugars

Sucrose synthase is used for the preparation of activated glucoses by the combination of sucrose synthase with kinases (myokinase) using inexpensive nucleoside monophosphate and sucrose. In this, myokinase catalyzes the synthesis of nucleoside diphosphate which is later used by sucrose synthase (Zervosen et al., 1998). The UDP-glucose produced by sucrose synthase can be a starting material for various important oligosaccharides. It can be converted to UDP-galactose by UDP-Gal-4-epimerase. This UDP-galactose in the presence of myoinositol can be converted to galactinol by galactinol synthase. Galactinol is the precursor for synthesis of the raffinose family of oligosaccharides (Wakiuchi et al., 2003).

2.2.3.2 For synthesizing sucrose analogues

Sucrose analogues can be synthesized using the sucrose synthase enzyme. The ability of the sucrose synthase to have wide specificity in its substrates in sucrose synthesis direction was utilized in the synthesis of various sucrose analogues. In such an effort, sucrose synthase 1 from potato expressed in *S. cerevisiae* is used for the preparative synthesis of 1'-deoxy-1'-fluoro-β-Dfructofuranosyl-α-D-glucopyranoside, $[^{13}C_1]$-β-D-fructofuranosyl-α-D-glucopyranoside, α-D-glucopyranosyl-α-L-sorbofuranoside and α-D-glucopyranosyl-α-D-lyxopyranoside, using 1-deoxy-1-fluoro-D-fructose, $[^{13}C_1]$-β-D-fructofuranoside, L-sorbose and D-lyxose respectively as substrates (Römer et al., 2001).

Susy1 from potato also accepts a wide spectrum of D/L ketoses (e.g., 1-deoxy-1-fluoro-D-fructose, L-sorbose and D-xylulose) and D/L aldoses (e.g., D-talose, D-idose, D-lyxose, L-arabinose and D-ribose) in the synthesis direction. These non-natural substrates to the enzyme are poor substrates when compared with the natural substrate-D-fructose (Römer et al., 2001, 2004).

2.2.3.3 For enhancing carbohydrate production

Cellulose production in *Acetobacter xylinum* was found to be enhanced by incorporating plant sucrose synthase. Expression of sucrose synthase in *Acetobacter xylinum* changed the microbial sucrose metabolism, thus efficiently increasing the UDP-glucose production. The low level of cellular UDP enhanced the incorporation of glucose to growing 1, 4-β-glucan increasing cellulose production. Sucrose synthase channeled carbon directly from sucrose to cellulose in the bacteria (Nakai et al., 1999). The over-expression of sucrose synthase gene in crops improves yield and starch content. It has been reported that sucrose synthase is involved in cellulose and starch biosynthesis in *Arabidopsis* (Baroja-Fernández et al., 2012).

2.3 Inulinases

Inulinases belong to glycoside hydrolase (GH) 32 family that mostly hydrolyzes inulin, a fructose polymer. Besides inulin, it can also hydrolyze sucrose, raffinose and levan (de Paula et al., 2008; Kim et al., 2008). Inulinases can be inulin-specific endo-inulinase (2,1–β-D-fructan fructanohydrolase EC 3.2.1.7) that hydrolyze inulin by breaking the bonds between fructose away from the ends; or exo-inulinases (fructan-β-fructosidase EC 3.2.1.80) that hydrolyze terminal, non-reducing 2,1 linked and 2,6-linked β-D fructofuranose residues in inulin, levan and sucrose, releasing β-D-fructose (Nagem et al., 2004; Sokolenko and Karpechenko, 2015). Inulinases are usually thermo stable and commercially available for industrial applications (Ettalibi and Baratti, 2001). Exo-inulinase, which can hydrolyze sucrose, is used for the production of high purity fructose syrups.

2.3.1 Inulinases catalytic mechanism

Exo inulinase is a monomeric protein encoded by exo-inulinase gene (inuD) which is present as a single copy in the genome (Moriyama et al., 2002). The three-dimensional structure and catalytic mechanism are similar to other members of GH 32 family. The high resolution X-ray study of exo-inulinase from *Aspergillus awamori* showed that the enzyme has two domains—a longer N-terminal catalytic domain of the rare five-bladed β-propeller fold and smaller C-terminal domain folded as a β-sandwich (Table 3).

The catalytic reaction of exo-inulinase follows double displacement/ping-pong mechanism and involves overall retention of anomeric configuration of the substrate. The catalytic amino acids in exo-inulinase from *Aspergillus awamori* are Glu 241 (proton donor) and Asp 41 (nucleophile) (Nagem et al., 2004).

2.3.2 Inulinases sources

Inulinases are found in plants, filamentous fungi, yeasts (de Paula et al., 2008) and bacteria (Kim et al., 2008). Microorganisms are the best sources for commercial inulinase production as they are easy to culture and facilitate high production of the

Table 3. Microbial production of Inulinases.

Producer	Substrate	Nature of Enzyme	Vm	Km	References
Aspergillus awamori	Inulin	Free	175±5 µmol/min/mg	0.003 ± 0.0001 mM	Arand et al. (2002)
Aspergillus awamori	Levan	Free	2.08 ± 0.04 mg/ml	1.2 ±0.02 µmol/min/mg	Arand et al. (2002)
Kluyveromyces marxianus var. *bulgaricus*	Sucrose	Free	37.60 IU/mg protein	61.83 mM	de Paula et al. (2008)
Kluyveromyces marxianus var. *bulgaricus*	Sucrose	Immobilized	31.45 IU/mg	149.28 mM	de Paula et al. (2008)
Aspergillus niger	Sucrose	Immobilized			Silva et al. (2013)
Aspergillus ficuum	Sucrose	Free	192 ± 5.9 IU/mg	0.060 ± 0.005 M	Ettalibi and Baratti (2001)
Aspergillus ficuum	Sucrose	Immobilized	163 ± 5.4 IU/mg	0.148 ± 0.010 M	Ettalibi and Baratti (2001)

enzyme (Chi et al., 2009). Among the bacterial, fungal and yeast genera of inulinase producers, yeasts are found to be better producers of the enzyme (Sokolenko and Karpechenko, 2015). But inulinases from fungi is advantageous for the sucrose hydrolysis as they are often extracellular and exo-acting (Pandey et al., 1999). Among filamentous fungi *Penicillium* sp. (Onodera and Shiomi, 1992), *Aspergillus* sp. (Arand et al., 2002), and *Rhizopus* sp. are common inulinase producers (Mohamed et al., 2015).

2.3.3 Inulinases applications

Inulinases have the following applications in industries:

2.3.3.1 In food industry

Exo-inulinase is a good alternative for invertase for the production of fructose syrups by having ability to act on more number of substrates including sucrose, inulin, levan and raffinose. The hydrolysation of these substrates by exo-inulinase is advantageous over the other methods of fructose production, like acid hydrolysis of substrates and starch hydrolysis followed by glucose isomerization (Zhou et al., 2015). Fructose can be used as a sweetener in the food industry and it is significantly sweeter than table sugars, like sucrose and glucose. Using inulinase, fructose can be produced by a single-step direct hydrolysis of fructan polysaccharides (Sirisansaneeyakul et al., 2007). Fructose is used instead of sucrose in food, pharmaceuticals and beverages. It has beneficial effects on diabetic patients, enhances iron absorption in children, stimulates calcium absorption, prevents colon cancer, has prebiotic effects and is used as dietary fiber because of its fat-like texture (Chi et al., 2009). Fructooligosaccharides, produced with inulinase, are finding application in the food industry because of its prebiotic and beneficial effects on the human health (Chen et al., 2009).

2.3.3.2 In fermentation technology

Ethanol can be produced from inulin-rich feedstock, like *Jerusalem artichoke* tubers through the consolidated bioprocessing strategy using inulinase-producing recombinant yeasts, like *Kluyveromyces marxianus* (Yuan et al., 2013) and *Saccharomyces* sp. (Li et al., 2013). Inulinases can hydrolyze inulin/sucrose-rich substrates to fermentable sugars, like glucose and fructose. Inefficient inulinase production under ethanol production may prolong the production. This can be overcome by overexpressing the inulinase gene (Yuan et al., 2013).

2.3.3.3 For measuring fructans

The exo-inulinase along with endo-inulinase is used for measuring fructans (inulin or fructo-oligosaccharides) in plant and food products. The fructans extracted from the samples should be treated with a mixture of sucrase (invertase), α-amylase, pullulanase and maltase to hydrolyze sucrose and starch to their constituents. These interfering sugar units present in the sample should be removed by borohydride reduction through treatment with alkaline borohydride solution that converts reducing sugars to sugar alcohols. It also reduces the terminal-reducing fructose in the fructan chain. The solution is neutralized and excess borohydride is removed, using dilute acetic acid. The fructan is then hydrolyzed to fructose and glucose using fructanase (exo- and endo-inulinases). The reducing sugar formed is spectrophotometrically measured after reaction with para-hydroxybenzoic acid hydrazide (McCleary et al., 2000).

3. Sucrose Isomerase

The natural sucrose isomers have glycosidic bonds between anomeric carbon of glucose and non-anomeric carbon of fructose moiety, resulting in five reducing disaccharides with individual properties having advantages over sucrose for some applications. The structural isomers of sucrose include trehalulose (α-1, 1), turanose (α-1, 3), maltulose (α-1, 4), leucrose (α-1, 5) and isomaltulose (α-1, 6). Turanose, maltulose and leucrose are formed as occasional side-products of polyglucan synthesis or starch hydrolysis. Trehalulose and isomaltulose are produced by the action of sucrose isomerase (Goulter et al., 2012).

Sucrose isomerase (EC 5.4.99.11) hydrolyzes the α-1,2 bond between glucose and fructose to form an α-1,6 bond to produce isomaltulose (palatinose™, α-D-glucopyranosyl-1,6-D-fructofuranose) or an α-1,1 bond to produce trehalulose (α-D-glucopyranosyl-1,1-D-fructofuranose). Based on the product formed, sucrose isomerase is known as isomaltulose synthase or trehalulose synthase (Contesini et al., 2013; Nam et al., 2014; Wu et al., 2015). It is believed that the microorganisms utilize this enzyme as a system to irreversibly sequester the source of carbon and energy into a form that is inaccessible to the host plant and other competing microorganisms (Nam et al., 2014). In addition to these products, the enzyme also hydrolyzes sucrose into small amounts of glucose and fructose (Mu et al., 2014).

Sucrose isomerases can convert sucrose to its isomers without any cofactors and the conversion is complete because of the multi-step mechanism and low free energy of the isomers (Goulter et al., 2012). Purified sucrose isomerase from several microorganisms exhibit differences in percentage of palatinose (isomaltulose) and trehalulose production. Sucrose isomerase from *Serratia plymuthica*, *Erwinia rhapontici*, *Klebsiella* sp., and *Pantoea dispersa* mainly produce palatinose (75–85 per cent) while sucrose isomerase from *Pseudomonas mesoacidophila* and *Agrobacterium radiobacter* produces trehalulose (91 per cent) as a major product (Aroonnual et al., 2007). The ratio of isomers produced not only depends on the microbial strain but also on the environmental conditions, such as pH, temperature and other reaction conditions in which the culture is maintained. A lower temperature increases the trehalulose ratio in the product (Nam et al., 2014).

3.1 Sucrose isomerase catalytic mechanism

Sucrose isomerase belongs to glycoside hydrolase (GH) 13 family based on homology. The isomaltulose synthase Pal I from *Klebsiella* sp. (Zhang et al., 2002) and trehalulose synthase Mut B from *Pseudomonas mesoacidophila* (Ravaud et al., 2005) having been extensively studied for their crystalline structure and catalytic mechanisms. Both Pal I and Mut B display quite a high sequence similarity with a common fold consisting of three domains–N-terminal catalytic domain with $(\alpha/\beta)_8$, barrel, a subdomain and C-terminus domain. The catalytic mechanism of sucrose isomerase, similar to other enzymes in this group, is a double displacement reaction with products retaining the α-anomeric configuration of the substrate *via* a covalent glycosyl-enzyme substrate intermediate (Table 4).

Table 4. Microbial production of sucrose isomerases.

Producer	Activity	Isomaltulose %	Trehalulose %	References
Erwinia sp.	38.75 U/mg	63.00	30.00	Contesini et al. (2013)
Pectobacterium carotovorum SIase expressed in *Echerichia coli*	47.6 U/mg dcw	20.72	79.16	Nam et al. (2014)
Klebsiella pneumoniae expressed in *E. coli*	2362 U/mg	73.00	25.00	Aroonnual et al. (2007)
Serratia plymuthica	120 IU/mg	72.60	6.60	Véronèse and Perlot (1999)

The general acid-basis catalysis of sucrose isomerase involves two steps—the hydrolysis of α-(1→2) glycosidic linkage followed by a simultaneous protonation of glycosidic bond and the nucleophilic attack of the anomeric carbon of the glucose moiety, leading to the formation of an enzyme-substrate intermediate (Ravaud et al., 2005). The glycosyl moiety can be transferred to a water molecule for hydrolysis and to the fructose moiety for sucrose isomers synthesis. In the isomerization step, the structure of the fructose determines the balance in the formation of two sucrose isomers—isomaltulose and trehalulose (Véronèse and Perlot, 1999).

Crystal structure of isomaltulose synthase (Pal I) from *Klebsiella* sp. showed a catalytic triad (Asp 241, Glu 295 and Asp 369) and two histidines (His 145 and His 368) present in the catalytic groove. Glu 295 acts as a general acid catalyst by protonating sucrose glycosidic bond; Asp 241 acts as the attacking nucleophile on the anomeric carbon of glucose moiety; and Asp 369 form hydrogen bonds with O_2 and O_3. His 145 and His 368 form hydrogen bonds with O_6 and O_2 respectively (Aroonnual et al., 2007). The two histidines are highly conserved in α-amylases and glucosyl transferases, which may participate in the stabilization of the substrate-binding transition state.

3.2 Sucrose isomerases applications

Sucrose isomerase has the following applications in industries:

3.2.1 Applications of isomaltulose

It is a potential disaccharide sugar substitute in food, beverages and medicines. It is known for its function as a food ingredient due to its superior acid stability, lower hygroscopicity and various other health benefits. It is found to prevent tooth decay, attenuate insulin level and glycemic index in the bloodstream (Wu et al., 2015). It may also reduce accumulation of visceral fat. Its low glycemic properties allow the sugar to enter the bloodstream at a slower rate. Since the hydrolysis of isomaltulose is slow, it is recommended for diabetics and athletes in their food (Contesini et al., 2013).

3.2.2 Applications of trehalulose

Similar to isomaltulose; trehalulose is considered as a non-cariogenic sweetener. The sweetness of trehalulose is approximately 60 per cent of that of sucrose. It can prevent dental cavities, mitigate diabetes mellitus and help maintain body weight (Wei et al., 2013). It is used in highly sweetened foods, like jellies and jams (Mu et al., 2014).

4. Conclusion

Enzymatic hydrolysis and isomerization of sucrose lead to production of fructo-oligosaccharides and other valuable products. These have been used widely as sweeteners as well as prebiotics. Their relative utilization in food and pharmaceutical fields on a commercial scale is very high. Various sucrose hydrolyzing and isomerizing enzymes are available in Nature. The production and catalytic mechanism of these enzymes depends on a lot of factors. Unfortunately only few researches have been carried out to reveal the secrets behind the maximum production/extraction and catalytic activity of these enzymes. Hence, giving the importance of their utility in the commercial sector, these enzymes require much more attention for effective employment on a wide scale.

Acknowledgments

The financial support from Department of Science (SB/FTP/ETA-212/2012), India is highly appreciated.

References

Andjelković, U., Milutinović-Nikolić, A., Jović-Jovičić, N., Banković, P., Bajt, T., Mojović, Z., Vujčić, Z. and Jovanović, D. 2015. Efficient stabilization of *Saccharomyces cerevisiae* external invertase by immobilisation on modified beidellite nanoclays. Food Chemistry, 168: 262–269.

Anwar, M.A., Kralj, S., van der Maarel, M.J. and Dijkhuizen, L. 2008. The probiotic *Lactobacillus johnsonii* NCC 533 produces high-molecular-mass inulin from sucrose by using an inulosucrase enzyme. Applied and Environmental Microbiology, 74(11): 3426–3433.

Arand, M., Brandao, N., Polikarpov, I. and Wattiez, R. 2002. Purification, characterization, gene cloning and preliminary X-ray data of the exo-inulinase from *Aspergillus awamori*. Biochemistry Journal, 362: 131–135.

Aroonnual, A., Nihira, T., Seki, T. and Panbangred, W. 2007. Role of several key residues in the catalytic activity of sucrose isomerase from *Klebsiella pneumoniae* NK33-98-8. Enzyme and Microbial Technology, 40(5): 1221–1227.

Awad, G.E., Amer, H., El-Gammal, E.W., Helmy, W.A., Esawy, M.A. and Elnashar, M.M. 2013. Production optimization of invertase by *Lactobacillus brevis* Mm-6 and its immobilization on alginate beads. Carbohydrate Polymers, 93(2): 740–746.

Badotti, F., Dário, M.G., Alves, S.L., Cordioli, M.L., Miletti, L.C., de Araujo, P.S. and Stambuk, B.U. 2008. Switching the mode of sucrose utilization by *Saccharomyces cerevisiae*. Microbial Cell Factories, 7(1): 4.

Bagal-Kestwal, D., Karve, M.S., Kakade, B. and Pillai, V.K. 2008. Invertase inhibition based electrochemical sensor for the detection of heavy metal ions in aqueous system: Application of ultra-microelectrode to enhance sucrose biosensor's sensitivity. Biosensors and Bioelectronics, 24(4): 657–664.

Baier, M.C., Barsch, A., Küster, H. and Hohnjec, N. 2007. Antisense repression of the *Medicago truncatula* nodule-enhanced sucrose synthase leads to a handicapped nitrogen fixation mirrored by specific alterations in the symbiotic transcriptome and metabolome. Plant Physiology, 145(4): 1600–1618.

Baroja-Fernández, E., Muñoz, F.J., Li, J., Bahaji, A., Almagro, G., Montero, M., Etxeberria, E., Hidalgo, M., Sesma, M.T. and Pozueta-Romero, J. 2012. Sucrose synthase activity in the sus1/sus2/sus3/sus4 *Arabidopsis* mutant is sufficient to support normal cellulose and starch production. Proceedings of the National Academy of Sciences, 109(1): 321–326.

Catana, R., Ferreira, B., Cabral, J. and Fernandes, P. 2005. Immobilization of inulinase for sucrose hydrolysis. Food Chemistry, 91(3): 517–520.

Chen, H.-Q., Chen, X.-M., Li, Y., Wang, J., Jin, Z.-Y., Xu, X.-M., Zhao, J.-W., Chen, T.-X. and Xie, Z.-J. 2009. Purification and characterisation of exo-and endo-inulinase from *Aspergillus ficuum* JNSP5-06. Food Chemistry, 115(4): 1206–1212.

Chi, Z., Chi, Z., Zhang, T., Liu, G. and Yue, L. 2009. Inulinase-expressing microorganisms and applications of inulinases. Applied Microbiology and Biotechnology, 82(2): 211–220.

Contesini, F.J., de Oliveira Carvalho, P., Grosso, C.R.F. and Sato, H.H. 2013. Single-step purification, characterization and immobilization of a sucrose isomerase from *Erwinia* sp. Biocatalysis and Agricultural Biotechnology, 2(4): 322–327.

de Paula, F.C., Cazetta, M.L., Monti, R. and Contiero, J. 2008. Sucrose hydrolysis by gelatin-immobilized inulinase from *Kluyveromyces marxianus* var. *bulgaricus*. Food Chemistry, 111(3): 691–695.

Esmon, P., Esmon, B., Schauer, I., Taylor, A. and Schekman, R. 1987. Structure, assembly, and secretion of octameric invertase. Journal of Biological Chemistry, 262(9): 4387–4394.

Ettalibi, M. and Baratti, J. 2001. Sucrose hydrolysis by thermostable immobilized inulinases from *Aspergillus ficuum*. Enzyme and Microbial Technology, 28(7): 596–601.

Falconer, D.J., Mukerjea, R. and Robyt, J.F. 2011. Biosynthesis of dextrans with different molecular weights by selecting the concentration of *Leuconostoc mesenteroides* B-512FMC dextransucrase, the sucrose concentration, and the temperature. Carbohydrate Research, 346(2): 280–284.

Figueroa, C.M., Diez, M.D.A., Kuhn, M.L., McEwen, S., Salerno, G.L., Iglesias, A.A. and Ballicora, M.A. 2013. The unique nucleotide specificity of the sucrose synthase from *Thermosynechococcus elongatus*. FEBS Letters, 587(2): 165–169.

Goulter, K.C., Hashimi, S.M. and Birch, R.G. 2012. Microbial sucrose isomerases: Producing organisms, genes and enzymes. Enzyme and Microbial Technology, 50(1): 57–64.

Kim, K.-Y., Nascimento, A.S., Golubev, A.M., Polikarpov, I., Kim, C.-S., Kang, S.-I. and Kim, S.-I. 2008. Catalytic mechanism of inulinase from *Arthrobacter* sp. S37. Biochemical and Biophysical Research Communications, 371(4): 600–605.

Kolman, M.A., Torres, L.L., Martin, M.L. and Salerno, G.L. 2012. Sucrose synthase in unicellular cyanobacteria and its relationship with salt and hypoxic stress. Planta, 235(5): 955–964.

Kulshrestha, S., Tyagi, P., Sindhi, V. and Yadavilli, K.S. 2013. Invertase and its applications—A brief review. Journal of Pharmacy Research, 7(9): 792–797.

Lammens, W., Le Roy, K., Van Laere, A., Rabijns, A. and Van den Ende, W. 2008. Crystal structures of *Arabidopsis thaliana* cell-wall invertase mutants in complex with sucrose. Journal of Molecular Biology, 377(2): 378–385.

Lee, S.S., Hong, S.Y., Errey, J.C., Izumi, A., Davies, G.J. and Davis, B.G. 2011. Mechanistic evidence for a front-side, SNi-type reaction in a retaining glycosyltransferase. Nature Chemical Biology, 7(9): 631–638.

Li, Y., Liu, G.-L. and Chi, Z.-M. 2013. Ethanol production from inulin and unsterilized meal of Jerusalem artichoke tubers by *Saccharomyces* sp. W0 expressing the endo-inulinase gene from *Arthrobacter* sp. Bioresource Technology, 147: 254–259.

McCleary, B.V., Murphy, A. and Mugford, D.C. 2000. Measurement of total fructan in foods by enzymatic/spectrophotometric method: collaborative study. Journal of AOAC International, 83(2): 356–364.

Mohamed, S.A., Salah, H.A., Moharam, M.E., Foda, M. and Fahmy, A.S. 2015. Characterization of two thermostable inulinases from *Rhizopus oligosporus* NRRL 2710. Journal of Genetic Engineering and Biotechnology, 13(1): 65–69.

Moriyama, S., Akimoto, H., Suetsugu, N., KawasakI, S., Nakamura, T. and Ohta, K. 2002. Purification and properties of an extracellular exoinulinase from *Penicillium* sp. strain TN-88 and sequence analysis of the encoding gene. Bioscience, Biotechnology and Biochemistry, 66(9): 1887–1896.

Mu, W., Li, W., Wang, X., Zhang, T. and Jiang, B. 2014. Current studies on sucrose isomerase and biological isomaltulose production using sucrose isomerase. Applied Microbiology and Biotechnology, 98(15): 6569–6582.

Myers, D., Lawlor, D. and Attfield, P. 1997. Influence of invertase activity and glycerol synthesis and retention on fermentation of media with a high sugar concentration by *Saccharomyces cerevisiae*. Applied and Environmental Microbiology, 63(1): 145–150.

Nagem, R., Rojas, A., Golubev, A., Korneeva, O., Eneyskaya, E., Kulminskaya, A., Neustroev, K. and Polikarpov, I. 2004. Crystal structure of exo-inulinase from *Aspergillus awamori*: the enzyme fold and structural determinants of substrate recognition. Journal of Molecular Biology, 344(2): 471–480.

Nakai, T., Tonouchi, N., Konishi, T., Kojima, Y., Tsuchida, T., Yoshinaga, F., Sakai, F. and Hayashi, T. 1999. Enhancement of cellulose production by expression of sucrose synthase in *Acetobacter xylinum*. Proceedings of the National Academy of Sciences, 96(1): 14–18.

Nam, C.-H., Seo, D.-H., Jung, J.-H., Koh, Y.-J., Jung, J.-S., Heu, S., Oh, C.-S. and Park, C.-S. 2014. Functional characterization of the sucrose isomerase responsible for trehalulose production in plant-associated *Pectobacterium* species. Enzyme and Microbial Technology, 55: 100–106.

Ohara, A., de Castro, R.J.S., Nishide, T.G., Dias, F.F.G., Bagagli, M.P. and Sato, H.H. 2015. Invertase production by *Aspergillus niger* under solid state fermentation: Focus on physical-chemical parameters, synergistic and antagonistic effects using agro-industrial wastes. Biocatalysis and Agricultural Biotechnology, 4(4): 645–652.

Onodera, S. and Shiomi, N. 1992. Purification and subsite affinities of exo-inulinase from *Penicillium trzebinskii*. Bioscience, Biotechnology and Biochemistry, 56(9): 1443–1447.

Paine, H., Walton Jr, C. and Badollet, M. 1925. Industrial applications of invertase. Industrial & Engineering Chemistry, 17(5): 445–450.

Pandey, A., Soccol, C.R., Selvakumar, P., Soccol, V.T., Krieger, N. and Fontana, J.D. 1999. Recent developments in microbial inulinases. Applied Biochemistry and Biotechnology, 81(1): 35–52.

Pito, D., Fonseca, I., Ramos, A., Vital, J. and Castanheiro, J. 2012. Hydrolysis of sucrose over composite catalysts. Chemical Engineering Journal, 184: 347–351.

Plascencia-Espinosa, M., Santiago-Hernández, A., Pavón-Orozco, P., Vallejo-Becerra, V., Trejo-Estrada, S., Sosa-Peinado, A., Benitez-Cardoza, C.G. and Hidalgo-Lara, M.E. 2014. Effect of deglycosylation on the properties of thermophilic invertase purified from the yeast *Candida guilliermondii* MpIIIa. Process Biochemistry, 49(9): 1480–1487.

Porchia, A.C., Curatti, L. and Salerno, G.L. 1999. Sucrose metabolism in cyanobacteria: sucrose synthase from *Anabaena* sp. strain PCC 7119 is remarkably different from the plant enzymes with respect to substrate affinity and amino-terminal sequence. Planta, 210(1): 34–40.

Qi, P., You, C. and Zhang, Y.-H.P. 2014. One-pot enzymatic conversion of sucrose to synthetic amylose by using enzyme cascades. ACS Catalysis, 4(5): 1311–1317.

Ravaud, S., Watzlawick, H., Mattes, R., Haser, R. and Aghajari, N. 2005. Towards the three-dimensional structure of a sucrose isomerase from *Pseudomonas mesoacidophila* MX-45. Biologia, Bratislava, 60(16): 89–95.

Reddy, V.A. and Maley, F. 1990. Identification of an active-site residue in yeast invertase by affinity labeling and site-directed mutagenesis. Journal of Biological Chemistry, 265(19): 10817–10820.

Röhrig, H., John, M. and Schmidt, J. 2004. Modification of soybean sucrose synthase by S-thiolation with ENOD40 peptide A. Biochemical and Biophysical Research Communications, 325(3): 864–870.

Römer, U., Nettelstroth, N., Köckenberger, W. and Elling, L. 2001. Characterization of recombinant sucrose synthase 1 from potato for the synthesis of sucrose analogues. Advanced Synthesis & Catalysis, 343(6-7): 655–661.

Römer, U., Schrader, H., Günther, N., Nettelstroth, N., Frommer, W.B. and Elling, L. 2004. Expression, purification and characterization of recombinant sucrose synthase 1 from *Solanum tuberosum* L. for carbohydrate engineering. Journal of Biotechnology, 107(2): 135–149.

Sainz-Polo, M.A., Ramírez-Escudero, M., Lafraya, A., González, B., Marín-Navarro, J., Polaina, J. and Sanz-Aparicio, J. 2013. Three-dimensional structure of *Saccharomyce*s invertase-Role of a non-catalytic domain in oligomerization and substrate specificity. Journal of Biological Chemistry, 288(14): 9755–9766.

Sauerzapfe, B., Engels, L. and Elling, L. 2008. Broadening the biocatalytic properties of recombinant sucrose synthase 1 from potato (*Solanum tuberosum* L.) by expression in *Escherichia coli* and *Saccharomyces cerevisiae*. Enzyme and Microbial Technology, 43(3): 289–296.

Silva, M.F., Rigo, D., Mossi, V., Golunski, S., de Oliveira Kuhn, G., Di Luccio, M., Dallago, R., de Oliveira, D., Oliveira, J.V. and Treichel, H. 2013. Enzymatic synthesis of fructooligosaccharides by inulinases from *Aspergillus niger* and *Kluyveromyces marxianus* NRRL Y-7571 in aqueous-organic medium. Food Chemistry, 138(1): 148–153.

Sirisansaneeyakul, S., Worawuthiyanan, N., Vanichsriratana, W., Srinophakun, P. and Chisti, Y. 2007. Production of fructose from inulin using mixed inulinases from *Aspergillus niger* and *Candida guilliermondii*. World Journal of Microbiology and Biotechnology, 23(4): 543–552.

Sokolenko, G. and Karpechenko, N. 2015. Expression of inulinase genes in the yeasts *Saccharomyces cerevisiae* and *Kluyveromyces marxianus*. Microbiology, 84(1): 23–27.

Sturm, A. and Tang, G.-Q. 1999. The sucrose-cleaving enzymes of plants are crucial for development, growth and carbon partitioning. Trends in Plant Science, 4(10): 401–407.

Uma, C., Gomathi, D., Ravikumar, G., Kalaiselvi, M. and Palaniswamy, M. 2012. Production and properties of invertase from a *Cladosporium cladosporioides* in SmF using pomegranate peel waste as substrate. Asian Pacific Journal of Tropical Biomedicine, 2(2): S605–S611.

Véronèse, T. and Perlot, P. 1999. Mechanism of sucrose conversion by the sucrose isomerase of *Serratia plymuthica* ATCC 15928. Enzyme and Microbial Technology, 24(5): 263–269.

Vetting, M.W., Frantom, P.A. and Blanchard, J.S. 2008. Structural and enzymatic analysis of MshA from *Corynebacterium glutamicum* substrate-assisted catalysis. Journal of Biological Chemistry, 283(23): 15834–15844.

Vorlova, L. and Pridal, A. 2002. Invertase and diastase activity in honeys of Czech provenience. Acta Univ. Agric. Et Silvic. Mendel Brun., pp. 57–66.

Wakiuchi, N., Shiomi, R. and Tamaki, H. 2003. Production of galactinol from sucrose by plant enzymes. Bioscience, Biotechnology and Biochemistry, 67(7): 1465–1471.

Wei, Y., Liang, J., Huang, Y., Lei, P., Du, L. and Huang, R. 2013. Simple, fast, and efficient process for producing and purifying trehalulose. Food Chemistry, 138(2): 1183–1188.

Wildberger, P., Aish, G.A., Jakeman, D.L., Brecker, L. and Nidetzky, B. 2015. Interplay of catalytic subsite residues in the positioning of α-D-glucose 1-phosphate in sucrose phosphorylase. Biochemistry and Biophysics Reports, 2: 36–44.

Wu, L., Liu, Y., Chi, B., Xu, Z., Feng, X., Li, S. and Xu, H. 2015a. An innovative method for immobilizing sucrose isomerase on ε-poly-l-lysine modified mesoporous TiO_2. Food Chemistry, 187: 182–188.

Wu, R., Diez, M.D.A., Figueroa, C.M., Machtey, M., Iglesias, A.A., Ballicora, M.A. and Liu, D. 2015b. The crystal structure of *Nitrosomonas europaea* sucrose synthase reveals critical conformational changes and insights into the sucrose metabolism in prokaryotes. Journal of Bacteriology, 197(17): 2734–2746.

Yuan, W., Zhao, X., Chen, L. and Bai, F. 2013. Improved ethanol production in Jerusalem artichoke tubers by overexpression of inulinase gene in *Kluyveromyces marxianus*. Biotechnology and Bioprocess Engineering, 18(4): 721–727.

Zervosen, A., Römer, U. and Elling, L. 1998. Application of recombinant sucrose synthase-large scale synthesis of ADP-glucose. Journal of Molecular Catalysis B: Enzymatic, 5(1): 25–28.

Zhang, D., Li, X. and Zhang, L.-H. 2002. Isomaltulose synthase from *Klebsiella* sp. strain LX3: gene cloning and characterization and engineering of thermostability. Applied and Environmental Microbiology, 68(6): 2676–2682.

Zheng, Y., Anderson, S., Zhang, Y. and Garavito, R.M. 2011. The structure of sucrose synthase-1 from *Arabidopsis thaliana* and its functional implications. Journal of Biological Chemistry, 286(41): 36108–36118.

Zhou, J., Lu, Q., Peng, M., Zhang, R., Mo, M., Tang, X., Li, J., Xu, B., Ding, J. and Huang, Z. 2015. Cold-active and NaCl-tolerant exo-inulinase from a cold-adapted *Arthrobacter* sp. MN8 and its potential for use in the production of fructose at low temperatures. Journal of Bioscience and Bioengineering, 119(3): 267–274.

Zittan, L. 1981. Enzymatic hydrolysis of inulin—an alternative way to fructose production. Starch-Stärke, 33(11): 373–377.

Sucrose Transforming Enzymes
Transfructosylation and Transglycosylation
Kiran Babu Uppuluri and Harish B.S. *

1. Introduction

Fructose and glucose units produced from sucrose can be transformed to various acceptors by the transfructosylation and transglycosylation reactions, respectively using various sucrose-transforming enzymes. Sucrose transfructosylation is catalyzed by various fructosyl transferases including inulosucrases, levansucrases and sucrose-sucrose fructosyl transferases whereas sucrose transglycosylation is aided by sucrose phosphorylases and dextransucrases.

2. Sucrose Transforming Enzymes for Producing Fructo-Oligosaccharides and Fructans

Fructooligosaccharides (FOs) are oligomers of fructose containing a single glucose moiety. They are mainly 1-kestose (GF_2), nystose (GF_3), ^{1}F-fructofuranosyl nystose (GF_4) or higher oligomers in which fructosyl units (F) are bound at the β-2,1 position of sucrose (Yun, 1996; Sánchez et al., 2008). They are used as food ingredients and have beneficial effects on human health and have prebiotic effects by improving the growth of Bifidobacteria in intestinal flora (Kurakake et al., 2009). They are calorie-

Bioprospecting Laboratory, School of Chemical and Biotechnology, SASTRA University, Thanjavur. Tamil Nadu-613401, India.
E-mail: kinnubio@gmail.com
* Corresponding author: harish.bs74@gmail.com

free, non-cariogenic and considered as soluble dietary fibers which can be used as alternative sweeteners (Zambelli et al., 2016). They are only one-third as sweet as sucrose and thus can be used where high sweetness of sucrose is not preferred. Since they are calorie free and not used by the body as energy sources they can be used by diabetic patients. Other notable property of FOs is that they reduce the level of serum cholesterol, phospholipid and triglyceride. Interestingly, the use of FOs instead of sucrose does not remove beneficial bulking properties that sucrose possess (Yun, 1996). These excellent physico, chemical and biological properties of fructose and FOs encourage the use of fructo-oligosaccharides in the food industry.

Enzymes from different microbes were identified to produce fructo-oligosaccharides and fructo-polymers from sucrose-rich substrates. These transferases have the ability to synthesize low molecular weight fructo-oligosaccharides or fructo-polymers from sucrose. The size and structure of the fructo-oligosaccharides depend upon the source of the enzyme (Olivares-Illana et al., 2002). Fructo-oligosaccharides can be inulin-type fructo-oligosaccharides (^{1}F-FOs), levan-type fructo-oligosaccharides (^{6}F-FOs) or neo-fructo-oligosaccharides (^{6}G-FOs). Inulin-type fructo-oligosaccharides are fructo-oligomers with terminal glucose in which 2–4 fructo-furanosyl moieties are linked by β-(2→1) bonds. These are produced by controlled hydrolysis of inulin or other fructans (by inulinases, EC 3.2.1.7); or by enzymatic trans-fuctosylation of sucrose by β-fructo-furanosidases (Invertase EC 3.2.1.26); or by fructosyl-transferases (inulosucrase EC 2.4.1.9). Levan-type fructo-oligosaccharides, containing two fructose units linked by β (2→6) linkages and neo-fructo-oligosaccharides, having glucose and fructose linked together, were reported to exhibit improved prebiotic properties and chemical stability. They are commonly produced by enzymes from yeasts. Levan-type fructo-oligosaccharides are produced by the action of fructosyl-transferases (levansucrase EC 2.4.1.10) (Zambelli et al., 2014).

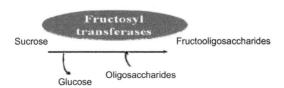

Figure 1. Sucrose transformation by fructosyl transferases.

2.1 Inulinases for fructo-oligosaccharide synthesis

Inulinases which are responsible for the hydrolysis of various fructans can also be used for the synthesis of fructo-oligosaccharides. The fructosyl-transferases activity of inulinase is responsible for the production of fructo-oligosaccharides from sucrose, in which fructosyl units are bound at the β (2→1) position of the sucrose (Santos and Maugeri, 2007; Silva et al., 2013). This reaction releases glucose as a by-product (Risso et al., 2012).

2.2 Invertase for fructo-oligosaccharide synthesis

Typically β-fructo-furanosidases hydrolyzes sucrose to glucose and fructose, but a few enzymes can transfer fructosyl residues to the sucrose molecule at high sucrose concentration in which the fructosyl residues are linked to sucrose by β-2→1 glycosidic bond (Kurakake et al., 2009). The fructosyl-transferring ability of β-fructo-furanosidase differs with the source of the enzyme and is characterized by the ratio of fructosyl transferring activity (U_t) to the hydrolyzing activity (U_h), denoted by U_t/U_h. Two forms of β-fructo-furanosidases have been separated from various strains of *Aspergillus* sp., *Penicillium* sp. and *S. cerevisiae*. These two forms are encoded by different genes. When the sucrose concentration is high, the production of β-fructofuranosidase with higher fructosyl-transferring activity increases and that of the enzyme with hydrolyzing activity reduces (Wallis et al., 1997; Kurakake et al., 2007; Kurakake et al., 2009).

2.3 Fructosyl transferases for fructo-oligosaccharide/fructan synthesis

Fructo-oligosaccharides are commercially produced by transfructosylation of sucrose by microbial enzymes (Sangeetha et al., 2004). Bacterial fructosyl-transferases (fructan-sucrase) catalyze transfer of fructosyl residues from sucrose either by (i) transfructosylation by using growing fructan chain (polymerization), sucrose, gluco- and fructo-saccharides (oligosaccharide synthesis); or by (ii) hydrolysis of sucrose using water as substrate.

The reaction specificity (i.e., hydrolysis or transferases activity) and the product specificity (type and size of fructans/fructo-oligosaccharides) of fructosyl-transferases depends on the reaction conditions (such as substrate concentration and temperature), the form in which the enzyme is used (free/immobilized), the presence of organic solvents or co-solvents and the source of the enzyme. The bacterial fructosyl transferases belong to glycoside hydrolase family (GH68). Most of the bacterial fructosyl transferases are inulosucrases and levansucrases (Ozimek et al., 2006; Bivolarski et al., 2013). As for most of the enzymes, Ca^{2+} stabilizes fructosyl transferases (del Moral et al., 2008).

The fructosyl transferases are β-retaining enzymes employing a double displacement mechanism that involves formation and subsequent hydrolysis of a covalent enzyme-glycosyl intermediate. When sucrose is bound to the enzyme, the glycosidic bond is cleaved and glucose is released with the formation of enzyme-glycosyl intermediate. Depending on the acceptor molecules, the enzyme can catalyze hydrolysis (water as acceptor) or transfructosylation (oligosaccharides or growing polymer chain as acceptor) (Ozimek et al., 2006).

2.4 Inulosucrase

Inulosucrases (sucrose: 2,1-β-D fructan 6-β-D fructosyl tranferase EC 2.4.1.9) are fructosyl transferases/fructansucrases produced by few gram positive bacteria which catalyze the production of fructan polymers/fructo-oligosaccharides with β(2→1) linked fructosyl units, by transferring fructosyl units from sucrose to an acceptor molecule (Van Hijum et al., 2003; Ozimek et al., 2005). The fructosyl unit can be transferred to the water molecule, resulting in the hydrolysis of sucrose (Peña-Cardeña et al., 2015). High sucrose concentration in conjugation with low inulosucrase concentration helps in the efficient synthesis of low molecular weight fructo-oligosaccharides, such as 1-kestose, 6-kestose, neokestose, nystose and f-nystose (Peña-Cardeña et al., 2015).

2.4.1 Inulosucrase sources

Inulosucrases have been isolated from plants, fungi and bacteria. Fungal inulosucrases have high specificity to low molecular weight inulin, low hydrolytic activity and high stability. Bacterial inulosucrases are used for synthesizing high molecular weight inulin with low specificity towards fructo-oligosaccharides. Bacterial strains, such as *Leuconostoc* sp. (Olivares-Illana et al., 2002), *Streptococcus mutans* and *Lactobacillus* sp. (Walter et al., 2008) are also used for inulosucrase production (Olivares-Illana et al., 2002; Catana et al., 2005; Peña-Cardeña et al., 2015).

2.4.2 Inulosucrase catalytic activity

They are multi-domain enzymes and have five-bladed β-propeller topology that enclose a funnel like central cavity where the conserved catalytic residues are located. Mature inulosucrases have three domains: (i) an N-terminal variable domain, (ii) a catalytic core domain and (iii) C-terminal variable regions that sometimes contain cell anchoring domain. Three dimensional crystal structure of a truncated, active bacterial inulosucrase of *Lactobacillus johnsonii*, in the apo form, with bounded substrate (sucrose) and with transfructosylation product, shows that the sucrose-binding product is similar to levansucrase. Catalytic amino acids founded Asp 272 as nucleophile, Asp 425 as transition state stabilizer and Glu 524 acts as catalytic acid/base (Pijning et al., 2011).

2.4.3 Inulosucrase applications

Inulosucrase has various applications in food, pharmaceuticals, medicines, etc.:

2.4.3.1 In food industry

Fructan produced by inulosucrase is important for its food and nutrition applications. Inulin is used as a fat substitute and to provide texture and stability to several food products, such as desserts, baked food and fermented dairy products. Inulin and inulin-type fructo-oligosaccharides are also prebiotic and repress the growth of pathogens. They support the growth of *Bifidobacteria* which are essential for the prevention of many diseases and for maintaining good health. They also provide relief from constipation usually encountered in elderly persons (Kaur and Gupta, 2002).

2.4.3.2 In medicine

Inulin, produced by the action of inulosucrase, also has the potential to reduce osteoporosis by increasing mineral absorption, especially of calcium. Inulin can also reduce the risk of atherosclerosis by lowering the synthesis of triglycerides and fatty acids in the liver and decreasing their level in serum (Kaur and Gupta, 2002). Inulin clearance test is one of the major tests used to measure the glomular filtration rate (GFR) for assessing the kidney function. The measurement is done in patients after overnight hospital stay with standardized metabolic conditions. A continuous intravenous hydration with 200 mL/h was given to assure water diuresis during the measurement. An inulin loading dose of 90 mg/Kg body weight was administered for 15 minutes. Urine and blood samples were collected at different time intervals for measurement. Either the Cockcroft–Gault (CG) equation or the MDRD formula is used commonly for estimating the renal functions on the basis of serum creatinine (Tsinalis and Thiel, 2009).

2.4.3.3 In pharma industry

Inulin increases the shelf-life of proteins and reduces the rate of denaturation of proteins when used as lyoprotectant. Inulin is widely used in the pharmaceutical industry as a protectant during drying and storage of protein-containing formulations (Malik et al., 2015). It has been studied on several therapeutically proteins like bovine plasma proteins (Hinrichs et al., 2001; Furlán et al., 2010), viruses like influenza virus (de Jonge et al., 2007) and several other therapeutic agents, like lipoplexes (Hinrichs et al., 2005).

The inulin type oligosaccharides have high glass transition temperature (Tg), less number of reducing groups and low rate of crystallization that extends the product's shelf-life. During drying, the water molecules around the protein are replaced by sugar molecules in the polysaccharide. Being a polyol, it will form multiple external hydrogen bonds with the protein that maintains the structural integrity of the protein. Low rate of crystallization reduces the chance of phase separation, thus maintaining the protective effect. For various reasons, it is preferred for these oligosaccharides to be in a glassy state during handling and storage. This reduces the molecular mobility and protects against the denaturing effects during drying. Less number of reducing sugars reduce the chance of its reaction with the amine group of proteins to form Schiff's base which may initiate Maillard reaction to severely affect the protein activity (Hinrichs et al., 2001). Use of inulin as a protectant during freeze drying of viruses, which can be used as vaccines or carrier systems for cellular delivery of therapeutic molecules, helps to maintain the structural integrity and fusogenic activity of virosomes (Hinrichs et al., 2005).

2.4.3.4 In bioplastic synthesis

Inulin can be also used as a potential surfactant by modifying the polymer. To produce amphipathic graft, copolymer alkyl group is introduced to the poly-fructose backbone of inulin. This alkyl group can attach to hydrophobic surfaces like carbon black or an oil drop and the sugar chain forms, stabilizing the chain which is highly water-soluble. Carbamoylated inulin has the ability to lower the surface tension of water and interfacial tension at the oil-water interface, thus providing a biodegradable surface-active agent (Tadros et al., 2004).

2.5 Levansucrase

Levansucrase (sucrose: 2,6-β-D-fructan 6-β-D-fructosyltransferase; EC 2.4.1.10) belongs to the glycoside hydrolase family (GH) 68, catalyzes conversion of sucrose to fructo-oligosaccharides and high molecular weight β-(2→6)-levan. It can directly use the free energy from the cleavage of sucrose to transfer the fructosyl groups to a variety of acceptor molecules. The acceptors can be monosaccharide (exchange), oligosaccharide (fructo-oligosaccharide synthesis) or a growing fructan chain (polymerization) (Tian et al., 2014). The fructosyl unit can be transferred to the water molecule resulting in the hydrolysis of sucrose. The reaction specificity (i.e., hydrolysis/transfructosylation) of levansucrase is modulated by several factors, such as sucrose concentration and the temperature or presence of organic solvents in the reaction medium. Usually a high sucrose concentration promotes transfructosylation reaction and reduces hydrolysis of sucrose (Porras-Domínguez et al., 2015).

It has a wide range of product spectrum (i.e., levan/FOs) depending on its reaction specificity and oligo-/polymerization ratio (Tian et al., 2014). Several bacteria, such as *Microbacterium laevaniformans, Zymomonas mobilis, Arthrobacter* sp., *Pseudomonas syringae, Rahnella aquatilis* and some *Bacillus* and *Erwinia* species are known to produce extracellular levansucrase (Sangiliyandi et al., 1999; Goldman et al., 2008). Levansucrases from some bacteria, like *Bacillus subtilis*, catalyze the formation of high molecular mass levan without accumulation of fructo-oligosaccharide molecules, suggesting the processive type of enzymatic reaction in which the enzyme catalyzes consecutive reactions without releasing the substrate. But the levansucrases from *Gluconacetobacter diazotrophicus, Zymomonas mobilis* and *Lactobacillus sanfranciscensis* mainly synthesize short fructo-oligosaccharides from sucrose, employing non-processive type reaction that, involves release of fructan chain after each fructosyl transfer (Ozimek et al., 2006).

The production of short-chain fructo-oligosaccharides with controlled molecular size (GF_2-GF_3) is significant as they are mostly absorbed in the small intestine and increase their prebiotic effect. The yields of short-chain fructo-oligosaccharides by levansucrase catalyzed transfructosylation reactions are low. The use of bi-enzymatic system of levansucrase and fructanase (endo-inulinase) regulate the molecular size of fructo-oligosaccharides, in which levansucrase catalyzes synthesis of levan and fructo-oligosaccharides from sucrose, while fructanase hydrolyzes fructan polymer into short-chain fructo-oligosaccharides and oligo-levan (Tian et al., 2014).

The structure and catalytic mechanism of levansucrase is similar to that of the inulosucrase. Levansucrase and inulosucrase from *Lactobacillus reuteri* are closely related to each other at amino acid level, showing 86 per cent similarity. The crystal structure of levansucrase from *Bacillus subtilis* displayed a five-bladed β-propellar topology. Catalytic amino acids were founded Asp 135 as nucleophile, Asp 309 as transition state stabilizer and Glu 401 as catalytic acid/base (Martínez-Fleites et al., 2005).

2.5.1 Levansucrase applications

Using fructosyl donor like sucrose and a series of mono-, di-, tri-, oligo-saccharides, as well as sugar and aliphatic alcohols as fructosyl acceptors, levan sucrase can produce a wide range of products with different applications (Lu et al., 2014).

2.5.1.1 In food industry

A fructose polymer levan, linked mostly by the β (2→6) fructo-furanosidic bond, is a non-toxic and non-mutagenic dietary fiber with a wide range of applications in food and beverage industries as viscosifier, stabilizer, emulsifier or as water-binding agent (Jung et al., 1998; Vaidya and Prasad, 2012). The levansucrase-catalyzed transfructosylation produces fructo-oligosaccharides from levan with potential health benefits, such as low caloric value, non-cariogenic properties and prebiotic effects (Vaidya and Prasad, 2012; Tian et al., 2014a; Tian et al., 2014b). Levansucrase is also used for synthesizing sucrose analogues, using monosaccharides as acceptors. Levansucrase from *B. subtilis* is found to have higher fructosyl transfer efficiency than monosaccharides like D-galactose, D-xylose and D-fucose.

2.5.1.2 In medicine

Levan, like many other polysaccharides, has immune-stimulating, anti-diabetic, anti-tumor and anti-inflammatory activities. Anti-tumor activity of levan can be due to many mechanisms. It can have direct inhibitory effect on tumor cells, modulate host immune response and can also augment other anti-tumor compounds. The anti-tumor activity of levan is dependent on molecular weight of levan (Yoo et al., 2004). As the branching degree of levan decreases, it is found to show lesser anti-tumor activity (Yoon et al., 2004). It can also lower cholesterol levels. So levan or levan derivatives can be used as hypocholesterolemic agents or antiobesity agents (Combie, 2006). Levan has application in the drug delivery. It is a promising material in the preparation of nanometric carriers of peptide and protein drug delivery (Sezer et al., 2011).

2.5.1.3 In cosmetics

Levan is found to have many relevant cosmoceutical properties, including its moisturizing effect, cell cytotoxicity, cell proliferation effect and anti-inflammatory effect. Despite the high molecular weight of levan, it still shows cell proliferation and anti-inflammatory effects. It might be due to much smaller particle size distribution when compared to other polysaccharides; and may also improve the penetrative ability. Levan from *Zymomonas mobilis* has been studied for its cosmoceutical properties and was found to be a safe and useful cosmoceutical agent (Kim et al., 2006).

2.5.1.4 In bioseparation

PEG-Levan aqueous two phase system has been demonstrated for the purification of biological compounds (Chung et al., 1997).

2.6 Sucrose-sucrose fructosyl tranferase

Plant fructan biosynthesis initiated by sucrose, sucrose 1-fructosyl transferase (SST, EC 2.4.1.99) irreversibly transfers fructosyl group from one sucrose to another, resulting in the formation of glucose and 1-kestose. Two distinct SSTs are found in plants: 1-SST and 6-SST, based on the fructosyl linkage in their products—β2→1 and β2→6 respectively. But 1-SST does not polymerize glucose above the trisaccharide level with sucrose as substrate. Chain elongation is made possible by another enzyme, called fructan 1-fructosyl transferase (EC 2.4.1.100), which catalyzes reversible transfer of fructosyl moiety from one fructan molecule to another (Puebla et al., 1999; Wada et al., 2003). Exposure of the plants to stress enhances SST production and thus increases fructan production (De Roover et al., 2000). The enzyme could be inhibited by Zn^{2+}, Ag^+ and Cu^{2+} ions (Hellwege et al., 1997). 6-SST can transfer fructosyl moiety of sucrose not only to sucrose (producing 6-kestose) but to a wide variety of fructans. So the enzyme initially designated as 6-SST can be called as 6-SFT (levansucrase) (Duchateau et al., 1995; Sprenger et al., 1995). The gene responsible for SST is expressed in microbes for the production of useful fructans. 1-SST from *Aspergillus foetidus* has been produced in yeast, using recombinant technology for the production of 1-kestose (Rehm et al., 1998).

Plant fructosyl transferases are highly homologous in primary sequence and typically consist of two subunits—a large N-terminal subunit and a small C-terminal subunit. The large subunit determines the catalytic specificity for sucrose: sucrose 1-fructosyl transferase (Altenbach et al., 2004).

3. Enzymes Involved in Transglycosylation or Glucan Synthesis from Sucrose

The following enzymes are involved in transglycosylation of sucrose:

- Sucrose phosphorylase
- Dextransucrases

3.1 Sucrose phosphorylase

Glycosylation will improve the solubility, stability, flavor and pharmacokinetic behavior of non-carbohydrate molecules, like flavanoids, steroids, alkaloids and benzoic acid. Enzymatic glycosylation is advantageous over chemical methods which often involve the use of toxic catalysts and solvents, which make the process undesirable. Enzymatic glycosylation can be done using glycosyl transferases, glycoside phosphorylases and glycoside hydrolases. Transfer of glucose units from cheap, readily available sucrose sources to a variety of acceptor compounds makes sucrose phosphorylases an attractive enzyme for the purpose (Pang et al., 2013).

Sucrose phosphorylase belongs to the large family of glycoside hydrolases and transglycosylases (GH-13 family, often referred as α-amylase family) that act on glucosidic oligosaccharides and polysaccharides. Sucrose phosphorylase (EC 2.4.1.7) catalyzes reversible phosphorolysis of sucrose to α-D-glucose-1-phosphate (α-Glc 1-P) and D-fructose. In this sucrose is glucosyl donor and phosphate is glucosyl acceptor substrate (De Winter et al., 2011; Wildberger et al., 2015). The α-D-glucose-1-phosphate (α-Glc 1-P), so produced is a key intermediate in the synthesis of nucleotide sugars and is an activated donor of glycosyl transferases. Moreover it is also involved in the simulation of active calcium transport as a substitute for inorganic phosphate in parenteral nutrition (De Winter et al., 2011).

Sucrose phosphorylase activity and synthesis are inhibited by the glucose. Increase in sucrose concentration compensates the activity loss, but the phosphate has not much effect. It is believed that sucrose and glucose bind to the same active site of the enzyme and thus the presence of glucose competitively inhibits the sucrose phosphorylase activity (Vandamme et al., 1987).

Sucrose phosphorylase is an attractive enzyme for glycosylation due to its broad acceptor promiscuity. It glycosylates a wide variety of mono, di and trisaccharides and non-carbohydrate molecules, like phenolic compounds and furanoses. Although the enzyme has broad acceptor promiscuity, it is highly structural specific to the glucopyranosyl ring. The enzyme requires sucrose, α-Glc 1-P or α-D-Glc 1-fluoride as glucosyl donor to the enzyme. Due to this property, sucrose phosphorylase can be used for glycosylation of many compounds, which may influence the properties like solubility, stability and pharmacokinetic behavior of the molecule (Aerts et al., 2011; De Winter et al., 2011).

3.1.1 Sucrose phosphorylase sources

The enzyme is produced from some bacterial species including *Bifidobacterium adolescentis* (Aerts et al., 2011), *Leuconostoc mesenteroides* (Morimoto et al., 2015), *Pseudomonas saccharophila* (Silverstein et al., 1967; Trethewey et al., 2001) and *Streptococcus mutans* (Russell et al., 1988). The production of the enzyme is enhanced by expressing the genes responsible for sucrose phosphorylase in host organisms, like *E. coli.*

Table 1. Microbial enzymes for fructo-oligosaccharide production.

Enzyme	Producer	Nature of Enzyme	Yield	References
Inulinase	*Aspergillus niger*	Free/Immobilized/ Pretreated immobilized		Silva et al. (2013)
Inulinase	*Kluyveromyces marxianus*	Free	16.7 ± 1.1 wt.%	Risso et al. (2012)
β-fructofuranosidase	*Aspergillus oryzae*	Immobilized	51.90%	Kurakake et al. (2009)
β-fructofuranosidase	*Aspergillus niger*			Wallis et al. (1997)
Inulosucrase	*Leuconostoc citreum*		65%	Peña-Cardeña et al. (2015)
Fructosyl transferases	*Aspergillus* sp.		69%	Sánchez et al. (2008)
Levansucrase/ endoinulinase bienzymatic system	*Bacillus amyloliquefaciens/ Aspergillus niger*		57–65%	Tian et al. (2014)

3.1.2 Sucrose phosphorylase catalytic mechanism

The three-dimensional structural studies of sucrose phosphorylase from *Bifidobacterium adolescentis* revealed a (β/α)8 barrel fold and found the catalytic amino acids as Asp 192 (nucleophile) and Glu 232 (general acid/base) (Sprogøe et al., 2004). The reaction mechanism of sucrose phosphorylase, like other members of the GH-13 family, is a double displacement mechanism/ping-pong mechanism. The reaction of sucrose phosphorylase involves two configurationally inverting steps with net retention of anomeric configuration. In the first step, carbon-oxygen bond of the glucosyl donor is cleaved with the formation of a covalent β-glucosyl-enzyme (β-Glc-E) intermediate and the release of fructose. This step is initiated by simultaneous protonation of sucrose glycosidic bond by Glu 232 (proton donor) and a nucleophilic attack on anomeric carbon of glucose moiety by Asp 192. In the second step β-Glc-E intermediate reacts with phosphate to yield α-Glc-1-P and this is termed as phosphorolysis. Sucrose phosphorylase catalyzes hydrolysis if β-Glc-E intermediate reacts with water, but this hydrolysis reaction is much slower than the phosphorolytic reaction. Sucrose phosphorylase can also catalyze the transglucosylation reaction when β-Glc-E intermediate is attacked by external nucleophiles and new α-D-glucosides are produced (Sprogøe et al., 2004; Goedl et al., 2007).

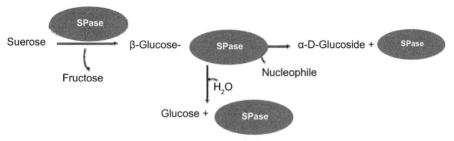

Figure 2. Sucrose phosphorylase catalytic mechanism.

Table 2. Sucrose phosphorylase producers which are expressed in *Escherichia coli* for the production.

Producer	Glucose Acceptor	Glucose Donor	Nature of Enzyme	References
Leuconostoc mesenteroides	8 different ketohexoses	α-D-glucose-1-phosphate (G1P)	Free	Morimoto et al. (2015)
Bifidobacterium adolescentis	Different monomeric sugars	sucrose and a-D-glucose-1-phosphate (G1P)	Free	van den Broek et al. (2004)
Bifidobacterium adolescentis	Phosphate	sucrose	Immobilized in Sepabeads EC-HFA	De Winter et al. (2011)
Streptococcus mutans	Phosphate	sucrose	Free	Fujii et al. (2006)
Bifidobacterium longum	Ascorbic acid	Sucrose	Free	Kwon et al. (2007)

3.1.3 Sucrose phosphorylase applications

Sucrose phosphorylase has the following applications:

3.1.3.1 In synthesis of glycosides and oligosaccharides

It can be used for functional diversification of the compounds such that entirely novel fields of applications can be opened. In such an effort, 1,2 propanediol and 3-aryloxy/alkyloxy derivatives thereof, which are bulk products obtained through immediate follow-up chemistry on glycerol (a by-product of biodiesel production) can be transformed into useful forms by glycosylation, using sucrose phosphorylase with sucrose as glucosyl donor (Luley-Goedl et al., 2010).

Qi et al. (2014) have described one-pot enzymatic conversion of sucrose to synthetic amylose, by using sucrose phosphorylase and α-glucan phosphorylase which were supplemented with glucose isomerase, glucose oxidase and catalase. The sucrose phosphorylase produced α-Glc 1-P, which is an efficient donor for chemical and enzymatic glycosylation reaction and was utilized by α-glucan phosphorylase to transfer glucose molecule to the non-reducing end of the primer-maltodextrin. Fructose, an inhibitor to sucrose phosphorylase, was removed by the action of glucose isomerase, glucose oxidase and catalase. Amylose is a very important compound as a precursor for biodegradable plastic films with low oxygen diffusion, a healthy food additive and a potential high-density hydrogen carrier (Qi et al., 2014).

3.1.3.2 In inorganic phosphate assay

Sucrose phosphorylase is used for the enzymatic assay of inorganic phosphate. Inorganic phosphate content in the sample is enzymatically measured using sucrose phosphorylase, phosphoglucomutase (PGM), glucose 6 phosphate dehydrogenase and 6-phosphogluconic dehydrogenase. Sucrose phosphorylase transfers inorganic phosphate to sucrose, producing α-D-glucose-1-phosphate and α-D-fructose. Glucose-1-phosphate is transphosphorylated by PGM in the presence of α-D-glucose-1, 6-bisphosphate to form α-D-glucose-6-phosphate. Glucose-6-phosphate is oxidized by NAD^+ and glucose 6 phosphate dehydrogenase to form 6-phophogluconate and NADH. Further 6-phophogluconate is oxidized by NAD^+ and 6-phosphogluconic dehydrogenase to form D-ribulose-5-phopshate and NADH. So, effectively two molecules of NADH are formed per molecule of inorganic phosphate, and that can be monitored by the change in absorbance at 340 nm (Tedokon et al., 1992).

3.1.3.3 In cosmetic applications

2-O-α-D-glucopyranosyl-sn-glycerol and 2-O-α-D-glucopyranosyl-R-glycerate, the two glycosyl glycerols are powerful osmolytes having promising cosmetic applications as moisturizing agents and therapeutic applications as protein stabilizers. These are easily synthesized using sucrose phosphorylase with sucrose as glucosyl donor (Luley-Goedl et al., 2010; Pang et al., 2013).

3.2 Dextransucrases

Various lactic acid bacteria genera, like *Leuconostoc mesenteroides*, *Streptococcus* sp., *Lactobacillus* sp. and *Weissella* produce extracellular α-glucans from glucopyranosyl residues of sucrose to protect them from environmental stress; they get catalyzed by glucansucrases. Glucansucrases, in addition to glucan synthesis can also catalyze transglycosylation reaction referred to as acceptor reaction (Seo et al., 2007; Falconer et al., 2011; Shi et al., 2016). In the presence of suitable hydroxyl group containing acceptors, such as low molecular weight carbohydrates, glucansucrases can transfer D-glucopyranosyl unit from sucrose to the acceptor, at the expense of glucan synthesis. Similar to sucrose phosphorylase enzyme, glucansucrases have broad acceptor promiscuity. Based on the structure of glucans, they synthesize glucansucrases which are classified into dextransucrases (EC 2.1.4.5), mutansucrases (EC 2.1.4.5), reuteransucrases (EC 2.1.4.5) and alternansucrases (2.1.4.140) (Bivolarski et al., 2013).

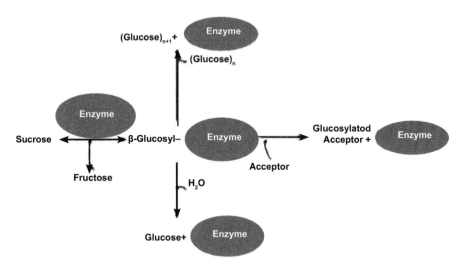

Figure 3. Dextransucrase catalytic mechanism.

Table 3. Microbial production of dextransucrase.

Producer	Glucose Acceptor	Glucose Donor	Products	Nature of Enzyme	References
Leuconostoc mesenteroides	Growing dextran chain	Sucrose	Dextrans	Native	Falconer et al. (2011)
Weissella confusa	Maltose	Sucrose	Gluco-oligosaccharides	Native	Shukla et al. (2014)
Weissella confusa dextransucrase expressed in *L. lactis*	Lactose and cellobiose	Sucrose	Trisaccharides	Recombinant	Shi et al. (2016)
Streptococcus strain	Growing dextran chain	Sucrose	Dextrans	Native free	Chludzinski et al. (1974)

Dextrans are polysaccharides of D-glucans with contiguous α (1→6) glycosidic linkages in the main chain and α-(1→2), α-(1→3) or α-(1→4) branch glycosidic linkages, depending on the specificity of particular dextransucrase (Seo et al., 2007). Dextransucrases (EC 2.1.4.5) are extracellular enzymes that synthesize dextrans from sucrose releasing fructose. In addition to the synthesis of dextrans, the enzyme is also capable of transglycosylation reaction/acceptor reaction in which other carbohydrates or even the dextran chain itself acts as the acceptor (Falconer et al., 2011; Leemhuis et al., 2013). Higher sucrose concentration will increase the microbial dextransucrase production, but very high concentration in the medium makes removal of cells very difficult. Optimum sucrose concentration for dextransucrase production by *Leuconostoc mesenteroides* was found to be 2 per cent (Santos et al., 2000). The acceptor reaction is also found to be enhanced by higher concentration of sucrose (Seo et al., 2007).

3.2.1 Dextransucrase catalytic action

All the glucansucrases belong to the glycoside hydrolase (GH) family 70 based on the structure analogy and amino acid sequence similarity (Leemhuis et al., 2013). Glucansucrase-producing strains form slimy colonies in sucrose containing media which are similar to fructansucrase-producing strains. But unlike fructansucrase-producing bacteria, it cannot use tri-saccharide-raffinose and this property can be used for screening glucansucrase strains from fructansucrase-producing ones (Leemhuis et al., 2013).

Similar to the GH 13 enzymes, the reaction mechanism of GH70 glucansucrases is a double displacement reaction. In the first stage, the glycosidic linkage of the sucrose is cleaved, resulting in β-glucosyl-enzyme intermediate and in the second stage glucosyl moiety is transferred to an acceptor with retention of the α-configuration (Robyt et al., 2008; Leemhuis et al., 2013).

3.2.2 Dextransucrase sources

Basically produced by carbohydrate fermenting gram-positive lactic acid bacteria genera of *Lactobacillus, Leuconostoc, Streptococcus* and *Weissella*, the strains are typically identified on solid or liquid media supplemented with sucrose by the appearance of slimy/ropy colonies or viscous liquids (Leemhuis et al., 2013).

3.2.3 Dextransucrase applications

Dextransucrase has the following applications in food, medicine and other industries:

3.2.3.1 In food industry

Dextran produced by dextransucrase is widely used as a food thickener and viscosifier (Leemhuis et al., 2013). Protein-dextran conjugate, which is prepared by covalently attaching proteins to the biopolymer, may improve the stability and solubility of the protein and shows good emulsifying properties. Hence such conjugates can be used as emulsifiers and as well as protein food additives where heat stability is required (Kato et al., 1990).

3.2.3.2 In medicine

Dextran 40 (MW around 40000) is known for antithrombotic effect and provides a prophylactic treatment for deep venous thrombosis and post-operative fatal pulmonary emboli. Dextran 70 (MW around 70000) is used as plasma volume expander in the treatment of hemorrhage, burns, surgery or trauma (de Belder, 1990).

3.2.3.3 In chromatography

Sephadex gels prepared by emulsion polymerization of dextran are one of the most important gel filtration media in analytical techniques (de Belder, 1990).

3.2.3.4 In the synthesis of oligosaccharides

Oligosaccharides can be synthesized using dextransucrase by adding carbohydrate in the reaction mixture which acts as an acceptor. If the added acceptor carbohydrate is D-glucose, maltose, isomaltose or compounds with free hydroxyl group(s), then a series of oligosaccharides and glycosides are formed. The resultant product contains varying number of D-glucose attaching to the non-reducing end of the acceptor by α (1→6) linkage. Higher sucrose concentration reduces the production of high molecular weight dextran with concomitant enhancement of low molecular weight dextran (Seo et al., 2007).

3.2.3.5 In cell culture studies

For culturing anchorage dependent animal cells cross linked dextran substituted with cationic substituents (cytodex) are used commercially. Human melanocytes (Smit et al., 1989), vero cells (Lindskog et al., 1986) and rat anterior pituitary cells (Smith and Vale, 1980) were cultured, using cytodex.

3.2.3.6 In bioseparation

PEG-dextran aqueous two-phase system was used for biomolecule purification (Diamond and Hsu, 1990).

4. Conclusion

Sucrose, the most abundant disaccharide in Nature can be converted to a wide range of useful products by the action of various sucrose-transforming enzymes produced by both plants and microbes. The enzymes can convert sucrose to its monosaccharide constituents—glucose and fructose; or to its isomers; or it can be used for synthesizing sucrose analogues and fructo-oligosaccharides. Some enzymes can transfer it monosaccharide units to various acceptors to produce a wide range of products that possess many industrial applications.

Acknowledgments

The financial support from Science and Technology (SB/FTP/ETA-212/2012), India is greatly acknowledged.

References

Aerts, D., Verhaeghe, T.F., Roman, B.I., Stevens, C.V., Desmet, T. and Soetaert, W. 2011. Transglucosylation potential of six sucrose phosphorylases toward different classes of acceptors. Carbohydrate Research, 346(13): 1860–1867.

Altenbach, D., Nüesch, E., Meyer, A.D., Boller, T. and Wiemken, A. 2004. The large subunit determines catalytic specificity of barley sucrose: Fructan 6-fructosyltransferase and fescue sucrose: sucrose 1-fructosyltransferase. FEBS Letters, 567(2): 214–218.

Bivolarski, V., Vasileva, T., Dzhambazov, B., Momchilova, A., Chobert, J.-M., Ivanova, I. and Iliev, I. 2013. Characterization of glucansucrases and fructansucrases produced by wild strains *Leuconostoc mesenteroides* URE13 and *Leuconostoc mesenteroides* LM17 grown on glucose or fructose medium as a sole carbon source. Biotechnology & Biotechnological Equipment, 27(3): 3811–3820.

Catana, R., Ferreira, B., Cabral, J. and Fernandes, P. 2005. Immobilization of inulinase for sucrose hydrolysis. Food Chemistry, 91(3): 517–520.

Chludzinski, A.M., Germaine, G.R. and Schachtele, C.F. 1974. Purification and properties of dextransucrase from *Streptococcus mutans*. Journal of Bacteriology, 118(1): 1–7.

Chung, B.H., Kyung Kim, W., Song, K.-B., Kim, C.-H. and Rhee, S.-K. 1997. Novel polyethylene glycol/levan aqueous two-phase system for protein partitioning. Biotechnology Techniques, 11(5): 327–329.

Combie, J. 2006. Properties of Levan and Potential Medical Uses. ACS symposium series. Oxford University Press. pp. 263–269.

de Belder, A.N. 1990. Dextran. Pharmacia.

de Jonge, J., Amorij, J.-P., Hinrichs, W.L., Wilschut, J., Huckriede, A. and Frijlink, H.W. 2007. Inulin sugar glasses preserve the structural integrity and biological activity of influenza virosomes during freeze-drying and storage. European Journal of Pharmaceutical Sciences, 32(1): 33–44.

De Roover, J., Vandenbranden, K., Van Laere, A. and Van den Ende, W. 2000. Drought induces fructan synthesis and 1-SST (sucrose: sucrose fructosyltransferase) in roots and leaves of chicory seedlings (*Cichorium intybus* L.). Planta, 210(5): 808–814.

De Winter, K., Cerdobbel, A., Soetaert, W. and Desmet, T. 2011. Operational stability of immobilized sucrose phosphorylase: Continuous production of α-glucose-1-phosphate at elevated temperatures. Process Biochemistry, 46(10): 2074–2078.

del Moral, S., Olvera, C., Rodriguez, M.E. and Munguia, A.L. 2008. Functional role of the additional domains in inulosucrase (IslA) from *Leuconostoc citreum* CW28. BMC Biochemistry, 9(1): 6.

Diamond, A.D. and Hsu, J.T. 1990. Protein partitioning in PEG/dextran aqueous two-phase systems. AIChE Journal, 36(7): 1017–1024.

Duchateau, N., Bortlik, K., Simmen, U., Wiemken, A. and Bancal, P. 1995. Sucrose: fructan 6-fructosyltransferase, a key enzyme for diverting carbon from sucrose to fructan in barley leaves. Plant Physiology, 107(4): 1249–1255.

Falconer, D.J., Mukerjea, R. and Robyt, J.F. 2011. Biosynthesis of dextrans with different molecular weights by selecting the concentration of *Leuconostoc mesenteroides* B-512FMC dextransucrase, the sucrose concentration, and the temperature. Carbohydrate Research, 346(2): 280–284.

Fujii, K., Iiboshi, M., Yanase, M., Takaha, T. and Kuriki, T. 2006. Enhancing the thermal stability of sucrose phosphorylase from *Streptococcus mutans* by random mutagenesis. Journal of Applied Glycoscience, 53(2): 91–97.

Furlán, L.T.R., Padilla, A.P. and Campderrós, M.E. 2010. Inulin like lyoprotectant of bovine plasma proteins concentrated by ultrafiltration. Food Research International, 43(3): 788–796.

Goedl, C., Schwarz, A., Minani, A. and Nidetzky, B. 2007. Recombinant sucrose phosphorylase from *Leuconostoc mesenteroides*: characterization, kinetic studies of transglucosylation, and application of immobilised enzyme for production of α-D-glucose 1-phosphate. Journal of Biotechnology, 129(1): 77–86.

Goldman, D., Lavid, N., Schwartz, A., Shoham, G., Danino, D. and Shoham, Y. 2008. Two active forms of zymomonas mobilis levansucrase—an ordered microfibril structure of the enzyme promotes levan polymerization. Journal of Biological Chemistry, 283(47): 32209–32217.

Hellwege, E.M., Gritscher, D., Willmitzer, L. and Heyer, A.G. 1997. Transgenic potato tubers accumulate high levels of 1-kestose and nystose: functional identification of a sucrose sucrose 1-fructosyltransferase of artichoke (*Cynara scolymus*) blossom discs. The Plant Journal, 12(5): 1057–1065.

Hinrichs, W., Prinsen, M. and Frijlink, H. 2001. Inulin glasses for the stabilization of therapeutic proteins. International Journal of Pharmaceutics, 215(1): 163–174.

Hinrichs, W., Sanders, N., De Smedt, S., Demeester, J. and Frijlink, H. 2005. Inulin is a promising cryo-and lyoprotectant for PEGylated lipoplexes. Journal of Controlled Release, 103(2): 465–479.

Jung, H.-C., Lebeault, J.-M. and Pan, J.-G. 1998. Surface display of *Zymomonas mobilis* levansucrase by using the ice-nucleation protein of Pseudomonas syringae. Nature Biotechnology, 16(6): 576–580.

Kato, A., Sasaki, Y., Furuta, R. and Kobayashi, K. 1990. Functional protein–polysaccharide conjugate prepared by controlled dry-heating of ovalbumin–dextran mixtures. Agricultural and Biological Chemistry, 54(1): 107–112.

Kaur, N. and Gupta, A.K. 2002. Applications of inulin and oligofructose in health and nutrition. Journal of Biosciences, 27(7): 703–714.

Kim, K., Chung, C., Kim, Y.H., Kim, K., Han, C. and Kim, C. 2006. Cosmeceutical properties of levan produced by *Zymomonas mobilis*. International Journal of Cosmetic Science, 28(3): 231.

Kurakake, M., Masumoto, R., Maguma, K., Kamata, A., Saito, E., Ukita, N. and Komaki, T. 2009. Production of fructooligosaccharides by β-fructofuranosidases from *Aspergillus oryzae* KB. Journal of Agricultural and Food Chemistry, 58(1): 488–492.

Kurakake, M., Ogawa, K., Sugie, M., Takemura, A., Sugiura, K. and Komaki, T. 2007. Two types of β-fructofuranosidases from *Aspergillus oryzae* KB. Journal of Agricultural and Food Chemistry, 56(2): 591–596.

Kwon, T., Kim, C.T. and Lee, J.-H. 2007. Transglucosylation of ascorbic acid to ascorbic acid 2-glucoside by a recombinant sucrose phosphorylase from *Bifidobacterium longum*. Biotechnology Letters, 29(4): 611–615.

Leemhuis, H., Pijning, T., Dobruchowska, J.M., van Leeuwen, S.S., Kralj, S., Dijkstra, B.W. and Dijkhuizen, L. 2013. Glucansucrases: Three-dimensional structures, reactions, mechanism, α-glucan analysis and their implications in biotechnology and food applications. Journal of Biotechnology, 163(2): 250–272.

Lindskog, U., Lundgren, B., Wergeland, I. and Billig, D. 1986. Microcarrier cell culture: Vero cells on cytodex® 1. Journal of Tissue Culture Methods, 9(4): 205–210.

Lu, L., Fu, F., Zhao, R., Jin, L., He, C., Xu, L. and Xiao, M. 2014. A recombinant levansucrase from *Bacillus licheniformis* 8-37-0-1 catalyzes versatile transfructosylation reactions. Process Biochemistry, 49(9): 1503–1510.

Luley-Goedl, C., Sawangwan, T., Brecker, L., Wildberger, P. and Nidetzky, B. 2010a. Regioselective O-glucosylation by sucrose phosphorylase: a promising route for functional diversification of a range of 1, 2-propanediols. Carbohydrate Research, 345(12): 1736–1740.

Luley-Goedl, C., Sawangwan, T., Mueller, M., Schwarz, A. and Nidetzky, B. 2010b. Biocatalytic process for production of a-glucosylglycerol using sucrose phosphorylase. Food Technology and Biotechnology, 48: 276–283.

Malik, A., Hapsari, M.T., Ohtsu, I., Ishikawa, S. and Takagi, H. 2015. Cloning and heterologous expression of the ftfCNC-2 (1) gene from *Weissella confusa* MBFCNC-2 (1) as an extracellular active fructansucrase in *Bacillus subtilis*. Journal of Bioscience and Bioengineering, 119(5): 515–520.

Martínez-Fleites, C., Ortíz-Lombardía, M., Hernandez, L., Pons, T., Tarbouriech, N., Taylor, E., Arrieta, J. and Davies, G.X.A.J. 2005. Crystal structure of levansucrase from the Gram-negative bacterium *Gluconacetobacter diazotrophicus*. Biochemistry Journal, 390: 19–27.

Morimoto, K., Yoshihara, A., Furumoto, T. and Takata, G. 2015. Production and application of a rare disaccharide using sucrose phosphorylase from *Leuconostoc mesenteroides*. Journal of Bioscience and Bioengineering, 119(6): 652–656.

Olivares-Illana, V., Wacher-Rodarte, C., Le Borgne, S. and López-Munguía, A. 2002. Characterization of a cell-associated inulosucrase from a novel source: A *Leuconostoc citreum* strain isolated from Pozol, a fermented corn beverage of Mayan origin. Journal of Industrial Microbiology and Biotechnology, 28(2): 112–117.

Ozimek, L., Euverink, G., Van Der Maarel, M. and Dijkhuizen, L. 2005. Mutational analysis of the role of calcium ions in the *Lactobacillus reuteri* strain 121 fructosyltransferase (levansucrase and inulosucrase) enzymes. FEBS Letters, 579(5): 1124–1128.

Ozimek, L.K., Kralj, S., Van der Maarel, M.J. and Dijkhuizen, L. 2006. The levansucrase and inulosucrase enzymes of *Lactobacillus reuteri* 121 catalyse processive and non-processive transglycosylation reactions. Microbiology, 152(4): 1187–1196.

Pang, H., Du, L., Pei, J., Wei, Y., Du, Q. and Huang, R. 2013. Sucrose hydrolytic enzymes: Old enzymes for new uses as biocatalysts for medical applications. Current Topics in Medicinal Chemistry, 13(10): 1234–1241.

Peña-Cardeña, A., Rodríguez-Alegría, M.E., Olvera, C. and Munguía, A.L. 2015. Synthesis of fructooligosaccharides by IslA4, a truncated inulosucrase from *Leuconostoc citreum*. BMC Biotechnology, 15(1): 2.

Pijning, T., Anwar, M.A., Böger, M., Dobruchowska, J.M., Leemhuis, H., Kralj, S., Dijkhuizen, L. and Dijkstra, B.W. 2011. Crystal structure of inulosucrase from *Lactobacillu*s: insights into the substrate specificity and product specificity of GH68 fructansucrases. Journal of Molecular Biology, 412(1): 80–93.

Porras-Domínguez, J.R., Ávila-Fernández, Á., Miranda-Molina, A., Rodríguez-Alegría, M.E. and Munguía, A.L. 2015. *Bacillus subtilis* 168 levansucrase (SacB) activity affects average levan molecular weight. Carbohydrate Polymers, 132: 338–344.

Puebla, A.F., Battaglia, M.E., Salerno, G.L. and Pontis, H.G. 1999. Sucrose-sucrose fructosyl transferase activity: A direct and rapid colorimetric procedure for the assay of plant extracts. Plant Physiology and Biochemistry, 37(9): 699–702.

Qi, P., You, C. and Zhang, Y.-H.P. 2014. One-pot enzymatic conversion of sucrose to synthetic amylose by using enzyme cascades. ACS Catalysis, 4(5): 1311–1317.

Rehm, J., Willmitzer, L. and Heyer, A.G. 1998. Production of 1-kestose in transgenic yeast expressing a fructosyltransferase from *Aspergillus foetidus*. Journal of Bacteriology, 180(5): 1305–1310.

Risso, F.V., Mazutti, M.A., Treichel, H., Costa, F., Maugeri, F. and Rodrigues, M.I. 2012. Comparison between systems for synthesis of fructooligosaccharides from sucrose using free inulinase from *Kluyveromyces marxianus* NRRL Y-7571. Food and Bioprocess Technology, 5(1): 331–337.

Robyt, J.F., Yoon, S.-H. and Mukerjea, R. 2008. Dextransucrase and the mechanism for dextran biosynthesis. Carbohydrate Research, 343(18): 3039–3048.

Russell, R., Mukasa, H., Shimamura, A. and Ferretti, J. 1988. *Streptococcus mutans* gtfA gene specifies sucrose phosphorylase. Infection and Immunity, 56(10): 2763–2765.

Sánchez, O., Guio, F., Garcia, D., Silva, E. and Caicedo, L. 2008. Fructooligosaccharides production by *Aspergillus* sp. N74 in a mechanically agitated airlift reactor. Food and Bioproducts Processing, 86(2): 109–115.

Sangeetha, P., Ramesh, M. and Prapulla, S. 2004. Production of fructo-oligosaccharides by fructosyl transferase from *Aspergillus oryzae* CFR 202 and *Aureobasidium pullulans* CFR 77. Process Biochemistry, 39(6): 755–760.

Sangiliyandi, G., Raj, K.C. and Gunasekaran, P. 1999. Elevated temperature and chemical modification selectively abolishes levan forming activity of levansucrase of *Zymomonas mobilis*. Biotechnology Letters, 21(2): 179–182.

Santos, A.M. and Maugeri, F. 2007. Synthesis of fructooligosaccharides from sucrose using inulinase from *Kluyveromyces marxianus*. Food Technology and Biotechnology, 45(2): 181.

Santos, M., Teixeira, J. and Rodrigues, A. 2000. Production of dextransucrase, dextran and fructose from sucrose using *Leuconostoc mesenteroides* NRRL B512 (f). Biochemical Engineering Journal, 4(3): 177–188.

Seo, E.-S., Nam, S.-H., Kang, H.-K., Cho, J.-Y., Lee, H.-S., Ryu, H.-W. and Kim, D. 2007. Synthesis of thermo-and acid-stable novel oligosaccharides by using dextransucrase with high concentration of sucrose. Enzyme and Microbial Technology, 40(5): 1117–1123.

Sezer, A.D., Kazak, H., Öner, E.T. and Akbuğa, J. 2011. Levan-based nanocarrier system for peptide and protein drug delivery: optimization and influence of experimental parameters on the nanoparticle characteristics. Carbohydrate Polymers, 84(1): 358–363.

Shi, Q., Juvonen, M., Hou, Y., Kajala, I., Nyyssölä, A., Maina, N.H., Maaheimo, H., Virkki, L. and Tenkanen, M. 2016. Lactose-and cellobiose-derived branched trisaccharides and a sucrose-containing trisaccharide produced by acceptor reactions of *Weissella confusa* dextransucrase. Food Chemistry, 190: 226–236.

Shukla, S., Shi, Q., Maina, N.H., Juvonen, M. and Goyal, A. 2014. *Weissella confusa* Cab3 dextransucrase: Properties and *in vitro* synthesis of dextran and glucooligosaccharides. Carbohydrate Polymers, 101: 554–564.

Silva, M.F., Rigo, D., Mossi, V., Golunski, S., de Oliveira Kuhn, G., Di Luccio, M., Dallago, R., de Oliveira, D., Oliveira, J.V. and Treichel, H. 2013. Enzymatic synthesis of fructooligosaccharides by inulinases from *Aspergillus niger* and *Kluyveromyces marxianus* NRRL Y-7571 in aqueous–organic medium. Food Chemistry, 138(1): 148–153.

Silverstein, R., Voet, J., Reed, D. and Abeles, R.H. 1967. Purification and mechanism of action of sucrose phosphorylase. Journal of Biological Chemistry, 242(6): 1338–1346.

Smit, N.P., Westerhof, W., Asghar, S.S., Pavel, S. and Siddiqui, A.H. 1989. Large-scale cultivation of human melanocytes using collagen-coated sephadex beads (cytodex 3). Journal of Investigative Dermatology, 92(1): 18–21.

Smith, M.A. and Vale, W.W. 1980. Superfusion of rat anterior pituitary cells attached to cytodex beads: Validation of a technique. Endocrinology, 107(5): 1425–1431.

Sprenger, N., Bortlik, K., Brandt, A., Boller, T. and Wiemken, A. 1995. Purification, cloning, and functional expression of sucrose: fructan 6-fructosyltransferase, a key enzyme of fructan synthesis in barley. Proceedings of the National Academy of Sciences, USA, 92(25): 11652–11656.

Sprogøe, D., van den Broek, L.A., Mirza, O., Kastrup, J.S., Voragen, A.G., Gajhede, M. and Skov, L.K. 2004. Crystal structure of sucrose phosphorylase from *Bifidobacterium adolescentis*. Biochemistry, 43(5): 1156–1162.

Tadros, T.F., Vandamme, A., Levecke, B., Booten, K. and Stevens, C. 2004. Stabilization of emulsions using polymeric surfactants based on inulin. Advances in Colloid and Interface Science, 108: 207–226.

Tedokon, M., Suzuki, K., Kayamori, Y., Fujita, S. and Katayama, Y. 1992. Enzymatic assay of inorganic phosphate with use of sucrose phosphorylase and phosphoglucomutase. Clinical Chemistry, 38(4): 512–515.

Tian, F., Karboune, S. and Hill, A. 2014a. Synthesis of fructooligosaccharides and oligolevans by the combined use of levansucrase and endo-inulinase in one-step bi-enzymatic system. Innovative Food Science & Emerging Technologies, 22: 230–238.

Tian, F., Khodadadi, M. and Karboune, S. 2014b. Optimization of levansucrase/endo-inulinase bi-enzymatic system for the production of fructooligosaccharides and oligolevans from sucrose. Journal of Molecular Catalysis B: Enzymatic, 109: 85–93.

Trethewey, R., Fernie, A., Bachmann, A., Fleischer-Notter, H., Geigenberger, P. and Willmitzer, L. 2001. Expression of a bacterial sucrose phosphorylase in potato tubers results in a glucose-independent induction of glycolysis. Plant, Cell & Environment, 24(3): 357–365.

Tsinalis, D. and Thiel, G.T. 2009. An easy to calculate equation to estimate GFR based on inulin clearance. Nephrology Dialysis Transplantation, gfp193.

Vaidya, V.D. and Prasad, D.T. 2012. Thermostable levansucrase from *Bacillus subtilis* BB04, an isolate of banana peel. Journal of Biochemical Technology, 3(4): 322–327.

van den Broek, L.A., van Boxtel, E.L., Kievit, R.P., Verhoef, R., Beldman, G. and Voragen, A.G. 2004. Physico-chemical and transglucosylation properties of recombinant sucrose phosphorylase from *Bifidobacterium adolescentis* DSM20083. Applied Microbiology and Biotechnology, 65(2): 219–227.

Van Hijum, S., Van Der Maarel, M. and Dijkhuizen, L. 2003. Kinetic properties of an inulosucrase from *Lactobacillus reuteri* 121. FEBS Letters, 534(1): 207–210.

Vandamme, E.J., Van Loo, J., Machtelinckx, L. and De Laporte, A. 1987. Microbial sucrose phosphorylase: fermentation process, properties, and biotechnical applications. Advances in Applied Microbiology, 32: 163–201.

Wada, T., Ohguchi, M. and Iwai, Y. 2003. A novel enzyme of *Bacillus* sp. 217C-11 that produces inulin from sucrose. Bioscience, Biotechnology and Biochemistry, 67(6): 1327–1334.

Wallis, G., Hemming, F. and Peberdy, J. 1997. Secretion of two β-Fructofuranosidases by *Aspergillus niger* growing in sucrose. Archives of Biochemistry and Biophysics, 345(2): 214–222.

Walter, J., Schwab, C., Loach, D.M., Gänzle, M.G. and Tannock, G.W. 2008. Glucosyltransferase A (GtfA) and inulosucrase (Inu) of *Lactobacillus reuteri* TMW1. 106 contribute to cell aggregation, *in vitro* biofilm formation, and colonization of the mouse gastrointestinal tract. Microbiology, 154(1): 72–80.

Wildberger, P., Aish, G.A., Jakeman, D.L., Brecker, L. and Nidetzky, B. 2015. Interplay of catalytic subsite residues in the positioning of α-d-glucose 1-phosphate in sucrose phosphorylase. Biochemistry and Biophysics Reports, 2: 36–44.

Yoo, S.-H., Yoon, E.J., Cha, J. and Lee, H.G. 2004. Antitumor activity of levan polysaccharides from selected microorganisms. International Journal of Biological Macromolecules, 34(1): 37–41.

Yoon, E.J., Yoo, S.-H., Cha, J. and Lee, H.G. 2004. Effect of levan's branching structure on antitumor activity. International Journal of Biological Macromolecules, 34(3): 191–194.

Yun, J.W. 1996. Fructooligosaccharides—occurrence, preparation, and application. Enzyme and Microbial Technology, 19(2): 107–117.

Zambelli, P., Fernández-Arrojo, L., Romano, D., Santos-Moriano, P., Gimeno-Perez, M., Poveda, A., Gandolfi, R., Fernández-Lobato, M., Molinari, F. and Plou, F. 2014. Production of fructooligosaccharides by mycelium-bound transfructosylation activity present in *Cladosporium cladosporioides* and *Penicilium sizovae*. Process Biochemistry, 49(12): 2174–2180.

Zambelli, P., Tamborini, L., Cazzamalli, S., Pinto, A., Arioli, S., Balzaretti, S., Plou, F.J., Fernandez-Arrojo, L., Molinari, F. and Conti, P. 2016. An efficient continuous flow process for the synthesis of a non-conventional mixture of fructooligosaccharides. Food Chemistry, 190: 607–613.

β-Galactosidases for Lactose Hydrolysis and Galactooligosaccharide Synthesis

Francisco J. Plou,[1,]* *Julio Polaina,*[2] *Julia Sanz-Aparicio*[3] *and María Fernández-Lobato*[4]

1. Introduction

β-Galactosidases (β-D-galactoside galactohydrolases, EC 3.2.1.23), also called lactases, catalyze the hydrolysis of the non-reducing-end galactosyl moiety of various oligosaccharides. These enzymes can be found naturally in a variety of sources including microorganisms (bacteria, fungi and yeasts), plants (especially in almonds, peaches and apricots) and animals (Husain, 2010). Microbial enzymes show great biotechnological interest and commercial value by offering advantages, such as easy handling, ensured availability and high production yield. A large number of microorganisms have been assessed as potential sources of β-galactosidases, and properties of the characterized enzymes differ markedly (Oliveira et al., 2011; Panesar et al., 2010).

β-Galactosidases have attracted attention from food industries as they can be used in different applications, mostly hydrolysis of lactose into the monosaccharide glucose and galactose (Ansari and Husain, 2010) and the production of galactosylated

[1] Instituto de Catálisis y Petroleoquímica, CSIC, 28049 Madrid, Spain.
[2] Instituto de Agroquímica y Tecnología de Alimentos, CSIC, 46980 Paterna, Valencia, Spain.
[3] Departamento de Cristalografía y BiologíaEstructural, Instituto de Química-FísicaRocasolano, CSIC, 28006 Madrid, Spain.
[4] Centro de Biología Molecular Severo Ochoa (CSIC-UAM), Universidad Autónoma de Madrid, 28049 Madrid, Spain.
* Corresponding author

derivatives, basically galactooligosaccharides (GOS). GOS show ability to induce a bifidogenic microbiota in the consumer, which is referred as prebiotic effect (Lamsal, 2012).

The use of immobilization technology has a significant economic impact in the β-galactosidases market because it allows reutilization of the enzyme and continuous operation; in some cases it can also help to improve the enzyme stability. Different enzyme immobilization methods have been applied to these enzymes, including binding to a support (adsorption or covalent attachment), cross-linking, matrix entrapment or encapsulation (Konar et al., 2011; Torres and Batista-Viera, 2012).

2. Structure and Mechanism of β-Galactosidases

β-Galactosidases are retaining glycosidases, thus resulting in the net retention of the anomeric configuration of the released galactose. The hydrolysis occurs through a double displacement mechanism, i.e., an initial nucleophilic substitution, provided by an enzyme nucleophile, forms an inverted covalent enzyme-carbohydrate intermediate which is, in turn, attacked by the acceptor molecule, leading to a net retention of anomeric stereochemistry. Two glutamates are essential for the enzymatic reaction—one acting as the nucleophile while the other is the acid/base catalyst. This second residue assists in providing the proton to the leaving group in the first step of the reaction, while giving an electron pair and activating the acceptor in the second. In general, the galactose molecule is recognized at the glycon (subsite-1) in a highly precise mode characteristic to each family, while a broad specificity is allowed at the aglycon portion of the active site (subsites +1, +2, ...). This trait results in a high variability of the potential substrates for both the hydrolytic and the synthetic reaction.

β-Galactosidases are members of the GH1, GH2, GH35, GH42, and GH59 families of the glycoside hydrolases (Cantarel et al., 2009). All of them contain a catalytic domain folded into a $(\beta/\alpha)_8$ barrel. This folding allocates the active center in the axis of the barrel surrounded by the loops connecting the end of the β-strands to the helices. These loops are the most variable regions in the sequence within each family, being mostly responsible for the different specificities encountered. Furthermore, and excluding enzymes within GH1 family—to which human lactase belongs, among others—a common characteristic of β-galactosidases is their multidomain character with all of them containing from 2 to 6 additional domains showing β secondary structure, with generally an unknown function. Most studied families are GH2 and GH35 while those from eukaryotic organisms are grouped into GH35 with the exception of *Kluyveromyces lactis* and *Kluyveromyces marxianus* β-galactosidases, which belong to the GH2 family together with prokaryotic enzymes, such as β-galactosidases from *Escherichia coli*, *Bifidobacterium bifidum* and *Thermotoga maritima*.

Among β-galactosidases, *Kluyveromyces lactis* β-galactosidase is one of the most important and widely used enzymes in the industry. Several studies report the existence of two active forms that were attributed to the presence of dimers and tetramers. The crystal structure (Pereira-Rodriguez et al., 2012) showed that the tetramer is, in fact, a dimer of dimers, explaining that equilibrium exists between the associated and dissociated dimers (Fig. 1). The subunit is a large polypeptide of 1032 amino acids

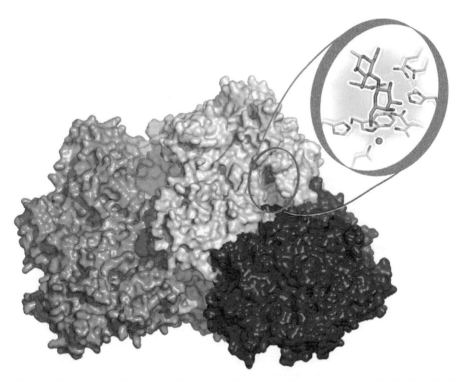

Figure 1. Structure 3D of the tetrameric β-galactosidase from *Kluyveromyces lactis*. Each dimer (red/beige, cyan/orange) locates two active sites faced in a connected narrow channel. A lactose molecule is shown in spheres representation at one of them. Zoom: The crystal analysis of complexes illustrates how the substrate lactose (green) is accommodated by interacting with key residues of the enzyme (beige) conferring specificity. A sodium (purple) and a magnesium ion (green) are involved in substrate binding [Courtesy: Julia Sanz-Aparicio (IQFR, CSIC)].

folded into five domains, with a long solvent exposed linker connecting domains 4 and 5. The special quaternary arrangement of the enzyme locates the catalytic pocket in a very narrow cavity surrounded not only by residues from the catalytic domain 3, but also from domains 1 and 5. Interestingly, two catalytic pockets are faced at the dimer interface but no cooperativity is detected between them. As it has been reported for other GH2 enzymes, *K. lactis* β-galactosidase contains a magnesium and sodium ion that are directly involved in accommodating the galactose at the catalytic pocket and seems essential for catalysis.

The first structural studies on a GH2 β-galactosidase carried out on the enzyme from *E. coli* delineated the main features explaining its function, essentially the ability to hydrolyze lactose or allolactose with equal catalytical efficiency, while being able to produce allolactose by transglycosylation (Juers et al., 2001). Also, the analysis of different complexes led to the proposals of a reaction mechanism that involves a movement of the galactosyl moiety from a shallow mode binding (proper of the substrates and the product allolactose) into a deep position (proper of intermediates and the product galactose), in which there is a conformational change in a loop that was stated to be responsible for selecting allolactose as the sole transglycosylating product. However, this loop is not conserved in *K. lactis* β-galactosidase and does not show conformational changes upon galactose binding. Consequently, the proposed mechanism may rule only for the GH2 enzymes regulated by the lac operon. By contrast, a marked distinguishing feature in *K. lactis* β-galactosidase active site is an insertion in one of the loops of the catalytic domain surrounding the catalytic pocket that protrudes from the channel and partially buries the entrance to the binding site. Interestingly, some residues from this loop make many links with the aglycon moiety of the lactose and consequently, this may account for the high affinity of *K. lactis* β-galactosidase for this substrate and therefore, might explain its unusually high hydrolytic activity. Moreover, this loop may be involved in selecting different acceptor molecules during transglycosylation, in particular, in conferring the unique affinity for acceptors glucose or galactose to yield disaccharides, as will be described below.

In contrast with *K. lactis* β-galactosidase–forming dimer/tetramer and other GH2 members, e.g., those from *E. coli* (tetramer) or *Arthrobacter* (hexamer), most GH35 β-galactosidases are monomeric or dimeric enzymes. They are isolated from microorganisms as fungi, bacteria, yeast and, also, from higher organisms as plants and mammals. Commonly, they show the highest activity in acidic conditions, with several of them having broad specificity and the ability to hydrolyze Gal-β(1→3), β(1→4) and β(1→6) substrates. The first structurally depicted GH35 β-galactosidase was that from *Penicillium* sp. (Rojas et al., 2004), which showed that it is a large protein composed of an N-terminal catalytic domain surrounded by five β-structure domains in a horseshoe shape. The later reported crystal structures of the β-galactosidases from *Trichoderma reesei* and *Aspergillus oryzae* (Maksimainen et al., 2013) confirmed the same domain compositions for all the three fungal enzymes. In contrast, the reported structures of β-galactosidases from *Streptococcus pneumonia* (Cheng et al., 2012) and *Bacillus circulans* (Henze et al., 2014) showed that they are composed of only three distinct domains, the fungal second and third domains being missing—a trait that is shared with the human β-galactosidase (Ohto et al., 2012). Interestingly, the human and the *S. pneumonia* β-galactosidases are dimeric and, thus, the oligomerization might compensate the functionality of the missing domains. The structure of many complexes has been reported, which has allowed the identification of the determinants involved in substrate recognition. Some conformational changes upon substrate binding have been found in the β-galactosidase from *T. reseei* (Maksimainen et al., 2011), but these changes have not been observed in other reported complexes.

3. Biotechnological Use of β-Galactosidases for Lactose Hydrolysis

3.1 Lactose intolerance

Intolerance to lactose is the inability to digest the milk sugar, caused by insufficient or nil production of lactase in the intestine. Lactose is not absorbed by enterocytes and the disaccharide must be hydrolysed into its two constituent monosaccharides—glucose and galactose which can be permeated into the cells. The most common type of lactase intolerance, called adult type hypolactasia (AHT) since it manifests itself in adults, is indeed a very common problem that affects about two-thirds of the human population. The symptoms caused by undigested lactose vary between individuals and depend on the amount ingested. The most common is diarrhoea often accompanied by abdominal pain, gas and nausea.

In the strict sense, adult lactose intolerance cannot be considered a disease but a normal condition. While congenital lactase deficiency is a rare defect with fatal consequences for newborns (if not averted), in most humans the production of lactase-phlorizin hydrolase—the specific type of intestinal enzyme responsible for lactose digestion—starts to decrease after weaning, when suckling is no longer necessary, and disappears in adulthood—a normal process known as lactase non-persistence. About 7,500 years ago, human groups which had adopted cattle herding as a way of life developed the capability of digesting lactose in adulthood (lactose persistency). This trait appeared independently in different populations, in Europe and Africa, caused by point mutations (single nucleotide replacement) at the regulatory (promoter) region of the LCT gene that encodes LPH (Enattah et al., 2002; Olds and Sibley, 2003). These mutations were transmitted to their descendants and as a consequence, of the selective advantage that it conferred, became highly predominant (> 90 per cent) among northwest European and Tutsi populations (Itan et al., 2009; Itan et al., 2010; Swallow, 2003; Tishkoff et al., 2007). In addition to the referred ATH, the other less frequent causes of lactose intolerance can also be damage of the small intestine due to surgery, chemotherapy, infection, etc. (Amiri et al., 2015; Deng et al., 2015).

3.2 Use of microbial lactases for the treatment of lactose intolerance

Milk and milk products are rich in nutrients and functional compounds that exert a positive effect on health. Therefore, they are generally considered important components of a healthy diet. Additionally, these products are essential ingredients in desserts, confectionary, ice-creams, etc. which are very dear to many consumers. To overcome the adverse effects associated with lactose intake by lactose-sensitive potential consumers, the addition of lactase to specific milk products and ingestion of lactase pills have become extended practices. Lactase is an ill-defined term to designate β-galactosidases with high specific activity on lactose (Adam et al., 2004). Lactases are synthesized by many microorganisms and it is relatively simple to produce them in large quantities by using genetic engineering techniques (Husain, 2010; Oliveira et al., 2011). Lactase preparations commercially available as food additives or complement to alleviate lactose deficiency, are generally obtained from GRAS (generally recognized as safe) microorganisms, specific strains of the yeast *Kluyveromyces* and fungus *Aspergillus*. *Kluyveromyces* is easy to cultivate and produces quite a large amount of highly active lactase and is relatively easy to purify (Mahoney et al., 1975). Although the β-galactosidase produced by *Aspergillus* is less efficient for lactose hydrolysis than the *Kluyveromyces* enzyme, industrial cultures of closely related *Aspergillus* species (either *A. niger* or *A. oryzae*) can be used for the simultaneous production of different valuable enzymes which make the process more profitable. Additionally, while *Kluyveromyces* lactase is an intracellular enzyme that requires cell disruption and purification, *Aspergillus* lactase is secreted to the culture medium from where it can be more easily purified by ultrafiltration (Nakayama and Amachi, 1999).

Despite the extended use of commercial lactase preparations, the enzymes currently used have some limitations that stimulate the search for new types of lactases. For instance, the enzymes commonly sold as pills to prevent lactose deficiency are not well suited to stand the conditions of pH and the attack by digestive enzymes that they confront in the stomach, where they become inactivated within a short time. An enzyme particularly suitable to be active under such conditions has now been characterized (O'Connell and Walsh, 2010). Protein-engineering techniques have been applied to improve the efficiency of *Aspergillus* β-galactosidase for lactose hydrolysis by making the enzyme more resistant to inhibition by its final product, galactose (Hu et al., 2010). Screening of metagenomic libraries has become a routine wide-range technique to search for enzymes with new capabilities. A specific search for genes isolated from metagenomic soil samples encoding β-galactosidase activity led to the characterization of new enzymes which proved more effective for lactose hydrolysis than commercial lactase (Erich et al., 2015).

3.3 Lactose-free milk

In the last decade, lactose-free milk and milk products have become easily available in supermarkets all over the world. An easy analysis of a random sample of lactose-free milk from different brands marketed in Spain (Fig. 2) showed that all of them contained *Kluyveromyces* lactase.

Figure 2. Lactose-free milk from different brands marketed in Spain, which were employed to analyze the strain of β-galactosidase involved in lactose removal [Courtesy: David Talens-Perales (IATA, CSIC)].

Figure 3 compares the pattern of proteins observed in milk samples (regular milk and milk without lactose of the same brand) with a commercial sample of *K. lactis* lactase. A protein band with the same mass than the lactase is present in the milk without lactose. The same result was obtained with all the other samples analyzed. In most of the lactose-free milk samples, the added lactase resisted the process of pasteurization since β-galactosidase activity was detectable. Among the enzymes used on an industrial scale, *Kluyveromyces* lactase is best suited to remove lactose from milk. As an intracellular enzyme, its optimum pH is about neutral, close to the slightly acid pH of milk, whereas the enzyme for *Aspergillus* works best at lower pH. The use of immobilised lactase offers an interesting alternative for the production of lactose-free milk without any exogenous unrequired constituent (Marin-Navarro et al., 2014).

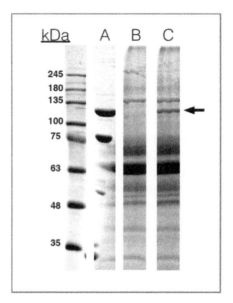

Figure 3. SDS-PAGE analysis of the proteins present in milk samples–regular milk (lane B) and lactose-free milk (lane C) of the same brand—compared with a commercial sample of *K. lactis* lactase (lane A). [Courtesy of David Talens-Perales (IATA, CSIC)].

3.4. Ethanol production from lactose

Worldwide, the cheese industry generates huge amounts of whey or the aqueous milk fraction obtained as a by-product once the coagulated casein has been separated. One kilogram of cheese produces nearly 10 liters of whey. Once other valuable proteins and fat have been recovered, the main component of whey is lactose at a concentration of about 5 per cent. Although lactose is a valuable sugar added to different food-related products, its recovery from whey by concentration or desiccation is expensive (Adam et al., 2004; Illanes, 2011). An alternative use of whey lactose is its conversion into ethanol. For this purpose, the yeast *Saccharomyces cerevisiae* is the microorganism of choice because of its powerful fermentative capability. However, *S. cerevisiae* is unable to hydrolyze lactose and lactase supplementation is too expensive. The technical solution is the use of genetically-modified strains to which the lactose metabolizing capability of *Kluyveromyces lactis* has been transferred. This requires the expression of two genes that code for lactose permease and β-galactosidase from *K. lactis* (Sreekrishna and Dickson, 1985). Efficient expression and stability of the *K. lactis* genes in *S. cerevisiae* can be achieved by multiple gene insertion (Rubio-Texeira et al., 1998; Rubio-Texeira et al., 2000) or by forced evolution (Guimarães et al., 2008). Strains with these characteristics are well suited for the production of ethanol from whey supplemented with corn steep liquor (Silva et al., 2010).

4. Application of β-Galactosidases for Galactooligosaccharides (GOS) Synthesis

4.1 Transgalactosylation activity of β-galactosidase

Another application of β-galactosidases is based on transgalactosylation reactions in which lactose or the released monosaccharides (glucose and galactose) serve as galactosyl acceptors, forming a series of disaccharides, trisaccharides (and eventually higher oligosaccharides) called galactooligosaccharides (GOS) (Park and Oh, 2010; Torres et al., 2010). In the presence of lactose, β-galactosidases catalyze both lactose hydrolysis and GOS synthesis (Fig. 4). The transferase/hydrolase ratio, which determines the maximum yield of GOS, depends basically on two parameters: (a) the concentration of lactose and (b) the intrinsic enzyme properties, that is, its ability to exclude H_2O from the acceptor binding site and to bind a sugar nucleophile (to which thegalactosyl moiety is transferred).

Maximum GOS production for any particular enzyme thus depends on the relative rates of the transgalactosylation and hydrolysis reactions (Plou et al., 2007). As the

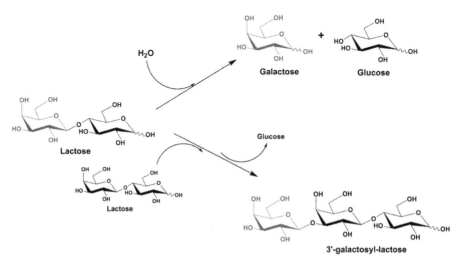

Figure 4. Hydrolytic and transglycosylation activity of β-galactosidases. [Courtesy: Francisco J. Plou (ICP, CSIC).]

lactose is consumed, the concentration of GOS increases until it reaches a maximum. At this point, the rate of synthesis of GOS products equals its rate of hydrolysis. Subsequently, product hydrolysis becomes the major process until thermodynamic equilibrium is reached. The existence of this kinetic maximum explains why transglycosylation results in higher yields of condensation products compared with equilibrium-controlled processes. It has been widely reported that, working under kinetic control conditions, enzyme concentration has no effect on the maximum GOS yield as long as no enzyme inactivation takes place. It only exerts a marked influence on the reaction time at which the maximum oligosaccharide concentration is achieved (Buchholz et al., 2005; Chockchaisawasdee et al., 2005).

GOS synthesis has been studied with different microbial β-galactosidases from Archea, Bacteria and Eukaryota super kingdoms. Table 1 summarizes the main reaction parameters and the GOS yield obtained with a series of β-galactosidases. In

Table 1. Examples of β-galactosidases employed for preparation of galactooligosaccharides (Adapted from Torres et al., 2010).

Organism	GH Family	Optimum pH	Optimum T (°C)	[Lactose][a] (g/L)	GOS Yield[b] (%)	Reference
Archaea						
Sulfobolus solfataricus	GH1	5.5–6.5	70–80	70–800	53	Petzelbauer et al. (2000)
Bacteria						
Bifidobacterium bifidum	GH2	6.0	37	100–400	44	Jorgensen et al. (2001)
Escherichia coli	GH2	6.5–7.2	30–37	22–240	56	Huber et al. (1976)
Bacillus circulans	GH2	5.5	40	400	49	Rodriguez-Colinas et al. (2012)
Geobacillus stearothermophilus	GH42	6.5	37	170	2.4	Placier et al. (2009)
Lactobacillus acidophilus	GH2	6.5	30	210	39	Nguyen et al. (2007)
Saccharopolyspora rectivirgula	GH2	7.0	70	600	44	Nakao et al. (1994)
Thermotoga maritima	GH2	6.0	80	200–500	19	Kim et al. (2004)
Eukaryota						
Aspergillus aculeatus	–	6.5	45–60	210–240	24	Del Val and Otero (2003)
Kluyveromyces lactis	GH2	6.8	40	400	44	Rodriguez-Colinas et al. (2011)
Penicillium expansum	GH35	5.4	50	50–480	29	Li et al. (2008)
Aspergillus oryzae	GH35	4.5	40	400	27	Urrutia et al. (2013)
Aspergillus niger	GH2	7.0	40	20–300	16	Toba and Adachi (1978)
Sterigmatomyces elviae	–	5.0	60	20	38	Onishi and Tanaka (1995)

[a] Initial lactose concentration
[b] Maximum GOS yield (%) = $([GOS]_{max}/[Lactose]_o) \times 100$

general, the upper range of GOS yield reported is close to 40–50 per cent (Hansson and Adlercreutz, 2001; Rabiu et al., 2001; Splechtna et al., 2006)—a value significantly lower than that obtained with other prebiotics, such as fructooligosaccharides (60–65 per cent) (Alvaro-Benito et al., 2007; Ghazi et al., 2006).

4.2 Properties of GOS

Infant food formulas are often supplemented with GOS to mimic the multiple benefits of human milk oligosaccharides (HMOs) (Moro and Arslanoglu, 2005; Shadid et al., 2007). GOS have been employed in the food industry as functional ingredients for more than 35 years, especially in Japan and Europe (Tzortzis and Vulevic, 2009).

HMOs are constituted by more than one hundred carbohydrates that exert positive effects on the health of breast-fed infants (Sela and Mills, 2010). The amount of HMOs in human milk (5–15 g/L) is nearly 100-fold higher than in cow's milk. In the colostrum the concentration of HMOs is even higher, representing nearly a quarter of total colostrum carbohydrates (Bode, 2006). HMOs contain lactose at their reducing end, normally elongated with N-acetyl-lactosamine or lacto-N-biose, which is further fucosylated or sialylated (Bode, 2012). The different combinations of monosaccharides and glycosidic linkages between the sugar moieties result in a complex family of linear and branched glycoderivatives, which from the practical point of view are difficult to synthesize (however, in the last few years a notable progress has been made in the preparation of HMOs by chemical and biotechnological methodologies (Zeuner et al., 2014). The formation of a beneficial bifidus microbiota in the intestine of milk-fed babies seems to be related with the prebiotic effect of HMOs present in the breast milk (Locascio et al., 2007).

GOS share structural features with HMOs and exhibit prebiotic properties (Gosling et al., 2010). A prebiotic is a non-digestible food ingredient that beneficially affects the host by selectively stimulating the growth and/or the activity of certain types of bacteria in the colon, basically of the genera *Bifidobacterium* and *Lactobacillus* (Gibson et al., 2004). Prebiotics produce positive effects on human health as the metabolism of these bacteria releases short-chain fatty acids (acetate, propionate and butyrate) and L-lactate (Roberfroid, 2007). Such fatty acids exert protective effects against colorectal cancer and bowel infectious diseases by inhibiting putrefactive and pathogen bacteria and improve the bioavailability of essential minerals. Prebiotics enhance the glucid and lipid metabolism and also reduce the level of cholesterol in the serum (Tuohy et al., 2005).

The chemical structure and composition of GOS (number of hexose units and types of linkages between them) may affect their fermentation pattern by probiotic bacteria in the gut (Cardelle-Cobas et al., 2011; Martinez-Villaluenga et al., 2008; Rodriguez-Colinas et al., 2013). Varying the source of β-galactosidase, the yield and composition of GOS can be modified (Iqbal et al., 2010; Maischberger et al., 2010; Rodriguez-Colinas et al., 2011; Splechtna et al., 2006). In general, complex mixtures of GOS of various chain lengths and glycosidic bonds are commonly obtained. The most studied β-galactosidases are those from *Kluyveromyces lactis* (Chockchaisawasdee et al., 2005; Martinez-Villaluenga et al., 2008; Maugard et al., 2003), *Aspergillus oryzae*

(Albayrak and Yang, 2002c; Guerrero et al., 2011; Iwasaki et al., 1996), *Bacillus circulans* (Gosling et al., 2011) and *Bifidobacterium* sp. (Hsu et al., 2007).

The complete identification of the GOS synthesized by a particular enzyme is a difficult task. In fact, considering the formation of Gal-β(1→2), β(1→3), β(1→4) and β(1→6) linkages, the theoretical number of synthesized GOS accounts for 7 disaccharides, 32 trisaccharides, 128 tetrasaccharides and so on. A combination of hydrophilic interaction chromatography (HPLC-HILIC), high performance anion-exchange chromatography coupled with pulsed amperometric detection (HPAEC-PAD), mass spectrometry and bidimensional NMR are typically required to unequivocally determine the structure of the synthesized oligosaccharides.

In the following sections we will describe the structural features of the GOS synthesized by three representative β-galactosidases.

4.3 Bacillus circulans β-galactosidase

At least three isoforms of the β-galactosidase from *Bacillus circulans* have been characterized for GOS production: isoform 1 showed major hydrolytic activity (Mozaffar et al., 1984), isoform 2 displayed higher transferase activity (Mozaffar et al., 1986), and isoform 3 was able to produce GOS with β(1→3) linkages (Fujimoto et al., 1998). The transgalactosylation activity of β-galactosidase from *B. circulans* has been also utilized in the synthesis of N-acetyl-lactosamine (Bridiau et al., 2010) or lactosucrose (Wei et al., 2009). This enzyme is also able to catalyze the galactosylation of different acceptors in the presence of several organic cosolvents (Usui et al., 1993). The *B. circulans* β-galactosidase has been immobilized on different supports (Mozaffar et al., 1986; Torres and Batista-Viera, 2012). The analysis of the GOS formed in the transglycosylation reaction with lactose has been scarcely described, probably due to the complexity of the reaction mixture derived from the presence of several isoforms with different regiospecificity (Mozaffar et al., 1986).

In a recent work, our group studied in detail the synthesis of GOS with a novel commercial preparation of β-galactosidase from *B. circulans* (Biolactase) starting with 400 g/L lactose (Rodriguez-Colinas et al., 2012). Figure 5 shows the chromatogram obtained by high performance anion-exchange chromatography coupled with pulsed amperometric detection (HPAEC-PAD) of the reaction mixture with this enzyme, close to the time of maximum GOS concentration. The two main GOS in the reaction mixture were the trisaccharide 4'-galactosyl-lactose [Gal-β(1→4)-Gal-β(1→4)-Glc] (peak *11*) and the tetrasaccharide Gal-β(1→4)-Gal-β(1→4)-Gal-β(1→4)-Glc (peak *17*). Other GOs were identified in the chromatograms: the disaccharides allolactose [Gal-β(1→6)-Glc] (peak *3*), 3-galactobiose [Gal-β(1→3)-Gal] (peak 5), 4-galactobiose [Gal-β(1→4)-Gal] (peak *6)* and Gal-β(1→3)-Glc (peak 8), as well as the trisaccharides 6'-galactosyl-lactose [Gal-β(1→6)-Gal-β(1→4)-Glc] (peak 7) and Gal-β(1→4)-Gal-β(1→3)-Glc (peak *15*). Peaks *9, 10, 12, 13, 14* and *16* remained unknown.

It is noteworthy that two of the major products synthesized by this enzyme (peaks *11* and *17*) contain only β-(1→4) bonds. Yanahira et al. (1995) were the first in performing structural analysis of the GOS formed by *B. circulans* β-galactosidase employing Biolacta (Daiwa Kasei). They reported that the main GOS was 4'-galactosyl-lactose. Although the presence of tetrasaccharides was not mentioned, they identified

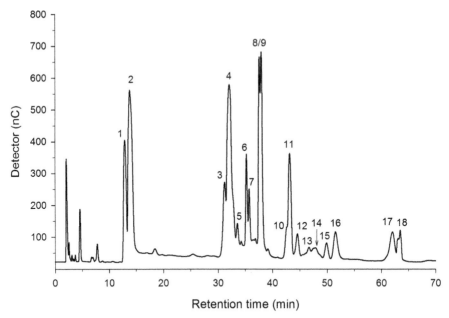

Figure 5. Chromatogram of the reaction of lactose with β-galactosidase from *Bacillus circulans* (Biolactase). Peaks: (1) Galactose; (2) Glucose; (3) Allolactose; (4) Lactose; (5) 3-Galactobiose; (6) 4-Galactobiose; (7) 6'-Galactosyl-lactose; (8) 3-Galactosyl-glucose; (11) 4'-Galactosyl-lactose; (15) Gal-β(1→4)-Gal-β(1→3)-Glc; (17) Gal-β(1→4)-Gal-β(1→4)-Gal-β(1→4)-Glc; (9, 10, 12, 13, 14, 16) Unknown GOS. Conditions: 400 g/L lactose, 0.1 M sodium acetate buffer (pH 5.5), 40°C (Adapted with permission from Rodriguez-Colinas et al., 2012).

several disaccharides and trisacccharides, some of them containing β-(1→2) linkages. Such products might correspond to some of the unidentified peaks in Fig. 5. Song et al. (2011) also analyzed the GOS production by different isoforms of *B. circulans* β-galactosidase; however, the structural analysis of the synthesized compounds was not reported.

We performed the GOS synthesis at 400 g/L lactose using 15 U/mL biocatalyst. Figure 6 shows that the maximum GOS concentration was achieved in 6.5 hours, with a yield of 198 g/L. This value corresponds to 49.4 per cent (w/w) of total sugars in the reaction mixture and it was higher than the value obtained at 1.5 U/mL (165 g/L, 41.3 per cent). It is worth noting that the remaining lactose at the equilibrium (10 g/L, 2.5 per cent of total carbohydrates) is significantly lower than that obtained at 1.5 U/mL. This effect probably indicated that the stability of *B. circulans* β-galactosidase was only moderate under typical reaction conditions. It is interesting to highlight that the GOS production with *B. circulans* β-galactosidase is one of the highest values reported to date.

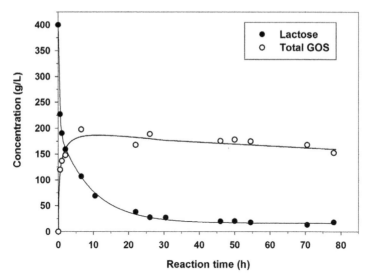

Figure 6. Kinetics of GOS formation at 15 U/mL using 400 g/L lactose catalyzed by β-galactosidase from B. circulans (biolactase). Reaction conditions: 0.1 M sodium acetate buffer (pH 5.5), 40°C (Adapted with permission from Rodriguez-Colinas et al., 2012).

4.4 Kluyveromyces lactis β-galactosidase

The mesophile yeast *Kluyveromyces lactis* is the major commercial source of β-galactosidase by far (Chockchaisawasdee et al., 2005; Martinez-Villaluenga et al., 2008; Maugard et al., 2003; Pal et al., 2009). Due to its intracellular nature, production of cell-free *K. lactis* β-galactosidase is hampered by the high cost associated with enzyme extraction and downstream processing as well as the low stability of the enzyme (Park and Oh, 2010; Pinho and Passos, 2011). Several approaches for immobilization of this enzyme on different carriers have been described (Maugard et al., 2003; Zhou et al., 2003; Zhou and Chen, 2001).

The production of GOS in batch and continuous bioreactors of *K. lactis* β-galactosidase has been reported (Chockchaisawasdee et al., 2005; Martinez-Villaluenga et al., 2008). Figure 7 shows the HPAEC-PAD chromatogram of the reaction of lactose with *K. lactis* β-galactosidase (Lactozym 3000L) close to the maximum GOS yield, under conditions that promote the transgalactosylation activity (Rodriguez-Colinas et al., 2011). The most abundant GOS synthesized by this enzyme corresponded to the disaccharides 6-galactobiose [Gal-β(1→6)-Gal] (peak 3) and allolactose [Gal-β(1→6)-Glc] (peak *4*) and the trisaccharide 6'-galactosyl-lactose [Gal-β(1→6)-Gal-β(1→4)-Glc] (peak *8*).

Interestingly the three major products synthesized by *K. lactis* β-galactosidase contained a β-(1→6) bond between two galactoses or between one galactose and one glucose. This product selectivity is of biological significance, as it is accepted

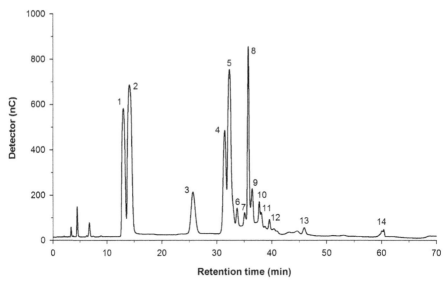

Figure 7. HPAEC-PAD chromatogram of the reaction of lactose with *K. lactis* β-galactosidase (Lactozym 3000L). Peaks: (1) Galactose; (2) Glucose; (3) 6-Galactobiose; (4) Allolactose (6-Galactosyl-glucose); (5) Lactose; (6) 3-Galactobiose; (7) 4-Galactobiose; (8) 6'-Galactosyl-lactose; (10) 3-Galactosyl-glucose; (9, 11, 12, 13, 14) Unknown GOS. Conditions: 400 g/L lactose, 0.1 M sodium phosphate buffer (pH 6.8), 40°C. Courtesy: Barbara Rodriguez-Colinas (ICP, CSIC).

that β(1→6) linkages are cleaved very fast by β-galactosidases from bifidobacteria (Martinez-Villaluenga et al., 2008), which represents a key factor in the prebiotic properties of GOS. The first in performing a preliminary structural analysis of the GOS synthesized by *K. lactis* β-galactosidase were Chockchaisawasdee et al. (2005) using the commercial preparation Maxilact L2000. They concluded that the major formed bonds were β(1→6). Maugard et al. (2003) and Cheng et al. (2006) detected the formation of disaccharides by *K. lactis* but no structural identification was performed. Martinez-Villaluenga et al. (2008) reported the formation of products with β(1→6) bonds with *K. lactis* β-galactosidase in accordance with our findings, but the formation of other GOS was not mentioned.

Therefore, the product specificity of *K. lactis* β-galactosidase is notably different to that of its *B. circulans* counterpart—the former exhibits a tendency to synthesize β(1→6) bonds, whereas the latter prefers the formation of β(1→4) bonds. In addition, both enzymes display a different behavior towards the formation of disaccharides: *K. lactis* enzyme is able to use efficiently free galactose and glucose as acceptors, yielding 6-galactobiose and allolactose, respectively, with notable yields (Rodriguez-Colinas et al., 2011) in contrast to the *B. circulans* β-galactosidase which only produces a moderate amount of allolactose, 4-galactobiose and 3-galactosyl-glucose.

The stability of *K. lactis* β-galactosidase is quite low. In such cases, the use of whole cells may offer advantages over soluble enzymes, allowing the reuse of the biocatalyst and continuous processing. However, the permeability barrier of the cell envelope for substrates and products often generates slow conversion rates, especially in yeasts. In order to increase the volumetric activity, permeabilization of yeast cells is a simple, economical and safe process that usually facilitates substrate access to the intracellular enzymes (Fontes et al., 2001; Kondo et al., 2000; Siso et al., 1992). The permeabilizing agent acts on the membrane phospholipids, thus facilitating the diffusion of low molecular weight compounds in and out of the cells (Manera et al., 2010). Several strategies of permeabilizing cells that do not cause a substantial reduction in enzymatic activity have been reported, including mechanical disruption, drying, treatment with solvents or surfactants, lyophilisation or ultrasonic treatment (Panesar et al., 2007; Panesar et al., 2011).

In this context, we assessed the use of *K. lactis* permeabilized cells for the GOS synthesis. Ethanol incubation followed by lyophilization was applied to harvested *K. lactis* cells. Although cells treated with short-chain alcohols may die in many cases, intracellular enzymes usually resist such treatment and, in consequence, are not inactivated. As a consequence, the permeability barriers of cell walls are lowered, so that the cells themselves can be repeatedly utilized as a biocatalyst.

The maximum GOS yield obtained with permeabilized cells (177 g/L, equivalent to 44 per cent of the total carbohydrates in the reaction mixture) was observed in 6 hours. At this point, the composition of the mixture was as follows: monosaccharides (32 per cent), lactose (24 per cent), 6-galactobiose (6 per cent), allolactose (10 per cent), 6'-galactosyl-lactose (16 per cent) and other GOS (12 per cent). Therefore GOS containing β-(1→6) bonds contributed substantially to total GOS throughout the process, as occurred with the soluble enzyme. On obtaining this maximum concentration, the amount of GOS diminished by the hydrolytic action of β-galactosidase until GOS concentration reached 105 g/L, which corresponded to the equilibrium position. Interestingly, lactose disappeared almost completely at the end of the process. Permeabilized *K. lactis* cells have not been only used for GOS synthesis, as they have been successfully applied to lactose hydrolysis (Fontes et al., 2001; Genari et al., 2003; Panesar et al., 2007; Siso et al., 1992) and for the transformation of fructose and lactose into disaccharide lactulose (Lee et al., 2004).

In summary, maximum GOS yield was slightly lower for soluble β-galactosidase (160 g/L for Lactozym, which represent 42 per cent w/w of the total carbohydrates in the mixture) compared with permeabilized cells (177 g/L, 44 per cent w/w). These results indicate that with the soluble enzyme the reaction finishes before reaching the equilibrium, probably due to the low stability of this β-galactosidase. In contrast, the enhanced stability of the enzyme embedded in the permeabilized cells is reflected in the typical reaction profile of the transglycosylation, showing a maximum of transglycosylated products followed by a progressive decrease in favor of the hydrolysis reaction. This behavior is called 'kinetic control of transglycosylation processes' (Rodriguez-Alegria et al., 2010).

4.5 Aspergillus oryzae β-galactosidase

The fungus *Aspergillus oryzae* produces a monomeric β-galactosidase of 105 kDa (Tanaka et al., 1975) and isoelectric point 4.6 (Ansari and Husain, 2010; Yang et al., 1994). Its optimum temperature is in the range 45–55°C (Guidini et al., 2010; Guleç et al., 2010) and shows optimum pH values of 4.5 with ONPG and 4.8 with lactose (Tanaka et al., 1975), two units lower than the optimum pH of *K. lactis* β-galactosidase.

The *A. oryzae* β-galactosidase has been successfully applied to the synthesis of GOS (Albayrak and Yang, 2002b; Iwasaki et al., 1996; Vera et al., 2012), lactulose (Guerrero et al., 2011; Mayer et al., 2004) and galactosyl-polyhydroxy alcohols (Irazoqui et al., 2009; Klewicki, 2007), using free or immobilized enzyme. In addition, the β-galactosidase from *A. oryzae* has been immobilized by different methodologies, including covalent attachment on to various carriers (Gaur et al., 2006; Huerta et al., 2011; Neri et al., 2011; Sheu et al., 1998), ionic adsorption followed by crosslinking (Guidini et al., 2011) and entrapment in alginate (Freitas et al., 2011).

In our laboratory, we assessed the synthesis of GOS catalyzed by a commercially available preparation of β-galactosidase from *A. oryzae* (Enzeco fungal lactase) using 400 g/L lactose and 15 U/mL. We analyzed by HPAEC-PAD the product specificity of *A. oryzae* β-galactosidase. The main oligosaccharide synthesized by this enzyme was trisaccharide 6'-galactosyl-lactose. Various disaccharides containing different linkages were also identified: Gal-β(1→6)-Gal, Gal-β(1→6)-Glc, Gal-β(1→3)-Gal and Gal-β(1→3)-Glc, as well as a series of tetrasaccharides.

The characterization of several GOS synthesized by *A. oryzae* β-galactosidase was carried out by Toba et al. (1985). In particular, they identified three trisaccharides: [6'-galactosyl-lactose, 3'-galactosyl-lactose and Gal-β(1→4)-Gal-β(1→6)-Glc], two tetrasaccharides [Gal-β(1→6)-Gal-β(1→6)-Gal-β(1→4)-Glc, Gal-β(1→6)-Gal-β(1→3)-Gal-β(1→4)-Glc] and one pentasaccharide [Gal-β(1→6)-Gal-β(1→6)-Gal-β(1→6)-Gal-β(1→4)-Glc]. On the other hand, Neri et al. (2011) covalently immobilized *A. oryzae* β-galactosidase on to a hydrazide-Dacron-magnetite composite and characterized several of the GOS synthesized. They identified three of the GOS previously characterized by Toba et al. (1985): the trisaccharides 6'-galactosyl-lactose and Gal-β(1→4)-Gal-β(1→6)-Glc, and the tetrasaccharide Gal-β(1→6)-Gal-β(1→6)-Gal-β(1→4)-Glc. The authors also reported the existence of a disaccharide containing a β(1→6) linkage, although no structural information was given. In general, the specific features of this enzyme indicated a tendency to form β(1→6) bonds followed by β(1→3), with a minor contribution of β(1→4) linkages.

Figure 8 depicts the kinetics of GOS formation using an enzyme concentration of 15 U/mL and 400 g/L lactose. GOS reached a maximum concentration of 107 g/L at 8 h of reaction, corresponding to 26.8 per cent of total carbohydrates in the mixture. Once the point of maximum yield was achieved, the GOS concentration slowly decreased to values of approximately 60 g/L. Regarding the effect of pH and temperature on GOS synthesis, it has been found that both the parameters affect the reaction rate but they do not modify maximum GOS concentration (Albayrak and Yang, 2002b; Neri et al., 2011).

It is well reported that the increase of lactose concentration [up to 30–40 per cent (w/v)] has a strong positive effect on the maximum amount of GOS obtained (Iwasaki et al., 1996; Matella et al., 2006). For that reason, initial lactose concentration must be accounted for in adequate comparison with other studies. Albayrak and Yang (2002a) used 500 g/lactose concentration and obtained a maximum GOS concentration of 27 per cent (w/w), of which more than 70 per cent corresponded to trisaccharides. In agreement with these results, Neri et al. (2011) obtained 20.2 per cent trisaccharides

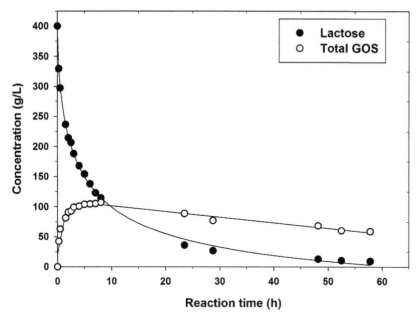

Figure 8. Reaction progress of GOS formation at 15 U/mL using 400 g/L lactose catalyzed by β-galactosidase from *A. oryzae*. Reaction conditions: 0.1 M citrate-phosphate buffer (pH 4.5), 40°C. Adapted with permission from Urrutia et al. (2013).

and 5.9 per cent tetrasaccharides (total yield of 26.1 per cent, 130 g/L) using 500 g/L lactose and 40°C. No significant differences were observed in comparison to the free or immobilized enzyme. Gaur et al. (2006), using 200 g/L lactose, reported that *A. oryzae* β-galactosidase formed only trisaccharides, with a maximum GOS yield of 22.6 per cent (w/w) for the soluble enzyme and 25.5 per cent (w/w) for the chitosan-immobilized β-galactosidase. Guleç et al. (2010) reported a GOS yield of 20.8 per cent (w/w)—mainly trisaccharides—using 320 g/L lactose and 55°C.

Significant differences were found in the behavior of β-galactosidases from *Bacillus circulans*, *Kluyveromyces lactis* and *Aspergillus oryzae* during the synthesis of GOS. First of all, the product specificity was dependent on the enzyme origin. Thus, *K. lactis* β-galactosidase exhibits a tendency to synthesize β(1→6) bonds, whereas the *B. circulans* counterpart prefers the formation of β(1→4) bonds. The β-galactosidase from *A. oryzae* displays a marked preference to form β(1→6) linkages followed by β(1→3).

In conclusion, the maximum yield and the bond specificity of the synthesized GOS are highly dependent on the origin of β-galactosidase. As the properties of GOS may depend significantly on the degree of polymerization and chemical structure, the selection of the appropriate enzyme could exert a considerable effect on the bioactivity of the resulting product. For example, to get a GOS enriched in disaccharides, *K. lactis* β-galactosidase is one of the best choices. However, the *B. circulans* enzyme would be preferred if a GOS with a high tri- and tetrasaccharides content is desirable.

5. Summary and Conclusion

β-Galactosidases (EC 3.2.1.23) are retaining glycosidases that belong to the GH1, GH2, GH35, GH42, and GH59 families of the glycoside hydrolases, and catalyze the release of the galactosyl moiety from the non-reducing end of different carbohydrates. Structurally, they are characterized by the presence of a catalytic domain folded into a $(\beta/\alpha)_8$ barrel. To overcome the adverse effects associated to lactose intake by lactose-sensitive consumers, the addition of lactase to specific milk products has become one of the most important food processes performed with enzymes to date. In particular, *Kluyveromyces lactis* β-galactosidase is commonly used to remove lactose from milk.

β-Galactosidases also catalyze transgalactosylation reactions in which lactose, as well as the glucose and galactose released by hydrolysis, serve as galactosyl acceptors, yielding galactooligosaccharides (GOS), which are widely employed in infant foods. GOS share structural features with human milk oligosaccharides (HMOs) and exhibit prebiotic properties. The yield of GOS (the highest value does not exceed 50 per cent) and the linkages present in the synthesized products strongly depend on the enzyme source and reaction conditions. Due to their availability and outstanding properties, β-galactosidases will continue attracting the attention not only from the food industry but also from other sectors, such as the biofuels manufacturers or the biosensor devices.

References

Adam, A.C., Rubio-Texeira, M. and Polaina, J. 2004. Lactose: The milk sugar from a biotechnological perspective. Critical Reviews in Food Science and Nutrition, 44: 553–557.

Albayrak, N. and Yang, S.T. 2002a. Immobilization of *Aspergillus oryzae* beta-galactosidase on to sylated cotton cloth. Enzyme and Microbial Technology, 31: 371–383.

Albayrak, N. and Yang, S.T. 2002b. Immobilization of beta-galactosidase on fibrous matrix by polyethyleneimine for production of galacto-oligosaccharides from lactose. Biotechnology Progress, 18: 240–251.

Albayrak, N. and Yang, S.T. 2002c. Production of galacto-oligosaccharides from lactose by *Aspergillus oryzae* beta-galactosidase immobilized on cotton cloth. Biotechnology and Bioengineering, 77: 8–19.

Alvaro-Benito, M., de Abreu, M., Fernandez-Arrojo, L., Plou, F.J., Jimenez-Barbero, J., Ballesteros, A., Polaina, J. and Fernandez-Lobato, M. 2007. Characterization of a β-fructofuranosidase from *Schwanniomyces occidentalis* with transfructosylating activity yielding the prebiotic 6-kestose. Journal of Biotechnology, 132: 75–81.

Amiri, M., Diekmann, L., von Köckritz-Blickwede, M. and Naim, H.Y. 2015. The diverse forms of lactose intolerance and the putative linkage to several cancers. Nutrients, 7: 7209–7230.

Ansari, S.A. and Husain, Q. 2010. Lactose hydrolysis by β-galactosidase immobilized on concanavalin A-cellulose in batch and continuous mode. Journal of Molecular Catalysis B: Enzymatic, 63: 68–74.

Bode, L. 2006. Recent advances on structure, metabolism, and function of human milk oligosaccharides. Journal of Nutrition, 136: 2127–2130.

Bode, L. 2012. Human milk oligosaccharides: Every baby needs a sugar mama. Glycobiology, 22: 1147–1162.

Bridiau, N., Issaoui, N. and Maugard, T. 2010. The effects of organic solvents on the efficiency and regioselectivity of N-acetyl-lactosamine synthesis, using the β-galactosidase from *Bacillus circulans* in hydro-organic media. Biotechnology Progress, 26: 1278–1289.

Buchholz, K., Kasche, V. and Bornscheuer, U.T. 2005. Biocatalysts and Enzyme Technology. Wiley-VCH Verlag, Weinheim.

Cantarel, B.L., Coutinho, P.M., Rancurel, C., Bernard, T., Lombard, V. and Henrissat, B. 2009. The Carbohydrate-Active EnZymes database (CAZy): An expert resource for Glycogenomics. Nucleic Acids Research, 37: D233–D238.

Cardelle-Cobas, A., Corzo, N., Olano, A., Pelaez, C., Requena, T. and Avila, M. 2011. Galactooligosaccharides derived from lactose and lactulose: Influence of structure on *Lactobacillus*, *Streptococcus* and *Bifidobacterium* growth. International Journal of Food Microbiology, 149: 81–87.

Cheng, C.C., Yu, M.C., Cheng, T.C., Sheu, D.C., Duan, K.J. and Tai, W.L. 2006. Production of high-content galacto-oligosaccharide by enzyme catalysis and fermentation with *Kluyveromyces marxianus*. Biotechnology Letters, 28: 793–797.

Cheng, W., Wang, L., Jiang, Y.L., Bai, X.H., Chu, J., Li, Q., Yu, G., Liang, Q.L., Zhou, C.Z. and Chen, Y. 2012. Structural insights into the substrate specificity of *Streptococcus pneumoniae* b(1,3)-galactosidase BgaC. Journal of Biological Chemistry, 287: 22910–22918.

Chockchaisawasdee, S., Athanasopoulos, V.I., Niranjan, K. and Rastall, R.A. 2005. Synthesis of galacto-oligosaccharide from lactose using beta-galactosidase from *Kluyveromyces lactis*: Studies on batch and continuous UF membrane-fitted bioreactors. Biotechnology and Bioengineering, 89: 434–443.

Del Val, M.I. and Otero, C. 2003. Biphasic aqueous media containing polyethylene glycol for the enzymatic synthesis of oligosaccharides from lactose. Enzyme and Microbial Technology, 33: 118–126.

Deng, Y., Misselwitz, B., Dai, N. and Fox, M. 2015. Lactose intolerance in adults: Biological mechanism and dietary management. Nutrients, 7: 8020–8035.

Enattah, N.S., Sahi, T., Savilahti, E., Terwilliger, J.D., Peltonen, L. and Järvelä, L. 2002. Identification of a variant associated with adult-type hypolactasia. Nature Genetics, 30: 233–237.

Erich, S., Kuschel, B., Schwarz, T., Ewert, J., Böhmer, N., Niehaus, F., Eck, J., Lutz-Wahl, S., Stressler, T. and Fischer, L. 2015. Novel high-performance metagenome β-galactosidases for lactose hydrolysis in the dairy industry. Journal of Biotechnology, 210: 27–37.

Fontes, E.A.F., Passos, F.M.L. and Passos, F.J.V. 2001. A mechanistic mathematical model to predict lactose hydrolysis by ß-galactosidase in a permeabilized cell mass of *Kluyveromyces lactis*: Validity and sensitivity analysis. Process Biochemistry, 37: 267–274.

Freitas, F.F., Marquez, L.D.S., Ribeiro, G.P., Brandao, G.C., Cardoso, V.L. and Ribeiro, E.J. 2011. A comparison of the kinetic properties of free and immobilized *Aspergillus oryzae* β-galactosidase. Biochemical Engineering Journal, 58-59: 33–38.

Fujimoto, H., Miyasato, M., Ito, Y., Sasaki, T. and Ajisaka, K. 1998. Purification and properties of recombinant β-galactosidase from *Bacillus circulans*. Glycoconjugate Journal, 15: 155–160.

Gaur, R., Pant, H., Jain, R. and Khare, S.K. 2006. Galacto-oligosaccharide synthesis by immobilized *Aspergillus oryzae* beta-galactosidase. Food Chemistry, 97: 426–430.

Genari, A.N., Passos, F.V. and Passos, F.M.L. 2003. Configuration of a bioreactor for milk lactose hydrolysis. Journal of Dairy Science, 86: 2783–2789.

Ghazi, I., Fernandez-Arrojo, L., Gomez de Segura, A., Alcalde, M., Plou, F.J. and Ballesteros, A. 2006. Beet sugar syrup and molasses as low-cost feedstock for the enzymatic production of fructo-oligosaccharides. Journal of Agricultural and Food Chemistry, 54: 2964–2968.

Gibson, G.R., Probert, H.M., Van Loo, J., Rastall, R.A. and Roberfroid, M.B. 2004. Dietary modulation of the human colonic microbiota: Updating the concept of prebiotics. Nutrition Research Reviews, 17: 259–275.

Gosling, A., Stevens, G.W., Barber, A.R., Kentish, S.E. and Gras, S.L. 2010. Recent advances refining galactooligosaccharide production from lactose. Food Chemistry, 121: 307–318.

Gosling, A., Stevens, G.W., Barber, A.R., Kentish, S.E. and Gras, S.L. 2011. Effect of the substrate concentration and water activity on the yield and rate of the transfer reaction of β-galactosidase from *Bacillus circulans*. Journal of Agricultural and Food Chemistry, 59: 3366–3372.

Guerrero, C., Vera, C., Plou, F. and Illanes, A. 2011. Influence of reaction conditions on the selectivity of the synthesis of lactulose with microbial beta-galactosidases. Journal of Molecular Catalysis B: Enzymatic, 72: 206–212.

Guidini, C.Z., Fischer, J., Resende, M.M.D., Cardoso, V.L. and Ribeiro, E.J. 2011. b-Galactosidase of *Aspergillus oryzae* immobilized in an ion exchange resin combining the ionic-binding and crosslinking methods: Kinetics and stability during the hydrolysis of lactose. Journal of Molecular Catalysis B: Enzymatic, 71: 139–145.

Guidini, C.Z., Fischer, J., Santana, L.N.S., Cardoso, V.L. and Ribeiro, E.J. 2010. Immobilization of *Aspergillus oryzae* b-galactosidase in ion exchange resins by combined ionic-binding method and cross-linking. Biochemical Engineering Journal, 52: 137–143.

Guimarães, P.M.R., François, J., Parrou, J.L., Teixeira, J.A. and Domingues, L. 2008. Adaptive evolution of a lactose-consuming *Saccharomyces cerevisiae* recombinant. Applied and Environmental Microbiology, 74: 1748–1756.

Guleç, H.A., Gürdaç., Albayrak, N. and Mutlu, M. 2010. Immobilization of *Aspergillus oryzae* β-galactosidase on low-pressure plasma-modified cellulose acetate membrane using polyethyleneimine for production of galactooligosaccharide. Biotechnology and Bioprocess Engineering, 15: 1006–1015.

Hansson, T. and Adlercreutz, P. 2001. Optimization of galactooligosaccharide production from lactose using beta-glycosidases from hyperthermophiles. Food Biotechnology, 15: 79–97.

Henze, M., You, D.J., Kamerke, C., Hoffmann, N., Angkawidjaja, C., Ernst, S., Pietruszka, J., Kanaya, S. and Elling, L. 2014. Rational design of a glycosynthase by the crystal structure of β-galactosidase from *Bacillus circulans* (BgaC) and its use for the synthesis of N-acetyllactosamine type 1 glycan structures. Journal of Biotechnology, 191: 78–85.

Hsu, C.A., Lee, S.L. and Chou, C.C. 2007. Enzymatic production of galactooligosaccharides by beta-galactosidase from *Bifidobacterium longum* BCRC 15708. Journal of Agricultural and Food Chemistry, 55: 2225–2230.

Hu, X., Robin, S., O'Connell, S., Walsh, G. and Wall, J.G. 2010. Engineering of a fungal β-galactosidase to remove product inhibition by galactose. Applied Microbiology and Biotechnology, 87: 1773–1782.

Huber, R.E., Kurz, G. and Wallenfels, K. 1976. A quantitation of the factors which affect the hydrolase and transgalactosylase activities of β-galactosidase (*E. coli*) on lactose. Biochemistry, 15: 1994–2001.

Huerta, L.M., Vera, C., Guerrero, C., Wilson, L. and Illanes, A. 2011. Synthesis of galacto-oligosaccharides at very high lactose concentrations with immobilized β-galactosidases from *Aspergillus oryzae*. Process Biochemistry, 46: 245–252.

Husain, Q. 2010. β-Galactosidases and their potential applications: A review. Critical Reviews in Biotechnology, 30: 41–62.

Illanes, A. 2011. Whey upgrading by enzyme biocatalysis. Electronic Journal of Biotechnology, 14.

Iqbal, S., Nguyen, T.H., Nguyen, T.T., Maischberger, T. and Haltrich, D. 2010. ß-galactosidase from *Lactobacillus plantarum* WCFS1: Biochemical characterization and formation of prebiotic galacto-oligosaccharides. Carbohydrate Research, 345: 1408–1416.

Irazoqui, G., Giacomini, C., Batista-Viera, F., Brena, B.M., Cardelle-Cobas, A., Corzo, N. and Jimeno, M.L. 2009. Characterization of galactosyl derivatives obtained by transgalactosylation of lactose and different polyols using immobilized β-galactosidase from *Aspergillus oryzae*. Journal of Agricultural and Food Chemistry, 57: 11302–11307.

Itan, Y., Jones, B.L., Ingram, C.J., Swallow, D.M. and Thomas, M.G. 2010. A worldwide correlation of lactase persistence phenotype and genotypes. BMC Evolutionary Biology, 10.

Itan, Y., Powell, A., Beaumont, M.A., Burger, J. and Thomas, M.G. 2009. The origins of lactase persistence in Europe. PLoS Computational Biology, 5.

Iwasaki, K.I., Nakajima, M. and Nakao, S.I. 1996. Galacto-oligosaccharide production from lactose by an enzymic batch reaction using β-galactosidase. Process Biochemistry 31: 69–76.

Jorgensen, F., Hansen, O.C. and Stougaard, P. 2001. High-efficiency synthesis of oligosaccharides with a truncated beta-galactosidase from *Bifidobacterium bifidum*. Applied Microbiology and Biotechnology, 57: 647–652.

Juers, D.H., Heightman, T.D., Vasella, A., McCarter, J.D., Mackenzie, L., Withers, S.G. and Matthews, B.W. 2001. A structural view of the action of *Escherichia coli* (lacZ) β-galactosidase. Biochemistry, 40: 14781–14794.

Kim, C.S., Ji, E.S. and Oh, D.K. 2004. Characterization of a thermostable recombinant beta-galactosidase from *Thermotoga maritima*. Journal of Applied Microbiology, 97: 1006–1014.

Klewicki, R. 2007. The stability of gal-polyols and oligosaccharides during pasteurization at a low pH. LWT-Food Science and Technology, 40: 1259–1265.

Konar, E., Sarkar, S. and Singhal, R.S. 2011. Galactooligosaccharides: Chemistry, production, properties, market status and applications—A review. Trends in Carbohydrate Research, 3: 1–16.

Kondo, A., Liu, Y., Furuta, M., Fujita, Y., Matsumoto, T. and Fukuda, H. 2000. Preparation of high activity whole cell biocatalyst by permeabilization of recombinant flocculent yeast with alcohol. Enzyme and Microbial Technology, 27: 806–811.

Lamsal, B.P. 2012. Production, health aspects and potential food uses of dairy prebiotic galactooligosaccharides. Journal of the Science of Food and Agriculture, 92: 2020–2028.

Lee, Y.J., Kim, C.S. and Oh, D.K. 2004. Lactulose production by beta-galactosidase in permeabilized cells of *Kluyveromyces lactis*. Applied Microbiology and Biotechnology, 64: 787–793.

Li, Z., Xiao, M., Lu, L. and Li, Y. 2008. Production of non-monosaccharide and high-purity galactooligosaccharides by immobilized enzyme catalysis and fermentation with immobilized yeast cells. Process Biochemistry, 43: 896–899.

Locascio, R.G., Ninonuevo, M.R., Freeman, S.L., Sela, D.A., Grimm, R., Lebrilla, C.B., Mills, D.A. and German, J.B. 2007. Glycoprofiling of bifidobacterial consumption of human milk oligosaccharides demonstrates strain specific, preferential consumption of small chain glycans secreted in early human lactation. Journal of Agricultural and Food Chemistry, 55: 8914–8919.

Mahoney, R.R., Nickerson, T.A. and Whitaker, J.R. 1975. Selection of strain, growth conditions and extraction procedures for optimum production of lactase from *Kluyveromyces fragilis*. Journal of Dairy Science, 58: 1620–1629.

Maischberger, T., Leitner, E., Nitisinprasert, S., Juajun, O., Yamabhai, M., Nguyen, T.H. and Haltrich, D. 2010. ß-Galactosidase from *Lactobacillus pentosus*: Purification, characterization and formation of galacto-oligosaccharides. Biotechnology Journal, 5: 838–847.

Maksimainen, M., Hakulinen, N., Kallio, J.M., Timoharju, T., Turunen, O. and Rouvinen, J. 2011. Crystal structures of *Trichoderma reesei* β-galactosidase reveal conformational changes in the active site. Journal of Structural Biology, 174: 156–163.

Maksimainen, M.M., Lampio, A., Mertanen, M., Turunen, O. and Rouvinen, J. 2013. The crystal structure of acidic β-galactosidase from *Aspergillus oryzae*. International Journal of Biological Macromolecules, 60: 109–115.

Manera, A.P., De Almeida Costa, F.A., Rodrigues, M.I., Kalil, S.J. and Maugeri Filho, F. 2010. Galacto-oligosaccharides production using permeabilized cells of *Kluyveromyces marxianus*. International Journal of Food Engineering, 6.

Marin-Navarro, J., Talens-Perales, D., Oude-Vrielink, A., Cañada, F.J. and Polaina, J. 2014. Immobilization of thermostable β-galactosidase on epoxy support and its use for lactose hydrolysis and galactooligosaccharides biosynthesis. World Journal of Microbiology and Biotechnology, 30: 989–998.

Martinez-Villaluenga, C., Cardelle-Cobas, A., Corzo, N., Lano, A. and Villamiel, M. 2008. Optimization of conditions for galactooligosaccharide synthesis during lactose hydrolysis by beta-galactosidase from *Kluyveromyces lactis* (Lactozym 3000 L HP G). Food Chemistry, 107: 258–264.

Matella, N.J., Dolan, K.D. and Lee, Y.S. 2006. Comparison of galactooligosaccharide production in free-enzyme ultrafiltration and in immobilized-enzyme systems. Journal of Food Science, 71: C363–C368.

Maugard, T., Gaunt, D., Legoy, M.D. and Besson, T. 2003. Microwave-assisted synthesis of galacto-oligosaccharides from lactose with immobilized beta-galactosidase from *Kluyveromyces lactis*. Biotechnology Letters, 25: 623–629.

Mayer, J., Conrad, J., Klaiber, I., Lutz-Wahl, S., Beifuss, U. and Fischer, L. 2004. Enzymatic production and complete nuclear magnetic resonance assignment of the sugar lactulose. Journal of Agricultural and Food Chemistry, 52: 6983–6990.

Moro, G.E. and Arslanoglu, S. 2005. Reproducing the bifidogenic effect of human milk in formula-fed infants: Why and how? Acta Paediatrica, 94: 14–17.

Mozaffar, Z., Nakanishi, K. and Matsuno, R. 1984. Purification and properties of β-galactosidases from *Bacillus circulans*. Agricultural and Biological Chemistry, 48: 3053–3061.

Mozaffar, Z., Nakanishi, K. and Matsuno, R. 1986. Continuous production of galacto-oligosaccharides from lactose using immobilized beta-galactosidase from *Bacillus circulans*. Applied Microbiology and Biotechnology, 25: 224–228.

Nakao, M., Harada, M., Kodama, Y., Nakayama, T., Shibano, Y. and Amachi, T. 1994. Purification and characterization of a thermostable β-galactosidase with high transgalactosylation activity from *Saccharopolyspora rectivirgula*. Applied Microbiology and Biotechnology, 40: 657–663.

Nakayama, T. and Amachi, T. 1999. β-Galactosidase. pp. 1291–1305. *In*: M.C. Flickinger and S.W. Drew (eds.). Encyclopedia of Bioprocess Technology, Fermentation, Biocatalysis and Bioseparation. Wiley, New York.

Neri, D.F.M., Balcão, V.M., Cardoso, S.M., Silva, A.M.S., Domingues, M.D.R.M., Torres, D.P.M., Rodrigues, L.R.M., Carvalho, L.B. and Teixeira, J.A.C. 2011. Characterization of galactooligosaccharides produced by β-galactosidase immobilized on to magnetized Dacron. International Dairy Journal, 21: 172–178.

Nguyen, T.H., Splechtna, B., Krasteva, S., Kneifel, W., Kulbe, K.D., Divne, C. and Haltrich, D. 2007. Characterization and molecular cloning of a heterodimeric beta-galactosidase from the probiotic strain *Lactobacillus acidophilus* R22. FEMS Microbiology Letters, 269: 136–144.

O'Connell, S. and Walsh, G. 2010. A novel acid-stable, acid-active β-galactosidase potentially suited to the alleviation of lactose intolerance. Applied Microbiology and Biotechnology, 86: 517–524.

Ohto, U., Usui, K., Ochi, T., Yuki, K., Satow, Y. and Shimizu, T. 2012. Crystal structure of human β-galactosidase: Structural basis of G M1 gangliosidosis and morquio B diseases. Journal of Biological Chemistry, 287: 1801–1812.

Olds, L.C. and Sibley, E. 2003. Lactase persistence DNA variant enhances lactase promoter activity *in vitro*: Functional role as a cis regulatory element. Human Molecular Genetics, 12: 2333–2340.

Oliveira, C., Guimaraes, V.M. and Domingues, L. 2011. Recombinant microbial systems for improved β-galactosidase production and biotechnological applications. Biotechnology Advances, 29: 600–609.

Onishi, N. and Tanaka, T. 1995. Purification and properties of a novel thermostable galacto-oligosaccharide-producing beta-galactosidase from *Sterigmatomyces elviae* CBS8119. Applied and Environmental Microbiology, 61: 4026–4030.

Pal, A., Pal, V., Ramana, K.V. and Bawa, A.S. 2009. Biochemical studies of ß-galactosidase from *Kluyveromyces lactis*. Journal of Food Science and Technology, 46: 217–220.

Panesar, P.S., Kumari, S. and Panesar, R. 2010. Potential applications of immobilized β-galactosidase in food-processing industries. Enzyme Research 2010, Article number 473137.

Panesar, R., Panesar, P.S., Singh, R.S. and Kennedy, J.F. 2011. Hydrolysis of milk lactose in a packed bed reactor system using immobilized yeast cells. Journal of Chemical Technology and Biotechnology, 86: 42–46.

Panesar, R., Panesar, P.S., Singh, R.S., Kennedy, J.F. and Bera, M.B. 2007. Production of lactose-hydrolyzed milk using ethanol permeabilized yeast cells. Food Chemistry, 101: 786–790.

Park, A.R. and Oh, D.K. 2010. Galacto-oligosaccharide production using microbial ß-galactosidase: Current state and perspectives. Applied Microbiology and Biotechnology, 85: 1279–1286.

Pereira-Rodriguez, A., Fernandez-Leiro, R., Gonzalez-Siso, M.I., Cerdan, M.E., Becerra, M. and Sanz-Aparicio, J. 2012. Structural basis of specificity in tetrameric *Kluyveromyces lactis* β-galactosidase. Journal of Structural Biology, 177: 392–401.

Petzelbauer, I., Zeleny, R., Reiter, A., Kulbe, K.D. and Nidetzky, B. 2000. Development of an ultra-high-temperature process for the enzymatic hydrolysis of lactose: II. Oligosaccharide formation by two thermostable beta-glycosidases. Biotechnology and Bioengineering, 69: 140–149.

Pinho, J.M.R. and Passos, F.M.L. 2011. Solvent extraction of β-galactosidase from *Kluyveromyces lactis* yields a stable and highly active enzyme preparation. Journal of Food Biochemistry, 35: 323–336.

Placier, G., Watzlawick, H., Rabiller, C. and Mattes, R. 2009. Evolved β-galactosidases from *Geobacillus stearothermophilus* with improved transgalactosylation yield for galacto-oligosaccharide production. Applied and Environmental Microbiology, 75: 6312–6321.

Plou, F.J., Gomez de Segura, A. and Ballesteros, A. 2007. Application of glycosidases and transglycosidases for the synthesis of oligosaccharides. pp. 141–157. *In*: J. Polaina and A.P. MacCabe (eds.). Industrial Enzymes: Structure, Function and Application. Springer, New York.

Rabiu, B.A., Jay, A.J., Gibson, G.R. and Rastall, R.A. 2001. Synthesis and fermentation properties of novel galacto-oligosaccharides by beta-galactosidases from *Bifidobacterium* species. Applied and Environmental Microbiology, 67: 2526–2530.

Roberfroid, M. 2007. Prebiotics: The concept revisited. Journal of Nutrition, 137: 830S–837S.

Rodriguez-Alegria, M.E., Enciso-Rodriguez, A., Ortiz-Soto, M.E., Cassani, J., Olvera, C. and Munguia, A.L. 2010. Fructooligosaccharide production by a truncated *Leuconostoc citreum* inulosucrase mutant. Biocatalysis and Biotransformation, 28: 51–59.

Rodriguez-Colinas, B., De Abreu, M.A., Fernandez-Arrojo, L., De Beer, R., Poveda, A., Jimenez-Barbero, J., Haltrich, D., Ballesteros, A.O., Fernandez-Lobato, M. and Plou, F.J. 2011. Production of galacto-oligosaccharides by the b-galactosidase from *Kluyveromyces lactis*: Comparative analysis of permeabilized cells versus soluble enzyme. Journal of Agricultural and Food Chemistry, 59: 10477–10484.

Rodriguez-Colinas, B., Kolida, S., Baran, M., Ballesteros, A.O., Rastall, R.A. and Plou, F.J. 2013. Analysis of fermentation selectivity of purified galacto-oligosaccharides by *in vitro* human faecal fermentation. Applied Microbiology and Biotechnology, 97: 5743–5752.

Rodriguez-Colinas, B., Poveda, A., Jimenez-Barbero, J., Ballesteros, A.O. and Plou, F.J. 2012. Galacto-oligosaccharide synthesis from lactose solution or skim milk using the β-galactosidase from *Bacillus circulans*. Journal of Agricultural and Food Chemistry, 60: 6391–6398.

Rojas, A.L., Nagem, R.A.P., Neustroev, K.N., Arand, M., Adamska, M., Eneyskaya, E.V., Kulminskaya, A.A., Garratt, R.C., Golubev, A.M. and Polikarpov, I. 2004. Crystal structures of β-galactosidase from *Penicillium* sp. and its complex with galactose. Journal of Molecular Biology, 343: 1281–1292.

Rubio-Texeira, M., Arevalo-Rodriguez, M., Luis Lequerica, J. and Polaina, J. 2000. Lactose utilization by *Saccharomyces cerevisiae* strains expressing *Kluyveromyces lactis* LAC genes. Journal of Biotechnology, 84: 97–106.

Rubio-Texeira, M., Castrillo, J.I., Adam, A.C., Ugalde, U.O. and Polaina, J. 1998. Highly efficient assimilation of lactose by a metabolically engineered strain of *Saccharomyces cerevisiae*. Yeast, 14: 827–837.

Sela, D.A. and Mills, D.A. 2010. Nursing our microbiota: Molecular linkages between bifidobacteria and milk oligosaccharides. Trends in Microbiology, 18: 298–307.

Shadid, R., Haarman, M., Knol, J., Theis, W., Beermann, C., Rjosk-Dendorfer, D., Schendel, D.J., Koletzko, B.V. and Krauss-Etschmann, S. 2007. Effects of galactooligosaccharide and long-chain fructooligosaccharide supplementation during pregnancy on maternal and neonatal microbiota and immunity—a randomized, double-blind, placebo-controlled study. American Journal of Clinical Nutrition, 86: 1426–1437.

Sheu, D.C., Li, S.Y., Duan, K.J. and Chen, C.W. 1998. Production of galactooligosaccharides by beta-galactosidase immobilized on glutaraldehyde-treated chitosan beads. Biotechnology Techniques, 12: 273–276.

Silva, A.C., Guimarães, P.M.R., Teixeira, J.A. and Domingues, L. 2010. Fermentation of deproteinized cheese whey powder solutions to ethanol by engineered *Saccharomyces cerevisiae*: Effect of supplementation with corn steep liquor and repeated-batch operation with biomass recycling by flocculation. Journal of Industrial Microbiology and Biotechnology, 37: 973–982.

Siso, M.I.G., Cerdán, E., Picos, M.A.F., Ramil, E., Belmonte, E.R. and Torres, A.R. 1992. Permeabilization of *Kluyveromyces lactis* cells for milk whey saccharification: A comparison of different treatments. Biotechnology Techniques, 6: 289–292.

Song, J., Abe, K., Imanaka, H., Imamura, K., Minoda, M., Yamaguchi, S. and Nakanishi, K. 2011. Causes of the production of multiple forms of β-galactosidase by *Bacillus circulans*. Bioscience, Biotechnology and Biochemistry, 75: 268–278.

Splechtna, B., Nguyen, T.H., Steinbock, M., Kulbe, K.D., Lorenz, W. and Haltrich, D. 2006. Production of prebiotic galacto-oligosaccharides from lactose using beta-galactosidases from *Lactobacillus reuteri*. Journal of Agricultural and Food Chemistry, 54: 4999–5006.

Sreekrishna, K. and Dickson, R.C. 1985. Construction of strains of *Saccharomyces cerevisiae* that grow on lactose. Proceedings of the National Academy of Sciences of the United States of America, 82: 7909–7913.

Swallow, D.M. 2003. Genetics of lactase persistence and lactose intolerance. Annual Review of Genetics, 37: 197–219.

Tanaka, Y., Kagamiishi, A., Kiuchi, A. and Horiuchi, T. 1975. Purification and properties of β-galactosidase from *Aspergillus oryzae*. Journal of Biochemistry, 77: 241–247.

Tishkoff, S.A., Reed, F.A., Ranciaro, A., Voight, B.F., Babbitt, C.C., Silverman, J.S., Powell, K., Mortensen, H.M., Hirbo, J.B., Osman, M., Ibrahim, M., Omar, S.A., Lema, G., Nyambo, T.B., Ghori, J., Bumpstead, S., Pritchard, J.K., Wray, G.A. and Deloukas, P. 2007. Convergent adaptation of human lactase persistence in Africa and Europe. Nature Genetics, 39: 31–40.

Toba, T. and Adachi, S. 1978. Hydrolysis of lactose by microbial β-galactosidases. Formation of oligosaccharides with special reference to 2-O-β-D-galactopyranosyl-D-glucose. Journal of Dairy Science, 61: 33–38.

Toba, T., Yokota, A. and Adachi, S. 1985. Oligosaccharide structures formed during the hydrolysis of lactose by *Aspergillus oryzae* β-galactosidase. Food Chemistry, 16: 147–162.

Torres, D.P., Goncalves, M., Teixeira, J.A. and Rodrigues, L.R. 2010. Galacto-Oligosaccharides: Production, properties, applications, and significance as prebiotics. Comprehensive Reviews in Food Science and Food Safety, 9: 438–454.

Torres, P. and Batista-Viera, F. 2012. Immobilization of β-galactosidase from *Bacillus circulans* onto epoxy-activated acrylic supports. Journal of Molecular Catalysis B: Enzymatic, 74: 230–235.

Tuohy, K.M., Rouzaud, G.C.M., Bruck, W.M. and Gibson, G.R. 2005. Modulation of the human gut microflora towards improved health using prebiotics—Assessment of efficacy. Current Pharmaceutical Design, 11: 75–90.

Tzortzis, G. and Vulevic, J. 2009. Galacto-oligosaccharide prebiotics. *In*: D. Charalampopoulos and R.A. Rastall (eds.). Prebiotics and Probiotics Science and Technology.: Springer, New York.

Urrutia, P., Rodriguez-Colinas, B., Fernandez-Arrojo, L., Ballesteros, A.O., Wilson, L., Illanes, A. and Plou, F.J. 2013. Detailed analysis of galactooligosaccharides synthesis with β-galactosidase from *Aspergillus oryzae*. Journal of Agricultural and Food Chemistry, 61: 1081–1087.

Usui, T., Kubota, S. and Ohi, H. 1993. A convenient synthesis of β-D-galactosyl disaccharide derivatives using the β-D-galactosidase from *Bacillus circulans*. Carbohydrate Research, 244: 315–323.

Vera, C., Guerrero, C., Conejeros, R. and Illanes, A. 2012. Synthesis of galacto-oligosaccharides by β-galactosidase from *Aspergillus oryzae* using partially dissolved and supersaturated solution of lactose. Enzyme and Microbial Technology, 50: 188–194.

Wei, L., Xiaoli, X., Shufen, T., Bing, H., Lin, T., Yi, S., Hong, Y. and Xiaoxiong, Z. 2009. Effective enzymatic synthesis of lactosucrose and its analogues by β-galactosidase from *Bacillus circulans*. Journal of Agricultural and Food Chemistry, 57: 3927–3933.

Yanahira, S., Kobayashi, T., Suguri, T., Nakakoshi, M., Miura, S., Ishikawa, H. and Nakajima, I. 1995. Formation of oligosaccharides from lactose by *Bacillus circulans* β-galactosidase. Bioscience Biotechnology and Biochemistry, 59: 1021–1026.

Yang, S.T., Marchio, J.L. and Yen, J.W. 1994. A dynamic light scattering study of β-galactosidase: Environmental effects on protein conformation and enzyme activity. Biotechnology Progress, 10: 525–531.

Zeuner, B., Jers, C., Mikkelsen, J.D. and Meyer, A.S. 2014. Methods for improving enzymatic trans-glycosylation for synthesis of human milk oligosaccharide biomimetics. Journal of Agricultural and Food Chemistry, 62: 9615–9631.

Zhou, Q.Z., Chen, X.D. and Li, X. 2003. Kinetics of lactose hydrolysis by ß-galactosidase of *Kluyveromyces lactis* immobilized on cotton fabric. Biotechnology and Bioengineering, 81: 127–133.

Zhou, Q.Z.K. and Chen, X.D. 2001. Effects of temperature and pH on the catalytic activity of the immobilized ß-galactosidase from *Kluyveromyces lactis*. Biochemical Engineering Journal, 9: 33–40.

8

Pectinases and their Biotechnological Applications

Anuja Vohra[a] and *Reena Gupta* *

1. Introduction

Enzymes are highly efficient catalysts from biological sources that catalyze all synthetic and degradative reactions of living organisms. They are used industrially because of their high catalytic power, specific mode of action, stereo-specificity, eco-friendly and reduced energy requirements. There are many advantages of using microorganisms as enzyme sources, such as easy culturing and propagation, less time and space requirement, etc. Above all, majority of the microbial enzymes are extracellular, heat stable and active over a wide pH range (Kumar et al., 2015). There is an ever-increasing demand to replace traditional chemical processes with advanced biotechnological processes involving microorganisms and enzymes, such as pectinases, cellulases and ligninases.

The commercial application of pectinases was first observed in 1930 in wine and fruit juice preparation. It was only in the 1960s that when the chemical nature of plant tissues became apparent that scientists began to use a greater range of enzymes more efficiently. As a result, pectinases are today one of the upcoming enzymes in the commercial sector (Zeni et al., 2010).

2. Pectin

The plant cell wall is composed of four major complex polysaccharides (celluose, hemicellulose, lignin and pectin), which constitute the most abundant polymers

Department of Biotechnology, Himachal Pradesh University, Summerhill, Shimla-171005, India.
[a] E-mail: anujavohra77@gmail.com
* Corresponding author: reenagupta_2001@yahoo.com

in Nature to be broken down into their assimilable form *via* cell-wall degrading enzymes, such as cellulases, ligninases and pectinases. Pectin-degrading enzymes are known as pectinases (pectinolytic or pectolytic enzymes). Pectic substances are widely distributed in fruits and vegetables (in turnips, peels of orange and in pulp of tomato, pineapple and lemon). Contrary to proteins, lipids and nucleic acids being polysaccharides, pectic substances do not have a defined molecular weight. The relative molecular masses of pectic substances range from 25 to 360 kDa (Table 1).

Table 1. Molecular weight of some pectic substances (Sakai et al., 1993).

Source	Molecular Weight (kDa)
Apple and lemon	200–360
Pear and prune	25–35
Orange	40–50
Sugar beet pulp	40–50

Pectic substances or essential components of plant cell wall with high molecular weight are negatively charged and acidic complex biomolecules in Nature to give shape to the soft non-woody parts of the plant. Pectins are composed of polymer of D-galacturonic acid formed by α-1, 4-glycosidic bonds which are cross-linked by carboxyl groups with divalent cations, such as Ca^{2+} and Mg^{2+} (Jayani et al., 2005). They are commonly produced during the initial stages of primary cell growth and make about one-third of the cell-wall dry substances. The highest concentration of pectins in the cell wall is seen in the middle lamella, with a gradual decrease from the primary cell wall towards the plasma membrane in dicots and nongraminaceous monocots.

Pectins are present in the middle lamella (the cementing material) between the cells of plant cell-wall in the form of calcium pectate and magnesium pectate. Pectins are among the most prevalent colloids in both must and wine. These substances have an elevated solvation capacity and thus increase the solubility of other substances, including salts (sodium bitartrate), proteins, etc. Thus, pectins impede both stabilization and clarification.

2.1 Structure and classification

Pectic substances mainly consist of galacturonans and rhamnogalacturonans in which the C-6 carbon of galactate is oxidized to a carboxyl group (Whitaker, 1990). These substances are a group of complex colloidal polymeric materials composed largely of a backbone of anhydrogalacturonic acid units as illustrated in Fig. 1 (Cho et al., 2001; Codner, 2001).

The carboxyl groups of galacturonic acid are partially esterified by methyl groups and partially or completely neutralized by sodium, potassium or ammonium ions (Kashyap et al., 2001). Some of the hydroxyl groups on C_2 and C_3 may be acetylated (Alkorta et al., 1998). The side chains of arabinan, galactan, arabinogalactan, xylose or fucose are connected to the main chain through their C_1 and C_2 atoms (Blanco et al., 1999; Van der Vlugt-Bergmans et al., 2000; Gummadi and Panda, 2003). Mostly

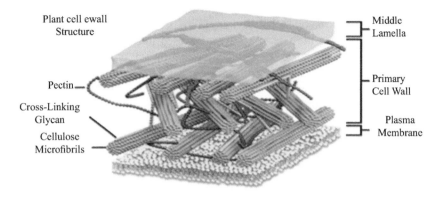

Figure 1. Primary structure of pectic substances.

the extended side chains tend to be homogenous polymers of either D-galacturonic acid units or of L-arabinose units.

The above description indicates that the pectic substances are present in various forms in plant cells and this is the probable reason for the existence of various forms of pectinolytic enzymes.

The American Chemical Society classified pectic substances into four main types (Alkorta et al., 1998):

1. *Protopectin*: It is water insoluble pectic substance present in the parenchyma of fruits and vegetables. Protopectin on restricted hydrolysis yields pectin or pectic acids.
2. *Pectic acid*: It is a water-soluble polymer of galacturonans that contains negligible amounts of methoxy groups. Normal or acid salts of pectic acids are called pectates.
3. *Pectinic acids*: It is the polygalacturonan chain that contains 0 and 75 per cent of methylated galacturonate units. Normal or acid salts of pectinic acid are referred to as pectinates. Pectinic acids alone have the unique property of forming a gel with sugar or acid or if suitably low in methyl content with certain other compounds, such as Ca^{2+} salts (Kashyap et al., 2001).
4. *Pectin* (polymethyl galacturonate): It is the polymeric material in which, at least 75 per cent of the carboxyl groups of the galacturonate units are esterified with methanol. It confers rigidity on the cell wall when bound to cellulose in the cell wall.

3. Pectinases

The pectinases or pectinolytic enzymes are a group of complex enzymes widely distributed in higher plants and microorganisms where pectin serves as the substrate. These are groups of enzymes that attack pectin and depolymerise it by hydrolysis and transelimination as well as by de-esterification reactions, which hydrolyse the ester bond between carboxyl and methyl groups of pectin (Ceci and Loranzo, 1998). Pectinases were among the first enzymes to be used in homes. They are of great significance, with a wide range of biotechnological applications (Muthezhilan et al., 2008; Raghuwanshi et al., 2013). Pectic enzymes have been used for the clarification of wines since the beginning of the 19th century. They are industrially useful enzymes for extraction, clarification and liquefaction of fruit juices and wines (Chauhan et al., 2001). The enzyme preparations used in the food industry are of mostly fungal origin because fungi are potent producers of pectic enzymes and the optimum pH of fungal enzymes is very close to the pH of many fruit juices, which range from 3.0 to 5.5 (Angelova, 2008). They are also used in the fabric industry to soak plant fibers and in the papermaking industry to solve the retention problems by de-clogging the pulps (Sakai et al., 1993). Pectinases are also found important in oil extraction, flavors and pigments from plant materials (Phugare et al., 2011).

3.1 Classification of pectinases

Pectin degradation or pectinolysis is a very important natural process for plants. Pectinolysis is a necessary process for cell elongation and growth and fruit ripening. Plant pathogenicity and spoilage of fruits and vegetables are signs of action of pectinolytic enzymes (Jayani et al., 2005). There are many systems of classification of pectinases but a recent classification based on the mode of action and substrate used is as follows (Saranraj and Naidu, 2014):

3.1.1 Protopectinases

Theses enzymes degrade insoluble protopectin and solubilize it in the form of highly polymerized pectin. Pectinosinase is also synonymous with protopectinase (PPase). They are further classified into two types on the basis of their reaction mechanism:

- *A Type PPases*: It reacts with the inner site, i.e., the polygalacturonic acid region of protopectin. A-type PPase is found in the culture filtrates of yeast and yeast-like fungi
- *B Type PPases*: It reacts on the outer site, i.e., on the polysaccharide chains that may connect the polygalacturonic acid chain and cell wall constituents. B-type PPases have been reported in *Bacillus subtilis* IFO 12113, *Bacillus subtilis* IFO 3134 and *Trametes* sp.

3.1.2 Esterases/pectinesterases

Pectinesterases catalyse the hydrolysis of methylated carboxylic ester groups in pectin into pectic acid and methanol.

3.1.3 Depolymerases

On the basis of the mechanism of action, depolymerases are further classified into two groups:

- *Hydrolases:* These enzymes degrade the pectic substances *via* hydrolysis, which may be random or specific
- *Lyases:* These enzymes degrade pectic substances *via* the elimination reaction

Another type of classification includes three criteria such as substrate used-whether pectin, pectic acid or oligo-D-galacturonic acid, type of cleavage—trans-elimination or hydrolysis and mode of action—whether cleavage is random (endo-liquifying or depolymerising enzymes) or end wise (exo or saccharifying enzymes). Based on these criteria, enzymes can be briefly classified (Table 2).

Table 2. Classification of pectinases (Alkorta et al., 1998; Kapoor et al., 2001; Hoondal et al., 2002).

Types of Pectinases	E.C. No.	Substrate	Mode of Action	Product
Esterase				
PME	3.1.1.11	Pectin	Hydrolysis	Pectic acid + Methanol
PAE	3.1.1.6	Pectin	Hydrolysis	Pectic acid + Ethanol
Depolymerase				
a. Hydrolases				
i. Endo PG	3.2.1.15	Pectic acid	Hydrolysis	Oligogalacturonates
ii. Exo PG	3.2.1.67	Pectic acid	Hydrolysis	Monogalacturonates
b. Lyases				
i. Endo PL	4.2.2.2	Pectic acid	Transelimination	Unsaturated oligogalacturonate
ii. Endo PNL	4.2.2.10	Pectic acid	Transelimination	Unsaturated methyloligogalacturonates

PME, pectin methylesterase; PAE, pectinacetylesterase; PG, polygalacturonases; PL, pectate lyase; PNL, pectin lyase

3.1.4 Pectin methyl esterase

The pectin methyl esterase (EC 3.1.1.11) is a hydrolytic enzyme removing the methoxy group of pectin-yielding polygalacturonic acid (PGA). Both extracellular and membrane-bound pectin methyl esterases have been reported in bacteria. The enzyme activity is important in fruit ripening of plants and in facilitating polygalacturonase activity on pectin.

3.1.5 The lyases

Two kinds of lyases can be identified from their substrate specificity:

- The Pectate Lyase (PL)
- The Pectin Lyase (PNL)

a) The pectate lyase

Pectate lyases cleave internal glycosidic bonds in PGA *via* β-elimination. Exo-pectate lyases (EC 4.2.2.9) and endo-pectate lyases have been reported both in bacteria and fungi. Exoenzymes generate an unsaturated digalacturonates at the non-reducing end. Membrane-bound exoenzyme present in the periplasm of bacteria utilize pectin as a carbon source. These membrane-bound enzymes are reportedly more active on oligogalacturonides than on polygalacturonic acid itself (Barras et al., 1994; Hugouvieux-Cotte-Pattat et al., 1996). Endo pectate lyases (EC 4.2.2.2) are only extracellular in nature and act only on polygalacturonic acid. They cleave the polymer producing 4, 5-unsaturated oligomers at the non-reducing end (Hugouvieux-Cotte-Pattat et al., 1996).

b) The pectin lyase

Pectin lyase (EC 4.2.2.10) cleaves natural pectin and highly (98 per cent) methyl esterified PGA through β-elimination, but it otherwise has no activity towards polygalacturonic acid (Hugouvieux-Cotte-Pattat et al., 1996). These enzymes are absent in fruit and they are present only in microorganisms.

3.1.6 Polygalacturonases

Polygalacturonases (PG) are pectinolytic enzymes that degrade polygalacturonan present in the cell walls of plants by hydrolysis of the glycosidic bonds that link galacturonic acid residues. These enzymes display a right-handed parallel β-helix topology. They are the most extensively studied among the family of pectinolytic enzymes.

Polygalacturonases are classified as Endo-polygalacturonase (EC 3.2.1.15) and Exo-polygalacturonase (EC 3.2.1.67). Endo-polygalacturonase (Endo-PG) is one of the well-studied pectic enzymes and brings about random hydrolysis of the polymer to yield oligomers of galacturonides whereas exo-polygalacturonase (Exo-PG) hydrolyses α-1, 4-glycosidic bonds and attacks the non-reducing end of PGA to release digalacturonate.

Accepted Name: Polygalacturonase; Other Names: Pectin depolymerase; pectinase; endopolygalacturonase; pectolase; pectin hydrolase; pectin polygalacturonase; endo-polygalacturonase; poly-α-1,4-galacturonide glycanohydrolase; endogalacturonase; endo-D–galacturonase; SYSTEMATIC NAME: Poly (1,4-α-D-galacturonide) glycanohydrolase.

4. Pectinases and Plant Pathogenicity

Microorganisms responsible for various plant diseases produce pectinases. The pathogenicity results from the secretion of pectinolytic enzymes responsible for disorganization of the plant cell-wall. The degraded polymer provides the essential carbon source to the microorganism. It also facilitates the entry of the microorganisms into the plant through its macerating action on the plant cell-wall (Barras et al., 1994; Chang et al., 1995; Ugwuanyi and Obeta, 1997).

Pectinases are one of the upcoming enzymes of the commercial sector.

5. Microorganisms Producing Commercially Viable Pectinases

Microbial pectinases have proved to be the single large economical source of pectinases. Microbes are the only industrially viable source for pectinolytic enzymes except for pectin methyl esterase. Pectinases are produced by prokaryotic microorganisms, i.e., *Bacillus* sp.*, Agrobacterium tumefaciens*, *Bacteroides thetaiotamicron*, *Ralstonia solanacearum* and also by eukaryotic microorganisms such as fungi and yeast (Hoondal et al., 2000; Jayani et al., 2005; Jayani et al., 2010; Priya and Sashi, 2014). Fungus *Aspergillus*, bacteria such as *Erwinia* and *Xanthomonas* (Ladjama et al., 2007; Chatterjee et al., 2011) and actinomycetes (Kasuo et al., 2004) have been extensively studied for their pectic enzymes (Table 3).

Table 3. Microorganism producing commercially viable pectinases.

Microorganism	Enzyme
Aspergillus niger, Penicillum viricatum	Polygalacturonase
Thermoascus aurantiacus	Polygalacturonase, pectin lyase
Bacillus sp.	Pectinase
Aspergillus niger, Penicillum, Tricoderma	Polygalacturonase, pectinlyase, Pectin esterase
Aspergillus awamori	Endo and exo polygalacturonase
Aspergillus foetidus	polygalacturonase, Polymethylgalacturonases
Aspergillus fumigates	pectinase, polygalacturonase

6. Mechanism of Action of Pectinases

Pectin-degrading enzymes are classified, based on their mode of action on pectin and pectic substances into PG, PME, PL, PNL (Sakai et al., 1993; Wong, 1995; Chauhan et al., 2001). The action mechanism of pectinolytic enzymes (Fig. 2) is as follows:

The reactions catalyzed by these enzymes are (Whitaker, 1990).

a. Pectin methylesterase

Pectin + H_2O ⟶ Polygalacturonate + methanol

b. Endopolygalacturonases: [Poly (1, 4-α-D-galacturonide) glucan hydrolase]

Polygalacturonate + nH_2O ⟶ Oligogalacturonates + galacturonate

c. Exopolygalacturonases: [Poly (1, 4-α-D-galacturonide) galacto hydrolase]

(Polygalacturonate)$_n$ + H_2O ⟶ (Polygalacturonate)n-1 + galacturonate

d. Exopolygalacturonate lyase: [Poly (1, 4-α-D-galacturonide) lyase]

Polygalacturonate ⟶ 4:5 unsaturated galacturonates

e. Endopolygalacturonate lyase: [Poly (1, 4-α-D-galacturonide) lyase]

Polygalacturonate ⟶ 4:5 unsaturated oligogalacturonates

f. Endopolymethylgalacturonase

(Pectin)n + H₂O ⟶ Oligomethylgalacturonates

g. Endopolymethylgalacturonate lyase: [Poly (methoxygalacturonide) lyase

Pectin ⟶ U saturated methyl oligogalacturonates

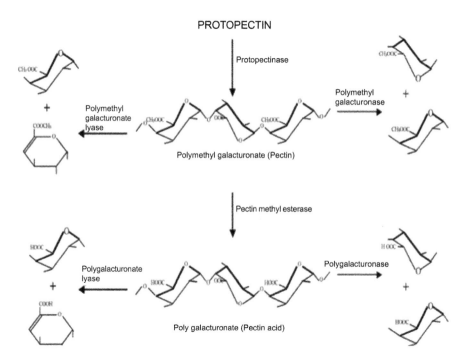

Figure 2. Action mechanism of pectinolytic enzymes.

7. Bioprocessing for Production of Pectinases

Pectinases can be produced either by submerged or solid-state fermentation. Solid-state fermentation are those processes in which microbial growth and products formation occur on the surfaces of solid substrates in the near absence of free water whereas submerged fermentation is dominant method of producing microbial products such as antibiotics, vitamins, amino acids and organic acids.

7.1 Advantages and disadvantages of solid state fermentation over submerged fermentation

Advantages

- Lower medium cost
- Easier downstream processing
- Higher productivity
- Better aeration
- Reduced energy and cost requirement
- Simple technology
- Minimum operational problems
- High recovery of product

Disadvantages

- Difficulties on scale up
- Contamination chances with unwanted fungal species
- Process parameter, such as pH, moisture content, substrate, oxygen and biomass concentration becomes a problem because of solid nature of the substrate

Culture conditions for maximum pectinase production in solid state fermentation are:

- Screening of agro-industrial wastes
- Variation in moisture content
- Inoculum size and type
- pH and temperature
- Period of cultivation
- Carbon and nitrogen supplements

Microorganisms and substrate used for the production of pectinases in solid state fermentation systems are:

Bacteria, yeasts and fungi can grow on solid substrates. Filamentous fungi are best adapted for solid state fermentation, such as *Aspergillus niger, Penicillum, Tricoderma*. Substates used for the cultivation of microorganisms to produce pectinases are sugar cane bagasse, wheat bran, rice bran, soy bran, banana waste, tea waste, apple pomace, starch, orange bagasse, coffee pulp, grape pomace, etc.

The future trend of solid state fermentation may improve as a result of continuous research to scale-up the process to industrial level for efficient realization of desired products in an eco-efficient and sustainable manner.

8. Novel Strategies for Improved Enzyme Production

In spite of a large number of applications, enzymes do not exhibit satisfactory characteristics in terms of selectivity, stability and activity. Today there is a need for new, improved and versatile enzymes in order to develop more sustainable, novel and economically competitive production processes. Microbial diversity and molecular techniques, such as metagenomics and genomics, are being used to discover microbial enzymes whose catalytic properties can be modified by different strategies based on rational, semi-rational (includes site-directed mutagenesis to target amino acid substitutions) and directed evolution (finding variants of existing enzymes). Most industrial enzymes are recombinant forms produced in bacteria and fungi. Rapid development of enzymes was due to the recent advances in large-scale genome sequencing, directed evolution, protein expression, metabolic engineering, high throughput screening and structural biology.

9. Pectinase Applications in Foods, Beverages and Other Industries

Pectinolytic enzymes are commercially important in a number of industrial processes and are almost exclusively derived from fungal species, especially *Mucor* and *Aspergillus*. Biochemical and thermal characterization of crude exo-polygalacturonase produced by *Aspergillus sojae* has also been reported (Canan et al., 2008). Recently, the enzyme polygalacturonase has been targeted to develop non-fungicidal control strategies in order to avoid postharvest losses due to a fungal pathogen (Jurick et al., 2009). Each application requires unique properties with respect to specificity, stability, temperature and pH dependence. Pectinases offer tremendous industrial applications and are used in various environment-friendly and economic industrial sectors with low production cost and pollution control (Tariq and Latif, 2012).

9.1 Fruit juice extraction

The largest industrial application of pectinases is in fruit juice extraction and clarification. Pectins contribute to fruit juice viscosity and turbidity. The apple juice faces the problems of turbidity and suspended particles like pectin, cellulose and hemicellulose. Due to these factors, the juice fails to match the international standards. The quality of juice can be improved by appropriate treatment with pectinolytic enzymes, leading to increased market potential; thereby, providing a significant boost to the economic state. Treatment of fruit pulps with pectinases also showed an increase in fruit juice volume from banana, grapes and apples (Kaur et al., 2004). In another study, combined use of pectinolytic enzyme from *Aspergillus niger* van Tieghem and gelatin led to 1.5–2.0 times more clarification as compared to that containing enzyme alone (Singh and Gupta, 2003). Pectinases in combination with other enzymes, *viz.* cellulases, arabinases and xylanases have been used to increase the pressing efficiency of fruits for juice extraction (Gailing et al., 2000). Vacuum infusion of pectinases has a commercial application to soften the peel of citrus fruits for removal. The effect of pectinases on juice extraction and clarification of plum, peach, pear and apricot juices was studied by Joshi et al. (2011). The juice recovery of enzymatically treated pulps increased significantly from 52 to 78 per cent in plum, 38 to 63 per cent in peach, 60 to 70 per cent in pear and 50 to 80 per cent in apricot. Other applications of pectinolytic enzymes in food processing are in the manufacture of better quality purees from peaches, prunes, apricots, strawberries, etc. Pectinases are also used in canning of orange segments and also in sugar extraction process from date fruits.

9.2 Improvement of chromaticity and stability of red wines

Pectinolytic enzymes added to macerated fruits before the addition of wine yeast in the process of producing red wine resulted in improved visual characteristics (color and turbidity) as compared to untreated wines. Enzymatically treated red wines presented chromatic characteristics, which are considered better than the control wines. These wines also showed greater stability as compared to the controlled ones (Revilla and Ganzalez-San Jose, 2003).

9.3 Coffee and tea fermentation

Pectinase treatment accelerates tea fermentation and also destroys the foam-forming property of instant tea powders by destroying pectins (Carr, 1985). They are also used in coffee fermentation to remove mucilaginous coat from coffee beans. Angayarkanni et al. (2002) studied the effect of both crude enzyme preparation and purified pectinase enzymes on the improvement of tea leaf fermentation in terms of theaflavin, thearubigin, high polymerized substances, total liquor color, dry matter content and total soluble solids of the tea produced. The crude enzyme preparations obtained from ethanol precipitation were found to be more effective in improving tea-leaf fermentation than the purified pectinase enzymes.

9.4 Retting of plant fibers

Pectinases have been used in the treatment of flux to separate the fibers and eliminate pectin. The fibers contain gum, which must be removed before use in textile making (Hoondal et al., 2000; Evans et al., 2002). Research revealed that the application of these enzymes results in shortening the process time by 33 per cent as compared to water retting (Kozlowski et al., 2006). Woody bast fibers are long, strong and usually stiff, derived from the phloem of plants and used for making cordage, matting and various fabrics.

Pectinolytic enzymes can also cause maceration of these fibers in addition to xylanolytic and cellulolytic enzymes. Enzymatic retting is faster, easily controlled and produces less odor as compared to traditional retting methods. Pulps prepared with pectinolytic enzymes produce bulkier paper with higher capacity and better printability than pulps prepared by an alkaline chemical process.

9.5 Degumming of plant bast fibers

Bast fibers are soft fibers formed in groups outside the xylem, phloem or pericycle, e.g., Ramie and sunn hemp. Biotechnological degumming using pectinases in combination with xylanases presents an ecofriendly and economic alternative to the above problem (Kapoor et al., 2001). Enzymatic mixtures are applied for quicker and simultaneous degumming of fiber from the woody core and pectin dispersion between the bundles of the fiber, which contributes to shortening of the degumming process, quality enhancement, mainly due to increase of divisibility and better fineness of fiber strands (Kozlowski et al., 2006).

9.6 Textile processing and bioscouring of cotton fibers

Bioscouring is a novel process for removal of non-cellulosic impurities from the fiber with specific enzymes. Pectinases are used for this purpose without any negative side effects on cellulose degradation (Guebitz et al., 2006). Bioscouring of fabrics with pectinases results in enhancement of various physical properties of fabrics, *viz.* whiteness, tensile strength and tearness over conventional alkaline-scoured fabrics (Ahlawat et al., 2009). Pectinases are most suitable for cotton bioscouring as they perform combined action of bioscouring and bleaching of cotton-based fabrics in the textile industry (Hebeish et al., 2013).

Pectinases have been used in combination with amylases, lipases, cellulases and hemicellulases to remove sizing agents from cotton in a safe and ecofriendly manner, thus replacing toxic caustic soda used for the purpose earlier (Hoondal et al., 2000). Ossola and Galante (2004) tested the scouring of flax roving with enzymes. They demonstrated the advantages of scouring with enzymes used under mild reaction conditions in comparison with traditional chemical scouring. In the laboratory, the following decreasing order of effectiveness was proved: Pectinase > xylanase = galactomannanase = protease > lipase > or = laccase. Enzyme preparation of roving was better than that prepared through a traditional chemical scouring with respect to higher yarn count, increased strength and decreased number of imperfections in the yarn after spinning.

9.7 Waste water treatment

Vegetable food-processing industries release pectin-containing wastewaters as byproduct. Pre-treatment of these wastewaters with pectinolytic enzymes facilitates removal of pectinaceous material and renders it suitable for decomposition by activated sludge treatment (Hoondal et al., 2002). They also aid in maintaining ecological balance by causing decomposition and recycling of waste plant materials (Jayani et al., 2005). Conventionally, the treatment of wastewater from citrus-processing industries containing pectic substances used to be carried out in multiple steps, including physical dewatering, spray irrigation, chemical coagulation, direct activated sludge treatment and chemical hydrolysis, which led to the formation of methane. This has several disadvantages, such as high cost of treatment and longer treatment time in addition to environmental pollution through use of chemicals. Thus an alternative, cost-effective and environment-friendly method is the use of pectin lyase from bacteria, which selectively removes pectic substances from waste water. The pre-treatment of pectic waste water from vegetable food-processing industries with alkaline pectinase and alkalophilic pectinolytic microbes facilitates removal of pectinaceous material and renders it suitable for decomposition by activated sludge treatment (Tanabe et al., 1987).

9.8 Paper and pulp industry

During papermaking, pectinase can depolymerise pectins and subsequently lower the cationic demand of pectin solutions and the filtrate from peroxide bleaching (Viikari et al., 2001; Reid and Richard, 2004). With advancement of biotechnology and increased reliance of paper and pulp industries on the use of microorganisms and their enzymes for bioleaching and papermaking, the use of enzymes other than xylanases and ligninases, such as mannanase, pectin lyase and α-galactosidase is increasing in the paper and pulp industries in many countries (Kirk et al., 1996; Bajpai, 1999). An overall bleach-boosting of eucalyptus Kraft pulp was obtained when pectinase from *Streptomyces* sp. QG-11–3 was used in combination with xylanase from the same organism for bioleaching (Beg et al., 2000). Pectin lyases depolymerise polygalacturonic acids and consequently decrease the cationic demand in the filtrate from peroxide bleaching of thermo-mechanical pulp (Viikari et al., 2001).

9.9 Oil extraction

Citrus oils, such as lemon oil, can be extracted with pectinases. They destroy the emulsifying properties of pectin, which interferes with the collection of oils from citrus peel extracts (Scott, 1978). Pectinases are found important in extraction of oils, flavors and pigments from plant materials (Phugare et al., 2011).

9.10 Animal feed

Intensive research into the use of various enzymes in animal and poultry feeds started in the early 1980s (Raghuwanshi et al., 2013). Pectinases are used in the enzyme cocktail for production of animal feeds. This reduces the feed viscosity, which increases absorption of nutrients, liberates nutrients either by hydrolysis of non-biodegradable fibers, or by liberating nutrients blocked by these fibers and reduces the amount of faeces (Hoondal et al., 2002).

9.11 Purification of plant viruses

In cases where the virus particle is restricted to phloem, alkaline pectinases and cellulases are used to liberate the virus from the tissues to give very pure preparations of the virus (Salazar and Jayasinghe, 1999).

10. Commercial Pectinases in Market

Biotechnology is gaining rapid growth as it has many advantages over conventional technologies. The global market for industrial enzymes was estimated at dollar 3.3 billion in 2010 and was expected to reach dollar 4.4 billion by 2015. The largest segment of industrial enzymes include food and beverage enzymes with revenues of nearly dollar 1.2 billion in 2011 and has been expected to grow by dollar 1.8 billion in 2016, at a composed annual growth rate of 10.4 per cent.

The market segmentation for various areas of application shows 11 per cent market for paper and pulp, 17 per cent by textile and 34 per cent for food and animal feed industries. At present pectinases are playing an important role in the commercial sector of biotechnology. In future more and novel applications will be displayed by these enzymes.

11. Perspective and Future Thrust

There are many drawbacks of chemical transformation processes used in industries, such as specially-designed equipments and high capital investment. Consumption of chemicals and harmful by-products has an adverse impact on the environment. Enzymatic reactions are the solution to these problems as these originate from biological system and are easily controlled.

The future thrust of enzymes will be development of more effective systems that use less energy, fewer chemicals, is environment-friendly, gives maximum product yield and requires minimum water. New enzymes through modern biotechnology, such as genomics, metagenomics, efficient expression and proteomics will produce better enzyme products. Immobilization is an active research area of pectinases. Efforts to achieve clean industrial products and processes will prove beneficial to the industry in future.

12. Conclusion

Pectolytic enzymes are widespread in plants, fungi and bacteria. They constitute a unique group of enzymes that are responsible for degradation of pectin and pectic substances in plant cell-walls. Over the years pectinases have been used in several conventional industrial processes, where the elimination of pectin is essential—fruit juice processing, coffee and tea processing, macerating of plants and vegetable tissue, degumming of plant fibers, wastewater treatment, extracting vegetable oil, bleaching of paper, alcoholic beverages and food industries. They also aid in maintaining the ecological balance by causing decomposition and recycling of waste plant materials. Purified pectinases have also been developed specifically for use in plant protoplast culture studies. When used with cellulase, purified pectinases have been found to be very useful in generating good yields of viable protoplast in several plant systems, e.g., corn, soybean, red beet, sunflower, tomato, citrus, etc. Purification and characterization of enzyme can lead to a better understanding of its contribution in these processes. Pectinases are today one of the upcoming enzymes of the commercial sector.

Acknowledgements

The financial support from Department of Biotechnology, Ministry of Science and Technology, Government of India to Department of Biotechnology, Himachal Pradesh University, Shimla is thankfully acknowledged. Fellowship granted to Miss Anuja Vohra from UGC in the form of Project Fellow is also thankfully acknowledged.

References

Ahlawat, S., Dhiman, S.S., Sharma, J., Mandhan, R.P. and Battan, B. 2009. Pectinase production by *Bacillus subtilis* and its potential application in biopreparation of cotton and micropoly fabric. Process Biochemistry, 10: 521–526.

Alkorta, I., Garbisu, C., Llama, M.J. and Serra, J.L. 1998. Industrial applications of pectic enzyme: A review. Process Biochemistry, 33: 21–28.

Angayarkanni, J., Palaniswamy, M., Murugesan, S. and Swaminathan, K. 2002. Improvement of tea leaves fermentation with *Aspergillus* spp. pectinase. Journal of Bioscience and Bioengineering, 94: 299–303.

Angelova, M.B. 2008. Microbial Pectinases: Application in Horticultural Industries, Microbial Biotechnology in Horticulture. pp. 101–179. *In:* R.C. Ray and O.P. Ward (eds.). Science Publishers, Enfield, NH.

Bajpai, P. 1999. Application of enzymes in the pulp and paper industry. Biotechnology Progress, 15: 147–157.

Barras, F., Gijsegem, F.V. and Chatterjee, A.K. 1994. Extracellular enzymes and pathogenesis of soft-rot *Erwinia*. Annual Review of Phytopathology, 32: 201–204.

Beg, Q.K., Bhushan, B., Kapoor, M. and Hoondal, G.S. 2000. Production and characterization of thermostable xylanase and pectinase from a *Streptomyces* sp. QG-11-3. Journal of Industrial Microbiology and Biotechnology, 24: 396–402.

Blanco, P., Sieiro, C. and Villa, T.G. 1999. Production of pectic enzymes in yeast. FEMS Microbiology Letters, 175: 1–9.

Canan, T., Dogan, N. and Gogus, N. 2008. Biochemical and thermal characterization of exo-polygalacturonase produced by *Aspergillus sojae*. Food Chemistry, 111: 824–829.

Carr, J.G. 1985. Tea, coffee and cocoa. pp. 133–154. *In*: B.J.B. Wood (ed.). Microbiology of Fermented Foods, Elsevier Science Ltd. London.

Ceci, L. and Loranzo, J. 1998. Determination of enzymatic activities of commercial pectinases for the clarification of apple juice. Food Chemistry, 61: 237–241.

Chang, T.S., Siddiq, M., Sinha, N.K. and Cash, J.N. 1995. Commercial pectinases and the yield and quality of Stanley plum juice. Journal of Food Processing and Preservation, 19: 89–101.

Chatterjee, A.K., Buchanan, G.E. and Behrens, M.K. 2011. Synthesis and excretion of polygalacturonic acid trans-eliminase in *Erwinia*, *Yersinia* and *Klebsiella* species. Canadian Journal of Microbiology, 25: 94–102.

Chauhan, S.K., Tyagi, S.M. and Singh, D. 2001. Pectinolytic liquefaction of apricot, plum and mango pulps for juice extraction. International Journal of Food Properties, 4: 103–109.

Cho, S.W., Lee, S. and Shin, W. 2001. The X-ray structure of *Aspergillus aculeatus* polygalacturonase octagalacturonate complex. Journal of Molecular Biology, 314: 863–878.

Codner, R.C. 2001. Pectinolytic and cellulolytic enzymes in the microbial modification of plant tissues. Journal of Applied Bacteriology, 34: 147–160.

Evans, J.D., Akin, D.E. and Foulk, J.A. 2002. Flax-retting by polygalacturonases containing enzyme mixtures and effects on fibre properties. Journal of Biotechnology, 97: 223–231.

Gailing, M.F., Guibert, A. and Combes, D. 2000. Fractional factorial designs applied to enzymatic sugar beet pulps pressing improvement. Bioprocess Engineering, 22: 69–74.

Guebitz, G., Valle, L.J., Onos, M., Grriga, P., Calafell, M. and Schnitzhefer, W. 2006. Bioscouring of cotton fiber with polygalacturonase induced in *Sclerotium rolfsii* using cellulose and glucose pectin. Textile Research J., 76: 400–405.

Gummadi, S.N. and Panda, T. 2003. Purification and biochemical properties of microbial pectinases: A review. Process Biochemistry, 38: 987–996.

Hebeish, A., Ramadan, M.A., Hashem, M., Sadek, B. and Abdel-Hady, M. 2013. New development for combined bioscouring and bleaching of cotton-based fabrics. Research Journal of Textile and Apparel, 17: 94–103.

Hoondal, G.S., Tiwari, R.P., Tewari, R., Dahiya, N. and Beg, Q.K. 2002. Microbial alkaline pectinases and their industrial applications: A review. Applied Microbiology Biotechnology, 59: 409–418.

Hoondal, G.S., Tiwari, R.P., Tiwari, R., Dahiya, N. and Beg, Q.K. 2000. Microbial alkaline pectinases and their applications: A review. Applied Microbiology and Biotechnology, 59: 409–418.

Hugouvieux-Cotte-Pattat, N., Condemine, G., Nasser, W. and Reverchon, S. 1996. Regulation of pectinolysis in *Erwinia chrysanthemi*. Annual Review of Microbiology, 50: 213–257.

Jayani, R.S., Saxena, S. and Gupta, R. 2005. Microbial pectinolytic enzymes: A review. Process Biochemistry, 40: 2931–2944.

Jayani, R.S., Shukla, S.K. and Gupta, R. 2010. Screening of bacterial strains for polygalacturonase activity: Its production by *Bacillus sphaericus* (MTCC 7542). Enzyme Research, 2010: 1–5.

Joshi, V.K., Parmar, M. and Rana, N. 2011. Purification and characterization of pectinases produced from Apple pomace and evaluation of its efficacy in fruit juice extraction and clarification. Indian Journal of Natural Products and Resources, 2: 189–197.

Jurick, W.M., Vico, I., Mc Evoy, J.L., Whitaker, B.D., Janisiewicz, W. and Conway, W.S. 2009. Isolation, purification, and characterization of a polygalacturonase produced in *Penicillium solitum*-decayed 'Golden Delicious' apple fruit. Phytopathology, 99: 636–641.

Kapoor, M., Beg, Q.K., Bhushan, B., Singh, K., Dadhich, K.S. and Hoondal, G.S. 2001. Application of an alkaline and thermostable polygalacturonase from *Bacillus* sp. MG-cp-2 in degumming of ramie (*Boehmeria nivea*) and sun hemp (*Crotalaria juncea*) bastfibres. Process Biochemistry, 36: 803–807.

Kashyap, D.R., Vohra, P.K., Chopra, S. and Tewari, R. 2001. Applications of pectinase in commercial sector: A review. Bioresource Technology, 77: 215–227.

Kasuo, M., Ai, O., Akio, T., Kouzou, H., Makiko, O. and Toshikatsu, O. 2004. Low molecular weight pectates lyase from *Streptomyces thermocarboxylus*. Journal of Applied Glycoscience, 51: 1–7.

Kaur, G., Kumar, S. and Satyanarayana, T. 2004. Production, characterization and application of a thermostable polygalacturonase of a thermophilic mould *Sporotrichum thermophile apinis*. Bioresource Technology, 94: 239–243.

Kirk, T.K., Jefferies, T.W. and Viikari, L. 1996. Role of microbial enzymes in pulp and paper processing. pp. 1–14. *In*: T.K. Kirk, T.W. Jefferies and L. Viikari (eds.). Enzymes for Pulp and Paper Processing. ACS Symposium Series. American Chemical Society, Washington D.C.

Kozlowski, R., Batog, J., Konczewicz, W., Mackiewicz-Talarczyk, M., Muzyczek Sedelnik, M. and Tanska, B. 2006. Enzyme in bast fibrous plant processing. Biotechnology Letters, 28: 761–765.

Kumar, S., Jain, N.K., Sharma, K.C., Paswan, R., Mishra, B.K., Srinivasan, R. and Mandhania, S. 2015. Optimization, purification and characterization of pectinases from pectinolytic strain, *Aspergillus foetidus* MTCC 10559. Journal of Environmental Biology, 36: 483–489.

Ladjama, A., Taibi, Z. and Meddour, A. 2007. Production of pectinolytic enzymes using *Streptomyces* strains isolated from palm grove soil in Biskra area (Algeria). African Crop Science Society, 8: 1155–1158.

Muthezhilan, R., Newtonraja, A., Abirami, D. and Jayalakshmi, S. 2008. Pectinase from an estuarine strain of *Aspergillus fumigatus*. Research Journal of Biological Sciences, 1: 64–70.

Ossola, M. and Galante, Y.M. 2004. Scouring of flax rove with the aid of enzymes. Enzyme and Microbiol. Technology, 34: 177–186.

Phugare, S.S., Kalyani, D.C., Patil, A.V. and Jadhav, J.P. 2011. Textile dye degradation by bacterial consortium and subsequent toxicological analysis of dye and dye metabolites using cytotoxicity, genotoxicity and oxidative stress studies. Journal of Hazardous Material, 186: 713–723.

Priya, V. and Sashi, V. 2014. Pectinase enzyme production by using agrowastes. International Journal of Engineering Sciences and Research Technology, 3: 8041–8046.

Raghuwanshi, S., Gupta, A., Srivastava, A., Singh, D., Nema, R., Jawre, A.K., Pratap, B., Verma, S., Shrivastva, H., Dadse, D., Binjhade, D., Ghidode, S. and Thakre, S. 2013. Pectolytic enzyme and its global applications: A review. CMBT Journal of Science and Technology, 1: 1–9.

Reid, I. and Richard, M. 2004. Purified pectinase lowers cationic demand in peroxide-bleached mechanical pulp. Enzyme and Microbial Technology, 34: 499–504.

Revilla, I. and Ganzalez-San Jose, M.L. 2003. Addition of pectolytic enzymes: An enological practice which improves the chromaticity and stability of red wines. International Journal of Food Science and Technology, 38: 29–36.

Sakai, T., Sakamoto, T., Hallaert, J. and Vandamme, E.J. 1993. Pectin, pectinases and propectin: Production, properties and applications. Advances in Applied Microbiology, 39: 231–294.

Salazar, L. and Jayasinghe, U. 1999. Fundamentals of Purification of Plant Viruses. Techniques in Plant Virology. CIP training Manual JO, Virus Purification, International Potato Centre, Peru, pp. 1–10.

Saranraj, P. and Naidu, M.A. 2014. Microbial pectinases: A review. Global Journal of Traditional Medicinal System, 3: 1–9.

Scott, D. 1978. Enzymes industrial: Encyclopedia of chemical technology. pp. 173–224. *In*: M. Grayon, D. Ekarth and K. Othmer (eds.). Wiley, New York.

Singh, S. and Gupta, R. 2003. Apple juice clarification using fungal pectinolytic enzyme and gelatine. Industrial Biotechnology, 3: 573–576.

Tanabe, H., Yoshihara, Y., Tamura, K., Kobayashi, Y. and Akamatsu, T. 1987. Pretreatment of pectic wastewater from orange canning process by an alkalophilic *Bacillus* sp. Journal of Fermentation Technology, 65: 243–246.

Tariq, A. and Latif, Z. 2012. Isolation and biochemical characterization of bacterial isolates producing different levels of polygalacturonases from various sources. African Journal of Microbiological Research, 6: 7259–7264.

Ugwuanyi, O. and Obeta, J.A.N. 1997. Some pectinolytic and celluolytic enzyme activities of fungi causing rots of cocoyams. Journal of the Science of Food and Agriculture, 73: 432–436.

Van der Vlugt-Bergmans, C.J.B., Meeuwsen, P.J.A., Voragen, A.G.J. and Van Oogen, A.J.J. 2000. Endo-xylogalacturonan hydrolase, a novel pectinolytic enzyme. Applied and Environment Microbiology, 66: 36–41.

Viikari, L., Tenakamen, M. and Suurnakki, A. 2001. Biotechnology in pulp and paper industry. Biotechnology. pp. 523–546. *In*: H.J. Rehm (ed.). VCH-Wiley, Washington.

Whitaker, J.R. 1990. New and future uses of enzymes in food processing. Food Biotechnology, 4: 669–697.

Wong, D.W.S. 1995. Pectic Enzymes. Food Enzymes. Structure and Mechanisms. New York: Chapman and Hall, pp. 212–236.

Zeni, J., Cence, K., Grando, C.E., Tiggermann, L., Colet, R., Lerin, L.A., Cansian, R.L., Toniazzo, G., Oliveira de, D. and Valduga, E. 2010. Screening of pectinase-producing microorganisms with polygalacturonase activity. Applied Biochemistry and Biotechnology, 163: 383–392.

Proteases

Olga Luisa Tavano

1. Introduction

Proteases, or peptidases, constitute the largest group of enzymes with a wide range of uses in the industrial process, especially in detergents and in pharmaceutical and food processing. These applications are associated with the proteases performing functions that focus on the effects of cleavage of peptide bonds. Proteolysis is a powerful tool in the modification of the properties of proteins and proteic systems, including changes in solubility, gelation, emulsifying and foaming characteristics, which make this process ideal in food processes. Among the main achieved effects in food modification, there are aspects relating to the nutritional value and functional properties of foods, including modifications of sensory quality (such as texture or flavor) and health benefits, such as the improvement of food digestibility, bioactive properties or reduction in allergenic responses (Panyam and Kilara, 1996; Van Boekel et al., 2010; Tavano, 2013).

The most important characteristic of proteases relates directly to their applications. Although protein hydrolysis can be carried out by enzymatic or chemical processes, including alkaline or acid hydrolysis, the processes tend to be difficult to control and yield products with modified amino acids. Besides, enzymatic hydrolysis can be performed under mild conditions, avoid amino acid chain side reactions and do not decrease the nutritional value of the protein source (Maldonado et al., 1998). Enzyme processes present substrate specificity which permits the development of protein hydrolysates with better defined results and characteristics (Tsugita and Scheffler, 1982). And in seeking to achieve a specific product or due process, it must have the knowledge that proteases perform to a more appropriate choice of the enzyme for best results in the desired application so that development of protein hydrolysates with better defined chemical and nutritional characteristics occurs.

Alfenas Federal University, Nutrition Faculty, 700 Gabriel Monteiro da Silva St, Alfenas, MG 37130-000, Brazil.
E-mail: tavanool@yahoo.com.br

2. Proteases Classifications

Due to their kinetic characteristics, proteases are classified in Class 3 in the international system for classification and nomenclature of enzymes (EC number) (Webb, 1993), as hydrolases, subclass 3.4 (hydrolysis of peptide bonds) divide proteases between 13 sub-classes on the basis of the catalytic reaction. Moreover, MEROPS databases (a peptidase database), considering principally its tertiary structure and catalytic sites, classified proteases in clans, and clans divided into families. Every clan provides information about the catalytic structure of the proteases. The group names take into account the iconic amino acid or metal present in the active site: Aspartic peptidases (A), Cysteine peptidases (C), Metallo peptidases (M), Serine peptidases (S), Mixed catalytic type (P) and Unknown type (U) (Rawlings and Barrett, 1993; Rawlings et al., 2008; Rawlings et al., 2012; Polgár, 2005). Serine proteases, for example, are known for their classical catalytic triad of amino acids. The first structure reported for a peptidase, a chymotrypsin, showed a very well-documented relationship between Asp102, His57 and Ser195 (Perona and Craik, 1997; Polgár, 2005). This classification method is based on the knowledge of the protease chemical structure, and it is important to emphasise that the structure around the active site of the protease determines how the substrate can bind to the sites of the protease. The enzyme active site has a characteristic arrangement of amino acid residues which define the enzyme–substrate interaction (Neurath, 1984). The surface of the protease that is able to accommodate the chain of the substrate is called the sub-site which determines the substrate specificity of a given protease (Turk, 2006). The specificity of a protease determines the position at which the enzyme will catalyze the peptide bond hydrolysis (Table 1).

Their mode of action can also define proteases as endopeptidases or exopeptidases. The peptide substrate runs through the entire length of the active site of an endopeptidase framework and is cleaved in the middle of the molecule. Exopeptidases that act near the end of the polypeptide chains are termed as aminopeptidases if they act at the N-terminus and as carboxypeptidases, if acting on peptide bonds from the C-terminus.

Moreover, proteases are also distinguished by their pH optima of action, such as acidic (pH < 7.0), neutral (around pH 7.0) and alkaline proteases (pH > 7.0), or even high alkaline proteases (pH > 10.0) (Sumantha et al., 2006; Gupta and Ayyachamy, 2012). It is important to remember that the proteases, as well as other enzymes, have their preferences on catalysis conditions, making an enzyme more or less suitable for processing a food, for example in the function of its pH. Acidic proteases are found

Table 1. Examples of protease activities.*

Protease	Microorganism	Preferential Cleavage Site
Subtilisin (EC 3.4.21.62)	*Bacillus amyloliquefaciens*	Preference for a large uncharged residue in P1
Thermolysin (EC 3.4.24.27)	*Bacillus thermoproteolyticus*	Preferably -/-Leu > -/-Phe
Protease from *Aspergillus oryzae*	Non-toxicogenic and non-pathogenic strains of *Aspergillus oryzae*	Hydrolysis at the N-terminal, serine and aspartic acid bonds

*Adapted from BRENDA (BRaunschweig ENzyme DAtabase), accessible free at <http://www.brenda-enzyme.org>

in animal cells, moulds and yeasts, but sometimes in bacteria too. Microbial rennin-like proteases have been found in *Mucor pusillus* and *Endothia parasitica*. But other species of *Mucor* produce interesting proteases, such as *M. hiemalis, M. racemosus*, and *M. bacilliformis* (Escobar and Barnett, 1993; Fernandez-Lahore et al., 1997; Fernandez-Lahore et al., 1998; Fernandez-Lahore et al., 1999; Tremacoldi et al., 2004; Kumar et al., 2005). *Bacillus cereus* and *Aspergillus* spp. also produce proteases with commercial value, especially pepsin-like acid proteases, as shown in Table 2. Many of them contain aspartate as the active amino acid and their characteristics are defined by the presence of aromatic amino acids or bulky side chains at both sides of the cleaving bond, and their content of carbohydrate confers heat stability to the protease chain (Kovaleva et al., 1972; Tsujita and Endo, 1978). The most significant property of acidic proteases is their ability to coagulate proteins, which reflects their application in the dairy industry thanks to their ability to coagulate milk casein to form cheese curds.

Neutral proteases, especially fungal neutral proteases are important components of commercial enzymes preparations, which have applications in baking, food processing, protein modification and also in the leather, animal feeds and pharmaceutical industries. Its affinity for hydrophobic amino acids provides beneficial in its utilization as a debittering agent. *Aspergillus oryzae* is the predominant fungal source of the neutral enzymes (Sumantha et al., 2006). In fact, *Aspergillus* species are highlighted in the proteases production, such as *A. sojae*, which produces a very interesting metalloprotease (Sumantha et al., 2006). Neutral metalloproteases present high specificity towards peptide linkages that contain hydrophobic amino acids at the N-terminal. Their principal applications include soy protein hydrolysis, soy sauce production, gelatin hydrolysis, casein and whey protein hydrolysis, meat protein recovery, fish protein hydrolysis and meat tenderization. Another protease from *Bacillus subtilis* is used to deproteinize crustacean waste to produce chitin (Yang et al., 2000). A very interesting, and one that is attracting many interests, is the thermostable protease *Thermolysin*, produced by *Bacillus stearothermophilus* which can act at 80°C. Other *Bacillus* spp., *Pseudomonas aeruginosa, Streptomyces griseus* and *Penicillium* are also reported to be producers of neutral proteases, especially neutral metalloprotease (Germano et al., 2003; Sumantha et al., 2006).

Alkaline proteases are industrially very significant in view of their activity and stability at alkaline pH. Subtilisin Carlsberg produced by *Bacillus licheniformis*, Subtilisin BPN and Subtilisin Novo are some serine alkaline proteases widely used. These can be employed in the dehairing and bating of hides, meat tenderizing, cheese flavor development, treatment of flour in the manufacture of baked goods, improvement of dough texture, flavor, and colour in cookies, improving digestibility of animal feeds (Guntelberg and Otteson, 1952; Smith et al., 1968; Kumar and Takagi, 1999; Sumantha et al., 2006).

3. Proteases Sources

Since proteolytic processes are intimately associated with vital biological pathways, it is not difficult to imagine importance and enormous range of structures and specificities of proteases with a very large repertory of functions. Therefore, proteases can

Table 2. Microbial protease preparations used in food processing following general specifications of Joint FAO/WHO Expert Committee on Food Additives (JECFA). *

Enzyme Preparations	Action	Sources	Applications in Food Processing
Acid protease from *Bacillus cereus** (Rennet from *Bacillus cereus*)	Aspartic proteinase preparation (EC 3.4.23.6), which hydrolyze polypeptides such as casein.	Produced by controlled fermentation of *Bacillus cereus*.	Used in cheese making.
Aspartyl protease* (Chymosin A from *Escherichia coli* K-12 containing the Prochymosin A gene)	Chymosin (EC 3.4.23.4) which cleaves a single bond in kappa-casein.	Produced intracellularly by the controlled fermentation of *Escherichia coli* K-12 containing the bovine Prochymosin A gene.	Used in clotting of milk cheese production.
Aspartyl protease* (Chymosin B from *Kluyveromyces lactis* containing the Prochymosin B gene)	Chymosin (EC 3.4.23.4) which cleaves a single bond in kappa-casein.	Produced extracellularly by the controlled fermentation of non-toxicogenic and non-pathogenic strains of *Kluyveromyces lactis* containing the bovine Prochymosin B gene.	Used in clotting of milk cheese production.
Aspartyl protease* (Chymosin B from *Aspergillus niger* var. *awamori* containing the Prochymosin B gene)	Chymosin (EC 3.4.23.4) which cleaves a single bond in kappa-casein.	Produced extracellularly by the controlled fermentation of non-toxicogenic and non-pathogenic strains of *Aspergillus niger* var. *awamori* containing the bovine Prochymosin B gene	Used in clotting of milk cheese production.
Mixed Microbial Carbohydrase and Protease from *Bacillus subtilis*, var.	Usually contain following two proteases (besides carbohydrase): serine proteinase (EC 3.4.21.14) and metalloproteinase (EC 3.4.24.4). Metalloproteinase cleavage preferentially bonds adjacent to a hydrophobic amino acid.	Produced by controlled fermentation of *Bacillus subtilis*; var.	Used in preparations of start syrups, alcohol, beer, bakery products, fish meal, tenderizing meat and protein hydrolysates.
Protease from *Aspergillus oryzae*	Hydrolysis at the N-terminal, serine and aspartic acid bonds.	Non-toxicogenic and non-pathogenic strains of *Aspergillus oryzae*.	Preparation of meat and fish products, beverages, soup and broths, dairy and bakery products.
Protease from *Streptomyces fradiae*	Hydrolyzes proteins and peptides with no clear specificity.	*Streptomyces fradiae.*	Preparation of beverages.

Table 2. contd....

Table 2. contd....

Enzyme Preparations	Action	Sources	Applications in Food Processing
Serine Protease with chymotrypsin specificity from *Nocardiopsis prasina* expressed in *Bacillus licheniformis*	Serine protease with chymotrypsin specificity. Preferencial cleavage: Tyr, Trp, Phe and Leu.	Produced by submerged fermentation of a genetically modified non-pathogenic and non-toxigenic strain of *Bacillus licheniformis* which contains a gene coding for serine protease with chymotrypsin specificity from *Nocardiopsis prasina*.	Used in the hydrolysis of proteins like casein, soy isolate or concentrate, wheat or corn gluten, for hydrolysates production.
Serine Protease with Trypsin specificity from *Fusarium oxysporum* Expressed in *Fusarium venenatum*	Serine protease with trypsin activity (EC 3.4.21.4). Preferential cleavage: Arg and Lys.	Produced by submerged fermentation of a genetically modified non-pathogenic and non-toxigenic strain of *Fusarium venenatum* which contains a gene coding for serine protease with trypsin specificity from *Fusarium oxysporum*.	Used in the production of partially or extensively hydrolysed proteins of vegetable or animal origin. This hydrolysate may be used for protein fortification or functional effects (emulsification or flavor enhancement) in food or beverages.

*Adapted from online database for food additives evaluated by JECFA, accessible free at <http://www.http://www.fao.org/food-safety-quality/scientific-advice/jecfa/jecfa-additives/en/>

usually recover from animal, plant or microbial sources and even the most primitive organism. Microorganism sources are especially interesting for many reasons. One prime reason is its path to culture. They can be cultured on a large scale, for example, over a short time-process of fermentation (Devasena, 2010). Both submerged and solid state fermentation are often used in the production of proteases from cultures of microorganisms. Microbial proteases are often produced as extracellular enzymes in nature; therefore, they can directly secreted into the fermentation broth which simplify downstream processing (understood as recovery and purification steps of the products from their sources) of the enzyme production as compared to proteases obtained from plants and animals (Savitha et al., 2011; Souza et al., 2015).

Submerged fermentation has advantages in process control and can present an easy recovery when extracellular enzymes are produced. Some species of filamentous fungi, such as *Aspergillus, Penicillium* and *Paecylomices* have been identified as great producers of extracellular protease in submerged fermentation. However, the produced enzymes are dilute in culture medium, which becomes a problem in industrial production, where the higher water amount allows the production of metabolites in a less concentrated form, making the downstream processing time-consuming and expensive. Solid-state fermentation has attracted the interest of many researchers. Its advantages include simplicity, lower production costs, low waste-water output and its static process avoids mechanical energy expenditures while low moisture content minimizes problems with bacterial contamination during fermentation, although problems such as temperature and pH control have been encountered (Sandhya et al., 2005). Whatever the chosen enzyme source or methodology of production, downstream processes are really crucial for enzymes applications. Considering this fact, production methods which avoid many downstream steps can be very advantageous as they check the expenses and are less time-consuming in the production of proteases.

Fungal cultures are particularly applied and have relative advantages depending on the type of medium used in their culture. Fungal proteases have attracted the attention of environmental biotechnologists because fungi can grow on low-cost substrates and secrete large amounts of enzymes into the culture medium which could ease downstream processing (Anitha and Palanivelu, 2013; Souza et al., 2015).

Different mechanisms are described to regulate the synthesis and secretion of protease, especially extracellularly. In general, the presence of an adequate substrate can induce protease secretion, which may increase when insufficient levels of amino acids and nitrogen that are easily metabolizable are available. Complex nitrogen sources are usually used for protease production, which can be induced at low levels, regardless of the availability of a substrate (Geisseler and Horwath, 2008). As described by Kucera (1981), fungi produce more proteolytic enzymes in more complex proteinaceous nitrogen sources than on low molecular weight or inorganic nitrogen sources. Enzyme synthesis was found to be repressed by rapidly metabolizable nitrogen sources, such as amino acids or ammonium ion concentrations in the medium (Kumar and Takagi, 1999).

Besides the relative ease of the use of culture of microorganisms, the enormous number and variety of microorganisms available help to seek the most appropriate source, using the traditional culture 'screening' process. The enormous range of different proteases with a large repertoire of structures and specificities improves the choice in acting on a specific substrate or intended products. Moreover, the application of proteases in food processes often requires them to maintain high activity in non-physiological conditions, such as high temperatures and pH, intensive calcium chelating agents or salt or others ingredients in a high concentration. The proteolysis, like any other enzymatic process, meets limitations in protease utilization, such as enzyme chain instability in denaturation conditions. The absence of such selection pressures during the natural evolution of living beings, makes most of the proteases unstable or inactive in many food environmental conditions. Since proteases are linked to the vital cycle of life, their variety reflects the living conditions. Although much variation and evolutionary changes exist, most living proteases reflect mild action, according to survival conditions (temperature, pH, pressure), especially with regard to animals and vegetables. Among microorganisms, this situation can be different, since it finds a wide range of livelihood extreme, with a variety of pH, temperature, media components concentrations. This is an important advantage when seeking enzymes in microorganism sources, due to their enormous metabolic variety (Li et al., 2013).

The isolation of protease variants from organisms living in extreme environments, for example, can be used to obtain microbial enzyme sources variants whose stability may be very attractive (Eijsink et al., 2005). These microorganisms, called extremophiles, naturally produce enzymes that are functional under extreme conditions, such as pressure, radiation, salinity, pH and temperature (Van den Burg, 2003).

Thermostable enzymes, for example, have been isolated from thermophilic microorganisms, which, at high environmental temperatures, can acquire enhanced mobility to act, but remain rigid enough to resist denaturation, while the mesophilic enzymes can suffer unfolding (Fontana et al., 1998). Their application in processes development at high temperatures is a real advantage. For treatment of industrial wastes, such as hard-to-degrade animal proteins generated in the meat industry, thermostable proteases can be used at high temperatures which must induce protein (as a substrate) thermal denaturation and result in greater proteolysis susceptibility (Suzuki et al., 2006). In this case, the high temperature may benefit not only the function of the enzyme but also the availability of the substrate to hydrolysis, making it more susceptible to the action of the enzyme and consequently increasing the speed of the reaction.

At the opposite thermal extreme, psychrophilic (cold-adapted) microorganisms are capable of proliferating at $\leq 0°C$, and even at $-20°C$ and these include mechanisms that provide the ability to adapt to these particular thermal environment, and obviously, the production of many cold-adapted enzymes (Siddiqui and Cavicchioli, 2006). Although among the isolated and cold-adapted enzymes may be seen fish proteases—the most cold-adapted enzymes that have been sourced from microorganisms. Its proteases present peculiarities that could explain its great adaptability to cold temperatures. Several studies convey that psychrophilic protease character is accompanied by an overall increase in flexibility (Tindbaek et al., 2004). A number of cold-adapted proteases present a higher Arg/Lys ratio compared to thermostable enzymes. Arg residues can form interactions with water and these residues generally enhance enzyme thermostability more than Lys residues by facilitating a greater number of very efficient electrostatic interactions (Cupo et al., 1980; Mrabet et al., 1992; Feller et al., 1997; Georlette et al., 2000; D'Amico et al., 2001; Siddiqui and Cavicchioli, 2006).

Halophilic microorganisms are considered important sources of proteases, since enzymes capable of supporting hypersaline conditions may provide distinct advantages for some processes. Extremely saline conditions may allow food processing in the absence of strictly sterile conditions due the growth inhibitory effects of other microorganisms. Structural and biochemical characterization of several halophilic enzymes shows that enhancing solvatation is the key requirement essential for maintaining solubility and activity of halophilic enzymes in low water activity, which can approach values as low as 0–75 in a saturated NaCl solution. A serine protease from *Natrinema* sp. is capable of deproteinizing chitin-containing biomass at high salinity, including shrimp shell power. Interestingly, low water-activity conditions mimic aqueous-organic solvent mixtures, and consequently halophilic enzymes generally retain considerable activity in organic media, making them potentially useful as industrial biocatalysts (DasSarma and DasSarma, 2015).

On the other hand, while the natural evolution is the largest source of variation among living beings, it is time consuming and does not necessarily reflect the industrial needs. Protease engineering by site-directed mutagenesis, or directed-evolution, can generate proteases with improved functions to meet the requirements of commercial applications (Li et al., 2013). Microorganisms can be 'engineered' to direct the generation of new enzymes, often with desirable characteristics (including overall stability, solubility and especially new specificities), using 'classical' protein engineering, such as rationale-based mutagenesis, or directed evolution. This is a further advantage of the use of microorganisms as sources of enzymes. Directed evolution is based on generation of diversity followed by selection/screening. As described by Cobb et al. (2012), directed evolution refers to the application of selective pressure to a library of variants of a target biological entity with the intent of identifying those with desired properties. Although 'laboratory evolution' accelerates the individual steps of natural evolution, this procedure is restricted by the size of possible libraries and can be very time-consuming at every cycle of library generation and screening/selection (Cobb et al., 2012). Classical protein engineering utilizes information on enzyme structure and on the molecular basis of stability to rationally predict stabilizing mutations (Eijsink et al., 2005), applying techniques for site-directed mutagenesis (including insertions, deletions or/and recombination of amino acids). One of the first successful trials in altering proteases was achieved and described by Craik et al. (1985). They modified eukaryotic genes by site-specific mutagenesis and expressed in mammalian cells, replacing the glycine residues at positions 216 and 226 in the binding cavity of trypsin by alanine residues and resulting in three trypsin mutants, which presented reduced catalytic rate but increased specificity. Other proteases have been successfully engineered for the enhancement of their innate properties. A number of other studies show that microbial subtilisin has been successfully engineered for various purposes (Li et al., 2013). Narhi et al. (1991) described studies on the effect of point mutations on the stability of aprA-subtilisin, which contains Asn-Gly pairs at positions 109–110 and 218–219. The substitution of N109S showed a 3°C increase in transition temperature, while the substitution of N219S led to a 4°C increase in transition temperature. Combination of these two mutations (N109S, N219S) exhibited stability gains that had a cumulative beneficial effect on subtilisin stability, with a 7°C higher transition temperature.

4. Food Applications

4.1 As enzyme

It is important to consider that different processes of food modification can use the isolated enzyme or even the cell itself, which does not always control the effect, but has the advantage to rely on the effect of the mixture of enzymes released by the microorganism. For example, in many types of cheese, proteolysis is not limited to the action of added enzymes, but also to the enzymes present from microorganisms. A classic example is the blue-veined cheese category, where the presence of blue moulds gives a characteristic appearance and the enzymes of these moulds produce typical aroma and taste. The same goes for the mould surface in cheese, such as brie and Camembert (Souza et al., 2001; Seratlic et al., 2011). Proteolysis can be extremely complex. For Camembert at least five proteases are active during ripening: chymosin and bovine pepsin (in rennet), plasmin, aspartylprotease and metalloproteinases from *Penicillium caseicolum,* each having a complementary and sequential role (Trieu-Cuot and Gripon, 1982). That is, when microorganisms are used for proteolysis, it is unlikely that only one enzyme acts. Thus, hydrolysis will be performed by a mixture of enzymes and, after liberation of the first peptides, these will be susceptible to the action of carboxypeptidases or aminopeptidases. Depending on the specificity of the enzyme that released the first peptides, distinct amino acids will be present at the extremities of the fragments and, as a result, distinct amino acid residues will be liberated over time by the exopeptidases. This will result in a wide variety of hydrolysed species (Castro and Martín-Hernandez, 1994; Otte et al., 1997; Milesi et al., 2009; Seratlic et al., 2011).

The use of specific enzymes can produce specific changes and much more controlled action. Many processes included in the production of food can make use of specific proteases to obtain products with controlled characteristics, but also to make use of some proteases that would have been naturally released by present microorganisms in the environment, in order to enhance a particular effect on the final product. Commercial proteinase preparations have been used to accelerate ripening in Cheddar and Dutch-type cheeses, to help alleviate bitterness. The inclusion of combinations of proteases with lipases has been reported to result in increased proteolysis suggesting a possible synergistic effect (Lin et al., 1987; Kheadr-Ehab et al., 2003). Exogenous proteases can be added before or with the starter culture and the coagulant protease, or even at dry salting and or directly into the cheese block (Wilkinson and Kilcawley, 2005).

Major industrial applications of these biocatalysts are in the brewing processes, avoiding protein effects on chill haze of beer, or tenderizing meat, accelerating the natural tenderization process via additional proteolytic tenderization by sprinkling a meat piece with commercial preparations of proteases (Fig. 1). Proteases are also very useful in the bread industry, which can produce a partial gluten hydrolysis, especially in cracker or wafer production with little or weak gluten. In these food processes, fungal or bacterial proteases are produced by controlled fermentation of *Streptomyces fradiae* or the endo- and exo-peptidases from non-toxicogenic and non-pathogenic strains of *Aspergillus.* But some proteases have aroused great interest among researchers

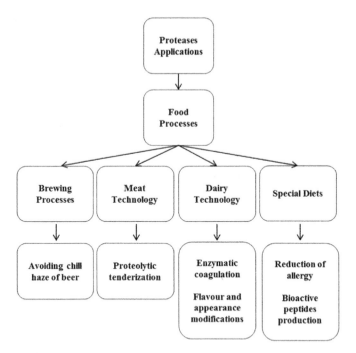

Figure 1. Some protease applications in food processes.

seeking its various applications in processed food modifications, especially aspartyl- and serine-proteases, such as Flavourzyme, Alcalase, Thermolysin, and even some rennet substitutes (Li et al., 2012).

4.2 As food additives

But, the key item in evaluating enzyme preparations applied to food processes is the safety assessment of the production strain—not from the point of view of their specificity and/or stability, but because in the case of food processing it has to be safe for health. The Joint FAO/WHO Expert Committee on Food Additives (JECFA) is an international expert scientific committee that is administered jointly by the Food and Agriculture Organization of the United Nations (FAO) and the World Health Organization (WHO). It has been meeting since 1956, initially to evaluate the safety of food additives. As described under 'General Specifications and Considerations for Enzyme Preparations used in Food Processing': *"Enzyme preparations consist of biologically active proteins, at times combined with metals, carbohydrates and/or lipids. They are obtained from animal, plant or microbial sources and may consist of whole cells, parts of cells, or cell-free extracts of the source used. They may contain one or more active components as well as carriers, solvents, preservatives, antioxidants and other substances consistent with good manufacturing practice. They may be liquid, semi-liquid, dry or in an immobilized form (immobilized enzyme preparations are preparations which have been made insoluble in their intended food matrix by physical and/or chemical means). Their colour may vary from virtually colourless to dark brown.*" And further, with regard to microbial sources: "*Microbial sources used in the production of enzyme preparations may be native strains or variants of microorganisms, or be derived from native strains or variants by the processes of selective serial culture or genetic modification. Production strains for food enzyme preparations must be non-pathogenic and non-toxigenic...*" Protease preparations may be 'food additives' if approved by FDA for specific use or GRAS (generally recognized as safe) substances. A substance may be GRAS only if its general recognition of safety is based on the views of experts qualified to evaluate the safety of the substance. Table 2 presents a compilation of microbial proteases listed as food additives (enzyme preparations) in JECFA website database, which provides the most recent specifications for food additives evaluated by JECFA.

5. Some Highlighted Proteases

5.1 Serine family

Serine proteases are a detachable family of proteases and have been grouped into six clans (Rawlings and Barrett, 1993), of which trypsin-like and subtilisin-like proteases present a detachable set of enzymes from the standpoint of their applications. Subtilisins are important industrial enzymes, found in microorganisms and constitute an important class of microbial serine proteases, especially those secreted from *Bacillus* species *amyloliquefaciens* (BPNP), *subtilis* (*subtilisin* E) and *lentus* (*savinase*), which are among the most mutagenized proteases (Bryan et al., 2000). Subtilisins can be divided into families like Subtilases, Thermitases, Proteinase K or Lantibioric peptidases.

Subtilases are members of the clan of subtilisin-like serine proteases, which are quite common in gram-positive bacteria, such as *Bacillus* (Schmidt et al., 1995). In most bacteria, archaea, and lower eukaryotes are extracellular; rather, unspecific enzymes which are required either for defense or for growth on proteinaceous substrates. Thermitase family is also found only in microorganisms, including some thermophiles and halophiles. Proteinase K family presents endopeptidases which are found only in fungi, yeasts and gram-negative bacteria. Lantibioric peptidase family constitutes a small number of highly specialized enzymes for cleavage of leader peptides from precursors of lantibiotics—a unique group of post-translationally modified, antimicrobial peptides (Sahl et al., 1995). These endopeptidases have been found only in gram-positive bacteria and several are intracellular (Siezen and Leunissen, 1997).

5.2 Flavourzyme

Flavourzyme is a commercial peptidase preparation (Novozymes) containing endo- and exopeptidases from *Aspergillus oryzae* and have a wide range of applications in industries and research. Seven principal proteases of the Flavourzyme preparation were recently identified and purified—leucine aminopeptidase A, leucine aminopeptidase 2, dipeptidyl peptidase 4, a dipeptidyl peptidase 5, a neutral protease 1, a neutral protease 2, and alkaline protease 1 (Merz et al., 2015a, 2015b). Although this combination is interesting, since this synergy of endopeptidases and exopeptidases may be identified as the key for an efficient hydrolysis of proteins, the presence of more than one enzyme can present some disadvantages, such as reduced control over the exact action of the preparation on the substrate. The variation of the Flavourzyme composition also affects the reproducibility of a casein-batch hydrolysis process, and this should be taken into account for any future research and industrial application. And, in addition, each enzyme can react differently to changing environmental conditions or the storage time. However, Novozymes consider that key enzyme activity is provided by exopeptidase, which liberates amino acids by hydrolysis of the N-terminal peptide bond and standardizes the enzyme preparation (Flavourzyme® 1000L) using the marker substrate H-leucine-para-nitroanilide (Leu-pNA), i.e., Flavourzyme® 1000L contains at least 1000 leucine aminopeptidase units (LAPU) per gram preparation. But this substrate mainly covers one aminopeptidase while other enzymes are not quantified nor specified, i.e., the variation of other enzymes is unclear. Variation of the Flavourzyme composition can affect the reproducibility of a casein-batch hydrolysis process and this should be taken into account for any future research and industrial application (Merz et al., 2015a, 2015b). Although the name 'Flavourzyme' suggests production of flavor-active compounds from various protein sources, its application in the production of bioactive peptides has been studied. Thamnarathip et al. (2016) produced a riceberry-bran protein hydrolysate using enzymatic hydrolysis. When Flavourzyme was used, a higher antioxidant activity was noted. Hrcková et al. (2002) showed that the enzymatic treatment of defatted soy flour with three different proteases (Flavourzyme 1000 L, Novozym FM 2.0 L and Alcalase 2.4 L FG) improved the foaming and gelling properties. In spite of the fact that the degrees of hydrolysis are not so different for the three enzymes but the latter have different specificity and produced different hydrolysates, especially when Flavourzyme was used.

5.3 Alcalase

Alcalase was initially obtained from *Bacillus subtilis*. Several strains of *Bacillus subtilis* produce an extracellular proteolytic enzyme of broad specificity with an alkaline pH optimum. The first of these enzymes, now called subtilisin Carlsberg, was discovered by Linderstrom-Lang and Ottesen and purified by Gtintelberg and Ottesen (DeLange and Smith, 1968). This enzyme is also called subtilisin A, subtilopeptidase A and finally Alcalase. A number of properties of subtilisin Carlsberg have been reported when comparing this strain of the enzyme with homologous subtilisins BPN' and Novo Alcalase, which show similar properties but differ markedly in amino acid composition and sequence (DeLange and Smith, 1968). Novo Nordisk (Bagsvaerd, Denmark) Alcalase was produced via fermentação submersa using *Bacillus licheniformis*. Although it has been developed for the detergent industry, many studies show a great application in the modification of food. Boza et al. (1994) studied hydrolysates of casein and whey proteins using a protease mix containing *Bacillus licheniformis* enzymes, obtaining hydrolysates of high nutritional values and with reduced the potential antigenicity of the whey protein (van Beresteijn et al., 1994). Roasted peanut protein when hydrolysed by Alcalase and Flavourzyme, presented a 65 per cent decrease in IgE reactivity when noted after 300 min. of hydrolysis with use of Flavourzyme.

However, at 30 min. of treatment with Flavourzyme, an increase in IgE reactivity was detected by ELISA, whereas a 100 per cent reduction in IgE reactivity was observed for Alcalase treatment (Cabanillas et al., 2012). Angiotensin I converting enzyme (ACE) produced inhibitory peptides from salmon byproduct proteins via enzymatic hydrolysis using Alcalase, flavourzyme, neutrase, pepsin, protamex and trypsin, where Alcalase hydrolysate showed the highest ACE inhibitory activity. The purified ACE inhibitory peptides were identified as Val-Trp-Asp-Pro-Pro-Lys-Phe-Asp, Phe-Glu-Asp-Tyr-Val-Pro-Leu-Ser-Cys-Phe, and Phe-Asn-Val-Pro-Leu-Tyr-Glu (Ahn et al., 2012). When sweet sorghum grain, protein was hydrolysed using Alcalase, after hydrolysis, the hydrolysate, with hydrolysis degree of 19 per cent, exhibited the strongest ACE inhibitory activity (Wu et al., 2016). Enzymatic hydrolysis of chickpea protein concentrate by Alcalase improved the antioxidant activity of the hydrolysate (Ghribi et al., 2015). Goat whey hydrolysates, produced from goat whey, using Alcalase® from *Bacillus licheniformis*, exhibited an interesting high antibacterial activity against *Escherichia coli* and *Bacillus cereus* (Osman et al., 2016).

5.4 Thermolysin

Thermolysin is a known thermostable zinc endopeptidase traditionally isolated from thermophile *Bacillus thermoproteolyticus*, but, several studies show that neutral proteases from *Bacillus subtilis, Bacillus stearothermophilus, Bacillus amyloliquefaciens* and *Bacillus cereus* are closely related to this thermolysin (Matthews, 1988). Thermolysin protein hydrolisates show very potent bioactive activities. Dried bonito, a Japanese traditional seasoning made out of bonito muscle, was hydrolysed by various proteases by Yokoyama et al. (1992) and the inhibitory activity of the hydrolysates for angiotensin I-converting enzyme (ACE) of thermolysin hydrolysate among the digests showed the most potent inhibitory activity, which is due to thermolysin specificity reaching peptides with Ile, Leu, Phe, or Ala at the amino terminus. When thermolysin catalysed hydrolysates of α-lactalbumin and β-casein, many peptides were responsible for the high ACE inhibitory activity (Otte et al., 2007). Whey protein concentrate enriched with β-lactoglobulin was hydrolysed using Corolase PP (a proteolytic enzyme preparation from pig pancreas) and thermolysin to produce hydrolysates with antioxidant activity, the highest radical scavenging activity was found in WPC hydrolysed with thermolysin after eight hours of digest (Contreras et al., 2011).

5.5 Rennet-like microbial proteases

Calf rennet (chymosin and pepsin extract as major components) is the protease traditionally used in cheese making. Cheese production usually starts via controlled protein hydrolysis (cleavage of Phe 105-Met 106 bond of the κ-casein) leading to milk protein coagulation. Bovine chymosin, an aspartyl protease extracted from the abomasum of suckling calves, is the principal component due to its specificity. Therefore, a suitable rennet substitute must have high specific caseinolytic activity and small-generalized proteolytic activity (Mohanty et al., 1999; Kumar et al., 2005). Several proteases have been investigated as likely rennet substitutes (animal, microbial or plant enzymes) (Ghorai et al., 2009). The aspartyl protease from *Rhizomucor miehei* has been used as a chymosin substitute in cheese making. A number of strains of *Rhizomucor miehei* was exploited for milk-clotting (MC) enzyme production. A recombinant strain of *Aspergillus niger* var. *awamori* has been widely used due to its high productivity. In order to develop a commercial process, a strain of *A. niger* var. *awamori* (GCAAP4), from which the native aspartic proteinase gene had been deleted (Berka et al., 1990), was transformed with pGAMpR. The best condition obtained (GC4-1) produced approximately 280 mg/1 of extracellular chymosin in shake-flask cultures (Ward et al., 1993; Olempska-Beer et al., 2006). In this way, chymosin is produced from recombinant microorganisms that contain the bovine prochymosin B gene (Table 2).

6. Conclusions

Proteases constitute the largest group of enzymes with a wide range of uses in food processes considering proteolysis as a powerful tool in food protein modifications, including changes in solubility, gelation, emulsifying and foaming characteristics, and also nutritional and bioactive properties. Microorganism sources are especially interesting for many reasons, such as relatively easy culture on a large scale, enormous range of variety of microorganisms producing many species of enzymes and the possibility of engineering by site-directed mutagenesis. Fungal cultures are particularly applied to protease production. Although many interesting proteases are highlighted in their ability to transform food, many of them are not listed as GRAS (generally recognized as safe) substances by FDA or could be considered as food additives following general specifications of Joint FAO/WHO Expert Committee on Food Additives (JECFA). Thus, the study of protease applications to food processes is still a large open field, asking for more innovative studies and directed at actual application possibilities.

References

Ahn, C., Jeon, Y., Kim, Y. and Je, J. 2012. Angiotensin I converting enzyme (ACE) inhibitory peptides from salmon byproduct protein hydrolysate by alcalase hydrolysis. Process Biochemistry, 47: 2240–2245.

Anitha, T.S. and Palanivelu, P. 2013. Purification and characterization of an extracellular keratinolytic protease from a new isolate of *Aspergillus parasiticus*. Protein Expression and Purification, 88: 214–220.

Berka, R.M., Ward, M., Wilson, L.J., Hayenga, K.J., Kodama, K.H., Carlomagno, L.P. and Thompson, S.A. 1990. Molecular cloning and deletion of the gene encoding aspergillopepsin A from *Aspergillus awamori*. Gene, 86: 153–162.

Boza, J.J., Jimenez, J., Martinez, O., Suarez, M.D. and Gil, A. 1994. Nutritional value and antigenicity of two milk protein hydrolysates in rats and guinea pigs. Journal of Nutrition, 124: 1978–1986.

Bryan, P.N. 2000. Protein engineering of subtilisin. Biochimica et Biophysica Acta, 1543: 203–222.

Cabanillas, B., Pedrosa, M.M., Rodriguez, J., Muzquiz, M., Maleki, S.J., Cuadrado, C., Burbano, C. and Crespo, J.F. 2012. Influence of enzymatic hydrolysis in the allergenicity of roasted peanut protein extract. International Archives of Allergy and Immunology, 157: 41–50.

Castro, M.A.M. and Martín-Hernández, M.C. 1994. Determination of aminopeptidase activity in cheese. European Journal of Food Research Technology, 198: 20–23.

Cobb, R.E., Si, T. and Zhao, H. 2012. Directed evolution: An evolving and enabling synthetic biology tool. Current Opinion in Chemical Biology, 16: 285–291.

Contreras, M.M., Hernández-Ledesma, B., Amigo, L., Martín-Álvarez, P.J. and Recio, I. 2011. Production of antioxidant hydrolyzates from a whey protein concentrate with thermolysin: Optimization by response surface methodology. LWT—Food Science and Technology, 44: 9–15.

Craik, C.S., Largman, C., Fletcher, T., Roczniak, S., Barr, P.J., Fletterick, R. and Rutter, W.J. 1985. Redesigning trypsin: Alteration of substrate specificity. Science, 228: 291–297.

Cupo, P., El-Deiry, W., Whitney, P.L. and Awad, W.M. 1980. Stabilization of proteins by guanidination. Journal of Biological Chemistry, 255: 10828–33.

D'Amico, S., Gerday, C. and Feller, G. 2001. Structural determinants of cold adaptation and stability in a large protein. Journal of Biological Chemistry, 276: 25791–25796.

DasSarma, S. and DasSarma, P. 2015. Halophiles and their enzymes: Negativity put to good use. Current Opinion in Microbiology, 25: 120–126.

Delange, R.J. and Smith, E. 1968. Subtilisin Carlsberg I. Amino acid composition, isolation and composition of peptides from the tryptic hydrolysate. Journal of Biological Chemistry, 243: 2134–2143.

Devasena, T. (ed.). 2010. Enzymology. Oxford University Press, Oxford.

Eijsink, V.G.H., Gaseidnes, S., Borchert, T.V. and Van den Burg, B. 2005. Directed evolution of enzyme stability. Biomolecular Engineering, 22: 21–30.

Escobar, J. and Barnett, S.M. 1993. Effect of agitation speed on the synthesis of *Mucor miehei* acid protease. Enzyme and Microbial Technology, 15: 1009–1013.

Feller, G., Zekhnini, Z., Lamotte-Rasseur, J. and Gerday, C. 1997. Enzymes from cold-adapted microorganisms. European Journal of Biochemistry, 244: 186–91.

Fernandez-Lahore, H.M., Auday, R.M., Fraile, E.R., Cascone, O., Bonino, M.B.J., Pirpignani, L. and Machalinski, C. 1999. Purification and characterization of an acid proteinase from mesophilic *Mucor* sp. solid-state cultures. Journal of Peptide Research, 53: 599–605.

Fernandez-Lahore, H.M., Fraile, E.R. and Cascone, O. 1998. Acid protease recovery from a solid-state fermentation system. Journal of Biotechnology, 62: 83–93.

Fernandez-Lahore, H.M., Gallego Duaigues, M.V., Cascone, O. and Fraile, E.R. 1997. Solid state production of a *Mucor bacilliformis* acid protease. Revista Argentina de Microbiología, 29: 1–6.

Fontana, A., De Filippis, V., Laureto, P.P., Scaramella, E. and Zambonin, M. 1998. Rigidity of thermophilic enzymes. Progress in Biotechnology, 15: 277–294.

Food and Agricultural Organization. 2014. Available: http://www.fao.org/biotech/fao-statement-on-biotechnology/en/: [02 Jan. 16].

Geisseler, D. and Horwath, W.R. 2008. Regulation of extracellular protease activity in soil in response to different sources and concentrations of nitrogen and carbon. Soil Biology and Biochemistry, 40: 3040–3048.

Georlette, D., Jonsson, Z.O., VanPetegem, F., Chessa, J. and VanBeeumen, J. 2000. A DNA ligase from the psychrophile *Pseudoalteromonas haloplanctis* gives insights into the adaptation of proteins to low temperatures. European Journal of Biochemistry, 267: 3502–12.

Germano, S., Pandey, A., Osaku, C.A., Rocha, S.N. and Soccol, C.R. 2003. Characterization and stability of proteases from *Penicillium* sp. produced by solid-state fermentation. Enzyme and Microbial Technology, 32: 246–251.

Ghorai, S., Banik, S.P., Verma, D., Chowdhury, S., Mukherjee, S. and Khowala, S. 2009. Fungal biotechnology in food and feed processing. Food Research International, 42: 577–587.

Ghribi, A.M., Sila, A., Przybylski, R., Nedjar-Arroume, N., Makhlouf, I., Blecker, C., Attia, H., Dhulster, P., Bougatef, A. and Besbes, S. 2015. Purification and identification of novel antioxidant peptides from enzymatic hydrolysate of chickpea (*Cicer arietinum* L.) protein concentrate. Journal of Functional Foods, 12: 516–525.

Guntelberg, A.V. and Otteson, M. 1952. Preparation of crystals containing the plakalbumin-forming enzyme from *Bacillus subtilis*. Nature, 170: 802.

Gupta, K. and Ayyachamy, M. 2012. Biotechnology of Fungal Genes, CRC Press, Boca Raton, pp. 400.

Hrčková, M., Rusňáková, M. and Zemanovi, J. 2002. Enzymatic hydrolysis of defatted soy flour by three different proteases and their effect on the functional properties of resulting protein hydrolysates. Czech Journal of Food Science, 20: 7–14.

Kheadr-Ehab, E., Vuillemard, J.C. and El-Deeb, S.A. 2003. Impact of liposome-encapsulating enzyme cocktails on Cheddar cheese ripening. Food Research International, 36: 241–252.

Kovaleva, G.G., Shimanskaya, M.P. and Stepanov, V.M. 1972. Site of diazoacetyl inhibitor attachment to acid proteinase of *A. awamori*, an analog of penicillopepsin and pepsin. Biochemical and Biophysical Research Communications, 49: 1075–1082.

Kucera, M. 1981. The production of toxic protease by the entomopathogenous fungus Metarhizium anisopliae in submerged culture. Journal of Invertebrate Pathology, 38: 33–38.

Kumar, C.G. and Takagi, H. 1999. Microbial alkaline proteases: From a bioindustrial viewpoint. Biotechnology Advances, 17: 561–594.

Kumar, S., Sharma, N.S., Saharan, M.R. and Singh, R. 2005. Extracellular acid protease from *Rhizopus oryzae*: Purification and characterization. Process Biochemistry, 40: 1701–1705.

Li, Q., Yi, L., Marek, P. and Iverson, B.L. 2013. Commercial proteases: Present and future. FEBS Letters, 587: 1155–1163.

Li, S., Yang, X., Yang, S., Zhu, M. and Wang, X. 2012. Technology prospecting on enzymes: Application, marketing and engineering computational and structural. Biotechnology Journal, 2: 1–11.

Lin, J.C., Jeon, I.J., Roberts, H.A. and Milliken, G.A. 1987. Effects of commercial food grate enzymes on proteolysis and textural changes in granular Cheddar cheese. Journal of Food Science, 52: 620–625.

Maldonado, J., Gil, A., Narbona, E. and Molina, J.A. 1998. Special formulas in infant nutrition: A review. Early Human Development, 53: S23–S32.

Matthews, B.W. 1988. Structural basis of the action of thermolysin and related zinc peptidases. Accounts of Chemical Research, 21: 333–340.

Merz, M., Appel, D., Berends, P., Rabe, S., Blank, I., Stressler, T. and Fischer, L. 2015a. Batch-to-batch variation and storage stability of the commercial peptidase preparation Flavourzyme in respect of key enzyme activities and its influence on process reproducibility. European Food Research and Technology, 241: 1–8.

Merz, M., Eisele, T., Berends, P., Appel, D., Rabe, S., Blank, I., Stressler, T. and Fischer, L. 2015b. Flavourzyme, an enzyme preparation with industrial relevance: Automated nine-step purification and partial characterization of eight enzymes. Journal of Agricultural and Food Chemistry, 63: 5682–93.

Milesi, M.M., Vinderola, G., Sabbag, N., Meinardi, C.A. and Hynes, E. 2009. Influence on cheese proteolysis and sensory characteristics of non-starter lactobacilli strains with probiotic potential. Food Research International, 42: 1186–1196.

Mohanty, A.K., Mukhopadhyay, U.K., Grover, S. and Batish, V.K. 1999. Bovine chymosin: Production by rDNA technology and application in cheese manufacture. Biotechnology Advances, 17: 205–217.

Mrabet, N.T., Van den Broeck, A., Van den Brande, I., Stanssens, P. and Laroche, Y. 1992. Arginine residues as stabilizing elements in proteins. Biochemistry, 31: 2239–2253.

Narhi, L., Stabinsky, Y., Levitt, M., Miller, L., Sachdev, R., Finley, S., Park, S., Kolvenbach, C., Arakawa, T. and Zukowski, M. 1991. Enhanced stability of subtilisin by three point mutations. Biotechnology and Applied Biochemistry, 13: 12–24.

Neurath, H. 1984. Evolution of proteolytic enzymes. Science, 224: 350–357.

Olempska-Beer, Z.S., Merker, R.I., Ditto, M.D. and DiNovi, M.J. 2006. Food-processing enzymes from recombinant microorganisms—A review. Regulatory Toxicology and Pharmacology, 45: 144–158.

Osman, A., Goda, H.A., Abdel-Hamid, M., Badran, S.M. and Otte, J. 2016. Antibacterial peptides generated by Alcalase hydrolysis of goat whey. LWT—Food Science and Technology, 65: 480–486.

Otte, J., Shalaby, S.M., Zakora, M., Pripp, A.H. and El-Shabrawy, S.A. 2007. Angiotensin-converting enzyme inhibitory activity of milk protein hydrolysates: Effect of substrate, enzyme and time of hydrolysis. International Dairy Journal, 17: 488–503.

Otte, J., Shalaby, S.M.A., Zakora, M. and Nielsen, M.S. 2007. Fractionation and identification of ACE-inhibitory peptides from α-lactalbumin and β-casein produced by thermolysin-catalysed hydrolysis. International Dairy Journal, 17: 1460–1472.

Panyam, D. and Kilara, A. 1996. Enhancing the functionality of food proteins by enzymatic modification. Trends in Food Science and Technology, 7: 120–125.

Perona, J.J. and Craik, C.S. 1997. Evolutionary divergence of substrate specificity within the chymotrypsin-like serine protease fold. The Journal of Biological Chemistry, 48: 29987–29990.

Polgár, L. 2005. The catalytic triad of serine peptidases. Cellular and Molecular Life Science, 62: 2161–2172.

Rawling, N.D. and Barrett, A.J. 1993. Evolutionary families of peptidases. Biochemical Journal, 290: 205–218.

Rawlings, N.D., Barrett, A.J. and Bateman, A. 2012. *MEROPS*: the database of proteolytic enzymes, their substrates and inhibitors. Nucleic Acid Research, 40: D343–D350.

Rawlings, N.D., Morton, F.R., Kok, C.Y., Kong, J. and Barrett, A.J. 2008. MEROPS: the peptidase database. Nucleic Acids Research, 36: D320–D325.

Sahl, H.G., Jack, R.W. and Bierbaum, G. 1995. Biosynthesis and biological activities of lantibiotics with unique post-translational modifications. European Journal of Biochemistry, 230: 827–853.

Sandhya, C., Sumantha, A., Szakacs, G. and Pandey, A. 2005. Comparative evaluation of neutral protease production by *Aspergillus oryzae* in submerged and solid-state fermentation. Process Biochemistry, 40: 2689–2694.

Savitha, S., Sadhasivam, S. and Swaminathan, K. 2011. Fungal protease: Production, purification and compatibility with laundry detergents and their wash performance. Journal of the Taiwan Institute of Chemical Engineers, 42: 298–304.

Schmidt, B.F., Woodhouse, L., Adams, R.M., Ward, T., Mainzer, S.E. and Lad, P.J. 1995. Alkalophilic *Bacillus* sp. strain LC I2 has a series of serine protease genes. Applied and Environmental Microbiology, 61: 4490–4493.

Seratlić, S.V., Miloradović, Z.N., Radulović, Z.T. and Maćej, O.D. 2011. The effect of two types of mould inoculants on the microbiological composition, physicochemical properties and protein hydrolysis in two Gorgonzola-type cheese varieties during ripening. International Dairy Technology, 64: 408–416.

Siddiqui, K.S. and Cavicchioli, R. 2006. Cold-adapted enzymes. Annual Review of Biochemistry, 75: 403–433.

Siezen, R.J. and Leunissen, J.A.M. 1997. Subtilases: The superfamily of subtilisin-like serine proteases. Protein Science, 98: 6301–6523.

Smith, E.L., DeLange, R.J., Evans, W.H., Landon, M. and Markland, F.S. 1968. Subtilisin Carlsberg. V. The complete sequences: comparison with Subtilisin BPN; Evolutionary relationship. Journal of Biological Chemistry, 243: 2184–2191.

Souza, M.J., Ard, Y. and McSweeneya, P.L.H. 2001. Advances in the study of proteolysis during cheese ripening. International Dairy Journal, 11: 327–345.

Souza, P.M., Bittencourt, M.L.A., Caprara, C.C., Freitas, M., Almeida, R.P.C., Silveira, D., Fonseca, Y.M., Ferreira Filho, E.X., Pessoa Junior, A. and Magalhães, P.O. 2015. A biotechnology perspective of fungal proteases. Brazilian Journal of Microbiology, 46: 337–346.

Sumantha, A., Larroche, C. and Pandey, A. 2006. Microbiology and industrial biotechnology of food-grade proteases: A perspective. Food Technology and Biotechnology, 44: 211–220.

Suzuki, Y., Tsujimoto, Y., Matsui, H. and Watanabe, K. 2006. Decomposition of extremely hard-to-degrade animal proteins by Thermophilic Bacteria. Journal of Bioscience and Bioengineering, 102: 73–81.

Tavano, O.L. 2013. Protein hydrolysis using proteases: An important tool for food biotechnology. Journal of Molecular Catalysis B: Enzimatic, 90: 1–11.

Thamnarathip, P., Jangchud, K., Jangchud, A., Nitisinprasert, S., Tadakittisarn, S. and Vardhanabhuti, B. 2016. Extraction and characterisation of Riceberry bran protein hydrolysate using enzymatic hydrolysis. International Journal of Food Science and Technology, 51: 194–202.

Tindbaek, N., Svendsen, A., Oestergaard, P.R. and Draborg, H. 2004. Engineering a substrate-specific cold-adapted subtilisin. Protein Engineering, Design and Selection, 17: 149–56.

Trieu-Cuot, P. and Gripon, J. 1982. A study of proteolysis during Camembert cheese ripening using isoelectric focusing and two-dimensional electrophoresis. Journal of Dairy Research, 49: 501–510.

Tsugita, A. and Scheffler, J.J. 1982. A rapid method for acid hydrolysis of protein with a mixture of trifluoroacetic acid and hydrochloric acid. European Journal of Biochemistry, 124: 585–588.

Tsujita, Y. and Endo, A. 1978. Presence and partial characterization of internal acid protease of *Aspergillus oryzae*. Applied and Environmental Microbiology, 36: 237–242.

Turk, B. 2006. Targeting proteases: successes, failures and future prospects. Nature Reviews, 5: 785–799.

Van Beresteijn, E.C.H., Peeters, R.A., Kaper, J., Meijer, R.J.G., Robbeb, A.J.P.M. and Schmidt, D.G. 1994. Molecular mass distribution, immunological properties and nutritive value of whey protein hydrolysates. Food Protection, 57: 619–625.

van Boekel, M., Fogliano, V., Pellegrini, N., Stanton, C., Scholz, G., Lalljie, S., Somoza, V., Knorr, D., Rao, P. and Eisenbrand, J.G. 2010. Molecular Nutrition and Food Research, 54: 1215–1247.

Van den Burg, B. 2003. Extremophiles as a source for novel enzymes. Current Opinion in Microbiology, 6: 213–218.

Ward, M., Wilson, L.J. and Kodama, K.H. 1993. Use of *Aspergillus* overproducing mutants, cured for integrated plasmid, to overproduce heterologous proteins. Applied Microbiology and Biotechnology, 39: 738–743.

Webb, E.C. 1993. Enzyme nomenclature: A personal retrospective. The FASEB Journal, 7: 1192–1194.

Wilkinson, M.G. and Kilcawley, K.N. 2005. Mechanisms of incorporation and release of enzymes into cheese during ripening. International Dairy Journal, 15: 817–830.

Wu, Q., Du, J., Jia, J. and Kuang, C. 2016. Production of ACE inhibitory peptides from sweet sorghum grain protein using alcalase: Hydrolysis kinetic, purification and molecular docking study. Food Chemistry, 199: 140–149.

Yang, J.K., Shih, I.L., Tzeng, Y.M. and Wang, S.L. 2000. Production and purification of a protease from a *Bacillus subtilis* that can deproteinize Crustacean wastes. Enzyme Microbial Technology, 26: 406–413.

Yokoyama, K., Chiba, H. and Yoshikawa, M. 1992. Peptide inhibitors for angiotensin I-converting enzyme from thermolysin digest of dried bonito. Bioscience, Biotechnology and Biochemistry, 56: 1541–1545.

10

Microbial Transglutaminase and Applications in Food Industry

Marek Kieliszek* and Stanisław Błażejak

1. Introduction

Transglutaminase (TGase) belongs to the transferase class of enzymes (EC 2.3.2.13) and is the common name of protein-glutamine γ-glutamyltransferase (Ando et al., 1989). The term TGase was first introduced by Mycek et al. (1959) to differentiate the enzyme that catalyzes transfer reactions involving the glutamine carboxamide group from other enzymes catalyzing transfer or hydrolysis reactions of amide groups, wherein free glutamine, such as γ-glutamyltransferase participates.

Transglutaminase participates in the reaction that catalyzes the formation of isopeptide bond between glutamine residue of γ-carboxamide (donor) and primary ε-amine groups of various compounds, for example, proteins (acyl residue acceptor) (Fig. 1a). The result of this reaction is the transfer of acyl group from one substrate to the other, accompanied by the release of ammonia. This reaction enables attachment of hydrophobic or hydrophilic groups to a protein particle. When the acyl acceptor is lysine, the protein becomes enriched with this amino acid. Acyl transfer to the lysine residue bonded in a polypeptide chain results in cross-linking, that is, formation of inter- or intramolecular cross-links of ε-(γ-glutamylo) lysine (Fig. 1b) (Kieliszek and Misiewicz, 2014). The isopeptide bonds thus formed contribute to the formation of stable protein networks, which enable the formation of inter- and intramolecular cross-links. These inter-protein bonds lead to the formation of high-molecular-weight polymers, resulting in the occurrence of the process of aggregation or even gelation. Intramolecular bonds are characterized by a compact structure with reduced hydrodynamic radius (Djoullah et al., 2015).

Faculty of Food Sciences, Department of Biotechnology, Microbiology and Food Evaluation, Warsaw University of Life Sciences – SGGW, Nowoursynowska 159 C, 02-776 Warsaw, Poland.
* Corresponding author: marek-kieliszek@wp.pl; marek_kieliszek@sggw.pl

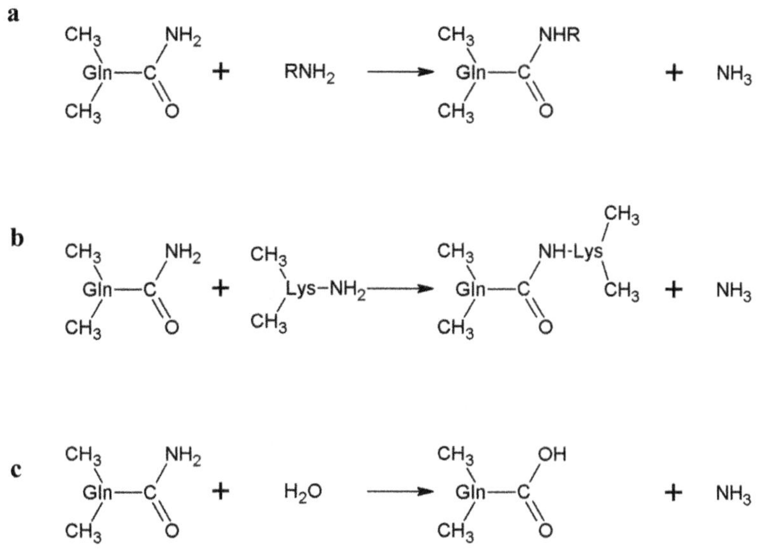

Figure 1. Reactions catalyzed by transglutaminase (TGase): (a) acyl-transfer reaction; (b) cross-linking reaction between Gln and Lys residues of proteins or peptides (c) deamidation.

When free amine groups are lacking, TGase participates in a deamination reaction (Fig. 1c) (Kuraishi et al., 2001). The reactions catalyzed by this enzyme bring about significant changes in the physical and chemical characteristics of proteins. These changes consist in the change of viscosity, thermal stability, elasticity and flexibility of a protein (Kieliszek and Misiewicz, 2014). Independent of the deamination or polymerization processes, the use of TGase affects the structure of protein systems. This causes changes in their functional properties during the process of production of different food products characterized by innovative textural properties (Gaspar and de Góes-Favoni, 2015).

2. Transglutaminase Characteristics

Transglutaminase (EC 2.3.2.13) is a natural enzyme commonly found in animal tissues and intercellular fluid (Fig. 2). Initially, it was presumed that a single enzyme is responsible for TGase activity; but today it is known that such activity is exhibited by a whole group of enzymes that show similarity in the type of the reactions catalyzed, but possibly differ in substrate specificity, expression method and physiological regulation. Microbial transglutaminase (mTGase) was isolated for the first time in 1989 from a strain of *Streptoverticillium* sp. The enzyme is a single-chain protein with molecular weight of approximately 38 kDa built of approximately 331 amino acids (Duran et al., 1998; Pasternack et al., 1998; Yokoyama et al., 2004). Active center of TGase is constituted by the residue of cysteine, histidine and aspartic acid or asparagine. Numerous factors affect the enzymatic activity of TGase. Maximum activity is exhibited at a temperature of approximately 50°C and acidic pH of 5–6. Enzyme inactivation takes place after 5 min. at a temperature of 75°C. The optimal

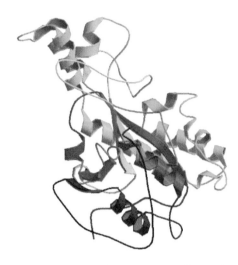

Figure 2. Structure of microbial transglutaminase from *Streptoverticillium mobaraense* strain (Kashiwagi et al., 2002; Kieliszek and Misiewicz, 2014; www.rcsb.org, with permission).

temperature favoring catalytic activity of TGase isolated from the *Streptoverticillium ladakanum* strain is 40°C (Ho et al., 2000). One of the exceptions is TGase isolated from *Streptomyces* sp., which acts most efficiently at slightly higher temperature of 45°C. At temperatures close to zero, the enzyme maintains its full activity (Kieliszek and Misiewicz, 2014; Yokoyama et al., 2004).

Presence of maltodextrin, sucrose, mannose, trehalose and reduced glutathione (GSH) in the media increases the thermal stability of the enzyme (Cui et al., 2006). Casein can protect TGase against degradation of extracellular proteolytic enzymes (Junqua et al., 1997). One activity unit of mTGase is defined as the amount of enzyme necessary for the formation of 1 μM hydroxylamine acid in 1 min. at a temperature of 37°C.

Enzymes synthesized by bacteria are stable in a broad range of pH from 4,5 to 8,0. In addition, contrary to TGases of animal origin, mTGase does not need calcium ions for activation. This characteristic is particularly desirable in practical use for an enzymatic preparation. TGase activity is higher in the presence of Co^{2+}, Ba^{2+}, and K^+ ions. Inhibitors of TGase activity include Zn^{2+}, Cu^{2+}, Hg^{2+}, and Pb^{2+} ions, which bind to the thiol group of cysteine found in the active center (Kieliszek and Misiewicz, 2014; Macedo et al., 2010).

Transglutaminase catalyzes the cross-linking reactions in proteins, leading to a stabilizing effect due to the formation of covalent bonds, which possess characteristics different to peptide bonds. Under the effect of this enzyme's activity, disulfide bridges stabilize the structure and improve the stiffness of a molecule. They also participate in cross-linking of neighboring chains or in the formation of polypeptide chain loop. This leads to changes in protein conformation; hence in modification of structure, gelation stability and water-binding ability, which, as a consequence, results in changes of rheological properties of protein products. These characteristics, along with the fact that the scientific community recognized the enzyme in 1998 as a safe substance (it has the GRAS status—Generally Recognized as Safe) by the FDA (Food and Drugs Administration), makes it a very attractive product for the food industry (Gaspar and de Góes-Favoni, 2015).

3. Transglutaminase Production Technologies

Developments of biotechnology in recent years enabled biosynthesis, isolation and improvement of enzymes that have a large significance in different branches of the food industry. Solutions for many previous technological problems have been found and ways for exciting possibilities have been established. In the 1970s, in Guinea, a technology of TGase production from the liver of guinea pig was developed (from 1 kg of liver only 230 mg of pure enzyme was obtained) (Folk and Chung, 1973; Kuraishi et al., 2001). However, due to the source of the enzyme and relatively expensive isolation and purifying method of the enzyme, it constituted an obstacle for its wide use in the industry. Modern science provides a wide spectrum of possibilities in the areas of biotechnology, enzymology and enzyme use at industrial scale. Researchers (Liu et al., 2014; Yu et al., 2008) were searching for new sources of the enzyme, which would facilitate the process of production of various food products. Scientists hope to develop better and less expensive sources of the enzyme with the use of microorganisms that will help to utilize their biotechnological abilities in human economy.

Transglutaminases are commonly used in the modern food technology. Their properties can be used to the benefit of both the food industry as well as the consumer. Selectivity of their effect improves control of product formation, while high efficiency and low energy requirements are beneficiary for the environment.

Microbiological media used for *Streptoverticillium* cultivation are not attractive from the economic point of view due to the requirement of high amount of expensive nutritional components, such as yeast extract and peptone (Téllez-Luis et al., 2002, 2004). Glucose, sucrose, starch and glycerol in the role of carbon source can be used for TGase biosynthesis. Literature provides information on the possibility to use raw materials of plant origin as nitrogen source, such as NH_4NO_3, $(NH_4)_2SO_4$, urea, $NaNO_3$, NH_4Cl, soybean, corn, wheat or wheat flour, rice, bran, corn steep liquor, essential minerals, and trace elements like phosphate, magnesium, potassium, iron, copper, zinc, and vitamins (Zhu et al., 1995). Numerous studies (Herrera et al., 2003; Tellez-Luis et al., 2002) demonstrate the issues involved in the use of agricultural waste as carbon source in TGase production. Xylose is a hemicellulose sugar, which can be used as a potential carbon and energy source for the growth of microorganisms. The interest in the use of xylose as a source of carbon for bacteria proliferation can be even greater, if the media are prepared using inexpensive raw materials, such as hemicellulose hydrolysates like sorghum straw (Téllez-Luis et al., 2002; Kieliszek and Misiewicz, 2014).

Peptone, yeast extract and casein are common nitrogen sources used in TGase biosynthesis (Ando et al., 1989; Gerber et al., 1994; Zhu et al., 1995). Ammonium salts turned out to be less favorable sources of nitrogen (Zhu et al., 1995). Bourneow et al. (2012) demonstrated that peptone turned out to be the best nitrogen source for TGase production from *Streptomyces* sp. P20 and *Streptoverticillium mobaraense* strains. Zhu et al. (1996), conducting a study on the optimization of medium composition, determined that introduction of additional compounds in the form of a proper set of amino acids to the medium containing peptone as the main nitrogen source caused a significant increase of TGase production by *Sv. mobaraense* strain (Kieliszek and Misiewicz, 2014).

Amino acids play an important role in the synthesis of mTGase. Precise mechanism of biosynthesis has not yet been entirely understood and this stage requires further study. Use of non-modified peptides limits synthesis of the enzyme by microorganisms (Zhu et al., 1995).

Microbial transglutaminase is an extracellular enzyme exhibiting the ability to dissolve in the culture medium. Methods commonly used in the process of enzyme purification are also used for the purification of mTGase from the medium. Examples include ethanol, acetone and isopropyl alcohol. Methods used for the purification of mTGase include ammonium sulfate, sodium chloride, dialysis processes, ultrafiltration, ion exchange chromatography, gel filtration, absorption and a method based on the isoelectric point analysis. Oftentimes, enzyme purification requires the use of more than one method and sometimes there is a need for combining some of the processes, which, as a consequence, may increase the mTGase recovery effect.

The process of expression and purification of TGase was performed using the following strains: *Streptomyces lividans* (Lin et al., 2004; Washizu et al., 1994), *Escherichia coli* (Yokoyama et al., 2000) and *Corynebacterium glutamicum* (Date et al., 2004). Currently, the enzyme is produced using *Streptomyces mobaraense* or *Bacillus subtilis* (Liu et al., 2014). Recently, a study was conducted on the production of recombinant TGase from the strain of *E. coli* (Yu et al., 2008). However, it should be emphasized that the processes of TGase translation as an active enzyme can have disastrous effects on microorganisms. Induction of mTGase in *E. coli* cells containing synthetic gene coding for the enzyme can influence inhibition of cell growth, followed by cell lysis. Cell death is apparently caused by the occurrence of cross-linking reactions between cytosol proteins in the presence of TGase (Pasternack et al., 1998).

The enzyme thus obtained can be bonded with enzyme stabilizers, such as various salts, sugars, proteins, lipids or surfactants (Zhu et al., 1995) in the subsequent stages. In the context of the above applications, it is justified to create more effective TGase production systems, which would be used in the food industry. The published literature does not provide many studies on elaboration of a commercial production process, improvement of the procedure, or efficiency of the whole enterprise related to the production of TGase.

In order to improve TGase production efficiency, studies were conducted on heterologous expression of recombinant genes responsible for mTGase production in the following microbes: *E. coli*, *S. lividans*, *C. glutamicum* and methylotrophic strains of yeast. mTGase isolated from the strain *Streptomyces* is naturally synthesized in the form of a zymogen (pro-TGase) which is then processed to obtain its active form by deleting its N-terminal peptide. Liu et al. (2015) conducted a study on the transformation of cloned gene from the strain *Streptomyces hygroscopicus* responsible for pro-TGase production to yeast strain *Yarrowia lipolytica* Po1h. The study demonstrated that the strain *Y. lipolytica* Po1h, which has vector pINA1297, when cultivated on a medium with glycerol in a 3 L bioreactor, exhibited enzymatic activity of TGase (N355Q) of 35,3 U/mL, whereas mTGase activity of the enzyme obtained from the wild strain amounted to only 5,3 U/mL. This study provides new possibilities for the production of more efficient mTGase with increased enzymatic activity, which can be used in food processing.

Lin et al. (2007) elaborated a rapid and relatively simple system for purifying recombinant TGase from the strain *S. lividans* 25-2. Purification of TGase was conducted on a laboratory scale, resulting in the enzyme with a purity of 90–95 per cent and with the enzymatic activity of 61–65 U/mg. The technique followed for enzyme purification in this study ensured high recovery. Data presented by these authors proved that the recombinant mTGase activity was 3,3 times higher than that obtained from the strain *Streptomyces platensis* M5218 (Lin et al., 2006). In other strains, such as *S. lividans* 25-2 (78,2 mg/L), the obtained efficiency was significantly better than that observed for *S. lividans* 3131-TS (< 0,1 mg/L) (Kieliszek and Misiewicz, 2014; Washizu et al., 1994).

A study by Junqua et al. (1997) demonstrated that the addition of casein to the medium containing 38,4 g/L and 31,2 g/L of glycerol increased mTGase production by three times (0,331 ± 0,038 U/mL) by the strain *Streptomyces cinnamoneum* CBS 683.68. On the other hand, Itaya and Kikuchi (2008) demonstrated that strains of *Corynebacterium ammoniagenes* were able to synthesize more amounts of the enzyme TGase than the bacteria *C. glutamicum*, which are commonly used for the biosynthesis of the enzyme (Yokoyama and Kikuchi, 2004). These species are widely used in commercial production of amino acids, such as lysine and glutamate, which are used in the food industry.

Transglutaminases, due to their ability to modify physical and chemical properties of proteins, are used in a variety of ways in different branches of the industry. For large-scale production of TGases, the bacterial expression system of *S. mobaraense* is primarily used. However, the system has its disadvantages, that is, it has problems related to the post-translational protein modification (Griffin et al., 2002). In the context of the above limitations, studies on a considerably less expensive and more efficient system should be conducted to allow a reduction in the costs related to distribution, storage, extraction and purification of the recombinant proteins (Kieliszek and Misiewicz, 2014).

Microbial transglutaminase is used as a biological glue in many fields devoted to biomedicine and biotechnology. For continuous development in this field, continuous enrichment of the knowledge on future aspects of the use of the enzyme in other industrial branches is indispensable; most likely it is one of the most challenging areas of study in the field of biochemistry and biotechnology. Constant development of new technologies for biosynthesis of the enzyme and in the field of molecular biology can lead to increased TGase production, which eventually may lead to direct distribution of the enzyme through many food products. Such efforts enable adjustment and creation of the needs of the market and economy. Among a variety of factors that contribute to the development, a very important role is played by the consumer and his needs. By 2020, a further increase in TGase consumption is anticipated from 0,22 to 0,32 times a year, where daily mTGase dosage may amount to 15 mg. TGase use in the restructuring or dosage amounts to approximately 50–100 mg of the enzyme per one kilogram of food (Lerner and Matthias, 2015). The great scale of interest in mTGase is reflected in the increasing number of patent applications on biosynthesis and use of the enzyme in various food products.

4. Use of Transglutaminase in the Food Industry

Transglutaminase preparations have potentially wide spectra of use (Fig. 3). They are very popular due to their use in the food industry for protein cross-linking (Buettner et al., 2012; Giosafatto et al., 2012; Kashiwagi et al., 2002; Pinterits and Arntfield, 2008; Zheng et al., 2002). The other field of use is represented by protein production, for example, casein films (Buettner et al., 2012; Dong et al., 2008).

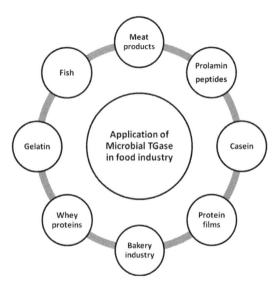

Figure 3. Transglutaminase and its use in food processing.

Transglutaminase can affect various food characteristics, such as texture, solubility, viscosity, gelation capacity and water retention capacity, which are reflected in various textural properties of the obtained food products (Table 1). TGase turned out to be very useful in this technology as it acts in a different direction than the majority of enzymes used in the food technology—it does not divide the media into smaller subunits but creates larger particles from the protein units *via* the cross-linking reaction and bond formation. Strong cross-linking in food products (i.e., pieces of meat) is maintained even during the process of cutting, freezing, or heating (Gaspar and Goes-Favoni, 2015; Lerner and Matthias, 2015).

The obtainment of products with characteristic quality is possible due to modifications of the food components, with both chemical as well as enzymatic methods. TGase of microbial origin (mTGase) belongs to the group of enzymes on which the food industry is currently focused. TGase forms larger particles from small

Table 1. Application of microbial transglutaminase in food processing.

Source	Product	Effect	Reference
Casein	Cross linked protein, mineral absorption promoters	Improved mineral absorption in intestine	Noguchi et al. (1992) Lauber et al. (2000) Ozer et al. (2007)
Milk	Cream, desserts, milk drinks, dressings	Improved quality and texture	Lauber et al. (2000) Şanlı et al. (2011)
Meat	Restructured meat, hamburger, meatballs dumplings canned meat frozen meat	Restructured meat texture, appearance, increased hardness	Zhu et al. 1995 Kuraishi et al. (1997) Motoki and Seguro (1998) Trespalacios and Pla (2007)
Wheat	Baked foods	Improved texture and high volume	Gerrard et al. (2001)
Gelatin	Sweet foods	Low calorie foods with good texture and elasticity	Giosafatto et al. (2012)
Fish	Fish paste restructured product	Increased hardness	Téllez-Luis et al. (2002)
Fat, oil	Solid fats	Pork fat substitute with good taste, texture and flavor	Zhu et al. (1995)
Soya bean	Mapo doufu tofu	Improved shelf-life and texture	Zhu et al. (1995) Kato et al. (1991)
Collagens	Shark-fin imitation	Imitation of delicious food	Zhu et al. (1995)
Plant proteins	Protein powders	Gel formation with good texture and taste	Zhu et al. (1995)

protein units *via* the reaction of cross-linking and bond formation. Use of TGase in the production of meat products influences the improvement of texture and color and some of the sensory quality factors of the food product.

4.1 Increasing shelf-life of fish products

For commercial purposes, the enzyme is produced through microbiological culturing of *S. mobaraensis*. mTGase is significantly cheaper, with activity comparable to that obtained from the animal origin. Examples of food products obtained with the use of mTGase are fish balls and fish fingers, surimi and products imitating crab meat and shark fins (Tani et al., 1990). Fish paste products are obtained from material containing primarily fish meat and 0,1–700 U TGase/g of fish proteins. The final product can be kamaboko, a Japanese fish paste, characterized by good texture and white color (Zhu et al., 1995).

One of the methods of improving shelf-life of the food product is by increasing the number of covalent bonds of the food matrix. In the case of products containing proteins, it can be obtained using TGase. This enzyme turned out to be useful in the production of various protein components and food products. Proteins are good substrates for TGase. In the production of surimi, stiff gels are produced by cross-linking of fish proteins (Benjakul et al., 2001). Gels can be produced from dissolved and dispersed proteins, colloidal systems, proteins covering interface surfaces, such as oil–water and air–water interfaces (Dłużewska and Florowska, 2014). Furthermore,

Kołodziejska et al. (2004) determined a positive effect of mTGase on gelating properties of gelatin from cod skin. The influence of mTGase on the durability of fish meat gel depends on the denaturation and degradation of the myofibrillar proteins (Jiang et al., 2000). Therefore, further studies on determining the efficiency of mTGase bonding in the products of fish muscles are necessary, an example of which can be Nile tilapia (*Oreochromis niloticus*) (Monteiro et al., 2015).

4.2 In production of tofu

The use of mTGase in the production of tofu is an example of practical use of the enzyme (Kato et al., 1991). Tofu is soybean curd produced in the process of soy milk coagulation. TGase added to soy milk with a coagulant (nigari) produces tofu with more compact texture and slows down the process of coagulation, facilitating its control. The ability to gelate tofu from old beans gets worsened by adding TGase. It is possible to obtain the same gel as with the use of young soy beans. It was determined that soy protein hydrolysates when subjected to enzymatic polymerization maintained their solubility in spite of being built of fractions with higher molecular mass as compared to the native protein. Furthermore, emulsifying and foaming properties of these proteins were improved with simultaneous elimination of sour flavor of the hydrolysate (Gaspar and de Góes-Favoni, 2015). Gels produced with the participation of TGase are more elastic and tougher than the thermally induced gels. Thus, a possibility exists to control their texture *via* the change of enzyme dozing method and production conditions. The addition of sodium chloride to soy protein isolate gels significantly decreases the value of breaking stress, whereas the addition of even 3 per cent salt to gels produced by TGase does not change their properties (Dłużewska and Florowska, 2014). Mapo-doufu (tofu, traditional meal of the Sichuan province in China) produced with the addition of mTGase exhibits good texture, taste and appearance even after six months of storage at 25°C and relative humidity of 60 per cent. Other methods to improve taste, texture, appearance and shelf-life were proposed by Nonaka et al. (1990).

In the food industry, endogenous animal TGase or TGase preparations of animal and microbial origin can be used. Enzymes of animal origin depend on the presence of metal ions, for example, calcium. In the absence of metal ions, the enzyme acts inefficiently or does not exhibit any activity at all. Metal ions not only influence the enzyme activity but also stabilize its structure; they may also be involved in the activation or inhibition of the activity. However, it is not feasible to add calcium salts to all the food products, particularly at concentrations required by the enzyme; mTGase does not require the presence of calcium ions. Therefore, production of commercial mTGase preparations significantly broadened the possibilities of its use in the food industry. Since 1989, TGase isolated from *Streptoverticillium* sp. has been used.

4.3 Production of restructured meat

Transglutaminase is used in the production of restructured meat (Monteiro et al., 2015). The use of TGase preparations—apart from positive effect on the texture of the finished product—enables strong binding of a meat block without the need for heat treatment and without the addition of common salt or phosphates (Kuraishi et al., 1997). Technological creation of characteristic quality of meat products depends mainly on the type and amount of raw components used and on the addition of different functional substances, like, salts (Hong et al., 2014). Phosphates are responsible in increasing water absorption by the muscle tissue to minimize fat leakage and to improve the binding properties of the finished product. A decrease in salt or phosphate addition to sausages leads to obtaining a product with lower water absorption and decreased texture and consistency. Addition of TGase is an alternative for the use of these additives in the meat product technology. Regardless of favorable technological properties, the use of phosphates in meat processing may raise many health concerns. The main issue may be their effect on human calcium metabolism (Pyrcz et al., 2014). TGase enzyme preparations produce similar effects on phosphates, particularly in the area of improved binding properties and consistency in meat preparations. The addition of TGase, at lower concentrations, to sausages eliminates unfavorable changes because bonds formed by the enzyme are thermostable, which as a result leads to the improvement in texture of preserved food products.

Sakamoto and Soeda (1991) elaborated a method for manufacturing food products containing minced meat with TGase. Minced meat and other food components were mixed with TGase and then packed in pressure-resistant containers. Hamburgers, meat balls and dumplings can be the final products. Manufactured food exhibits improved elasticity, texture and more intense taste and smell. Similar methods of TGase use in meat products can be found in the literature (Seguro and Motoki, 1991; Zhu et al., 1995).

The use of TGase in the meat-processing industry allows improvement in the texture of the finished product, which is exhibited by improved hardness. The texture of poultry sausages with addition of TGase improved and was comparable to that of pork sausages (Cegiełka and Geryk, 2007). Spatial bonds formed as an effect of TGase action strengthen the meat protein structure, improve their water absorption and facilitate sausage slicing. The use of TGase in the meat industry has made possible the use of less valuable raw material, that is, collagen, blood proteins and mechanically separated meat, for the manufacture of meat products with increased nutritional value via enriching with the lack in the amino acids (e.g., exogenous lysine) (Kowalski and Pyrcz, 2009). These components can be subjected to enzymatic modifications; even other substances introduced into meat products can be subjected to similar processes. In order to give the texture typical to meat (fibrousness, juiciness), textured soy proteins were modified with the use of TGase (Gaspar and de Góes-Favoni, 2015; Weng and Zheng, 2015).

The use of mTGase created new technological possibilities on the production of fine and coarse minced sausages, wieners and smoked meats. To obtain these products, low-grade meat as well as additives, such as skimmed-milk powder, soy and wheat flour can be used. As a result of TGase effect on the proteins of these raw materials, products which do not differ from the analogous products manufactured from high-quality meat in their appearance, consistency, taste, smell and nutritional value (Kuraishi et al., 1998) are obtained. In order to improve meat value, salt extracts of muscle proteins or alkaline reagents are used, but heat treatment is necessary. An alternative can be found in the use of TGase, which does not require basic salting, freezing, or heating processes. TGase also contributes to maintaining the original taste and aroma of raw materials with improved efficiency. A better texture of a product is obtained by simultaneous use of TGase and sodium caseinate (Mirzaei, 2011). However, the bonds formed were not strong enough to protect the raw, restructured meat from disintegration during heat treatment (Dłużewska and Florowska, 2014). Use of TGase allowed production of some sausage types with reduced fat content; for example, salami type. The products obtained with the addition of filler do not differ organoleptically from traditional meat products (Kieliszek and Misiewicz, 2014; Nielsen, 1995).

Transglutaminase has been used for the manufacture of a whole range of poultry meat products. Possible benefits related to this, apart from modification of the texture of meat, are increase of firmness, effect on water retention by the finished product and ability to change the appearance and taste qualities of products manufactured from the most valuable parts of poultry carcass, that is, the breast muscles. In order to use small meat pieces and to improve their commercial value, they may be restructured. The conventional method consists in binding meat by using salt extracts of the muscle proteins or alkaline reagents, but such preparations require heat treatment. Restructured poultry meat products manufactured with the use of TGase preparations can at the same time be marinated and then offered to the customers in various forms (de Almeida et al., 2015; Trespalacios and Pla, 2007; Tseng et al., 2000).

4.4 Improving functional characteristics of milk products

The study of Monogioudi et al. (2011) demonstrated that β-casein enzymatic cross-linking is more resistant to pepsin digestion as compared to noncross-linked β-casein form. Giosafatto et al. (2012) demonstrated that the obtained results may have a significant effect on the development of new food structures with better characteristics (Giosafatto et al., 2012). Polymerization of milk proteins *via* the effect of TGase may lead to the formation of protein film, which increases functional characteristics of milk products (Rossa et al., 2011). According to Hiller and Lorenzen (2009), cross-linking is a dominant process, which leads to the formation of ε-(γ-glutamyl)-lysine within and between isopeptide chains. The degree of cross-linking depends on the protein structure. More effective cross-linking takes place in proteins containing glutamine residues in the elastic areas of a protein. Thus casein constitutes a better substrate than ovalbumin and β-lactoglobulin. Furthermore, denaturation and adsorption on the oil-water interface increase the protein reactivity (Dłużewska and Florowska, 2014; Kieliszek and Misiewicz, 2014).

In the dairy industry, TGase has been used in a variety of products, for example, in yogurts, to prevent syneresis or improve structure (more firm with soft consistency) (Lorenzen et al., 2002). As a result of casein modification with TGase, it is possible to produce a range of new dairy products with their characteristic structure and consistency. TGase reactivity with casein decreases in the following order: κ-casein > α-casein > β-casein (Tang et al., 2005). However, it should be emphasized that the reactivity depends on the source of the enzyme. Cross-linked casein is characterized by better gelating and emulsifying properties and may be widely used as a functional additive in the food industry (de Kruif et al., 2002).

A recent study by Prakasan et al. (2015) showed that the addition of TGase to *paneer* (cottage cheese) results in improvement of textural properties and water retention with a simultaneous minimization of heat treatment. *Paneer* is a traditional Indian dairy product that is popular among vegetarians. It is valued for its nutritional and health advantages. It primarily consists of solid milk substances. The study suggests that *paneer* exhibiting improved functional characteristics can be manufactured using enzymatic cross-linking method with mTGase.

The use of cross-linking enzymes allows for addition of proteins and polysaccharides or reduction of dry substances without changes in texture and the property of water retention. Yogurts obtained from milk incubated with TGase are manufactured using this method (Iličić et al., 2014; Ozer et al., 2007). They are characterized by a uniform, compact and very creamy consistency and smooth, dry surface of the coagulum. It is a result of the phenomenon of syneresis (Lorenzen et al., 2002). Yogurts manufactured by these methods are used in the preparation of creams, various frozen desserts, ice creams, milk drinks and dressings (Lauber et al., 2000; Şanlı et al., 2011). The enzyme can be added to milk before its heat treatment, which inactivates TGase, than adding a starter culture to initiate the fermentation process. The result of TGase action is increase in compactness and viscosity of a gel, improvement of water retention, and eventually, reduction of syneresis.

Currently, protein cross-linking processes by TGase are drawing a lot of attention. A study showed (Liu et al., 2014) that formation of ε-(γ-glutamyl)-lysine may improve water retention and stabilize the three-dimensional network of yogurt gel (viscosity), reducing separation of whey proteins. However, on the other hand, the pore size of the milk gel catalyzed by TGase is decreased and this may affect the rate and extent of the syneresis process and result in the release of liquid from it (Li et al., 2015). It should also be noted that whey proteins in their native structure are globular and less susceptible to cross-linking reaction, mainly due to the stabilization of globular conformation with disulfide bonds. Cross-linking of whey proteins can be improved by a prior denaturation step followed with heat treatment (Iličić et al., 2014).

Study by Mahmood et al. (2009) on the improvement of efficiency and properties of a soft cheese showed that TGase addition prior to the addition of rennet prevents milk coagulation. On the contrary, by adding the enzyme and rennet at the same time, the durability and hardness of cheese, as well as protein and fat level in whey, decrease considerably. Cheeses with TGase addition are characterized by a less dry structure. Furthermore, such cheese contains more whey proteins. Milk ice creams, particularly low caloric and sugar-free, modified with TGase have smoother consistency. The effect of the enzyme action is similar to that caused by the addition of ice-cream stabilizer—lower syneresis is observed, better water retention and lower susceptibility to excessive formation of ice crystals (Dłużewska and Florowska, 2014).

4.5 In baking industry

Transglutaminases are currently used in the bakery technology to create bonds between prolamin polypeptide chains. First observations of baking a dough with the use of TGase were performed by Gottmann et al. (1992). As a result of the study, it turned out that TGase affected the stability and volume of the dough while improving the quality of baking flours of poor quality, leading to the improvement of bread texture (Kieliszek and Misiewicz, 2014; Marco and Rosell, 2008). Losche (1995) noted that TGase improve the rheological properties of the dough, ensuring proper size of pores and elasticity of bread after baking. Addition of TGase to dough improves the mechanical and water absorption properties and reduces the production costs. In fried cereal products, it is possible to obtain lower content in the finished product.

Addition of TGase strengthens the dough structure, which may be used in the production of frozen doughs, because the dough then is more resistant to degradation associated with the freezing process (Kim et al., 2008). In the bakery industry, TGase is used to improve the volume and texture of bread and durability of flours (Moore et al., 2006). The objective of TGase addition to gluten-free bread, with the participation of proteins other than gluten, is to obtain the spatial structure produced by this protein. Recently, numerous studies have been published on the effect of TGase on the quality of gluten-free bread (Dłużewska and Florowska, 2014). It was shown that the effect of this enzyme depends not only on its concentration in the dough, but also on the type of proteins added (Kuraishi et al., 2001). Various studies on modification of gluten proteins with the use of TGase, the objective of which is to reduce the allergenicity of gluten, were also conducted. As a result of TGase modification of wheat flour proteins, there was an improvement in the dough elasticity and flexibility. At the same time, a 14 per cent increase in the volume of the bread was comparable to the bread baked from the traditional dough (Gerrard et al., 2001). TGases are also used in enriching prolamins with lysine or other exogenous amino acids or fructooligosaccharides. From the nutritional viewpoint, rice flour contains many valuable nutrients, such as protein, dietary fiber and vitamins E and B. However, its use is limited to non-fermented bakery products. The study by Gujral and Rosell (2004) showed that the addition of TGase to rice flour improves the rheological characteristics of the dough. Changes in the amount of TGase added facilitated the control of pasta texture. The bonds formed as a result of cross-linking reactions in the presence of TGase are thermally stable, which in consequence leads to compactness and elasticity of pastas long after their cooking. Texture of such pastas does not change as a result of heat treatment (Yamazaki and Nishimura, 2002). In dough strengthened with TGase, starch remains to a greater extent on an expanded gluten network. Therefore, in the process of heat treatment, lesser amounts of dry mass components are released. As a result, the amount of starch lost is low, whereas the surface of pastas exhibits lower viscosity (Mariniello et al., 2008). In instant noodles, TGase addition may cause the product to absorb less fat during frying.

Kuraishi et al. (2001) studied the effect of TGase addition on fat absorption during donut frying. Donuts without the addition of TGase contained 18,2 per cent fat, whereas donuts with TGase contained 13,8 per cent. The value of fat absorption was reduced by 25 per cent. Similarly, use of TGase as a component of breadcrumbs may cause deep fried cutlets to absorb less fat and keep their energy value at a lower level. The addition of TGase strengthens dough structure, which may be used for production of doughs that are frozen prior to baking. This makes the dough less susceptible to freezing-related damage (Marapana and Jiang, 2004). The results presented by Li et al. (2015) demonstrated that recombinant TGase isolated from *Pichia pastoris* GS115 can be used in the food industry. It was determined that enzymatic cross-linking in the dairy proteins by the enzyme significantly improves the functional characteristics of fat-free yogurts. In addition, it was noted that properties of products with lower fat content can be modified with the use of the enzyme.

4.6 In fruits and vegetable processing

Transglutaminase enabled the creation of new products, for example, protein coatings used to coat fresh vegetables, fruits and processed food products in order to improve their shelf-life and freshness (Porta et al., 2015). One of the most important edible coating functions is the protection of fruits against moisture losses, improvement of shelf-life and reduction of losses related to transport, distribution and sale. Takagaki et al. (1991) described a method of coating vegetables and fruits with the addition of TGase. Freshness of vegetables and fruits could be preserved by coating with a film containing proteins and TGase. As an example, celery was treated with an aqueous solution containing TGase, gelatin and natural antibacterial substances, and then heated at a temperature of 50°C for 5 min. to form the coatings. Coated celery was stored at 20°C for three days. The obtained celery was characterized by a lower number of microbes than the product without coating (Zhu et al., 1995).

For the production of such coatings, whey protein modified with TGase is used. These coatings are edible and can be consumed with a product. In addition, depending on the technology used, they exhibit various properties such as water permeability, elasticity, flexibility, tensile strength, and resistance to mechanical damage (Yildirim and Hettiarachy, 1998). Natural coatings produced with the addition of TGase slow down the vital processes in plants, thus contributing to their longer shelf-life. They have protective functions, regulate gas exchange, protect against penetration of harmful substances from the environment (pathogens, contaminations) and protect the tissue against damage. Use of TGase in the production of coatings aims at reduction of technological losses, enrichment of products with protein (improving nutritional value) and ensuring the consistency of the products. When producing edible coatings with the use of various proteins, it is very important to understand the course of the gelating process and the composition and properties of the obtained gels, which may retain water molecules, lipids, or other substances favoring coating formation.

4.7 As protein modifier

In numerous branches of the food industry, TGase is used as a protein modifier to improve nutritional values of substandard proteins *via* embodying them with desired amino acids and peptides. Noguchi et al. (1992) elaborated a method for increasing the absorption of mineral components in humans. It is connected with the process of casein deamination *via* treatment with TGase. The obtained product increases absorption of mineral components in the intestines and can be used in the food industry or pharmaceutical industry for the production of medicines or mineral supplements for adults, children and infants. Casein structure can store mineral components or other chemical elements before they are transported to the intestine and dissolved. TGase is also used to block allergenic, proteolysis-resistant peptides of soy proteins (Babiker et al., 1998). Products of protein modifications formed as a result of the action of the enzyme are used in the cosmetic, pharmaceutical and leather industries (Nielsen, 1995; Zhu et al., 1995).

5. Conclusion

Discovery of a non-expensive TGase source in the form of microbes, which are able to biosynthesize the enzyme, and practical applications of the enzyme in many branches of the food industry are important for meeting customer expectations. The essential step in enzyme preparation is its production from the culture medium in a purified form. TGase acts in mild conditions and is considered an entirely safe food additive. TGase is listed as GRAS (Generally Recognized As Safe). On the background of the majority of enzymes, TGase is far from the standard due to its cross-linking properties. The present study on microbe modification and a constant development toward profitable production of the enzyme may result in the manufacture of more accessible products with a wider spectrum of applications. TGase use in the processes of food manufacture has numerous advantages: enables better raw material use, prevents unfavorable quality changes (color, taste, smell, or texture), improves sensory attractivity, increases manufacturing efficiency of the products, as well as allows for obtaining new products. Great application possibilities of mTGase encourage the search for new sources of this biocatalyst (of microbes), which are able to synthesize considerable amounts of the enzyme using the least expensive culture media.

References

Ando, H., Adachi, M., Umeda, K., Matsuura, A., Nonaka, M., Uchio, R., Tanaka, H. and Motoki, M. 1989. Purification and characteristics of a novel transglutaminase derived from microorganisms. Agricultural and Biological Chemistry, 53: 2613–2617.

Babiker, E.F.E., Matsudomi, N. and Kato, A. 1998. Masking of antigen structure of soybean protein by conjugation with polysaccharide and crosslinkage with microbial transglutaminase. Nahrung, 42: 158–159.

Benjakul, S., Visessanguan, W. and Srivilai, J. 2001. Porcine plasma protein as gel enhancer in bigeye snapper (*Priacanthus tayenus*) surimi. Journal of Food Biochemistry, 25: 285–308.

Bourneow, C., Benjakul, S. and H-Kittikun, A. 2012. Hydroxamate-based colorimetric method for direct screening of transglutaminase producing bacteria. World Journal of Microbiology and Biotechnology, 28: 2273–2277.

Buettner, K., Hertel, T.C. and Pietzsch, M. 2012. Increased thermostability of microbial transglutaminase by combination of several hot spots evolved by random and saturation mutagenesis. Amino Acids, 42: 987–996.

Cegiełka, A. and Geryk, A. 2007. Zastosowanie preparatu transglutaminazy i wybranych hydrokoloidow w procesie produkcji restrukturyzowanej szynki z miesa kurczat. Roczniki Instytutu Przemysłu Mięsnego i Tłuszczowego, 1(45): 163–171.

Cui, L., Zhang, D.-X., Huang, L., Liu, H., Du, G.-C. and Chen, J. 2006. Stabilization of a new microbial transglutaminase from *Streptomyces hygroscopicus* WSH03–13 by spray drying. Process Biochemistry, 41: 1427–1431.

Date, M., Yokoyama, K., Umezawa, Y., Matsui, H. and Kikuchi, Y. 2004. High-level expression of *Streptomyces mobaraensis* transglutaminase in *Corynebacterium glutamicum* using a chimeric pro-region from *Streptomyces cinnamoneus* transglutaminase. Journal of Biotechnology, 110: 219–226.

de Almeida, M.A., Villanueva, N.D.M., Gonçalves, J.R. and Contreras-Castillo, C.J. 2015. Quality attributes and consumer acceptance of new ready-to-eat frozen restructured chicken. Journal of Food Science and Technology, 52(5): 2869–2877.

de Kruif, C.G., Tuinier, R., Holt, C., Timmins, P.A. and Rollema, H.S. 2002. Physicochemical study of κ- and β-casein dispersions and the effect of cross-linking by transglutaminase. Langmuir, 18(12): 4885–4891.

Djoullah, A., Krechiche, G., Husson, F. and Saurel, R. 2015. Size measuring techniques as tools to monitor pea proteins intramolecular crosslinking by transglutaminase treatment. Food Chemistry, 190: 197–201.

Dłużewska, E. and Florowska, A. 2014. Wykorzystanie transglutaminazy w przemyśle spożywczym. Przemysł Spożywczy, 68: 21–24.

Dong, Z.J., Xia, S.Q., Hua, S., Hayat, K., Zhang, X.M. and Xu, S.Y. 2008. Optimization of cross-linking parameters during production of transglutaminase-hardened spherical multinuclear microcapsules by complex coacervation. Colloids and Surfaces B: Biointerfaces, 63(1): 41–47.

Duran, R., Junqua, M., Schmitter, J.M., Gancet, C. and Goulas, P. 1998. Purification, characterisation and gene cloning of transglutaminase from *Streptoverticillium cinnamoneum* CBS 683.68. Biochimie, 80: 313–319.

Folk, J.E. and Chung, S.I. 1973. Molecular and catalytic properties of transglutaminases. Advances in Enzymology and Related Areas of Molecular Biology, 38: 109–191.

Gaspar, A.L.C. and de Góes-Favoni, S.P. 2015. Action of microbial transglutaminase (MTGase) in the modification of food proteins: A review. Food Chemistry, 171: 315–322.

Gerber, U., Jucknischke, U., Putzien, S. and Fuchsbauer, H.L. 1994. A rapid and simple method for the purification of transglutaminase from *Streptoverticillium mobaraense*. Biochemical Journal, 299: 825–829.

Gerrard, J.A., Fayle, S.E., Brown, P.A., Sutton, K.H., Simmons, L. and Rasiah, I. 2001. Effects of microbial transglutaminase on the wheat proteins of bread and croissant dough. Journal of Food Science, 66: 782–786.

Giosafatto, C.V.L., Rigby, N.M., Wellner, N., Ridout, M., Husband, F. and Mackie, A.R. 2012. Microbial transglutaminase-mediated modification of ovalbumin. Food Hydrocolloids, 26: 261–267.

Gottmann, K. and Sprossler, B. 1992. Backmittel oder Backmehl, sowie Verfahren zur Herstellung von Backteigen und Backwaren. Europe Patent # 0,492,406.

Griffin, M., Casadio, R. and Bergamini, C.M. 2002. Transglutaminases: Nature's biological glues. Biochemical Journal, 368: 377–396.

Gujral, S.H. and Rosell, M.C. 2004. Functionality of rice flour modified with a microbial transglutaminase. Journal of Cereal Science, 39: 225–230.

Herrera, A., Téllez-Luis, S.J., Ramírez, J.A. and Vázquez, M. 2003. Production of xylose from sorghum straw using hydrochloric acid. Journal of Cereal Science, 37: 267–274.

Hiller, B. and Lorenzen, P.C. 2009. Functional properties of milk proteins as affected by enzymatic oligomerisation. Food Research International, 42(8): 899–908.

Ho, M.L., Leu, S.Z., Hsieh, J.F. and Jiang, S.T. 2000. Technical approach to simplify the purification method and characterization of microbiol transglutaminase produced form *Streptoverticillium ladakanum*. Journal of Food Science, 65: 76–80.

Hong, G.P., Chun, J.Y., Jo, Y.J. and Choi, M.J. 2014. Effects of pH-shift processing and microbial transglutaminase on the gel and emulsion characteristics of porcine myofibrillar system. Korean Journal of Food Science and Technology, 34(2): 207–213.

Iličić, M.D., Milanović, S.D., Carić, M.D., Dokić, L.P. and Kanurić, K.G. 2014. Effect of transglutaminase on texture and flow properties of stirred probiotic yoghurt during storage. Journal of Texture Studies, 45(1): 13–19.

Itaya, H. and Kikuchi, Y. 2008. Secretion of *Streptomyces mobaraensis* protransglutaminase by coryneform bacteria. Applied Microbiology and Biotechnology, 78: 621–625.

Jiang, S.T., Hsieh, J.F., Ho, M.L. and Chung, Y.C. 2000. Microbial transglutaminase affects gel properties of golden threadfin-bream and pollack surimi. Journal of Food Science, 65(4): 694–699.

Junqua, M., Duran, R., Gancet, C. and Goulas, P. 1997. Optimization of microbial transglutaminase production using experimental designs. Applied Microbiology and Biotechnology, 48: 730–734.

Kashiwagi, T., Yokoyama, K., Ishikawa, K., Ono, K., Ejima, D., Matsui, H. and Suzuki, E. 2002. Crystal structure of microbial transglutaminase from *Streptoverticillium mobaraense*. Journal of Biological Chemistry, 277: 44252–44260.

Kato, A., Wada, T., Kobayashi, K., Seguro, K. and Motoki, M. 1991. Ovomucin-food protein conjugates prepared through the transglutaminase reaction. Agricultural and Biological Chemistry, 55(4): 1027–1031.

Kieliszek, M. and Misiewicz, A. 2014. Microbial transglutaminase and its application in the food industry: A review. Folia Microbiologica, 59(3): 241–250.

Kim, Y.S., Huang, W., Du, G., Pan, Z. and Chung, O. 2008. Effects of trehalose, transglutaminase and gum on rheological, fermentation, and baking properties of frozen dough. Food Research International, 41(9): 903–908.

Kołodziejska, I., Kaczorowski, K., Piotrowska, B. and Sadowska, M. 2004. Modification of the properties of gelatin from skins of Baltic cod (*Gadus morhua*) with transglutaminase. Food Chemistry, 86(2): 203–209.

Kowalski, R. and Pyrcz, J. 2009. Innowacyjne dodatki technologiczne w przemyśle mięsnym. Przemysł Spożywczy, 63: 28–32.

Kuraishi, C., Sakamoto, J. and Soeda, T. 1998. Application of transglutaminase for meat processing. Fleischwirtschaft, 78: 656–662.

Kuraishi, C., Sakamoto, J., Yamazaki, K., Susa, Y., Kuhara, C. and Soeda, T. 1997. Production of restructured meat using microbial transglutaminase without salt or cooking. Journal of Food Science, 62: 488–490.

Kuraishi, C., Yamazaki, K. and Susa, Y. 2001. Transglutaminase: Its utilization in the food industry. Food Reviews International, 17: 221–246.

Lauber, S., Henle, T. and Klostermeyer, H. 2000. Relationship between the crosslinking of caseins by transglutaminase and the gel strength of yoghurt. European Food Research and Technology, 210: 305–309.

Lerner, A. and Matthias, T. 2015. Changes in intestinal tight junction permeability associated with industrial food additives explain the rising incidence of autoimmune disease. Autoimmunity Reviews, 14(6): 479–489.

Li, H., Cui, Y., Zhang, L., Luo, X., Fan, R., Xue, Ch., Wang, S., Liu, W., Zhang, S., Jiao, Y., Du, M., Yi, H. and Han, X. 2015. Production of a transglutaminase from *Zea mays* in *Escherichia coli* and its impact on yoghurt properties. International Journal of Dairy Technology, 68(1): 54–61.

Lin, S.-J., Hsieh, Y.-F., Wang, P.-M. and Chu, W.-S. 2007. Efficient purification of transglutaminase from recombinant *Streptomyces platensis* at various scales. Biotechnology Letters, 29: 111–115.

Lin, Y.-S., Chao, M.-L., Liu, C.-H., Tseng, M. and Chu, W.-S. 2006. Cloning of the gene coding for transglutaminase from *Streptomyces platensis* and its expression in *Streptomyces lividans*. Process Biochemistry, 41: 519–524.

Lin, Y.-S., Chao, M.-L., Liu, CH.-H., Chu, W.-S. 2004. Cloning and expression of the transglutaminase gene from *Streptoverticillium ladakanum* in *Streptomyces lividans*. Process Biochemistry, 39: 591–598.

Liu, S., Wan, D., Wang, M., Madzak, C., Du, G. and Chen, J. 2015. Overproduction of pro-transglutaminase from *Streptomyces hygroscopicus* in *Yarrowia lipolytica* and its biochemical characterization. BMC Biotechnology, 15(1): 1–9.

Liu, Y., Lin, S., Zhang, X., Liu, X., Wang, J. and Lu, F. 2014. A novel approach for improving the yield of *Bacillus subtilis* transglutaminase in heterologous strains. Journal of Industrial Microbiology and Biotechnology, 41(8): 1227–1235.

Lorenzen, P.C., Neve, H., Mautner, A. and Schlimme, E. 2002. Effect of enzymatic cross-linking of milk proteins on functional properties of set-style yoghurt. Journal of Dairy Technology, 55: 152–157.

Losche, I.K. 1995. Enzymes in baking. The World of Ingredients, May–June, 22–25.

Macedo, J.A., Cavallieri, A.L.F., da Cunha, R.L. and Sato, H.H. 2010. The effect of transglutaminase from *Streptomyces* sp. CBMAI 837 on the gelation of acidified sodium caseinate. International Dairy Journal, 20: 673–679.

Mahmood, W.A. and Sebo, N.H. 2009. Effect of microbial transglutaminase treatment on soft cheese properties. Mesopotamia Journal of Agriculture, 37, http://www.iasj.net/iasj?func=fulltext&aId=27525.

Marapana, R.A.U.J. and Jiang, B. 2004. Protein cross-linking in food by microbial transglutaminase (MTGase) and its application and usefulness in food industry. Tropical Agricultural Research and Extension, 7: 49–61.

Marco, C. and Rosell, C.M. 2008. Breadmaking performance of protein enriched, gluten-free breads. European Food Research and Technology, 227: 1205–1213.

Mariniello, L., Di Pierro, P., Giosafatto, C.V.L., Sorrentino, A. and Porta, R. 2008. Transglutaminase in food biotechnology. pp. 185–211. *In*: R. Porta, L. Mariniello and P. Di Pierro (eds.). Recent Developments in Food Biotechnology: Enzymes as Additives or Processing Aids. Research Signpost, Trivandum, Kerala.

Mirzaei, M. 2011. Microbial transglutaminase application in food industry. International Conference on Food Engineering and Biotechnology. IPCBEE, Vol. 9. IACSIT Press, Singapoore, pp. 267–271.

Monogioudi, E., Faccio, G., Lille, M., Poutanen, K., Buchert, J. and Mattinen, M.-L. 2011. Effect of enzymatic cross-linking of β-casein on proteolysis by pepsin. Food Hydrocolloids, 25: 71–81.

Monteiro, M.L.G., Mársico, E.T., Lázaro, C.A., da Silva, A.C.V.C., da Costa Lima, B.R.C., da Cruz, A.G. and Conte-Júnior, C.A. 2015. Effect of transglutaminase on quality characteristics of a value-added product tilapia wastes. Journal of Food Science and Technology, 52(5): 2598–2609.

Moore, M.M., Heinbockel, M., Dockery, P., Ulmer, M.H. and Arendt, E.K. 2006. Network formation in gluten-free bread with application of transglutaminase. Cereal Chemistry, 83: 28–36.

Mycek, M.J., Clarke, D.D., Neidle, A. and Waelsch, H. 1959. Amine incorporation into insulin as catalyzed by transglutaminase. Archives of Biochemistry and Biophysics, 84: 528–540.

Nielsen, P.M. 1995. Reactions and potential industrial applications of transglutaminase. Review of literature and patents. Food Biotechnology, 6: 119–156.

Noguchi, T., Tanimoto, H., Motoki, M. and Mori, M. 1992. A Promoting Material for Absorption of Minerals and Compositions Containing It. Japanese Kokai Tokkyo Koho JPatent # 04349869.

Nonaka, M., Soeda, T., Yamagiwa, K., Kobata, H., Motoki, M. and Toiguchi, S. 1990. Tofu for Long-term Storage and Its Manufacture Rusing a Novel Enzyme. Japanese Kokai Tokkyo Koho JPatent #0269155.

Ozer, B., Kirmaci, H.A., Oztekin, S., Hayaloglu, A. and Atamer, M. 2007. Incorporation of microbial transglutaminase into non-fat yogurt production. International Dairy Journal, 17: 199–207.

Pasternack, R., Dorsch, S., Otterbach, J.T., Robenek, I.R., Wolf, S. and Fuchsbauer, H.L. 1998. Bacterial pro-transglutaminase from *Streptoverticillium mobaraense*. European Journal of Biochemistry, 257(3): 570–576.

Pinterits, A. and Arntfield, S.D. 2008. Improvement of canola protein gelation properties through enzymatic modification with transglutaminase. LWT-Food Science and Technology, 41(1): 128–138.

Porta, R., Di Pierro, P., Rossi-Marquez, G., Mariniello, L., Kadivar, M. and Arabestani, A. 2015. Microstructure and properties of bitter vetch (*Vicia ervilia*) protein films reinforced by microbial transglutaminase. Food Hydrocolloids, 50: 102–107.

Prakasan, V., Chawla, S.P. and Sharma, A. 2015. Effect of transglutaminase treatment on functional properties of *paneer*. International Journal of Current Microbiology and Applied Sciences, 4(5): 227–238.

Pyrcz, J., Kowalski, R., Kostecki, A. and Danyluk, B. 2014. Technologiczna przydatność enzymu transglutaminazy i soli fosforanowej w produkcji kutrowanych kiełbas parzonych. Aparatura Badawcza i Dydaktyczna, 19: 315–320.

Rossa, P.N., de Sá, E.M.F., Burin, V.M. and Bordignon-Luiz, M.T. 2011. Optimization of microbial transglutaminase activity in ice cream using response surface methodology. LWT—Food Science and Technology, 44(1): 29–34.

Sakamoto, H. and Soeda, T. 1991. Minced Meat Products Containing Transglutaminase. Japanese Kokai Tokkyo Koho JPatent # 03175929.

Şanli, T., Lezgin, E., Deveci, O., Şenel, E. and Benli, M. 2011. Effect of using transglutaminase on physical, chemical and sensory properties of set-type yoghurt. Food Hydrocolloids, 25: 1477–1481.

Seguro, K. and Motoki, M. 1991. Manufacture of Canned Meats Containing Transglutaminase. Japanese Kokai Tokkyo Koho JPatent # 03210144.

Takagaki, Y., Narakawa, K. and Uchio, R. 1991. Coating of Vegetables and Fruits with Transglutaminase and Proteins for Preservation. Japanese Kokai Tokkyo Koho JPatent # 03272639.

Tang, C., Yang, X.Q., Chen, Z., Wu, H. and Peng, Z.Y. 2005. Physicochemical and structural characteristics of sodium caseinate biopolymers induced by microbial transglutaminase. Journal of Food Biochemistry, 29(4): 402–421.

Tani, T., Iwamoto, K., Motoki, M. and Toiguchi, S. 1990. Manufacture of Shark Fin Imitation Food. Japanese Kokai Tokkyo Koho JPatent # 02171160.

Téllez-Luis, S.J., González-Cabriales, J.J., Ramírez, J.A. and Vázquez, M. 2004. Production of transglutaminase by *Streptoverticillium ladakanum* NRRL-3191 grown on media made from hydrolysates of sorghum straw. Food Technology and Biotechnology, 42: 1–4.

Téllez-Luis, S.J., Uresti, R.M., Ramírez, J.A. and Vázquez, M. 2002. Low-salt restructured fish products using microbial transglutaminase. Journal of the Science of Food and Agriculture, 82: 953–959.

Trespalacios, P. and Pla, R. 2007. Simultaneous application of transglutaminase and high pressure to improve functional properties of chicken meat gels. Food Chemistry, 100: 264–272.

Tseng, T.-F., Liu, D.-C. and Chen, M.-T. 2000. Evaluation of transglutaminase on the quality of low-salt chicken meat-balls. Meat Science, 55(4): 427–431.

Washizu, K., Ando, K., Koikeda, S., Hirose, S., Matsuura, A., Takagi, H., Motoki, M. and Takeuchi, K. 1994. Molecular cloning of the gene for microbial transglutaminase from *Streptoverticillium* and its expression in *Streptomyces lividans*. Bioscience, Biotechnology and Biochemistry, 58: 82–87.

Weng, W. and Zheng, H. 2015. Effect of transglutaminase on properties of tilapia scale gelatin films incorporated with soy protein isolate. Food Chemistry, 169: 255–260. www.rcsb.org.

Yamazaki, K. and Nishimura, Y. 2002. Method for Producing Noodles. U.S. Patent # 6,403,127. Washington, DC: U.S. Patent and Trademark Office.

Yildirim, M. and Hettiarachchy, N.S. 1998. Properties of films produced by cross-linking whey proteins and 11S globulin using transglutaminase. Journal of Food Science, 63(2): 248–252.

Yokoyama, K., Nakamura, N., Seguro, K. and Kubota, K. 2000. Overproduction of microbial transglutaminase in *Escherichia coli, in vitro* refolding, and characterization of the refolded form. Bioscience, Biotechnology and Biochemistry, 64: 1263–1270.

Yokoyama, K., Nio, N. and Kikuchi, Y. 2004. Properties and applications of microbial transglutaminase. Applied Microbiology and Biotechnology, 64: 447–454.

Yu, Y.J., Wu, S.C., Chan, H.H., Chen, Y.C., Chen, Z.Y. and Yang, M.T. 2008. Overproduction of soluble recombinant transglutaminase from *Streptomyces netropsis* in *Escherichia coli*. Applied Microbiology and Biotechnology, 81: 523–532.

Zheng, M.Y., Du, G.C., Chen, J. and Lun, S.Y. 2002. Modelling of temperature effects on batch microbial transglutaminase fermentation with *Streptoverticillium mobaraense*. World Journal of Microbiology and Biotechnology, 18: 767–771.

Zhu, Y., Rizeman, A., Tramper, J. and Bol, J. 1995. Microbial Transglutaminase—A review of its production and application in food processing. Applied Microbiology and Biotechnology, 44: 277–282.

Zhu, Y., Rizeman, A., Tramper, J. and Bol, J. 1996. Medium design based on stoichiometric analysis of microbial transglutaminase production by *Streptoverticillium mobaraense*. Biotechnology and Bioengineering, 50: 291–298.

11

Microbial Milk Coagulants

Ekaterini Moschopoulou

1. Introduction

Enzymatic coagulation of milk is a critical phase of cheese-making process. Traditionally, the enzymes involved are gastric aspartic proteinase, i.e., chymosin (EC 3.4.23.4, formerly rennin), pepsin (EC 3.4.23.1) and gastriscin (EC 3.4.23.3), which are secreted in the abomasa of ruminants in the form of inactive precursors called prochymosin, pepsinogen and progastriscin, respectively. Precursors of enzymes are converted to active enzymes automatically in the acid conditions of the stomach. Substitutes of chymosin, such as pepsins of different animal origin, vegetable acid proteinases, microbial proteinases and bovine chymosin from cloned microorganisms referred to as 'fermentation produced chymosin', are also used in the cheese industry due to limited slaughtering of young animals, increased cheese production, diet and religious matters. The enzymatic products of ruminant origin are classified as rennet, while those of different origin are classified as milk coagulants (Harboe et al., 2010; Kumar et al., 2010). The use of calf rennet in cheese industry currently represents only 20–30 per cent of milk-clotting enzymes (Feijoo-Siota et al., 2014). The characteristics of bovine chymosin (Foltman, 1981, 1993; Chitpinityol and Crabbe, 1998; Crabbe, 2004) as well as of lamb (Baudys et al., 1988) or goat kid chymosin (Kumar et al., 2006; Moschopoulou, 2006, 2011) have been extensively reviewed.

This chapter deals with the characteristics, production and use of microbial milk coagulants, especially those from fungi that are used in the cheese industry since 1960s.

Laboratory of Dairy Research, Department of Food Science and Human Nutrition, Agricultural University of Athens, Iera Odos 75, Athens 118 55, Greece.
E-mail: catmos@aua.gr

2. Enzymatic Coagulation of Milk

Enzymatic coagulation of milk takes place in two separate phases—the primary, enzymatic phase, in which the proteinase cleaves the Phe_{105}-Met_{106} bond of κ-casein to destabilize the casein micelles and the secondary, non-enzymatic phase, in which the destabilized micelles precipitate in the presence of calcium ions. Cleavage of κ-casein yields an insoluble part, called para-κ-casein (1–105 amino acids fraction), which remains in the cheese curd and a soluble part called caseino-macropeptide (106–169 amino acids fraction), which passes into the whey. Therefore, the cleavage of sensitive to chymosin Phe_{105}-Met_{106} bond of κ-casein is the 'first step' for a new proteinase to be characterized as milk-clotting enzyme.

3. Microorganisms and Main Molecular Characteristics of Their Proteinases

In general, milk-clotting enzymes belong to the small group of aspartic (acid) proteinases (EC 2.4.23), which contain two aspartic acid residues at their active site, act at pH range 2.5–7 and are inhibited by pepstatin. Aspartic proteinases molecule has bilobal conformation with the catalytic centre being between the two lobs, each containing the aspartic acid residue. Furthermore, the molecule of these enzymes is heterogeneous because of N-glycosylation, phosphorylation, deamination or partial proteolysis process (Chitpinityol and Crabbe, 1998; Claverie-Martin and Vega-Hernandez, 2007; Harboe et al., 2010; Feijoo-Siota et al., 2014).

Microbial milk-clotting enzymes are usually bacterial or fungal in origin. More than 30 different bacteria, mainly from genus *Bacillus* spp. have been investigated for their suitability in producing such enzymes, but all of them are too proteolytic. Regarding fungal sources, more than 100 have been reported (Garg and Johri, 1994). The aspartic proteinases from fungi are generally divided into two groups:

- pepsin-like enzymes derived from *Aspergillus*, *Penicillium*, *Rhizopous* and *Neurospora* and
- chymosin (formerly rennin)-like enzymes derived from *Cryphonectria* and *Rhizomucor* spp. (Rao et al., 1998; Sumantha et al., 2006; Theron and Divol, 2014)

Only three filamentous fungi, i.e., *Rhizomucor miehei* (formerly *Mucor miehei*), *Rhizomucor pusillus* (formerly *Mucor pusillus*), which are related species of Zygomecetes and *Cryphonectria parasitica* (formerly *Entothia parasitica*) are used for commercial production of milk-clotting enzymes (Table 1).

R. miehei proteinase (3.4.23.23), known as *Mucor* rennin, is most commonly used and exists in four types, i.e., 'Type L' that is very heat stable, 'Type TL' that is modified by oxidation in order to be heat labile, more pH-dependent and slightly less proteolytic than Type L, 'Type XL' that is more heat labile, more pH-dependent and even less proteolytic than Type XL and 'Type XLG or XP' which is a chromatographically purified form of Type XL, containing thus less non-enzymatic impurities. *R. pusillus* proteinase (3.4.23.23) is similar to those of *R. miehei* and is usually commercially

Table 1. Commercial microbial milk coagulants.

Microbial Source	Name of Product	Producer
Rhizomucor miehei	Fromase L/XL/XLG	DSM
	Hannilase L/XL	CHR HANSEN
	Marzyme	Danisco
	Meito Rennet Super	Meito Sangyo Co.
	Milase XQL/TQL/Premium	CSK food enrichment
	Enzymaks	Enzymaks
	Reniplus/Proquiren	Proquiga Biotech
	Microclerici	Clerici-Sacco Group
	Rennilase	Iran Industrial Enzymes
R. pusillus, Lindt/*R. miehei*	Meito Microbial Rennet	Meito Sangyo Co.
Cryphonectria parasitica	Suparen	DSM
	Thermolase	CHR HANSEN

available as a mixture with the latter. *C. parasitica* proteinase (3.4.23.22), known as endothiapepsin, is very heat labile and for this reason it is recommended to be used only for highly cooked cheeses like Emmental (Harboe et al., 2010).

Fungal asparic proteinases are synthesized as pre-proenzymes, which after cleavage of the signal peptide the proenzyme is secreted and automatically activated. In general, the active enzymes consist of a single polypeptide chain containing about 320–360 amino acids residues (Claverie-Martín and Vega-Hernández, 2007). Amino acids composition and characteristics of purified commercial microbial proteinases have been reviewed by Garg and Johri (1994).

The structure of the *R. miehei* proteinase and *R. pusillus* proteinase are very similar. They are immunologically cross-reactive and display at least 83–85 per cent sequence homology, although cleavage of their peptide chain exhibits some differences (Yang et al., 1997; Harboe et al., 2010). *R. miehei* proteinase is secreted as precursor having a signal peptide of 22 residues and a pro-peptide of 47 residues (position 23–69). The active protease consists of a single polypeptide chain having 361 amino acids (Feijoo-Siota et al., 2014), has molecular weight 40.5 kDa and tertiary structure similar to those of bovine chymosin, is stable at acidic pH and preferentially cleaves the Phe_{105}-Met_{106} bond of κ-casein (Preetha and Boopathy, 1997; Feijoo-Siota et al., 2014). It is most glycosylated among the aspartic proteinases (Claverie-Martin and Vega-Hernandez, 2007), displaying 2 glycosylation sites, i.e., at Asn 79 and Asn 188, active protease numbering (Yang et al., 1997).

R. pusillus proteinase has the same specificity towards κ-casein, has optimum pH around 4 and is secreted as zymogen with a signal peptide of 22 residues, a pro-peptide of 44 amino acids (position 23–66). The active protease consists of a single polypeptide chain having 361 amino acids (Feijoo-Siota et al., 2014), having also glycosylation sites, i.e., at Asn 79, Asn 113 and Asn 188, active protease numbering (Aikawa et al., 1990). Molecular weight of *R. pusillus* proteinase is 49 kDa (Nouani et al., 2009).

C. parasitica proteinase is secreted as zymogen consisting of 419 amino acids, with a potential signal peptide of 20 amino acids, a pro-peptide of 69 amino acids (position 21–89). The single polypeptide chain of active proteinase consists of 330 amino acids having a molecular weight of about 34–39 kDa, pI 5.5 and does not contain carbohydrates (Feijoo-Siota et al., 2014).

4. Main Technological Properties

In general, aspartic proteinases have mainly endopeptidase activity and very low exopeptidase activity. Like rennet, microbial coagulants have two distinct activities in cheesemaking, i.e., the milk-clotting activity (MCA) and the proteolytic activity (PA), the latter being one of the main proteolytic agents during ripening of certain cheese types. The higher the ratio MCA/PA the best milk-clotting enzyme and this is an important criterion for the choice of a new strain or microorganism for producing new microbial coagulant.

Technological properties of microbial coagulants compared with those of calf rennet are summarized in Table 2.

4.1 Specificity and proteolytic activity

Specificity of aspartic acid proteinases is not very narrow because of the wide binding cleft in their molecule, which is able to host at least seven amino acids (Harboe et al., 2010). During the enzymatic phase of milk clotting, all commercial microbial coagulants cleave the same bond as chymosin, i.e., Phe_{105}-Met_{106} bond of bovine κ-casein, except *C. parasitica* proteinase that cleaves the Ser_{104}-Phe_{105} bond (Drøhse and Foltmann, 1989). *Rhizomucor* sp. and *Cryphonectria* proteinases also cleave several other bonds in κ-casein (Shammet et al., 1992a,b), but in general, bond specificity, active site and milk-clotting mechanism of *R. miehei* proteinase are similar to those of calf chymosin (Sternberg, 1972).

Proteolytic activity of commercial microbial coagulants has been studied using different substrates, i.e., milk, reconstituted skim milk, casein solution and purified casein fractions. In any case, proteolytic activity of these enzymes is significantly higher than those of traditional calf rennet or fermentation-produced chymosin (Jacob et al., 2011a) and affects the cheese yield and proteolysis during cheese ripening. The optimum proteolytic activity of *R. miehei* proteinase is at pH 4.1, while 50 per cent of it is lost after heating at 45°C for 30 min. (Preetha and Boopathy, 1997) or can be reduced to 50 per cent by acid treatment (Hubble and Mann, 1984). *C. parasitica* proteinase has higher proteolytic activity than those of *Rhizomucor miehei*, mainly towards ovine β-casein (Trujillo et al., 2000), and in general, is more proteolytic than the latter and chymosin on various casein substrates and at most pH values (Tam and Whitaker, 1972; Paquet and Alais, 1978; Harhoe et al., 2010). Regarding the degradation of whey proteins, commercial proteinases of *R. miehei* as well as of *R. pusillus* are also more proteolytic, especially on β-lactoglobulin than pure porcine, bovine pepsin and chymosin (Candioti et al., 2002).

4.2 Thermal stability

Thermal stability of milk-clotting enzyme is very important, especially when after cheese making the whey is going to be processed. It increases with decreasing pH

Table 2. Technological properties of microbial coagulants and calf rennet (Hyslop et al., 1979; Garnot, 1985; Ramet, 1997).

A. Sensitivity in different factors (increased from 1→4)			
	Temperature	*pH*	*Ca^{2+}*
Rhizomucor miehei	1	2	2
R. pusillus	2	1	1
Cryphonectria parasitica	4	4	4
Calf rennet	3	2	3
B. Proteolytic activity towards the casein fractions			
	$α_s$-casein	*β-casein*	*κ-casein*
R. miehei	++	+	++
R. pusillus	++	++	++
C. parasitica	+++	++	+
Calf rennet	++	+	++
C. Residual MCA (expressed as % of initial) in curds with different pH			
pH	*5.2*	*6.0*	*6.6*
R. miehei	19	19	19
R. pusillus	11	12	14
Calf rennet	83	70	30
D1. Heat stability at different pH (residual MCA as % of initial) after heating at 63,3°C for 1 min.			
pH	*5.2*	*6.0*	*6.6*
R. miehei	99	60	3
R. pusillus	33	0	0
C. parasitica	3	1	0
Calf rennet	10	0	0
D2. Heat stability at pH 6.0 (residual MCA as % of initial) after heating at different temperatures for 15 s.			
T *R. miehei* *R. pusillus* Calf rennet	*67* 10 10 10	*71* 10 10 10	*77* 10 10 10
D3. Heat stability at pH 6.0 (residual MCA in whey as % of initial) after heating at different temperatures for 15 sec.			
T *R. miehei* Calf rennet	*72* 70 0.5	*74* 40 0.5	

(Table 2) and, depending on the cooking temperatures, affects the proteinase amount retained in the curd. Fungal aspartic proteinases are generally heat stable. Compared to different milk-clotting enzymes, *C. parasitica* proteinase is the most thermally sensitive, while proteinases from *R. miehei* are the most thermally stable (Walsh and Li, 2000; Harboe et al., 2010).

The very high thermal stability of *R. miehei* proteinase is attributed to its high glycosylation degree (Yang et al., 1997). Thermal stability can be reduced by chemical modifications, i.e., oxidation of Met with oxidizing agents such as peracetic acid, hydrogen peroxide and sodium hypochlorite, so that at 74°C the oxidized type has ≤ 0.5 per cent of the initial clotting activity (Branner et al., 1984; Garnot, 1985). Alternatively, thermal stability of these proteinases can be modified by genetic approaches. For example, gene of *R. pusillus* proteinase has been successfully expressed in *Saccharomyces cerevisiae*. Substitution of both Thr_{101} with Ala and Asp_{186} with Gly resulted in the lowest heat stability (Yamashita et al., 1994). In another study, removal of the N-linked carbohydrates from *R. pusillus* proteinase by genetic engineering decreased not only its thermal stability but also its proteolytic activity, while increasing its milk-clotting activity (Aikawa et al., 1990).

In general, several methods, both chemical and genetic, have been developed for improving technological properties of microbial coagulants. Most of them are described in patents (Feijoo-Siota et al., 2014) and concern the decrease of high thermal stability and increase of MCA/PA ratio.

5. Production Process

All commercial microbial coagulants are produced by fermentation of fungi. At the beginning of fermentation they are secreted as inactive precursors which are activated automatically by the end of fermentation because of the acid pH of fermentation conditions.

When a fermentation process is designed, the main objective is to get a high yield of the product in a short time running. In addition, the production strain and the kind of an upstream or downstream processing must be taken into consideration especially in the case of viscous, with high level of fine particulate matter and of high proteinase activity product (Hjort, 2007). The strains used for the production of microbial coagulants are selected and improved in order to produce as less as possible of undesirable secondary enzymes, such as lipase. For example, the *Rhizomucor* spp. coagulants usually contain starch-degrading enzymes, which for the product 'Type XL' must be removed by an additional step (Harboe et al., 2010).

The fermentation process used for biosynthesis of Food-Grade proteinases can be carried out either as solid state fermentation or as submerged fermentation (Sumantha et al., 2006). The latter is almost exclusively used today in industrial microbial enzyme production since surface fermentation is labor-intensive and the downstream process is more complicated.

5.1 Solid state fermentation (SSF)

In SSF microbial growth takes place on a solid undissolved substrate in which there is little or no free water. The culture media that are used in SSF for producing all kinds of industrial enzymes have been reviewed by Pandey et al. (1999) and Sumantha et al. (2006). Literature concerns production on the laboratory scale.

R. pusillus Lindt (F-27) proteinase was successfully produced many years ago by solid culture on wheat bran at 30°C for about 70 hours followed by water extraction. The ratio MCA/PA of this enzyme was similar to those of calf rennet (Arima et al., 1967). It was also better obtained from semi-solid fermentation containing 50 per cent wheat bran (Neelakantan et al., 1999). SSF of *Rhizomucor* sp. on a ratio of wheat bran to moisture 1:0.6 (w/v) at room temperature for 5 days resulted in the highest MCA/PA ratio of *R. miehei* proteinase, while addition of 4 per cent (w/v) skim milk powder to the media improved only the MCA/PA ratio of *R. pusillus* proteinase. Moreover, Preetha and Boopathy (1994) showed that in the case of co-cultivation of both microorganisms, the ratio MCA/PA did not change.

Purified proteinase from *R. miehei* cultured by SSF in small-scale experiments had high ratio MCA/PA (6.6:1) and this ratio was comparable to those from large-scale production (Thakur et al., 1990). In SSF of *R. miehei,* increase of glucose concentration affected the enzyme synthesis, while casein was the prime factor in the enzyme synthesis induction (Silveira et al., 2005). Glucose concentration has also been found to affect enzyme biosynthesis of *R. mucedo* KP736529 (Ayana et al., 2015). SSF of *Rhizomucor* sp. (M-105), followed by two step purification of crude extract resulted in a proteinase, which exhibited high and similar to that of commercial microbial coagulant MCA/PA ratio, molecular weight 33 kDa and full inhibition by pepstatin (Fernandez-Lahore et al., 1999).

Downstream processing of microbial proteinase from *Rhizomucor* spp. cultivated by SSF may involve extraction from the moldy bran using a semicontinuous multiple contact-forced percolation method and further treatment of the extract using filtration through 5 per cent R16 clay to eliminate lipases and other proteinases (Thakur et al., 1993) or ultrafiltration of the crude extract and enzyme purification by ion-exchange followed by size-exclusion chromatography (Fernandez-Lahore et al., 1999). Methods for removal of undesirable enzymes from microbial proteinases have been reviewed by Garg and Johri (1994).

Different filamentous fungi and different strains will be always in search of milk-clotting enzyme production. Thus, on the laboratory scale SSF, *Thermomucor indicae-seudaticae* N31 is shown to be a good producer of microbial coagulant because its proteolytic activity is similar to those obtained with commercial *R. miehei* proteinase (Merheb-Dini et al., 2010). Also, a newly isolated thermophilic fungus *Rhizomucor nainitalensis* cultivated under SSF conditions has produced milk-coagulating enzyme which is free of toxin and may be commercially viable (Khademi et al., 2013). An ochratoxin-free extracellular acid proteinase has been produced by SSF of *Aspergillus niger* FFB1 and compared to calf rennet in fresh cheese manufacture. Results based on cheese pH, color and taste showed the potential use of this enzyme (Fazouane-Naimi et al., 2010). In addition, a milk-clotting enzyme with high MCA/PA ratio and low thermal stability has been produced by SSF of *A. oryzae* MTCC 5341, showing that it could be a good substitute of calf rennet (Vishwanatha et al., 2010).

In an attempt to produce new microbial proteinases, a milk-clotting enzyme was produced by culture of *Bacillus subtilis* var. *(natto) takahashi*, a commercial starter for the traditional Japanese food natto, i.e., soybeans fermented with this microorganism and has been proved to be comparable with *Rhizomucor* proteinase in MCA and MCA/PA ratio (Shieh et al., 2009). Purified proteinases derived from soybeans fermented by *B. subtilis* var. *natto* or by *Rhizopous oligosporus* have been also compared to the commercial product from *R. miehei* Type II and have been showed to be inferior to the commercial one (Chen et al., 2010). Moreover, good probability of successful use in cheese production has been showed for a proteinase produced from an indigenously isolated *Bacillus subtilis* growing on a medium containing fructose and casein (Dutt et al., 2008).

5.2 Submerged fermentation (SmF)

Submerged fermentation can be batch, fed-batch or continuous, but microbial coagulants are most often produced in submerged fed batch mode (Harboe et al., 2010).

On laboratory scale, several studies were made for producing microbial proteinases by SmF, with emphasis on the culture media. *R. miehei* and *C. parasitica* are well cultivated in SmF, giving good yields of proteinases, especially when *R. miehei* grows in a medium containing 4 per cent potato starch, 3 per cent soybean meal and 10 per cent barley (Neelakantan et al., 1999). The significant role of concentration of D-glucose on milk-clotting activity of microbial proteinase has been showed. In a continuously fed SmF of *R. miehei* NRRL 3420, a commercial strain, the produced enzyme showed the maximum MCA when D-glucose concentration was 7.5 g/dm^3, while, after concentrating the fermentation medium, the ratio MCA/PA was similar to those of a commercial coagulant (Seker et al., 1998; Beyenal et al., 1999a). Furthermore, in model simulation studies, the maximum milk-clotting activity of proteinase from the same fungus was obtained when a multiple linear function of parameters, such as fermentation medium pH, D-glucose, dissolved oxygen concentration and dilution rate was applied (Seker et al., 1999), while the enzyme production rate was rather affected by the fermentation medium than the operational parameters of the bioreactor (Beyenal et al., 1999b). Osorio et al. (2008) showed that whey as both a carbon and nitrogen source or only as a carbon source together with corn flour as nitrogen source was the best substrate for improving the enzymatic activity of *R. miehei* proteinase produced in a laboratory batch reactor. *R. pusillus* QM 436 also produced a proteinase of high level of milk-clotting activity coupled with a low level of thermal stability when grew on salted whey (El-Tanboly et al., 2013).

In batch shake flask fermentation of *R. miehei* ATCC 3420, the highest concentration of MCA was achieved at initial D-glucose concentration 29.6 g/dm^3, initial pH 6.8, temperature 37.6°C, with agitation rate at 81 strokes per min. and inoculum ratio 5.2 per cent (v/v) (Ayhan et al., 2001).

Other fungi or bacteria have been cultivated on laboratory scale for producing milk-clotting proteinases (Jacob et al., 2011b). Static flask fermentation of *Penicillium oxalicum* using sucrose 5 per cent as carbon source and a mixture of yeast extract and peptone or baker's yeast as nitrogen source resulted in a proteinase having the highest yield of milk-clotting activity (Hashem, 1999). A proteinase from *Myxococcus xanthus* strain 422, a gram negative bacterium, was produced in flask fermentation and characterized as a true milk-clotting enzyme, which has a molecular weight 40 kDa, optimum milk clotting activity at pH 6 and 37°C and is completely inactivated by heating at 65°C for 12 min. (Poza et al., 2003). On the other hand, an extracellular serine proteinase having molecular weight of 34 kDa and showing typical milk-clotting kinetics has been produced by SmF of *Bacillus licheniformis* strain USC13

(Ageitos et al., 2007). Recently, a milk-clotting proteinase from the thermophilic fungus *Thermomucor indicae-seudaticae* N31 was produced by SmF in a medium containing 4 per cent wheat bran as substrate in 0.3 per cent saline solution. This proteinase, which was suggested for industrial-scale production, was relatively heat resistant, i.e., milk-clotting activity stable up to 55°C for one hour, and exhibited MCA/PA ratio 510 (Silva et al., 2014).

5.3 SmF on industrial scale

In such a production, the selected fungal strain is initially inoculated into a flask containing an agar or liquid medium to sporulate and then transferred to a seed fermenter in which the cells adapt to the environment and the nutrients used in the rest of the process (Hjort, 2007). After seeding, material is inoculated into a sterile medium for the main fermentation process, which may last for days. The sterile medium accounts for the major part of the fermentation costs (Harboe et al., 2010). During fermentation, parameters like temperature, air flow, pressure, agitation, pH, oxygen tension, concentrations of important ingredients in the medium, level of enzyme and by-products are controlled in order to optimize the proteinase production. In fed-batch fermentation the production strain is fed with an additional medium. After main fermentation, the mixture of microbial cells, medium and enzyme products are downstream processed (Hjort, 2007). Microbial milk coagulants are extracellular proteinases and thus the enzyme is recovered by filtration or centrifugation to remove the fungus; it is then concentrated by ultrafiltration or evaporation before being filtered.

Industrial microbial milk coagulants are not usually subjected to any purification, being thus crude fermentates (Harboe et al., 2010), which, however, may contain not only the coagulating enzymes, but also an associated non-coagulating enzyme fraction, i.e., lipases, cellulases, unspecific proteinases. The latter can be removed by various treatments like salt precipitation, filtration, chromatography, etc. (Garg and Johri, 1994). Analysis of the contaminant fraction in preparation from calf rennet, fermentation-produced chymosin, artisanals lamb/kids rennets and different commercial microbial coagulants showed that those from *R. miehei* had the highest NaCL and pH values 5.3–5.7 and those from *C. parasitica* had the highest ammonia and small peptides content, high free NH_2 content and a mean pH value 4.4 (Rolet-Répécaud et al., 2013). On laboratory scale, purification of commercial proteinase preparations may isolate 20 mg of pure *R. miehei* proteinase and more than 20 mg of pure *C. parasitica* proteinase from 3 ml and 16 ml of commercial enzyme solutions respectively (Kobayashi et al., 1982).

The final steps in commercial microbial coagulant production is formulation, standardization and quality control of the milk-clotting activity. Formulation involves addition of sodium chloride as stabilizer, a buffer substance and often a preservative like sodium benzoate. Commercial milk coagulant preparations may be powdered, granular, tablets or liquid. Standardization concerns the milk-clotting activity and enzymes content (Harboe et al., 2010). Total milk clotting activity of microbial rennet is determined by an ISO/IDF standard method using reference standard powder from *R. miehei* and is expressed as IMCU/ml or IMCU/g (ISO-IDF, 2012).

6. Application in Cheese-Making Process

Although only 2–3 per cent of the coagulants added to the cheese milk are retained in the curd (Fox and McSweeney, 1997), microbial proteinases of fungal origin show usually higher proteolytic activity during cheese manufacture and ripening than the other milk-clotting enzymes. This, depending on the cheese type, may lead to decreased cheese yield and flavor defects, i.e., bitter flavor. Much research has been done on the use of these coagulants and especially of *R. miehei* proteinase in the manufacture of the most popular cheeses worldwide with, however, contradictory results. On the other hand, the use of *R. miehei* coagulant increased the antioxidant activity of Burgos-type cheese, showing the highest values for DPPH inhibition and chelating effect compared to those obtained with using calf rennet (Timón et al., 2014).

As far as brined cheese is concerned, Domiati cheese made with *R. mucedo* KP736529 enzyme exhibited higher proteolysis level and higher organoleptical scores than the cheese made with calf rennet (Ayana et al., 2015). Also, Teleme cheese from goat milk showed higher proteolysis level than the cheese made with calf rennet (Yetismeyen et al., 2003). *A. niger* proteinase and *R. miehei* proteinase were used in Tulum cheese and compared to calf rennet showed that α_{s1}-casein and β-casein degradation was higher in the cheese manufactured with *R. miehei* proteinase and both the cheeses were significantly softer than those made with calf rennet (Şengül et al., 2014). In contrary, Feta cheese made with commercial preparations of *Rhizomucor* sp. and *C. parasitica* proteinases did not differ significantly in proteolysis level, yield and organoleptic properties from the typical cheese (Alichanidis et al., 1984). Turkish white brined cheese made with *R. miehei* proteinase exhibited similar concentrations of free amino acids and sensory properties to those of cheese made with calf rennet (Eren-Vapur and Ozcan, 2012).

In mature Cheddar cheese made with Rennilase, the degradation of β-casein was higher than in cheese made with calf rennet, but the type of coagulant had little effect on the sensory scores (Johnston et al., 1994). In low fat Cheddar cheese, made with commercial *R. miehei* coagulant at three different concentrations, the ratio of pH 4.6 soluble nitrogen/total nitrogen increased significantly with increasing coagulant concentration and this resulted in a thinner cheese protein network (Soodam et al., 2015).

Buffalo mozzarella cheese made with *R. miehei* coagulant showed higher proteolysis rate and firmness and the same meltability, fat leakage and oiling off, compared to the cheese made with calf rennet (Ahmed et al., 2011). The same coagulant provided a higher level of thermal stability, proteolysis and meltability values in Malatya, a Halloumi type cheese, than calf rennet (Hayaloglu et al., 2014). In Kashkaval, a popular Mediterranean pasta filata cheese, the use of the enzyme product Rennilase resulted in a coagulum which contained larger casein particles compared to those produced by calf chymosin (Milanović et al., 1998).

Gouda cheese made at laboratory and at a pilot and commercial scale using *R. miehei* coagulant, exhibited lower solid transfer from milk to curds and higher level of proteolysis and bitterness at 12 weeks of ripening than the cheeses using calf rennet (Jacob et al., 2010). Also, Gouda made with microbial coagulant showed not only higher proteolysis level but also higher lipolysis level compared to the control cheese made with calf rennet (El-Tanboly et al., 2000). In a semi-hard ovine cheese made with microbial coagulant, the concentration of bitter peptides (those with a molecular size of 165–6500 g.mol^{-1}) was also the highest (Agboola et al., 2004).

Fresh goat's cheese manufactured with *R. miehei* coagulant did not show any statistically significant difference in physicochemical, proteolysis, sensory and texture profiles, compared to the cheese manufactured with calf rennet (García et al., 2012).

As for Spanish cheeses, *R. miehei* coagulant caused lower proteolysis and hydrophobic peptides level in ovine Manchego cheese than in cheese made with vegetable coagulant or neutral proteinase from *Bacillus subtilis* (Gaya et al., 1999) and this also happened with bovine Hispánico, a semi-hard cheese, using the same coagulants (Carrera et al., 1999).

Furthermore, experiments using non-commercial milk-clotting enzymes derived from bacteria also showed that the produced cheese is subject to a high proteolysis degree. The use of bacterial coagulants derived from *Bacillus* spp. and especially *B. polymyxa*, *B. subtilis* and *B. mesentericus* in different cheese types has been reviewed by Garg and Johri (1994). In general, bacterial proteinases are too proteolytic and result in reduced yield and poor quality of mature cheese. Recently, miniature Cheddar-type cheese made with proteinase from *B. amyloliquefaciens* exhibited higher proteolytic rate and softer texture than the cheese made with calf rennet (An et al., 2014).

7. Conclusions and Future Trends

Microbial milk-clotting enzymes have advantages over animal rennet because (a) they are easily produced, (b) their availability is inexhaustible and (3) their characteristics can be improved, using biotechnology tools. On the other hand, there are technological constraints regarding their suitability for cheesemaking and these derive from their high general proteolytic activity and thermal stability.

Consequently, the first stage in developing an industrially-produced milk-clotting enzyme is to isolate a microbial strain that is characterized by high MCA/PA ratio, low thermostability and also has the potential to produce high yields. Furthermore, the selected strain must neither produce toxins and nor have any pathogenic potential against human, animals or plants, i.e., only risk Class I can be used for commercial microbial coagulant. Till now, several methods, both chemical and genetic, have been developed for improving technological properties of microbial coagulants and new knowledge will continue to be produced in this direction especially with the aid of biotechnology, i.e., expression of genes of known microbial aspartic proteinases in other bacteria, fungi or yeasts.

References

Agboola, S., Chen, S. and Zhao, J. 2004. Formation of bitter peptides during ripening of ovine milk cheese made with different coagulants. Lait, 84: 567–578.

Ageitos, J.M., Vallejo, J.A., Sestelo, A.B.F., Poza, M. and Villa, T.G. 2007. Purification and characterization of a milk-clotting protease from *Bacillus licheniformis* strain USC13. Journal of Applied Microbiology, 103: 2205–2213.

Ahmed, N.S., El-Gawad, M.A.A., El-Abd, M.M. and Abd-Rabou, N.S. 2011. Properties of buffalo mozzarella cheese as affected by type of coagulant. Acta Scientiarum Polonorum, Technologia Alimentaria, 10: 339–357.

Aikawa, J., Yamashita, T., Nishiyama, M., Horinouchi, S. and Beppu, T. 1990. Effects of glycosylation on the secretion and enzyme activity of Mucor rennin, an aspartic proteinase of *Mucor pusillus*, produced by recombinant yeast. Journal of Biological Chemistry, 265: 13955–13959.

Alichanidis, E., Anifantakis, E.M., Polychroniadou, A. and Nanou, N. 1984. Suitability of some microbial coagulants for Feta cheese manufacture. Journal of Dairy Research, 51: 141–147.

An, Z., He, X., Gao, W., Zhao, W. and Zhang, W. 2014. Characteristics of miniature Cheddar-type cheese made by microbial rennet from *Bacillus amyloliquefaciens*: A comparison with commercial calf rennet. Journal of Food Science, 79: M214–M221.

Arima, K., Iwasaki, S. and Tamura, G. 1967. Milk-clotting enzyme from microorganisms. Agricultural and Biological Chemistry, 31: 540–551.

Ayana, I.A.A.A., Ibrahim, A.E. and Saber, W.I.A. 2015. Statistical optimization of milk-clotting enzyme biosynthesis by *Mucor mucedo KP736529* and its further applications in cheese production. International Journal of Dairy Science, 10: 61–76.

Ayhan, F., Çelebi, S.S. and Tanyolaç, A. 2001. The effect of fermentation parameters on the production of *Mucor miehei* acid protease in a chemically defined medium. Journal of Chemical Technology and Biotechnology, 76: 153–160.

Baudys, M., Erdene, T.G., Kostka, V., Pavlik, M. and Foltmann, B. 1988. Comparison between prochymosin and pepsinogen from lamb and calf. Comparative Biochemistry and Physiology, 89B: 385–391.

Beyenal, H., Şeker, S., Salih, B. and Tanyolaç, A. 1999a. The effect of D-glucose on milk-clotting activity of *Mucor miehei* in a chemostat with biomass retention. Journal of Chemical Technology and Biotechnology, 74: 527–532.

Beyenal, H., Şeker, S., Ayhan, F., Tanyolaç, A. and Salih, B. 1999b. Production of microbial rennin in a continuous system fermenter and modelling of the activity [Surekli sistem fermentorde mikrobiyal rennin uretimi ve aktivitenin modellenmesi]. Turkish Journal of Engineering and Environmental Sciences, 23: 83–91.

Branner, S., Eigtved, P., Christensen, M. and Thogersen, H. 1984. Amino acid and NMP analysis of oxidized *Mucor miehei* rennet. Enzyme Engineering: Proceedings of the International Enzyme Engineering Conference, 340–342.

Candioti, M.C., Hynes, E.R., Perotti, M.C. and Zalazar, C.A. 2002. Proteolytic activity of commercial rennets and pure enzymes on whey proteins. Milchwissenschaft, 57: 546–550.

Carrera, E., Gaya, P., Medina, M. and Nuñez, M. 1999. Effect of milk coagulant on the formation of hydrophobic and hydrophilic peptides during the manufacture of cows' milk hispánico cheese. Milchwissenschaft, 54: 146–148.

Chen, M.T., Lu, Y.Y. and Weng, T.M. 2010. Comparison of milk-clotting activity of proteinase produced by *Bacillus subtilis* var, *natto* and *Rhizopus oligosporus* with commercial rennet. Asian-Australasian Journal of Animal Sciences, 23: 1369–1379.

Chitpinityol, S. and Crabbe, M.J.C. 1998. Review: Chymosin and aspartic proteinases. Food Chemistry, 61: 395–418.

Claverie-Martín, F. and Vega-Hernández, M. 2007. Aspartic proteases used in cheese making. pp. 207–219. *In*: J. Polaina and A.P. MacCabe (eds.). Industrial Enzymes. Springer, New York.

Crabbe, M.J.C. 2004. Rennets: General and molecular aspects. pp. 19–45. *In*: P.F. Fox, P.L.H. McSweeney, T.M. Cogan and T.P. Guinee (eds.). Cheese: Chemistry, Physics and Microbiology. Volume 1. General Aspects. 3rd ed. Elsevier Academic Press, London.

Drøhse, H.B. and Foltmann, B. 1989. Specificity of milk-clotting enzymes towards bovine κ-casein. Biochimica et Biophysica Acta (BBA)/Protein Structure and Molecular, 995: 221–224.

Dutt, K., Meghwanshi, G.K., Gupta, P. and Saxena, R.K. 2008. Role of casein on induction and enhancement of production of a bacterial milk clotting protease from an indigenously isolated *Bacillus subtilis*. Letters in Applied Microbiology, 46: 513–518.

El-Tanboly, E., El-Hofi, M.A. and Ismail, A. 2000. Changes of proteolytic and lipolytic activities during ripening of Gouda cheese prepared with fungal rennet substitute. Milchwissenschaft, 55: 624–62.

El-Tanboly, E.-S., El-Hofi, M., Youssef, Y.B., El-Desoki, W. and Ismail, A. 2013. Utilization of salt whey from egyptian ras (cephalotyre) cheese in microbial milk clotting enzymes production. Acta Scientiarum Polonorum, Technologia Alimentaria, 12: 9–19.

Eren-Vapur, U. and Ozcan, T. 2012. Determination of free amino acids in whole-fat Turkish White Brined Cheese produced by animal and microbial milk-clotting enzymes with and without the addition of starter culture [Odred{stroke}ivanje slobodnih aminokiselina u punomasnom turskom bijelom siru u salamuri proizvedenom od životinjskih i mikrobnih enzima grušanja s dodatkom ili bez dodatka starter kulture]. Mljekarstvo, 62: 241–250.

Fazouane-Naimi, F., Mechakra, A., Abdellaoui, R., Nouani, A., Daga, S.M., Alzouma, A.M., Gais, S. and Penninckx, M.J. 2010. Characterization and cheese-making properties of rennet-like enzyme produced by a local Algerian isolate of *Aspergillus niger*. Food Biotechnology, 24: 258–269.

Feijoo-Siota, L., Blasco, L., Rodríguez-Rama, J.L., Barros-Velázquez, J., Miguel, T.D., Sánchez-Pérez, A. and Villa, T.G. 2014. Recent patents on microbial proteases for the dairy industry. Recent Advances in DNA and Gene Sequences, 8: 44–55.

Fernandez-Lahore, H.M., Auday, R.M., Fraile, E.R., Biscoglio De Jimenez Bonino, M., Pirpignani, L., Machalinski, C. and Cascone, O. 1999. Purification and characterization of an acid proteinase from mesophilic *Mucor* sp. solid-state cultures. Journal of Peptide Research, 53: 599–605.

Foltmann, B. 1981. Gastric proteinases-structure, function, evolution and mechanism of action. Essays in Biochemistry, 17: 53–84.

Foltmann, B. 1993. General and molecular aspects of rennets. pp. 37–68. *In*: P.F. Fox (ed.). Cheese: Chemistry, Physics and Microbiology, Vol. 1, 2nd ed. Chapman & Hall, London.

Fox, P.F. and McSweeney, P.L.H. 1997. Rennets: Their role in milk coagulation and cheese ripening. pp. 1–49. *In*: B.A. Law (ed.). Microbiology and Biochemistry of Cheese and Fermented Milk. Blackie Academic & Professional, London.

García, V., Rovira, S., Teruel, R., Boutoial, K., Rodríguez, J., Roa, I. and López, M.B. 2012. Effect of vegetable coagulant, microbial coagulant and calf rennet on physicochemical, proteolysis, sensory and texture profiles of fresh goats cheese. Dairy Science and Technology, 92: 691–707.

Garg, S.K. and Johri, B.N. 1994. Rennet: Current trends and future research. Food Reviews International, 10: 313–355.

Garnot, P. 1985. Heat—Stability of Milk-clotting Enzymes. Technological Consequences. IDF Bulletin 194, International Dairy Federation, Brussels, pp. 2–7.

Gaya, P., Carrera, E., Medina, M. and Nuñez, M. 1999. Formation of hydrophobic and hydrophilic peptides during the manufacture of ewes' milk Manchego cheese using different milk coagulants. Milchwissenschaft, 54: 556–558.

Harboe, M., Broe, M.L. and Qvist, K.B. 2010. The production, action and application of rennet and coagulants. pp. 98–129. *In*: B.A. Law and A.Y. Tamime (eds.). Technology of Cheesemaking, 2nd ed. Wiley-Blackwell, Oxford.

Hashem, A.M. 1999. Optimization of milk-clotting enzyme productivity by *Penicillium oxalicum*. Bioresource Technology, 70: 203–207.

Hayaloglu, A.A., Karatekin, B. and Gurkan, H. 2014. Thermal stability of chymosin or microbial coagulant in the manufacture of Malatya, a Halloumi type cheese: Proteolysis, microstructure and functional properties. International Dairy Journal, 38: 136–144.

Hjort, C. 2007. Industrial enzyme production for food applications. pp. 43–59. *In*: R. Rastall (ed.). Novel Enzyme Technology for Food Applications. CRC Press, Boca Raton.

Hubble, J. and Mann, P. 1984. Destabilization of microbial rennet. Biotechnology Letters, 6: 341–344.

Hyslop, D.B., Swanson, A.M. and Lund, D.B. 1979. Heat inactivation of milk-clotting enzymes at different pH. Journal of Dairy Science, 62: 1227–1232.

ISO-IDF. 2012. Milk and Milk Products—Microbial Coagulants—Determination of Total Milk-clotting Activity. ISO15174/IDF176 International Standard, International Dairy Federation, Brussels.

Jacob, M., Jaros, D. and Rohm, H. 2010. The effect of coagulant type on yield and sensory properties of semihard cheese from laboratory-, pilot- and commercial-scale productions. International Journal of Dairy Technology, 63: 370–380.

Jacob, M., Jaros, D. and Rohm, H. 2011a. Proteolytic activity of traditional calf rennet and selected milk coagulant substitutes. Milchwissenschaft, 66: 240–244.

Jacob, M., Jaros, D. and Rohm, H. 2011b. Recent advances in milk-clotting enzymes. International Journal of Dairy Technology, 64: 14–33.

Johnston, K.A., Dunlop, F.P., Coker, C.J. and Wards, S.M. 1994. Comparisons between the electrophoretic pattern and textural assessment of aged Cheddar cheese made by using various levels of calf rennet or microbial coagulant (Rennilase 46L). International Dairy Journal, 4: 303–327.

Khademi, F., Abachi, S., Mortazavi, A., Ehsani, M.A., Tabatabaei, M.R. and Malekzadeh, F.A. 2013. Optimization of fungal rennet production by local isolate of *Rhizomucor nainitalensis* under solid substrate fermentation system. Journal of Pharmacy and Biological Sciences, 5: 115–121.

Kobayashi, H., Kusakabe, I. and Murakami, K. 1982. Rapid isolation of microbial milk-clotting enzymes by N-acetyl (or N-isobutyryl)-pepstatin-aminohexylagarose. Analytical Biochemistry, 122: 308–312.

Kumar, A., Sharma, J., Mohanty, A.K., Grover, S. and Batish, V.K. 2006. Purification and characterization of milk-clotting enzyme from goat (*Carpa hircus*). Comparative Biochemistry and Physiology-B. Biochemistry and Molecular Biology, 145: 108–113.

Kumar, A., Grover, S., Sharma, J. and Batish, V.K. 2010. Chymosin and other milk coagulants: Sources and biotechnological interventions. Critical Reviews in Biotechnology, 30: 243–258.

Merheb-Dini, C., Gomes, E., Boscolo, M. and da Silva, R. 2010. Production and characterization of a milk-clotting protease in the crude enzymatic extract from the newly isolated *Thermomucor indicae-seudaticae N31* (Milk-clotting protease from the newly isolated *Thermomucor indicae-seudaticae* N31). Food Chemistry, 120: 87–93.

Milanović, S., Kaláb, M. and Carić, M. 1998. Structure of Kashkaval curd manufactured from milk or UF retentate using enzymes of various origin. LWT—Food Science and Technology, 31: 377–386.

Moschopoulou, E., Kandarakis, I., Alichanidis, E. and Anifantakis, E. 2006. Purification and characterization of chymosin and pepsin from kid. Journal of Dairy Research, 73: 49–57.

Moschopoulou, E. 2011. Characteristics of rennet and other enzymes from small ruminants used in cheese production. Small Ruminant Research, 101: 188–195.

Neelakantan, S., Mohanty, A.K. and Kaushik, J.K. 1999. Production and use of microbial enzymes for dairy processing. Current Science, 77: 143–148.

Nouani, A., Belhamice, N., Slamani, R., Belbraouet, S., Fazouane, F. and Bellal, M.M. 2009. Extracellular protease from *Mucor pusillus*: Purification and characterization. International Journal of Dairy Technology, 62: 112–117.

Osorio, A., Gómez, N. and Sánchez, C. 2008. Evaluation of different sources of carbon and nitrogen in the production of rennet from *Mucor miehei* [Evaluación de diferentes fuentes de carbono y de nitrógeno para la producción de renina a partir del moho *Mucor miehei*]. Revista Facultad de Ingenieria, 45: 17–26.

Pandey, A., Selvakumar, P., Soccol, C.R. and Nigam, P. 1999. Solid state fermentation for the production of industrial enzymes. Current Science, 77: 149–162.

Paquet, D. and Alais, C. 1978. Action of fungal proteases on bovine casein and its constituents. Milchwissenschaft, 33: 87–90.

Poza, M., Sieiro, C., Carreira, L., Barros-Velázquez, J. and Villa, T.G. 2003. Production and characterization of the milk-clotting protease of *Myxococcus xanthus* strain 422. Journal of Industrial Microbiology and Biotechnology, 30: 691–698.

Preetha, S. and Boopathy, R. 1994. Influence of culture conditions on the production of milk-clotting enzyme from *Rhizomucor*. World Journal of Microbiology and Biotechnology, 10: 527–530.

Preetha, S. and Boopathy, R. 1997. Purification and characterization of a milk clotting preotease from *Rhizomucor miehei*. World Journal of Microbiology and Biotechnology, 13: 573–578.

Ramet, J.P. 1997. La presure et les enzymes coagulantes. pp. 165–173. *In*: A. Eck and J.-C. Gillis (eds.). Le Fromage, 3rd ed. Lavoisier Tec and Doc, Paris.

Rao, M.B., Tanksale, A.M., Ghatge, M.S. and Deshpande, V.V. 1998. Molecular and biotechnological aspects of microbial proteases. Microbiology and Molecular Biology Reviews, 62: 597–635.

Rolet-Répécaud, O., Berthier, F., Beuvier, E., Gavoye, S., Notz, E., Roustel, S., Gagnaire, V. and Achilleos, C. 2013. Characterization of the non-coagulating enzyme fraction of different milk-clotting preparations. LWT—Food Science and Technology, 50: 459–468.

Şeker, S., Beyenal, H., Ayhan, F. and Tanyolaç, A. 1998. Production of microbial rennin from *Mucor miehei* in a continuously fed fermenter. Enzyme and Microbial Technology, 23: 469–474.

Seker, S., Beyenal, H. and Tanyolaç, A. 1999. Modelling milk-clotting activity in the continuous production of microbial rennet from *Mucor miehei*. Journal of Food Science, 64: 525–529.

Şengül, M., Erkaya, T., Dervişoglu, M., Aydemir, O. and Gül, O. 2014. Compositional, biochemical and textural changes during ripening of Tulum cheese made with different coagulants. International Journal of Dairy Technology, 67: 373–383.

Shammet, K.K., Brown, R.J. and McMahon, D.J. 1992a. Proteolytic activity of some milk-clotting enzymes on κ-casein. Journal of Dairy Science, 75: 1373–1379.

Shammet, K.K., Brown, R.J. and McMahon, D.J. 1992b. Proteolytic activity of proteinases on macripeptide isolated from κ-casein. Journal of Dairy Science, 75: 1380–1388.

Shieh, C.-J., Phan Thi, L.-A. and Shih, I.-L. 2009. Milk-clotting enzymes produced by culture of *Bacillus subtilis natto*. Biochemical Engineering Journal, 43: 85–91.

Silva, B.L., Geraldes, F.M., Murari, C.S., Gomes, E. and Da-Silva, R. 2014. Production and characterization of a milk-clotting protease produced in submerged fermentation by the thermophilic fungus *Thermomucor indicae-seudaticae N31*. Applied Biochemistry and Biotechnology, 172: 1999–2011.

Silveira, G.G., de Oliveira, G.M., Ribeiro, E.J., Monti, R. and Contiero, J. 2005. Microbial rennet produced by *Mucor miehei* in solid-state and submerged fermentation. Brazilian Archives of Biology and Technology, 48: 931–937.

Soodam, K., Ong, L., Powell, I.B., Kentish, S.E. and Gras, S.L. 2015. Effect of rennet on the composition, proteolysis and microstructure of reduced-fat Cheddar cheese during ripening. Dairy Science and Technology, 95: 665–686.

Sternberg, M. 1972. Bond specificity, active site and milk clotting mechanism of the *Mucor miehei* protease. BBA—Protein Structure, 285: 383–392.

Sumantha, A., Larroche, C. and Pandey, A. 2006. Microbiology and industrial biotechnology of food-grade proteases: A perpective. Food Technology and Biotechnology, 44: 211–220.

Tam, J.J. and Whitaker, J.R. 1972. Rates and extents of hydrolysis of several caseins by pepsin, rennin, *Endothia parasitica* protease and *Mucor pusillus* protease. Journal of Dairy Science, 55: 1523–1531.

Thakur, M.S., Karanth, N.G. and Nand, K. 1990. Production of fungal rennet by *Mucor miehei* using solid state fermentation. Applied Microbiology and Biotechnology, 32: 409–413.

Thakur, M.S., Karanth, N.G. and Nand, K. 1993. Downstream processing of microbial rennet from solid state fermented moldy bran. Biotechnology Advances, 11: 399–407.

Theron, L.W. and Divol, B. 2014. Microbial aspartic proteases: Current and potential applications in industry. Applied Microbiology and Biotechnology, 98: 8853–8868.

Timón, M.L., Parra, V., Otte, J., Broncano, J.M. and Petrón, M.J. 2014. Identification of radical scavenging peptides (<3kDa) from Burgos-type cheese. LWT—Food Science and Technology, 57: 359–365.

Trujillo, A.J., Guamis, B., Laencina, J. and Lopez, M.B. 2000. Proteolytic activities of some milk-clotting enzymes on ovine casein. Food Chemistry, 71: 449–457.

Vishwanatha, K.S., Appu Rao, A.G. and Singh, S.A. 2010. Production and characterization of a milk-clotting enzyme from *Aspergillus oryzae* MTCC 5341. Applied Microbiology and Biotechnology, 85: 1849–1859.

Walsh, M.K. and Li, X. 2000. Thermal stability of acid proteinases. Journal of Dairy Research, 67: 637–640.

Yamashita, T., Higashi, S., Higashi, T., Machida, H., Iwasaki, S., Nishiyama, M. and Beppu, T. 1994. Mutation of a fungal aspartic proteinase, *Mucor pusillus* rennin, to decrease thermostability for use as a milk coagulant. Journal of Biotechnology, 32: 17–28.

Yang, J., Teplyakov, A. and Quail, J.W. 1997. Crystal structure of the aspartic proteinase from *Rhizomucor miehei* at 2.15-Å resolution. Journal of Molecular Biology, 268: 449–459.

Yetismeyen, A., Gencer, N., Gürsoy, A., Deveci, O., Karademir, E., Senel, E. and Öztekin, S. 2003. The effect of some technological processes on the properties of Teleme from goat milk. Milchwissenschaft, 58: 286–28.

12

Lipase

Properties, Functions and Food Applications

Maria Antonia Pedrine Colabone Celligoi, Cristiani Baldo,
Marcelo Rodrigues de Melo, Fabiana Guillen Moreira Gasparin,
Thiago Andrade Marques* and *Márcio de Barros*

1. Introduction

The utilization of enzymes in industrial processes arouses great interest due to its availability and advantages in relation to chemical catalysts, more specificity, lower energy consumption and increased reaction speed. In addition, the enzymatic catalysis allows an increase in the quality of products, reduction of production costs and less environmental hazardous impact. Consequently, the enzyme industry has progressively grown. In 2010, the global market for industrial enzymes was estimated at $3.3 billion and is expected to reach $8.0 billion by 2015. Lipase currently ranks third among the most currently commercialized enzymes after proteases and carboxylases (Gupta et al., 2015).

Lipases (triacylglycerol ester hydrolase, EC 3.1.1.3) are enzymes that catalyze the total or partial hydrolysis of fats and oils, releasing free fatty acids, diacylglycerols, monoacylglycerols and glycerol (Villeneuve et al., 2000). These enzymes differ from esterases (carboxylesterases, EC 3.1.1.1) that act only on water-soluble carboxylic ester molecules (Verger, 1997). Under specific conditions, lipases also catalyze synthesis reactions, such as esterification, transesterification (interesterification, acidolysis and alcoholysis), aminolysis (the synthesis of amides) and lactonization (intermolecular

Department of Biochemistry and Biotechnology, Centre of Exact Science, State University of Londrina, 86057-970-Londrina, Parana, Brazil.
* Corresponding author: macelligoi@uel.br

esterification) (Gupta et al., 2015). In general, lipases do not require cofactors, operate in a wide pH range, are stable at high temperatures, have high specificity and exhibits regio-, chemo- and enantioselectivity (Villeneuve et al., 2000).

Owing to their unique properties, lipases are widely used in various industrial sectors, such as food, pharmaceutics, biofuels, oleochemical, textile, agro-chemical, paper manufacturing, cosmetics and many others. In the food industry, lipases can be used as flavor modifiers by synthesis of short chain fatty acids esters and alcohols, and to obtain products of increased nutritional value by modifying the triacylglycerol structure for inter- or transesterification (Verma et al., 2012). In bakery, lipases are potential candidate substitutes for emulsifiers (Colakoglu and Ozkaya, 2012). In the wine industry, these enzymes are used to produce characteristic wine esters. In addition, lipases can be used in many processes, such as to synthesize structural lipids, low calorie lipids and milk fat and in ripening cheese (Gupta et al., 2015; Reetz et al., 2002; Macedo et al., 2003). This chapter describes the main sources of microbial lipases, production by fermentation and enzymatic properties with emphasis on lipases which are used in different sectors of food industry.

2. Microbial Sources of Lipases

Lipases are ubiquitous in nature and are produced by plants, animals and microorganisms (Barros et al., 2010). However, microbial lipases, native or recombinant, are mostly used in different biotechnological applications (Choudhury and Bhunia, 2015).

Nature provides a wide biodiversity of microbial resources. Microorganisms have great adaptive capabilities, even in inhospitable environments, such as the Dead Sea, Antarctica, alkaline lakes, hot springs, volcanic vents and contaminated soils, which offer remarkable potential for production of enzymes with specific characteristics (Borrelli and Trono, 2015; Bonugli-Santos et al., 2015). In this sense, the proportion of enzymes used in the food industry is constantly growing, with perspects of growing more in the coming years, due to the demand for new applications in the fields of dairy and baking, among others. Studies report that marine microorganism capacity in the production of active compounds, including proteins and enzymes (Basheer et al., 2011; Dewapriya and Kim, 2014), reveal new sources of industrial enzymes, some producing microorganism of lipases and their application in the food industry, as shown in Table 1.

3. Production of Lipases by Fermentation

The production of microbial lipases is developed mainly by submerged fermentation (Singh and Mukhopadhyay, 2012), using bench-scale or industrial bioreactors (Gulati et al., 2001; Alonso et al., 2005; Oliveira et al., 2014; Fickers et al., 2006; Asih et al., 2014). Generally, submerged fermentation could be conducted by batch, but productivity is increased by the fed batch or continuous processes (Singh and Mukhopadhyay, 2012). Literature also describes the production of lipase by solid-state fermentation (Singh and Mukhopadhyay, 2012).

Table 1. Microorganisms producers of lipases and their mainly application in food industry.

Source	Food applications	Reference
Aspergillus oryzae	Synthesis of saturated fatty acids	Toida et al., 1998
Bacillus subtilis	Bread making	Sanchez et al., 2002
Candida antarctica	Oils and fats enriched	Anobom et al., 2014
Chromo bacterium Viscosum	Synthesis of aroma and flavour compounds	Carlile et al., 1996
Candida rugosa	Human Milk fat substitute	Ray, 2006
Candida lipolytica	Cheese ripening, fatty acid production	Ray, 2006
Humicola lanuginose,	Non-hydrogenated solid fats	Grillitsch et al., 2011
Penicillium camembertii	Production of glycerolglycolipids	Nakano et al., 1995
Thermomyces lanuginose	Non-hydrogenated solid fats	Grillitsch et al., 2011
Pseudomonas sp.	Food processing and oil Manufacture	Gilbert, 1993
Penicillium camembertii	Synthesis of saturated triacyl glycerides	Fan et al., 2012
Rhizomucor javanicus (meih)	Non-hydrogenated solid fats	Carlile et al., 1996
Penicillum roquefortii	Fragrance development Indairy production	Ha and Lindsay, 1993
Rhizomucor miehei	Oils and fats enriched, cocoa butter substitutes, synthesis of bioactive molecules	Marion and Oliver, 2013
Mucor miehei	Oils and fats enriched, cocoa butter substitutes, synthesis of bioactive molecules	Marion and Oliver, 2013
Penicillium roquefortii	Cheese ripening	Mhetras et al., 2009
Candida rugosa	Cheese ripening	Kosikowski, 1977

Microbial growth occurs in response to different environmental conditions and gene expression. Each microorganism may respond differently to the same environment, as well as the adverse environmental conditions may cause changes in the behavior of the same microbial strain. Usually, the lipase production occurs during the late logarithmic phase or stationary phase. Thus, the cultivation period may vary according to the microorganism and its growth rate (Sharma et al., 2001; Schimidt-Dannert et al., 1994). The majority of lipases are inducible by oils (Sharma et al., 2001), though, other inducers also induce increased production of lipases, such as free fatty acids, hydrolyzable esters, bile salts and glycerol (Ghosh et al., 1996). Lipase producers normally grow in complex growth media, which include carbon sources (oils, sugars), nitrogen sources and salts. Compounds such as olive oil, oleic acid and Tween 80 are also important in enzyme synthesis. Other important factors that significantly influence the lipase production include initial pH, temperature and dissolved nitrogen concentration (Sharma et al., 2001; Elibol and Ozer, 2000). Thus, several studies focus on optimizing the bioprocess parameters using statistical tools for simultaneous optimization of several variables.

Production costs are a major obstacle in industrial application of lipases. An interesting alternative is the immobilization of lipases and thus reducing the production costs due to reutilization of the enzyme (Narwal et al., 2014). Moreover, modern genetic recombination systems are used for insertion of fungal genes in bacteria to increase production. Interestingly, a bacterial lipase gene from *Bacillus subtilis* was expressed in *Saccharomyces cerevisiae* and a significant increase in lipase production was observed. According to the authors, this recombinant strain has wide application in the manufacture of bread (Sanchez et al., 2002).

Another alternative for decreasing lipase production costs is the use of alternative substrates, such as industrial waste or by-products. Marques and co-workers (2014a) showed that dairy effluent constitutes a useful medium for lipase production by *Trichoderma atroviride* 676 and thus reducing the fermentation cost and helping to minimize environmental problems caused by the dairy industry. Residual chicken fat and soybeen oil by-products were also used in the lipase production by a strain of *Pseudomonas* sp. (data not published). Ramani and colleagues (2010) also showed the production of lipase by *Pseudomonas gessardii* using beef tallow as a substrate. Furthermore, Salihu and co-workers (2011) reported the production of lipase by *Candida cylindracea*, using effluent oil obtained in the refining industry. Recently, Asih and co-workers (2014) reported the production of lipase by *Candida cylindracea* using palm oil mill effluent.

3.1 Effect of physical and chemical factors

The biosynthesis lipase may be influenced by various physical and chemical factors, such as agitation rate, temperature and initial pH. The agitation rate is considered an important factor in the production of lipase (Liu et al., 2011). An increase in the agitation rate improves the oxygen transfer rate and mixing efficiency, allowing an increase of cell growth and lipase production. However, high agitation rates may cause shear stress, which leads to negative effects on cell growth and enzyme activity. Liu and co-workers (2011) evaluated different agitation speeds (50 to 400 rpm) on lipase production by *Burkholderia* sp. and observed agitation speed showed significant effects on the overall lipase productivity and cell growth and the maximum overall lipase productivity occurred at 100 rpm. In other study, Cihangir and Sarikaya (2004) investigated the effect of agitation on the lipase activity produced by *Aspergillus* sp. The cultures conducted with agitation (150 rpm) showed enzymatic activity of 16.50 U/mL, while under static conditions, it reached 6.32 U/mL. The authors reported that the agitation speed also influenced the biomass production. Lipase production by *Aspergillus oryzae* was also influenced by the agitation rate and the best results were found on using 300 rpm in 10-liter fermentor (Toida et al., 1998).

Aeration also plays an important role on lipase production by different microorganisms. Lipase production by *Pseudomonas fragi*, *Pseudomonas aeruginosa* and *Rhizopus oligosporus* was reduced by vigorous aeration (Ghosh et al., 1996). For lipase production by *Penicillum roqueforti* IAM 7268, cultivation in a 30 liter jar-fermentors was done in 20 liters of medium at 25°C for 72 hours with agitation speed at 300 rpm and aeration at 20 liters per minute (Mase et al., 1995).

The temperature of incubation is also a critical factor in the production of lipase and may vary in different microorganisms (Ramani et al., 2010). Sharma et al. (2009) optimized the physical parameters for lipase production by using *Arthrobacter* sp. and studying the effects of different pH values (8.0, 9.0 and 10.0), temperatures

(40, 60 and 80°C) and incubation period (12, 30 and 48 hours). The optimum condition for lipase production was 40°C, pH 10.0 and 48 hours of incubation, archiving of 13.75 U/ml of enzymatic activity. The three variables significantly influenced the production of lipase. Yadav and co-workers (2011) evaluated the effects of temperature and initial pH on lipase production by *Yarrowia lipolytica* and found that the best conditions for enzyme production were 20°C and pH 6.0. Studying the production of lipase by *Kluyveromyces marxianus*, Stergiou et al. (2012) reported that temperature of 32.5°C and pH of 6.4 were the best conditions for enzyme production. According to Toida et al. (1998), the high production of lipase by *Aspergillus oryzae* was pH 5.5 and 30°C. For lipase production by *Penicillium roqueforti* IAM 7268, an initial pH of 6.0 and temperature of 25°C was employed (Mase et al., 1995). In another study, pH of 6.0 and 28°C resulted in the best production of lipase by *Trichoderma atroviride* 676 (Marques et al., 2014b).

3.2 Effect of nutritional conditions

Microorganisms are exposed to a variety of nutritional and physical conditions that can modify the yield of lipase production, inducing or suppressing the enzymatic synthesis. Thus, to obtain an economically feasible culture medium it is necessary to select the carbon and nitrogen sources to promote the growth of the microorganisms to obtain the maximum yield in the product (Hasan et al., 2006).

Carbon and nitrogen source deeply influence the fermentative production of lipase. Kantak and co-workers (2011) studied lipase production by *Rhizopus* sp. using various carbon (glucose, sucrose, maltose, olive oil and glycerol) and nitrogen sources (peptone, tryptone, amino acids, casein, yeast extract and corn steep liquor). The results indicated that the best carbon and nitrogen sources are glucose and corn steep liquor, respectively, resulting in a 16-fold increase in lipase activity as compared to the basal medium without the source of carbon and nitrogen.

As already mentioned, lipases are inductive enzymes produced in the presence of lipids as a source of carbon, such as oils (Kumar et al., 2011). Lima et al. (2003) evaluated various oils (olive, corn, soybean and sunflower oil) and glucose as carbon sources in the production of lipase by *Penicillium aurantiogriseum*. The authors reported that best lipase production occurs in a medium containing olive oil that encloses high oleic acid content (C18:1), while corn, soybean and sunflower oils have higher content of linoleic acid (C18:2). There was no lipase production in a medium containing only glucose as the carbon source, confirming that oil is required for lipase production. Similar results were reported by Wang and co-workers (2008) that tested the effect of different oils on lipase production by *Rhizopus chinensis.* Olive oil was also the best carbon source for production of lipase by *Burkholderia multivorans* (Dandavate et al., 2009).

Messias et al. (2009) reported the effect of different plant seed oils and glycerol in the production of lipase by nine isolates of *Botryosphaeria* spp. According to the authors, *Botryosphaeria ribis* EC-01 revealed the highest lipase activity on soybean oil and glycerol, while eight isolates of *Botryosphaeria rhodina* produced significantly lower enzyme activity. For production of lipase by *Debaryomyces hansenii*, the combination of glucose (1.0 g/L) and olive oil (19 g/L olive oil) showed 7.44 U/ml of enzymatic activity (Papagora et al., 2013).

It is well known that surfactants can increase cell permeability, thereby increasing the secretion of various molecules across the cell membrane (Silva et al., 2005). These compounds may be used as the only carbon source for the production of lipase, since they have fatty acids in their chemical composition (Messias et al., 2009). According to Ramani et al. (2010), the addition of surfactant in the culture medium can increase both the activity and the stability of the enzymes. In this sense, Silva et al. (2005), studying the lipase production by *Metarhizium anisopliae,* reported that the highest lipase activity was observed in the presence of sodium dodecyl sulfate and Tween 80.

The influence of nitrogen sources on lipase production has been considerably studied (Hasan-Beikdashti et al., 2012; Bora and Bora, 2012; Liu et al., 2012). Evaluating the effect of nitrogen sources on production of lipase by *Candida* sp., Tan and co-workers (2003) observed that casein, soybean meal and $(NH_4)_2SO_4$ or ammonium sulphate had the positive effect on lipase production. Gupta et al. (2004) reported the effect of nitrogen source on lipase production by *Bacillus* sp. The authors concluded that organic sources of nitrogen significantly stimulated the production of lipase by this microorganism. In another study, yeast extract had a positive effect on lipase production through the use of *Bacillus* sp. (Hasan et al., 2006). Turki et al. (2009) studied the effect of organic nitrogen sources on lipase production by *Yarrowia lipolytica* and reported that peptone and tryptone demonstrated a greater stimulatory effect on lipase production. On the other hand, Ramani and co-workers (2010) reported the adverse effect of peptone, NH_4Cl (ammonium chloride), $(NH_4)_2SO_4$ and urea on lipase production by *Pseudomonas gessardii.* According to the authors, the proteases released by the addition of nitrogen sources in the culture medium may have caused the denaturation of the lipase produced. Peptone and yeast extract significantly influenced the synthesis of lipase by *Stenotrophomonas maltophilia* (Hasan-Beikdashti et al., 2012).

Mineral salts may stimulate or inhibit lipase production (Zong and Li, 2010). Lin et al. (2006) investigated the effects of $MgSO_4$ (magnesium sulphate), $FeSO_4$ (iron sulphate), NaCl (sodium chloride), KCl (potassium chloride), $CaCl_2$ (calcium chloride) and $FeCl_3$ (ferrous chloride) on the production of lipase by *Antrodia cinnamomea*, and observed that only KCl stimulate the lipase production. On the contrary, $MgSO_4$ had a positive effect on lipase production by *Pseudomonas aeruginosa* (Ruchi et al., 2008). Lipase production by *Pseudomonas gessardii* was slightly inhibited KH_2PO_4 and $CaCl_2$ had a stimulating effect on enzyme production (Ramani et al., 2010). It is reported that the production of lipase by *Penicillium camembertii* was significantly inhibited by Hg^{2+} and Fe^{3+} (Chahinian et al., 2000). In another study, Mase et al. (1995) demonstrated that lipase from *Penicillium roqueforti* IAM 7268 displayed high tolerance towards NaCl (sodium chloride).

4. Kinetic and Physicochemical Characteristics of Microbial Lipases

Knowledge of kinetic and physicochemical characteristics of microbial lipases is of fundamental importance in the choice and application of lipases in food production. Moreover, these characteristics are important parameters in differentiating between lipases and esterases. Both enzymes catalyze the hydrolysis of carboxylic esters, belong to the group of α/β-hydrolases and are used in food production (Anobon et al., 2014; Ghosh et al., 1996). Two characteristics initially used to differentiate lipases from esterases were interfacial activation and the presence of an amphiphilic domain covering the active site (lid). Interfacial activation is the increased hydrolytic activity which is observed when lipase is at water/lipid interface. This phenomenon results from enzyme conformational change when oligopeptidic sequence of the lid is at water/lipid interface. The movement of the lid allows access of lipid to the active site of lipase (Jaeger et al., 1999). However, with the discovery of new lipases and elucidation of their three-dimensional structures since the 1990s, it was demonstrated that these two features are not universally present. Lipases from *Pseudomonas aeruginosa* (Jaeger et al., 1993; Jaeger et al., 1994) and *Candida antarctica* (CALB) (Uppmberg et al., 1994) do not show the interfacial activation phenomenon despite having the lid (Verger, 1997), while lipase from *Bacillus subtilis* (Lesuisse et al., 1993) and cutinase from *Fusarium solani* (Martinez et al., 1992) do not present either interfacial activation or the lid. Thus, the main characteristic used to differentiate between lipases from esterases is the ability of the former hydrolyze carboxylic esters of long chain (> 10 carbons) from triglycerides, while esterases activity is limited to carboxylic esters of short chain (< 10 carbons) (Jaeger et al., 1999; Verger, 1997).

Although interfacial activation is not present in all the lipases, lipolysis occurs primarily at water/lipid interfaces so that their kinetic cannot be described by the Michaelis-Menten model, which describes reactions that occur in a homogeneous medium and considers that the enzyme and substrate are soluble in the medium. In this way, kinetic models for reactions catalyzed by lipases should consider the need for the water/lipid interface to occur (Jaeger et al., 1994). Verger and Haas (1976) have proposed a kinetic model to lipolysis (Fig. 1). This model consists of two successive steps—in the first stage occurs the adsorption equilibrium of the enzyme

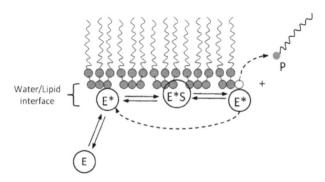

Figure 1. Kinetic model for lipase activity.

at the interface (E ↔ E*); in the second step, the balance follows the kinetic model of Michaelis-Menten, wherein the enzyme adsorbed in the interface (E*) binds to the substrate (S) for forming the enzyme-substrate complex (E*+S ↔ E*S), proceeds to form the product (E*S ↔ E*+P) and to enzyme regeneration in their adsorbed form. However, since the substrate concentration in the adsorbed lipase neighborhood is a concentration on the surface of water/lipid interface, the substrate concentration must be expressed per area unit and not per volume unit (Jaeger et al., 1994). This kinetic model also helps explain the effect of lid in the lipase interfacial activation. Once lid is an amphiphilic sequence, their interaction with a hydrophobic surface at water/lipid interface triggers a conformational change that releases the access of the active site of enzyme to hydrophobic substrates. In turn, the opening of the active site to hydrophobic substrates at the interface results in considerable increase of the hydrolysis activity, which responds by interfacial activation (Brzozowsk et al., 1991; Van Tilbeurgh et al., 1993).

In addition to hydrolysis of the ester linkage between long-chain fatty acids and glycerol lipases exhibit chemo and region selectivity, acting on specific fatty acids and in specific positions on the glycerol molecule, respectively (Rogalska et al., 1997). The specificities regarding the type and position of fatty acid make lipases important tools for modifying physical, chemical and nutritional characteristics of the lipids present in food, allowing manipulation of sensory attributes and nutritional quality of food. Lipase from *Rhizomucor miehie* is used in the food industry for modification of cocoa butter, replacing a palmitic acid molecule by another of stearic acid and resulting in a triglyceride containing stearic-oleic-estearic acids. This modification increases the melting temperature of the natural cocoa butter (37°C) and removes the unpleasant sensory characteristic of melting in the mouth (Coleman and Macrae, 1980). Lipases can also be used to selectively enrich oils and fats with polyunsaturated acids, which are essential for human nutrition (Gill and Valivety, 1997).

Considering the chemo and regio selectivity of lipases in general, these enzymes are grouped according to their ability to hydrolyze fatty acids of short, medium or long chain and for the ability to hydrolyze fatty acids linked to Position 1 and 3 of glycerol or attached to Position 2, respectively. From combination of these properties emerge lipases with varied characteristics. Using hierarchical clustering analysis, Song et al. (2008) grouped lipases of 9 microorganisms into 4 groups. Of these, lipases with a specificity for esters of medium-chain fatty acids and no hydrolysis at position 2 of glycerol (*Mucor racemosus*, *Rhizopus oryzae* and *Trichosporon capitatum*), and lipases with specificity for esters of short-chain fatty acids and hydrolysis at position 2 of glycerol (*Aspergillus niger*, *Aspergillus oryzae* and *Bacillus coughing*) were considered the most important. Table 2 shows the main enzymes used in food processing and production, all of which are of fungal origin. Among these enzymes, the largest group is of lipases used in the dairy industry and whose main activity is the hydrolysis of short- and medium-chain fatty acids.

Regardless of lipase substrate specificity, all microbial lipases share the same mechanism of action for hydrolysis of fatty acids from glycerol and the same basic structure in the active site consisting of a catalytic triad containing the amino acids serine, aspartic or glutamic acid and histidine, organized into a motif β-turn-α that comprises the conserved pentapeptide GX_1SX_2G (G: glycine, X1: histidine,

Table 2. Lipases used in food industry.

Industry	Application	Enzyme	Microorganism	Supplier	pH	T (°C)	Properties
Dairy	EMC	CheeseMax®	*Rhizopus oryzae*	Amano	6.0	45	Hydrolyzes triacylglyceride short, medium and long-chain fatty acids with a preference for short and medium chain fatty acids at the 1 and 3 positions of triacylglycerides.
	EMC	Lipase A "Amano" 12	*Aspergillus niger*	Amano	5.0–6.0	45	Hydrolyzes medium and long fatty acids from 1 and 3 positions of triacylglycerol.
	EMC	Lipase AY "Amano" 30SD	*Candida cylindracea*	Amano	7.0	50	Hydrolyzes short, medium and long fatty acids from 1, 2 and 3 positions of triacylglycerol.
	EMC	Lipase AY "Amano" 30G	*Candida rugosa*	Amano	7.0	45	Hydrolyzes short, medium and long fatty acids from 1, 2 and 3 positions of triacylglycerol.
	EMC	Lipase R "Amano"	*Penicillium roqueforti*	Amano	7.0	40	Hydrolyzes short and medium in preference to long fatty acids from 1 and 3 positions of triacylgricerol.
	EMC	Piccantase	*Mucor miehei*	DSM	7.0	37–40	Mixed fungal esterase.
	EMC (american cheddar type flavour)	Lipomod™ 187P-L187P	mixed fungal esterasc, including *Penicillium roqueforti*	Biocatalysts	6.0–7.0	40–50	Mixed fungal esterase.
	EMC (blue-type flavors)	Lipomod™ 338P-L338P	*Penicillium roqueforti*	Biocatalysts	5.0–7.0	40–50	Strong preference for hydrolysis of short chain fatty acids from triglycerides.
	EMC (mild buttery flavours)	Lipomod™ 34P-L034P	*Candida cylindracea [rugosa]*	Biocatalysts	5.0–8.0	40–55	Broadly active but more specific for short chain fatty acids. It is active against all 3 positions on the triglyceride molecule.
	Cheese flavor (cheddar-type flavors)	Lipomod™ 621P-L621	*Penicillium* sp./ *Aspergillus* sp./ *Candida* sp.	Biocatalysts	5.0–7.0	40–50	Mixed fungal esterase and protease.
	EMC (cheddar-type flavors)	Lipomod™ 29P-L029P	*Candida cylindracea* + porcine pancreas	Biocatalysts	6.0–8.0	40–50	Blend of animal/fungal lipases. It contains both lipase and esterase activity and is active against all 3 positions on the triglyceride molecule.

	Use	Trade name	Source	Manufacturer	pH	Temp	Properties
	EMC (rounded cream notes)	Lipomod™ 691P-L691P	*Candida* sp./*Rhizopus* sp.	Biocatalysts	6.0–7.0	40–50	Mixed fungal lipase.
	EMC (subtle sweet undertones)	Lipomod™ 768P-L768P	Mixed Fungal	Biocatalysts	5.0–8.0	40–55	Mixed fungal lipase. It is active against all 3 positions on triacylglycerol substrates (fat); hydrolysing short, medium and long chain fatty acids.
	Cheese-flavor enhancement	Palatase®	*Rhizomucor miehei*	Novozymes	7.0–10.0	30–50	Hydrolysis of small fatty acids on the carboxylic acid part.
Oil and fat	Fats and oils processing	Lipase M "Amano" 10	*Mucor javanicus*	Amano	7.5	35	Hydrolyze short, medium and long chain fatty acids at 1, 2, and 3 positions of tri, di and monoglycerides.
	Fats and oils processing	Lipase G "Amano" 50	*Penicillium camembertii*	Amano	5.0	40	High sterifying activity and hydrolyzes glycerides and patial glycerides more rapidly than triglyceride, producing glycerol and fatty acid.
	Fats and oils processing	Lipase DF "Amano" 15	*Rhizopus oryzae*	Amano	6–7	35–40	Positional specificity for the 1 and 3 positions of glycerides, and hydrolyzes ester bonds of 1 and 3 positions of triglycerides. Relatively specific to fatty acids with long and medium chain length.
	Interesterification of vegetable oil	Lipozyme® TL IM	*Thermomyces lanuginosus*	Novozymes	7.0–10.0	20–50	Transesterification, interesterification, ester hydrolysis and desymmetrization os esteres.
Baking	Emulsifier	Lipopan® F	*Thermomyces lanuginosus* (donor)/*Aspergillus oryzae* (host)	Novozymes	6.5–7.0	40–45	Specifically acts on the fatty acid in position 1 in both triglyceride substrates and, like other known lipases', in phospholipid substrates.
Noodles/pasta	Improvement of quality of noodles and wheat-based pasta products	Noopazyme®	*Thermomyces lanuginosus* (donor)/*Aspergillus oryzae* (host)	Novozymes			A purified 1,3 specific lipase.

EMC: Enzyme modified cheese.

Information related to the mentioned enzymes were obtained from the manufacturers' websites.

S: serine, X_2: aspartic or glutamic acid) (Bornscheuer, 2002). The hydrolysis of ester bond between the fatty acid and glycerol begins with the binding of the catalytic serine to carbonyl in the ester bond of lipid, forming a tetrahedral intermediate. This intermediate is stabilized by the other catalytic residues, His and Asp/Glu and the ester bond is cleaved with fatty acid esterified to catalytic serine. Then, one molecule of water hydrolyzes the ester bond between the serine and fatty acid, forming another tetrahedral intermediate, from which the fatty acid is released (Jaeger et al., 1999).

Despite using the same basic catalytic mechanism, the physicochemical properties of lipases vary considerably, with temperature and pH being some of the most important characteristics for application of these enzymes in processing and production of foods. In the food industry, for example, processing temperatures and pHs are relatively mild and these enzymes should present optimum activity within the processing conditions (López-López et al., 2014). As shown in Table 2, most of the lipases used in food production show thermophilic features, with optimum activity at temperatures above 40°C and pHs of action around neutrality. Food processing at elevated temperatures reduces the risk of contamination with mesophilic microorganisms, reduces the viscosity of the feedstock and increases the reaction rate (Haki et al., 2003). Other important factors to lipase activity are the presence of cations in the reactional medium. In general, it is observed that Ca^{2+} stimulates lipase activity, which explains, in part, the success of lipases in the dairy industry. On the other hand, cations like Co^{2+}, Ni^{2+}, Sn^{2+}, and Hg^{2+} are strong lipase inhibitors, whereas Zn^{2+}, Mg^{2+}, EDTA and SDS result in moderate inhibition of these enzymes (Ghosh et al., 1996).

5. Lipases in Food Industry

5.1 Lipase in aromatic compounds

The production of low-molecular-weight esters, such as aromatic compounds has great importance in the food industry (Ferreira-Dias et al., 2013). In this context, the demand for esters and aromatic fragrance for food, cosmetics and pharmaceutical industry rank fourth among the most currently commercialized food additives (Ahmed et al., 2010). Low-molecular-weight ester may be obtained by direct extraction of fruit, enzymatic synthesis or chemical synthesis by esterification reactions in the presence of inorganic catalysts at elevated temperatures. However, esters obtained by chemical synthesis are considered 'unnatural' and have lower market value when compared to natural (Mendes et al., 2012). The use of enzymes for ester synthesis is an important alternative for obtaining the 'natural' compounds that retain the odor compared to the esters prepared by chemical synthesis (Ahmed et al., 2010). The main flavors synthesized by microbial lipases are shown in Table 3.

The 2-phenylethyl acetate (2-PEAc), a rose aromatic ester, is widely used in cosmetics, soaps, food and drinks. Kuo and co-workers (2012) reported the enzymatic synthesis of 2-PEAc by vinyl acetate transesterification with 2-phenethyl alcohol catalyzed by an immobilized lipase (Novozym ® 435) of *Candida antartica*. Terpene esters of fatty acids are also essential oils applied in pharmaceuticals, cosmetics and the food industry. Among them, the acyclic terpene alcohols—geraniol and citrollenol—are the most important (Serri, 2010). Interestingly, the use of a lipase from *Aspergillus*

Table 3. A summary of some aromatic esters in food industry.

Esters	Flavors	Source	Reference
Ethyl acetate	Pineaple, apple, banana, apricot and butter	*Candida rugosa* lipase (CRL)	Ozyilmaz and Gezer, 2010
Hexyl-acetate or hexanoate	Pear	*Candida rugosa* lipase (CRL)	
ethyl valerate	apple	*Candida rugosa* lipase (CRL)	
Butyl acetate isoamyl acetatoe	Banana	*Candida rugosa* lipase (CRL)	
2-Phenylethyl acetate (2-PEAc)	Rose	(Novozym ® 435) *Candida antarctic*	Kuo et al. (2012)
Citronellyl esters (acetate, propionate, butyrate, oleyc	Feeling of freshness, fragrant, bitter taste	(Novozym 435) de *Aspergillus niger* recombinante	Paroul et al., 2012; Smaniotto et al., 2014
Geranyl propionate	Aroma of rose, sweet	*Penicillium crustosum*	Ferraz et al. (2015)

niger to synthesize terpene alcohol esters of short chain (C3-C6) and fatty acid esters (methyl acetate, methyl propionate, methyl butyrate, 2,2-dimethyl methyl valerate) has already been reported (Contesini et al., 2010). Indeed, several citronelil esters (acetate, propionate, butyrate, oleic) have been synthesized using immobilized lipase (Novozym 435) from *Aspergillus niger* (Paroul et al., 2012; Smaniotto et al., 2014). The production of geranyl propionate using non-commercial immobilized lipase of *Penicillium crustosum* was also studied by Ferraz and co-workers (2015). According to Joseph and co-workers (2008), lipase from *Pseudomonas* strains P38 has been used for synthesis of n-heptane butyl caprylate flavoring compounds. In summary, these results demonstrate the great potential of lipases for obtaining several aromatic compounds frequently used in the food industry.

5.2 Lipase in dairy products

Lipases are used in the dairy industry to break the milk fat, modifying the length of the fatty acids and increasing the production of flavors during ripening of cheese (Sharma et al., 2001; Ghosh et al., 1996; Ray, 2012). Traditional sources of lipases for enhancing the cheese flavor are pancreatic enzymes (bovine and porcine) and gastric tissue lipases from young ruminants (calf, lamb, kid). Additionally, microbial lipase are also employed to produce cheese of good quality (Aravidan et al., 2007). Some examples of microbial lipases used in the production of cheese are shown in Table 4.

Lipases from *Yarrowia lipolytica* are considered important for cheese maturation due to the release of fatty acids (propionic, myristic, palmitic, palmitoleic, butanoic, stearic and oleic acid) responsible for the sensory characteristics of the cheese. The butanoic acid, for instance, is responsible for the flavor of Cheddar and Camembert cheese (Zinjarde, 2014). Lipase from *Apergillus niger* was evaluated in an encapsulated form for accelerating the ripening of cheese. The sensory evaluation results showed that

Table 4. Microbial lipases used in cheese production.

Cheese	Lipase source	Reference
Romano, Domiati, Feta, Fontina, Ras, Romi, Lighvan	*Mucor miehei*	El-Hofi et al., 2011 Aminifar and Emam-Djomeh, 2014
Camembert, Parmesan, Provolone, Brie	*Penicillium camemberti*	El-Hofi et al., 2011 El-Fadaly et al., 2015
Roquefort, Cheddar, Manchego, Blue cheese Gorgonzola, Stilton, Danish blue	*Penicillium roqueforti* *Aspergillus oryzae/* *Aspergillus niger*	El-Hofi, 2011 Walker and Mills, 2014 El-Fadaly, 2015
Chedar, Dutch	Lactic Bacteria	El-Fadaly et al., 2015
Mozzarella	*Fusarium venenatum*	Nielsen and Hoier, 2009

the encapsulated form showed the best appearance, odor and texture when compared with other treatments (Yazdanpanah, 2014).

Changes of texture, microstructure and free fatty acid contents of lighvan cheese during accelerated ripening with lipase from *Rhizomucor miehei* were also reported (Aminifar and Eman-Djomeh, 2014). The main changes resulting from the addition of lipase were accumulation of long, medium and short-chain fatty acids, increase of hardness in cheese and a reduction in the average diameter of the fat globules trapped in the clot network. After 90 days of maturation, the fat globules completely disappear and the protein matrix increases, filling the empty spaces organized by the fermentation process.

The *Rhizomucor miehei* lipases are commercially available in soluble and immobilized form and may be applied in dairy processing. The genome of *R. miehei* CAU432 has 40-glycerol-ester hydrolase genes including the group of true lipases (triacylglycerol lipases). Furthermore, the fungus has 29 genes belonging to the group of true phospholipases (lisofosfolipases) and 24 genes being thiolester and sulfuric-ester hydrolases (Zhou et al., 2014).

In dairy processing, phospholipases represent an important group of enzymes useful for increasing the retention of fat in the rennet curd. Libaek and co-workers (2006) studied the effect of phospholipase A1 on the production and functional properties of mozzarella cheese and found that the enzymatic treatment with phospholipase increased the retention of fat in cheese due to an increase in lysophospholipids in the cheese curd.

Concerning the microbial lipase used in dairy processing, the *Rhizopus oryzae* is a major zygomycetes fungi approved by the human consumer since is considered safe by the FDA (Gosh and Ray, 2011). Enzyme-modified cheese (EMC) is, when incubated, in the presence of lipases or proteases at elevated temperature, producing a concentrated flavor with an intensity ranging from 10 to 30 times higher than normal cheese (Aravidan et al., 2007; Beermann and Hartung, 2012). The cheese flavor is very complex and differs between different types of cheeses (El-Hofi et al., 2011). The flavor and its intensity depends on the enzymes, starter microorganisms, fermentation and emulsifiers added to the production process. Indeed, in order to control the production it is necessary to know the activity and specificity of each enzyme (Beermann and Hartung, 2012).

The numerous compounds involved in the aroma and flavor of cheeses are derived from lactate, protein and lipids catabolism (Molimard and Spinnler, 1996; El-Hofi, 2011). Hydrolysis of lipids into fatty acids may contribute directly to the flavor of cheese. The free fatty acids could also indirectly act as precursors for the production of volatile flavor compounds, such as esters, lactones and ketones (El-Hofi, 2011). Importantly, the type of flavor depends on the chain size of fatty acids, whereas a short-chain, a medium-chain and long-chain fatty acids result in cheese, butter and soap flavor, respectively (Beermann and Hartung, 2012).

In blue cheese, specific methylketones produced by *Penicillum* in lipidic medium, contribute to its characteristic flavor. The production of methylketones in blue cheese by lipase from *Penicillum roquefort* was evaluated by Cao and co-workers (2014), using gas chromatography and sensory methods to quantify the volatiles products. Blue cheese treated with lipase showed three times more methylketone than the non-treated blue cheese. The production of methylketones by *P. roquefort* has been exploited in the cheese industry to provide the characteristic flavor to blue cheese. Methylketones appear to be produced by a futile cycle associated with incomplete β-oxidation of fatty acids, by formation of medium-chain 3-oxoacyl-CoA intermediates that are converted by thioesterases, releasing coenzyme A and the 3-oxoacids that are decarboxylated to methylketones. The process possibly occurs due to the peroxisomal β-oxidation pathway (Walker and Mills, 2014).

5.3 Lipase in oils and fats industry

Fats and oil modification is one of the primary areas in the food-processing industry that requests novel economic and green technologies. Due to the versatility to specifically catalyze hydrolysis reactions, such as esterification and interesterification, lipases are extensively used in the field (Rodrigues and Lafuente, 2010; Singh and Mukhopadhyay, 2012; Garcia-Galan et al., 2013; Sharma and Kanwar, 2014).

Lipases are enzymes which are able to modify the properties of the lipid, changing the location of the fatty acid chains in the triglyceride molecule and/or exchanging one or more fatty acid chains by a new molecule (Barros et al., 2010; Garcia-Galan et al., 2013). Lipases could be used to produce oils with specific compositions, without the use of chemical catalysts that might contaminate the final product, getting a higher value-added product and without harming the environment (Ray, 2012; Garcia-Galan et al., 2013).

After evidences of deleterious effects of trans-fats on human health, the partial hydrogenation process used to produce margarine, for example, has been replaced by other modification methods of oils and fats, such as fractionation, interesterification and full hydrogenation. The interesterification that involves rearranging the fatty acid chains in a triacylglycerol may be promoted by catalysis—alkaline or enzymatic. The alkaline catalysis produces a mixture of triglycerides in which the fatty acids are randomized while catalysis enzyme with lipase is specific when changing the positions 1 and 3, without altering the binding of the fatty acid in the position 2 of glycerol (Messias et al., 2011).

The interesterification of vegetable oil is used to obtain the 'hard stock', that is the main component in margarine. The conventional interesterification process occurs at high temperatures and it is catalyzed by an inorganic catalyst of sodium methoxide. The process uses a great quantity of energy and produces by-products which are removed in a series of bleaching steps (Jegannathan and Nielsen, 2013). Alternately, an immobilized lipase may be used as a catalyst in the interesterification of oils. The process is highly specific, runs at low temperature and generates less by-product (Holm and Cowan, 2008).

Lipase from *Thermomyces lanuginosus*, commercially available as TLL (IM® Lipozyme TL), is frequently used in oil and fat due to its high stability when immobilized. The TLL has been used in the hydrolysis of oils and fats to obtain fatty acids and free glycerol (Fernandez-Lafuente, 2010).

Other applications of lipases include hydrolysis of fish oil for obtaining polyunsaturated fatty acids. Lyberg and Adlercreutz (2008) evaluated the specificity of five lipases in the hydrolysis of eicosapentaenoic acid (EPA) and docosapentaenoic acid (DHA) present in fish oil in a system containing methyl ester of EPA, DHA and palmitic acid. The authors reported a higher hydrolysis activity of both EPA and DHA in fish oil as compared to the methyl esters. Among the lipases tested, the TLL and one lipase from *Candida rugosa* showed greater efficiency.

In addition, lipase from *Rhizomucor miehei* (RML) that is commercial available, has been reported in the hydrolysis of fatty acids. Although RML is a sn-1,3-specific lipase, the spontaneous acyl migration from position 2 to positions 1 or 3 allow the full hydrolysis of triglycerides. Studies have shown that the enzyme showed high efficiency in the hydrolysis of different types of oils as peanut oil, which has the same high concentration of polyunsaturated fatty acids in position sn-1,3 (Rodrigues and Ferandez-Lafuente, 2010). In another study, Sovavá and co-worker (2008) evaluated the hydrolysis activity of RML on blackcurrant seed oil (rich in linoleic and linolenic acids) in order to obtain different levels of α- and/or γ-linolenic acid in the mixture of liberated fatty acids and in the fraction of di- and monoacylglycerols, making them suitable for special dietary needs. The results showed that the liberated fatty acids contained more α-linolenic, palmitic and stearic acids, while di- and monoacylglycerols contained increased levels of γ-linolenic and stearidonic acids.

The esterification reaction is considered the inverse of reaction of lipase in Nature. This reaction involves direct condensation of a fatty acid and an alcohol that may be glycerin, a mono- or diglyceride, or other alcohol (Rodrigues and Ferandez-Lafuente, 2010). One example, is the process of palm oil deacidification by *Rhizomucor miehei* lipase (Kumar and Krishna, 2014). Crude red palm oil of 8.7 per cent free fatty acid content was deacidified using lipase, solvent (ethanol) and sodium hydroxide. Deacidification of oil, using enzyme, showed nearly 100 per cent product yield. The enzyme deacidified oil showed a higher value in unsaponifiable matter (0.91 per cent), monoacylglycerols (2.8 per cent) and diacylglycerols (18.7 per cent) contents as compared to the other two methods of deacidification.

Milk fat can be considered as a good source of essential fatty acids and fat-soluble vitamins. Due to its versatility and complexity, the milk fat provides both opportunities and challenges for modifying its composition for different applications (Soumanou et al., 2013; Kontkanen et al., 2011). Tecelão and co-workers (2012) studied the potential of *Rhizopus oryzae* lipase (rROL) to form substitutes of triglycerides in the human breast milk. The rROL showed a percentage of incorporation of oleic acid of about 30 and 22 mol.per cent of oleic acid incorporation, when rROL was immobilized in lewatit and accurel, respectively. The lipase Commercial RML-Lipozyme RM IM (*R. miehei*) also was successfully used in the incorporation of oleic acid in triglyceride fat milk (Zhang et al., 2013).

In the dairy industry, lipases from yeast and filamentous fungi with high specificity have multiple applications as in low-calorie milk fat synthesis. Lipases from *Candida antarctica* (Novozyme 435) and from *Rhyzomucror miehei* (1M RM) have been described as important enzymes for synthesis of di- or monoglycerides of special composition (Gupta, 2015). The literature also describes the insertion of linoleic acid in dairy foods by immobilized lipases (Ray, 2012).

Omega-3 fatty acids revealed many benefits in human health. The two most important omega-3 long chains are EPA and DHA, the last one being essential for brain development and fetal retina (Swanson et al., 2012). Milk fat can be enriched with omega 3 (linolenic acid) which is the precursor in the metabolism for the formation of EPA and DHA. Faustino et al. (2015) analyzed the efficiency of non-commercial heterologous lipase from *Rhizopus oryzae* (rROL), immobilized on Lewatit VP OC 1600 or on Relizyme OD403/S, and immobilized commercial lipase from *Rhizomucor miehei* (Lipozyme RM IM) in the incorporation of omega-3 linolenic acid in tripalmin. The authors observed 48.9, 43.6, and 18.3 mol. percentage of fatty acid incorporation in triacylglycerols (TAG) by using Lipozyme RM IM and rROL immobilized on Lewatit or on Relizyme, respectively. The rROL immobilized on Lewatit was selected as an alternative to the commercial immobilized lipases.

Cocoa butter fat is an important co-product of chocolate production, but it is often is short supply and the price can fluctuate widely (Andaluema et al., 2012). Therefore, the fat substituents of cocoa butter are very important in the food industry. Cocoa butter consists mainly of triglycerides esterified with saturated fatty acids (palmitic and stearic acid) at position sn-1,3 and monounsaturated fatty acid (oleic acid) at position sn-2. This particular structure is responsible for the rheological and sensorial characteristics found in cocoa butter (Mohamed, 2012; Ferreira-Dias et al., 2013). Lipases are able to catalyze the transesterification reactions of inexpensive oils and one fraction of palm oil (pal mid fraction) and produce cocoa butter. The Unilever and Fuji Oil industries have developed an enzymatic process for obtaining a substituent lipid matrix cocoa butter on a large scale (Coleman and Macrae, 1977; Matsuo et al., 1981). Cocoa butter equivalent was also produced by acidolysis of refined palm oil and palmitic, stearic acids (Mohamed, 2012) using the commercial immobilized lipase IM Lipozyme from *Mucor miehei* (Novo Nordisk A/S Corp.). The process resulted in production of cocoa butter equivalent with melting temperature ranging between 34.7 and 39.6°C, which is similar to that of cocoa butter.

Monoglycerides (MG) are amphipathic molecules widely used as emulsifying agents in the food industry. They are obtained from the chemical catalysis of glycerolysis of fats or oils to high temperatures (220–250°C) that consume high energy and result in a dark-colored product due to formation of by-products. Moreover, the chemical process yields a mixture of MG (5–60 per cent), diacylglycerols (DG) (35–50 per cent); triglycerides (1–20 per cent); free fatty acids (1–10 per cent) and alkali metal salts. However, the ideal would be to obtain a content of 80–90 per cent of MG. Then, several studies were conducted to produce MG from glycerol esterification with palmitic acid. Freitas et al. (2010) reported that 46 per cent of MG, 6 per cent, 3 per cent of diacylglycerols and 3 per cent triglycerides were obtained using a lipase from *Penicillium camemberti* immobilized on Epoxy-PVA-SiO2 composite. Lipases from *Penicillium* genus exhibit great skill in modifying oils and fats, such as enzymes from *P. roqueforti, P. camembertti, P. expansum* and *P. abeanum* (Li and Zhong, 2010). Kapoor and Gupta (2012) studied different formulations of lipase B from *Candida antarctica* (CLEAS: cross-linked enzyme agregates; PCMS: protein coated microcrystals; CLPCMCs: cross linked protein coated microcrystals) and found 81 per cent of MG and 4.5 per cent DG using the CLEAS formulation; 82 per cent of MG and 4 per cent of DG with PCMCs; and 87 per cent de MG e de 3,3 per cent de DG using the CLPCMs formulation.

5.4 Lipases in Bakery

Bakery products, such as bread, cake, sweets (pastries), crackers, cookies, pies and tortillas are composed of carbohydrate (flour and sugar), protein (wheat flour and eggs) and lipids (flour, eggs, emulsifiers, margarine and butter). Lipids are important in providing benefits to the process of baking and storage. Furthermore, they can be enzymatically converted into useful emulsifiers for bread dough (Miguel et al., 2013). In this context, lipases reinforce stability of the mass, increase loaf volume and improve the texture and shelf-life (Primo-Martin et al., 2006; Moayedallaie et al., 2010).

Wheat flour lipids are subdivided into amide lipids that are located in the starch granules and non-starchy lipids that may be found free or bound to other constituents of flour as proteins. The starch granules are called polar and can be extracted with polar solvents, such as butanol, while the non-starchy lipids can be extracted with non-polar solvents, such as n-hexane. The surfactants used in bakery are diacetyl tartaric esters of mono- and diglycerides (DATEM), sodium stearoyl lactylate (SSL) or monoacylglycerols. These surfactants are important to stabilize the gas-liquid interfacial interaction during the fermentation process of backing (Gerits et al., 2014a).

The use of lipase in bakery is quite recent as compared to other enzymes, like α-amylase and proteases (Colakoglu and Özkaya, 2012). The first generation of commercial lipase preparations was introduced in the market in the year 1990 and was used to improve the dough stability and increase the dough volume (Qi Si, 1997; Miguel et al., 2013). The second generation of lipases produces more polar lipids, provided a greater increase in volume, better stability to mechanical stress on the dough and a fine, uniform bread crumb structure compared to the first generation lipases. Recently, a third generation lipase was found to increase expansion of the gluten network, increase the wall thickness and reduce cell density, enhancing the volume and crumb structure of high-fiber white bread (Miguel et al., 2013).

In this sense, lipase functionality in bread manufacturing is related to the hydrolysis of one or more fatty acids from nonpolar triglycerides and/or polar lipids (phospholipids and glycolipids) to form the polar mono- and diacyl-forms. These products have been investigated due to their significant emulsifier characteristics and equivalent functionality to DATEM that have been extensively used as dough conditioners and/or crumb softeners in breadmaking (Colakoglu e Özkaya, 2012).

Three reasons may explain the recent interest in lipases in the baking industry. First, the lipases may replace the traditional surfactants by *in situ* generation of surfactants through hydrolysis of endogenous lipids. This process would reduce the cost of production, storage and transport. Second, lipases are not detected in the final product due to the denaturation during the baking process. Third, lipases are easily produced at ambient conditions and can easily be heterologously expressed (Gerits et al., 2014b).

The use of microbial lipases to improve dough properties was described by Olesen et al., 2000 (Patent 6110508) that discloses a method for improving dough properties through the addition of microbial lipases from *Humicola lanuginosu, Rhizomucor miehei, Pseudomonas cepacia,* and *C. Pastase, Rhizomucor miehei.*

The influence of a fungal lipase on bread quality was reported by Paslaru and co-workers (2008) after studying the commercial enzyme Belpan type B from *Rhizopus oryzae.* The enzyme was capable of increasing the dough stability and the loaf volume, making the bread fresh for a long time due to the increase in porosity and elasticity.

Studying the role of two commercial lipases, Lipopan F-BG (Novozyme, Denmark) and Grindamyl Powerbake 4101 (Danisco, Denmark), Colakoglu and colleagues (2012), reported a decrease in the softening degree, stickiness and increasing stability and maximum resistance to extension and hardness. The results were comparable to those obtained for DATEM. Moayedallaie and co-workers (2010) studied the effects of Lipopan F-BG, GRINDAMYL Exel-16, Lipopan Xtra, Lipopan 50-BG and DATEM emulsifier on white flour bread. Except for the enzyme Lipopan 50-BG, all enzymes and emulsifiers induced an increase in loaf volume with no significant difference between them. Lipopan F-BG is a lipase from *Apergillus oryzae* which is able to modify the apolar and polar lipids of flour (Sinkuniene et al., 2011). Lipopan Xtra is a phospholipase from *Thermomyces* sp. expressed in *Apergillus. oryzae* (Patent WO2014161876 A1, Bellido et al., 2014). GRINDAMYL Exel-16 is a *Thermomyces lanuginosus* lipase expressed in *Apergillus oryzae* (WO 2003/099016, Olsen et al., 2005).

The effect of four commercial lipases of microbial sources: YieldMAX (phospholipase A1 from *Fusarium venenatum*), Lipolase (lipase from *Thermomyces lanuginosus*), Lecitase Ultra (resulting from the combination of phospholipase *T. lanuginosus* gene and *Fusarium oxysporum*) and Lipopan F (an enzyme preparation of *Fusarium oxysporium*), on bread volume was evaluated (Gerits et al., 2014b). The increase in loaf volume by lipases was attributed to stabilized gas cells during fermentation and requires a balance between the different types of lipids. Enzymes increased the volume of bread by changing the composition of wheat flour lipids by decreasing the levels of galactolipids and phospholipids, and increasing the lysolipids and free fatty acids (Gerits et al., 2014b). Schaffarczyk and co-workers (2014) confirmed the importance of lipid composition and performance of bread flour during the baking process after the treatment of flour with two different commercial microbial lipases (Lipopan F and Lipopan Xtra -BG -BG). The changes induced by lipases in the lipid fraction caused increase in the bread volume by 56–58 per cent, depending on the type and concentration of added lipase.

As far as we know, there are a few bacterial lipases used in the bakery industry. A study shows the use of a lipase from *Bacillus subtilis* expressed in *Saccharomyces cerevisiae* as a cell-wall-immobilized enzyme (Paciello et al., 2015).

5.5 Lipases in wine

Wine is a complex mixture of thousands of compounds that contribute to its color, taste and aromatic properties (Sumby et al., 2010). Studies on wine aroma have received much attention in recent decades and numerous sensory compounds have been identified. Among these compounds, the ethyl esters have received great attention due to its pronounced influence on taste (Sumby et al., 2010; Lee et al., 2012; Esteban-Torres et al., 2014). The ethyl acetate ester is the most common compound found in wines. However, other esters have been reported in literature—such as, 2-ethyl hidroxpropionato, diethyl butanediato, ethyl butanoate, ethyl hexanoate, octanoate, ethyl decanoate, ethyl 2-methyl-propionate, ethyl 3-methyl-propionate, ethyl 3-methylbutanoate, ethyl cinnamate, methyl-butyl acetate, 2-phenyl-ethyl acetate and hexyl acetate (Sumby et al., 2010; Esteban-Torres et al., 2014).

Wine esters can be classified into two groups—those formed enzymatically during the aging process of the beverage, and those formed by chemical esterification between an alcohol and an acid. Together with esterases, lipases with specific properties, such as high activity on pH 5–7, stability in the presence of ethanol, sodium metabisulfate, malic, tartaric, citric and lactic acid (Sumby et al., 2010), can be used to produce ethyl acetate, ethyl butanoate, ethyl hexanoate and ethyl octanotato (Sumby et al., 2010; Esteban-Torres et al., 2014).

Esteban-Torres and colleagues (2014) evaluated the characteristics of the lipase/esterase produced by *Escherichia coli* BL21 genetically modified by insertion of gene encoding the lipase/esterase enzyme derived from *Lactobacillus plantarum* WCFS. The enzyme produced by the microorganism presented a high potential for application in the wine production process and showed high activity at low pH and stability in the presence of ethanol, sodium metabisulfite and tartaric, lactic and citric acids.

In another study, Lee and co-workers (2012) evaluated the effect of sequential inoculation of yeasts, *Williopsis saturnus* var. mrakii NCYC2251 and *Saccharomyces cerevisiae* var. bayanus R2, on ester production during the processing of papaya wine. The authors suggest that sequential inoculation of non-*Saccharomyces* and *Saccharomyces* yeasts may be a useful tool to manipulate yeast succession and to modulate the volatile profiles and organoleptic properties of papaya wine.

5.6 Other applications

Lipases could also be used for the removal of fat from meat and fish products by a procedure called biolipolysis. In addition, they are also important in the fermentation processes of sausage and are used to determine changes in the long-chain fatty acids during its ripening. The utilization of lipases in the refining of rice flour, modification of soy milk, and improving and accelerating the fermentation process of rice wine also was described. Immobilized lipases were also used for esterification of phenols from functionalized vegetable sunflower oil for synthesis of antioxidants (Andualema et al., 2012). The application of lipases as antioxidants for incorporation in oils was also cited (Messias et al., 2011). Other interesting applications of lipases include the processing of black tea for degrading membrane lipids and initiating the formation of volatile flavoring of any product (Verma et al., 2012; Latha and Ramarethianam, 1999; Ramarethianam et al., 2002).

6. Conclusion and Future Perspectives

Lipases present remarkable potential applications in different fields of the food industry, as in dairy products, synthesis of aromas, bakery, alcoholic beverages, modification of oils and fats, among others. Therefore, it is essential to develop new cost-effective technologies for increased production, scaling up and purification of this useful enzyme. It is undeniable that genetic engineering is the main tool to obtain lipases with specific characteristics, but we should not forget the wide existing biodiversity which certainly provides microorganisms having lipases with desired characteristics for use in various biotechnology processes.

Acknowledgment

The authors thank Coordenação de Aperfeiçoamento de Pessoal de Nível Superior (CAPES/PNPD) and Fundação Araucária of Brazil for their financial support.

References

Ahmed, E.H., Raghavendra, T. and Madamwar, D. 2010. An alkaline lipase from organic solvent tolerant *Acinetobacter* sp. EH 28: Application for ethyl caprylate synthesis. Bioresource Technology, 101: 3628–3634.

Alonso, F.O.M., Oliveira, E.B.L., Dellamora-Ortiz, G.M. and Pereira-Meirelles, F.V. 2005. Improvement of lipase production at different stirring speeds and oxygen levels. Brazilian Journal of Chemical Engineering, 22: 9–18.

Aminifar, M. and Emam-Djomeh, Z. 2014. Changes of texture, microstructure and free fatty acid contents of lighvan cheese during accelerated ripening with lipase. Journal of Agricultural Science and Technology, 16: 113–123.

Andualema, B. and Gessesse, A. 2012. Microbial lipase and their industrial applications: Review. Biotechnology, 3: 100–118.

Anobon, C.D., Pinheiro, A.S., De Andrade, R.A., Aguieiras, E.C.G., Andrade, G.C., Moura, M.V., Almeida, R.V. and Freire, D.M. 2014. From structure to catalysis: Recent developments in the biotechnological applications of lipases. BioMed Research International, 684506. doi.org/10.1155/2014/684506.

Aravidan, R., Anbumathi, P. and Viruthagiri, T. 2007. Lipase applications in food industry. Indian Journal of Biotechnology, 6: 141–158.

Asih, D.R., Alam, M.Z., Salleh, M.N. and Salihu, A. 2014. Pilot-scale production of lipase using palm oil mill effluent as a basal medium and its immobilization by selected materials. Journal of Oleo Science, 63(8): 779–85.

Barros, M., Fleuri, L.F. and Macedo, G.A. 2010. Seed lipase: Sources, applications and properties a review. Brazilian Journal of Chemical Engineering, 27: 15–29.

Basheer, S.M., Chellappan, S., Beena, P.S., Sukumaran, R.K., Elyas, K.K. and Chandrasekaran, M. 2011. Lipase from marine *Aspergillus awamori* BTMFW032: Production, partial purification and application in oil effluent treatment. New Biotechnology, 28: 627–638.

Bellido, G., Gazzola, G. and Matveeva, I. 2014. Method of producing a baked product with alpha-amylase, lipase and phospholipase. Patent No. WO2014161876 A1.

Beermann, C. and Hartung, J. 2012. Current enzymatic milk fermentation procedures. European Food Research and Technology, 235: 1–12.

Bonugli-Santos, R.C., Vasconcelos, M.R.S., Passarini, M.R.Z., Vieira, G.A.L., Lopes, V.C.P., Mainardi, P.H., dos Santos, J.A., Duarte, L.A., Otero, I.V.R., Yoshida, A.M.S., Feitosa, V.A., Pessoa Jr., A. and Sette, L.D. 2015. Marine-derived fungi: Diversity of enzymes and biotechnological applications. Frontiers in Microbiology, 6: 1–15.

Bora, L. and Bora, M. 2012. Optimization of extracellular thermophilic highly alkaline lipase from thermophilic *Bacillus* sp. isolated from hot spring of Arunachal Pradesh, India. Brazilian Journal of Microbiology, 43(1): 30–42.

Bornscheuer, U.T. 2002. Microbial carboxyl esterases: Classification, properties and application in biocatalysis. FEMS Microbiology Reviews, 26: 73–81.

Borrelli, G. M. and Trono, D. 2015. Recombinant lipases and phospholipases and their use as biocatalysts for industrial applications. International Journal of Molecular Sciences, 16: 20774–20840.

Brzozowski, A.M., Derewenda, U., Derewenda, Z.S., Dodson, G.G., Lawson, D.M., Turkenburg, J.P., Bjorkling, F., Huge-Jensen, B., Patkar, S.A. and Thim, L. 1991. A model for interfacial activation in lipases from the structure of a fungal lipase inhibitor complex. Nature, 351: 491–494.

Cao, M., Fonseca, L.M., Schoenfuss, T.C. and Rankin, S.A. 2014. Homogenization and lipase treatment of milk and resulting methyl ketone generation in blue cheese. Journal of Agricultural and Food Chemistry, 62(25): 5726–5733.

Carlile, K., Rees, G.D., Robinson, B.H., Steer, T.D. and Svenson, M. 1996. Lipase-catalyzed interfacial reactions in reverse micellar systems. Journal of the Chemical Society, Faraday Transactions, 92: 4701–4708.

Chahinian, H., Vanot, G., Ibrik, A., Rugani, N., Sarda, L. and Comeau, L.C. 2000. Production of extracellular lipases by *Penicillium cyclopium* purification and characterization of a partial acylglycerol lipase. Bioscience Biotechnology and Biochemistry, 64(2): 215–222.

Choudhury, P. and Bhunia, B. 2015. Industrial application of lipase: A review. Biopharm Journal, 1(2): 41–47.

Cihangir, N. and Sarikaya, E. 2004. Investigation of lipase production by a new isolate of *Aspergillus* sp. World Journal of Microbiology and Biotechnology, 20: 193–197.

Colakoglu, A.S. and Özkaya, H. 2012. Potential use of exogenous lipases for DATEM replacement to modify the reological and thermal properties of wheat flour dough. Journal of Cereal Science, 55: 397–404.

Coleman, M.H. and Macrae, A.R. 1977. Rearrangement of fatty in fat reaction reactants. German Patent DE 2 705608 (Unilever N. V.).

Coleman, M.H. and Macrae, A.R. 1980. Fat process and composition. UK Patent No. 1 577 933.

Contesini, F.J., Lopes, D.B., Macedo, G.A., Nascimento, M.G. and Carvalho, P.O. 2010. *Aspergillus* sp. lipase: Potential biocatalyst for industrial use. Journal of Molecular Catalysis: Enzymatic, 67: 163–171.

Dandavate, V., Jinjala, J., Keharia, H. and Madamwar, D. 2009. Production, partial purification and characterization of organic solvent tolerant lipase from *Burkholderia multivorans* V2 and its application for ester synthesis. Bioresource Technology, 100: 3374–3381.

Dewapriya, P. and Kim, S.-K. 2014. Marine microorganisms: An emerging avenue in modern nutraceuticals and functional foods. Food Research International, 56: 115–125.

El-Fadaly, H.A., El-Kadi, S.M., Hamad, M.N. and Habib, A.A. 2015. Role of fungal enzymes in the biochemistry of egyptian ras cheese during ripening period. Open Access Library Journal, 2: e1819. http://dx.doi.org/10.4236/oalib.1101819.

El-Hofi, M., El-Tanboly, E.-S. and Abd-Rabou, N.S. 2011. Industrial application of lipases in cheese making: A review. Internet Journal of Food Safety, 13: 293–302.

Elibol, M. and Ozer, D. 2000. Influence of oxygen transfer on lipase production by *Rhizopus arrhizus*. Process Biochemistry, 36: 325–329.

Esteban-Torres, M., Barcenilla, J.M., Mancheño, J.M., de las Rivas, B. and Munõz, R. 2014. Characterization of a versatile arylesterase from *Lactobacillus plantarum* active on wine esters. Journal of Agricultural and Food Chemistry, 62: 5118−5125.

Fan, X., Niehus, X. and Sandoval, G. 2012. Lipases as biocatalyst for biodiesel production. Methods in Molecular Biology, 861: 471–483.

Faustino, A.R., Osório, N.M., Tecelão, C., Canet, A., Valero, F. and Ferreira-Dias, S. 2015. Camelina oil as a source of polyunsaturated fatty acids for the production of human milk fat substitutes catalyzed by a heterologous *Rhizopus oryzae* lipase. European Journal of Lipid Science and Technology, doi: 10.1002/ejlt.201500003.

Fernandez-Lafuente, R. 2010. Lipase from *Thermomyces lanuginosus*: Uses and prospects as an industrial biocatalyst. Journal of Molecular Catalysis B: Enzymatic, 62: 197–212.

Ferraz, L.I.R., Possebom, J., Alvez, E.V., Cansian, R.L., Paroul, N., de Oliveira, D. and Treichel, H. 2015. Application of home-made lipase in the production of geranyl propionate by esterification of geraniol and propionic acid in solvent-free system. Biocatalysis and Agricultural Biotechnology, 4: 44–48.

Ferreira-Dias, S., Sandoval, G., Plou, F. and Valero, F. 2013. The potential use of lipase in the production of fatty acid derivatives for the food and nutraceutical industries. Electronic Journal of Biotechnology, 16: 1–33.

Fickers, P., Ongena, M., Destain, J., Weekers, F. and Thonart, P. 2006. Production and down-stream processing of an extracellular lipase from the yeast *Yarrowia lipolytica*. Enzyme and Microbial Technology, 38: 756–759.

Freitas, L., Paula, A.V., dos Santos, J.C., Zanin, G.M. and de Castro, H.F. 2010. Enzymatic synthesis of monoglycerides by esterification reaction using *Penicillium camembertii* lipase immobilized on epoxy SiO_2-PVA composite. Journal of Molecular Catalysis B: Enzymatic, 65: 87–90.

Garcia-Galan, C., Barbosa, O., Ortiz, C., Torres, R., Rodrigues, R.C. and Fernandez-Lafuente, R. 2013. Biotechnological prospects of the lipase from *Mucor javanicus*. Journal of Molecular Catalysis B: Enzymatic, 93: 34–43.

Gerits, L.R., Pareyt, B., Decamps, K. and Delcour, J.A. 2014a. Lipases and their functionality in the production of wheat-based food systems. Comprehensive Reviews in Food Science and Food Safety, 13: 978–989.

Gerits, L.R., Pareyt, B. and Delcour, J.A. 2014b. A lipase-based approach for studying the role of wheat lipids in breadmaking. Food Chemistry, 156: 190–196.

Ghosh, P.K., Saxena, R.K., Gupta, R., Yadav, R.P. and Davidson, S. 1996. Microbial lipases: Production and applications. Science Progress, 79: 119–157.

Gilbert, E.J. 1993. *Pseudomonas* lipases: Biochemical properties and molecular cloning. Enzyme and Microbial Technology, 15: 634–645.

Gill, I. and Valivety, R. 1997. Polyunsaturated fatty acids, Part 1: Occurrence, biological activities and applications. Trends in Biotechnology, 15: 401–409.

Gosh, B. and Ray, R.R. 2011. Current commercial perspective of *Rhizopus oryzae*: A review. Journal of Applied Sciences, 11(14): 2470–2486.

Grillitsch, K. and Daum, G. 2011. Triacylglycerol lipases of the yeast. Frontiers in Biology, 3: 219–230.

Gupta, R., Kumari, A., Syal, P. and Singh, Y. 2015. Molecular and functional diversity of yeast and fungal lipases: Their role in biotechnology and cellular physiology. Progress in Lipid Research, 57: 40–54.

Gupta, R., Gupta, N. and Rathi, P. 2004. Bacterial lipases: An overview of production, purification and biochemical properties. Applied Microbiology and Biotechnology, 64(6): 763–781.

Gulati, R., Bhattacharya, A., Prasad, A.K., Gupta, R., Parmar, V.S. and Saxena, R.K. 2001. Biocatalytic potential of Fusarium globulosum lipase in selective acetylation/deacetylation reactions and in ester synthesis. Journal of Applied Microbiology, 90(4): 609–613.

Ha, J.K. and Lindsay, R.C. 1993. Release of volatile branched-chain and other fatty acids from ruminant milk fats by various lipases. Journal of Dairy Science, 76: 677–690.

Haki, G.D. and Rakshit, S.K. 2003. Developments in industrially important thermostable enzymes: A review. Bioresource Technology, 89: 17–34.

Hasan, F., Shah, A.A. and Hameed, A. 2006. Influence of culture conditions on lipase production by *Bacillus* sp. FH5. Annals of Microbiology, 56: 247–252.

Hasan-Beikdashti, M., Forootanfar, H., Safiarian, M.S., Ameri, A., Ghahremani, M.H., Khoshayand, M.R. and Faramarzi, M.A. 2012. Optimization of culture conditions for production of lipase by a newly isolated bacterium, *Stenotrophomonas maltophilia*. Journal of the Taiwan Institute of Chemical Engineers, 43: 670–677.

Holm, H.C. and Cowan, W.D. 2008. The evolution of enzymatic interesterification in the oils and fats industry. European Journal of Lipid Science and Technology, 110: 679–691.

Jaeger, K.-E., Dijkstra, B.W. and Reetz, M.T. 1999. Bacterial biocatalysts: Molecular biology, three-dimensional structures, and biotechnological applications of lipases. Annual Review of Microbiology, 53: 315–351.

Jaeger, K.-E., Ransac, S., Dijkstra, B.W., Colson, C., Vanheuvel, M. and Misset, O. 1994. Bacterial lipases. FEMS Microbiology Reviews, 15: 29–63.

Jaeger, K.-E., Ransac, S., Koch, H.B., Ferrato, F. and Dijkstra, B.W. 1993. Topological characterization and modelling of the 3D structure of lipase from *Pseudomonas aeruginosa*. FEBS Letters, 332: 143–149.

Jegannathan, K.R. and Nielsen, P.H. 2013. Environmental assessment of enzyme use in industrial production—A literature review. Journal of Cleaner Production, 42: 228–240.

Joseph, B., Ramteke, P.W. and Thomas, G. 2008. Cold active microbial lipases: Some hot issues and recent developments. Biotechnology Advances, 26: 457–470.

Kantak, J.B., Bagade, A.V., Mahajan, S.A., Pawar, S.P., Shouche, Y.S. and Prabhune, A.A. 2011. Isolation, identification and optimization of a new extracellular lipase producing strain of *Rhizopus* sp. Applied Biochemistry and Biotechnology, 164: 969–978.

Kapoor, M. and Gupta, M.N. 2012. Obtaining monoglycerides by esterification of glycerol with palmitic acid using some high activity preparations of *Candida antarctica* lipase B. Process Biochemistry, 47(3): 503–508.

Kontkanen, H., Rokka, S., Kemppinen, A., Miettinen, H., Hellström, J., Kruus, K., Marnila, P., Alatossava, T. and Korhonen, H. 2011. Enzymatic and physical modification of milk fat: A review. International Dairy Journal, 21: 3–13.

Kumar, P.K.P. and Krishna, A.G. 2014. Impact of different deacidification methods on quality characteristics and composition of olein and stearin in crude red palm oil. Journal of Oleo Science, 63: 1209–1221.

Kumar, R., Mahajan, S., Kumar, A. and Singh, V. 2011. Identification of variables and value optimization for optimum lipase production by *Bacillus pumilus* RK31 using statistical methodology. New Biotechnology, 28(1): 187–192.

Kuo, C.H., Chiang, S.H., Ju, H.Y., Chen, Y.M., Lião, M.Y., Liu, Y.C. and Shie, C.J. 2012. Enzymatic synthesis of rose aromatic ester (2-phenylethyl acetate) by lipase. Journal of the Science of Food and Agriculture, 92(10): 2041–2047.

Latha, K. and Ramarethinam, S. 1999. Studies on lipid acyl during tea processing. Annals of Plant Physiology, 3: 73–78.

Lee, P.R., Chong, I.S.M., Yu, B., Curran, P. and Liu, S.Q. 2012. Effects of sequentially inoculated *Williopsis saturnus* and *Saccharomyces cerevisiae* on volatile profiles of papaya wine. Food Research International, 45(1): 177–183.

Lesuisse, E., Schanck, K. and Colson, C. 1993. Purification and preliminary characterization of the extracellular lipase of *Bacillus subtilis*168, an extremely basic pH tolerant enzyme. European Journal of Biochemistry, 216: 155–160.

Li, N. and Zong, M.H. 2010. Lipases from the genus *Penicillium*: Production, purification, characterization and applications. Journal of Molecular Catalysis B: Enzymatic, 66: 43–54.

Lilbæk, H.M., Broe, M.L., Høier, E., Fatum, T.M., Ipsen, R. and Sørensen, N.K. 2006. Improving the yield of Mozzarella cheese by phospholipase treatment of milk. Journal of Dairy Science, 89(11): 4114–4125.

Lima, V.M.G., Krieger, N., Sarquis, M.I.M., Mitchell, D.A., Ramos, L.P. and Fontana, J.D. 2003. Effect of nitrogen and carbon sources on lipase production by *Penicillium aurantiogriseum*. Food Technology Biotechnology, 41(2): 105–110.

Lin, E.-S., Wang, C.-C. and Sung, S.-C. 2006. Cultivating conditions influence lipase production by the edible Basidiomycete *Antrodia cinnamomea* in submerged culture. Enzyme and Microbial Technology, 39: 98–102.

Liu, C.-H., Huang, C.-C., Wang, Y.-W. and Chang, J.-S. 2012. Optimizing lipase production from isolated *Burkholderia* sp. Journal of the Taiwan Institute of Chemical Engineers, 43: 511–516.

Liu, C.-H., Chen, C.-Y., Wang, Y.-W. and Chang, J.-S. 2011. Fermentation strategies for the production of lipase by an indigenous isolate *Burkholderia* sp. C20. Biochemical Engineering Journal, 58(1): 96–102.

López-López, O., Cerdán, M.E. and González-Siso, M.I. 2014. New extremophilic lipases and esterases from metagenomics. Current Protein and Peptide Science, 15: 445–455.

Lyberg, A.M. and Adlercreutz, P. 2008. Lipase specificity towards eicosapentaenoic acid and docosahexaenoic acid depends on substrate structure. Biochimica et Biophysica Acta (BBA)—Proteins and Proteomics, 1784(2): 343–350.

Macedo, G.A., Lozano, M.M.S. and Pastore, G.M. 2003. Enzymatic synthesis of short chain citronellyl esters by a new lipase from *Rhizopus* sp. Electronic Journal of Biotechnology, 6(1): 72–75.

Marion, B.A.S. and Oliver, T. 2013. Review article: Immobilised lipases in the cosmetics industry. Chemical Society Reviews, 42: 6475–6490.

Marques, T.A., Baldo, C., Borsato, D., Buzato, J.B. and Celligoi, M.A.P.C. 2014a. Utilization of dairy effluent as alternative fermentation medium for microbial lipase production. Romanian Biotechnological Letters, 19(1): 9042–9050.

Marques, T.A., Baldo, C., Borsato, D., Buzato, J.B. and Celligoi, M.A.P.C. 2014b. Production and partial characterization of a thermostable, alkaline and organic solvent tolerant lipase from *Trichoderma atroviride* 676. International Journal of Scientific & Technology Research, 3(5): 77–83.

Martinez, C., de Geus, P., Lauwereys, M., Matthyssens, M. and Cambillau, C. 1992. *Fusarium solani* cutinase is a lipolytic enzyme with a catalytic serine accessible to solvent. Nature, 356: 615–618.

Mase, T., Matsumiya, Y. and Matsuura, A. 1995. Purification and characterization of *Penicillium roqueforti* IAM 7268 lipase. Bioscience, Biotechnology, and Biochemistry, 59(2): 329–330.

Matsuo, T., Sawamura, N., Hashimoto, Y. and Hashida, W. 1981. The enzyme and method for enzymatic transesterifiction of lipid. European Patent No. 0035 883 (Fuji Oil Co.).

Mendes, A.A., Oliveria, P.C. and Castro, H.F. 2012. Properties and biotechnological application of porcine pancreatic lipase. Journal of Molecular Catalysis B: Enzymatic, 78: 119–134.

Messias, J.M., da Costa, B.Z., Lima, V.M.G., Dekker, R.F.H., Rezende, M.I., Krieger, N. and Barbosa, A.M. 2009. Screening *Botryosphaeria* species for lipases: Production of lipase by *Botryosphaeria ribis* EC-01 grown on soybean oil and other carbon sources. Enzyme and Microbial Technology, 45: 426–431.

Messias, J.M., da Costa, B.Z., De Lima, V.M.G., Giese, E.C., Dekker, R.F.H. and Barbosa, A.M. 2011. Microbial lipases: Production, properties and biotechnological applications. Semina: Ciências Exatas e Tecnológicas, 32(2): 213–234.

Mhetras, N.C., Bastawde, K.B. and Gokhale, D.V. 2009. Purification and characterization of acidic lipase from *Aspergillus niger* NCIM 1207. Bioresource Technology, 3: 1486–1490.

Miguel, A.S.M., Martins-Meyer, T.S., Figueiredo, E.V.C., Lobo, B.W.P. and Dellamora-Ortiz, G.M. 2013. Enzymes in bakery: Current and future trends. pp. 287–321. *In*: Food Industry, 2013.

Moayedallaie, S., Mirzaei, M. and Paterson, J. 2010. Bread improvers: Comparison of a range of lipases with a traditional emulsifier. Food Chemistry, 122: 495–499.

Mohamed, I.O. 2012. Lipase-catalyzed synthesis of cocoa butter equivalent from palm olein and saturated fatty acid distillate from palm oil physical refinery. Applied Biochemistry and Biotechnology, 168(6): 1405–1415.

Molimard, P. and Spinnler, H.E. 1996. Review: Compounds involved in the flavour of the surface mould-ripened cheeses: Origins and properties. Journal of Dairy Science, 79(2): 169–184.

Nakano, H., Kitahata, S., Shimada, Y., Nakamura, M. and Tominaga, Y. 1995. Esterification of glycosides by a mono- and diacylglycerol lipase from *Penicillium camembertii* and comparison of the products with *Candida cylindracea* lipase. Journal of Fermentation and Bioengineering, 80: 24–29.

Narwal, S.K., Saun, N.K. and Gupta, R.R. 2014. Characterization and catalytic properties of free and silica-bound lipase: A comparative study. Journal of Oleo Science, 63(6): 599–605.

Nielsen, P.H. and Høier, E. 2009. Environmental assessment of yield improvements btained by the use of the enzyme phospholipase in mozzarella cheese production. The International Journal Life Cycle Assessment, 14: 137–143.

Olesen, V.T., Qi Si, J. and Donelyan, V. 2000. Use of Lipases in Baking. United States Patent No. 6.110.508.

Oliveira, A.C.D., Fernandes, M.L. and Mariano, A.B. 2014. Production and characterization of an extracellular lipase from *Candida guilliermondii*. Brazilian Journal of Microbiology, 45(4): 1503–1511.

Olsen, T.S. and Povlsen, I.L. 2005. A Method of Improving the Rheological Properties of a Flour Dough. European patent No. 1 608 227 B1.

Ozyilmaz, G. and Gezer, E. 2010. Production of aroma esters by immobilized *Candida rugosa* and porcine pancreatic lipase into calcium alginate gel. Journal of Molecular Catalysis B: Enzymatic, 64: 140–145.

Paciello, L., Landi, C., Orilio, P., Di Matteo, M., Zueco, J. and Parascandola, P. 2015. Bread making with *Saccharomyces cerevisiae* CEN.PK113-5D expressing lipase A from *Bacillus subtilis*: Leavening characterisation and aroma enhancement. International Journal of Food Science and Technology, 50: 2120–2128.

Pandey, A. 2003. Solid-state fermentation. Biochemical Engineering Journal, 13: 81–84.

Papagora, C., Roukas, T. and Kotzekidou, P. 2013. Optimization of extracellular lipase production by *Debaryomyces hansenii* isolates from dry-salted olives using response surface methodology. Food and Bioproducts Processing, 91(4): 413–420.

Paroul, N., Grzegozeski, L.P., Chiaradia, V., Treichel, H., Cansian, R.L., Oliveira, J.V. and de Oliveira, D. 2012. Solvent-free production of bioflavors by enzymatic esterification of citronella (*Cymbopogon winterianus*) essential oil. Applied Biochemistry and Biotechnology, 166: 13–21.

Paslaru, V., Niculita, I. and Leahu, A. 2008. Lipases influence on bread's quality. Journal of Agroalimentary Processes and Technologies, 14: 309–311.

Primo-Martin, C., Hamer, R.J. and de Jongh, H.H.J. 2006. Surface layer properties of dough liquor components: Are they key parameters in gas retention in bread dough? Food Biophysics, 1: 83–93.

Qi Si, J. 1997. Synergistic effect of enzymes for breadmaking. Cereal Foods World, 42(10): 802–807.

Ramani, A., Kennedy, L.J., Ramakrishnan, M. and Sekaran, G. 2010. Purification, characterization and application of acidic lipase from *Pseudomonas gessardii* using beef tallow as a substrate for fats and oil hydrolysis. Process Biochemistry, 45(10): 1683–1691.

Ramarethinam, S., Latha, K. and Rajalakshmi, N. 2002. Use of a fungal lipase for enhancement of aroma in black tea. Food Science and Technology Research, 8(4): 328–332.

Ray, A. 2012. Application of lipase in industry. Asian Journal of Pharmacy and Technology, 2(2): 33–37.

Reetz, M.T. 2002. Lipases as practical biocatalysts. Current Opinion in Chemical Biology, 6(2): 145–150.

Rodrigues, R.C. and Fernandez-Lafuente, R. 2010. Lipase from *Rhizomucor miehei* as a biocatalyst in fats and oils modification. Journal of Molecular Catalysis B: Enzymatic, 66(1): 15–32.

Rogalska, E., Douchet, I. and Verger, R. 1997. Microbial lipases: Structures, function and industrial applications. Biochemical Society Transactions, 25(1): 161–164.

Ruchi, G., Anshu, G. and Khare, S.K. 2008. Lipase from solvent tolerant *Pseudomonas aeruginosa* strain: Production optimization by response surface methodology and application. Bioresource Technology, 99(11): 4796–4802.

Salihu, A., Alam, M.Z., Abdulkarim, M.I. and Salleh, H.M. 2011. Optimization of lipase production by *Candida cylindracea* in palm-oil mill effluent based medium using statistical experimental design. Journal of Molecular Catalysis B: Enzymatic, 69: 66–73.

Sánchez, M., Prim, N., Rández-Gil, F., Pastor, F.I.J. and Diaz, P. 2002. Engineering of baker's yeasts, *Escherichia coli* and *Bacillus* hosts for the production of *B. subtilis* lipase A. Biotechnology and Bioengineering, 78(3): 339–345.

Schaffarczyk, M., Ostdal, H. and Koehler, P. 2014. Lipases in wheat breadmaking: Analysis and functional effects of lipid reaction products. Journal of Agricultural and Food Chemistry, 36: 8229–8237.

Schimidt-Dannert, C., Sztajer, H., Stocklein, W., Menge, W. and Schimid, R.D. 1994. Screening purification and properties of thermophilic lipase from *Bacillus thermocatenulatus*. Biochimica et Biophysica Acta (BBA)—Lipids and Lipid Metabolism, 1214: 43–53.

Serri, N.A., Kamaruddin, A.H. and Tau Len, K.Y. 2010. A continuous esterification of malonic acid with citronellol using packed bed reactor: Investigation of parameter and kinetics study. Food and Bioproducts Processing, 88(2): 327–332.

Sharma, A., Bardhan, D. and Patel, R. 2009. Optimization of physical parameters for lipase production from *Arthrobacter* sp. BGCC#490. Indian Journal of Biochemistry and Biophysics, 46(2): 178–183.

Sharma, R., Chisti, Y. and Banerjee, U.C. 2001. Production, purification, characterization, and applications of lipases. Biotechnology Advances, 19: 627–662.

Sharma, S. and Kanwar, S.S. 2014. Organic solvent tolerant lipases and applications. The Scientific World Journal, 625258. doi.org/10.1155/2014/625258.

Silva, W.O.B., Matidieri, S., Schrank, A. and Vainstein, M.H. 2005. Production and extraction of an extracelular lipase from the entomopathogenic fungus *Metarhizium anisopliae*. Process Biochemistry, 40: 321–326.

Singh, A.K. and Mukhopadhyay, M. 2012. Overview of fungal lipase: A review. Applied Biochemistry and Biotechnology, 166(2): 486–520.

Sinkuniene, D., Bendikiene, V. and Juodka, B. 2011. Response surface methodology based optimization of lipase-catalyzed triolein hydrolysis in hexane. Romanian Biotechnological Letters, 16(1): 5891–5901.

Smaniotto, A., Skovronski, A., Rigo, E., Tsai, S.M., Durrer, A., Foltran, L.L., Paroul, N., Di Luccio, M., Oliveira, J.V., de Oliveira, D. and Treichel, H. 2014. Concentration, characterization and application of lipases from *Sporidiobolus pararoseus* strain. Brazilian Journal of Microbiology, 45: 294–302.

Song, X., Qi, X., Bao, B. and Qu, Y. 2008. Studies of substrate specifities of lipases from different sources. European Journal of Lipid Science and Technology, 110: 1095–1101.

Soumanou, M., Pérignon, M. and Villenueve, P. 2013. Lipase-catalyzed interesterification reactions for human milk fat substitutes production: A review. European Journal of Lipid Science and Technology, 115(3): 270–285.

Sovová, H., Zarevúcka, M., Bernášek, P. and Stamenic´, M. 2008. Kinetics and specificity of Lipozyme-catalyzed oil hydrolysis in supercritical CO_2. Chemical Engineering Research Design, 86(7): 673–681.

Stergiou, P.-Y., Foukis, A., Sklivaniti, H., Zacharaki, P., Papagianni, M. and Papamichael, E.M. 2012. Experimental investigation and optimization of process variables affecting the production of extracellular lipase by *Kluyveromyces marxianus* IFO 0288. Applied Biochemistry and Biotechnology, 168(3): 672–680.

Sumby, K.M., Grbin, P.R. and Jiranek, V. 2010. Microbial modulation of aromatic esters in wine: Current knowledge and future prospects. Food Chemistry, 121: 1–16.

Swanson, D., Block, R. and Mousa, S.A. 2012. Omega-3 fatty acids EPA and DHA: Health benefits throughout life. Advances in Nutrition, 3: 1–7.

Tan, T., Zhang, M., Wang, B., Ying, C. and Deng, L. 2003. Screening of high lipase producing *Candida* sp. and production of lipase by fermentation. Process Biochemistry, 39(4): 459–465.

Tecelão, C., Guillén, M., Valero, F. and Ferreira-Dias, S. 2012. Immobilized heterologous *Rhizopus oryzae* lipase: A feasible biocatalyst for the production of human milk fat substitutes. Biochemical Engineering Journal, 67: 104–110.

Toida, J., Arikawa, Y., Kondou, K., Fukuzawa, M. and Sekiguchi, J. 1998. Purification and characterization of triacylglycerol lipase from *Aspergillus oryzae*. Bioscience, Biotechnology and Biochemistry, 62(4): 759–763.

Turki, S., Kraeim, I.B., Weeckers, F., Thonart, P. and Kallel, H.M. 2009. Isolation of bioactive peptides from tryptone that modulate lipase production in *Yarrowia lipolytica*. Bioresource Technology, 100(10): 2724–2731.

Uppenberg, J., Hansen, M.T., Patkar, S. and Jones, T.A. 1994. The sequence, crystal structure determination and refinement of two crystal forms of lipase B from *Candida antarctica*. Structure, 2(4): 293–308.

Van Tilbeurgh, H., Egloff, M.P., Martinez, C., Rugani, N., Verger, R. and Cambillau, C. 1993. Interfacial activation of the lipase-procolipase complex by mixed micelles revealed by X-ray crystallography. Nature, 362: 814–820.

Verger, R. 1997. Interfacial activation of lipases: facts and artifacts. Trends in Biotechnology, 15: 32–38.

Verger, R. and de Haas, G.H. 1976. Interfacial enzyme kinetics of lipolysis. Annual Review of Biophysics and Bioengineering, 5: 77–117.

Verma, N., Thakur, S. and Bhatt, A.K. 2012. Microbial lipases: Industrial applications and properties (a review). International Research Journal of Biological Sciences, 1(8): 88–92.

Villeneuve, P., Muderhwa, J.M., Graille, J. and Haas, M.J. 2000. Customizing lipases for biocatalysis: A survey of chemical, physical and molecular biological approaches. Journal of Molecular Catalysis B: Enzymatic, 9: 113–148.

Walker, V. and Mills, G.A. 2014. 2-Pentanone production from hexanoic acid by *Penicillium roqueforti* from Blue Cheese: Is this the Pathway Used in Humans? The Scientific World Journal, 215783 doi 10.1155/2014/215783.

Wang, D., Xu, Y. and Shan, T. 2008. Effects of oils and oil-related substrates on the synthetic activity of membrane-bound lipase from *Rhizopus chinensis* and optimization of the lipase fermentation media. Biochemical Engineering Journal, 41: 30–37.

Yadav, K.N.S., Adsul, M.G., Bastawde, K.B., Jadhav, D.D., Thulasiram, H.V. and Gokhale, D.V. 2011. Differential induction, purification and characterization of cold active lipase from *Yarrowia lipolytica* NCIM 3639. Bioresource Technology, 102: 10663–10670.

Yazdanpanah, S., Ehsani, M.R. and Mizani, M. 2014. Modelling lipolysis in acceleration of ripening of ultrafiltered-feta cheese. International Journal of Biosciences, 4(2): 309–315.

Zhang, J.-H., Jiang, Y.-Y., Lin, Y., Sun, Y.-F., Zheng, S.-P. and Han, S.-Y. 2013. Structure-guided modification of *Rhizomucor miehei* lipase for production of structured lipids. Plos One, 8(7): e67892. doi: 10.1371/journal.pone.0067892.

Zhou, P., Zhang, G., Chen, S., Jiang, Z., Tang, Y., Henrissat, B., Yan, Q., Yang, S., Chen, C.F., Zhang, B. and Du, Z. 2014. Genome sequence and transcriptome analyses of the thermophilic zygomycete fungus *Rhizomucor miehei*. BMC Genomics, 15: 294. doi: 10.1186/1471-2164-15-294.

Zinjarde, S.S. 2014. Food-related applications of *Yarrowia lipolytica*. Food Chemistry, 152: 1–10.

13

Catalases

Types, Structure, Applications and Future Outlook

Balwinder Singh Sooch,[1,a] *Baljinder Singh Kauldhar*[1,b] *and
Munish Puri*[1,2,]*

1. Introduction

Biocatalysts are the most proficient biological macromolecules produced by living organisms, offering more competitive processes as compared to chemical catalysts (Choi et al., 2015). These are basically proteins that speed up chemical reactions and may cause transfer of electrons or functional groups to produce a specific biochemical reaction. They are responsible for all important chemical inter-conversions that are required to sustain life (Gurung et al., 2013). Applications of tailor-made biocatalysts with desired properties in different industries have surged in recent times due to the development of recombinant technology, protein engineering and their computational designing.

Industrial demand for microbial enzymes is escalating with each passing year and catalases are one of them. The industrial enzyme market across the globe is growing fast with a total turnover of about $3.3 billion in 2010 and which is expected to touch $7.1 billion in 2018 (BCC Research Report, 2014). Catalases (EC 1.11.1.6) are among

---

[1] Enzyme Biotechnology Laboratory, Department of Biotechnology, Punjabi University, Patiala, Punjab, India.

[a] E-mail: soochb@pbi.ac.in

[b] E-mail: baljinderbiotech@gmail.com

[2] Centre for Chemistry and Biotechnology (CCB), Faculty of Science Engineering and Built Environment, Deakin University, Waurn Ponds, Victoria 3217, Australia.

* Corresponding author: munish.puri@deakin.edu.au

the most widespread industrial enzymes with highest turnover and having numerous industrial, diagnostic and therapeutic applications (Sooch et al., 2014b). Catalase is a haem-containing enzyme belonging to Class Oxidoreductases. Catalases were among the first enzymes to be characterized in biochemistry (Loew, 1900). They scavenge hydrogen peroxide (H_2O_2) to oxygen and water and constitute a significant component of the cell defence mechanism against oxidative stress (Foyer and Noctor, 2000). They are present in a wide range of plants, animals and microorganisms. Besides the ability to catalyze the decomposition of hydrogen peroxide (H_2O_2), catalases are reported to cause several other redox reaction types. However, the exact mechanisms for oxidase and peroxidase activities are yet to be determined experimentally. While mechanistic studies are under progress, catalases are further explored and exploited for a variety of applications. Catalases are localized in peroxisomes, in eukaryotes and in prokaryotes (Wolf et al., 2010; Williams et al., 2012). However, some evidences of cytosolic catalases from eukaryotes, such as *Saccharomyces cerevisiae* (Wieser et al., 1991), *Caenorhabditis elegans* (Taub et al., 1999) and *Neurospora crassa* (Schliebs et al., 2006) have also been reported.

Promising applications of catalases have emerged in recent years. Microbial catalases are preferred due to their economic feasibility, high yield, ease of product modification and optimization, regular supply due to absence of seasonal fluctuations and rapid growth of microbes on an inexpensive media. However, it is worth noting that most commercial catalases have optimal activity at 20°C to 50°C and at neutral pH which makes them unable to withstand the adverse conditions in some industrial processes (Spiro and Griffith, 1997). Despite great potential of catalases, their industrial applications have been hampered mainly due to their undesirable properties in terms of low stability, less catalytic efficiency and specificity. To overcome such shortcomings, a variety of approaches have been attempted, including screening of enzymes from natural sources, designing microbes with recombinant technologies, random mutations and immobilization.

Catalases produced from alkalothermophilic microorganisms have the ability to withstand high temperature and pH conditions. These extraordinary life forms have been extensively applied in sectors, such as wastewater treatment (Gudelj et al., 2001), therapeutics (Nishikawa et al., 2005), baking and brewing (Burg, 2003), pulp and paper industry (Ishida et al., 1997). It is also believed that the ability of alkalothermophiles to synthesize tough enzymes contributes in bridging the gap between chemical and biological processes. In the present write-up, information about structure and uses of catalases with respect to various industries have been discussed.

2. Classification and Structural Aspect of Microbial Catalases

Depending on the physical and biochemical properties of catalases, these encompass four different types: mono-functional haem catalases (classical catalase), catalase-peroxidases (atypical catalase), non-haem catalases (pseudocatalases) and minor catalases (Sooch et al., 2014b). The ribbon representation of different types of catalases is presented in Fig. 1.

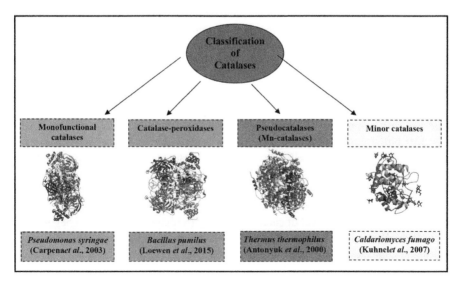

Figure 1. Types of catalases.

Mono-functional haem catalases belong to the original class of catalases found ubiquitously in animals, plants and microorganisms. The structure of mono-functional catalases varies with respect to the number and identity of domains in different species of organisms. Typically, catalases exist as a dumbbell-shaped tetramer of four identical subunits (Carpena et al., 2003). Molecular mass of these catalases varies between 200–340 kDa with a haem prosthetic group at the catalytic center. The haem group is responsible for catalase enzymatic activity and is located between the internal walls of the beta barrel and several helices (Zamocky and Koller, 1999; Lee et al., 2007). The classification of haem catalases into three clades has evolved from at least two gene duplication events (Klotz et al., 1997). Clade 1 catalases have about 500 residues per subunit and are mainly of plant origin, including a subgroup of bacterial origin. Clade 2 catalases, with approximately 750 residues per subunit, are mostly of bacterial or fungal origin (Nicholls et al., 2000; Diaz et al., 2012). Clade 3 catalases with nearly 500 residues per subunit are found in archaea, bacteria, fungi and some eukaryotes. The absence of these older taxonomic groups of catalases in organisms suggests that they arose later in evolution (Diaz et al., 2012).

Catalase-peroxidase is a less prevalent class and belongs to the second group of catalases (Chelikani et al., 2004; Loewen et al., 2015). These are found only in aerobic bacteria and their molecular weight ranges from 120–340 kDa (Obinger et al., 1997). Catalase-peroxidases resemble plant and fungal peroxidases but are not found in plants and animals (Chelikani et al., 2004; Zamocky et al., 2008). These bi-functional catalases are also haem-containing catalases though they are distinct in having peroxidative activity in addition to the catalytic activity (Welinder, 1991; Kapetanaki et al., 2007). A distinctive feature in most catalase-peroxidase structures is the presence of a covalent adduct in which tyrosine is attached at its ortho position with methionine on one side and a tryptophan is linked on the other side (Donald et al., 2003).

Third class is the non-haem catalases and which is a minor bacterial protein family with manganese in the active site rather than a haem molecule and is also called pseudo-catalase or manganese catalase (Antonyuk et al., 2000). These exhibit molecular mass of 170–210 kDa. The highly conserved essential ligands (glutamate, histidine) form typical signatures for the manganese catalase sequence. The classification of these non-haem catalases into five distinct clades occurs due to lateral gene transfer between various bacterial taxa. Clade 1 is spread among archaebacteria and at least one Firmicutes, while Clade 2 includes Actinomycetes and Firmicutes. Clade 3 is distributed among Proteobacteria and Firmicutes. Clade 4 comprises mainly the *Bacteroides* genus as well as many Cyanobacteria, whereas Clade 5 includes Proteobacteria. It has been observed that Clades 1 and 2 share a common ancestor, while, Clade 5 is the most distantly related to the other clades (Zamocky et al., 2008). The fourth minor class of catalases includes several haem-containing proteins and these exhibit a very low level of catalytic activity (Kuhnel et al., 2007). They include bi-functional enzymes, such as chloroperoxidases, bromoperoxidase (Nicholls et al., 2000) and catalase-phenol oxidases (Vetrano et al., 2005). Chloroperoxidase occurs as a 42 kDa monomer, whereas, catalase-phenol oxidase is a tetrameric haem protein with molecular mass of 320 kDa (Kocabas et al., 2008). Catalase-phenol oxidase has the unique ability to oxidize a number of phenolic compounds in the absence of hydrogen peroxide and has catalase activity (Kocabas et al., 2008).

3. Applications of Microbial Catalases

Relevance of enzymes for producing diverse types of chemical and biological products has become an established technology in food and other industries because enzyme-based processes usually lead to reduction in the process time, amount of waste and number of reaction steps. In other words, they provide a more potent way of producing a pure product with less wastage mainly through high stereo-selectivity, region-selectivity and chemo-selectivity. Catalases are used in a wide range of industrial processes (Table 1) and some specific applications of the enzyme are described in detail below (Fig. 2).

4. Catalases in Food Industry

Biocatalysis in the food industry has been used to produce varied products since a long time. A number of catalases used in several industries are available in the market under the brand name of Fermcolase and CAT-IG (Genencor DuPont, Rochester, USA), Enzeco (Enzyme Development, New York, USA) and Terminox (Novozymes, Denmark). The use of catalase in food items has already been approved by the FDA and there are numerous patents that illustrate the use of catalase in the food industry (Loncar and Fraaije, 2015).

Table 1. Applications of catalases in different sectors.

Industry	Source	Application	Reference
Food	*Aspergillus niger*	Removal of H_2O_2 from pasteurized milk and dairy effluent	Akertek and Tarhan, 1995; Yildiz et al., 2004
	ND	Preservative	Dondero et al., 1993
	Bovine liver	Detection of calcium in milk and water samples	Akyilmaz and Kozgus, 2009
	Bovine liver	Detection of infected milk	Futo et al., 2012
	Scytalidium thermophilum	Antioxidant properties	Koclar et al., 2013
Bioremediation	*Bacillus* sp.	Degradation of H_2O_2	Gudelj et al., 2001
	Comamonas terrigena N3H	Phenolic compounds degradation	Zamocky et al., 2001
	Bacillus halodurans LBK 261	Whole cell-based system for removal of H_2O_2 from effluents	Oluoch et al., 2006
Medical	Human	Immunomodulatory potential	Shi et al., 2013
	Human	Removal of H_2O_2 from blood	Gaetani et al., 1994
	ND	Contact lens	Cook and Worsley, 1996
	Human	Inhibition of tumor growth	Nishikawa et al., 2005
	Human	Artificial blood substitute	Bian et al., 2012
Pharmaceutical	ND	Biotransformation reactions	Magner and Klibanov, 1995
	Bacillus pumilus	Preparation of sulfoxides of β-lactams	Sangar et al., 2012
	Escherichia coli	Enzymatic polymerization reaction	Di Gennaro et al., 2012
Others	ND	Artificial enzyme-powered microfishes	Orozco et al., 2012
	E. coli	Prevention of biocorrosion	Baeza et al., 2013
	ND	Catalase-based micromotors	Simmchen et al., 2014

ND: Not determined

4.1 As a biosensor

An immobilized catalase based Clark-type electrode for the detection of highly toxic chemical azide in fruit juices, such as black cherry juice, orange juice and apricot juice, was developed by Sezginturk and co-workers (2005) which can detect azide in the range 25 μM to 300 μM in fruit juices. Similarly, a polyaniline-based catalase biosensor was also constructed for the estimation of hydrogen peroxide and azide in various biological samples (Singh et al., 2009).

The use of immobilized catalase from *Aspergillus niger* on to modified SiO_2 support in a batch-type reactor for removing traces of hydrogen peroxide used during pasteurization of milk, was investigated by Akertek and Tarhan (1995). A catalase based

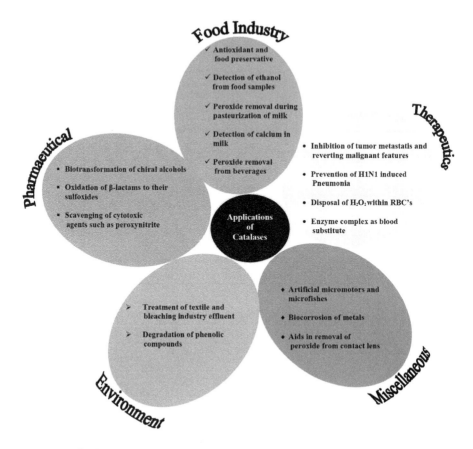

Figure 2. Applications of catalases.

biosensing system was fabricated by immobilizing enzyme on a dissolved oxygen probe membrane with the help of gelatin to determine the decomposition level of hydrogen peroxide in milk samples in dairy industries (Yildiz et al., 2004). Another catalase-based amperometric bio-device for detection of calcium in milk and water samples was designed by immobilizing enzyme on teflon membrane by Akyilmaz and Kozgus (2009). The detection limit of 1 mM to 10 mM and response time of one min. was recorded for this bio-device. A quick and straightforward biosensor-based method for detection of mastitis infection in milk was constructed, based on enzyme activity of catalase in infected milk samples (Futo et al., 2012). Hnaien et al. (2010) pioneered the construction of conductimetric biosensor by co-immobilizing catalase and ethanol oxidase for the detection of ethanol in foods. Aruldoss and Kalaichelvan (2014) also developed an amperometric catalase-based biosensor for the estimation of alcohol concentration in alcoholic drinks. This bio-device was also competent to determine calcium levels within 3 min. in cow milk samples.

Kroll and co-workers (1989) designed a catalase based method for assessing contamination of bacterial colonies in food samples with detection limit of 10^3 cells/

ml. Jasass and Fung (1998) suggested that catalase activity can be used as an index of microbial load for the determination of bacterial contamination in food samples. Similar catalase-based detection methods for food-borne pathogens, like *Listeria monocytogenes* and *Staphylococcus aureus* have also been developed to eliminate microbial population from food items (Patel and Beuchat, 1995; Majumdar et al., 2013).

4.2 As a preservative

An increase in color stabilization along with improved aroma and taste of white grape juice by treatment of high hydrostatic pressure and further with combination of glucose oxidase-catalase enzymes was achieved by Castellari et al. (2000). Sodium azide is a pisonous, white, odourless compound used in agriculture fields to control pests. It is widely used in laboratories as a preservative and also in detonators and explosives.

With increasing awareness on nutritional needs, a significant amount of attention has been paid to the bifunctional catalases having additional peroxidase activity. The use of catalase coupled with oxidase helps to scavenge free radicals, thereby avoiding oxidation which is responsible for deterioration of food items (Loncar and Fraaije, 2015). The preservative effect of catalase (Cat) and glucose oxidase (Gox) in 4 per cent (w/v) glucose (Glu) solution on fishes and shrimps has also been investigated by Dondero et al. (1993). It was observed that holding the shrimp and fish in Cat/Gox/ Glu solution retarded the microbial spoilage and in addition, it increased the freshness of product. It was then postulated on the basis of the above observations that the active oxygen species generated by these enzymes acted as bactericides which aid in increasing the shelf-life of food items.

4.3 As an antioxidant

A recent trend in the food industry is the exploitation of catalytic enzymes exhibiting phenol oxidase activity in addition to catalase activity (CATPO) as an antioxidant has been observed. Catalase-phenol oxidase has been found to show activity towards some phenolic substrates, such as catechin, catechol, caffeic acid, etc. The analysis of end products of these substrates revealed the formation of numerous metabolites and these phenolic secondary metabolites which serve various physiological roles, such as anti-oxidants, pro-oxidants and signal transduction molecules (Perron and Brumaghim, 2009; Maqsood et al., 2014). CATPO may also be developed as a valuable biocatalytic tool for food and biotechnological applications aimed at changing the antioxidant potential of phenolic compounds. Further exhaustive work focused on their exploitation (CATPO) in human consumption as part of the diet or a supplement is in progress (Koclar et al., 2013).

4.4 As an anti-corrosive agent

Stainless steel is used in a wide range of industries, ranging from food industry to sewage tanks, all due to its strength and resistance to corrosion but it is prone to microbial corrosion. Microbiologically-triggered corrosion is responsible for a large portion of the corrosion on machines in food and fermentation industries as biofilm formation alters the electrochemical conditions on the metal surface (Little and Lee, 2014). Of late, catalases are also targeted for corrosion prevention strategies in the food and fermentation industry tools and utensils which are prone to microbial corrosion during fermentation processes (Busalmen et al., 2002). Bacterial cultures of *E. coli* and a catalase deficient *E. coli* strain were also used for immersion of stainless steel to demonstrate the inability of the later cells to cause corrosion, which further indicates that catalases are an interesting target for corrosion-prevention strategies (Baeza et al., 2013).

4.5 Removal of H_2O_2 from beverages

Hydrogen peroxide (H_2O_2) is widely used for the removal of microbiological contaminants (bacteria, fungi) present in beverages and this process is commonly referred to as cold pasteurization. However, the removal of H_2O_2 prior to application in the food industry is important. An efficient method was developed for removal of H_2O_2 from beverages before packaging using catalase entrapped in alginate capsules (Trusek-Holownia and Noworyta, 2015).

4.6 As functional food

Catalase is a food enzyme known for promoting health and protecting against many age-related diseases. Proteinases of the gastrointestinal tract and cellular lysosome degrade catalase enzyme and limit their effectiveness and utility. However, Qi et al. (2012) developed an immobilized system with catalase enzyme loaded on to soybean phosphatidylcholine (SPC)-based polymeric support for its potential application in delivery and protection of this functional food enzyme in gastrointestinal tract. Kwak (2010) also found that the supplementation of a combination of vitamin C and catalase helps in reducing Anaphylaxis in mice models.

5. Catalase-based Pharmaceuticals

In recent times, the quest for green technologies for synthesis of pharmaceutical substances has increased. Consequently, the industry is looking for low-cost and greener biocatalytic pathways as alternatives to traditional chemical processes (Tomsho et al., 2012; Huisman and Collier, 2013). The use of catalases in the pharmaceutical industry has been known for numerous redox reactions. The purified catalase-peroxidase from *Bacillus pumilus* was exploited for oxidation of pharmacophore for antibiotics (i.e., β-lactams) into their sulfoxides in less time than the chemical route with enantio pure building blocks. A biocatalytic method for stereo-selective oxidation of pharmacophores with potential applications in the treatment of bacterial infections was also developed (Sangar et al., 2012). Magner and Klibanov (1995) investigated the catalase-mediated enantio-selectively biotransformation of some chiral alcohols (2, 3-butanediol, *trans*-1, 2-cyclohexanediol, *trans*-3-methylcyclohexanol, menthol, 2-methyl-1-butanol, etc.) in organic solvents. The catalytic role of catalase in decay reaction of peroxynitrite in the synthetic polymer industry has also been demonstrated by Gebicka and Didik (2009).

In polymer synthesis, catalase also serves as a catalyst to degrade residual H_2O_2 after a peroxide-triggered polymerization reaction (Loncar and Fraaije, 2015). Residual H_2O_2 has to be destroyed after the polymerization reaction as it can compromise long-term storage of polyacrylate preparations or may interfere with polyacrylate formulations (Di Gennaro et al., 2014). As these reactions are carried out at low pH, acidophilic catalase-producing microorganisms are needed that can work efficiently under acidic conditions (Sooch and Kauldhar, 2014a).

6. Catalases in Bioremediation

Stringent environmental legislations control disposal of H_2O_2-containing industrial effluents into open streams (Correia et al., 1994; Jensen, 2000). H_2O_2 is commonly used as a bleaching agent in the textile and paper industries. Traditional methods to remove the unused H_2O_2 involve extensive washing of bleached fabrics, resulting in the generation of large volumes of alkaline waste water. Alternatively, H_2O_2 is removed with chemicals, such as sodium bisulphite that also lead to high salt levels in the process streams (Switala and Loewen, 2002). The high catalytic rates render catalases as attractive eco-friendly alternatives for H_2O_2 removal. Various researchers have successfully exploited the use of catalase enzyme to remove hydrogen peroxide from the textile and bleaching industry effluents to reduce the pollution load.

An improvement in dyeing quality of cotton fabrics was also observed after the elimination of hydrogen peroxide residues with catalase (Amorim et al., 2002). Paar et al. (2001) used immobilized thermoalkalistable catalase from *Bacillus* sp. for degradation of hydrogen peroxide in textile bleaching effluent which enabled the reuse of treated water for dyeing. In another study, catalase-producing alginate entrapped *Bacillus* sp. TE-7 cells were used to degrade H_2O_2 in a packed bed reactor (Sooch and Kauldhar, 2015). The degradation of phenolic compounds by the action of catalase-peroxidase from *Comamonas terrigena* N_3H has been reported (Zamocky et al., 2001). A life cycle analysis study of textile manufacturing aimed at replacing traditional washing steps with an industrially produced catalase suggested water savings of 20 m^3/t of yarn when using 1 kg of Novozymes' Terminox Ultra® 50-L catalase preparation. Use of catalase reduces the water consumption and waste of energy by half, and costs of chemicals up to 83 per cent, and saves 33 per cent of time required for completion of process (Blackburn, 2009).

7. Other Applications

Catalase is reported to play an important role in the decomposition of hydrogen peroxide within the red blood cells, with half of the hydrogen peroxide generated in cells being disposed by this enzyme (Gaetani et al., 1994). Catalase-based therapy for inhibiting tumor in various parts of the body has been developed by Nishikawa et al. (2005). A novel enzyme complex (containing catalase, superoxide dismutase, carbonic anhydrase and polyhemoglobin) developed as a blood-substitute transports oxygen and carbon dioxide (Bian et al., 2012). The complex also acts as an antioxidant and therapeutic agent against ischemia-reperfusion injuries. Shi et al. (2013) also demonstrated the application of an immunomodulatory recombinant catalase for fighting H1N1 pneumonia. Aside from its use in biomedical and clinical diagnoses, the application of catalase for the removal of hydrogen peroxide from contact lenses has been patented (Cook and Worsley, 1996).

A novel use of catalase is the construction of artificial enzyme-powered microfishes which are solely based on catalase activity and used for toxicity testing of water samples by visual inspection. Propulsion of these artificial biocatalytic micro-swimmers is dependent on the presence of pollutants which affect catalase activity (Orozco et al., 2012). Artificial micro-fishes as detecting devices may avoid the use of live fish. Enzyme-based micro-propulsion is being further investigated for construction of more robust micro-motors (Simmchen et al., 2014).

8. Conclusion and Future Outlook

Biocatalytic processes have continuously substituted traditional chemical processes in many areas over the past few decades. Research efforts for exploitation of catalase in the food industry in the form of analytical devices, food preservative, functional food, antioxidant agent, etc. have been strengthened in the last few years. The conventional analytical techniques for food analysis are very tedious and time consuming; hence there is a need to develop quick, sensitive and reliable enzyme-based techniques for quick monitoring of food quality and safety. The improvement in stability of catalases achieved through various strategies would result in a vast array of new applications, like in the form of portable food biosensors to detect deterioration in food products. New catalase-based detection methods as an index of microbial load in food materials are up-coming with improved sensitivity and reliability. Some recent developments on catalase as a new functional food for promoting health have opened new opportunities for exploring this interesting biomolecule in the food industry. Further detailed research into catalases at structural level will aid in the design of new protein nano-chip-based devices for food industries. Their potential for diagnosis in the form of novel analytical devices will materialize with pressing advances in the field of nano-biotechnology.

Acknowledgements

Author BSK thanks University Grants Commission, New Delhi for providing Rajiv Gandhi National Fellowship to pursue doctoral studies.

References

Akertek, E. and Tarhan, L. 1995. Characterization of immobilized catalases and their applications in pasteurization of milk with H_2O_2. Applied Biochemistry and Biotechnology, 50(3): 291–303.
Akyilmaz, E. and Kozgus, O. 2009. Determination of calcium in milk and water samples by using catalase enzyme electrode. Food Chemistry, 115(1): 347–351.

Amorim, A.M., Gasques, M.D.G., Andreaus, J. and Scharf, M. 2002. The application of catalase for the elimination of hydrogen peroxide residues after bleaching of cotton fabrics. Annals of the Brazilian Academy of Sciences, 74(3): 433–436.

Antonyuk, S.V., Melik-Adamyan, V.R., Popov, A.N., Lamzin, V.S., Hempstead, P.D., Harrison, P.M., Artymyuk, P.J. and Barynin, V.V. 2000. Three-dimensional structure of the enzyme di-manganese catalase from *Thermus thermophilus* at 1 Å resolution. Crystallography Reports, 45(1): 105–116.

Aruldoss, V. and Kalaichelvan, P.T. 2014. Biosensor application of extracellular catalase from *Ganoderma lucidum* AVK-1 and *Acinetobacter calcoaceticus* AV-6 and comparative study of commercial catalase from bovine liver catalase. IOSR Journal of Pharmacy and Biological Sciences, 9(1): 51–54.

Baeza, S., Vejar, N., Gulppi, M., Azocar, M., Melo, F. and Monsalve, A. 2013. New evidence on the role of catalase in *Escherichia coli* mediated bio-corrosion. Corrosion Science, 67: 32–41.

BCC Research Report. 2014. Global markets for enzymes in industrial applications. http://www.bccresearch.com.

Blackburn, R.S. 2009. Sustainable Textiles: Life Cycle and Environmental Impact. Woodhead, Cambridge, UK.

Bian, Y., Chang, T.S. and Rong, Z. 2012. Polyhemoglobin-superoxide dismutase-catalase-carbonic anhydrase: A novel biotechnology-based blood substitute that transports both oxygen and carbon dioxide and also acts as an antioxidant. Artificial Cells Blood Substitutes and Immobilization Biotechnology, 40(1-2): 28–37.

Burg, B. 2003. Extremophiles as a source for novel enzymes. Current Opinion in Microbiology, 6: 213–218.

Busalmen, J.P., Vazquez, M. and de Sanchez, S.R. 2002. New evidences on the catalase mechanism of microbial corrosion. Electrochimica Acta, 47: 1857–1865.

Carpena, X., Soriano, M., Klotz, M.G., Duckworth, H.W., Donald, L.J., Melik-Adamyan, W., Fita, I. and Loewen, P.C. 2003. Structure of the Clade 1 catalase, CatF of *Pseudomona ssyringae*, at 1.8 Å resolution. Proteins, 50: 423–436.

Castellari, M., Arfelli, G., Carpi, G. and Galassi, S. 2000. Effects of high hydrostatic pressure processing and of glucose oxidase-catalase addition on the color stability and sensorial score of grape juice. Food Science and Technology International, 6(1): 17–23.

Chelikani, P., Fita, I. and Loewen, P.C. 2004. Diversity of structures and properties among catalases. Cellular and Molecular Life Sciences, 61: 192–208.

Choi, J.M., Han, S.S. and Kim, H.S. 2015. Industrial applications of enzyme biocatalysis: Current status and future aspects. Biotechnology Advances, 33(7): 1443–1454.

Cook, J.N. and Worsley, J.L. 1996. Compositions and Method for Destroying Hydrogen Peroxide on Contact Lens. US Patent No. 5,521,091.

Correia, V.M., Stephenson, T. and Judd, S.J. 1994. Characterization of textile wastewaters: A review. Environmental Technology, 15(10): 917–929.

Di Gennaro, P., Bargna, A., Bruno, F. and Sello, G. 2014. Purification of recombinant catalase-peroxidase HPI from *E. coli* and its application in enzymatic polymerization reactions. Applied Microbiology and Biotechnology, 98(3): 1119–1126.

Diaz, A., Loewen, P.C., Fita, I. and Carpena, X. 2012. Thirty years of heme catalases structural biology. Archives of Biochemistry and Biophysics, 525(2): 102–110.

Donald, L.J., Krokhin, O.V., Duckworth, H.W., Wiseman, B., Deemagarn, T. and Singh, R. 2003. Characterization of the catalase–peroxidase KatG from *Burkholderia pseudomallei* by mass spectrometry. Journal of Biological Chemistry, 278: 35687–35692.

Dondero, M., Egana, W., Tarky, W., Cifuentes, A. and Torres, J.A. 1993. Glucose oxidase/catalase improves preservation of shrimp (*Heterocarpus reedi*). Journal of Food Science, 58(4): 774–779.

Futo, P., Markus, G., Kiss, A. and Adanyi, N. 2012. Development of a catalase-based amperometric biosensor for the determination of increased catalase content in milk samples. Electroanalysis, 24(1): 107–113.

Foyer, C.H. and Noctor, G. 2000. Oxygen processing in photosynthesis: Regulation and signalling. New Phytologist, 146(3): 359–388.

Gaetani, G.F., Kirkman, H.N., Mangerini, R. and Ferraris, A.M. 1994. Importance of catalase in the disposal of hydrogen peroxide within human erythrocytes. Blood, 84(1): 325–330.

Gebicka, L. and Didik, J. 2009. Catalytic scavenging of peroxynitrite by catalase. Journal of Inorganic Biochemistry, 103: 1375–1379.

Gudelj, M., Fruhwirth, G.O., Paar, A., Lottspeich, F., Robra, K.H. and Cavaco-Paulo, A. 2001. A catalase-peroxidase from a newly isolated thermoalkaliphilic *Bacillus* sp. with potential for the treatment of textile bleaching effluents. Extremophiles, 5(6): 423–429.

Gurung, N., Ray, S., Bose, S. and Rai, V. 2013. A broader view: Microbial enzymes and their relevance in industries, medicine and beyond. Biomed. Research International, 11: 329121/1-18.

Hnaien, M., Lagarde, F. and Jaffrezic-Renault, N. 2010. A rapid and sensitive alcohol oxidase/catalase conductometric biosensor for alcohol determination. Talanta, 81(1-2): 222–227.

Huisman, G.W. and Collier, S.J. 2013. On the development of new biocatalytic processes for practical pharmaceutical synthesis. Current Opinion in Chemical Biology, 17(2): 284–292.

Ishida, M., Yoshida, M. and Oshima, T. 1997. Highly efficient production of enzymes of an extreme thermophile, *Thermus thermophilus*: A practical method to over express GC-rich genes in *Escherichia coli*. Extremophiles, 1(3): 157–162.

Jasass, F.B. and Fung, D.Y.C. 1998. Catalase activity as an index of microbial load and end point cooking temperature of fish. Journal of Rapid Methods and Automation in Microbiology, 6: 159–197.

Jensen, N.P. 2000. Catalase enzyme. Textile Chemistry Color and Dyestuff Reporter, 32: 23–24.

Kapetanaki, S.M., Zhao, X., Yu, S., Magliozzo, R.S. and Schelvis, J.P. 2007. Modification of the active site of *Mycobacterium tuberculosis* KatG after disruption of the Met–Tyr–Trp cross-linked adduct. Journal of Inorganic Biochemistry, 101(3): 422–433.

Klotz, M.G., Klassen, G.R. and Loewen, P.C. 1997. Phylogenetic relationships among prokaryotic and eukaryotic catalases. Molecular Biology and Evolution, 14(9): 951–958.

Kocabas, D.S., Bakir, U., Phillips, S.E., McPherson, M.J. and Ogel, Z.B. 2008. Purification, characterization and identification of a novel bifunctional catalase-phenol oxidase from *Scytalidium thermophilum*. Applied Microbiology and Biotechnology, 79(3): 407–415.

Koclar, A.G., Coruh, N., Bolukbasi, U. and Ogel, Z.B. 2013. Oxidation of phenolic compounds by the bifunctional catalase-phenol oxidase (CATPO) from *Scytalidium thermophilum*. Applied Microbiology and Biotechnology, 97(2): 661–672.

Kroll, R.G., Frears, E.R. and Bayliss, A. 1989. An oxygen electrode-based assay of catalase activity as a rapid method for estimating the bacterial content of foods. Journal of Applied Microbiology, 66(3): 209–217.

Kuhnel, K., Derat, E., Terner, J., Shaik, S. and Schlichting, I. 2007. Structure and quantum chemical characterization of chloroperoxidase compound 0, a common reaction intermediate of diverse heme enzymes. Proceedings of the National Academy of Sciences of the United States of America, 104(1): 99–104.

Kwak, Y.S. 2010. Studies of exercise-induced allergy anaphylaxis mechanisms and the effects of Vitamin C and catalase supplementation in exercise-induced allergy anaphylaxis models. Journal of Life Science, 20(4): 511–518.

Lee, D.H., Oh, D.C., Oh, Y.S., Malinverni, J.C., Kukor, J.J. and Kahang, H.Y. 2007. Cloning and characterization of monofunctional catalase from photosynthetic bacterium *Rhodospirillum rubrum* S1. Journal of Microbiology and Biotechnology, 17(9): 1460–1468.

Little, B.J. and Lee, J.S. 2014. Microbiologically influenced corrosion: An update. International Materials Reviews, 59(7): 384–393.

Loew, O. 1900. A new enzyme of general occurrence in organism. Science, 11: 701–702.

Loncar, N. and Fraaije, M.W. 2015. Catalases as biocatalysts in technical applications: Current state and perspectives. Applied Microbiology and Biotechnology, 99: 3351–3357.

Loewen, P.C., Villanueva, J., Switala, J., Donald, L.J. and Ivancich, A. 2015. Unprecedented access of phenolic substrates to the heme active site of a catalase: Substrate binding and peroxidase-like reactivity of *Bacillus pumilus* catalase monitored by X-ray crystallography and EPR spectroscopy. Proteins, 83(5): 853–866.

Magner, E. and Klibanov, A.M. 1995. The oxidation of chiral alcohols catalyzed by catalase in organic solvents. Biotechnology and Bioengineering, 46(2): 175–179.

Majumdar, T., Chakraborty, R. and Raychaudhuri, U. 2013. Rapid electrochemical quantification of food-borne pathogen *Staphylococcus aureus* based on hydrogen peroxide degradation by catalase. Journal of the Electrochemical Society, 160(4): G75–G78.

Maqsood, S., Benjakul, S., Abushelaibi, A. and Alam, A. 2014. Phenolic compounds and plant phenolic extracts as natural antioxidants in prevention of lipid oxidation in seafood: A detailed review. Comprehensive Reviews in Food Science and Food Safety, 13(6): 1125–1140.

Nicholls, P., Fita, I. and Loewen, P.C. 2000. Enzymology and structure of catalases. Advances in Inorganic Chemistry, 51: 51–106.

Nishikawa, M., Hyoudou, K., Kobayashi, Y., Umeyama, Y., Takakura, Y. and Hashida, M. 2005. Inhibition of metastatic tumor growth by targeted delivery of antioxidant enzymes. Journal of Controlled Release, 109(1-3): 101–107.

Obinger, C., Regelsberger, G., Strasser, G., Burner, U. and Peschek, G.A. 1997. Purification and characterization of a homodimeric catalase-peroxidase from the cyanobacterium *Anacystis nidulans*. Biochemical and Biophysical Research Communications, 235(3): 545–552.

Oluoch, K.R., Welander, U., Andersson, M.M., Mulaa, F.J., Mattiasson, B. and Hatti-Kaul, R. 2006. Hydrogenperoxide degradation by immobilized cells of alkaliphilic *Bacillus halodurans*. Biocatalysis and Biotransformation, 24(3): 215–222.

Orozco, J., García-Gradilla, V., D'Agostino, M., Gao, W., Cortes, A. and Wang, J. 2012. Artificial enzyme-powered microfish for water-quality testing. ACS Nano, 7(1): 818–824.

Paar, A., Costa, S., Tzanov, T., Gudelj, M., Robra, K.H. and Cavaco-Paulo, A. 2001. Thermoalkalistable catalases from newly isolated *Bacillus* sp. for the treatment and recycling of textile bleaching effluents. Journal of Biotechnology, 8: 147–153.

Patel, J.R. and Beuchat, L.R. 1995. Enrichment in Fraser broth supplemented with catalase or Oxyrase ®, combined with the microcolony immunoblot technique, for detecting heat-injured *Listeria monocytogenes* in foods. International Journal of Food Microbiology, 26: 165–176.

Perron, N. and Brumaghim, J. 2009. A review of the antioxidant mechanisms of polyphenol compounds related to iron binding. Cell Biochemistry and Biophysics, 53(2): 75–100.

Qi, Ce., Chen, Y., Huang, J.H., Jina, Q.Z. and Wang, X.G. 2012. Preparation and characterization of catalase-loaded solid lipid nanoparticles based on soybean phosphatidylcholine. Journal of the Science of Food and Agriculture, 92(4): 787–793.

Sangar, S., Pal, M., Moon, L.S. and Jolly, R.S. 2012. A catalase-peroxidase for oxidation of beta-lactams to their (R)-sulfoxides. Bioresourse Technology, 115: 102–110.

Schliebs, W., Wurtz, C., Kunau, W.H., Veenhuis, M. and Rottensteiner, H.A. 2006. Eukaryote without catalase-containing microbodies: *Neurospora crassa* exhibits a unique cellular distribution of its four catalases. Eukaryotic Cell, 5(9): 1490–1502.

Sezginturk, M.K., Goktug, T. and Dinckaya, E. 2005. A biosensor based on catalase for determination of highly toxic chemical azide in fruit juices. Biosensors and Bioelectronics, 21: 684–688.

Shi, X., Shi, Z., Huang, H., Zhu, H., Ju, D. and Zhou, P. 2013. PEGylated human catalase elicits potent therapeutic effects on H1N1 influenza induced pneumonia in mice. Applied Microbiology and Biotechnology, 97(23): 10025–10033.

Simmchen, J., Baeza, A., Ruiz-Molina, D. and Vallet-Regi, M. 2014. Improving catalase-based propelled motor endurance by enzyme encapsulation. Nanoscale, 6(15): 8907–8913.

Singh, R.P., Kang, D.Y., Oh, B.K. and Choi, J.W. 2009. Polyaniline-based catalase biosensor for the detection of hydrogen peroxide and azide. Biotechnology and Bioprocess Engineering, 14(4): 443–449.

Sooch, B.S. and Kauldhar, B.S. 2014a. A Process for Hyperproduction of Catalase Enzyme from Novel Extremophilic Bacterium *Geobacillus extremocatsoochus* MTCC 5873 and Strain Thereof. Indian Patent Application No. 362/DEL/2014.

Sooch, B.S., Kauldhar, B.S. and Puri, M. 2014b. Recent insights into microbial catalases: Isolation, production and purification. Biotechnology Advances, 32(8): 1429–1447.

Sooch, B.S. and Kauldhar, B.S. 2015. Development of an eco-friendly whole cell-based continuous system for the degradation of hydrogen peroxide. Journal of Bioprocessing and Biotechniques, 5(6): 1–5.

Spiro, M.C. and Griffith, W.P. 1997. The mechanism of hydrogen peroxide bleaching. Textile Chemist and Colorist, 29: 12–13.

Switala, J. and Loewen, P.C. 2002. Diversity of properties among catalases. Archives of Biochemistry and Biophysics, 401(2): 145–154.

Taub, J., Lau, J.F., Ma, C., Hahn, J.H., Hoque, R. and Rothblatt, J. 1999. A cytosolic catalase is needed to extend adult lifespan in *Caenorhabditis elegans* daf-C and clk-1mutants. Nature, 399(6732): 162–166.

Tomsho, J.W., Pal, A., Hall, D.G. and Benkovic, S.J. 2012. Ring structure and aromatic substituent effects on the pK(a) of the benzoxaborole pharmacophore. ACS Medicinal Chemistry Letters, 3(1): 48–52.

Trusek-Holownia, A. and Noworyta, A. 2015. Catalase immobilized in capsules in microorganism removal from drinking water, milk and beverages. Desalination and Water Treatment, 55(10): 2721–2727.

Vetrano, A.M., Heck, D.E., Mariano, T.M., Mishin, V., Laskin, D.L. and Laskin, J.D. 2005. Characterization of the oxidase activity in mammalian catalase. Journal of Biological Chemistry, 280(42): 35372–35381.

Welinder, K.G. 1991. Bacterial catalase–peroxidases are gene duplicated members of the plant peroxidase superfamily. Biochimica et Biophysica Acta, 1080(3): 215–220.

Wieser, R., Adam, G., Wagner, A., Schuller, C., Marchler, G. and Krawiecs, Z. 1991. Heat shock factor: Independent heat control of transcription of the CTT1 gene encoding the cytosolic catalase T of *Saccharomyces cerevisiae*. Journal of Biological Chemistry, 266(19): 12406–12411.

Williams, C., Aksam, E.B., Gunkel, K., Veenhuis, M. and van der Klei, I.J. 2012. The relevance of the non-canonical PTS1 of peroxisomal catalase. Biochimica et Biophysica Acta, 1823(7): 1133–1141.

Wolf, J., Schliebs, W. and Erdmann, R. 2010. Peroxisomes as dynamic organelles: Peroxisomal matrix protein import. The FEBS Journal, 277(16): 3268–3278.

Yildiz, H., Akyilmaz, E. and Dinckaya, E. 2004. Catalase immobilization in cellulose acetate beads and determination of its hydrogen peroxide decomposition level by using a catalase biosensor. Artificial Cells Blood Substitutes and Immobilization Biotechnology, 32(3): 443–452.

Zamocky, M., Furtmuller, P.G. and Obinger, C. 2008. Evolution of catalases from bacteria to humans. Antioxidants and Redox Signaling, 10(9): 1527–1548.

Zamocky, M. and Koller, F. 1999. Understanding the structure and function of catalases: Clues from molecular evolution and *in vitro* mutagenesis. Progress in Biophysics and Molecular Biology, 72(1): 19–66.

Zamocky, M., Godocikova, J., Koller, F. and Polek, B. 2001. Potential application of catalase-peroxidase from *Comamonas terrigena* N3H in the biodegradation of phenolic compounds. Antonie Van Leeuwenhoek, 79(2): 109–117.

14

Microbial Laccases as Potential Eco-friendly Biocatalysts for the Food Processing Industries

Susana Rodríguez-Couto

1. Introduction

Laccase (*p*-diphenol oxygen oxidoreductases EC 1.10.3.2) enzymes are copper-containing oxidases, widely distributed in Nature. The interest in these almost-neglected enzymes has exploded after the discovery that their catalytic action can be extended by the use of the so-called redox mediators. This aspect, together with its broad substrate specificity and the fact that it only needs oxygen (easily available from air) to exert its catalytic function makes laccases very attractive for a great variety of industrial and biotechnological applications, including food processing.

White-rot fungi are the only microorganisms that degrade the bulky, recalcitrant and heterogeneous polymer lignin (Kirk and Fenn, 1982). This ability is due to the secretion of an extracellular and non-specific enzymatic complex during their secondary metabolism (Wesenberg et al., 2003), usually triggered by nitrogen depletion. This enzymatic complex mainly consists of lignin peroxidases (LiPs; EC 1.11.1.14), manganese-dependent peroxidases (MnPs; EC 1.11.1.13) and laccases (benzenediol: oxygen oxidoreductases; EC 1.10.3.2) together with other complementary enzymes (Ruiz-Dueñas and Martínez, 2009). Peroxidases are stronger oxidants than laccases but they need hydrogen peroxide to exert their catalytic action. Also, the MnP dependence on Mn^{2+} or the LiP dependence on veratryl alcohol limit their practical application.

CEIT-IK4, Unit of Environmental Engineering, Paseo Manuel de Lardizábal 15, 20018, San Sebastian, Spain; IKERBASQUE, Basque Foundation for Science, Díaz de Haro 3, 48013, Bilbao, Spain.
E-mail: srodriguez@ceit.es

In contrast, laccases only need oxygen (easily available from air) to perform their catalysis, producing water as the only by-product, due to which they are considered as 'green' catalysts. These properties make laccases promising biocatalysts for different industrial applications, such as bio-bleaching, bio-pulping, stabilizers in wine production, biodegradation of organic pollutants, textile decolouration, bio-fuel cells, biosensors, manufacture of antibiotics and anti-cancer drugs, polymer and fiber surface modifications, etc. (Rodriguez-Couto and Toca-Herrera, 2006; Medhavi and Lele, 2009). Among these applications, the ability of laccase to degrade lignin opens great opportunities for lignin valorisation which is presently a hot topic of research. Also, laccases are excellent candidates for food industry applications since many of the laccase substrates (e.g., unsaturated fatty acids, phenols, thiol-containing proteins) are present in different foods and beverages.

At present there are several laccase-based preparations commercially available in the market, mainly for textile applications (Table 1). Being energy-saving and biodegradable, laccase-based biocatalysts are especially appropriate for the development of sustainable and environment-friendly industries. However, the high commercial price of laccases (Table 2) limits their industrial exploitation.

Table 1. Commercial laccase-based preparations.

Brand Name	Main Application	Manufacturer
Lignozym®-process	Pulp bleaching	Lignozym GmbH (Germany) now Wacker Chemie
LaSOX	Pulp and paper industry (delignification and bleaching)	Bioscreen E.K. (Germany)
DeniLite®	Textile industry	Novozymes (Denmark)
Zylite	Denim bleaching	Zytex Pvt. Ltd. (India)
Novozym® 51003	Paper industry	Novozymes (Denmark)
NS51002	Deinking	Novozymes (Denmark)
NS51003	Deinking	Novozymes (Denmark)
LACCASE Y120	Colour enhancement in tea, etc.	Amano Enzyme USA Co. Ltd.
Flavourstar	Brewing	Advanced Enzyme Technologies Ltd. (India)
Bleach Cut 3-S	Denim bleaching	Season Chemicals (China)
Ecostone® LC10	Denim finishing	AB Enzymes GmbH (Germany)
Cololacc BB	Denim finishing	Colotex Biotechnology Co. Ltd. (Hong Kong)
Primagreen® Ecofade LT100	Denim bleaching and shading	Genencor International Inc. (USA)
IndiStar™	Denim finishing	Genencor, Danisco USA Inc.
Purizyme Laccase	Bleaching of indigo dyed garments	Puridet (Asia) Ltd. (Hong Kong)
Novoprime® Base 268	Denim bleaching	Novozymes (Denmark)
Suberase®	Cork modification; synthesis of phenolic colourants	Novozymes (Denmark)
MetZyme	Pulp and paper; biofuel; wastewater treatment	Metgen Oy (Finland)

Table 2. Price of some commercially available laccases.

Company	Source	Quantity	Price (EUR)
Sigma Aldrich	*Agaricus bisporus* (≥ 4 U/mg)	100 mg	90.5
(www.sigmaaldrich.com)	*Rhus vernicifera* (≥ 50 U/mg)	1 g	546
	Trametes versicolor (≥ 0.5 U/mg)	10 KU	221
		1 g	73.3
		10 g	465.5
Jena Biosciences GmbH	*Cerrena unicolor*	1000 U	162
(www.jenabioscience.com)		5000 U	648
NZYTech (www.nzytech.com)	*Escherichia coli* (recombinant)	1 mg	81
	Bacillus subtilis (recombinant)	3 × 1 mg	217
		1 mg	81
		3 × 1 mg	217

The methodology and expression of laccase activity are different among the companies

2. Laccase Producing Microorganisms

Laccases are ubiquitous enzymes in Nature. The first laccase was found in the exudates of the Japanese lacquer tree, *Toxicodendron vernicifluum* (formerly *Rhus vernicifera*) (Yoshida, 1883), from which the name laccase was derived. A few years later, it was also found in fungi (Bertrand, 1896). Since then, they have been found in many plants, fungi, bacteria (Dwivedi et al., 2011) and a few insects (Xu, 1999). The first bacterial laccase was found in the plant-root-associated bacterium, *Azospirillum lipoferum* (Givaudan et al., 1993). Later, laccase activity was found in other bacteria, such as *Marinomonas mediterranea, Mycobacterium tuberculosis* or in the coat protein A (cotA) endospores of *Bacillus sphaericus* and *B. subtilis* (Claus, 2004). Most of the studied bacterial laccases are intracellular although extracellular bacterial laccases have also been found (Arias et al., 2003). Among fungi, laccases are particularly abundant in the white-rot fungi (Arora and Sharma, 2010), which are the only microorganisms to degrade the whole wood components (i.e., cellulose, hemicellulose and lignin) so far (Kirk and Fenn, 1982). Almost all white-rot fungi species are reported to produce laccases to different extents (Hatakka, 2001). Among them, *Trametes versicolor* (formerly known as *Coriolus versicolor* or *Polyporus versicolor*) is the most studied fungus for the production of laccase (Fig. 1). Other laccase producers of the *Trametes* genus include *T. pubescens* (Galhaup et al., 2002a), *T. hirsuta* (Rodríguez-Couto et al., 2003) and *T. gallica* (Dong et al., 2005). In plants, laccase facilitates lignification whereas in fungi, it is involved in delignification, sporulation, pigment production, fruiting body formation and plant pathogenesis (Thurston, 1994; Yaver et al., 2001).

3. Catalysis Mechanism

Laccases are multi-copper oxidases. They contain four copper atoms at their active site distributed in one mono-nuclear termed T1, which is responsible for the blue colour of the enzyme and one tri-nuclear cluster T2/T3 (Mayer and Staples, 2002). Laccases lacking the T1 copper are called yellow or white laccases.

Figure 1. Fruiting bodies of the white-rot-fungus *Trametes versicolor* (Source: www.mykoweb.com).

In the substrate oxidation catalyzed by laccase, the four copper ions at the active site transfer the electrons involved in the reaction. The 'blue' copper (T1 site) is thought to accept electrons from the reducing substrates and the electrons are subsequently transferred to the tri-nuclear copper cluster (T2/T3 site), where oxygen is reduced to water. Thus, the net result is the oxidation of four substrate molecules into four radicals, while one oxygen molecule is reduced into two water molecules (Morozova et al., 2007). The redox potential of the T1 copper site is directly responsible for the catalytic capacity of the enzyme. In contrast to most enzymes, laccases have a low substrate specificity which allows them degrading compounds with a structure similar to lignin, such as polyaromatic hydrocarbons (PAHs), textile dyes and other xenobiotic compounds (Mayer and Staples, 2002), and, hence, their industrial interest. However the range of substrates susceptible to be oxidized by laccases is limited by their low redox potential, since the oxidation of a substrate depends on the redox potential difference between that substrate and the T1 copper of laccase (Xu, 1996). The redox potentials of the T1 copper of laccase range from 0.4 V to 0.8 V *versus* a normal hydrogen electrode (NHE) (Xu et al., 1996; Garzillo et al., 2001; Morozova et al., 2007). Therefore, if the substrate of interest has a particularly high redox potential, such as the non-phenolic lignin units, laccase is not able to oxidize it directly. Also, the substrate of interest can be a compound that due to its size or steric hindrance does not fit into the active center of laccase. These limitations can be overcome by the use of the so-called redox mediators. The redox mediators are organic compounds of low molecular weight to be oxidized by laccases, forming highly reactive cation radicals that oxidize compounds that laccases alone cannot. In Fig. 2, a scheme of laccase catalysis with and without redox mediators is presented. More than 100 redox mediators have been described but the most commonly used are 2,2'-azino-bis (3-ethylbenzothiazoline-6-sulfonic acid (ABTS) and 1-hydroxybenzotriazole (HBT). However, these synthetic mediators are toxic and not economical, which has impelled the search for alternative mediators of natural origin. Thus, Cañas and Camarero (2010) reported that several lignin-derived phenolic compounds are economical and environment-friendly alternatives to synthetic mediators. The chemical structures of several synthetic and natural redox mediators are shown in Fig. 3.

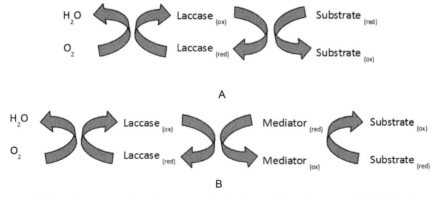

A

B

Figure 2. Schematic representation of laccase-catalysed redox cycles for substrates oxidation in the absence (A) or in the presence (B) of redox mediators (Riva, 2006).

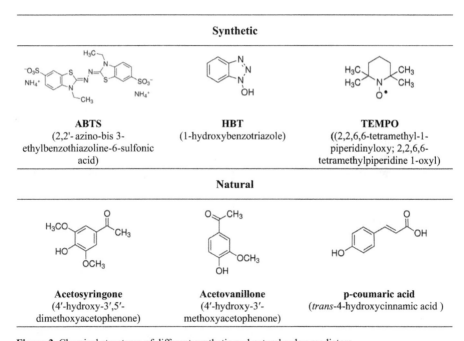

Synthetic

ABTS (2,2'- azino-bis 3- ethylbenzothiazoline-6-sulfonic acid)	**HBT** (1-hydroxybenzotriazole)	**TEMPO** ((2,2,6,6-tetramethyl-1- piperidinyloxy; 2,2,6,6- tetramethylpiperidine 1-oxyl)

Natural

Acetosyringone (4'-hydroxy-3',5'- dimethoxyacetophenone)	**Acetovanillone** (4'-hydroxy-3'- methoxyacetophenone)	**p-coumaric acid** (*trans*-4-hydroxycinnamic acid)

Figure 3. Chemical structures of different synthetic and natural redox mediators.

4. Properties

The biochemical properties of laccases have been reviewed in detail by Babu et al. (2012). Laccases often occur as isoenzymes or monomers that oligomerize to form multimeric complexes (Kunamneni et al., 2008a,b). Each isoenzyme has four atoms of copper and is able to individually perform the catalytic mechanism of laccases. The molecular mass of the laccase monomers ranges from 40 to 130 kDa and are glycosylated to an extent between 10 and 25 per cent in fungi and from 20 to 45 per cent in plants. This feature has been proposed to protect laccases against proteolysis and inactivation at high temperatures (Yoshitake et al., 1993).

Usually, fungal laccases have an optimum pH in the range of 3–5 and bacterial laccases in the range of 5–6, when the substrate is a hydrogen atom donor compound (e.g., ABTS). When the substrate is a phenolic compound (e.g., syringaldazine), the optimal pH is shifted to 6–7 as a result of the balance of redox potentials between the substrate and the inhibition of the T2/T3 copper site by the binding of a hydroxide anion (Xu, 1997). Fungal laccases have isoeletric points (pI) ranging from 3 to 7, whereas plant laccase pI is about 9 (Babu et al., 2012). Also, fungal laccases are more stable at higher acidic pH (Leonowicz et al., 1984), although exceptions exist (Baldrian, 2004).

The optimum temperature for laccase usually depends on the source. In general, laccases have optimum temperatures at 30–50°C and rapidly lose activity at temperatures above 60°C (Galhaup et al., 2002b; Jung et al., 2002; Palonen et al., 2003). However, the temperature stability varies considerably (Baldrian, 2006). On the other hand, the redox potential ranges from 0.4–0.5 V in plants and bacteria (Gianfreda et al., 1999; Durão et al., 2006) to 0.4–0.9 V in fungal laccases (Wesenberg et al., 2003).

5. Laccase Production

Submerged (SmF) and solid-state fermentation (SSF) techniques are used for laccase production. SmF involves the growth of microorganisms in a liquid medium which is rich in nutrients under high oxygen concentration (aerobic conditions). The uncontrolled growth of mycelium is the major problem associated with submerged fermentation as it results in mass and oxygen transfer limitations (Rodriguez-Couto and Toca-Herrera, 2007). This problem can be solved by using a pulsed system to control pellet growth (Lema et al., 2001) or by cell immobilization (Rodriguez-Couto and Toca-Herrera, 2007 and references therein).

SSF involves the growth of microorganisms in the absence or near absence of free-flowing liquid, using an inert substrate (synthetic material) or a natural substrate (organic material) as a solid support (Pandey et al., 1999a). SSF is shown to be particularly suitable for the production of enzymes by filamentous fungi (Moo-Young et al., 1983; Pandey et al., 1999a), since it reproduces the natural habitat (Pandey et al., 1999b). In Fig. 4, photographs of the white-rot fungus *T. pubescens* grown under SmF conditions (Fig. 4A) and under SSF conditions on an inert support (nylon sponge cubes) (Fig. 4B) are shown.

The main obstacles for laccase commercialization are its low yield and high production cost. Some attempts have been made over the last few years to overcome this problem and achieve low-cost overproduction of laccase. Thus, the use of inducers of the enzyme activity and agro- and forestry-wastes as substrates has led to promising results in both SmF and SSF conditions (Elisashvili et al., 2006; Toca-Herrera et al., 2007; Elisashvili et al., 2008; Rodriguez-Couto et al., 2009; Songulashvili et al., 2015). Also, the selection of a suitable host overproducing laccase is another approach that has gained the attention of many researchers either by isolating and screening new laccase producers (Kiiskinen et al., 2004; Dhakar et al., 2014) or by constructing gene engineering strains (Sun et al., 2012; Theerachat et al., 2012).

Figure 4. The white-rot fungus *Trametes pubescens* grown under SmF conditions (A) and under SSF conditions on an inert support (B).

6. Applications of Laccases to Food Industries

The potential application of laccases to the food industry has been reviewed by several researchers (Minussi et al., 2002; Brijwani et al., 2010; Osma et al., 2010) and more recently by Pezzela et al. (2015). In the following text the different actual and potential applications of laccase to the food industry are described.

6.1 Beverages

6.1.1 Wine

Polyphenols have undesirable effects on wine production and on its organoleptic characteristics; therefore they must be removed from wine (Rosana et al., 2002). Although different innovative treatments, such as enzyme inhibitors, complexing agents and sulphate compounds, have been developed to remove the above-mentioned compounds, the possibility of using laccase-based treatments as a specific and mild technology is seen as an appalling approach (Cantarelli et al., 1989; Arora and Sharma, 2010). Also, biosensors with immobilized laccases have been developed to determine the total phenolic content (i.e., tannins) in wines (Chawla et al., 2012; Lanzelloto et al., 2014).

6.1.2 Beer

Haze formation has been a persistent problem in the brewing industry (McMurrough et al., 1999). It is produced by small quantities of naturally-occurring pro-anthocyanidins that generate protein precipitation (Mathiasen, 1995). Different authors have investigated the use of laccases for polyphenol oxidation as an alternative to the traditional methods (Rossi et al., 1988; Giovanelli, 1989; Mathiasen, 1995). In addition, laccases have been used to remove the undesired oxygen at the end of the brewing processing (Mathiasen, 1995) and, thereby, increase the beer storage life. Moreover, Dhillon et al. (2012) showed the laccase potential for the clarification and flocculation of crude beer as a sustainable alternative to the traditional flocculants (e.g., stabfix, bentonite). Also, laccase-based biosensors have been constructed to determine the polyphenolic content in beers (ElKaoutit et al., 2008).

6.1.3 Juice clarification

The main purpose of juice clarification is to reduce the amount of phenolic compounds and decrease astringency (Alper and Acar, 2004). Ultrafiltration is the commonly used method to stabilize fruit juices (Sibert, 1999) but ultrafiltrated juices are not always stable due to reactive phenolic compounds cannot be retained by the membrane (Stuz, 1993). Thus, a pre-treatment technique with laccase before ultrafiltration was developed (Maier et al., 1994). Stanescu et al. (2012) reported that the addition of immobilized laccase from *T. pubescens* to apple juice stabilized its content in phenols, avoiding the deposit of solid polymers. More recently, Gasara-Chatti et al. (2013) showed that laccase encapsulated in hydrogels can replace the conventional ultrafiltration process.

6.1.4 Tea

Bowens et al. (1997) reported that treatment with a fungal laccase from a *Pleurotus* species enhanced the colour of a tea-based product. Also, Murugesan et al. (2002) showed that the addition of cellulase and laccase from *T. versicolor* at a ratio of 3:2 (v/v) increased the quality of black tea and, thus, its market value. On the other hand, Sabela et al. (2012) developed a laccase-based biosensor to evaluate the antioxidants properties of herbal tea samples.

6.1.5 Preventing taint in cork stoppers

Cork stoppers in bottled wine may impart an unpleasant flavour to the wine due likely to the phenolic compounds in the cork (Sponholz et al., 1994, 1997). These compounds can be removed by a laccase-based process (Conrad et al., 2000) and, in fact, a commercial laccase product named Suberase® has been developed and marketed by the company Novozymes (Denmark) to perform such a treatment.

6.2 Removal of aflatoxin B1 from foodstuffs

Aflatoxin B_1 is the most abundantly produced aflatoxin and has been shown to be highly mutagenic, toxic, carcinogenic and teratogenic to humans and animals (Eaton and Gallagher, 1994; Mishra and Das, 2003). Contamination of feed and foodstuffs by aflatoxin causes significant economic losses (Yabe and Nakajima, 2004) and health problems (Wogan, 2000). The different physical and chemical methods used to inactivate aflatoxin has resulted to be non-effective and costly (Mishra and Das, 2003). Therefore, an effective and low-cost method is required. Thus, Alberts et al. (2009) studied the degradation of aflatoxin B_1 by different laccases from different white-rot fungi. They found that aflatoxin B_1 was considerably degraded when treated with pure laccase from *T. versicolor* (87.34 per cent) and recombinant laccase from *Aspergillus niger* (55 per cent). In addition, a significant loss of its mutagenicity was obtained.

More recently, Scarpari et al. (2014) found that the treatment of contaminated maize with culture filtrates of *T. versicolor,* containing mainly laccase enzymes, reduced significantly the content of aflatoxin B_1. So, laccase treatment could be a promising approach to remove aflatoxin B_1 from foodstuffs.

6.3 Baking

The ability of laccases to cross-link biopolymers has made possible to use them in baking. Thus, it was shown that laccase addition to the dough of baking products increased volume, strength and stability, improved crumb structure and softness and reduced stickiness (Si, 1994).

In the last few years, increasing awareness of celiac disease has raised the interest of the food industry in marketing gluten-free cereal products. Thus, cereal flours, such as oat and corn, have been the focus for development of gluten-free baked products (Gallagher, 2009), but they lack the protein matrix responsible for dough formation and the physical characteristics found in wheat-based baked products. To overcome these issues, it was shown that the addition of laccases to oat flour improved the texture quality of oat bread (Renzetti et al., 2010).

6.4 Sugar beet pectin gelation

Sugar beet pectin is a functional food ingredient that can form thermo-irreversible gels through oxidative cross-linking of ferulic acid (Thibault et al., 1991). This type of gel is very interesting for the food industry as it can be heated while maintaining the gel structure. It was shown that laccase can cross-link beet pectin molecules (Kuuva et al., 2003; Littoz and McClements, 2008). Jung and Wicker (2012) reported that the conjugation of sugar beet pectin by laccase led to a high molecular weight and branched structure.

6.5 Other food applications

Other potential applications of laccase enzymes could be in the enhancement of taste and flavour in cacao (Takemori et al., 1992), deoxygenation of vegetal oils (Petersen et al., 1996), recovery of antioxidants from olive mill wastewater (Hamza et al., 2012), as an oxygen-scavenging system in active packaging (Johansson et al., 2012), development of a bio-electronic tongue for the analysis of grapes (Medina-Plaza et al., 2014), biosensors for quantification of pesticides in fruits (Ribeiro et al., 2014) and improvement of yogurt texture (Struch et al., 2015).

7. Conclusion and Future Outlook

Since laccase enzyme is not yet allowed as a food additive, the use of immobilized laccase might be a suitable method to overcome such legal barriers as in this form it may be classified as a technological aid. Furthermore, it would allow its re-utilization, making the process more economical. However, before laccases can be applied to the food industry, a method to produce laccase economically at an industrial scale should be available. In addition, issues related to laccase inactivation and the toxicity and cost of redox mediators also need to be solved.

References

Alberts, J.F., Gelderblom, W.C.A., Botha, A. and van Zyl, W.H. 2009. Degradation of aflatoxin B$_1$ by fungal laccase enzymes. International Journal of Food Microbiology, 135: 47–52.

Alper, N. and Acar, J. 2004. Removal of phenolic compounds in pomegranate juices using ultrafiltration and laccase-ultrafiltration combinations. Nahrung, 48: 184–187.

Arias, M.E., Arenas, M., Rodríguez, J., Soliveri, J., Ball, A.S. and Hernández, M. 2003. Kraft pulp biobleaching and mediated oxidation of a nonphenolic substrate by laccase from *Streptomyces cyaneus* CECT 3335. Applied and Environmental Microbiology, 69: 1953–1958.

Arora, D.A. and Sharma, R.K. 2010. Ligninolytic fungal laccases and their biotechnological applications. Applied Biochemistry and Biotechnology, 160: 1760–1788.

Babu, R.P., Pinnamaneni, R. and Koona, S. 2012. Occurrences, physical and biochemical properties of laccase. Universal Journal of Environmental Research and Technology, 2: 1–13.

Baldrian, P. 2004. Purification and characterization of laccase from the white-rot fungus *Daedalea quercina* and decolorization of synthetic dyes by the enzyme. Applied Microbiology and Biotechnology, 63: 560–563.

Baldrian, P. 2006. Fungal laccases-occurrence and properties. FEMS Microbiology Reviews, 30: 215–242.

Bertrand, G. 1896. Sur la presence simultanee de la laccase et de la tyrosinase dans le suc de quelques champignons. Comptes Rendus Hebdomadaires des Seances de l'Academie des Sciences, 123: 463–465.

Bowens, E.C.M., Trivedi, K., Van, V.C. and Winkel, K. 1997. Method of Enhancing Colour in a Tea Based Foodstuff. EP 0760213 A1.

Brijwani, K., Rigdon, A. and Vadlani, P.V. 2010. Fungal Laccases: Production, function and applications in food processing. Enzyme Research. DOI:10.4061/2010/149748.

Cañas, A.I. and Camarero, S. 2010. Laccases and their natural mediators: Biotechnological tools for sustainable eco-friendly processes. Biotechnology Advances, 28: 694–705.

Cantarelli, C., Brenna, O., Giovanelli, G. and Rossi, M. 1989. Beverage stabilization through enzymatic removal of phenolics. Food Biotechnology, 3: 203–214.

Chawla, S., Rawal, R., Kumar, D. and Pundir, C.S. 2012. Amperometric determination of total phenolic content in wine by laccase immobilized on to silver nanoparticles/zinc oxide nanoparticles modified gold electrode. Analytical Biochemistry, 430: 16–23.

Claus, H. 2004. Laccases: Structure, reactions, distribution. Micron, 35: 93–96.

Conrad, L.S., Sponholz, W.R. and Berker, O. 2000. Treatment of Cork with a Phenol Oxidizing Enzyme. US patent 6152966.

Dhakar, K., Jain, R., Tamta, S. and Pandey, A. 2014. Prolonged laccase production by a cold and pH tolerant strain of *Penicillium pinophilum* (MCC 1049) isolated from a low temperature environment. Enzyme Research. DOI:10.1155/2014/120708.

Dhillon, G.S., Kaur, S., Braur, S.K. and Verma, M. 2012. Flocculation and haze removal from crude beer using in-house produced laccase from *Trametes versicolor* cultured on brewer's spent grain. Journal of Agricultural and Food Chemistry, 60: 7895–7904.

Dong, J.L., Zhang, Y.W., Zhang, R.H., Huang, W.Z. and Zhang, Y.Z. 2005. Influence of culture conditions on laccase production and isozyme patterns in the white-rot fungus *Trametes gallica*. Journal of Basic Microbiology, 45: 190–198.

Durao, P., Bento, I., Fernandes, A.T., Melo, E.P., Lindley, P.F. and Martins, L.O. 2006. Perturbation of the T1 copper site in CotA-laccase from *Bacillus subtilis*: Structural, biochemical, enzymatic and stability studies. Journal of Biological Inorganic Chemistry, 11: 514–526.

Dwivedi, U.N., Singh, P., Pandey, V.P. and Kumar, A. 2011. Structure–function relationship among bacterial, fungal and plant laccases. Journal of Molecular Catalysis B: Enzymatic, 68: 117–128.

Eaton, D.L. and Gallagher, E.P. 1994. Mechanisms of aflatoxin carcinogenesis. Annual Review of Pharmacology and Toxicology, 34: 135–172.

Elisashvili, V., Penninckx, M., Kachlishvili, E., Asatiani, M. and Kvesitadze, G. 2006. Use of *Pleurotus dryinus* for lignocellulolytic enzymes production in submerged fermentation of mandarin peels and tree leaves. Enzyme and Microbial Technology, 38: 998–1004.

Elisashvili, V., Kachlishvili, E. and Pennickx, M. 2008. Effect of growth substrate, method of fermentation and nitrogen source on lignocellulose-degrading enzymes production by white-rot basidiomycetes. Journal of Industrial Microbiology and Biotechnology, 35: 1531–1538.

ElKaoutit, M., Naranjo-Rodriguez, I., Temsamani, K.R., Hernandez-Artiga, M.P., Bbellido-Milla, D. and Hidalgo-Hidalgo de Cisneros, J.L. 2008. A comparison of three amperometric phenoloxidase–Sonogel–carbon-based biosensors for determination of polyphenols in beers. Food Chemistry, 110: 1019–1024.

Galhaup, C., Wagner, H., Hinterstoisser, B. and Haltrich, D. 2002a. Increased production of laccase by the wood-degrading basidiomycete *Trametes pubescens*. Enzyme and Microbial Technology, 30: 529–536.

Galhaup, C., Goller, S., Peterbauer, C.K., Strauss, J. and Haltrich, D. 2002b. Characterization of the major laccase isoenzyme from *Trametes pubescens* and regulation of its synthesis by metal ions. Microbiology, 148: 2159–2169.

Gallagher, E. 2009. Improving gluten-free bread quality through the application of enzymes. European Journal of Nutraceuticals and Functional Foods, 20: 34–37.

Garzillo, A.M., Colao, M.C., Buonocore, V., Oliva, R., Falcigno, L., Saviano, M., Santoro, A.M., Zappala, R., Bonomo, R.P., Bianco, C., Giardina, P., Palmieri, G. and Sannia, G. 2001. Structural and kinetic characterization of native laccases from *Pleurotus ostreatus*, *Rigidoporus lignosus*, and *Trametes trogii*. Journal of Protein Chemistry, 20: 191–201.

Gasara-Chatti, F., Brar, S.K., Ajila, C.M., Verma, M., Tyagi, R.D. and Valero, J.R. 2013. Encapsulation of ligninolytic enzymes and its application in clarification of juice. Food Chemistry, 137: 18–24.

Gianfreda, L., Xu, F. and Bollag, J.M. 1999. Laccases: A useful group of oxidoreductive enzymes. Bioremediation Journal, 3: 1–26.

Giovanelli, G. 1989. Enzymic treatment of malt polyphenols for beer stabilization. Industrie delle Bevande, 18: 497–502.

Givaudan, A., Effosse, A., Faure, D., Potier, P., Bouillant, M.L. and Bally, R. 1993. Polyphenol oxidase in *Azospirillum lipoferum* isolated from rice rhizosphere: Evidence for laccase activity in non-motile strains of *Azospirillum lipoferum*. FEMS Microbiology Letters, 108: 205–210.

Hamza, M., Khoufi, S. and Sayadi, S. 2012. Fungal enzymes as a powerful tool to release antioxidants from olive mill wastewater. Food Chemistry, 131: 1430–1436.

Hatakka, A. 2001. Biodegradation of lignin. pp. 129–180. *In*: M. Hofrichter and A. Steinbüchel (eds.). Biopolymers, Vol. 1: Lignin, Humic Substances and Coal, Wiley-VCH, Weinheim.

Johansson, K., Winestrand, S., Johansson, C., Järnström, L. and Jönsson, L.J. 2012. Oxygen-scavenging coatings and films based on lignosulfonates and laccase. Journal of Biotechnology, 161: 14–18.

Jung, H., Xu, F. and Li, K. 2002. Purification and characterization of laccase from wood degrading fungus *Trichophyton rubrum* LKY-7. Enzyme and Microbial Technology, 30: 161–168.

Jung, J. and Wicker, L. 2012. Laccase mediated conjugation of sugar beet pectin and the effect on emulsion stability. Food Hydrocolloids, 28: 168–173.

Kiiskinen, L.L., Ratto, M. and Kruus, K. 2004. Screening for novel laccase-producing microbes. Journal of Applied Microbiology, 97: 640–646.

Kirk, T.K. and Fenn, P. 1982. Formation and action of the ligninolytic system in basidiomycetes. pp. 67–90. *In*: J.C. Franklin, J.N. Hedges and M.J. Swift (eds.). Decomposer Basidiomycetes, British Mycological Society Symposium 4, Cambridge University Press, Cambridge.

Kunamneni, A., Camarero, S., García-Burgos, C., Plou, F., Ballesteros, A. and Alcalde, M. 2008a. Engineering and applications of fungal laccases for organic synthesis. Microbial Cell Factories, 7: 32.

Kunamneni, A., Plou, F., Ballesteros, A. and Alcalde, M. 2008b. Laccases and their applications: A patent review. Recent Patents on Biotechnology, 2: 10–24.

Kuuva, T., Lantto, R., Reinikainen, T., Buchert, J. and Autio, K. 2003. Rheological properties of laccase-induced sugar beet pectin gels. Food Hydrocolloids, 17: 679–684.

Lanzelloto, C., Favero, G., Antonelli, M.L., Tortolini, C., Canistraro, S., Coppaei, E. and Mazzei, F. 2014. Nanostructured enzymatic biosensor based on fullerene and gold nanoparticles: Preparation, characterization and analytical applications. Biosensors and Bioelectronics, 55: 430–437.

Lema, J.M., Roca, E., Sanromán, A., Núñez, M.J., Moreira, M.T. and Feijoo, G. 2001. Pulsing bioreactors. pp. 309–329. *In:* J.M.S. Cabral, M. Mota and J. Tramper (eds.). Multiphase Bioreactor Design, Taylor & Francis, London.

Leonowicz, A., Edgehill, R.U. and Bollag, J.-M. 1984. The effect of pH on the transformation of syringic and vanillic acids by the laccases of *Rhizoctonia praticola* and *Trametes versicolor*. Archives of Microbiology, 137: 89–96.

Littoz, F. and McClements, D.J. 2008. Bio-mimetic approach to improving emulsion stability: Cross-linking adsorbed beet pectin layers using laccase. Food Hydrocolloids, 22: 1203–1211.

MacMurrough, I., Madigan, D., Kelly, R. and O'Rourke, T. 1999. Haze formation. Shelf-life prediction for lager beer. Food Technology, 53: 58–62.

Maier, G., Frei, M., Wucherpfennig, K., Dietrich, H. and Ritter, G. 1994. Innovative processes for production of ultrafiltrated apple juices and concentrates. Fruit Process, 5: 134–138.

Mathiasen, T.E. 1995. Laccase and beer storage. PCT International Application WO 9521240 A2.

Mayer, A.M. and Staples, R.C. 2002. Laccase: New functions for an old enzyme. Phytochemistry, 60: 551–565.

Medhavi, V. and Lele, S.S. 2009. Laccase: Properties and applications. Bioresources, 4: 1694–1717.

Medina-Plaza, C., de Saja, J.A. and Rodriguez-Mendez, M.L. 2014. Bioelectronic tongue based on lipidic nanostructured layers containing phenol oxidases and lutetium bisphthalocyanine for the analysis of grapes. Biosensors and Bioelectronics, 57: 276–283.

Minussi, C.R., Pastore, G.M. and Durán, N. 2002. Potential applications of laccase in the food industry. Trends in Food Science and Technology, 13: 205–216.

Mishra, H.N. and Das, C. 2003. A review on biological control and metabolism of aflatoxin. Critical Reviews in Food Science and Nutrition, 43: 245–264.

Moo-Young, M., Moreira, A.R. and Tengerdy, R.P. 1983. Principles of solid state fermentation. pp. 117–144. *In:* J.E. Smith, D.R. Berry and B. Kristiansen (eds.). The Filamentous Fungi. Edward Arnold Publishers, London.

Morozova, O., Shumakovich, G., Gorbacheva, M., Shleev, S. and Yaropolov, A. 2007. 'Blue' laccases. Biochemistry (Moscow), 72: 1136–1150.

Murugesan, G.S., Angayarkanni, J. and Swaminathan, K. 2002. Effect of tea fungal enzymes on the quality of black tea. Food Chemistry, 79: 411–417.

Osma, J.F., Toca-Herrera, J.L. and Rodriguez-Couto, S. 2010. Uses of laccases in the food industry. Enzyme Research. DOI:10.4061/2010/918761.

Palonen, H., Saloheimo, M., Viikari, L. and Kruus, K. 2003. Purification, characterization and sequence analysis of a laccase from the ascomycete *Mauginiella* sp. Enzyme and Microbial Technology, 33: 854–862.

Pandey, A., Selvakumar, P., Soccol, C.R. and Nigam, P. 1999a. Solid state fermentation for the production of industrial enzymes. Current Science, 77: 149–162.

Pandey, A., Azmi, W., Singh, J. and Banerjee, U.C. 1999b. pp. 383–426. *In:* V.K. Joshi and A. Pandey (eds.). Biotechnology: Food Fermentation, Educational Publishers & Distributors, New Delhi.

Petersen, B.R., Mathiasen, T.E., Peelen, B. and Andersen, H. 1996. Use of Laccase for Deoxygenation of Oil-containing Product such as Salad Dressing. PCT International Application WO 9635768 A.

Pezzella, C., Guarino, L. and Piscitelli, A. 2015. How to enjoy laccases. Cellular and Molecular Life Sciences, 72: 923–940.

Renzetti, S., Courtin, C.M., Delcour, J.A. and Arendt, E.K. 2010. Oxidative and proteolytic enzyme preparations as promising improvers for oat bread formulations: Rheological, biochemical and microstructural background. Food Chemistry, 119: 1465–1473.

Ribeiro, F.W.P., Barroso, M.F., Morais, S., Viswanathan, S., de Lima-Neto, P., Correia, A.N., Oliveira, M.B.P.P. and Delarue-Matos, C. 2014. Simple laccase-based biosensor for formetanate hydrochloride quantification in fruits. Bioelectrochemistry, 95: 7–14.

Riva, S. 2006. Laccases: Blue enzymes for green chemistry. Trends in Biotechnology, 24: 219–226.

Rodríguez-Couto, S., Moldes, D., Liébanas, A. and Sanromán, A. 2003. Investigation of several bioreactor configurations for laccase production by *Trametes versicolor* operating in solid-state conditions. Biochemical Engineering Journal, 15: 21–26.

Rodriguez-Couto, S. and Toca-Herrera, J.L. 2006. Industrial and biotechnological applications of laccases: A review. Biotechnology Advances, 24: 500–513.

Rodriguez-Couto, S. and Toca-Herrera, J.L. 2007. Laccase production at reactor scale by filamentous fungi. Biotechnology Advances, 25: 558–569.

Rodriguez-Couto, S., Osma, J.F. and Toca-Herrera, J.L. 2009. Removal of synthetic dyes by an eco-friendly strategy. Engineering in Life Sciences, 2: 116–123.

Rosana, C., Minussi, Y., Pastore, G.M. and Durany, N. 2002. Potential applications of laccase in the food industry. Trends in Food Science and Technology, 13: 205–216.

Rossi, M., Giovanelli, G., Cantarelli, C. and Brenna, O. 1988. Effects of laccase and other enzymes on barley wort phenolics as a possible treatment to prevent haze in beer. Bulletin de Liaison-Groupe Polyphenols, 14: 85–88.

Ruiz-Dueñas, F.J. and Martínez, A.T. 2009. Microbial degradation of lignin: How a bulky recalcitrant polymer is efficiently recycled in Nature and how we can take advantage of this. Microbial Biotechnology, 2: 164–177.

Sabela, M.I., Gumede, N.J., Singh, P. and Bisetty, K. 2012. Evaluation of antioxidants in herbal tea with a laccase biosensor. International Journal of Electrochemical Science, 7: 4918–4928.

Scarpari, M., Bello, C., Pietricola, C., Zaccaria, M., Bertocchi, L., Angelucci, A., Ricciardi, M.R., Scala, V., Parroni, A., Fabbri, A.A., Reverberi, M., Zjalic, S. and Fanelli, C. 2014. Aflatoxin control in maize by *Trametes versicolor*. Toxins, 6: 3426–3437.

Si, J.Q. 1994. Use of Laccase in Industry. International patent application PCT/DK94/ 00232.

Sibert, K.J. 1999. Protein–polyphenol haze in beverages. Food Technology, 53: 54–57.

Songulashvili, G., Spindler, D., Jiménez-Tobón, G.A., Jaspers, G., Kerns, G. and Penninckx, M.J. 2015. Production of a high level of laccase by submerged fermentation at 120-L scale of *Cerrena unicolor* C-139 grown on wheat bran. Comptus Rendus Biologies, 338: 121–125.

Sponholz, W.R. and Muno, H. 1994. Der Korkton-ein mikrobiologisches Problem 2. Die Wein-Wissenschaft, 49: 17–22.

Sponholz, W.R., Grossmann, M.K., Muno, H. and Hoffmann, A. 1997. The distribution of chlorophenols and chloroanisoles in cork and a microbiological method to prevent their formation. Industrie delle Bevande, 26: 602–607.

Stanescu, M.D., Gavrilas, S., Ludwig, R. Haltrich, D. and Lozinsky, V.I. 2012. Preparation of immobilized Trametes pubescens laccase on a cryogel-type polymeric carrier and application of the biocatalyst to apple juice phenolic compounds oxidation. European Food Research and Technology, 234: 655–662.

Struch, M., Linke, D., Mokoonlall, A., Hinrische, J. and Berger, R.G. 2015. Laccase-catalysed cross-linking of a yogurt-like model system made from skimmed milk with added food-grade mediators. International Dairy Journal, 49: 89–94.

Stutz, C. 1993. The use of enzymes in ultrafiltration. Fruit Process, 7: 248–252.

Sun, J. and Peng, R.H. 2012. Secretory expression and characterization of a soluble laccase from the *Ganoderma lucidum* strain 7071-9 in *Pichia pastoris*. Molecular Biology Reports, 39: 3807–3814.

Takemori, T., Ito, Y., Ito, M. and Yoshama, M. 1992. Flavor and Taste Improvement of Cacao Nib by Enzymatic Treatment. Japan Kokai Tokkyo Koho JP 04126037 A2.

Theerachat, M., Emond, S., Cambon, E., Bordes, F., Marty, A., Nicaud, J.M., Chulalaksananukul, W., Guieysse, D., Remaud-Simeon, M. and More, S. 2012. Engineering and production of laccase from *Trametes versicolor* in the yeast *Yarrowia lipolytica*. Bioresource Technology, 125: 267–274.

Thibault, J.F., Guillon, F. and Rombouts, F.M. 1991. Gelation of sugar beet pectin by oxidative coupling. pp. 119–133. *In:* R.H. Walter and S. Taylor (eds.). The Chemistry and Technology of Pectin, Academic Press Inc., San Diego.

Thurston, C.F. 1994. The structure and function of fungal laccases. Microbiology+, 140: 19–26.

Toca-Herrera, J.L., Osma, J.F. and Rodriguez-Couto, S. 2007. Potential of solid-state fermentation for laccase production. pp. 391–400. *In:* A. Mendez-Vilas (ed.). Communicating Current Research and Educational Topics and Trends in Applied Microbiology, Formatex, Badajoz, Spain.

Wesenberg, D., Kyriakides, I. and Agathos, S.N. 2003. White-rot fungi and their enzymes for the treatment of industrial dye effluents. Biotechnology Advances, 22: 161–187.

Wogan, G.N. 2000. Impacts of chemicals on liver cancer risk. Seminars in Cancer Biology, 10: 201–210.

Xu, F. 1996. Oxidation of phenols, anilines and benzenethiols by fungal laccases: Correlation between activity and redox potentials as well as halide inhibition. Biochemistry, 35: 7608–7614.

Xu, F., Shin, W., Brown, S.H., Wahleithner, J.A., Sundaram, U.M. and Solomon, E.I. 1996. A study of a series of recombinant fungal laccases and bilirubin oxidase that exhibit significant differences in redox potential, substrate specificity and stability. Biochimica et Biophysica Acta (BBA)—Protein Structure and Molecular Enzymology, 1292: 303–311.

Xu, F. 1997. Effects of redox potential and hydroxide inhibition on the pH activity profile of fungal laccases. The Journal of Biological Chemistry, 272: 924–928.

Xu, F. 1999. Recent progress in laccase study: Properties, enzimology, production and applications. *In*: M.C. Flickinger and S.W. Drew (eds.). The Encyclopedia of Bioprocessing Technology: Fermentation, Biocatalysis and Bioseparation, John Wiley & Sons, New York.

Yabe, K. and Nakajima, H. 2004. Enzyme reactions and genes in aflatoxin biosynthesis. Applied Microbiology and Biotechnology, 64: 745–755.

Yaver, D.S., Berka, R.M., Brown, S.H. and Xu, F. 2001. The Presymposium on Recent Advances in Lignin Biodegradation and Biosynthesis, 3-4 Vikki Biocentre, University of Helsinki, Helsinki, Finland.

Yoshida, H. 1883. Chemistry of lacquer (Urushi). Journal of the Chemical Society, 43: 472–486.

Yoshitake, A., Katayama, Y., Nakamura, M., Iimura, Y., Kawai, S. and Morohoshi, N. 1993. N-linked carbohydrate chains protect laccase III from proteolysis in *Coriolus versicolor*. Journal of General Microbiology, 139: 179–185.

PART 3

MICROBIAL ENZYMES IN FOOD FERMENTATION

15

Role of Intrinsic and Supplemented Enzymes in Brewing and Beer Properties

Sergio O. Serna-Saldivar[1,]* and *Monica Rubio-Flores*[2]

1. Introduction

Beer is the foremost alcoholic beverage consumed on this planet and has been manufactured for at least eight millennium. Roughly 1.89 billion hectoliters of barley beer were brewed throughout the world in the year 2013 (FAO, 2015). The top five producers—China, USA, Brazil, Russia and Germany with a yearly production of 506, 224, 135, 89, 86 million hectoliters of barley beer, accounted for more than 55 per cent of the total world production. Interestingly, during the past 50 years, beer production has more than quadrupled (FAO, 2015). The beer market is expected to continue to grow by approximately 1.2 per cent per year in the next few years mainly due to the consumers' increasing purchasing power as well as the their changing drinking habits.

Babylon is considered the birthplace of beer. Historically, it is well documented that malt was used 2400 years BC for the manufacture of opaque beers by the ancient Egyptian and Babylonian cultures. The Egyptians passed on their brewing techniques to the Greeks and Romans and from there it spread to all places around the world (Serna-Saldivar, 2010a).

[1] Professor, Centro de Biotecnología FEMSA, Escuela de Ingeniería y Ciencias, Tecnológico de Monterrey, Av. Eugenio Garza Sada 2501 Sur, Monterrey N.L. CP 64849, México.
[2] Graduate Student, Masters Program in Biotechnology, Centro de Biotecnología FEMSA, Escuela de Ingeniería y Ciencias Tecnológico de Monterrey, Av. Eugenio Garza Sada 2501 Sur, Monterrey N.L. CP 64849, México.
E-mail: monica.rubiof@gmail.com
* Corresponding author: sserna@itesm.mx

European beer production is divided into four main and sequential operations: malting, mashing, addition of hops and pitching or yeast fermentation. Three of these operations are based in the conversion of substrates by enzymes provided by malt and yeast. Thus the success of brewing operations greatly depends on the action of these important enzymes provided by malt and yeast or purposely supplemented as processing aids to modify the final beer properties.

The aim of this chapter is to review the critical role of intrinsic enzymes associated with malt and yeast and extrinsic or supplemented enzymes in the production of European beer.

2. Regular Barley Beer Production

European beer is usually produced from four major ingredients (barley malt, adjuncts, hops and yeast) and involves four chief operations (malting, mashing, addition of hops and yeast fermentation). Figure 1 depicts the major operations adopted to produce lager and ale beer.

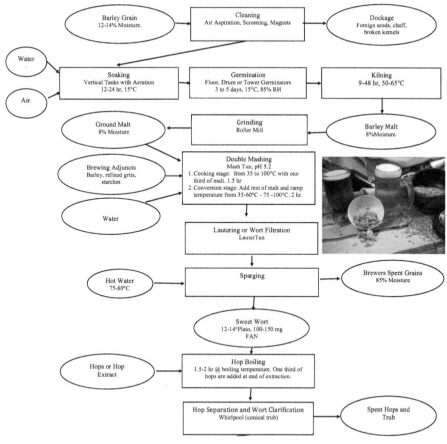

Figure 1. contd....

Figure 1. contd.

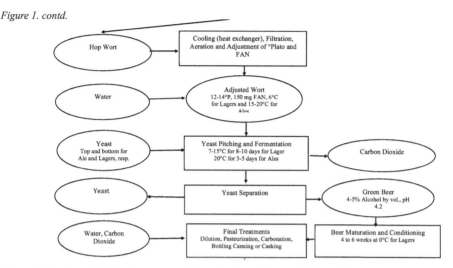

Figure 1. Flowchart of the brewing processes for production of lager and ale beers (Serna-Saldivar, 2010b).

2.1 Beer ingredients

2.1.1 Barley malt

Barley (*Hordeum vulgare*) has been the preferred raw material in brewing because of its high germination ability, presence of husk and mainly its capacity to synthesize critical enzymes during malting. Among the array of enzymes, undoubtedly the most relevant are α- and β-amylases, which are responsible of imparting the amylolytic or diastasic activity (DA). In order to optimize enzymes, the barley caryopsis is purposely germinated under controlled conditions. The three basic steps of germination or malting are grain soaking, germination and kilning (Fig. 1). The grain gets activated during soaking due to synthesis of gibberellic acid in the germ. This hormone controls germination and therefore the different enzymes produced mainly in the germ are *scutellum* and aleurone. The enzymes are sequentially produced—first in the germ and afterwards in the multi-layered aleurone. Generally, the first enzymes produced are lipases and lipooxygenases followed by cell wall degrading or fibrolitic enzymes, like celullases, β-glucanases, xylanases and hemicellulases which allow entry into the endosperm of the most relevant proteases and amylases. Both α- and β-amylases are produced by the aleurone cells and start degrading from the outer to the inner starchy endosperm. The ideal malting barley is relatively low in protein and yields sound and uniform kernels that upon germination at temperatures higher than 5°C synthesize high amounts of both α and β-amylases (Hough et al., 1993; Serna-Saldivar, 2010a,b).

2.2 Brewing adjuncts

Brewing adjuncts are feedstocks rich in carbohydrates that do not provide significant levels of enzymes. However, these are converted to fermentable sugars and dextrins during mashing in the presence of diastatic malt. The majority of brewing adjuncts consist of refined maize (*Zea mays*) or rice (*Oryza sativa*) grits or refined starches. Starches are practically pure starch whereas refined grits are also a source of endosperm proteins conformed by prolamins and glutelins. Glucose or maltose syrups are also used as adjuncts, especially for brewing light or dark beer. Starchy brewing adjuncts are hydrolyzed by the malt amylases into linear and branched dextrins, maltotriose, maltose and traces of glucose. Dextrins are responsible for imparting the typical beer body whereas simpler carbohydrates are yeast substrates, which are further converted into ethanol and other organic compounds. Most brewing grits have an average particle size of 40 to 60 US mesh that favors extraction and lautering. Refined grits are preferred because they contain more convertible carbohydrates, less pigments and oil and better color. The resulting beer is less prone to oxidation and has better flavor and color (Priest and Stewart, 2006; Serna-Saldivar, 2010a).

2.3 Hops

The chief feature of European beer is that it is brewed with hops (*Humulus lupulus*) in order to impart the typical flavor and aroma. Generally, 100 to 300 g hops are added for 1 hL sweet wort. Hops are obtained from the cones or female flower of the perennial plant, which adapts to cold or temperate climates. Hop plantations consist of female plants that grow in fields equipped with long posts that intercommunicate with wires, which support the vines. The female mature flowers have a green to yellow coloration before harvesting and provide the bittering and aromatic compounds enclosed in microscopic organelles named lupulins, which are rich in resins, phenolics, tannins and essential oils. After harvesting, the cones are carefully dehydrated at temperatures of 60 to 65°C for an average of 10 hours or until the moisture drops to about 8–10 per cent. The shelf-stable dried hops must be stored cold in an airtight condition to minimize oxidation and degradation of bitter and flavor compounds. Hops have several different functionalities in the beer system. They contribute to the typical flavor and aroma due to the presence of essential oils and resins. The tannins and other polyphenols help to clarify the beer because they bind soluble proteins that cause haze. Furthermore, hops contribute to the typical beer-flavor profile because they are rich in aldehydes, carboxylic acids and alcohols. The distinctive bitter flavor is due to resins classified as humulones or α-acids and lupulones or β-acids. The essential oils are a mixture of approximately 300 compounds, such as terpenes, aldehydes, ketones and alcohols. These organic compounds polymerize to form resin bodies. The essential oil imparts the characteristic aroma that is lost during boiling of wort. This is the reason why a portion of the total hops are commonly added when wort boiling is discontinued. Hops are classified into aroma and bittering types, depending on the ratio of oil to resin. Most commercial varieties are classified into mid-range, high and super alpha. Most hops used by the brewing industry are dehydrated and formed into dense pellets that are vacuum packaged. Hop extracts are produced by first treating the dehydrated cones with ethanol or methylene chloride and then removing the organic solvent by distillation to extract the spent hops with water in order to solubilize polyphenols and pectins (Bamforth, 2003; Hardwick, 1995; Hough et al., 1993; Priest and Stewart, 2006; Serna-Saldivar, 2010a).

2.4 Yeast

Yeast (*Saccharomyces cerevisiae*) plays an essential role in beer production because it is responsible for transforming the wort into beer (Fig. 1). The biochemical pathway of yeast is thoroughly covered by Ingledew (1995). In the brewing operations, yeast is selected according to the desired fermentation power, fermentation temperature and based on the flavorings and volatiles that are generated. Brewing yeasts are classed, according to the mode of action, into top and bottom. The basic difference is in the cell-wall structure. Comparatively, the cell walls of top yeast tend to be more hydrophobic; bottom yeasts do not sporulate and adapt to low temperature, thus being preferred for production of lagers. Floating brewing yeasts usually produce strong fermentation at high temperatures (15–22°C) and are almost always used for manufacturing pilsners. Regardless of the type of beer, natural yeast ferment soluble sugars into ethanol, carbon dioxide and intermediate organic metabolites that greatly affect beer flavor and aroma (Bamforth, 2009; Hardwick, 1995; Hough et al., 1993; Priest and Stewart, 2006; Serna-Saldivar, 2010b). Today, yeasts are being genetically engineered in order to generate unique strains capable of expressing enzymes that affect the brewing process and beer quality.

3. Major Brewing Operations

3.1 Malting

Malting is aimed at production of an array of enzymes critical for mashing and the rest for the brewing steps (Briggs, 1998; Steward, 2013; Zhou et al., 2013). The quality of malt greatly depends on growing conditions, postharvest grain management and malting conditions. Malting is divided into three sequential major operations: steeping or soaking, germination and kilning (Fig. 1). The typical process starts when cleaned barley kernels are steeped in cold water (15°C) for an average of 24 hours in order to activate the grain. The key step is germination of the hydrated kernels until achieving the maximum DA that usually peaks after three to five days followed by kilning in order to stop activity and develop important flavoring and colorful compounds (Serna-Saldivar, 2010b). There are different malting technologies, but the most widely practiced are floor, drum, rectangular or Saladin and tower. Briggs (1998) points out the advantages and disadvantages of these commercial malting systems.

3.2 Steeping

The aim of steeping is to hydrate the kernel under aerobic conditions in order to activate the synthesis of gibberellic acid that controls the germination process. The water enters the kernel through the mycropylar region of the germ and rapidly hydrates both the embryonic axis and *scutellum*. Then water is distributed to the rest of the kernel through the endocarp tube cells. The water diffuses inward because the endocarp cross cells act as a seal. The water first permeates to the aleurone layer and later, slowly hydrates from the outer to the inner starchy endosperm. The hydration process under ideal conditions lasts one to two days. The imbibing of the germ triggers respiration and the secretion of gibberellins both in the embryo and *scutellum* and the absorbed water catalyzes the synthesis of enzymes and germination weakening the grain structure (Serna-Saldivar, 2010a,b). According to Desai et al. (1997), gibberellins stimulate production of α-amylase by the aleurone layer only after eight hour exposure.

Generally, the soaking operation is simply performed by immersing or by spraying water on to the grains for 24 to 80 hours to increase the grain moisture to 42–48 per cent. Most operations take place in vertical tanks with a conic base or vertical cylinders. These containers have perforated pipelines to inject air. The water temperature plays a critical role because it affects the hydration rate and the microbiology of the grain. A relatively high water temperature allows excessive microbial growth, decreasing the oxygen availability for the developing embryo. The water temperature is usually adjusted at 15°C. The other important control factor is the application of air at time intervals which create aerobic conditions and take out the respiration heat and carbon dioxide. The frequency of the aeration cycles is less than an hour. Soaking is usually interrupted after 12–24 hours by first draining the water and then adding new water. During steeping, the barley kernels absorb water which catalyzes respiration, synthesis of enzymes and germination (Briggs, 1998; Hough et al., 1993; Hardwick, 1995; Palmer and Bathgate, 1976; Pyler and Thomas, 2000; Priest and Stewart, 2006; Serna-Saldivar, 2010b).

3.3 Germination

After steeping, the soaked barley kernels are allowed to germinate under special conditions so as to achieve the desired DA, cell division and the development of rootlets (primordial roots) and acrospires (first leaf). The process is usually carried on malting beds placed inside special rooms with strict temperature and relative humidity controls. This process lasts four to six days and is greatly affected by the malting-house temperature. The sprouting rate is controlled according to the initial moisture, germination temperature and airflow that circulate through the malting bed. The operation can be performed on malting floors or in rotative drums. The main goal of malting is to achieve the maximum DA with the lowest dry matter loss. Gibberellins are mainly responsible for triggering the synthesis of key enzymes that will degrade the nutrients stored in the first and second reserve tissues of the germ and endosperm, respectively. Two gibberellic acids (GA_1 and GA_3) are the major hormones produced by the embryo. These hormones are later on transported along with the absorbed water to the endosperm through the tube cells and induce aleurone cells to synthesize enzymes, such as amylases and proteases, which will catalyze the depolymerization of stored starch and endosperm proteins (Desai et al., 1997; Serna-Saldivar, 2010a,b; Steward, 2013; Zhou et al., 2013).

Table 1 summarizes the enzymes that are sequentially synthesized during germination (Serna-Saldivar, 2010a). The first enzymes generated during malting are oxidative and reducing, associated to oxygen absorption required for respiration followed by citases which hydrolyze cell walls and facilitate the entrance of other enzymes into the endosperm. Then, phytases, lipolytic, fibrolitic, proteolitic and amylolitic enzymes are produced. The main lipolytic enzymes are lipases A_1, A_2, phospholipases A_1, A_2 and lipooxygenases. Lipases release free fatty acids from

Table 1B-1. Main enzymes synthesized during malting.

Lipases	Attacks triglycerides releasing fatty acids. Lipase A1 hydrolyze fatty acids located on the ends of triglycerides whereas lipase A2 has specificity for the middle fatty acid.
Phospholipases	Attacks specifically phospholipids liberating fatty acids. Phospholipase A1 hydrolyzes the fatty acid located on the end of the phospholipid whereas phospholipase A2 has specificity for the middle fatty acid. Lysophospholipase frees the fatty acid form lysophospholipids or of phospholipids initially hydrolyzed by phospholipase A1.
Lipooxygenase	Enzyme that oxidizes polyunsaturated fatty acids to hydroxyperoxides. It is produced in the embryo and germ.
Phytases	Regulates phosphate release from phytic acid which is critical for germination.
Cellulases	Catalyze the hydrolysis of ß-glucosyl in ß-glucans linked by ß-1-4 glycosidic bonds (cellulose). It is mainly synthesized in the scutellum.
ß-Glucanases	Catalyze the hydrolysis of ß-glucosyl linkages in 1-3 or 1-4 ß-Glucans associated to cell walls where the glucosyl residue is substituted at the C(O) 3 position.
Xylanases (endo and exo)	Depolymerize arabinoxylanases associated to cell walls.
Arabinofuranosidases	Removes arabinosyl side chains allowing exoxylanase to release xylose.
Endopeptidase	Enzymes that cleave polypeptide chains within the chain at any susceptible point away from the N and C termini. They are subdivided according to the catalytic mechanism or preference for certain amino acids into serine, aspartic and cysteine. These enzymes attack proteins producing smaller peptides. Endopeptidases and carboxypeptidases act synergistically.
Carboxypeptidase	The different types of carboxypeptidases hydrolyze peptide bonds adjoining COOH terminal amino acids. These enzymes produce small peptides and free amino acids.
α-Amylase	Endohydrolyses that cleave internal α-1-4 glycosidic bonds of starch in an essentially random fashion. It is a calcium dependent enzyme. The enzyme hydrolyze amylose and amylopectin into linear and branched dextrins.
ß-Amylase	Exohydrolyses that cleave the penultimate α-glycosidic bonds from the nonreducing end of α-glucans or dextrins to release maltose and simpler dextrins. The activity ceases when it reaches a α-1-6 bond. It complements the action of α-amylase.
Limit Dextrinase	Hydrolyses α-1-6 bonds from starch or dextrins and increases the abundance of linear α-glucans chains. Also known as debranching enzyme which occurs in the endosperm and aleurone layer of mature barley kernels.
α-Glucosidase	Releases glucose from hydrolyzed starch, dextrins and maltose.

the triglycerides stored in the germ and aleurone, whereas the phospholipases from phospholipids. The resulting glycerol is metabolized through glycolysis after its oxidation and the fatty acids through the β-oxidation pathway by sequentially removing acetyl CoA that enters the TCA cycle for its complete oxidation to carbon dioxide and water. Most phosphorus associated to cereal grains exists in the form of phytic acid (mioinositol hexaphosphate) that is stored in phytic bodies mainly located in the aleurone cells. These molecules bind potassium, magnesium and other minerals. The phytases degrade these compounds to release phosphate, other minerals and myoinositol. The myoinositol is a known precursor of sugars associated with cell-wall polysaccharides and promoter of seedling growth. The free phosphorus is critically important for synthesis of nucleic acids and phospholipids for cellular membrane proliferation, and ATP and energy production. Fibrolitic or cell wall degrading enzymes are constituted by cellulases, hemicellulases, pectinases, xylanases, β-glucanases and others that synergistically attack cell walls and enhance the entrance of the other enzymes into the cells. The enzymes are mainly secreted from the aleurone or *scutellum* and therefore approach their substrates from outside the cells of the starchy endosperm. The simple sugars released from wall polysaccharides make a contribution to the total energy needed for seedling development (Serna-Saldivar, 2010b).

Proteolytic enzymes degrade the different types of protein fractions associated with the germ (globulins and albumins) and endosperm (prolamins and glutelins) to generate energy and enhance the susceptibility of starch granules to amylolytic enzymes. The group of enzymes consists of endopeptidases, carboxypeptidases, aminopeptidases and peptide hydrolases that degrade proteins into small peptides and free amino acids. The endopeptidases cleave proteins mainly associated with protein and aleurone bodies to yield lower molecular-weight polypeptides and peptides. The aleurone proteins are mainly mobilized to provide amino acids for the synthesis of important enzymes, such as amylases. The free amino acids are used for enzyme synthesis or oxidized for energy after deamination or formation of keto acids. They hydrolyze conjugated proteins associated with amylases, so helping to activate starch degrading enzymes. The free nitrogenous compounds, quantified as free amino nitrogen (FAN) are used as substrate by the developing embryo.

The most important enzymes are amylases. During germination, most of the bound β-amylases are liberated or activated whereas α-amylases are synthesized. The highest DA is achieved when the malt develops an acrospire of approximately two-thirds of the kernel length. The endosperm starch granules are degraded by α- and β-amylases into hydrosoluble maltose and dextrins. The dextrins are further degraded to glucose by limit dextrinases or pullulanase and amyloglucosidase. Alpha amylase randomly hydrolyzes α 1-4 glycosidic bonds of both amylose and amylopectin, yielding linear and branched dextrins whereas β-amylase cleaves successive maltose units, starting from the non-reducing end of amylose, amylopectin and large dextrins. The catalytic action of β-amylase ceases when it encounters an α 1-6 glycosidic bond. Other enzymes known as starch phosphorylase cleaves glucose units from the non-reducing end of both types of starches by introducing phosphate rather than water and thus producing the activated glucose-1-phosphate molecules. Maltose is hydrolyzed to glucose by α-glucosidase. The germinating barley kernel generates a more favorable and balanced α to β-amylase ratio compared to sorghum (*Sorghum bicolor*) or the rest of the cereals. Therefore, sorghum malt contains lesser amounts of maltose and higher amounts of dextrins compared to barley malt (Briggs, 1998; Fincher and Stone, 1993; Hough et al., 1993; Hardwick, 1995; Palmer and Bathgate, 1976; Pyler and Thomas, 2000; Priest and Stewart, 2006; Serna-Saldivar, 2010a,b; Steward, 2013; Zhou et al., 2013).

3.4 Kilning

This drying operation serves several purposes. The foremost is to halt germination and the botanical growth of the seedling so as to obtain a shelf-stable product that recuperates the enzymatic activity upon rehydration. The dehydration program lowers the moisture to a level that can be stored at room temperature for several months. Kilning conditions are critical to produce different sorts of malts with different enzyme activities, colors and flavors. Malts destined for lagers are generally less extensively modified (higher soluble sugars and FAN) than those aimed for ales that are kilned to a relatively mild regime. Lager malts therefore develop less color and produce pale or straw or amber-colored beer. Malts destined for ale production are kilned to higher temperature and thus resulting malts have darker colorations. The high temperature produces melanoidins from soluble sugars and amino acids. The higher kilning temperature also develops complex flavors that leach into the wort and beer. If the malt is kilned at high temperatures, it is possible to make special dark products and develop flavors described as burnt and smoky. This is characteristic of the color stouts.

The green malt is kilned using different methods. The most popular is drying the malt bed, using air that flows through perforated floors, rotary drums or vertical dryers. Floor drying consists of placing layers of malt for nine to 48 hours of kilning with forced air. The initial stages are carried out at temperatures lower than 50–65°C in order to prevent enzyme denaturation. When the moisture content of the malt drops to approximately 10 per cent, the temperatures can be increased up to 100°C. The application of a higher temperature program decreases DA but improves aroma and flavor. The malt is generally dehydrated to 4.5 per cent of moisture. After kilning, the malt is cooled down with fresh air and the rootlets stripped or deculmed (Fig. 1). The culms are separated because they impart bitter flavor, are high in nitrogenous compounds and contain high levels of nitrosamines. In the specific case of sorghum, the culms are rich in cyanogenic compounds (Serna-Saldivar, 2010b). The typical deculming devices include a slow rotating cylindrical screen equipped with a set of beaters that separate the rootlets from the malt or pneumatic equipment which impacts the malt to release the rootlets that are lifted by air aspiration (Briggs, 1998; Hough et al., 1993; Hardwick, 1995; Palmer and Bathgate, 1976; Pyler and Thomas, 2000; Priest and Stewart, 2006). The best quality brewing malts are obtained after storage or aging of the kilned malt for at least one month. During this phase, the moisture inside the kernel equilibrates and as a result, the malt is roll-milled more efficiently. During malting, the barley loses between 5–10 per cent of its weight. The main chemical changes that barley suffers during malting are: the assayable starch content gets reduced from 64 to about 59 per cent and hemicelluloses from 9 to 7 per cent, whereas reducing sugars increase from 0.2 to 4 per cent and sucrose from 2 to 5 per cent (Briggs, 1998; Serna Saldivar, 2010b).

3.5 Mashing

Mashing is an operation where the enzymes generated during germination are reactivated in order to convert the barley and adjunct starches and proteins into soluble compounds. The efficiency of this process greatly influences the yield and beer properties. Prior to mashing, the malt is coarsely roll-milled in order to enhance the release of glumes or husks as flakes. It is important that the glumes remain as intact as possible because they will act more efficiently as a filtrating bed during lautering (Fig. 1). Mashing is aimed at hydrolyzing the starch and proteins associated as the brewing adjuncts into fermentable carbohydrates/dextrins and soluble nitrogen. The fermentable carbohydrates and soluble nitrogen are important yeast substrates whereas dextrins impart the typical beer body. Independent of the mashing method, the raw materials used are water, usually containing controlled levels of soluble salts, ground malt and brewing adjuncts. In some mashing processes, supplementary enzymes are added to reactors. The most common mashing procedures are detailed by Briggs (1998). Infusion mashing simply consists of mixing the malt and brewing adjuncts that are treated at 63–67°C for about one hour stand. Then, the resulting mash is filtered until it becomes clear. The temperature-programmed mashing is carried in two vessels—a mash mixing and a wort separation device or lauter tun. In the mash mixing vessel, the mash is stirred and warmed through a carefully chosen program designed to allow optimal enzyme conversion. The temperature program usually starts at 35°C and after 30 min. increased to 50, 65 and 75°C. Next, the mash is transferred to the lauter tun in order to separate the sweet wort from brewers spent grains. The decocotion mashing employs a mash and decoction vessels and a lauter tun. The decoction vessel is used as a cereal cooker. The grist is mixed with water and heated to 35–40°C for enzyme conversion. After 1.5 to two hours, about one-third of the mash is transferred to the decoction cooker where it is first heated to 65°C for about 20 min. to enhance starch liquefaction and after that heated to boiling. Then, this material is pumped back to the stirring mash-mixing vessel, increasing the temperature of the combined mash to 52°C where it is maintained for about an hour. Once more, about one-third of the mash undergoes decoction and returns to the mash vessel where the temperature increases to 65°C. Finally a third decoction takes place, increasing the mash temperature to 76°C. The mash is then transferred to the lauter tun. The disadvantage of the decoction mashing process is that it takes about six hours (Briggs, 1998; Hardwick, 1995; Hough et al., 1993; Priest and Stewart, 2006; Serna Saldivar, 2010b; Steward, 2013).

Most lagers in the American continent are produced by the double-mashing procedure because the grist formulations usually contain high amounts of starchy adjuncts. These starchy materials require cooking in order to achieve complete starch gelatinization. The mashing consists of two distinctive stages. In the first stage, commonly known as adjunct mash, the brewing adjuncts are mixed with water and heated to 35°C. After a 30 to 60 min. stand, the contents are heated first to 70°C for about 30 min. and then to 100°C for 30–45 min. The aim is to first hydrate the adjuncts and malt and then promote starch gelatinization and conversion since most cereal starches gelatinize at temperatures higher than 65°C. Boiling is required to denature proteins and to inactivate the microbial load and enzymes. Once the adjunct mashing schedule is ongoing, the second mash is prepared by mixing and heating to 35°C most of the malt and water. Then the contents of the two vessels are mixed. The aim of the second phase is to hydrolyze most of the starch and proteins. This is optimally achieved by programming a gradual temperature increase, which starts at 35°C for approximately 30 min. The temperature is usually ramped 10–15°C until achieving 70°C. The sequential temperature increase favors proteases followed by β- and α-amylases (optimum temperatures of 60 and 70°C, respectively). The high temperature of the last mashing stage stops most enzymatic activity, reduces viscosity and improves the fluidity and filtering capacity of the wort. Brewing with other malts with less DA and/or brewing grits less prone to gelatinization may require changes in the temperature used, especially during the second phase of mashing (Bamforth,

2003; Hardwick, 1995; Hough et al., 1993; Priest and Stewart, 2006; Serna-Saldivar, 2010b; Steward, 2013).

3.6 Lautering/sparging

After completing the mashing step, the mash is transferred to the lauter tun where the sweet wort is separated from the brewers spent grains. This step is a common bottleneck in brewing operations. Poor lautering not only causes a loss in production capacity but can also lead to loss of extract yield. Furthermore, a slow lautering process negatively affects the quality of the wort, which may lead to problems with beer filtration, flavor and stability. A relevant factor influencing lautering is wort viscosity. In order to maximize performance, the starch and β-glucans should be properly hydrolyzed. The lauter tun is a relatively shallow apparatus with a double mesh floor which is equipped with rakes that loosen the filtrating bed structure, minimizing compaction. The mash contents are allowed to precipitate for 30 min. so as to enhance the formation of the natural filtration bed, rich in husks. The sweet wort is separated by filtration through the double mesh floor. Generally in the first stage of lautering, the wort is re-circulated through the bed of spent grains in order to decrease wort turbidity. The filtering is generally carried out at temperatures of 65–70°C and the spent grains sparged with hot water (75–80°C) (Bamforth, 2003; Hardwick, 1995; Hough et al., 1993; Priest and Stewart, 2006; Serna-Saldivar, 2010b; Steward, 2013).

3.7 Addition of hops

Hops (*Humulus lupulus*) are added to sweet wort in order to impart the distinctive European beer flavor and aroma. The process simply consists of adding the hops to the wort in order to promote the extraction of solubles by boiling and lixiviation for 1.5–2.5 hours (Fig. 1). Generally, half to two-thirds of the hops are added at the beginning of the program and the rest at the end of the process with the objective of keeping key volatiles that enhance the beer flavor and aroma. During boiling, the enzymes are inactivated, the wort becomes darker due to caramelization reactions and the hopped wort becomes sterile. During this thermal treatment from 4–12 per cent, water is lost due to evaporation. In addition, some soluble proteins will bind to tannins and precipitate, decreasing the haze. Spent hops are removed from the hopped in a large whirlpool equipped with a conical trub collector located in the central part of the base of the apparatus. The hopped wort is cooled in a heat-exchanger to about 6°C for traditional lagers and as high as 15–20°C for ales, adjusted to a predetermined Plato and aerated with sterile air in order to increase the oxygen which is important for yeast growth and budding, especially during the early phase of fermentation. During cooling, more proteins becomes insoluble and are removed by centrifugation (Bamforth, 2003; Hough et al., 1993; Palmer and Bathgate, 1976; Serna-Saldivar, 2010b).

3.8 Fermentation

The wort is fermented into beer in special reactors equipped with cooling coils or jackets. The most popular equipment consists of stainless steel conical and hermetically sealed tanks (150–500 hL capacity). The wort is pitched with 1.5–2.5 g dry yeast or 4.5 to 7.5 g fresh compressed yeast/L. During the first stage of fermentation, the yeast reproduces asexually by budding, increasing the biomass from four to five times and utilizes the available oxygen. Thus, the reactor conditions gradually switch from aerobic to anaerobic. It is considered that after 12–24 hours of fermentation, the conditions become anaerobic. During this phase, the yeast metabolizes fermentable carbohydrates and FAN, producing ethanol and fusel alcohols (isopropanol, amylic, isoamylic and butanol), respectively. During this stage, carbon dioxide is also produced and intermediate organic products that help to impart the characteristic beer flavor.

For production of lagers, fermentation is commonly carried with bottom yeast at 7–15°C during eight to 10 days and for ales at approximately 20°C for three to five days. Lagers are matured in closed tanks at a temperature of 0°C for four to six additional weeks in order to further reduce oxygen to a level of less than 0.2 ppm and enhance bouquet and aroma due to chemical changes, such as the generation of diacetyl, dimethylsulfide and hydrogen sulfide. During fermentation, yeast cells transform maltose and maltotriose into glucose that is further metabolized into carbon dioxide, energy, ethanol, organic acids and volatile compounds. The initial wort density is about 1.040 and the beer between 1.008 to 1.010 g/cm^3. The organic acid production decreases pH to a level of 4.2. The change in the acidity coagulates some proteins and decreases even more the solubility of some acidic hop resins. Also, during fermentation, important quantities of FAN are metabolized into fusel alcohols that affect the organoleptic properties of beer. The sweetness of beer is due to residual sugars that have not been fermented into alcohol (Bamforth, 2003, 2009; Hardwick, 1995; Hough et al., 1993; Ingledew, 1995; Priest and Stewart, 2006; Steward, 2013; Serna Saldivar, 2010a).

Many different types of chemical compounds besides ethanol, fusel alcohols and those derived from hops also influence beer flavor, aroma and stability. A wide array of esters present in concentrations of ppm affect sensory properties. The most relevant are ethyl acetate, butyl acetate, isoamyl acetate, ethyl valerate, isoamyl propionate, phenylethyl acetate, methyl caprate and methyl and isoamyl caprate, which impart different fruit flavors. The same applies for sulfur compounds, such as dimethyl sulfide or disulfide, ethyl or amyl mercaptan and methional. These sulfur containing volatiles affect flavor (onion, garlic, egg, rotting leek) at even lower concentrations (Bamforth, 2003, 2009).

4. Supplementary Enzymes Utilized in Brewing Operations

The major enzymes for brewing are provided by the malt. However, a wide array of exogenous enzymes might be supplemented in the different stages of the brewing process in order to expedite production, the ratio of fermentable to un-fermentable sugars and therefore the wort fermentability and beer characteristics. Exogenous enzymes are mainly used to improve extract recovery, the critical operation of lautering or wort filtration, and mainly to add to the quality, appearance (clarity and haze) and shelf-life of beers. These enzymes are mainly of microbial origin although some are extracted from plants, and even from animal sources. In most cases, these enzymes are commonly supplemented during mashing or wort fermentation and should be approved for use in foods (Godfrey and Reichelt, 1983; Power, 1993; Zhou et al., 2013).

Both intrinsic malt and supplementary enzymes have different optimum catalytic conditions especially in terms of temperature, pH and presence of metal ions; so brewers should intelligently give them the time and conditions to perform their intended task.

4.1 Starch-degrading enzymes

Undoubtedly these enzymes are the most supplemented in brewing operations, especially when mashing with high amounts of adjuncts or brewing light or gluten-free beer. The major categories of starch-degrading enzymes are α and β-amylases, amyloglucosidases or glucoamylases and limit dextrinases or pullulanases. Their chemistry, classification, catalytic action and purification, are described by Bryce (2003), MacGregor (2003), Reilly (2003) and Wong and Robertson (2003a,b).

4.1.1 α-amylases

These are classified as endoglucanases that catalyze the cleavage of internal α 1-4 glycosidic bonds in starch at random to yield linear and branched dextrins and oligosaccharides. The α 1-4 bonds near α 1-6 branches are resistant to hydrolysis (Wong and Robertson, 2003a). The process of converting starch into dextrins (average of 10 glucose units) with α-amylase is industrially known as liquefaction due to the high reduction in viscosity of the gelatinized starch. Most α-amylases optimally work at pH's in the range of 5.5 to 7 and in the presence of calcium ions (200 ppm $CaCl_2$ or 300 ppm $CaSO_4$). The α-amylase present in barley and other cereal grains work optimally at 60°C whereas most microbial counterparts widely used nowadays by both the sweetener and bioethanol industries are thermostable and act optimally at temperatures ranging from 85 to 95°C. These thermostable enzymes can resist temperatures close to 130°C. The advantages of these enzymes are the higher conversion rates that reduce the catalysis time and that the thermal treatment aimed towards starch gelatinization can be integrated into liquefaction. However, the disadvantage is the difficulty to inactivate the enzymes in order to prevent further undesirable catalysis.

α-Amylases are very effective in liquefying the starch and act synergistically with counterparts associated with barley malt or are used to strengthen the action of low DA barley malts. Both fungal and bacterial α-amylases are commonly used in the industry. Fungal α-amylases are usually obtained from *Aspergillus oryzae* whereas heat-stable bacterial α-amylases are generally produced by a genetically modified strain of the genus like *Bacillus licheniformis or Bacillus amyloliquefaciens* (Table 2). Other reported microorganisms are *Bacillus stearothermophilus* and *Bacillus subtilis.* Fungal enzymes from *Aspergillus* have pH and temperature optima in the range of 5.0 to 6.5 and 55 to 65°C. On the other hand, the α-amylase extracted from *Bacillus subtilis* has an optimal temperature and pH between 65 to 70°C and 6.0 and 7.5 (Hobbs, 2003; Teague and Brumm, 1992; Woods and Swinton, 1995; Wong and Robertson, 2003a). The heat-stable amylase from *Bacillus licheniformis* is not effective at temperatures below 70°C and is recommended to thin or hydrolyze the starch as it gelatinizes. This

Table 1B-2. Role and source of main exogenous enzymes.

Enzyme	Source	Major Role
α-Amylase	*Bacillus amyloliquefaciens* *Bacillus licheniformis* *Bacillus stearpothermophilus* *Bacillus subtilis* *Aspergillus oryzae*	Enzymes added when malts have poor diastatic power or when higher amounts of starchy adjuncts are used in the grist. Supplemented to enhance starch hydrolysis and thin down the adjunct starch as it gelatinizes during cooking and mashing. Added to decrease mash viscosity and improve runoff during lautering.
ß-Amylase	*Bacillus polymixa* *Bacillus cerus*	Used to increase the amount of maltose and wort fermentability. Used to reduce the amounts of dextrins and production of light beers.
Limit Dextrinase or Pullulanase	*Klebsiella pneumonia* *Klebsiella aerogens* *Bacillus acidopullylyticus* *Trichoderma* spp.	Generates more substrate for β-amylase and secures maximum fermentability of the wort. Used to reduce the amounts of dextrins, wort viscosity and production of light beers.
Glucoamylase or Amyloglucosidase	*Aspergillus* spp. *Rhizopus* spp.	Added to mashes to increase the fermentability of the wort and production of light beers (low in dextrins). Supplemented to mashes to improve runoff during lautering due to a reduced viscosity.
Proteases	*Bacillus subtilis* Plants such as papaya, pineapple and fig	Degrade proteins that surrounds starch granules improving its hydrolysis by amylases. Increase soluble protein or free amino-nitrogen which secures proper yeast fermentation. Supplemented to degrade polypeptide haze precursors in beer
Prolylendopeptidase	*Aspergillus niger*	Improve beer stability by decreasing hazing Used to degrade peptides that trigger celiac disease and possible production of gluten free or reduced gluten beers.
Cellulases and ß-Glucanases	*Bacillus subtilis* *Penicillium funiculosum* *Penicillum emersonii* *Aspergillus niger* *Trichodermareesei* *Rhizopus microsporus*	Enzymes used to reduce viscosity and improve runoff during lautering and as aid filtration. Added to prevent chill haze.
Xylanases	*Disporotrichum* spp. *Trichoderma* spp. *Aspergillus* spp	Improve extraction, reduce wort viscosity and improve lautering and beer filtration. Added to prevent haze.
α-Acetolactate decarboxylase	*Acetobacteraceti* *Enterobacteraerogenes*	Decrease fermentation time by avoiding formation of vicinal diketones (VDK). Prevent off flavor development in beer.

prevents a high wort viscosity and minimizes starch retrogradation when added to the colder malt mash. One of the major advantages of supplemental α-amylase is that it allows more adjunct to be used in the cooking vessel and prevents the formation of off flavor from the husk of the barley malt. Despite the very complete starch digestion, the hydrolysis with α-amylase yields very low amounts of fermentable carbohydrates (Morgan and Priest, 1981; Ryder and Power, 2006).

Adequate α-amylase activity during mashing is undeniably critical for acceptable runoff of the wort during mash filtration. Malt normally contains more than enough α-amylase to allow good runoff. However, if a problem with starch degradation occurs, the bacterial amylase of *Bacillus amyloliquefaciens* has been proposed to improve runoff. Alpha-amylases from *Aspergillus oryzae* have also been used to prevent starchy haze caused by the presence of high-molecular-weight dextrins. This enzyme will increase the level of fermentable sugars during fermentation when used at a high level. However, at lower concentration, they breakdown dextrins without significant increase in fermentable sugar levels. Preparation of this enzyme effectively reduces starchy hazes that form in beer during storage within a day, even at cold storage temperatures (Ryder and Power, 2006).

4.1.2 β-amylases

Beta-amylases (α 1-4 D glucan maltohydrolases) are exoenzymes present only in plants and bacteria which convert linear and branched dextrins into maltose and lower molecular-weight dextrins from the non-reducing end (Wong and Robertson, 2003b). The β-amylase activity ceases when it reaches an α-1-6 glycosidic bond. It is usually known as the complementary amylase because it hydrolyzes dextrins previously yielded by α-amylase. The optimum pH and temperature ranges for most β-amylases are 5.0–5.5 and 55–60°C, respectively (Hobbs, 2003; Kruger et al., 1987; Teague and Brumm, 1992; Woods and Swinton, 1995).

Supplemental β-amylases can be extracted from particular strains of bacteria, such as *Bacillus polymixa* or *Bacillus cerus* (Table 2). β-amylases are mainly employed to increase the amounts of fermentable maltose and thus, wort fermentability. In addition, they can be supplemented to decrease dextrins and the caloric content of light beers. Their optimum pH and temperature is 5.3 and 55°C; therefore, they work favorably at the typical acidity of mashes. Beta-amylases are readily denatured by heat and are fully deactivated before lautering (Briggs et al., 1981).

Norris and Lewis (1985) applied food-grade commercial barley β-amylase in mashing. When mash temperature was varied between 60 and 80°C and adjunct ratio between 0 and 98 per cent, wort fermentability was affected more than the extract yield. The supplemented enzyme markedly increased fermentability and may find special application in the production of reduced dextrin or light beers.

4.1.3 Limit dextrinases/pullulanases

Limit dextrinases, also known as pullulanases (α dextrin endo α 1-6 glucosidase) are debranching enzymes which selectively catalyze α 1-6 glycosidic linkages and have no action on α 1-4 glycosidic bonds (Bryce, 2003). These enzymes are considered complementary to amylases to achieve higher starch conversions into fermentable sugars mainly in terms of maltose concentration (Table 2, MacGregor, 2003). MacGregor et al. (1999) modeled the contribution of α-amylase, β-amylase and limit dextrinase to starch degradation during mashing and concluded that addition of limit dextrinase to the mashes resulted in a substantial increase in levels of fermentable carbohydrates.

The limit dextrinase showed a synergistic effect in increasing levels of maltose in the mash liquor when combined with high levels of β-amylase.

Regular and heat-stable pullulanases are produced today. Regular bacterial pullulanases are usually obtained from *Klebsiella aerogens, Klebsiella pneumoniae* or *Bacillus acidopullylyticus* or from a fungi *Trichoderma* (Hobbs, 2003; Kruger et al., 1987). These enzymes have optimum activity at pH and temperatures between 4 to 7.5 and 50 to 65°C, respectively so that they are readily inactivated by heat or beer pasteurization (MacGregor, 2003; Willox et al., 1977). According to Kunamneni and Singh (2006) and Saha et al. (1988) heat-stable pullulanases capable of working at temperatures of up to 90°C can be extracted from strains of *Clostridium thermohydrosulfuricum* or *Bacillus* sp.

4.1.4 Amyloglucosidases

Also known as glucoamylases or saccharifying enzymes are generally extracted from molds, such as *Aspergillus* spp. or *Rhizopus* spp. Amyloglucosidases attack the non-reducing ends of starch chains and dextrin-releasing glucose units (Reilly, 2003). Their hydrolysis on α 1-4 bonds is comparatively rapid relative to the attack on α 1-6 bonds so that the conversion of starch into glucose is accelerated by the addition of limit dextrinases. The optimal pH range is 4.0 to 5.5 and the enzyme will act for extended periods at 60 to 65°C. Recombinant yeasts capable of expressing glucoamylase have been successfully developed. Bui et al. (1996) cloned the glucoamylase gene of yeast *Arxula adeninivorans* in *Saccharomyces cerevisiae*. The transformants secreted 95 per cent of the enzyme into the culture medium and the glucoamylase activities observed promised to improve soluble sugars and alcohol content after fermentation.

Amyloglucosidases are added to mashes, particularly those containing large proportions of adjuncts, to increase the fermentability of the wort and have proved effective in production of lager beer with lower viscosity and reduced calories and sorghum beer (Table 2). The amyloglucosidase is deactivated during the last part of the mashing regime or during hop boiling.

Recently, sorghum has been viewed as a source of brewing adjuncts and diastatic malt for production of gluten-free beer. The main problems when brewing with sorghum are the lower DA of its malt especially deficient in β-amylase activity, and the comparatively higher gelatinization temperature of its starch compared to barley starch (Owuama, 1997; Serna-Saldivar, 2010b). Compared with barley malt, mashing with sorghum requires adjusted mashing in order to increase the solubility and hydrolysis of the starch (NicPhiarais and Arendt, 2008; Ogbonna, 2011). The addition of extrinsic β-amylase or microbial glucoamylase during mashing of sorghum worts significantly increased the extract and starch conversion and favored the ratio between fermentable sugars and dextrins. Therefore, the most favorable wort composition augmented the alcohol content and yield. Thus, the best technology to produce gluten-free sorghum beer is with the use of white sorghum genotypes adequate for the production of malt and refined grits supplemented with glucoamylase during mashing (Cortés-Ceballos et al., 2015; Espinosa Ramirez et al., 2013a,b, 2014; Urias Lugo and Serna Saldivar, 2005; Del Pozo Insfran et al., 2004).

4.2 Proteases

Protein is the second most abundant component in malt and starch-brewing adjuncts. These raw cereal-based materials contain about 10 per cent protein which is divided into albumins, globulins, prolamins and glutelins. The last two are associated with the starchy endosperm and are the main fractions. The malting procedure enhances the formation of proteases, like endopeptidases and carboxipeptidases (Table 1) that during mashing, break down prolamins and glutelin-yielding peptides that are commonly assayed with the FAN ninhydrin colorimetric assay. The water-soluble FAN compounds are important substrate for yeast during fermentation; so the lack of sufficient FAN in wort limits yeast activity and fermentation and the production of acceptable beer (Aastrup and Olsen, 2008; Barredo-Moguel et al., 2001; Button and Palmer, 1974; Ryder and Power, 2006).

Various supplementary proteases have been used to break down gluten proteins that hinder starch hydrolysis during mashing. This is especially important when using sorghum malt or refined sorghum adjuncts (Duodu et al., 2003; Odibo et al., 2002). As a result of the strong interaction between protein and starch granules and the weaker protease activity of sorghum malt, worts produced from sorghum malt and grits are usually deficient in FAN compounds. Therefore, supplementary plant (ficin, papain and bromelin) and microbe-derived proteases improve wort extraction and assayable FAN. These enzymes hydrolyze proteins at temperatures as high as 65°C. According to Ryder and Power (2006), the generation of excess levels of soluble nitrogenous compounds may lead to reduction in both foam formation and foam stability.

Supplementary proteases are also used to degrade polypeptide haze precursors in beer (Aastrup and Olsen, 2008; Nelson and Young, 1987). Papain, bromelin and ficin, plant protease-derived from papaya (*Carica papaya*), pineapple (*Ananas* spp.) and figs (*Ficus* spp.) respectively have been successfully added to beer as stabilizing agents. These proteases act similarly, degrading the haze precursor and are denatured by thermal treatment of pasteurization. Papain has also been added after the yeast has been removed from beer to prevent haze during cold storage.

Compared to plant proteases, microbial proteases have an additional advantage in terms of degrading foam precursors. Proteases extracted from *Bacillus subtilis* work adequately at temperatures ranging from 45 to 50°C and common mash pH values of around 5.6. Fungal proteases have an optimum pH in the range 3 to 6, and temperature of about 50°C (Briggs et al., 2004).

A proline endopeptidase extracted from *Aspergillus niger* has been used to improve beer stability. This enzyme specifically hydrolyzes sensitive polypeptides because of their enhanced proline content and promotes the physical stability of beer with no effect on foam stability. Lopez and Edens (2005) incubated beer wort with a proline-specific protease in order to hydrolyze proteins that promote haze. Pre-digestion of the proline-rich wheat gliadin with different proteases pointed towards a strong haze-suppressing effect by this specific enzyme. This finding was confirmed in small-scale brewing experiments using the proline-specific protease capable of hydrolyzing at acidic pH. Subsequent pilot plant trials demonstrated that upon its addition during the fermentation phase of beer brewing, even low levels of this enzyme effectively prevented chill-haze formation in bottled beer.

Recent research has also proposed the use of prolyl endopeptidases for gluten degradation in foods and beverages (Caputo et al., 2010; Tanner, 2014). Germinated cereal grains as well as bacteria and fungi are known to be suitable sources of these enzymes. Prolyl endoproteinase is normally derived from *Aspergillus niger*. In an attempt to produce gluten-free beer, Guerdrum and Bamforth (2012) researched at what point in the brewing process the gluten of barley was hydrolyzed when supplementing prolyl endoproteinase. Approximately half of the prolamin was lost during mashing and lautering presumably by proteolysis or precipitation. After fermentation, only 1.9 per cent of the original hordein from the malt remained in the unfiltered beer. These authors concluded that the prolamin in the beer after fermentation represented less than 2 per cent of the total prolamin in the malt rendering beer essentially free of gluten and without negatively impacting the foaming capacity.

4.3 Fiber-degrading enzymes

Plant cell walls are composed of three major polymers: cellulose, hemicellulose and lignin. Cellulose and hemicelluloses are the first and second most abundant polysaccharides in Nature. As a part of the carbon cycle, these two constituents are degraded by specialized enzyme systems (Biely, 2003). Fibrolytic enzymes play an important role as supplementary catalysts in brewing operations even though fiber is relatively low in malts and especially in brewing adjuncts. The industry utilizes an array of these enzymes in order to improve beer processability, especially in terms of mash runoff and viscosity and appearance. Addition to the mash of microbial enzymes capable of breaking down β-glucans and other hemicelluloses, such as xylans and pentosans, improve runoff especially with difficult malts or when barley is used as an adjunct (Ryder and Power, 2006).

4.3.1 Cellulases and β-glucanases

Cellulases belong to the group of β-glucan hydrolyses which attack β 1-4 linkages of cellulose (Tenkanen et al., 2003). Cellulases are not among the main enzymes used by the brewing industry but are added to enhance the action of other more relevant enzymes, such as β-glucanases and xylanases. Cellulase preparations from *Penicillium funiculosum* or *Penicillum emersonii* have been added to break down β-glucans, holocellulosic material and pentosans present in barley mashes. The first enzyme cocktail exhibited adequate activity in the pH range of 4.3 to 5.0 and temperatures of 65°C while the other at temperatures up to 80°C.

As cell-wall components, β-glucans make a relatively minor contribution to the total weight of cereals but have a great impact in processing and nutritional value because of the propensity of these carbohydrates to be extracted with water and thereafter to form viscous solutions (Hramova and Fincher, 2003). These soluble fiber components consist predominantly of long, linear chains of glucosyl residues which are linked with β 1-3 and 1-4 bonds. It is well known that undegraded wort β-glucans negatively influence beer filtration, reduce filter capacity and raise the consumption of filter aids, such as diatomaceous earth used to eliminate beer haze (Harmova and Fincher, 2003; Silveira Celestino et al., 2006).

β-glucanases are synthesized in barley malt or obtained from microbial sources (Tables 1 and 2). These enzymes have a broad activity because they attack β-1,3 and 1-4 linkages of glucans. The β-glucan exohydrolases hydrolyze β 1-4 glycosidic bonds more slowly than β 1,3 linkages. Brunswick et al. (1987) studied the development of β-D glucanases during the germination of barley and the effect of kilning on individual isoenzymes. Kilning and dry-milling of barley malt caused losses of 80–90 per cent in the specific endo-1, 3 and 1,4-glucanase activity. In contrast, endo-1,3-β-glucanase and cellobiase activities in malt extracts were less affected by kilning. Likewise, Loi et al. (1987) collected green malt and samples of kiln-dried malt from a commercial malt-house with the aim of monitoring 1-3, 1-4 glucan endohydrolase activity. The finished malt, which was dried at temperatures up to 80°C, retained 38 per cent of the activity detected in green malt. Immunological assays revealed that the activity in kiln-dried malt was attributable entirely to (1-3, 1-4)-β-glucanase isoenzyme II and that isoenzyme I disappeared during initial drying at 60 to 65°C. When the kiln-dried malt was subjected to laboratory-scale mashing at 45°C, approximately 50 per cent of the initial activity remained stable for at least 30 min. Although activity was lost more rapidly as mashing temperature increased, significant β-glucanase activity survived for up to 5 min. at mashing temperatures of 65°C.

The β-glucanase from *Bacillus subtilis* was first applied or supplemented during mashing. It has moderate heat resistance and an optimum activity at 50°C and pH of 6.5. This enzyme has a mode of action similar to that of barley glucanase, resulting in a quick size reduction with practically nil production of fermentable carbohydrates. Various types of fungal (*Aspergillus* and *Trichoderma*) β-glucanases have been effectively used to break down β-glucans during mashing. These enzymes show optimal activity at pH between 4 and 5. The fungal enzymes derived from *Aspergillus niger* and *Trichoderma reesei* work effectively at pH 4.2 of beer and showed good activity at temperatures above 60°C. The *Trichoderma* glucanase is especially effective in enhancing beer filtration (Ducroo and Delecourt, 1972; Ryder and Power, 2006). Silveira Celestino et al. (2006) researched the use of exogenous β-glucanases from *Rhizopus microsporus* to reduce the β-glucans already present in the barley malt. The optimal pH and temperature for hydrolysis of 1,3 and 1,4 β-glucans were in the ranges of 4–5 and 50–60°C, respectively. The purified enzyme was able to reduce both the viscosity and the filtration time or runoff of mashes. In comparison to the values determined for the mash treated with two commercial glucanases, the relative viscosity value for the mash treated with fungal 1,3-1,4-β-glucanase was determined to be consistently lower. The same β-glucanases can be supplemented to beer during storage. However, the relatively high pH optimum for these enzymes makes it less effective in beer than in the mash.

4.3.2 Xylanases

Hemicellulose consists of a series of matrix and cross-linking heteropolysaccharides which include glucans, mannans, arabinans, galactans and xylans (Biely, 2003). Gluco and arabinoxylans are hydrosoluble compounds associated with cell walls of cereals, such as maize, sorghum, wheat, rye or triticale. These compounds are degraded by many bacteria, fungi, yeast and protozoa (Biely, 2003). Xylanases are synthesized during germination where they catalyze hydrolysis of hemicellulose to remove the physical barrier imposed by cell walls on free diffusion of amylases and proteases (Biely, 2003). As same as β-glucanases, xylanases are used by brewers to enhance wort separation and speed up beer filtration and decrease beer haze. Sometimes these preparations contain complex mixtures of enzymes that act synergistically. Beg et al. (2001) reviewed the microbial xylanase complex involved in its complete break down and their potential industrial applications.

Du et al. (2013) recently characterized three novel thermophilic xylanases from *Humicola insolens* with potential for the brewing industry. Recombinant XynA-C produced in *Pichia pastoris* showed optimal activities at pH 6.0–7.0 and at temperatures of 70–80°C, besides exhibiting good stability over a broad pH range. The gene xyn C produced by *Humicola insolens* was similar in enzyme properties as compared to the one expressed by *Pichia*. XynA exhibited better alkaline adaptation and thermostability and had higher catalytic efficiency and wider substrate specificity. Under simulated mashing conditions, addition of XynA-C showed better performance on filtration acceleration (37.4 per cent) and viscosity reduction (13.5 per cent) than a commercial enzyme used by brewers.

4.4 Other enzymes

Many other enzymes with different functionalities have been proposed in brewing operations. These include α-acetolactate decarboxylase, tanninases, phosphatases oxidases, transglycosylases and glucosidases. These are used as processing aids or to degrade metabolites that negatively affect beer flavor and aroma (Briggs et al., 2004).

4.4.1 α-acetolactate decarboxylase

The management of diacetyl and other vicinal diketones (VDK) during beer fermentation and maturation is critical because these compounds affect beer taste and shelf-life. Brewers have used α-acetolactate decarboxylase in order to decarboxylate α-acetolactate into the flavor inactive acetoin, preventing the formation of diacetyl. Thus, this decarboxylase has been commercially used for removal of α-acetolactate and α-aceto-α-hydroxybutyrate to a level below the taste threshold of volatile VDK, without affecting other important beer features (Table 2). Commercial enzymes generally regarded as safe are usually obtained from *Acetobacter aceti*, which is usually supplemented to sweet wort prior to pitching or fermentation. Godtfredsen and Otresen (1982) studied the maturation of beer with α-acetolactate decarboxylase isolated from *Enterobacter aerogenes*. The enzyme was applied at 10°C during beer maturation in order to prevent off flavor development. The beer matured in the presence of the enzyme was judged to be of an equally satisfactory quality when compared with conventionally prepared beer. Yamano et al. (1994) successfully cloned a gene, encoding α-acetolactate decarboxylase from *Acetobacter aceti* ssp. into brewer's yeast. As expected, the transformed yeast grown at laboratory-scale considerably reduced the total diacetyl concentration.

5. Future Trends and Conclusion

Supplemented enzymes play a more relevant role in the brewing industry because they can help to design more efficient processes and improve beer features, organoleptic properties and shelf-life. With new biotechnological tools available nowadays, it will be feasible to produce new generation enzymes or genetically-modified yeast expressing specific enzymes for particular uses. Both enzymes and yeast will have to be graded as 'Generally Recognized as Safe' or GRAS in order to be readily adopted by the industry.

Recent beer trends demanded by consumers, such as gluten-free, low calorie or carbohydrate, low alcohol, high alcohol among others can successfully be carried out with the use of proper enzymes, as well as reducing or completely removing malt from the process. In short, the use of exogenous enzymes in the brewing industry is expected to expand in order to generate a wider array of beer with different properties and/or facilitate and lower costs of current processes. Undoubtedly, supplementary enzymes will allow processors to design new processes and innovate especially in terms of product development in the competitive and still growing beer market worldwide.

References

Aastrup, S. and Olsen, H.S. 2008. Enzymes in brewing. BioZoom, 2: 1–12.
Bamforth, C.W. 2003. Beer: Tap into the Art and Science of Brewing, 2nd Edition. Oxford University Press, New York.
Bamforth, C.W. 2009. Beer: A Quality Perspective, Academic Press, New York.
Barredo-Moguel, L.H., Rojas de Gante, C. and Serna-Saldivar, S.O. 2001. Alpha amino nitrogen and fusel alcohols of sorghum worts fermented into lager beer. Journal of the Institute of Brewing, 107(6): 367–372.
Beg, Q., Kapoor, M., Mahajan, L. and Hoondal, G.S. 2001. Microbial xylanases and their industrial applications: A review. Applied Microbiology and Biotechnology, 56: 326–338.
Biely, P. 2003. Xylanolytic enzymes. pp. 273–282. *In*: J.R. Whitaker, A.G.J. Voragen and D.W.S. Wong (eds.). Handbook of Food Enzymology. Marcel Dekker, New York.
Briggs, D.E. 1998. Malts and Malting. Blackie Academic & Professional, London.
Briggs, D.E., Boulton, C.A., Brookes, P.A. and Stevens, R. 2004. Malts, adjuncts and supplementary enzymes. pp. 11–51. *In:* D.E. Briggs, C.A. Boulton, P.A. Brookes and R. Stevens (eds.). Brewing Science and Practice. CRC Press, Boca Raton.
Briggs, D.E., Hough, J.S., Stevens, R. and Young, T.W. 1981. Malting and Brewing Science Malt and Sweet Wort, 2nd Edition. Chapman & Hall, London.
Brunswick, P., Manners, D.J. and Stark, J.R. 1987. The development of β-D glucanases during the germination of barley and the effect of kilning on individual isoenzymes. Journal of Institute of Brewing, 93: 181–186.
Bryce, J.H. 2003. Limit dextrinase. *In*: J.R. Whitaker, A.G.J. Voragen and D.W.S. Wong (eds.). Handbook of Food Enzymology, Marcel Dekker, New York.
Bui, D.M., Kunze, I., Forster, S., Wartmann, T., Horstmann, C., Manteuffel, R. and Kunze, G. 1996. Cloning and expression of an *Arxula adeninivorans* glucoamylasegene in *Saccharomyces cerevisiae*. Applied Microbiology and Biotechnology, 44: 610–619.
Button, A.H. and Palmer, J.R. 1974. Production scale brewing using high proportions of barley. Journal of Institute of Brewing, 80: 206–213.
Caputo, I., Lepretti, M., Martucciello, S. and Esposito, C. 2010. Enzymatic strategies to detoxify gluten: Implications for celiac disease. Enzyme Research, 2010: 174354.

Cortés-Ceballos, E., Nava-Valdez, Y., Pérez-Carrillo, E. and Serna-Saldívar, S.O. 2015a. Effect of the use of thermoplastic extruded corn or sorghum starches on the brewing performance of lager beers. Journal of the American Society of Brewing Chemists, 73(4): 318–322.

Del Pozo-Insfran, D., Urias-Lugo, D., Hernandez-Brenes, C., and Serna-Saldivar, S.O. 2004. Effect of amyloglucosidase on wort composition and fermentable carbohydrate depletion during fermentation of sorghum lager beer. Journal of Institute of Brewing, 110(2): 124–132.

Desai, B.B., Kotecha, P.M. and Salunke, D.K. 1997. Seed germination. pp. 51–87. *In*: B.B. Desai (ed.). Seeds Handbook: Biology Production, Processing and Storage. 2nd Edition, Marcel Dekker, Inc., New York.

Du, Y., Shi, P., Huang, H., Zhang, X., Luo, H., Wang, Y. and Yao, B. 2013. Characterization of three novel thermophilic xylanases from *Humicola insolens* Y1 with application potentials in the brewing industry. Bioresource Technology, 130: 161–167.

Ducroo, P. and Delecourt, R. 1972. Enzymic hydrolysis of barley beta-glucans, Wallerstien Laboratories Communications, 35: 219–226.

Duodu, K.G., Taylor, J.R.N., Belton, P.S. and Hamaker, B.R. 2003. Factors affecting sorghum protein digestibility. Journal of Cereal Science, 38: 117–131.

Espinosa-Ramírez, J., Pérez-Carrillo, E. and Serna-Saldívar, S.O. 2013a. Production of brewing worts from different types of sorghum malts and adjuncts supplemented with β-amylase or amyloglucosidase. Journal of the American Society of Brewing Chemists, 71(1): 49–56.

Espinosa-Ramírez, J., Pérez-Carrillo, E. and Serna-Saldívar, S.O. 2013b. Production of lager beers from different types of sorghum malts and adjuncts supplemented with β-amylase or amyloglucosidase. Journal of the American Society of Brewing Chemists, 71: 208–213.

Espinosa-Ramírez, J., Pérez-Carrillo, E. and Serna-Saldívar, S.O. 2014. Maltose and glucose during fermentation of barley and sorghum lager beers as affected by β-amylase or amyloglucosidase addition. Journal of Cereal Science, 60(3): 602–609.

FAO. 2015. Food Agriculture Organization Statistical database http://faostat.fao.org. Accessed Sept. 16, 2015, Rome.

Fincher, G.B. and Stone, B.A. 1993. Physiology and biochemistry of germination in barley. pp. 247–298. *In*: A.W. MacGregor and R.S. Bhatty (eds.). Barley Chemistry and Technology. American Association of Cereal Chemists, St. Paul.

Godfrey, T. and Reichelt, J. 1983. Industrial Enzymology. The Application of Enzymes in Industry, Macmillan, Basingstoke.

Godtfredsen, S.E. and Otresen, M. 1982. Maturation of beer with α-acetolactate decarboxylase. Carlsberg Research Communication, 47: 93–102.

Guerdrum, L.J. and Bamforth, C.W. 2012. Prolamin levels through brewing and the impact of prolyl endoproteinase. Journal of the American Society of Brewing Chemists, 70: 35–38.

Hardwick, W. 1995. Handbook of Brewing. Marcel Dekker, New York.

Hough, J.S., Briggs, D.E., Stevens, R. and Young, T.W. 1993. Malting and Brewing Science. Vols. I and II. Chapman & Hall, London.

Hobbs, L. 2003. Corn sweeteners. pp. 635–670. *In*: P.J. White and L.A. Johnson (eds.). Corn Chemistry and Technology, Second Edition. American Association of Cereal Chemists, St. Paul.

Hramova, M. and Fincher, G. 2003. Enzymic hydrolysis of cereal (1→ 3, 1→ 4) β-Glucans. pp. 879–916. *In*: J.R. Whitaker, A.G.J. Voragen and D.W.S. Wong (eds.). Handbook of Food Enzymology. Marcel Dekker, New York.

Ingledew, W.M. 1995. The biochemistry of alcohol production. pp. 55–79. *In*: T.P. Lyons, D.R. Kelsall and J.E. Murtagh (eds.). The Alcohol Textbook. Nottingham University Press, Nottingham.

Kruger, J.E., Lineback, D. and Stauffer, C.E. 1987. Enzymes and Their Role in Cereal Technology. American Association of Cereal Chemists, St. Paul, pp. 403.

Kunamneni, A. and Singh, S. 2006. Improved high thermal stability of pullulanase from a newly isolated thermophilic *Bacillus* sp. AN-7. Enzyme and Microbial Technology, 39: 1399–1404.

Loi, L., Barton, P.A. and Fincher, G.B. 1987. Survival of barley (1→3, 1→w4)-β-glucanase isoenzymes during kilning and mashing. Journal of Cereal Science, 5: 45–50.

Lopez, M. and Edens, L. 2005. Effective prevention of chill-haze in beer using an acid proline-specific endoprotease from *Aspergillus niger*. Journal of Agriculture and Food Chemistry, 53: 7944–7949.

MacGregor, E.A. 2003. Limit dextrinase/pullulanase. *In*: J.R. Whitaker, A.G.J. Voragen and D.W.S. Wong (eds.). Handbook of Food Enzymology. Marcel Dekker, New York, pp.

MacGregor, A.W., Bazin, S.L., Macri, L.J. and Babb, J.C. 1999. Modeling the contribution of alpha amylase, β-amylase and limit dextrinase to starch degradation during mashing. Journal of Cereal Science, 29: 161–169.

Morgan, F.J. and Priest, F.G. 1981. Characterization of a thermostable α-amylase from *Bacillus licheniformis* NCIB 6346. Journal of Applied Microbiology, 50(1): 107–114.

Nelson, G. and Young, T.W. 1987. The addition of proteases to the fermenter to control chill haze formation. Journal of Institute of Brewing, 93: 116–120.

NicPhiarais, B.P. and Arendt, E.K. 2008. Malting and brewing with gluten-free cereals. pp. 347–372. *In*: *E.K. Arendt and F. Dal Bello (eds.)*. Gluten-free Cereal Products and Beverages. Academic Press, Burlington.

Norris, K. and Lewis, M.J. 1985. Application of a commercial barley β-amylase in brewing. Journal of the American Society of Brewing Chemists, 43: 96–101.

Odibo, F.J.C., Nwankwo, L.N. and Agu, R.C. 2002. Production of malt extract and beer from Nigerian sorghum varieties. Process Biochemistry, 37: 851–855.

Ogbonna, A.C. 2011. Current developments in malting and brewing trials with sorghum in Nigeria: A review. Journal of Institute of Brewing, 117: 394–400.

Owuama, C.I. 1997. Sorghum: A cereal with lager beer brewing potential. World Journal of Microbiology and Biotechnology, 13: 253–260.

Palmer, G.H., and Bathgate, G.M. (1976). Malting and brewing. pp. 237–324. *In*: Y. Pomeranz (ed.). Advances in Cereal Science and Technology. American Association of Cereal Chemists, St. Paul.

Power, J. 1993. Enzymes in brewing. pp. 439–457. *In*: T. Nagodawithana and G. Reed (eds.). Enzymes in Food Processing. 3rd Edition. Academic Press, San Diego.

Priest, F.G. and Stewart, G.G. 2006. Handbook of Brewing. CRC Press, Taylor and Francis, Boca Raton.

Pyler, R.E. and Thomas, D.A. 2000. Malted cereals: Their production and use. pp. 685–696. *In*: K. Kulp and J.G. Ponte (eds.). Handbook of Cereal Science and Technology, 2nd Edition. Marcel Dekker, Inc. New York.

Reilly, P.J. 2003. Glucoamylase. *In*: J.R. Whitaker, A.G.J. Voragen and D.W.S. Wong (eds.). Handbook of Food Enzymology. Marcel Dekker, New York.

Ryder, D.S. and Power, J. 2006. Miscellaneous ingredients in aid of the process. pp. 334–381. *In*: F.G. Priest and G.G. Steward (eds.). Handbook of Brewing, 2nd Edition. CRC Press, Boca Raton.

Saha, B.C., Mathupala, S.P. and Zeikus, G. 1988. Purification and characterization of highly thermostable novel pulluanase from *Clostridium thermohydrosulfuricum*. Biochemistry Journal, 252: 343–348.

Serna-Saldivar, S.O. 2010a. Grain development, morphology and structure. pp. 109–128. *In*: S.O. Serna Saldivar (ed.). Cereal Grains Properties, Processing and Nutritional Attributes. CRC Press, Boca Raton.

Serna-Saldivar, S.O. 2010b. Production of malts, beers, alcoholic spirits and fuel ethanol. pp. 417–462. *In*: S.O. Serna Saldivar (ed.). Cereal Grains Properties, Processing and Nutritional Attributes. CRC Press, Boca Raton.

Silveira Celestino, K.R., Cunha, R.B. and Feliz, C.R. 2006. Characterization of a β-glucanase produced by *Rhizopus microsporus* var. *microsporus* and its potential for application in the brewing industry. BMC Biochemistry, 7: 23.

Steward, G.G. 2013. Biochemistry of brewing. pp. 291–318. *In*: N.A. Michael Eskin and F. Shahidi (eds.). Biochemistry of Foods, 3rd Edition. Academic Press, New York.

Tanner, G.J. 2014. Gluten, celiac disease, and gluten intolerance and the impact of gluten minimization treatments with prolylendopeptidase on the measurement of gluten in beer. Journal of the American Society of Brewing Chemists, 72(1): 36–50.

Teague, W.M. and Brumm, P.J. 1992. Commercial enzymes for starch hydrolysis products. pp. 45–78. *In*: F.W. Schenck and R.E. Hebeda (eds.). Starch Hydrolysis Products. VCH Publishers, New York.

Tenkanen, M., Niku-Paavola, M.-L., Linder, M. and Vikari, L. 2003. Cellulases in food processing. *In*: J.R. Whitaker, A.G.J. Voragen and D.W.S. Wong (eds.). Handbook of Food Enzymology. Marcel Dekker, New York.

Urias-Lugo, D. and Serna-Saldivar, S.O. 2005. Effect of amyloglucosidase on properties of lager beers produced from sorghum malt and waxy grits. Journal of the American Society of Brewing Chemists, 63(2): 63–68.

Willox, I.C., Rader, S.R., Riolo, J.M. and Stern, H. 1977. The addition of starch debranching enzymes to mashing and fermentation and their influence on attenuation. MBAA Tech. Q., 14: 105–110.

Wong, D.W.S. and Robertson, G.H. 2003a. α-Amylases. *In*: J.R. Whitaker, A.G.J. Voragen and D.W.S. Wong (eds.). Handbook of Food Enzymology. Marcel Dekker, New York.

Wong, D.W.S. and Robertson, G.H. 2003b. β-Amylases. *In*: J.R. Whitaker, A.G.J. Voragen and D.W.S. Wong (eds.). Handbook of Food Enzymology. Marcel Dekker, New York.

Woods, L.F.J. and Swinton, S.J. 1995. Enzymes in the starch and sugar industries. *In*: G.A. Tucker and L.F.J. Woods (eds.). Enzymes in Food Processing. 2nd Edition. Chapman & Hall, Blackie Academic & Professional. Glasgow.

Yamano, G., Tanaka, J. and Inoue, T. 1994. Cloning and expression of the gene encoding α-acetolactate decarboxylase from *Acetobacter aceti* ssp. in brewer's yeast. Journal of Biotechnology, 32: 165–171.

Zhou, K., Slavin, M., Lutterodt, H., Whent, M., Michael Eskin, N.A. and Yu, L. 2013. Cereals and legumes. pp. 3–48. *In*: N.A. Michael Eskin and F. Shahidi (eds.). Biochemistry of Foods, Third Edition, Academic Press, New York.

16

Enzymes in Baking

Angela Dura and *Cristina M. Rosell**

1. Introduction

Breadmaking is based in several fundamental complex processes of physical, chemical and biochemical changes like evaporation of water, volume expansion and formation of a porous structure, denaturation of protein, gelatinization of starch, crust formation and browning reaction that are essential for the quality of the final product (Eliasson and Larsson, 1993; Rosell, 2011). Bread has been one of the principal forms of food from earliest times and is one of the major food products across the world. Wheat bread is the most common and most consumed type of bread, being an important source of nutrients. Particularly, bread provides complex carbohydrates, dietary fiber, has low content of fat (without containing cholesterol), minerals, especially calcium, phosphorous, iron and potassium and B vitamins (Rosell and Garzon, 2015). A number of European countries recommend a daily bread intake of about 250 g, which corresponds to four to eight slices, depending on the national food habits (WHO, 2003). The main ingredient in baked goods is flour from cereal grains, which are the fruit of plants belonging to the grass family (*Gramineae*). Apart from wheat, other major cereals that are used for bread making are corn, rice, barley, sorghum, millet, oats and rye. They are grown on approximately 60 per cent of the cultivated land in the world. Wheat, corn and rice take up 90 per cent of cereal grain production worldwide, the largest quantities of cereal grains (FAO, 2012) and provide two-thirds of energy requirement for humans through diet (Bharath Kumar and Prabhasankar, 2014). Different types of grains affect the quality of the final product and display variability in nutritional value. Consumers' quality perceptions of bread from a health perspective have lately increased over time. It has led to a growing demand for healthier cereal-based products by incorporating

Department of Food Science. Institute of Agrochemistry and Food Technology. Spanish Research Council. IATA-CSIC. Avda.CatedráticoAgustínEscardino, 7. 46980 Paterna. Spain.
  E-mail-andudemi@iata.csic.es
* Corresponding author: crosell@iata.csic.es

into their diet ancient grains, such as millet, quinoa, sorghum, oat, barley, rye and teff, because consumption of whole grain foods and high dietary fiber, including bread, is related to prevention of the metabolic syndrome, obesity and associated chronic diseases, such as cardiovascular disease and Type 2 diabetes (Björck et al., 2012). Bread has become a good vehicle to incorporate from a nutritional point of view, a good supply of dietary fiber, an assortment of vitamins and minerals and a source of energy in the form of starch. Moreover, the total consumption of bread has increased over the past 20 years. Bread and bakery goods are widely consumed by world's population because of its low price and high nutritional value. Due to the consumers' demand, bakery industry is becoming very active regarding processing, recipes, shapes and in the production, wholesaling and retailing of baked goods including breads, cakes, pastries, breakfast cereals, snacks, cookies and crackers. Nowadays, the bakery industry is taking advantage of optimizing the quality of bread with respect to primary ingredients, advance technology design to monitor the process by means of approved methods, but also to identify and incorporate values that are attractive to different consumer segments. Research in the baking industry has largely focused on investigating methods to improve the final product by extending the shelf-life and preserving the product freshness.

In this trend, enzyme research has played an important role in cereal processing and especially in the baking industry as they play an important role as processing or technological aids. One additional benefit is that enzymes are considered clean label compounds. Progressively, chemical supplements that were used in the baking industry are being replaced by enzymes, which are the best and safest alternative owing to their GRAS (Generally Recognized as Safe) status. Enzymes are able to catalyze chemical reactions and are no longer active after breadmaking owing to the denaturation of their protein structure during baking. Baking enzymes have become an essential part of the industry and are able to modify and improve the functional, nutritional and sensorial properties of ingredients and products. Thus enzymes have found extensive applications in processing and production of bread products. Traditionally, enzymes were used as processing aids in the baking industry, but lately, owing to the nutrition-awareness, a more extensive concept is being applied and they are considered as healthy aids because of their role in improving the nutritious quality of the bakery products (Rosell, 2014).

Enzymes that contribute to the quality and processing characteristics involved in breadmaking can be derived from endogenous enzymes of grain raw materials or commercial enzymes that are usually of microbial origin, added during breadmaking process or supplemented to the flour. Several commercial enzyme preparations and enzyme-containing bread improvers are available. There are many enzymes used in industrial baking to accomplish the desired functionality (Rosell and Collar, 2008; Rosell and Dura, 2015). Enzymes are classified according to the reactions they catalyze and the substrate they act on. Each enzyme is given an EC number by the Nomenclature Committee of the International Union of Biochemistry and Molecular Biology (Cornish-Bowden, 2014). The most common enzymes used in the baking industry are amylases, proteases, lipases, hemicellulases and oxidoreductases. All these enzymes added individually or in combination enable better dough performance and stability, strengthening and bleaching of the dough, maintaining bread volume, crumb softness, crust coloring or browning and provision of anti-staling properties. Their action during breadmaking is largely dependent on the flour constituents, because of which becomes primordial to know them in order to understand the enzymes functionality in baking.

Rapid advances in biotechnology has allowed improved yields by fermentation, modifying specificity, selectivity and stability of enzymes, and also to increase the number of enzymes available for the baking industry.

2. Principles of Breadmaking

Since ancient time cereals have played an important role in human nutrition, particularly common wheat which is most commonly used for bread and baking goods. As mentioned before, many other grains have been used, like millet, quinoa, sorghum, oat, barley, rye and teff, to fulfill the consumer's demand or special nutritional needs. Cereal grains contain 66–76 per cent carbohydrates, where starch is the most abundant carbohydrate and the most relevant reserve constituent of cereals (55–70 per cent) followed by minor constituents, such as arabinoxylans (1.5–8 per cent), β-glucans (0.5–7 per cent), sugars (~3 per cent), cellulose (~2.5 per cent), and glucofructans (~1 per cent). Another important group of constituents is the proteins which fall within an average range of about 8–11 per cent. With the exception of oats (~7 per cent), cereal lipids belong to the minor constituents (2–4 per cent) along with minerals (1–3 per cent) (Rosell, 2012). Bread is the product of baking a mixture of flour, the most essential ingredient and key source of enzyme substrates, water, salt and yeast. Breadmaking is a traditional ability practiced wideworld. Despite the diverse range of cereal grains and different processed bread products, a number of central stages are common to all bread products and breadmaking processes. The basic principles of breadmaking involve mixing of ingredients, dough fermentation, rounding and molding of dough, proofing, baking of bread in oven and cooling operation. The aim of the breadmaking process is to convert flour into an aerate, edible and tasty bread for the consumer by a number of stages that can proceed differently, depending on the final product but that have a common purpose.

2.1 Mixing ingredients

The mixing of flour (mainly wheat), water, yeast, salt and other listed ingredients in appropriate ratios is the first stage to form an homogenous dough as it has a direct consequence on the quality of the end product. During mixing, fermenting and baking, the dough is exposed to different physio-chemical and biological transformations, such as evaporation of water, gelatinization of starch, volume expansion, crust formation, denaturation of protein, browning reactions and so on, which are largely affected by temperature and water content (Rosell and Collar, 2009). Mixing provides the necessary mechanical energy for developing the protein network and incorporates air bubbles into the dough. The real goal of vigorous and continued mixing and kneading is to improve the bread volume and crumb softness. This process produces changes in the physical properties of the dough, making it extensible enough to expand during proofing and elastic enough to improve its ability to retain the carbon dioxide gas, which will later be generated by yeast fermentation and be stable enough to hold its shape and cell structure (Cauvain and Young, 2007).

2.2 Dough fermentation

Dough is a macroscopically homogeneous three-dimensional mixture of starch, protein, fat, salt, yeast and other components that form a visco-elastic material that exhibits an intermediate rheological behavior between a viscous liquid and anelastic solid. Dough development begins by mixing the ingredients with the formation of gluten, which requires the hydration of flour proteins and the application of energy through the process of kneading. Gluten proteins (gliadin and glutenin) play an important role in determining the baking quality of wheat by conferring water absorption capacity, cohesivity, viscosity and elasticity on dough. It is widely recognized that gliadins confer viscous properties while glutenins impart strength and elasticity (Shewry et al., 1986). Rheological properties of gluten are a direct consequence of the wheat quality and affect the textural characteristics of the finished bread. Starch, as the major carbohydrate present in flour, is able to form a continuous network of particles together with the macromolecular network of hydrated gluten and therefore affects the rheological properties of doughs (Song and Zheng, 2007). Other constituents, such as non-starch polysaccharides and lipids, can affect the dough stability and hence the breadmaking performance of the flour (Goesaert et al., 2005). After mixing, during fermentation, yeast metabolism results in carbon dioxide production and expansion of the dough only if that carbon dioxide gas is retained in the dough. Fermentation occurs with the expansion of air bubbles previously incorporated during mixing and provides the characteristic aerated structure of bread. Enzymes present in yeast and flour or added intentionally in the process breaks down starch and oligosaccharides into carbon dioxide and alcohol during alcoholic fermentation. The growth of gas bubbles determines the expansion of the dough and therefore, the ultimate volume and texture of the baked product (He and Hoseney, 1991).

2.3 Other dough operations during breadmaking

Different stages take place before the final baking. The dough is divided to generate the shape and size of product required. The accuracy of the system depends on the homogeneity of the dough through the distribution of gas bubbles. After dividing, pieces are subjected to shaping, usually by rounding. Mechanical molding exposes the dough to stresses and strains that may lead to damage of the existing gas-bubble structure (Cauvain, 2012).

2.4 Proving and baking

Proving refers to the final rise that the dough undergoes in a resting period which takes place after being shaped and before it is baked. After proofing of the dough, baking takes place through stabilization of a porous structure by altering the molecular configuration of the polymeric components in the cell walls through the application of heat. At the temperatures and moisture contents typical for bread baking, thermal transitions occur. Baking temperatures could vary from different ovens and with the product. Baking time, oven temperature and source of heat are factors that decide the quality of bread during the baking process. Initial temperature leads to an increase in the dough volume and initiates rapid evaporation of water and release of carbon dioxide, ethanol and some aromatic compounds. The structural changes also occur during the bread-baking process, which this promotes the gelatinization of starch and denaturation of protein with the coagulation of gluten, which comprises solidification and expansion. The increase in temperature and lower moisture content induce non-enzymatic browning reaction, the rising of the top layer followed by crumb development and finally, gradual color development.

Due to the existing variability of bread products, breadmaking process can differ from one product to another. In breadmaking applications, replacement of wheat flour by significant amounts of other cereal or pseudo-cereal flours can vary the process in order to obtain a suitable product.

3. Enzymes Role in Breadmaking to Improve Process

Enzymes are proteins that catalyze chemical reactions and convert substrates into different molecules. Enzymes are heat-sensitive biological catalysts that have an optimum temperature and pH for activity and are also dependent upon the availability of water, the amount of enzyme used, the availability of substrate and the time allowed for the reaction. Once optimum temperature has reached its maximum tolerance, enzymes reach a denaturation point and will lose their functionality. Baking enzymes have been an essential part to the industry throughout the ages and are able to modify and improve the functional, nutritional and sensorial properties of ingredients and products; thus enzymes have found extensive applications in processing and production of bread products. Enzymes (α-amylases, xylanases, lipoxygenases, glucooxidases, transglutaminases, proteolytic enzymes and lipases) are added to improve the dough-handling properties, loaf volume, fresh bread stability and shelf-life of bread (Dunnewind et al., 2002; Haros et al., 2002a; Caballero et al., 2007; Steffolani et al., 2010; Rosell and Dura, 2015). In baking, the most commonly used enzyme's classes fall within the category of oxidoreductases (i.e., glucose oxidase and lipoxygenase) used for strengthening and bleaching of the dough, hydrolases, such as amylases (to convert starch to sugar and to produce dextrins), proteases and hemicellulases which affect wheat gluten, and lipases improve the dough's rheological properties and quality of the baked product. All these enzymes play an important role in maintaining bread volume, crumb softness, crust crispiness, crust coloring or browning and in maintaining its freshness.

Nowadays, the baking industry conforms to the higher consumer's demands, claiming more natural products and new products adapted to special nutritional needs. In this tendency, enzymes have a number of advantages, being used as an alternative to traditional chemical-based technology, lowering energy consumption levels and fewer by-products (Whitehurst and Oort, 2010) and does not require any changes in the processing line operation during the breadmaking process.

As a consequence of the research and development over the years, numerous new methods have been established to modify enzymes or increase their yield, making their use economically viable in breadmaking. The common way to use those enzymes is by mixing them with the rest of the ingredients in the initial stage of breadmaking. There is no special requirement when adding enzymes, which makes them very useful as processing aids. Enzymes can be added individually or in composite mixtures, which may act in a synergistic, additive, or antagonistic way in the production of bread. The supplementation of flour and dough with enzyme improvers is usual practice for flour standardization to achieve the desirable effects, to provide consistent product quality or to increase the yield in the process and their action will continue during all the breadmaking stages. Therefore, this section focuses on the enzymes used as processing aids in the breadmaking process added as a single enzyme or as a combination of them.

4. Application of Individual Enzymes

The baking industry is a large consumer of starch and starch-modifying enzymes. Many enzymes added individually in different breadmaking stages have been reported to improve dough characteristics and bread quality.

4.1 Amylases

Amylases comprise a variety of amylolytic enzymes capable of digesting glycosidic linkages from the starch in the flour into smaller dextrins, which are subsequently fermented by yeast. For a long time, enzymes, such as malt and fungal α-amylase, have been used in breadmaking. It is rather important to take into account that the source of enzymes determines their activity and stability (Rosell et al., 2001). Overall, fungal enzymes are less thermostable and most of their activity is lost during starch gelatinization (Bowles, 1996). Addition of amylases mainly aims at optimizing the amylase activity of the flour (flour standardization). By enzyme supplementation of flour, the fermentation rate is enhanced owing to the increase in fermentable sugars. The dough viscosity is reduced, resulting in improvement in the volume and texture of the product (Stefanis and Turner, 1981). It also generates additional sugars in the dough, which improves the taste, crust color and toasting qualities of the bread. The action of α-amylases becomes readily evident in the crumb microstructure, which has been associated with the structural changes promoted at dough level (Błaszczak et al., 2004). Initially, the effect of α-amylases was attributed to the increase in sugar content, but later on it has been confirmed that the effects are wider than sugar production. Patel et al. (2012) studied the effects of fungal α-amylase on chemically leavened doughs, finding that the baking qualities of chemically leavened dough improved in the presence of fungal α-amylases. Differences in rheological properties were attributed to the higher amount of free liquid and the lower viscosity in the dough containing α-amylases. Further on, Sahnoun et al. (2013) observed that the addition of a preparation at an adequate concentration of α-amylase from *Aspergillus oryzae* produces positive effects on dough properties and bread quality, increasing the dough resistance to extension and bread volume. It must be remarked that some bacterial α-amylases have high themostability and may survive the baking process. Consequently, they are difficult to control during storage and can result in progressively reduced crumb structure properties (Bowles, 1996). As reported by Bosmans et al. (2013), the individual addition of three different bacterial amylases to a bread recipe impact bread firming, leading to lower initial crumb resilience. Bread loaves that best retained their quality were those obtained with α-amylase from *Bacillus stearothermophilus*, whereas *Bacillus subtilis* α-amylase led to extensive degradation of the starch network, resulting in partial structure collapse and poor crumb resilience.

Upon storage, the shelf-life of fresh bread is rather short as a number of complex physico-chemical alterations occur, which are together referred to as staling, comprising the increase in crumb firmness. Retrogradation of the starch fraction in bread plays a very important role in staling. Retrogradation/recrystallisation of starch, specifically of the short amylopectin side chains, plays a major role in bread firming after the initial cooling process (Gray and BeMiller, 2003). Taste and freshness define to a great extent the quality of bread products. From the bakery industry's point of view, it is of considerable economic importance, since it limits the shelf-life of baked products. Several additives and different combinations have been used to delay staling and improve, texture, volume and flavor of bakery products. Enzymes active on starch, mainly amylases, have been suggested to act as anti-staling agents, particularly a maltogenic α-amylase being a very effective anti-staling agent (Goesaert et al., 2009). A combination of emulsifiers, hydrocolloids and α-amylase was proposed by Armero and Collar (1996) for retarding staling in white and wholemeal breads where different synergistic and antagonistic effects resulted from their combinations (DATEM*SSL, α-amylase*SSL, α-amylase*HPMC). Many authors have reported the anti-firming effects of amylases in bread staling (Goesaert et al., 2009; Gray and BeMiller, 2003; Hug-Iten et al., 2003; Fadda et al., 2014). More recently, Gomes-Ruffi et al. (2012) studied the effect of maltogenic amylase in combination with an emulsifier on the quality of the pan bread during storage resulting in an increase in the bread volume and a reduction of firmness and high sensory scores with multiple combinations of the emulsifier and the enzyme.

Although enzymes are usually added at the mixing stage, it has also reported the use of enzymes on to the dough (Primo-Martin et al., 2006) or even on to the surface of partially baked breads (Altamirano-Fortoul et al., 2014). Primo-Martin et al. (2006) reported the effect of different enzymes (endoprotease, transglutaminase, α-amylase) sprayed on to the dough surface as a possible strategy for extending crust crispiness, with the protease having the best effect. On the other hand, amyloglucosidase sprayed on to the partially baked loaf surface has been proposed for modulating the properties of the bread crust and increase its crispness by increasing the number of voids and texture fragility (Altamirano-Fortoul et al., 2014).

A less known enzyme in bakery is a branching enzyme that is a carbohydrate-active enzyme belonging to the amylases family. A gene encoding a glycogen branching enzyme was cloned and expressed in *Escherichia coli*. The purified enzyme was added to wheat breadmaking and the effect was studied by Wu et al. (2014). Addition of this enzyme to wheat bread resulted in increased specific volume, decrease in crumb firmness and simultaneously the retrogradation was significantly retarded in storage.

4.2 Non-starch hydrolases

Xylanases from family 8 and 11 have been found very effective as baking processing aids (Collins et al., 2006) and within the fungal xylanases, *Aspergillus oryzae* xylanase was the best bread improver and *Trichoderma reesei* xylanase the best anti-staling improver (Baskinskiene et al., 2007). Ahmad et al. (2012) studied the effect of *Aspergillus niger* xylanase either added during tempering of wheat kernels or mixing the dough. Both treatments changed the dough characteristics, improving the dough performance and resulting in bigger loaf volume and better sensory characteristics of bread. But, the overall development was more prominent in the samples receiving enzyme treatment during tempering. The hydrolytic activity of xylanases, at optimum level dosage, resulted in monomers and oligomers that affect the water balance, transforming protein-starch interaction; hence improving dough machinability, dough stability, oven spring, loaf volume, crumb structure and shelf-life as well as reducing the staling rate (Poutanen, 1997). Purification of an extracellular endo-β-1,4-xylanase and its application in breadmaking was studied by McPhillips et al. (2014). Optimal activity of the enzyme was measured at pH 6.0 and 65°C, which was suitable for breadmaking applications. Hydrolysis products from the action of the enzyme by cleaving 1,4-β-D-xylosidic linkages resulted in xylosaccharides, predominantly xylotriose and xylobiose. Basic wheat bread recipe at low dosages showed its suitability to increase both loaf volume and softness, while reducing bread staling up to four days of storage. The arabinoxylans (AX), the major fraction from non-starch polysaccharide (85–90 per cent) consists of a backbone of β-(1,4)-linked xylose residues. They are present in water-extractable (WE-AX) and water-unextractable (WU-AX) forms (Dornez et al., 2009). Their structure results in exclusive physicochemical properties, which strongly determine their functionality in breadmaking. Dornez et al. (2011) compared the individual action of three psychrophilic xylanases and a mesophilic xylanase. Xylanases, with a high capacity to solubilize WU-AX during mixing, increased the bread volume more than xylanases that mainly solubilized WU-AX during fermentation. The xylanase efficiency could be related to their substrate hydrolysis behavior and was mainly dependent on the enzyme's temperature-activity profile and its inhibition sensitivity. Xylanases have been useful also in different type of flour. The effect of xylanase with ascorbic and citric acid was also studied to improve the breadmaking performance of two spelt cultivars from organic production differing in dough rheological properties (Filipcev et al., 2014). The optimized formulations allowed substantial increase in specific volume and crumb softening for weaker spelt cultivar.

Pentosans present in flour have an important role in bread quality due to their water absorption capability and interaction with gluten, which is vital for the formation of the loaf structure. The 'short' gluten flours cause problems in breadmaking due to their high elasticity and low extensibility, which negatively affect the dough capacity to retain the gases released from fermentation. Stoica et al. (2013) demonstrated that the combination of a reducing agent (L-cysteine) and a fungal pentosanase resulted in an increase in bread volume, porosity and elasticity.

Other enzymes acting on non-starch cereal carbohydrates have been tested for anti-staling effects. A kinetic study of the firmness along storage using Avrami equation conducted by (Haros et al., 2002b) showed the potential anti-staling effect of different carbohydrases. Particularly, the simultaneous analysis of the hardening and starch retrogradation indicated that the anti-staling effect of xylanase might be due to the delay in starch retrogradation, having the greatest anti-staling effect. Conversely, cellulose and β-glucanase reduced the bread firming speed, but some other mechanism should be also involved. The use of enzymes like xylanases has been described to play an important role in increasing the shelf-life of bread and to reduce bread staling (Butt et al., 2008).

4.3 Tranglutaminase (TGase)

TGase is a protein-glutamine γ-glutamyl-transferase, which catalyzes an acyl-transfer reaction between the γ-carboxyamide group of peptide-bound glutamine residues and a variety of primary amines (Motoki and Seguro, 1998). TGase has been applied in breadmaking when aimed at decreasing the dough extensibility and increasing resistance to stretching due to reinforcement of the protein network induced by TGase (Autio et al., 2005). The use of enzyme increases dough stability and bread volume, and it also improves the bread crumb structure (Gerrard et al., 1998). This enzyme has also been used for improving the baking quality of weak wheat flours (Basman et al., 2002) and bug damaged flours (Rosell et al., 2003). TGase has also shown positive effects on non-wheat-based bread. Beck et al. (2011) showed that an optimal dosage of TGase improves rheological properties of rye dough, by creating a continuous protein network, and showed positive effects on loaf volume and crumb texture of rye bread. It was also reported that higher levels of the enzyme had detrimental effects on loaf volume, resulting in an increase of crumb springiness and hardness. Gluten-free bakery products have been one of the most grave concerns of the research community, due to the growing demand for these products. The demand for gluten-free products is encouraging the use of ingredients such as rice flour, corn flour and meal, ancient grains and tubers and pulses. Nevertheless, the proteins of those flours are unable to retain carbon dioxide during fermentation due to lack of viscoelastic properties, causing one to find technological alternatives to mimic gluten viscoelastic properties (Arendt, 2008). However, the protein network created by TGase is greatly dependent on the protein origin, its thermal compatibility and the dosage of the enzyme (Marco and Rosell, 2008b). In this respect, TGase has been the enzyme most extensively proposed for creating protein crosslinks in rice flour (Gujral et al., 2003). Authors studied the addition of increasing amounts of TGase (0.5, 1.0 or 1.5 per cent w/w) to rice flour to obtain a progressive increase of the viscous and elastic moduli, but higher bread volume and softer crumb was obtained with 1.0 per cent TGase and further improvement was obtained with the addition of 2 per cent HPMC. Furthermore, Gujral and Rosell (2004a) applied TGase to rice flour for promoting protein interactions, which lead to a better breadmaking performance. Renzetti et al. (2008) used TGase for network forming potential on flours from six different gluten-free cereals (brown rice, buckwheat, corn, oat, sorghum and teff) used in breadmaking. Three-dimensional confocal laser scanning micrographs confirmed the formation of protein complexes by TGase. Better fundamental rheological analysis and bread quality indicated an improving effect of 10U TGase on buckwheat and brown rice batters and breads, but a detrimental effect on corn batters with increased specific volume and decreased crumb hardness and chewiness on corn breads. A complete revision about the role of enzymes in gluten-free flours has been recently published (Renzetti and Rosell, 2015). Lately, some proteins (albumin, casein, pea protein and so on) have been added to enhance the action of transglutaminase with the aim to create a protein network in gluten-free matrixes (Marco and Rosell, 2008a; Storck et al., 2013).

4.4 Oxidases

Different oxidases have a beneficial effect on dough development and dough quality and are mostly attributed to the reinforcement or strengthening effect of dough (Oort, 1996), but also as dough-bleaching agents (Gelinas et al., 1998). Zhang et al. (2013) described the effect of adding recombinant lipoxygenase to wheat flour whiteness and the enzyme-treated bread showed higher volume and loaf height and better crumb color, resilience, chewiness and hardiness. Treatment of dough with glucose oxidase (GOX) increases the gluten macro-polymer content due to disulphide and non-disulphide cross-linking (Steffolani et al., 2010) and leads to protein aggregates and broken segments (Indrani et al., 2003). A study conducted by Bonet et al. (2006) from macroscopic to molecular level showed a reinforcement or strengthening of wheat dough and an improvement in bread quality, although excessive addition of GOX produced excessive cross-linking in the gluten network and was responsible for the negative effect on the breadmaking properties. Degrand et al. (2015) have recently reported the possible valuable effect as dough and as bread improver of a carbohydrate oxidase similar to *Aspergillus niger* GOX. Carbohydrate oxidase exhibits a similar efficiency towards maltose as GOX towards glucose, whatever be the oxygen supply. GOX has also been used to replace chemical oxidants when using weak flours in the breadmaking process, thus reducing the economic losses caused by plagues. This enzyme was able to restore the broken covalent bonds between glutenin subunits and forming dityrosine cross-links between the wheat proteins, which reinforced the gluten network and gave away the dough functionality (Bonet et al., 2007). The use of gluten-free flour, such as rice flour, makes the breadmaking process difficult to perform. Gujral and Rosell (2004b) confirmed the effects of GOX on rice-flour dough rheology and protein modification, obtaining rice bread with better specific volume and texture.

4.5 Lipases

Although, lipids are minor constituents in flour, the use of lipases in breadmaking has also been reported as individual effect. Lipase is mainly used to enrich the flavor content of bakery products by liberating short-chain fatty acids through esterification. This enzyme also participates in improving the dough's rheological properties, texture and softness. Lipase as a dough and bread improver reduces crumb pore diameter, increases crumb homogeneity and improves the gluten index in dough (Aravindan et al., 2007; Poulson et al., 2010). Specifically, a strengthening effect on the gluten network leads to an increase in dough stability, oven spring and specific volume (Moayedallaie et al., 2010), or even a change in the bread crust fracture behavior (Primo-Martin et al., 2008). Other applications of this enzyme was found in brewer's spent grain, as the main by-product of the brewing industry, with a high content of fibre and which could be treated with a mixture of lipases to increase expansion of the gluten network, to increase the wall thickness and to reduce cell density, thus enhancing the volume and crumb structure of high-fiber white bread (Stojceska and Ainsworth, 2008). Moreover, application of bioprocessing techniques (sourdough fermentation and technological aids) for brewer's spent grain breads were reported by Ktenioudaki et al. (2015). Buckwheat flour, as a highly nutritious pseudo-cereal containing dietary fiber, was added with phospholipase; the hydrolysis of the lipids improved the dough stability and quality (Ozer et al., 2010).

At final stage of baking, the increase in temperature and lower moisture content induce non-enzymatic browning reaction, which results in crust formation on the surface of the bread. Due to the presence of reducing sugars and L-asparagine, besides the high temperatures, acrylamide (chemical compound with probable human carcinogenic effect) is formed in many bakery products. Mohan et al. (2013) confirmed that asparaginase reduced the acrylamide formation in the crust and crumb regions by 97 per cent and 73 per cent, respectively, without changing the rheological properties and physico-sensory characteristics of bread. Similar results were observed by Kukurova et al. (2009), reducing acrylamide formation up to 90 per cent. This enzyme also converted glutamine into glutamic acid, but neither action affected the browning and Maillard reactions.

4.6 Enzymes combinations for enhancing bread quality

Numerous enzyme combinations have been applied for improving bread's industrial or instrumental quality. Nevertheless, it is necessary to understand the possible synergies or antagonisms within enzymes prior to selecting the right or optimal combination to reach better dough performance and quality of the finished baked product (Collar et al., 1998; Rosell et al., 2001). Further, a suitable application of enzyme combination offers a wide range of alternatives to improve bread quality. Combinations of the extensive variety of enzymes focus on improving dough performance in breadmaking and the final product characteristics. Some enzyme combinations are not appropriate for the intended purpose with antagonist or lack of effect. According to Collar et al. (2000), the incorporation of xylanase with GOX is not recommended because GOX counteracts the softening effect of xylanases. Moreover, limited effects on dough and bread properties were found by Bilgicli et al. (2014) when combining two cross-linking enzymes, TGase and GOX in conjunction with several emulsifiers and ascorbic acid as an oxidant in wheat-lupin flour blend. It has also been reported that GOX acts synergistically with phospholipase, ensuring rolls shape uniformity and tolerance to process changes, besides greater bread volume (Novozymes, 2001). Likewise, results obtained by Selinheimo et al. (2006) confirmed that dough hardening induced by GOX was counteracted when combined with laccase, because the later mediated the depolymerization of the cross-linked arabinoxylans network.

Conversely, ideal enzyme blends have been proposed, in which enzymatic activities are enhanced. Gil et al. (1999) reported the combined action of pentosanase, bacterial α-amylase and lipase in increasing bread crumb elasticity and reducing firmness during storage up to 72 hours. An optimal combination composed of 0.010

per cent (w/w) xylanase, 0.005 per cent (w/w) papain and 0.002 per cent (w/w) GOX was defined using an orthogonal experimental design by comparing browning index enhancement in 24 hours, which efficiency was confirmed when used as rheological improver for fresh whole wheat dough (Yang et al., 2014). Immobilized GOX on chitosan (CS)-sodium tripolyphosphate (TPP) combined with fungal α-amylase added to wheat flour resulted in a slightly increase of bread springiness and specific volume and a decrease in hardness, improving the bread's final quality (Tang et al., 2014). The combination of GOX and fungal α-amylase supplemented to weak wheat flour was also tested by Stoica et al. (2010). Both enzymes have a strengthening action on weak gluten network, thus increasing its capacity to retain the fermentation gases. Enzyme combination caused a substantial improvement in bread quality indicators, such as volume, porosity and elasticity, clearly superior to their individual use. In fact, flour with low α-amylase activity needs to be supplemented with additional α-amylase, but α-amylase added to weak flour can decrease the quality of the dough. Shafisoltani et al. (2014) proposed a combination of GOX and xylanase to improve the quality of wheat flour after being optimized by adding α-amylase. Results showed that this enzyme combination led to dough with low stickiness and bread with higher specific volume, better shape, lower crumb hardness and higher total score in sensory evaluation test, but those effects were dose dependent.

Non-starch carbohydrases are commonly used in combinations. Stoica et al. (2013) studied the effect of three hemicellulolytic enzymes—a fungal hemicellulase from *Aspergillus oryzae*, a fungal pentosanase from *Humicola insolens* and a bacterial xylanase from *Bacillus subtilis* in short gluten flours to improve characteristics of pan bread, resulting in better physical indices like bread volume, porosity and elasticity. The use of an enzymatic cocktail comprising xylanase, xylosidase and α-L-arabinofuranosidase, produced by a thermophilic fungus *Thermoascus aurantiacus*, on wheat bread confirmed that these enzymes mainly attack water-un-extractable arabinoxylan, with the subsequent increase in specific volume and reduction in crumb firmness and amylopectin retrogradation during bread storage (Oliveira et al., 2014). Similar results were obtained by Hsing et al. (2010) that recommended a combination of hemicellulase and/or endoxylanase with ascorbic acid, which delayed the staling and also improved the quality of the freshly baked bread.

Gluten-crosslinking enzymes can greatly improve the functional properties of dough. The effect of a number of gluten-crosslinking enzymes (transglutaminase, glucose oxidase and laccase), along with polysaccharide and gluten degrading enzymes (alpha-amylase, xylanase and protease) were analyzed by Caballero et al. (2007) in breadmaking systems. Results revealed significant interactions within enzymes and improvement of their anti-staling effects. In addition the effect between the pairs glucose oxidase-laccase, glucose oxidase-pentosanase, amylase-laccase, amylase-protease and pentosanase-protease showed significant synergistic effects on bread quality (Caballero et al., 2007).

Most recently, a combination of three lipases added to wheat flour was proposed to change flour lipid composition, specifically decreasing the level of galactolipids and phospholipids and in parallel increasing the level of 'lyso'-lipid as well as free fatty acids (Gerits et al., 2014).

5. Enzymes as Healthy Aids in Baking Goods

For baked goods selection the major drivers are product presentation, freshness and taste. These, in conjunction with health benefits, are top trends in today's in-store bakeries. The increasing consumer demand for healthy bread has prompted the development of breads that combine health benefits with good sensory properties (Rieder et al., 2012). Even in this context, enzymes play a significant role and, besides their role as processing aids, they can extend their application to improve the healthy impact of bakery goods. According to this, enzymes could also be coined as healthy aids (Rosell, 2014).

Consumers are seeking bakery products made with high fiber and whole grains, while avoiding high fructose corn syrup, trans-fats and hydrogenated fats. Bread, as a traditional food, plays an important role in daily diet and can easily be the vehicle for improving the nutritional profile from common wheat bread (Rosell, 2012). Supplementation, combination and/or substitution of wheat flour with whole grains improve the nutritional quality of wheat products. More accurately, the consumer's awareness of using healthy high-fibre foods is increasing, which is forcing enrichment with dietary fiber owing to their benefits, such as prevention of certain diseases including cancer, the blood cholesterol level, gastrointestinal disorders, diabetics and coronary heart disease (Fuentes-Zaragoza et al., 2010; Champ et al., 2003). In this development there are a group of enzymes that play a significant role as bread improvers. For these specific actions a group of starch and non-starch carbohydrate-hydrolyzing enzymes have been usually proposed to improve dough rheological properties, bread specific volume and crumb firmness. Particularly, xylanases added during mixing of the whole wheat flour lead to improvement in the bread sensory evaluation (Driss et al., 2013). In addition, some studies show prebiotic activity of cereal-derived arabinoxylan oligosaccharides produced by xylanases. Three mesophilic xylanases from *Bacillus subtilis*, *Aspergillus niger* and *Hypocrea jecorina* and one thermophilic xylanase from *H. jecorina* were used in different dosages in whole meal breads enriched with arabinoxylan rich materials (Damen et al., 2012). Chapati bread from whole wheat flour treated with amylases and xylanases had higher content of soluble dietary fiber and soluble polyphenols, besides high antioxidant properties (Hemalatha et al., 2012). Furthermore, the α-amylase, xylanase and lipase blend was found effective in high-fiber wheat bread, where enzymes and sourdough lower firmness, amylopectin crystallinity and rigidity of polymers (Katina et al., 2006). Consumption of bread containing whole grains or functional ingredients rich in dietary fiber and β-glucans has shown a dramatic increase because of its health benefits. Endo-glucanases have been also tested in rye breads, resulting in lower proofing times due to their action on β-glucans, but they do not influence the content or solubility of the arabinoxylans (Autio et al., 1996). Most recently, the effect of β-glucanase on wheat barley composite flour, despite some negative dough properties, overall indicated that dough rheological characteristics and bread quality can be improved by using an optimal enzyme concentration (Li et al., 2014).

Bran can previously be treated by baking enzymes to increase the amount of water-soluble dietary fiber during the baking process of wheat bread with added bran. The baking enzyme mixtures tested (xylanase and glucanase/cellulase, with and without lipase) increased the amounts of soluble arabinoxylan and protein resistant to digestion, which in turn can exert prebiotic effects on certain potentially beneficial microbes (Saarinen et al., 2012). The effect of wheat bran and enzyme addition to dough's functional performance and phytic acid levels was also studied by Sanz Penella et al. (2008). α-Amylase addition was suggested for improving the dough development and gassing power parameters during proofing. Researchers reported that a combination of bran with amylolytic and phytate-degrading enzymes is advisable for overcoming the detrimental effects of bran on the mineral availability (phytase) or on the technological performance of doughs (α-amylase). The combination of fungal phytase addition, breadmaking process and frozen storage for overcoming the detrimental effect of bran on the mineral bioavailability was reported by Rosell et al. (2009) when studying the effects of different breadmaking processes (conventional, frozen dough, frozen partially baked bread) and the effect of the storage period on the technological quality of fresh whole wheat breads. Results supported that combination of phytase addition and freezing process increased the bioavailability of phosphorus. Rice bran enzymatically treated with an enzyme combination of alcalase, cellulose and phytase was incorporated into wheat flour mixed by enhancing potential fiber content. Results showed that either raw or enzyme treated rice bran was fit for breadmaking and with expects better sensory and nutritional values (Teng et al., 2015).

Recent research shows that microbial TGase could be used to detoxify gluten either by selectively modifying glutamine residues of intact gluten by transamidation with lysine methyl ester or by crosslinking gluten peptides in beverages via isopeptide bonds, so that they can be removed by filtration (Wieser and Koehler, 2012). Other enzymes have been tested in gluten-free breads, like commercial preparations of laccase, glucose oxidase and protease (Renzetti et al., 2010). Protease added individually (Kawamura-Konishi et al., 2013) or in combination with glucose oxidase (Renzetti and Arendt, 2009) has also been studied in rice flour and different commercial gluten-free flours (buckwheat, corn, sorghum and teff), respectively, in order to evaluate their impact on the breadmaking performance. In this trend, the addition of different groups of enzymes in gluten-free bread production was investigated in order to improve the quality of the final product, mainly by crosslinking proteins (Rosell, 2009). Most recently, the potential use of cyclodextrin-glycosyltransferase (CGTase) enzyme produced by *Bacillus firmus* strain 37 in breadmaking and the development of gluten-free bread with pinion and corn flours were investigated by Basso et al. (2015). The addition of CGTase increased the specific volume, improved the texture and the sensory characteristics of the bread. Reminding the action of CGTase, cyclization is the specific enzymatic reaction that releases cyclodextrins (CDs), the most common being CDs α-, β-, and γ-CD consisting of six, seven and eight glucose monomers in cycles, respectively. Corn starch samples were modified with CGTase below the gelatinization temperature to obtain porous, partially hydrolyzed, granules with CDs that may be used as an alternative to modified corn starches in foods applications (Dura et al., 2016). Benefits of containing mainly β-CD influence pasting behavior and may impede the orientation of amylases slowing hydrolysis, which may help to maintain lower blood glucose levels in mice. CDs are granted GRAS (Generally Recognized as Safe), considering the presence of CDs in a food formulation to improve or modify the characteristics of the final product that may lead to starch samples which are more slowly digested.

CGTase has been also used for treatment of native and extruded wheat flours (Román et al., 2016). The level of released CDs revealed that γ-CD was predominant, followed by α-CD, whereas very low β-CD values were obtained probably due to the formation of CD–lipid complexes. CDs are non-toxic ingredients; they are not absorbed in the upper gastrointestinal tract and are completely metabolized by the colon microflora. They have been extensively used in the pharmaceutical and food industry (Del Valle, 2004; Szente and Szejtli, 2004). The presence of CDs in cereal-based products could reveal new alternatives to obtain a healthy product with an additional benefit.

6. Conclusion and Future Prospects

Enzymes are very useful in bakery as processing aids in improving the dough performance and bread quality properties. With new concerns over chemical additives, enzymes are acquiring even more importance because they can lead to green label bakery products. In addition, increasing awareness about nutrition, has prompted research to focus on the potential role of enzymes improving the health properties of baked goods. Therefore, the initial use of enzymes as processing aids is now being extended to healthy aids.

Acknowledgements

Authors acknowledge the financial support of the Spanish Ministry of Economy and Competitiveness (Project AGL2011-23802 and AGL2014-52928-C2-1-R), the European Regional Development Fund (FEDER) and GeneralitatValenciana (Project Prometeo 2012/064).

References

Ahmad, Z., Butt, M.S., Ahmed, A., Riaz, M., Sabir, S.M., Farooq, U. and Rehman, F.U. 2012. Effect of *Aspergillus niger* xylanase on dough characteristics and bread quality attributes. Journal of Food Science and Technology, 49(6): 737–744.

Altamirano-Fortoul, R., Hernando, I. and Rosell, C.M. 2014. Influence of amyloglucosidase in bread crust properties. Food Bioprocess Technology, 7(4): 1037–1046.

Aravindan, R., Anbumathi, P. and Viruthagiri, T. 2007. Lipase applications in food industry. Indian Journal of Biotechnology, 6 (2): 141–158.

Arendt, E.K., Morrissey, A., Moore, M.M. and Fabio Dal Bello, F. 2008. Gluten-free breads. pp. 289–319. *In*: E.K. Arendt and F. Dal Bello (eds.). Gluten-Free Cereal Products and Beverages. Academic Press.

Armero, E. and Collar, C. 1996. Antistaling additive effects on fresh wheat bread quality. Food Science and Technology International, 2(5): 323–333.

Autio, K., Harkonen, H., Parkkonen, T., Frigard, T., Poutanen, K., Siikaaho, M. and Aman, P. 1996. Effects of purified endo-β-xylanase and endo-β-glucanase on the structural and baking characteristics of rye doughs. LWT—Food Science and Technology, 29(1-2): 18–27.

Autio, K., Kruus, K., Knaapila, A., Gerber, N., Flander, L. and Buchert, J. 2005. Kinetics of transglutaminase-induced cross-linking of wheat proteins in dough. Journal of Agriculture and Food Chemistry, 53(4): 1039–1045.

Baskinskiene, L., Garmuviene, S., Juodeikiene, G. and Haltrich, D. 2007. Comparison of different fungal xylanases for wheat breadmaking. Getreidetechnologie, 61(4): 228–235.

Basman, A., Koksel, H. and Ng, P.K.W. 2002. Effects of increasing levels of transglutaminase on the rheological properties and bread quality characteristics of two wheat flours. European Food Research and Technology, 215(5): 419–424.

Basso, F.M., Mangolim, C.S., Aguiar, M.F.A., Monteiro, A.R.G., Peralta, R.M. and Matioli, G. 2015. Potential use of cyclodextrin-glycosyltransferase enzyme in breadmaking and the development of gluten-free breads with pinion and corn flours. International Journal of Food Science and Nutrition, 66(3): 275–281.

Beck, M., Jekle, M., Selmair, P.L., Koehler, P. and Becker, T. 2011. Rheological properties and baking performance of rye dough as affected by transglutaminase. Journal of Cereal Science, 54(1): 29–36.

Bharath Kumar, S. and Prabhasankar, P. 2014. Low glycemic index ingredients and modified starches in wheat-based food processing: A review. Trends in Food Science & Technology, 35(1): 32–41.

Bilgicli, N., Demir, M.K. and Yilmaz, C. 2014. Influence of some additives on dough and bread properties of a wheat-lupin flour blend. Quality Assurance and Safety of Crops & Foods, 6(2): 167–173.

Björck, I., Östman, E., Kristensen, M., Mateo Anson, N., Price, R.K., Haenen, G.R.M.M., Havenaar, R., Bach Knudsen, K.E., Frid, A., Mykkänen, H., Welch, R.W. and Riccardi, G. 2012. Cereal grains for nutrition and health benefits: Overview of results from *in vitro*, animal and human studies in the HEALTHGRAIN project. Trends in Food Science & Technology, 25(2): 87–100.

Błaszczak, W., Sadowska, J., Rosell, C.M. and Fornal, J. 2004. Structural changes in the wheat dough and bread with the addition of alpha-amylases. European Food Research and Technology, 219(4): 348–354.

Bonet, A., Rosell, C.M., Caballero, P.A., Gómez, M., Pérez-Munuera, I. and Lluch, M.A. 2006. Glucose oxidase effect on dough rheology and bread quality: A study from macroscopic to molecular level. Food Chemistry, 99(2): 408–415.

Bonet, A., Rosell, C.M., Pérez-Munuera, I. and Hernando, I. 2007. Rebuilding gluten network of damaged wheat by means of glucose oxidase treatment. Journal of Science of Food and Agriculture, 87(7): 1301–1307.

Bosmans, G.M., Lagrain, B., Fierens, E. and Delcour, J.A. 2013. The impact of baking time and bread storage temperature on bread crumb properties. Food Chemistry, 141(4): 3301–3308.

Bowles, L.K. 1996. Amylolytic enzymes. pp. 105–129. *In*: R.E. Hebeda and H.F. Zobel (eds.). Baked goods freshness: Technology, evaluation, and inhibition of staling. Marcel Dekker, New York.

Butt, M.S., Tahir-Nadeem, M., Ahmad, Z. and Sultan, M.T. 2008. Xylanases and their applications in baking industry. Food Technology and Biotechnology, 46(1): 22–31.

Caballero, P.A., Gomez, M. and Rosell, C.M. 2007. Bread quality and dough rheology of enzyme-supplemented wheat flour. European Food Research and Technology, 224(5): 525–534.

Caballero, P.A., Gómez, M. and Rosell, C.M. 2007. Improvement of dough rheology, bread quality and bread shelf-life by enzymes combination. Journal of Food Engineering, 81(1): 42–53.

Cauvain, S.P. (ed.). 2012. Breadmaking: Improving Quality, 2nd Edition. Woodhead Publishing, London.

Cauvain, S.P. and Young, L.S. 2007. Technology of Breadmaking, 2nd edition, Springer.

Champ, M., Langkilde, A.M., Brouns, F., Kettlitz, B. and Collet, Y.L. 2003. Advances in dietary fiber characterisation. 1. Definition of dietary fibre, physiological relevance, health benefits and analytical aspects. Nutrition Research Reviews, 16(1): 71–82.

Collar, C., Andreu, P. and Martinez-Anaya, M.A. 1998. Interactive effects of flour, starter and enzyme on bread dough machinability. Zeitschrift Fur Lebensmittel-Untersuchung Und-Forschung a-Food Research and Technology, 207(2): 133–139.

Collar, C., Martinez, J.C., Andreu, P. and Armero, E. 2000. Effects of enzyme associations on bread dough performance. A response surface analysis. Food Science and Technology International, 6(3): 217–226.

Collins, T., Hoyoux, A., Dutron, A., Georis, J., Genot, B., Dauvrin, T., Arnaut, F., Gerday, C. and Feller, G. 2006. Use of glycoside hydrolase family 8 xylanases in baking. Journal of Cereal Science, 43(1): 79–84.

Cornish-Bowden, A. 2014. Current IUBMB recommendations on enzyme nomenclature and kinetics. Perspectives in Science, 1(1–6): 74–87.

Damen, B., Pollet, A., Dornez, E., Broekaert, W.F., Van Haesendonck, I., Trogh, I., Arnaut, F., Delcour, J.A. and Courtin, C.M. 2012. Xylanase-mediated *in situ* production of arabinoxylan oligosaccharides with prebiotic potential in whole meal breads and breads enriched with arabinoxylan rich materials. Food Chemistry, 131(1): 111–118.

Degrand, L., Rakotozafy, L. and Nicolas, J. 2015. Activity of carbohydrate oxidases as influenced by wheat flour dough components. Food Chemistry, 181: 333–338.

Del Valle, E.M.M. 2004. Cyclodextrins and their uses: A review. Process Biochemistry, 39(9): 1033–1046.

Dornez, E., Gebruers, K., Delcour, J.A. and Courtin, C.M. 2009. Grain-associated xylanases: Occurrence, variability, and implications for cereal processing. Trends in Food Science & Technology, 20(11-12): 495–510.

Dornez, E., Verjans, P., Arnaut, F., Delcour, J.A. and Courtin, C.M. 2011. Use of psychrophilic xylanases provides insight into the xylanase functionality in bread making. Journal of Agriculture and Food Chemistry, 59(17): 9553–9562.

Driss, D., Bhiri, F., Siela, M., Bessess, S., Chaabouni, S. and Ghorbel, R. 2013. Improvement of breadmaking quality by xylanase gh11 from *Penicillium occitanis* Pol6. Journal of Texture Studies, 44(1): 75–84.

Dunnewind, B., van Vliet, T. and Orsel, R. 2002. Effect of oxidative enzymes on bulk rheological properties of wheat flour doughs. Journal of Cereal Science, 36(3): 357–366.

Dura, A., Yokoyama, W. and Rosell, C. 2016. Glycemic response to corn starch modified with cyclodextrin glycosyltransferase and its relationship to physical properties. Plant Foods for Human Nutrition (In Press).

Eliasson, A.C. and Larsson, K. 1993. Cereals in breadmaking: A Molecular Colloidal Approach, Marcel Dekker, New York, 373 pp.

Fadda, C., Sanguinetti, A.M., Del Caro, A., Collar, C. and Piga, A. 2014. Bread staling: Updating the view. Comprehensive Review in Food Science and Food Safety, 13(4): 473–492.

FAO. 2012. Food Outlook: Global Market Analysis. Food and Agriculture Organization of the United Nations. http://www.fao.org/worldfoodsituation/csdb/en/.

Filipcev, B., Simurina, O. and Bodroza-Solarov, M. 2014. Combined effect of xylanase, ascorbic and citric acid in regulating the quality of bread made from organically grown spelt cultivars. Journal of Food Quality, 37(3): 185–195.

Fuentes-Zaragoza, E., Riquelme-Navarrete, M.J., Sánchez-Zapata, E. and Pérez-Álvarez, J.A. 2010. Resistant starch as functional ingredient: A review. Food Research International, 43(4): 931–942.

Gelinas, P., Poitras, E., McKinnon, C.M. and Morin, A. 1998. Oxido-reductases and lipases as dough-bleaching agents. Cereal Chemistry, 75(6): 810–814.

Gerits, L.R., Pareyt, B. and Delcour, J.A. 2014. A lipase-based approach for studying the role of wheat lipids in breadmaking. Food Chemistry, 156: 190–196.

Gerrard, J.A., Fayle, S.E., Wilson, A.J., Newberry, M.P., Ross, M. and Kavale, S. 1998. Dough properties and crumb strength of white pan bread as affected by microbial transglutaminase. Journal of Food Science, 63(3): 472–475.

Gil, M.J., Callejo, M.J., Rodriguez, G. and Ruiz, M.V. 1999. Keeping qualities of white pan bread upon storage: Effect of selected enzymes on bread firmness and elasticity. European Food Research and Technology, 208(5/6): 394–399.

Goesaert, H., Brijs, K., Veraverbeke, W.S., Courtin, C.M., Gebruers, K. and Delcour, J.A. 2005. Wheat flour constituents: How they impact bread quality and how to impact their functionality. Trends in Food Science & Technology, 16(1-3): 12–30.

Goesaert, H., Leman, P., Bijttebier, A. and Delcour, J.A. 2009. Antifirming effects of starch degrading enzymes in bread crumb. Journal of Agriculture and Food Chemistry, 57(6): 2346–2355.

Goesaert, H., Slade, L., Levine, H. and Delcour, J. 2009. Amylases and bread firming—An integrated view. Journal of Cereal Science, 50(3): 345–352.

Gomes-Ruffi, C.R., da Cunha, R.H., Almeida, E.L., Chang, Y.K. and Steel, C.J. 2012. Effect of the emulsifier sodium stearoyl lactylate and of the enzyme maltogenic amylase on the quality of pan bread during storage. LWT—Food Science and Technology, 49(1): 96–101.

Gray, J.A. and BeMiller, J.N. 2003. Bread staling: Molecular basis and control. Comprehensive Review in Food Science and Food Safety, 2(1): 1–21.

Gujral, H.S., Guardiola, I., Carbonell, J.V. and Rosell, C.A. 2003. Effect of cyclodextrinase on dough rheology and bread quality from rice flour. Journal of Agriculture and Food Chemistry, 51(13): 3814–3818.

Gujral, H.S. and Rosell, C.M. 2004a. Functionality of rice flour modified with a microbial transglutaminase. Journal of Cereal Science, 39(2): 225–230.

Gujral, H.S. and Rosell, C.M. 2004b. Improvement of the breadmaking quality of rice flour by glucose oxidase. Food Research International, 37(1): 75–81.

Haros, M., Rosell, C.M. and Benedito, C. 2002a. Effect of different carbohydrases on fresh bread texture and bread staling. European Food Research and Technology, 215(5): 425–430.

Haros, M., Rosell, C.M. and Benedito, C. 2002b. Improvement of flour quality through carbohydrases treatment during wheat tempering. Journal of Agriculture and Food Chemistry, 50(14): 4126–4130.

He, H. and Hoseney, R.C. 1991. Gas retention of different cereal flours. Cereal Chemistry, 68(4): 334–336.

Hemalatha, M.S., Bhagwat, S.G., Salimath, P.V. and Rao, U. 2012. Enhancement of soluble dietary fibre, polyphenols and antioxidant properties of chapatis prepared from whole wheat flour dough treated with amylases and xylanase. Journal of Science of Food and Agriculture, 92(4): 764–771.

Hsing, I.L., Faubion, J.M. and Walker, C.E. 2010. Using enzyme-oxidant combinations to improve frozen-dough bread quality. Getreidetechnologie, 64(2): 124–134.

Hug-Iten, S., Escher, F. and Conde-Petit, B. 2003. Staling of bread: Role of amylose and amylopectin and influence of starch-degrading enzymes. Cereal Chemistry, 80(6): 654–661.

Indrani, D., Prabhasankar, P., Rajiv, J. and Rao, G.V. 2003. Scanning electron microscopy, rheological characteristics and bread-baking performance of wheat-flour dough as affected by enzymes. Journal of Food Science, 68(9): 2804–2809.

Katina, K., Salmenkallio-Marttila, M., Partanen, R., Forssell, P. and Autio, K. 2006. Effects of sourdough and enzymes on staling of high-fiber wheat bread. LWT—Food Science and Technology, 39(5): 479–491.

Kawamura-Konishi, Y., Shoda, K., Koga, H. and Honda, Y. 2013. Improvement in gluten-free rice bread quality by protease treatment. Journal of Cereal Science, 58(1): 45–50.

Ktenioudaki, A., Alvarez-Jubete, L., Smyth, T.J., Kilcawley, K., Rai, D.K. and Gallagher, E. 2015. Application of bioprocessing techniques (sourdough fermentation and technological aids) for brewer's spent grain breads. Food Research International, 73: 107–116.

Kukurova, K., Morales, F.J., Bednarikova, A. and Ciesarova, Z. 2009. Effect of L-asparaginase on acrylamide mitigation in a fried-dough pastry model. Molecular Nutrition & Food Research, 53(12): 1532–1539.

Li, Z., Dong, Y., Zhou, X.H., Xiao, X., Zhao, Y.S. and Yu, L.T. 2014. Dough properties and bread quality of wheat-barley composite flour as affected by β-glucanase. Cereal Chemistry, 91(6): 631–638.

Marco, C. and Rosell, C.M. 2008a. Breadmaking performance of protein-enriched, gluten-free breads. European Food Research and Technology, 227(4): 1205–1213.

Marco, C. and Rosell, C.M. 2008b. Functional and rheological properties of protein-enriched gluten-free composite flours. Journal of Food Engineering, 88(1): 94–103.

McPhillips, K., Waters, D.M., Parlet, C., Walsh, D.J., Arendt, E.K. and Murray, P.G. 2014. Purification and characterisation of a β-1,4-Xylanase from *Remersonia thermophila* CBS 540.69 and its application in breadmaking. Applied Biochemistry and Biotechnology, 172(4): 1747–1762.

Moayedallaie, S., Mirzaei, M. and Paterson, J. 2010. Bread improvers: Comparison of a range of lipases with a traditional emulsifier. Food Chemistry, 122(3): 495–499.

Mohan, N.S., Shimray, C.A., Indrani, D. and Manonmani, H.K. 2013. Reduction of acrylamide formation in sweet bread with l-asparaginase treatment. Food Bioprocess and Technology, 7(3): 741–748.

Motoki, M. and Seguro, K. 1998. Transglutaminase and its use for food processing. Trends in Food Science & Technology, 9(5): 204–210.

Novozymes, A.S. 2001. Combined use of glucose oxidase and phospholipase for baking. Research Disclosure (No. 442): 231.

Oliveira, D.S., Telis-Romero, J., Da-Silva, R. and Franco, C.M. 2014. Effect of a *Thermoascus aurantiacus* thermostable enzyme cocktail on wheat bread quality. Food Chemistry, 143: 139–46.

Oort, M.v. 1996. Oxidases in baking. A review of the uses of oxidases in breadmaking. International Food Ingredients (No. 4): 42–44.

Ozer, M.S., Kola, O. and Duran, H. 2010. Effects of buckwheat flour combining phospholipase or DATEM on dough properties. Journal of Food, Agriculture & Environment, 8(2): 13–16.

Patel, M.J., Ng, J.H.Y., Hawkins, W.E., Pitts, K.F. and Chakrabarti-Bell, S. 2012. Effects of fungal α-amylase on chemically leavened wheat flour doughs. Journal of Cereal Science, 56(3): 644–651.

Poulson, C.H., Soe, J.B., Rasmussen, P., Madrid, S.M. and Zargahi, M.R. 2010. Lipase and Use of Same for Improving Doughs and Baked Products. Danisco A/S.

Poutanen, K. 1997. Enzymes: An important tool in the improvement of the quality of cereal foods. Trends in Food Science & Technology, 8(9): 300–306.

Primo-Martin, C., de Beukelaer, H., Hamer, R.J. and Van Vliet, T. 2008. Fracture behaviour of bread crust: Effect of ingredient modification. Journal of Cereal Science, 48(3): 604–612.

Primo-Martin, C., de Pijpekamp, A.V., van Vliet, T., de Jongh, H.H.J., Plijter, J.J. and Hamer, R.J. 2006. The role of the gluten network in the crispness of bread crust. Journal of Cereal Science, 43(3): 342–352.

Renzetti, S. and Arendt, E.K. 2009. Effect of protease treatment on the baking quality of brown rice bread: From textural and rheological properties to biochemistry and microstructure. Journal of Cereal Science, 50(1): 22–28.

Renzetti, S., Courtin, C.M., Delcour, J.A. and Arendt, E.K. 2010. Oxidative and proteolytic enzyme preparations as promising improvers for oat bread formulations: Rheological, biochemical and microstructural background. Food Chemistry, 119(4): 1465–1473.

Renzetti, S., Dal Bello, F. and Arendt, E.K. 2008. Microstructure, fundamental rheology and baking characteristics of batters and breads from different gluten-free flours treated with a microbial transglutaminase. Journal of Cereal Science, 48(1): 33–45.

Renzetti, S. and Rosell, C.M. 2015. Role of enzymes in improving the functionality of proteins in non-wheat dough systems. Journal of Cereal Science. In Press, Corrected Proof, Available online 25 September *2015*.

Rieder, A., Holtekjolen, A.K., Sahlstrom, S. and Moldestad, A. 2012. Effect of barley and oat flour types and sourdoughs on dough rheology and bread quality of composite wheat bread. Journal of Cereal Science, 55(1): 44–52.

Román, L., Dura, Á., Martínez, M.M., Rosell, C.M. and Gómez, M. 2016. Combination of extrusion and cyclodextrin glucanotransferase treatment to modify wheat flours functionality. Food Chemistry, 199: 287–295.

Rosell, C. 2012. The nutritional enhancement of wheat flour. pp. 687–710. *In*: S. Cauvain (ed.). Breadmaking: Improving Quality, Second Edition. Woodhead Publishing, UK.

Rosell, C. 2014. Enzymes in Gluten-free Bread. AACC International Annual Meeting 2014, Providence, Rhode Island, US October 5–8, 2014.

Rosell, C. and Dura, A. 2015. Enzymes in bakeries. pp. 171–195. *In*: M. Chandrasekaran (ed.). Enzymes in Food and Beverages Processing. CRC Press. Florida.

Rosell, C. and Garzon, R. 2015. Chemical composition of bakery products. *In*: P.C.K. Cheung and B.M. Mehta (eds.). Handbook of Food Chemistry. Springer-Verlag Berlin Heidelberg.

Rosell, C.M. 2009. Enzymatic manipulation of gluten-free breads. pp. 83–98. *In*: E. Gallagher (ed.). Gluten-free Food Science and Technology. Wiley-Blackwell. London.

Rosell, C.M. 2011. The science of doughs and bread quality. pp. 3–14. *In*: R.R. Watson, V.R. Preedy and V.B. Patel (eds.). Flour and Breads and their Fortification in Health and Disease Prevention. Academic Press, London.

Rosell, C.M. and Collar, C. 2008. Effect of various enzymes on dough rheology and bread quality. pp. 165–183. *In*: R. Porta, P. Di Pierro and L. Mariniello (eds.). Enzymes as additives or processing aids. Research Signpost, Kerala, India.

Rosell, C.M. and Collar, C. 2009. Effect of temperature and consistency on wheat dough performance. International Journal of Food Science and Technology, 44(3): 493–502.

Rosell, C.M., Haros, M., Escrivá, C. and Benedito de Barber, C. 2001. Experimental approach to optimize the use of α-amylases in breadmaking. Journal of Agriculture and Food Chemistry, 49(6): 2973–2977.

Rosell, C.M., Rojas, J.A. and de Barber, C.B. 2001. Combined effect of different antistaling agents on the pasting properties of wheat flour. European Food Research and Technology, 212(4): 473–476.

Rosell, C.M., Santos, E., Sanz Penella, J.M. and Haros, M. 2009. Wholemeal wheat bread: A comparison of different breadmaking processes and fungal phytase addition. Journal of Cereal Science, 50(2): 272–277.

Rosell, C.M., Wang, J., Aja, S., Bean, S. and Lookhart, G. 2003. Wheat flour proteins as affected by transglutaminase and glucose oxidase. Cereal Chemistry, 80(1): 52–55.

Saarinen, M.T., Lahtinen, S.J., Sorensen, J.F., Tiihonen, K., Ouwehand, A.C., Rautonen, N. and Morgan, A. 2012. Treatment of bran containing bread by baking enzymes; effect on the growth of probiotic bacteria on soluble dietary fiber extract *in vitro*. Bioscience Biotechnology and Biochemistry, 76(6): 1135–1139.

Sahnoun, M., Naili, B., Elgharbi, F., Kammoun, R., Gabsi, K. and Bejar, S. 2013. Effect of *Aspergillus oryzae* CBS 819.72 α-amylase on rheological dough properties and bread quality. Biologia, 68(5): 808–815.

Sanz Penella, J.M., Collar, C. and Haros, M. 2008. Effect of wheat bran and enzyme addition on dough functional performance and phytic acid levels in bread. Journal of Cereal Science, 48(3): 715–721.

Selinheimo, E., Kruus, K., Buchert, J., Hopia, A. and Autio, K. 2006. Effects of laccase, xylanase and their combination on the rheological properties of wheat doughs. Journal of Cereal Science, 43(2): 152–159.

Shafisoltani, M., Salehifar, M. and Hashemi, M. 2014. Effects of enzymatic treatment using response surface methodology on the quality of bread flour. Food Chemistry, 148: 176–83.

Shewry, P.R., Tatham, A.S., Forde, J., Kreis, M. and Miflin, B.J. 1986. The classification and nomenclature of wheat gluten proteins—A reassessment. Journal of Cereal Science, 4(2): 97–106.

Song, Y. and Zheng, Q. 2007. Dynamic rheological properties of wheat flour dough and proteins. Trends in Food Science & Technology, 18(3): 132–138.

Stefanis, V.A.D. and Turner, E.W. 1981. Modified enzyme system to inhibit bread firming method for preparing same and use of same in bread and other bakery products. International Telephone & Telegraph Corp.

Steffolani, M.E., Ribotta, P.D., Pérez, G.T. and León, A.E. 2010. Effect of glucose oxidase, transglutaminase, and pentosanase on wheat proteins: Relationship with dough properties and breadmaking quality. Journal of Cereal Science, 51(3): 366–373.

Stoica, A., Barascu, E. and Hossu, A.-M. 2013. Improving the quality of bread made from 'short' gluten flours using a fungal pentosanase and L-cysteine combination. Revista De Chimie, 64(9): 951–954.

Stoica, A., Hossu, A.M., Barascu, E., Iordan, M. and Maria, M.F. 2010. Influence of the fungal alpha-amylase and glucose oxidase combinations on the physical properties of bread made from weak white flours. Revista De Chimie, 61(8): 724–727.

Stojceska, V. and Ainsworth, P. 2008. The effect of different enzymes on the quality of high-fiber enriched brewer's spent grain breads. Food Chemistry, 110(4): 865–872.

Storck, C.R., da Rosa Zavareze, E., Gularte, M.A., Elias, M.C., Rosell, C.M. and Guerra Dias, A.R. 2013. Protein enrichment and its effects on gluten-free bread characteristics. LWT—Food Science and Technology, 53(1): 346–354.

Szente, L. and Szejtli, J. 2004. Cyclodextrins as food ingredients. Trends in Food Science & Technology, 15(3-4): 137–142.

Tang, L.L., Yang, R.J., Hua, X., Yu, C.H., Zhang, W.B. and Zhao, W. 2014. Preparation of immobilized glucose oxidase and its application in improving breadmaking quality of commercial wheat flour. Food Chemistry, 161: 1–7.

Teng, Y.F., Liu, C.Y., Bai, J. and Liang, J.F. 2015. Mixing, tensile and pasting properties of wheat flour mixed with raw and enzyme treated rice bran. Journal of Food Science and Technology, 52(5): 3014–3021.

Whitehurst, R. and Oort, M. 2010. Enzymes in Food Technology, Wiley, VCH.

WHO. 2003. Food-Based Dietary Guidelines in the WHO European Region. World Health Organization.

Wieser, H. and Koehler, P. 2012. Detoxification of gluten by means of enzymatic treatment. Journal of AOAC International, 95(2): 356–363.

Wu, S.P., Liu, Y., Yan, Q.J. and Jiang, Z.Q. 2014. Gene cloning, functional expression and characterisation of a novel glycogen branching enzyme from *Rhizomucor miehei* and its application in wheat breadmaking. Food Chemistry, 159: 85–94.

Yang, T.Y., Bai, Y.X., Wu, F.F., Yang, N.J., Zhang, Y.J., Bashari, M., Jin, Z.Y. and Xu, X.M. 2014. Combined effects of glucose oxidase, papain and xylanase on browning inhibition and characteristics of fresh whole wheat dough. Journal of Cereal Science, 60(1): 249–254.

Zhang, C., Zhang, S., Lu, Z., Bie, X., Zhao, H., Wang, X. and Lu, F. 2013. Effects of recombinant lipoxygenase on wheat flour, dough and bread properties. Food Research International, 54(1): 26–32.

17

Enzymes in Winemaking

José Manuel Rodríguez-Nogales,[1,a,*]
Encarnación Fernández-Fernández[1,b] *and Josefina Vila-Crespo*[2]

1. Introduction

The biological process of winemaking is the result of a series of biochemical transformations brought about by the action of several enzymes. Many of these enzymes originate from the grape itself, the indigenous microflora on the grape and the microorganisms present during winemaking. Moreover, since the endogenous enzymes of grapes, yeasts and other microorganisms present in must and wine are often neither efficient nor sufficient under winemaking conditions, commercial enzyme preparations are also widely used as supplements (Moreno-Arribas and Polo, 2005).

This chapter focuses on the main commercial preparations of enzymes used in oenology, some of them are very commonly used by winemakers (i.e., pectinases) while some others are not so widely used (urease). Typical white and red winemaking processes (showing the stage at which enzymes are added) are illustrated in Figs. 1 and 2, respectively.

2. Enzymes to Improve Maceration

Color is one of the most important attributes in the initial assessment of red wine. High-colored wines are usually associated with high perceived quality scores (Parpinello

---

[1] Área de Tecnología de Alimentos, Universidad de Valladolid, Escuela Técnica Superior de Ingenierías Agrarias, Av. Madrid 44, 34071 Palencia, Spain.
[a] E-mail: rjosem@iaf.uva.es
[b] E-mail: effernan@iaf.uva.es
[2] Área de Microbiología, Universidad de Valladolid, Escuela Técnica Superior de Ingenierías Agrarias, Av. Madrid 44, 34071 Palencia, Spain.
E-mail: jvila@pat.uva.es
* Corresponding author

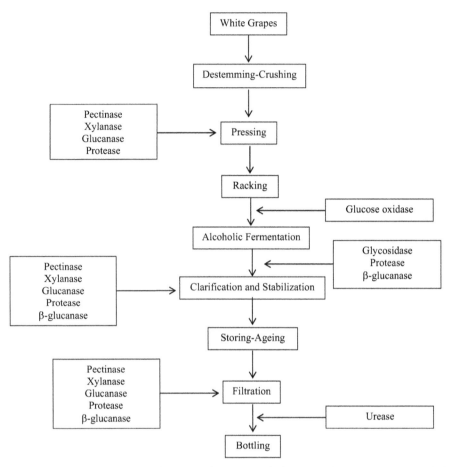

Figure 1. Scheme white wines. Typical production process of white wines showing the stages at which enzymes are added.

et al., 2009). Anthocyanins contribute strongly to the sensory quality of red wines because these molecules and their interactions with other phenolic compounds are responsible for the color and its stability during ageing (Boulton, 2001; Fulcrand et al., 2006; Monagas et al., 2006). Anthocyanins are mainly located inside the skin cell vacuoles and are partially extracted from the berry skin into the must/wine during winemaking (González-Neves et al., 2008). The anthocyanin content and composition of red wines depend on the amount of pigments in the berry skin at harvest and on the ease of their extraction. Although the qualitative and quantitative composition of anthocyanins in wine is directly related to the wine grape variety, ripening stage, culture practices, growing season and environmental conditions, the oenological practices also play an important role (González-Neves et al., 2012; González-Neves et al., 2008; Sacchi et al., 2005).

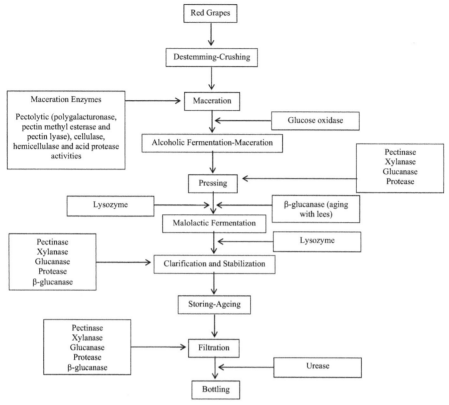

Figure 2. Scheme red wines. Typical production process of red wines showing the stages at which enzymes are added.

Exogenous enzymes are widely used in red winemaking, attempting to accelerate the extraction of anthocyanins from the berry skin and thus increasing the color intensity of the resulting wine (Bautista-Ortín et al., 2005; Gil-Muñoz et al., 2009; Ortega-Heras et al., 2012; Romero-Cascales et al., 2012; Soto Vázquez et al., 2010). The commercial enzyme preparations mainly show pectolytic (polygalacturonase, pectin methyl esterase and pectin lyase) cellulase, hemicellulase and acid protease activities (Romero-Cascales et al., 2008; Bautista-Ortín et al., 2005).

Maceration enzymes degrade the berry skin pecto-cellulosic cell walls by partial hydrolysis of structural polysaccharides. Therefore, the permeability of the cell wall increases facilitating the diffusion process of anthocyanins from the vacuoles into the must during fermentation (Romero-Cascales et al., 2008, 2012).

Contradictory results were reported about the impact of maceration enzymes on the anthocyanin content and color intensity in red wines (Sacchi et al., 2005). The discrepancies are probably due to the different nature and activities of the commercial enzyme preparations (Bautista-Ortín et al., 2005; Romero-Cascales et al., 2008), as well as to the varietal and vintage effects on the grape anthocyanin content and composition or on the skin cell wall morphology and composition (Bautista-Ortín et al., 2007; Ducasse et al., 2010; Ortega-Heras et al., 2012).

The differences in the mechanical properties of the berry skin are also linked to variations in the chemical composition of the cell walls, which determine the resistance of the skin to the anthocyanin release (Hernández-Hierro et al., 2014). In fact, berry skin hardness and berry skin thickness influence the rate and extent of the anthocyanin extractability (Río Segade et al., 2011; Rolle et al., 2012). Although it is well known that degradation of the cell walls causes the skin softening (Ortega-Regules et al., 2008), the effect of maceration enzymes on the skin mechanical properties has only been quantified to date by Río-Segade et al. (2015), who found that enzyme preparation influenced the mechanical properties of the berry skin, increasing the softening that naturally occurs during maceration as a result of the degradation process. That effect of enzymes on the skin hardness was instrumentally quantified for the first time and the mechanical properties of the skin may be considered predictors of the extraction yield of anthocyanins during maceration. Furthermore, the use of enzymes permitted increase in the extraction yield of anthocyanins, in shortening the maceration time, and in preventing the loss of anthocyanins released during maceration. However, a variety effect was found depending on the anthocyanin composition of wine grapes.

Maceration affects the anthocyanins release, chromatic characteristics and color stability in red wines. A longer maceration time usually contributes to a greater anthocyanin extraction from the skins and improves the color stability of the wine (González-Neves et al., 2008; Kelebek et al., 2009; Romero-Cascales et al., 2005; Romero-Cascales et al., 2012; Sacchi et al., 2005). Nevertheless, this relationship may be affected by the participation of extracted anthocyanins in oxidation and polymerization reactions occurring during the maceration process, their partial adsorption by the yeasts and their fixation on to the grape solid parts (Bautista-Ortín et al., 2007; González-Neves et al., 2008; Romero-Cascales et al., 2005, 2012).

Another factor to consider is the maceration temperature in red winemaking. Traditional red winemaking is characterized by a simultaneous development of alcoholic fermentation and maceration. New winemaking techniques have been proposed considering the selective effect of maceration conditions on the extraction of grape components and the reactions in which they participate (Martín and Morata de Ambrosini, 2014). Among these methods, Pre-fermentation Cold Maceration (PCM) is employed prior to fermentation to encourage extraction in an aqueous medium at a low temperature (5–15°C), which results in preferential solubility of water-soluble compounds and encourages selective extraction of anthocyanins and tannins of low molecular weight (Sacchi et al., 2005; Alvarez et al., 2006). Additionally, it results in a better-structured product—one that is richer in phenolic and aromatic compounds, thus preserving a strong relationship with the production area (Alvarez et al., 2006). On the other hand, fermentations conducted at low temperature (15–20°C) can increase the production and retention of volatile compounds, thus improving the aromatic profile of wines (Molina et al., 2007). White winemaking is traditionally handled at low temperature and it is currently gaining ground in red winemaking. However, low maceration temperatures can diminish color extraction in red wines—an effect that can be counteracted using cold-active pectinolytic enzymes as they ensure good color extraction at low temperature. Martín and Morata de Ambrosini (2013) reported a cold-active pectinolytic system obtained from a psychrotolerant *Bacillus* sp., which increased color extraction in short macerations with red grape skins at low temperature.

Recently the same authors (Martín and Morata de Ambrosini, 2014) showed that a pectinolytic enzymatic system obtained from *Bacillus* sp. CH15 bacterial strain (Cabeza et al., 2011) significantly accelerated color extraction by reducing the maceration time necessary for winemaking at low temperature and shortening the PCM stage. Enzyme-treated wines exhibited better chromatic parameters than their controls at devatting and after six months of storage. The cold-active enzyme compensated for the decrease in color extraction due to the low maceration temperature, achieving high-quality wines with chromatic characteristics, similar to those of traditional wines.

3. Enzymes to Improve Clarification, Filtration and Yield of Juice and Wine

The addition of these commercial enzymes (pectinases, xylanases, glucanases, proteases) to improve clarification and filtration (Forgaty et al., 1983; Cologrande et al., 1994; Ugliano, 2009) and to increase the pressing efficiency and juice extraction (Canal-Llaubères, 1993) is a common practice in winemaking since the 1970s. However, the number and variety of products available and the knowledge of their action mechanisms and effect on wine quality has evolved dramatically over the last few years.

Pectinases are the most important enzymes in winemaking. Together with cellulose, hemicellulose and lignin, pectic compounds form part of the grape cell wall and act as adherents between cells giving consistency to the cell wall (Bisson and Butzke, 1996; Vidal et al., 2003). The rupture of these cell structures favors the extraction of substances contained within the solid part of the grape, mainly in the pulp and in the skin. Pectolytic enzymes break down these compounds and improve the extraction and clarification processes of the must. They can also favor the extraction of substances that affect the aroma and color contained in the skin and in the pulp (Bakker et al., 1999; Pardo et al., 1999; Revilla and González-San José, 2001).

All these commercial enzyme preparations are obtained from microorganisms cultivated on substrates under conditions that optimize their production and facilitate their purification at a competitive cost (Van Rensburg and Pretorius, 2000).

Most existing commercial preparations are derived from different species of filamentous fungi, mainly *Aspergillus* spp., accepted as GRAS (Generally Recognised As Safe) and included in the International Code for Enological Practices of the *Office International of Vine and Wine* (O.I.V.). Many of them are mixtures of enzymes that fulfill more than one function in the process. Preparations based on pectinase, cellulase and hemicellulase enzymes are the most used in winemaking to improve clarification, filtration and yield of juice and wine. Often the commercial preparations available have, apart from desirable effects, other unwanted concomitant or side effects. These effects may include the oxidation of phenolic compounds during maceration (Fernández-Zurbano et al., 1999) and the production of methanol due to the pectin-methyl-esterase activity found in different pectolytic preparations (Revilla and González-San José, 1998), or even enhancement of the production of volatile phenols due to the presence of cinnamyl-esterase activity in most enzymatic preparations (Gerbaux et al., 2002). Other collateral effects refer to a considerable loss of proteins and glucidiccolloids (Lao et al., 1999). As a result, it would be useful to produce these enzymes in a cost-effective manner in order to develop innovative applications. The current trend consists of seeking alternative sources of pectinases, in particular from yeasts (Alimardani-Theuil et al., 2011). Yeasts have advantages over filamentous fungi in large-scale pectinase production. Some efforts have already been made to investigate whether yeast pectinases can be used in the food industry. *Saccharomyces* and *Kluyveromyces* yeasts possessing GRAS status are the most studied and seem to be the most promising. Therefore, their pectinases can be used to clarify fruit juices and wines while maintaining low methanol levels (Fernández-González et al., 2004, 2005).

In 1994, Gainvors et al. demonstrated that when added to grape must, a crude pectolytic extract from a *S. cerevisiae* strain had the same effect on the turbidity as the same quantity of a commercial pectinase preparation. Blanco et al. (1997) have shown that when wine fermentations are carried out using pectolytic strains from *S. cerevisiae*, the clarification process is greatly facilitated, with the filtration time being reduced up to 50 per cent in some cases.

Pretorius (2000) has predicted that pectolytic wine yeasts may improve liquefaction, clarification and filterability of grape must, releasing more color and flavor compounds entrapped in the grape skins and thereby making a positive contribution to the wine bouquet. In their work, Fernández-González et al. (2004) transformed a wine yeast strain with good enological qualities with the yeast PGU1 gene fused to the promoter of the PGK1 gene in order to enhance polygalacturonase (PG) expression during vinification (Fernández-González et al., 2005). When microvinification assays were performed on white and red musts treated with this transformed strain, there was a major improvement in the yield of must/wine extraction. However, Radoi et al. (2005) did not find the pectolytic recombinant strains to be any more efficient than pectolytic wild-type strains in winemaking.

3.1 β-glucanases to improve clarification and filtration

Of all the polysaccharides, β-glucans produced by *Botrytis cinerea* in botrytized grape juice can be regarded as the one with the greatest influence on the clarification and stabilization of must and wine. Fragmentation of β-glucans promoted by hydrolytic enzymes (i.e., β-glucanases) can enhance water solubility by reducing the degree of polymerization. β-D-glucans are constituted by a main chain formed by 3-D-glucose units joined together by β-1,3 glucosidic bonds in yeasts and fungi, such as *Saccharomyces cerevisiae, S. bayanus* and *Botrytis cinerea* or β-1,4 glucosidic bonds in cereals and by a variable number of lateral chains with different lengths that are linked to the main chain by β-1,6 or β-1,2 bonds. β-Glucanases represent a broad class of enzymes, which include both exo/endo forms of β-(1,6)-glucanases that are able to promote the hydrolysis of β-(1,6)-O-glycosidic bonds, connecting lateral branches with the main polymeric chain of β-glucans, as well as exo-β-(1,3) and endo-β-(1,3)-glucanases, which promote the lysis of terminal and internal β-(1,3)-bonds respectively (Venturi et al., 2013). Commercial preparations of these enzymes are utilized extensively in the brewing industry and in winemaking to facilitate the filtration of musts and wines, particularly those coming from grapes affected by *B. cinerea*.

β-Glucanase preparations are also employed at the end of the alcoholic fermentation to promote the lysis of the yeast cell walls, followed by release into the wine of their main constituents, including manno-proteins and oligomeric fragments of β-glucans (Parodi, 2002; Morata et al., 2003; Comuzzo et al., 2005; Palomero et al., 2007).

4. Assisted Wine Aging on Lees with β-Glucanase

Only white and sparkling wines were traditionally elaborated in contact with their lees. Nevertheless, this technology has recently been extended to other types of wine, like red ones since it is considered to improve the organoleptic perception of wines by the consumers. In these wines, once the alcoholic fermentation has finished, an aging on lees takes place for a variable period according to their particular features. Lees are mainly composed by microorganisms (yeasts and bacteria) and to a lesser extent by tartaric acid and inorganic matter (Perez-Serradilla and de Castro, 2008). During aging on lees, the autolysis of wine yeasts occurs, modifying the chemical and sensorial characteristics of the wine. Yeast autolysis involves the hydrolysis of biopolymers under the action of hydrolytic enzymes (β-glucanases and proteases among them) which release different compounds into the wine (Pozo-Bayon et al., 2009). Peptides, amino acids, fatty acids and nucleotides come from the yeast cytoplasm, while glucans and manno-proteins come from the yeast cell wall (Torresi et al., 2014). These compounds have a capital implication in the organoleptic properties and stability of the wines, and in the foaming of sparkling wines.

The period of aging on lees is variable according to the particular desired wine features, although a longer time positively contribute to the quality of the wine (Rodriguez-Nogales et al., 2012a). In order to reduce this long period, and therefore the risks of oxidation and microbial contamination and the high costs of production, enzymatic preparations rich in β-glucanases can be utilized. Exogenous glucanases can catalyze the hydrolysis of β-(1-3) and β-(1-6)-glycosidic bonds of the cell wall β-glucan chains, progressively degrading the cell wall and accelerating the yeast lysis (Zinnai et al., 2010).

Some studies have reported that this enzymatic treatment in wines aging on lees did not induce any major change either in their basic chemical composition (Masino et al., 2008; Rodriguez-Bencomo et al., 2010) or in the phenolic compounds, neutral and total polysaccharides, proteins and color in red wines (Del Barrio-Galan et al., 2011, 2012).

Little information is available in relation to the effect of the addition of exogenous β-glucanases to sparkling wine. A slight increase in the level of free amino was found, while this treatment did not substantially influence either the content of total proteins or the foam characteristics (Torresi et al., 2014). Rodriguez-Nogales et al. (2012a) showed that the enzyme treated-wines had a higher content of neutral polysaccharides and from a sensorial point of view, more volume, flavor intensity, yeast and dough smell and sweeter taste. Moreover, it has been proved that the addition of β-glucanases to sparkling wines increases their antioxidant properties (Rodriguez-Nogales et al., 2012b).

5. Enzymes for Enhancing Wine Aroma

Varietal aroma of grapes includes volatile free odor substances (mainly monoterpenes, C13-norisoprenoids, benzene derivatives and long-chain aliphatic alcohols) and odorless non-volatile glycoconjugates (Styger et al., 2011). Glycosidically bound volatiles are composed by glucose and also by rhamnose, arabinose or apiose in disaccharides glycosides (Hjelmeland and Ebeler, 2015). These glycosidic aroma precursors can give rise to odorous volatiles using exogenous glycosidases (Pogorzelski and Wilkowska, 2007). Usually, this process involves a two-step mode reaction. Firstly, α-rhamnosidase, α-arabinosidase or β-apiosidase release the terminal sugar (according to the sugar moieties of the substrates); secondly, a β-glucosidase liberates the aromatic aglycone and glucose (Palmeri and Spagna, 2007).

It has been demonstrated that the glycosidic aroma precursors remain stable during winemaking operations since the effect of grape and *S. cerevisiae* glycosidases is very limited (Maicas and Mateo, 2005). For that, the selection of active yeast, fungal and bacterial glycosidases to enhance the wine aroma has been extensively studied (Maicas and Mateo, 2005; Villena et al., 2006; Michlmayr et al., 2012). At present, commercial preparations, rich in glycosidases, are available from *Aspergillus niger* (Pogorzelski and Wilkowska, 2007). These fungal enzymes are active under winemaking conditions; however, because of their inhibition by glucose, they must be added at the end of fermentation.

In terms of industrial application, the commercial preparations must be free of cinnamate esterase activity in order to avoid the synthesis of off-aromas (like vinylphenols) in a synergistic reaction with a yeast decarboxylation. Moreover, application of these commercial preparations has to be limited to white wines to prevent loss of color from the hydrolysis of anthocyanins to anthocyanidins (Hjelmeland and Ebeler, 2015). Wang et al. (2013) have found a β-glucosidase from *Trichosporon asahii* with a minimal impact on the major anthocyanins, making possible its use in red wines.

The effectiveness of the treatment using commercial glycosidases, even with a broad range of enzymatic activities, is closely dependent on the aromatic potential of the grape. It has been suggested that the previous enrichment of wine in glycosides by skin-contact treatment of the must and the subsequent addition of glycosidic precursors significantly improved the wine's volatile composition (Cabaroglu et al., 2003).

Volatile thiols are also powerful odoriferous compounds identified in wine from different grape varieties (Sauvignon Blanc, Macabeo, Gewürztraminer, Riesling, Muscat, Verdejo, etc.). Mercapto-4-methylpentan-2-one (4MMP), 3-mercaptohexan-1-ol (3MH) and its acetate (3MHA) are the volatile thiols with more odorant impact in wines (Roland et al., 2011). It has been demonstrated that commercial yeast strains of *S. cerevisiae* release these volatile thiols through its β-lyase activity from the odorless precursors (cysteinylated and glutathionylated precursors) of the juices. However, their efficiency is very limited (Murat et al., 2001; Howell et al., 2004; Roncoroni, 2011). The use of exogenous β-lyase could, theoretically, enhance the volatile thiols of the wine; however, no commercial enzyme preparations are available yet.

6. Protein Haze Prevention Using Proteolytic Enzymes

Proteins are one of the main compounds of the grape juice and white wine, together with polysaccharides and polyphenols (Ferreira et al., 2001; Waters et al., 2005). Some of these proteins are responsible for the sediments and haze in white wines when affected by inadequate storage. The haze-forming proteins have been identified as pathogenesis-related (PR) proteins, involving mainly chitinases, thaumatin-like proteins (TLPs) and β-glucanases (Vincenzi et al., 2011). It has also been suggested that the wine pH plays an important role in this issue (Mesquita et al., 2001).

The mechanism of protein haze formation has been recently updated (Van Sluyter et al., 2015) and it involves a three-stage model. In the first stage, the increase of temperature provokes unfolding of the proteins and exposure of their hydrophobic binding sites. In the second stage, the unfolding proteins are self-aggregated *via* hydrophobic interactions. Finally, in the third stage, the protein aggregates gradually become crosslinked due to the actions of sulfates and polyphenols, flocculating into a hazy suspension and, in the end, formatting precipitates.

The treatment with bentonite is the most common strategy employed by winemakers to prevent wine haze (Ferreira et al., 2001). Nevertheless, a detrimental effect on wine aroma and flavor, loss of color, loss of wine volume, high handling costs and environmental impact of the residues have been attributed to the use of bentonite (Waters et al., 2005).

In order to reduce or eliminate the drawbacks of the bentonite, different strategies to prevent protein haze have been studied. However, the success of these treatments has not been completely reached, whether for their efficiency to eliminate haze or for their negative effects on the wine quality.

The use of proteolytic enzymes for the hydrolysis of the haze-forming proteins is an attractive alternative. Proteases catalyze the cleavage of hydrolytic bonds within proteins, thereby releasing peptides and/or amino acids (Tavano, 2013). It must be noted that proteases for oenological use have to work under extreme conditions, such as acid pH of 3–4, presence of inhibitors (alcohol, sulfur, etc.), and low temperatures. Moreover, the resistance of the haze-forming proteins to enzymatic hydrolysis is high due to their structure. Both chitinases and TLPs are small, compact and have globular structures (Theron and Divol, 2014). These particular circumstances make the attack of the proteases complex under winemaking conditions.

For that, one of the main issues is the selection of the adequate source of proteases. Potential sources of active proteases in winemaking environment could be grapes, yeast, bacteria and spoilage microbes. Endogenous proteases of the grapes are active at wine pH; however they show low degree of proteolysis (Exposito et al., 1991). Regarding the main wine yeast, *S. cerevisae*, it does not normally secrete external proteases, but several studies demonstrate that some strains secrete proteases (Rosi et al., 1987; Younes et al., 2011). *S. cerevisiae* PlR1 presented an exo-proteolytic activity during grape juice fermentation, though it seems to be too weak to reduce the effect of PR proteins (Younes et al., 2011, 2013). On the other hand, there are many examples of acid-tolerant proteases secreted by non-*Saccharomyces* yeasts (Dizy and Bisson, 2000; Reid et al., 2012) but these enzymes have not been sufficiently characterized under winemaking conditions.

Promising strategies for the degradation of haze-forming proteins based on a thermal denaturation of PR proteins together with an enzymatic treatment have been proposed (Marangon et al., 2012). Recently, a study reports an acid protease from *Botrytis cinerea*, a fungal pathogen responsible for grey rot, which presented a relevant proteolytic activity against chitinases at winemaking temperature of 17°C, without needing a thermal denaturation step (Van Sluyter et al., 2013).

7. Glucose Oxidase to Reduce Wine Alcohol Content

Wine is an alcoholic drink made from fermented grape juice with a high water and alcohol content. The alcohol concentration plays a positive effect on the aging, stability and organoleptic properties of the wine (Ribéreau-Gayon et al., 2006). However, from a sensorial point of view, high levels of alcohol can mask the sensorial perception of some volatile compounds, may induce the perception of hotness, body and viscosity and can also promote the perception of bitterness (Ozturk and Anli, 2014). Regarding the consumers, the high alcoholic wines can be perceived as unhealthy products (Pickering, 2000).

In later years, the climate change is causing a not well-balanced grape ripening. This fact leads to the production of wines with a high ethanol concentration (Vila-Crespo et al., 2010). Thus, the development of strategies to reduce excessive ethanol content in wine in order to improve the taste balance of wines is of great interest for winemakers. On the other hand, there is a growing market for the production of dealcoholized (> 0.5 per cent, v/v), low-(0.5–1.2 per cent v/v), and reduced-alcohol (1.2–6.5 per cent v/v) wines (Pickering, 2000).

Many different techniques have been tested to reduce or eliminate the wine alcohol content (Schmidtke et al., 2012; Ozturk and Anli, 2014). Some of them require the reduction of fermentable sugars through early grape harvest, juice dilution, membrane filtration techniques or using glucose oxidase enzyme (GOX). Some others focus on the fermentation process using modified yeast strains with lowered ethanol production. Finally, there is a third group of post-fermentation techniques to remove wine alcohol employing membrane transport processes (reverse osmosis, pervaporation and osmotic distillation) or non-membrane extractions (supercritical solvent extraction and spinning cone column).

As mentioned above, the application of GOX is an alternative to the depletion of the sugar level in grape must, resulting in wines with reduced alcohol after the fermentation of the must with native or selected yeasts, usually *S. cerevisiae*.

GOX is a flavoprotein which catalyzes the oxidation of β-D-glucose to D-glucono-d-lactone and H_2O_2 using molecular oxygen as an electron acceptor. D-glucono-d-lactone is non-enzymatically hydrolyzed to gluconic acid and subsequently the flavine adenine dinucleotide (FAD) ring of GOX is reduced to $FADH_2$ (reductive half reaction). The reduced GOD is reoxidized by oxygen to yield H_2O_2 (oxidative half reaction) (Bankar et al., 2009). A second enzymatic activity (catalase) is usually present in commercial GOX preparations to break the by-product of the reaction (H_2O_2).

The use of GOX has been explored in different grape varieties (Müller Thurgau, Muscat, Riesling, and Chasselas) (Heresztyn, 1987; Villettaz, 1987; Pickering et al.,

1999a). In grape must, the fermentable sugar fraction is about 50 per cent of glucose and 50 per cent of fructose. For that, theoretically, the total depletion of glucose using GOX could achieve wines with approximately half of the potential alcohol content. However, the reduction in the levels of alcohol reported in GOX-treated musts were from less than 4 per cent to 40 per cent, since the efficiency of the enzymatic process depends upon the enzyme concentration, must pH, concentration of dissolved oxygen and reaction time and temperature (Pickering et al., 1999a,b,c). Concerning the wine chemical and sensorial characteristics, the GOX-treated wines showed higher acidity than the control ones due to the presence of large amounts of gluconic acid. These wines also presented higher color intensity and less fruit aromas and length of flavor (Pickering et al., 1999a,c). Furthermore, higher concentration of esters and fatty acids and little change in the concentration of the other volatile compounds was reported (Pickering et al., 2001).

A refined method for the production of wines with lower content of alcohol, involving the treatment of the unfermented grape juice with GOX and glucose isomerase, has been recently reported (Van Den Brink and Bjerre, 2014).

Despite the above-mentioned advantages of the use of GOX to reduce the wine alcoholic degree, this technology is not currently recognized by the International Organization of Vine and Wine (IOV, 2015).

8. Lysozyme in Winemaking

Lysozyme is a muramidase enzyme, normally obtained from hen egg white, which can be used for the control of malolactic fermentation (MLF) during winemaking. This enzyme hydrolyzes the cell wall of susceptible bacteria, mainly Gram-positive bacteria, which increases their permeability and causes the cell to burst (Chassy and Giuffrida, 1980). Several methods have been tested in order to extend the antimicrobial spectrum of lysozyme to Gram-negative bacteria, such as denaturation, attachment of other compounds and the use of membrane-permeabilizing agents (Masschalck and Michiels, 2003). Hen egg white lysozyme (HEWL) is considered a natural antimicrobial and qualified as a food preservative in the European Union (EU) and several other countries (Liburdi et al., 2014).

MLF is a bioconversion process of malic acid into lactic acid by means of lactic acid bacteria that takes place in red and in some white wines. However, some of the species involved in MLF can also be responsible for biogenic amine formation, wine spoilage and sluggish or stuck fermentations. Several authors have studied the role of LAB in wine and the application of lysozyme against LAB in winemaking (Gerbaux et al., 1997; Liu et al., 1998; Pilatte et al., 2000; Costello et al., 2001; Gao et al., 2002; Isabel et al., 2009; Azzolini et al., 2010; Guzzo et al., 2011). In this sense, lysozyme is being used as an alternative to reduce the use of sulfur dioxide to control LAB proliferation in wine (Sonni et al., 2009); so regulation EC No. 2066/2001 allows up to 500 mg/L of HEWL to be added to wine or must. Lysozyme can be used in winemaking for different purposes: to inhibit MLF development, to reduce the competition between yeasts and bacteria, to decrease the occurrence of sluggish and stuck fermentations and for the microbiological stabilization of wine after MLF. No adverse effects on the wine composition parameters have been reported after HEWL wine treatment (Liburdi et al., 2014). Nevertheless, the same authors found different wine components to be responsible for the inhibition of several enzymes in wine, including lysozyme.

Lysozyme contains a well-known major allergen from hen's eggs. Thus several studies have demonstrated that HEWL can be considered as a possible allergen (Liburdi et al., 2014) and because of this allergenicity, lysozyme must be labeled on wine bottles if present at trace level in the final product. Taking this into account, Liburdi et al. (2014) focused on lysozyme immobilization and developed an integrated perspective on how to use and customize the enzymes for their specific wine application.

9. Urease in Wine

Urea in wine and other alcoholic beverages is a precursor of ethyl carbamate (urethane, EC) which has revealed potential carcinogenic activity when administered in high doses in animal tests (Schlatter and Lutz, 1990). The presence of urea is linked to the metabolic activity of the different microorganisms present in the winemaking process, both yeasts and bacteria.

Among the several preventive actions to reduce EC levels, the hydrolysis of urea by acid urease (urea amidohydrolase) appears to be a suitable process to prevent EC formation. Ureases are a group of highly proficient enzymes, widely distributed in Nature and whose catalytic function is to catalyze the hydrolysis of urea, its final products being carbonic acid and ammonia (Krajewska, 2009). Acid ureases are a distinct subgroup that, unlike typical (neutral) ureases, have the optimal pHs in the range 2–4.5; so, they can be used in wines. Oenological urease is produced by *Lactobacillus fermentum*. The International Oenological Codex establishes that it must be carefully incorporated and mixed in wine to be aged more than one year if it contains more than 3 mg/l of urea. After a noticeable decrease in urea, for example less than 1 mg/l, all enzymatic activity is eliminated by filtering the wine (diameter of pores under 1 μm).

Several authors have studied different technical aspects in order to improve the oenological use of urease by the optimization of the enzyme (Liu et al., 2012; Yang et al., 2015). Some others have focussed on the kinetics of urea degradation in different stirred bioreactors (Andrich et al., 2009; Andrich et al., 2010).

10. Conclusion and Future Trends

The use of commercial preparations of enzymes in oenology has gained popularity among winemakers in recent years, so that research in enzymes in order to improve the quality of the wines will hopefully increase. One line of action could be to improve the activity and stability of the available enzymes under the extreme and changing winemaking conditions (high sugar concentration, low temperature, presence of sulphur, alcohol production, acidic pH, etc.). Several strategies can be used as the search and selection of new microbial sources of enzymes, designing hyperproductive microbial strains through genetic engineering, the modification of existing enzymes by protein engineering technologies and/or the use of immobilizing technologies. Another possibility could be to design rich in C-S lyase and esterase enzymatic preparations to increase the aroma in wines. Finally, the need to reduce toxic products in the wine must also be addressed by enzymes, such as amine oxidases to remove biogenic amines.

References

Alimardani-Theuil, P., Gainvors-Claisse, A. and Duchiron, F. 2011. Yeasts: An attractive source of pectinases—From gene expression to potential applications: A review. Process Biochemistry, 46: 1525–1537.

Alvarez, I., Aleixandre, J.L., Garcia, M.J. and Lizama, V. 2006. Impact of pre-fermentation maceration on the phenolic and volatile compounds in Monastrell red wines. Analytica Chimica Acta, 563: 109–115.

Andrich, L., Esti, M. and Moresi, M. 2009. Urea degradation in model wine solutions by free immobilized acid urease in a stirred bioreactor. Journal of Agricultural and Food Chemistry, 57: 3533–3542.

Andrich, L., Esti, M. and Moresi, M. 2010. Urea degradation in some white wines by immobilized acid urease in a stirred bioreactor. Journal of Agricultural and Food Chemistry, 58: 6747–6753.

Azzolini, M., Tosi, E., Veneri, G. and Zapparoli, G. 2010. Evaluating the efficacy of lysozyme against lactic acid bacteria under different winemaking scenarios. South African Journal of Enology and Viticulture, 31: 99–105.

Bakker, J., Bellworthy, S.J., Reader, H.P. and Watkins, S.J. 1999. Effect of enzymes during vinification on color and sensory properties of Port wines. American Journal of Enology and Viticulture, 50: 271–276.

Bankar, S.B., Bule, M.V., Singhal, R.S. and Ananthanarayan, L. 2009. Glucose oxidase—An overview. Biotechnology Advances, 27: 489–501.

Bautista-Ortín, A.B., Fernández-Fernández, J.I., López-Roca, J.M. and Gómez-Plaza, E. 2007. The effects of enological practices in anthocyanins, phenolic compounds and wine color and their dependence on grape characteristics. Journal of Food Composition and Analysis, 20: 546–552.

Bautista-Ortín, A.B., Martínez-Cutillas, A., Ros-García, J.M., López-Roca, J.M. and Gómez-Plaza, E. 2005. Improving color extraction and stability in red wines: The use of maceration enzymes and enological tannins. International Journal of Food Science and Technology, 40: 867–878.

Bisson, L.F. and Butzke, C.E. 1996. Technical enzymes for wine production. Agro Food Industry Hi-Tec., May/June: 11–14.

Blanco, P., Sieiro, C., Diaz, A., Reboredo, N.M. and Villa, T.G. 1997. Grape juice biodegradation by polygalacturonases from *Saccharomyces cerevisiae*. International Biodeterioration and Biodegradation, 40: 115–118.

Boulton, R. 2001. The copigmentation of anthocyanins and its role in the color of redwine: A critical review. American Journal of Enology and Viticulture, 52: 67–87.

Cabaroglu, T., Selli, S., Canbas, A., Leproutre, J.P. and Gunata, Z. 2003. Wine flavor enhancement through the use of exogenous fungal glycosidases. Enzyme and Microbial Technology, 33: 581–587.

Cabeza, M.S., Baca, F.L., Muñoz Puntes, E., Loto, F., Baigorí, M.D. and Morata, V.I. 2011. Selection of psychrotolerant microorganisms producing cold-active pectinases for biotechnological processes at low temperature. Food Technology and Biotechnology, 49: 187–195.

Canal-Llaubères, R.M. 1993. Enzymes in winemaking. In: G.H. Fleet (ed.). Wine Microbiology and Biotechnology. Hardwood Academic Publishers: Chur, Switzerland, 886: 477–506.

Chassy, B.M. and Giuffrida, A. 1980. Method for the lysis of gram-positive, asporogenous bacteria with lysozyme. Applied Environmental Microbiology, 39: 153–158.

Colagrande, O., Silva, A. and Fumi, M.D. 1994. Recent applications of biotechnology in wine productions. Biotechnology Progress, 10: 2–18.

Comuzzo, P., Battistutta, F. and Tasso, A. 2005. Effet d'un lysat industriel de lévure sur l'évolution des vms rouges en bouteille. Journal International des Sciences de la Vigne et du Vin, 39: 83–90.

Costello, P.J., Lee, T.H. and Henschke, P. 2001. Ability of lactic acid bacteria to produce N-heterocycles causing mousy off-flavor in wine. Australian Journal of Grape and Wine Research, 7: 160–167.

Del Barrio-Galan, R., Perez-Magarino, S. and Ortega-Heras, M. 2011. Techniques for improving or replacing ageing on lees of oak aged red wines: The effects on polysaccharides and the phenolic composition. Food Chemistry, 127: 528–540.

Del Barrio-Galan, R., Perez-Magarino, S. and Ortega-Heras, M. 2012. Effect of the aging on lees and other alternative techniques on the low molecular weight phenols of tempranillo red wine aged in oak barrels. Analytica Chimica Acta, 732: 53–63.

Dizy, M. and Bisson, L.F. 2000. Proteolytic activity of yeast strains during grape juice fermentation. American Journal of Enology and Viticulture, 51: 155–167.

Ducasse, M.A., Canal-Llauberes, R.M., de Lumley, M., Williams, P., Souquet, J.M., Fulcrand, H., Doco, T. and Cheynier, V. 2010. Effect of macerating enzyme treatment on the polyphenol and polysaccharide composition of red wines. Food Chemistry, 118: 369–376.

Exposito, J.M., Gordillo, C.M., Marino, J.I.M. and Iglesias, J.L.M. 1991. Purification and characterization of a cysteine protease in *Vitis vinifera* grapes (Macabeo variety). Nahrung-Food, 35: 139–142.

Fernández-González, M., Ubeda, J.F., Cordero-Otero, R.R., Thanvanthri-Gururajan, V. and Briones, A.I. 2005. Engineering of an oenological *Saccharomyces cerevisiae* strain with pectinolytic activity and its effect on wine. International Journal of Food Microbiology, 102: 173–183.

Fernández-González, M., Ubeda, J.F., Vasudevan, T.G., Cordero Otero, R.R. and Briones, A.I. 2004. Evaluation of polygalacturonase activity in *Saccharomyces cerevisiae* wine strains. FEMS Microbiology Letters, 237: 261–266.

Fernández-Zurbano, P., Ferreira, V., Peña, C., Escudero, A. and Cacho, J. 1999. Effects of maceration time and pectolytic enzymes added during maceration on the phenolic composition of must. Food Science and Technology International, 5: 319–325.

Ferreira, R.B., Picarra-Pereira, M.A., Monteiro, S., Loureiro, V.B. and Teixeira, A.R. 2001. The wine proteins. Trends in Food Science and Technology, 12: 230–239.

Fogarty, W.M. and Kelly, C.T. 1983. Pectic enzymes. pp. 131–182. *In*: W.M. Fogarty (ed.). Microbial Enzymes and Biotechnology. Applied Science Publishers: London, U.K.

Fulcrand, H., Dueñas, M., Salas, E. and Cheynier, V. 2006. Phenolic reactions during winemaking and aging. American Journal of Enology and Viticulture, 57: 289–297.

Gainvors, A., Karam, N., Lequart, C. and Belarbi, A. 1994. Use of *Saccharomyces cerevisiae* for the clarification of fruit juices. Biotechnology Letters, 16: 1329–1334.

Gao, Y.C., Zhang, G., Krentz, S., Darius, S., Power, J. and Lagarde, G. 2002. Inhibition of spoilage lactic acid bacteria by lysozyme during wine alcoholic fermentation. Australian Journal of Grape and Wine Research, 8: 76–83.

Gerbaux, V., Villa, A., Monamy, C. and Bertrand, A. 1997. Use of lysozyme to inhibit malolactic fermentation and to stabilize wine after malolactic fermentation. American Journal of Enology and Viticulture, 48: 49–54.

Gerbaux, V., Vincent, B. and Bertrand, A. 2002. Influence of maceration temperature and enzymes on the content of volatile phenols in Pinot noirwines. American Journal of Enology and Viticulture, 53: 132–137.

Gil-Muñoz, R., Moreno-Pérez, A., Vila-López, R., Fernández-Fernández, J.I., Martínez- Cutillas, A. and Gómez-Plaza, E. 2009. Influence of low temperature prefermentative techniques on chromatic and phenolic characteristics of Syrah and Cabernet Sauvignon wines. European Food Research and Technology, 228: 777–788.

González-Neves, G., Gil, G. and Barreiro, L. 2008. Influence of grape variety on the extraction of anthocyanins during the fermentation on skins. European Food Research and Technology, 226: 1349–1355.

González-Neves, G., Gil, G., Favre, G. and Ferrer, M. 2012. Influence of grape composition and winemaking on the anthocyanin composition of red wines of Tannat. International Journal of Food Science and Technology, 47: 900–909.

Guzzo, F., Cappello, M.S., Azzolini, M., Tosi, E. and Zapparoli, G. 2011. The inhibitory effects of wine phenolics on lysozyme activity against lactic acid bacteria. International Journal of Food Microbiology, 148: 184–190.

Heresztyn, T. 1987. Conversion of glucose to gluconic acid by glucose oxidase enzyme in Muscat Gordo juice. Australian and New Zealand Grape Grower and Winemaker April, 25–27.

Hernández-Hierro, J.M., Quijada-Morín, N., Martínez-Lapuente, L., Guadalupe, Z., Ayestarán, B., Rivas-Gonzalo, J.C. and Escribano-Bailón, M.T. 2014. Relationship between skin cell-wall composition and anthocyanin extractability of VitisviniferaL cv. Tempranillo at different grape ripeness degree. Food Chemistry, 146: 41–47.

Hjelmeland, A.K. and Ebeler, S.E. 2015. Glycosidically bound volatile aroma compounds in grapes and wine: A review. American Journal of Enology and Viticulture, 66: 1–11.

Howell, K.S., Swiegers, J.H., Elsey, G.M., Siebert, T.E., Bartowsky, E.J., Fleet, G.H., Pretorius, I.S. and Lopes, M.A.D. 2004. Variation in 4-mercapto-4-methyl-pentan-2-one release by *Saccharomyces cerevisiae* commercial wine strains. FEMS Microbiology Letters, 240: 125–129.

[IOV] International Organization for Vine and Wine. International Code of Oenological Practices. 2015. OIV, Paris, France. Available http://www.oiv.int/oiv/info/enplubicationoiv#code.

Isabel, L., Santamaria, P., Tenorio, C., Garijo, P., Gutierrez, A.R. and Lopez, R. 2009. Evaluation of lysozyme to control vinification process and histamine production in Rioja wines. Journal of Microbiology and Biotechnology, 19: 1005–1012.

Kelebek, H., Canbas, A. and Selli, S. 2009. Effects of different maceration times and pectolytic enzyme addition on the anthocyanin composition of Vitisviniferacv Kalecik Karasi wines. Journal of Food Processing and Preservation, 33: 296–311.

Krajewska, B. 2009. UreasesI. Functional, catalytic and kinetic properties: A review. Journal of Molecular Catalysis B. Enzymatic, 59: 9–21.

Lao, C., López-Tamames, E., Buxaderas, S. and de la Torre-Boronat, M.C. 1999. Grape pectic enzyme treatment effect on white musts and wines composition. Journal of Food Science, 61: 553–556.

Liburdi, K., Benucci, I. and Esti, M. 2014. Lysozyme in wine: An overview of current and future applications. Comprehensive Reviews in Food Science and Food Safety, 13: 1062–1073.

Liu, J., Xu, Y., Nie, Y. and Zhao, G. 2012. Optimization production of acid urease by *Enterobacter* sp. in an approach to reduce urea in Chinese rice wine. Bioprocess and Biosystems Engineering, 35: 651–657.

Liu, S.Q. and Pilone, G.J. 1998. A review: Arginine metabolism in wine lactic acid bacteria and its practical significance. Journal of Applied Microbiology, 84: 315–327.

Maicas, S. and Mateo, J.J. 2005. Hydrolysis of terpenyl glycosides in grape juice and other fruit juices: A review. Applied Microbiology and Biotechnology, 67: 322–335.

Marangon, M., Van Sluyter, S.C., Robinson, E.M.C., Muhlack, R.A., Holt, H.E., Haynes, P.A., Godden, P.W., Smith, P.A. and Waters, E.J. 2012. Degradation of white wine haze proteins by aspergillopepsin I and II during juice flash pasteurization. Food Chemistry, 135: 1157–1165.

Martín, M.C. and Morata de Ambrosini, V.I. 2013. Cold-active acid pectinolytic system from psychrotolerant bacillus. Application in color extraction from red grape skin. American Journal of Enology and Viticulture, 64: 495–504.

Martín, M.C. and Morata de Ambrosini, V.I. 2014. Effect of a cold-active pectinolytic system on colour development of Malbec red wines elaborated at low temperatura. International Journal of Food Science and Technology, 49: 1893–1901.

Masino, F., Montevecchi, G., Arfelli, G. and Antonelli, A. 2008. Evaluation of the combined effects of enzymatic treatment and aging on lees on the aroma of wine from Bombinobianco grapes. Journal of Agricultural and Food Chemistry, 56: 9495–9501.

Masschalck, B. and Michiels, C.W. 2003. Antimicrobial properties of lysozyme in relation to foodborne vegetative bacteria. Critical Reviews in Microbiology, 29: 191–214.

Mesquita, P.R., Picarra-Pereira, M.A., Monteiro, S., Loureiro, V.B., Teixeira, A.R. and Ferreira, T.B. 2001. Effect of wine composition on protein stability. American Journal of Enology and Viticulture, 52: 324–330.

Michlmayr, H., Eder, R., Kulbe, K.D. and del Hierro, A. 2012. B-glycosidase activities of *Oenococcusoeni*: Current state of research and future challenges. Mitteilungen Klosterneuburg, 62: 87–96.

Molina, A., Swiegers, J.H., Varela, C., Pretorius, I.S. and Agosin, E. 2007. Influence of wine fermentation temperature on the synthesis of yeast-derived volatile aroma compounds. Applied Microbiology and Biotechnology, 77: 675–687.

Monagas, M., Martín-Álvarez, P.J., Bartolomé, B. and Gómez-Cordovés, C. 2006. Statistical interpretation of the color parameters of red wines in function of their phenolic composition during aging in bottle. European Food Research and Technology, 222: 702–709.

Morata, A., Gómez-Cordovés, M.C., Suberviola, J., Bartolomé, B., Colomo, B. and Suarez, J.A. 2003. Adsortion of anthocyanins by yeast cell walls during the fermentation of red wines. Journal of Agricultural of Food Chemistry, 51: 4084–4088.

Moreno-Arribas, M.V. and Polo, C. 2005. Winemaking biochemistry and microbiology: Current knowledge and future trends. Critical Reviews in Food Science and Nutrition, 45: 265–286.

Murat, M.L., Masneuf, I., Darriet, P., Lavigne, V., Tominaga, T. and Dubourdieu, D. 2001. Effect of *Saccharomyces cerevisiae* yeast strains on the liberation of volatile thiols in sauvignon blanc wine. American Journal of Enology and Viticulture, 52: 136–139.

Ortega-Heras, M., Pérez-Magariño, S. and González-Sanjosé, M.L. 2012. Comparative study of the use of maceration enzymes and cold pre-fermentative maceration on phenolic and anthocyanic composition and color of a Mencía red wine. LWT—Food Science and Technology, 48: 1–8.

Ortega-Regules, A., Ros-García, J.M., Bautista-Ortín, A.B., López-Roca, J.M. and Gómez-Plaza, E. 2008. Differences in morphology and composition of skin and pulp cell walls from grapes (Vitisvinifera L.): Technological implications. European Food Research and Technology, 227: 223–231.

Ozturk, B. and Anli, E. 2014. Different Techniques for Reducing Alcohol Levels in Wine: A Review. Proceedings of 37th World Congress of Vine and Wine, and 12th General Assembly of the OIV (Pt 1) 3.

Palmeri, R. and Spagna, G. 2007. B-glucosidase in cellular and acellular form for winemaking application. Enzyme and Microbial Technology, 40: 382–389.

Palomero, F., Morata, A., Benito, S., González, M.C. and Suarez-Lepe, J.A. 2007. Conventional and enzyme-assisted autolysis during ageing over lees in red wines: Influence on the release of polysaccharides from yeast cell walls and on wine nonnumeric anthocyanin content. Food Chemistry, 105: 838–846.

Pardo, F., Salinas, M.R., Alonso, G.L., Navarro, G. and Huerta, M.D. 1999. Effect of diverse enzyme preparations on the extraction and evolution of phenolic compounds in red wines. Food Chemistry, 67: 135–142.

Parodi, G. 2002. L'affinamento dei vini con l'ausilio di preparati enzimatici. Vignevini, 5: 54–58.

Parpinello, G.P., Versari, A., Chinnici, F. and Galassi, S. 2009. Relationship among sensory descriptors, consumer preference and color parameters of Italian Novello red wines. Food Research International, 42: 1389–1395.

Perez-Serradilla, J.A. and de Castro, M.D.L. 2008. Role of lees in wine production: A review. Food Chemistry, 111: 447–456.

Pickering, G J., Heatherbell, D.A. and Barnes, M.F. 1999a. The production of reduced-alcohol wine using glucose oxidase treated juice. Part I. Composition. American Journal of Enology and Viticulture, 50: 291–298.

Pickering, G.J. 2000. Low- and reduced-alcohol wines: A review. Journal of Wine Research, 11: 129–144.

Pickering, G.J., Heatherbell, D.A. and Barnes, M.F. 1999b. The production of reduced-alcohol wine using glucose oxidase-treated juice. Part II. Stability and SO$_2$-binding. American Journal of Enology and Viticulture, 50: 299–306.

Pickering, G.J., Heatherbell, D.A. and Barnes, M.F. 1999c. The production of reduced-alcohol wine using glucose oxidase-treated juice. Part III. Sensory. American Journal of Enology and Viticulture, 50: 307–316.

Pickering, G.J., Heatherbell, D.A. and Barnes, M.F. 2001. GC-MS analysis of reduced-alcohol Muller-Thurgau wine produced using glucose oxidase-treated juice. Lebensmittel-Wissenschaft Und-Technologie—Food Science and Technology, 34: 89–94.

Pilatte, E., Nygaard, M., Gao, Y., Krentz, S., Power, J. and Lagarde, G. 2000. Etude de l'effet du lysozyme sur differentes souches d'*Oenococcusoeni*. Applications dans la gestion de la fermentation malolactique. Revue Française d'œnologie, 185: 26–31.

Pogorzelski, E. and Wilkowska, A. 2007. Flavor enhancement through the enzymatic hydrolysis of glycosidic aroma precursors in juices and wine beverages: A review. Flavor and Fragrance Journal, 22: 251–254.

Pozo-Bayon, M.A., Martinez-Rodriguez, A., Pueyo, E. and Victoria Moreno-Arribas, M. 2009. Chemical and biochemical features involved in sparkling wine production: From a traditional to an improved winemaking technology. Trends in Food Science and Technology, 20: 289–299.

Pretorius, I.S. 2000. Tailoring wine yeast for the new millennium: Novel approaches to the ancient art of winemaking. Yeast, 16: 675–729.

Radoi, F., Kishida, M. and Kawasaki, H. 2005. Characteristics of wines made by Saccharomyces mutants which produce a polygalacturonase under wine-making conditions. Bioscience, Biotechnology and Biochemistry, 69: 2224–2226.

Reid, V.J., Theron, L.W., du Toit, M. and Divol, B. 2012. Identification and partial characterization of extracellular aspartic protease genes from *Metschnikowiapulcherrima* IWBT Y1123 and *Candida apicola* IWBT Y1384. Applied and Environmental Microbiology, 78: 6838–6849.

Revilla, I. and González-San José, M.L. 1998. Methanol release during fermentation of red grapes treated with pectolytic enzymes. Food Chemisty, 63: 307–312.

Revilla, I. and González-San José, M.L. 2001. Evolution during the storage of red wines treated with pectolytic enzymes: New anthocyanin pigment formation. Journal of Wine Research, 12: 183–197.

Ribéreau-Gayon, P., Dubourdieu, D., Donèche, B. and Lonvaud, A. 2006. Handbook of Enology, the Microbiology of Wine and Vinifications. John Wiley and Sons, New York.

Río Segade, S., Giacosa, S., Gerbi, V. and Rolle, L. 2011. Berry skin thickness as main texture parameter to predict anthocyanin extractability in wine grapes. LWT—Food Science and Technology, 44: 392–398.

Río Segade, S., Pace, C., Torchio, F., Giacosa, S., Gerbi, V. and Rolle, L. 2015. Impact of maceration enzymes on skin softening and relationship with anthocyanin extraction in wine grapes with different anthocyanin profiles. Food Research International, 71: 50–57.

Rodriguez-Bencomo, J., Ortega-Heras, M. and Perez-Magarino, S. 2010. Effect of alternative techniques to ageing on lees and use of non-toasted oak chips in alcoholic fermentation on the aromatic composition of red wine. European Food Research and Technology, 230: 485–496.

Rodriguez-Nogales, J.M., Fernandez-Fernandez, E. and Vila-Crespo, J. 2012a. Effect of the addition of β-glucanases and commercial yeast preparations on the chemical and sensorial characteristics of traditional sparkling wine. European Food Research and Technology, 235: 729–744.

Rodriguez-Nogales, J.M., Fernandez-Fernandez, E., Gomez, M. and Vila-Crespo, J. 2012b. Antioxidant properties of sparkling wines produced with β-glucanases and commercial yeast preparations. Journal of Food Science, 235: 729–744.

Roland, A., Schneider, R., Razungles, A. and Cavelier, F. 2011. Varietal thiols in wine: Discovery, analysis and applications. Chemical Reviews, 111: 7355–7376.

Rolle, L., Torchio, F., Ferrandino, A. and Guidoni, S. 2012. Influence of wine-grape skin hardness on the kinetics of anthocyanin extraction. International Journal of Food Properties, 15: 249–261.

Romero-Cascales, I., Fernández-Fernández, J.I., López-Roca, J.M. and Gómez-Plaza, E. 2005. The maceration process during winemaking extraction of anthocyanins from grape skins into wine. European Food Research and Technology, 221: 163–167.

Romero-Cascales, I., Fernández-Fernández, J.I., Ros-García, J.M., López-Roca, J.M. and Gómez-Plaza, E. 2008. Characterisation of the main enzymatic activities present in six commercial macerating enzymes and their effects on extracting color during winemaking of Monastrell grapes. International Journal of Food Science and Technology, 43: 1295–1305.

Romero-Cascales, I., Ros-García, J.M., López-Roca, J.M. and Gómez-Plaza, E. 2012. The effect of a commercial pectolytic enzyme on grape skin cell-wall degradation and color evolution during the maceration process. Food Chemistry, 130: 626–631.

Roncoroni, M., Santiago, M., Hooks, D.O., Moroney, S., Harsch, M.J., Lee, S.A., Richards, K.D., Nicolau, L. and Gardner, R.C. 2011. The yeast IRC7 gene encodes a β-lyase responsible for production of the varietal thiol 4-mercapto-4-methylpentan-2-one in wine. Food Microbiology, 28: 926–935.

Rosi, I., Costamagna, L. and Bertuccioli, M. 1987. Screening for extracellular acid protease(s) production by wine yeasts. Journal of the Institute of Brewing, 93: 322–324.

Sacchi, K.L., Bisson, L.F. and Adams, D.O. 2005. A review of the effect of winemaking techniques on phenolic extraction in red wines. American Journal of Enology and Viticulture, 56: 197–206.

Schlatter, J. and Lutz, W.K. 1990. The carcinogenic potential of ethyl carbamate (urethane): Risk assessment at human dietary exposure levels. Food and Chemical Toxicology, 28: 205–211.

Schmidtke, L.M., Blackman, J.W. and Agboola, S.O. 2012. Production technologies for reduced alcoholic wines. Journal of Food Science, 77: 25–41.

Sonni, F., Cejudo-Bastante, M.J., Chinnici, F., Natali, N. and Riponi, C. 2009. Replacement of sulphur dioxide by lysozyme and oenological tannins during fermentation: Influence on volatile composition of white wines. Journal of Science and Food Agriculture, 89: 688–696.

Soto Vázquez, E., Río Segade, S. and Orriols Fernández, I. 2010. Effect of the winemaking technique on phenolic composition and chromatic characteristics in young redwines. European Food Research and Technology, 231: 789–802.

Styger, G., Prior, B. and Bauer, F.F. 2011. Wine flavor and aroma. Journal of Industrial Microbiology and Biotechnology, 38: 1145–1159.

Tavano, O.L. 2013. Protein hydrolysis using proteases: An important tool for food biotechnology. Journal of Molecular Catalysis B-Enzymatic, 90: 1–11.

Theron, L.W. and Divol, B. 2014. Microbial aspartic proteases: Current and potential applications in industry. Applied Microbiology and Biotechnology, 98: 8853–8868.

Torresi, S., Frangipane, M.T., Garzillo, A.M.V., Massantini, R. and Contini, M. 2014. Effects of a β-glucanase enzymatic preparation on yeast lysis during aging of traditional sparkling wines. Food Research International, 55: 83–92.

Ugliano, M. 2009. Enzymes in winemaking. pp. 103–126. *In*: M.C. Polo and M.V. Moreno-Arribas (eds.). Wine Chemistry and Biochemistry. Springer: New York.

Van Den Brink, J.M. and Bjerre, K. 2014. Method for the Production of a Wine with Lower Content of Alcohol. U.S. Patent # 8,765,200.

Van Rensburg, P. and Pretorius, L.S. 2000. Enzymes in winemaking: Harnessing natural catalysis for efficient biotransformations—A review. South African Journal of Enology and Viticulture, 21: 52–73.

Van Sluyter, S.C., McRae, J.M., Falconer, R.J., Smith, P.A., Bacic, A., Waters, E.J. and Marangon, M. 2015. Wine protein haze: Mechanisms of formation and advances in prevention. Journal of Agricultural and Food Chemistry, 63: 4020–4030.

Van Sluyter, S.C., Warnock, N.I., Schmidt, S., Anderson, P., van Kan, J.A.L., Bacic, A. and Waters, E.J. 2013. Aspartic acid protease from *Botrytis cinerea* removes haze-forming proteins during white winemaking. Journal of Agricultural and Food Chemistry, 61: 9705–9711.

Venturi, F., Andrich, G., Quartacci, M.F., Sanmartin, C., Andrich, L. and Zinnai, A. 2013. A kinetic method to identify the optimum temperature for β-glucanase activity. South African Journal for Enology and Viticulture, 34: 281–286.

Vidal, S., Williams, P., Doco, T., Moutounet, M. and Pellerin, P. 2003. The polysaccharides of red wine: Total fractionation and characterization. Carbohydrate Polymers, 54: 439–447.

Vila-Crespo, J., Rodriguez-Nogales, J.M., Fernández-Fernández, E. and Hernanz-Moral, M. 2010. Strategies for the enhancement of malolactic fermentation in the new climate conditions. Current Research, Technology and Education Topics in Applied Microbiology and Microbial Biotechnology, Microbiology Series, Badajoz.

Villena, M.A., Iranzo, J.F.U., Gundllapalli, S.B., Otero, R.R.C. and Perez, A.I.B. 2006. Characterization of an exocellular β-glucosidase from *Debaryomyces pseudopolymorphus*. Enzyme and Microbial Technology, 39: 229–234.

Villettaz, J.C. 1987. Method for Production of a Low Alcoholic Wine. U.S. Patent # 4,675,191.

Vincenzi, S., Marangon, M., Tolin, S. and Curioni, A. 2011. Protein evolution during the early stages of white winemaking and its relations with wine stability. Australian Journal of Grape and Wine Research, 17: 20–27.

Wang, Y., Zhang, C., Li, J. and Xu, Y. 2013. Different influences of β-glucosidases on volatile compounds and anthocyanins of Cabernet Gernischt and possible reason. Food Chemistry, 140: 245–254.

Waters, E.J., Alexander, G., Muhlack, R., Pocock, K.F., Colby, C., O'Neill, B.K., Hoj, P.B. and Jones, P. 2005. Preventing protein haze in bottled white wine. Australian Journal of Grape and Wine Research, 11: 215–225.

Yang, Y., Kang, Z., Zhou, J. and Chen, J. 2015. High-level expression and characterization of recombinant acid urease for enzymatic degradation or urea in rice wine. Applied Microbiology and Biotechnology, 99: 301–308.

Younes, B., Cilindre, C., Jeandet, P. and Vasserot, Y. 2013. Enzymatic hydrolysis of thermo-sensitive grape proteins by a yeast protease as revealed by a proteomic approach. Food Research International, 54: 1298–1301.

Younes, B., Cilindre, C., Villaume, S., Parmentier, M., Jeandet, P. and Vasserot, Y. 2011. Evidence for an extracellular acid proteolytic activity secreted by living cells of *Saccharomyces cerevisiae* PlR1: Impact on grape proteins. Journal of Agricultural and Food Chemistry, 59: 6239–6246.

Zinnai, A., Venturi, F., Quartacci, M.F., Andreotti, A. and Andrich, G. 2010. The kinetic effect of some wine components on the enzymatic hydrolysis of β-glucan. South African Journal of Enology and Viticulture, 31: 9–13.

18

Microbial Linamarase in Cassava Fermentation

Sudhanshu S. Behera[1] and *Ramesh C. Ray*[2,*]

1. Introduction

The cassava (*Manihot esculenta* Crantz) is a major root crop in many tropical and subtropical countries, including Latin America, Asia and especially West Africa (Sayre et al., 2011). Cassava is variously named as manioc, mandioca, tapioca, or yucca and referred as a cyanogenic food crop (Abban et al., 2013). Cassava is also considered as a robust crop that grows well under harsh conditions in which few other crops can survive (Cach et al., 2006). The cassava plant is drought tolerant and offers resistance to most diseases and pests (Cach et al., 2006; Setter and Fregene, 2007). Furthermore, cassava is a form of staple food for over half a billion people in more than 90 countries and the third most important source of calories in the tropics, after rice and corn (Abban et al., 2013; Montagnac et al., 2009; Murugan et al., 2014). However, toxicity of cassava happens due to presence of the cyanogenic glucosides or cyanogens (naturally occurring substrates), namely linamarin (2-hydroxy-isobutyro-nitrile/~-D-glucopyranoside) and lotaustralin (2-hydroxyl-2metylbutyro-nitrile/~-D-glucopyranoside) (Fig. 1), which have fatal consequences when consumed in unprocessed food (Gleadow and Woodrow, 2002; Montagnac et al., 2009). Moreover, these substrates (cyanogens) may cause cyanide poisoning with symptoms of dizziness, stomach pain, headache, nausea, vomiting and occasionally death (Lambri et al., 2013; Mlingi et al., 1995; Muzanila et al., 2000; Sornyotha et al., 2010). In addition, daily consumption of cassava food products that still retain residual levels of these

[1] Department of Fisheries and Animal Resources Development, Government of Odisha, India.
[2] ICAR- Central Tuber Crops Research Institute (Regional Centre), Bhubaneswar 751 019, Orissa, India.
* Corresponding author: rc.ray6@gmail.com

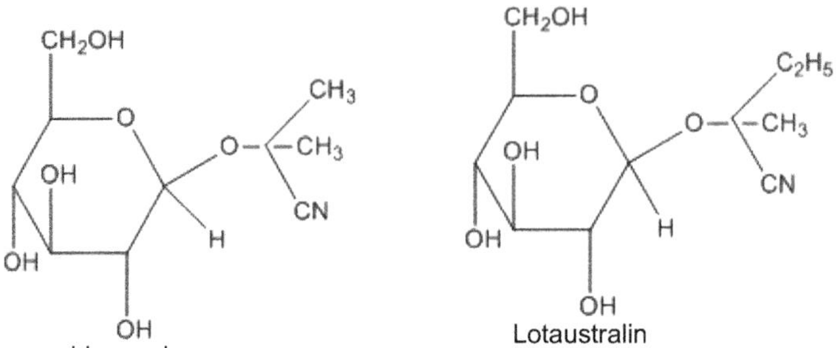

Figure 1. Linamarin (a) and lotaustralin (b), principal cyanogenic glucosides of cassava.

cyanogenic glucosides (cyanogens) can cause several chronic diseases, such as cretinism, goitre, tropical ataxic neuropathy and tropical diabetes (Ahaotu et al., 2013; Ogbonnaya, 2016). It has been recommended by the World Health Organization that 10 mg HCN/kg body weight is the maximum safe intake of cyanide-containing food/feed for humans and animals (Tritscher et al., 2013; Chikezie and Ojiako, 2013). Hence, considerable efforts are made to reduce/remove the cyanogenic glucosides by the use of endogenous or exogenous (microbial) linamarase (β-glucosidase) enzyme that releases the cyanohydric acid, which is also toxic (Kobawila et al., 2005). Traditional techniques, such as spontaneous fermentation are adopted to eliminate cyanohydric acids in cassava tubers so that they are suitable for human consumption (Murugan et al., 2014). These enzymes convert the cyanide-containing compounds into acetone cyanohydrins that further spontaneously decomposes to hydrogen cyanide (HCN) (Fig. 2). The compound, HCN is either released into the air or it readily dissolves in water (Murugan et al., 2014).

This chapter highlights the implications of microorganisms for fermentation of cassava and linamarase activity in detoxification of cyanogenic glucosides. The process, challenges and improvement achieved in reduction of toxic cyanogens in view of increasing nutritional aspects of cassava-based fermented foods are discussed.

2. Microbial Linamarase (β-Glucosidase)

Linamarase (EC 3.2.1.21; β-D-glucoside glucohydrolase) constitutes a group of well studied β-glucosidase (hydrolase/cellobiase) that catalyzes the hydrolysis of glycosidic linkages through transfer of glycosyl groups. Linamarase (native hydrolase) is obtained from crop-like cassava (endogenous linamarase) and microbial sources (exogenous linamrase). The endogenous linamarase is mainly responsible for the *in situ* detoxification of linamarin and lotaustralin in cassava.

Yeast and lactic acid bacteria (LAB) are the microbial strains that have been frequently associated with the production of linamarase during fermentation of cassava and in the development of flavor (Okafor et al., 1998; Lacerda et al., 2005; Nwokoro and Anya, 2011). The most predominant LAB-producing linamarase for the preparation of traditional cassava food are *Lactobacillus plantarum, Lb. fermentum, Lb. brevis, Leuconostoc pseudomesenteroides, Lc. fallax* and *Weissella paramesenteroides* (Kostinek et al., 2005; Nuraida, 2015). Other major LAB were

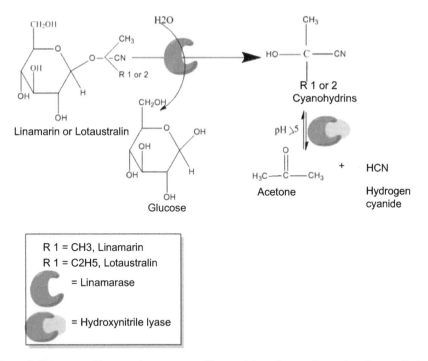

Figure 2. The enzyme (linamarase) converts cyanide containing substates, linamarin or lotaustralin into acetone cyanohydrins, which further spontaneously or enzymatic (hydroxynitrile lyase) decompose to acetone (CH_3COCH_3) and hydrogen cyanide (HCN).

some cocci including *Streptococcus* spp., *Corynebacterium* spp., whose numbers decreased with fermentation time (Amoa-Awua et al., 1996). However, a number of yeasts isolated from fermenting cassava was shown to possess linamarase enzyme (Lei et al., 1999; Kobawila et al., 2005) and these have been associated with detoxification of cyanogenic glucosides during cassava fermentation (Blagbrough et al., 2010).

Nwokoro and Anya (2011) investigated the degradation of cassava cyanide and studied the biochemical properties and purification of linamrase enzyme from *Lactobacillus delbrueckii* NRRL B-763. The crude enzyme showed a reduction from 2.1 mg HCN/10 g sample to 0.11 mg/ HCN/10 g sample (95 per cent reduction) after 20 hours of treatment. However, untreated control samples showed a reduction of only 5.7 per cent from 2.1 mg HCN/10 g sample to 1.98 mg HCN/10 g sample after 40 hours of incubation. The enzyme was purified 33 fold with a 40 per cent yield and showed maximum activity at pH 4.5.

Murugan et al. (2014) isolated cyanogenic glucoside using indigenous bacteria, *Bacillus subtilis* KM05 from cassava peel and evaluated its potential for cyanogen detoxification. The linamarase (53 KDa) enzyme was partially purified from *B. subtilis* KM05 strain and activity was found to be 9.6 U/ml. It was also inferred that the enzyme had the ability to degrade cynogens effectively.

Ogbonnaya (2016) investigated the abilities of cassava endogenous and microbial enzymes to hydrolyze cyanides in cassava flour. A total of eight linamarase-producing bacteria (*Lb. plantarum, Lb. fermentum, Lb. amylovorus, Lb. cellobiosus, Lc.mesenteroides, Pseudomonas stutzeri, Bacillus pumilus* and *B. subtilis*) were identified and isolated from cassava tubers and soil samples. All the isolates were grown in media containing surfactant, Tween 80 solution. The best enzyme activity (6.82 U/mL) was obtained by the *Lb. fermentum* in the medium containing the surfactant. The cassava flour containing residual cyanides treated with linamarase of *Lb. fermentum* was found undetected after 30 hours. However, the residual cyanides in cassava flour treated with endogenous cassava linamarase were found to be 0.39 mg/10 g cassava flour after 80 hours. The results revealed that fermenting cassava tubers with isolates had better control over the cassava fermentation process that could lead to production of non-toxic cassava food products.

3. Linamarase in Fermentation Processes

Fermentation aided linamarase is essential for improving the product quality as well as to provide safety, especially by reduction of toxic cyanogenic glycosides in cassava food products (Kostinek et al., 2005). Moreover, fermentation is an indispensable means for processing so as to improve the textural quality, palatability and to upgrade nutritive value by enriching with protein. Furthermore, the cassava fermentation process is widely used to transform and preserve the unique organoleptic qualities of final product, due to its low technology and energy requirements (Achi and Akomas, 2006).

Cassava processing methods cover different combinations of grating, drying, soaking, boiling and fermentation of whole or fragmented tubers (Obilie et al., 2004). However, the fermentation process can be broadly categorized into submerged (involving soaking in water, e.g., fufu) and solid state (without soaking, e.g., gari) (Ray and Swain, 2011).

The fermented foods from cassava are lafun, fufu, agbelima, chickwanghe, attiek, kivunde and gari in Africa, tape in Asia, and *cheese* bread and *coated* peanut in Latin America (Ray and Sivakumar, 2009).

3.1 Submerged fermentation (SmF)

SmF is a process in which the growth and anaerobic or partially anaerobic decomposition of the carbohydrates by action of microorganisms in a liquid medium with ample availability of free water (Ray et al., 2007) takes place.

Ezekiel et al. (2010) studied the performance of *Trichoderma viride* (ATCC 36316) in protein enriched cassava peels by SmF and also investigated the effect of enzyme pre-treatment prior to the fermentation process on the enriched product. The fermentation (for three to four days) showed eight fold (4.2 to 37.6 per cent) increase in crude protein content of cassava peels.

3.1.1 Cassava fermented food products through SmF

The cyanogenic glucosides content in tubers and freshly harvested leaves of cassava vary between 137 and 1515 ppm. After the traditional fermentation of tubers and leaves of cassava, the cyanogenic glucosides contents are reduced significantly by 70 to 75 per cent (Dhellot et al., 2015).

Lambri et al. (2013) investigated a drying procedure with and without the fermentation process for elimination of cyanogens in cassava tubers (pressed pulp). The process of fermentation was carried out in the presence of *Saccharomyces cerevisiae*; detoxification was found effective. In drying conditions, a temperature of 60°C, even for a shorter duration (say eight hours), lowered the cyanide content (> 90 per cent). However, the process of dehydration followed by fermentation showed maximum removal of cyanide content.

Several studies have identified the microorganisms associated with submerged cassava fermentation and the different traditional fermented products are listed below.

3.1.1.1 Fufu

Fufu is a traditional fermented cassava food product, indigenous for most Nigerians in the south and eastern zones (Achi and Akomas, 2006). It ranked next to gari (discussed in the later section) which comes as a wet mash or dry powder of fermented cassava tubers. It is formed by steeping whole or cut tubers in water to ferment for a period of three to five days, depending on the ambient temperature (Okpokiri et al., 1985).

During the production of fufu, a succession of organisms ferment the mash. Fufu is a fermented cassava food product which comes as a wet mash or dry powder (Umeh and Odibo, 2014). It is made by steeping whole or cut tubers in water to ferment for a period of three to five days, depending on the ambient temperature (Okpokiri et al., 1984).

3.1.1.2 Lafun

Lafun is a fermented cassava food product (fine powdery) commonly consumed in West Africa, mainly in the western states of Nigeria and Benin (Falade and Akingbala, 2010). It is produced by peeling and cutting cassava tubers into pieces followed by soaking in water (stationary water, or immersed in stream) for three to four days. The water soak-peeled tubers are processed by smF that leads to acidification and making them soft (Ferraro et al., 2015).

Padonou et al. (2009) investgated the microorganisms, such as LAB, aerobic bacteria and yeast associated with lafun fermentation using genotypic and phenotypic methods. The LAB including *Lb. fermentum, Lb. plantarum, Lb. fallax* and aerobic bacteria including *B. subtilis, B. cereus, Pantoea agglomerans;* and yeasts such as *S. cerevisiae, Hanseniaspora guilliermondii, Pichia scutulata, P. rhodanensis, P. kudriavzevii, Candida glabrata, Kluyveromyces marxianus* are the microorganisms that occur during lafun production.

3.1.1.3 Agbelima

Agbelima is one of the main fermented products processed from cassava (Amoa-Awua et al., 1997). It is widely consumed in countries like Ghana, Togo and Benin (Oduro et al., 2000). It is produced by grinding peeled cassava tubers together with an inoculum. The powdered mash is packed into pockets, dewatered and allowed to ferment for two to three days. The whole mass is fermented into a cassava meal called agbelima (Adamafio et al., 2010).

Mante et al. (2003) reported that LAB, *Bacillus* sp. and yeasts are mainly responsible for fermentation of cassava dough into agbelima.

3.1.1.4 Akyeke

Akyeke is steamed and sour granular cassava meal consumed in southwestern Ghana but more extensively in Ivory Coast. It is processed by fermenting grated cassava, then de-watering the mash, followed by sieving and streaming of granular product (Obilie et al., 2003).

Obilie et al. (2004) investigated the species of LAB responsible for the souring of cassava mash and examined the β-glucosidase activity during akyeke fermentation. The percentage of LAB, such as *Lb. plantarum, Lb. brevis, Lc. mesenteroides*, increased during five days of the fermentation process. Moreover, a reduction in cyanogens contents (about 98 per cent) occurred in cassava root during akyeke processing.

3.1.1.5 Attieke

Attieke is essentially a flavored starchy ingredient produced from fermented cassava tubers. It is a steamed granular cassava meal which is ready-to-eat, has slightly sour taste and whitish colour (Djeni et al., 2011). The attieke is a well-known product and widely consumed in Ivory Coast and neighboring countries (Daouda et al., 2012). Several studies highlight the preparation of attieke from cassava tubers. The cassava tubers are peeled, cut into pieces, washed and milled. During grinding, the cassava paste is mixed with inoculum or starter culture (Moorthy and Mathew, 1998).

3.1.1.6 Tape

Tape is a famous fermented food from Indonesia. It is formed from steamed cassava and then mixed with starter culture or with inoculum. The starter culture of tape consists of bacteria, moulds, and yeasts (Barus et al., 2013).

However, tape from steamed cassava, usually followed traditional methods which suffers from several drawbacks and neither is the manufacturing process standardized (Arihantana et al., 1987).

3.2 Solid state fermentation (SSF)

SSF is a process where microbial growth and product formation exists on the surface of solid surfaces. This process take place in the absence of 'free' water, where moisture is absorbed into the solid matrix (Pandey, 2003; Singhania et al., 2009). In recent years, many processing methods have been developed by the cassava-farming population. One of these involves SSF method of processing through application of microbial linamarase. This process not only reduces the linamarin content but even the microflora causes effective root softening of cassava flour (Essers et al., 1995).

Esser and Jurgens (1999) examined the effect of six individual strains (microflora) in SSF fermenting cassava on cyanogen levels. The six fungal strains of *Geotrichum candidum, Mucor racemosus, Neurospora sitophila, Rhizopus oryzae* and *Rhizopus stolonifer* and a *Bacillus* sp., were inoculated into cassava flour for 72 hours. The results found that both fermentation duration and microbial growth were involved in reducing cyangonic glucoside levels (62.7 per cent) and changing the cyanogen composition in cassava flour.

The most important mechanism of microbial linamarase in the cassava detoxification process is that the enzyme increases the cell-wall-degrading activity but improves the contact between endogenous linamrases and linamrain (Essers et al., 1995).

3.2.1 Cassava fermented food products through SSF

Cassava is fermented by SSF process to produce several products in different parts of Africa and Latin America. Such products include biomass, flour, starch, bread and gari (Ezekiel et al., 2010; Ezekiel and Aworh, 2013).

3.2.1.2 Cassava flour

Cassava tubers are highly perishable and begin to deteriorate after two days of harvesting. Moreover, suitable methods for storing the cassava tubers in fresh form are limited (Amoa-Awua et al., 2005). One method is to process the tubers into high quality cassava flour (Fig. 3). Therefore, it offers a means to preserve the cassava tubers, which are subsequently used for both industrial and traditional purposes (Defloor et al., 1993; Amoa-Awua et al., 2005).

The cassava tubers are grated, pressed and then dried or toasted into cassava flour (Demiate and Kotovicz, 2011). Akindahunsi (1999) investigated the pure strain of *Rhizopus oryzae* in the fermentation process of cassava pulp and the product was subsequently processed into flour. The protein content of flour showed an increase to a value of 97 per cent. However, the anti-nutrients factor, such as tannin (flour, 0.16 mg/100 g) and total cyanide (flour, 17.21 mg/kg) contents of flour and gari were considerably low.

Oboh and Akindahunsi (2003) reported that nutritional quality, such as protein and fat content, of cassava flour increased in the presence of *S. cerevisae*. The nutritional quality was determined in the SSF condition. The result revealed that protein (10.9 per cent) and fat (4.5 per cent) contents increased and conversely, cyanogens content decreased (9.5 mg/kg) in cassava flour.

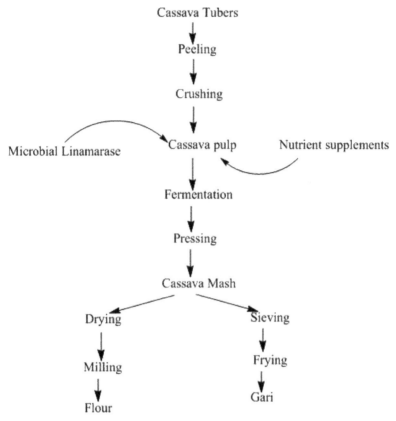

Figure 3. Production chart for solid media fermented (SSF) cassava products (flour and gari) (Oboh and Akindahunsi, 2003).

3.2.1.2.1 Retting

Retting is a spontaneous lactic acid fermentation of cassava tubers in the preparation of cassava-based foods, such as foo-foo and chickwaizgue (cassava bread) of Central Africa (Ampe and Brauman, 1995).

3.2.1.3 Chickwaizgue (Cassava bread)

One of the most popular food uses of cassava flour worldwide is the manufacture of baked products (Shittu et al., 2008). Recently, the Government of Nigeria made it mandatory for flour mills to include a minimum of 10 per cent high quality cassava flour in wheat flour for making composite flour used for baking purposes (Eduardo et al., 2013).

3.2.1.4 Cassava starch

Cassava starch is a valued raw material for producing many kinds of modified stanches (Demiate and Kotovicz, 2011). The cassava starch is the second most important botanical source of industrial starch, next to corn.

Cassava starch is generally prepared from peeling tubers and grinded and washed with water by using sieves. This is followed by centrifugation or decantation for recovery of cassava starch. In Brazil, the production of fermented and sun-dried cassava starch is also known as '*Polvilho azedo*' (Lacerda et al., 2005; Garrido et al., 2014).

3.2.1.5 Gari

Gari is one of the most popular foods derived from solid state fermentation of cassava (Kostinek et al., 2005; Udoro et al., 2014). Gari is consumed by nearly 200 million people in West Africa (Yao et al., 2009). Many authors have suggested that inoculation of fermenting cassava with extraneous linamarase has effectively reduced the cyanogen toxicity (Giraud et al., 1993).

Cassava fermentation is mainly due to LAB, *Lactobacillus* sp. and to a lesser extent *Streptococcus* sp. is responsible for acid production and gari flavour (Ngaba and Lee, 1979). However, Yao et al. (2009) reported that the strains *Lb. plantarum* VE36, G2/25, *Lb. fermentum* G2/10 and *Weissella parameseneroides* LC11 were selected to be developed as freeze-dried starter cultures in garri production.

Gari have been popularly consumed by people of Nigeria (Asegbeloyin and Onyimonyi, 2007). It is commercially prepared by several forms of consortia of microorganisms. Akindahunsi (1999) investigated the pure strain of *Rhizopus oryzae* in the fermentation process of cassava pulp and the product was subsequently processed into gari. The protein content of gari showed an increase in the value by 53 per cent. However, the anti-nutrients factors, such as tannin (0.08 mg/100 g) and total cyanides (14.85 mg/kg) contents of gari, were considerably low.

In order to increase the nutritional quality of gari, *S. cerevisae* was used in SSF of cassava pulp (Oboh and Akindahunsi, 2003; Aguoru et al., 2015). The resultant gari was analyzed with respect to protein and fat contents and also anti-nutrient availability. The results disclosed that there were notable increases in the protein (6.3 per cent) and fat (3.0 per cent) contents. Conversely, there was a significant decrease in the cyanide level (9.1 mg/kg) (Oboh and Akindahunsi, 2003).

Oguntoyinbo and Dodd (2010) investigated the microbial dynamics and diversity during SSF of cassava gari production in West Africa. Several advance molecular techniques, such as 16S rDNA gene sequence analysis, pulsed field gel electrophoresis (PFGF) analysis were used to monitor the bacterial dynamics during cassava fermentation. The lactic acid bacterial species and their close relatives, including *Lb. plantarum*, *Lb. fermentum*, *Lb. pentosus*, *Lb. acidophilus* and *Lb. casei* were identified.

Ahaotu et al. (2013) studied the fermentation of un-dewatered cassava pulp by microbial linamarase (*Alcaligenes faecalis*, *Lb. plantarum*, *Leuconostoc cremoris* and *Geotrichum candidum*) that was involved in the break down of cyanogenic glucosides and the extent of protein enrichment. The results claimed that the protein content of cassava mash fermented by a mixed culture of bacteria provided the highest value (14.60 mg/g dry matter) (Table 1).

More recently, Chikezie and Ojiako (2013) investigated the role of palm oil in conjugation with the duration of fermentation on cyanide and aflatoxin loads of gari. The 48-hour fermentation scheme of processed cassava gari showed significant reduction in cyanide content, whereas 48-hour fermentation with palm oil scheme increased the reduction of aflatoxin content.

Table 1. Various microbial strains in cassava fermentation process.

Microbiota Involved in Cassava Fermentation	Cassava Food/ Feed Product	References
Trichoderma viride (ATCC 36316)	Cassava peels	Ezekiel et al. (2010)
Saccharomyces cerevisiae	Cassava pulp	Lambri et al. (2013)
Lactobacillus fermentum, Lb. plantarum, Lb. fallax, Weisswella confuse, Bacillus substilis, B. cereus, Kluyveromyces pneumoniae, K. marxianus, Pantoea agglomerans, P. scutulata, P. rhodanensis, P. kudriavzevii, S. cerevisiae, Hansenia guilliermondii, Candida glabrata	Lafun	Padonou et al. (2009)
Bacillus sp. and yeasts	Agbelima	Mante et al. (2003)
Lb. plantarum, Lb. brevis, Leuconostoc mesentorides	Akyeke	Obilie et al. (2004)
Rhizopus oryzae	Cassava flour	Akindahunsi (1999)
S. cerevisae	Cassava Flour	Oboh and Akindahunsi (2003)
R. oryzae	Gari	Akindahunsi (1999)
Lb. plantarum VE36, G2/25, *Lb. fermentum* G2/10 and *Weissella parameseneroides* LC11	Gari	Yao et al. (2009)
Lb. plantarum, Lb. fermentum, Lb. pentosus, Lb. acidophilus and *Lb. casei*	Gari	Oguntoyinbo and Dodd (2010)
Alcaligenes faecalis, Lb. plantarum, Lc. cremoris and *Geotrichum candidum*	Gari	Ahaotu et al. (2013)
Aspergillus niger, Panus tigrinus	Poultry feed	Purwadaria (2014)

4. Biomass Conversion and Applications of Cassava Fermentation

Linamarases and cassava fermentation processes were initially investigated several decades ago for bioconversion of biomass (cassava), leading to research on the industrial applications of cassava as animal feed, food, biofuel and in the paper industry (Rattanachomsri et al., 2009; Poonpairoj and Chitradon, 2011).

4.1 Enriched poultry feed or animal feed

Cassava peels, which are mostly generated as waste during the processing of cassava, have been employed as an important source of carbohydrate in livestock feeds (Tijani et al., 2012). Besides the carbohydrate content, the cassava peels also have high level of crude fiber (low level of protein content) and facilitated its use as a potential source for animal feeds. However, cassava peels contain high amount of toxic cyanogens (Cardoso et al., 2005).

Purwadaria (2014) reported that the SSF process use of mould is beneficial in detoxification and degradation of toxic cyanogenic compounds. Moreover, the fiber content of cassava peels decrease due to mould degradation activity and hence used for poultry feed. The SSF process was carried out using cassava peels and/or mixed with wheat flour, with indigenous microbes, *Aspergillus niger* or *Panus tigrinus* as inoculums. The fermented products reported optimum substitution in poultry feed.

5. Conclusion and Future Prospects

Linamarase plays a pivotal role in the detoxification of toxic cyanogenic glucosides in cassava-based foods. Most of the fermenting microorganisms produce linamarase that helps in cyanogens detoxification. Although it is evident from cassava fermentation studies that the cyanogen level decreases during the course of fermentation, activity of microbial linamarase has not been fully explored. Further research and development needs to be conducted on solid state and in submerged fermentation for production of microbial linamarase in large-scale bioreactors for bulk production of the enzyme for utilization in the cassava-based food industry.

References

Abban, S., Brimer, L., Abdelgadir, W.S., Jakobsen, M. and Thorsen, L. 2013. Screening for *Bacillus subtilis* group isolates degrade cyanogens at pH 4. 5–5.0. International Journal of Food Microbiology, 161(1): 31–35.

Achi, O.K. and Akomas, N.S. 2006. Comparative assessment of fermentation techniques in the processing of fufu, a traditional fermented cassava product. Pakistan Journal of Nutrition, 5(3): 224–229.

Aguoru, C.U., Onda, M.A., Omoni, V.T. and Ogbonna, I.O. 2015. Characterization of moulds associated with processed garri stored for 40 days at ambient temperature in Makurdi, Nigeria. African Journal of Biotechnology, 13(5): 673–677.

Ahaotu, I., Ogueke, C.C., Owuamanam, C.I., Ahaotu, N.N. and Nwosu, J.N. 2013. Fermentation of un-dewatered cassava pulp by linamarase-producing microorganisms: Effect on nutritional composition and residual cyanide. American Journal of Food and Nutrition, 3(1): 1–8.

Akindahunsi, A.A., Oboh, G. and Oshodi, A.A. 1999. Effect of fermenting cassava with *Rhizopus oryzae* on the chemical composition of its flour and gari products. Rivista Italiana delle Sostanze Grasse, 76: 437–440.

Amoa-Awua, W.K., Frisvad, J.C., Sefa-Dedeh, S. and Jakobsen, M. 1997. The contribution of moulds and yeasts to the fermentation of 'agbelima' cassava dough. Journal of Applied Microbiology, 83(3): 288–296.

Amoa-Awua, W.K., Owusu, M. and Feglo, P. 2005. Utilization of unfermented cassava flour for production of an indigenous African fermented food, agbelima. World Journal of Microbiology and Biotechnology, 21: 1201–1207.

Ampe, F. and Brauman, A. 1995. Origin of enzymes involved in detoxification and root softening during cassava retting. World Journal of Microbiology and Biotechnology, 11(2): 178–182.

Arihantana, M.B. and Buckle, K.A. 1987. Cassava detoxification during tape fermentation with traditional inoculum. International Journal of Food Science and Technology, 22(1): 41–48.

Asegbeloyin, J.N. and Onyimonyi, A.E. 2007. The effect of different processing methods on the residual cyanide of 'gari'. Pakistan Journal of Nutrition, 6: 163–166.

Barus, T., Kristani, A. and Yulandi, A. 2013. Diversity of amylase-producing *Bacillus* spp. From 'Tape' (fermented cassava). HAYATI Journal of Biosciences, 20: 94–98.

Blagbrough, I.S., Bayoumi, S.A., Rowan, M.G. and Beeching, J.R. 2010. Cassava: An appraisal of its phytochemistry and its biotechnological prospects. Phytochemistry, 71(17): 1940–1951.

Cach, N.T., Lenis, J.I., Perez, J.C., Morante, N., Calle, F. and Ceballos, H. 2006. Inheritance of useful traits in cassava grown in sub-humid conditions. Plant Breeding, 125(2): 177–182.

Cardoso, A.P., Mirione, E., Ernesto, M., Massaza, F., Cliff, J., Haque, M.R. and Bradbury, J.H. 2005. Processing of cassava tubers to remove cyanogens. Journal of Food Composition and Analysis, 18(5): 451–460.

Chikezie, P.C. and Ojiako, O.A. 2013. Cyanide and aflatoxin loads of processed cassava (*Manihot esculenta*) tubers (Gari) in Njaba, Imo state, Nigeria. Toxicology International, 20(3): 261.

Daouda, N., Achille, T.F., Abodjo, K.C., Charlemagne, N. and Georges, A.N.G. 2012. Influence of traditional inoculum and fermentation time on the organoleptic quality of 'Attiéké'. Food and Nutrition Sciences, 3(10): 1335–1340.

Defloor, I., Nys, M. and Delcour, J.A. 1993. Wheat starch, cassava starch, and cassava flour impairment in the breadmaking potential of wheat flour. Cereal Chemistry, 70: 526–526.

Demiate, I.M. and Kotovicz, V. 2011. Cassava starch in the Brazilian food industry. Food Science and Technology (Campinas), 31(2): 388–397.

Dhellot, J.R., Mokemiabeka, S.N., Moyen, R., Kobawila, S.C. and Louembe, D. 2015. Volatile compounds produced in two traditional fermented foods of Congo: Nsamba (palm wine) and bikedi (retted cassava dough). African Journal of Biotechnology, 13: 4119–4123.

Djeni, N.T., N'guessan, K.F., Toka, D.M., Kouame, K.A. and Dje, K.M. 2011. Quality of attieke (a fermented cassava product) from the three main processing zones in Côte d'Ivoire. Food Research International, 44(1): 410–416.

Eduardo, M., Svanberg, U., Oliveira, J. and Ahrné, L. 2013. Effect of cassava flour characteristics on properties of cassava-wheat-maize composite bread types. International Journal of Food Science, 2013.

Esser, A.J.A. and Jurgens, C.M. 1999. Contribution of selected fungi to the reduction of cyanogen levels during solid-substrate fermentation of cassava. Journal of Food Technology in Africa, 4(2).

Essers, A.J.A., Bennik, M.H.J. and Nout, M.J.R. 1995. Mechanisms of increased linamarin degradation during solid-substrate fermentation of cassava. World Journal of Microbiology and Biotechnology, 11(3): 266–270.

Ezekiel, O.O. and Aworh, O.C. 2013, March. Solid state fermentation of cassava peel with *Trichoderma viride* (ATCC 36316) for protein enrichment. *In*: Proceedings of World Academy of Science, Engineering and Technology (No. 75, p. 667). World Academy of Science, Engineering and Technology (WASET).

Ezekiel, O.O., Aworh, O.C., Blaschek, H.P. and Ezeji, T.C. 2010. Protein enrichment of cassava peel by submerged fermentation with *Trichoderma viride* (ATCC 36316). African Journal of Biotechnology, 9: 177–194.

Falade, K.O. and Akingbala, J.O. 2010. Utilization of cassava for food. Food Review International, 27(1): 51–83.

Ferraro, V., Piccirillo, C. and Pintado, M.E. 2015. Cassava (*Manihot esculenta* Crantz) and yam (*Discorea* spp.) crops and their derived foodstuffs: Safety, security and nutritional value. Critical Reviews in Food Science and Nutrition (just-accepted), 00-00.

Garrido, L.H., Schnitzler, E., Zortéa, M.E.B., de Souza Rocha, T. and Demiate, I.M. 2014. Physicochemical properties of cassava starch oxidized by sodium hypochlorite. Journal of Food Science and Technology, 51(10): 2640–2647.

Giraud, E., Gosselin, L. and Raimbault, M. 1993. Production of a *Lactobacillus plantarum* starter with linamarase and amylase activities for cassava fermentation. Journal of the Science of Food and Agriculture, 62(1): 77–82.

Gleadow, R.M. and Woodrow, I.E. 2002. Mini-Review: Constraints on effectiveness of cyanogenic glycosides in herbivore defense. Journal of Chemical Ecology, 28(7): 1301–1313.

Kobawila, S.C., Louembe, D., Keleke, S., Hounhouigan, J. and Gamba, C. 2005. Reduction of the cyanide content during fermentation of cassava tubers and leaves to produce bikedi and ntoba mbodi, two food products from Congo. African Journal of Biotechnology, 4(7): 689–696.

Kostinek, M., Specht, I., Edward, V.A., Schillinger, U., Hertel, C., Holzapfel, W.H. and Franz, C.M. 2005. Diversity and technological properties of predominant lactic acid bacteria from fermented cassava used for the preparation of gari, a traditional African food. Systematic and Applied Microbiology, 28(6): 527–540.

Lacerda, I.C., Miranda, R.L., Borelli, B.M., Nunes, Á.C., Nardi, R.M., Lachance, M.A. and Rosa, C.A. 2005. Lactic acid bacteria and yeasts associated with spontaneous fermentations during the production of sour cassava starch in Brazil. International Journal of Food Microbiology, 105(2): 213–219.

Lambri, M., Fumi, M.D., Roda, A. and De Faveri, D.M. 2013. Improved processing methods to reduce the total cyanide content of cassava tubers from Burundi. African Journal of Biotechnology, 12(19): 2685–2691.

Lei, V., Amoa-Awua, W.K.A. and Brimer, L. 1999. Degradation of cyanogenic glycosides by *Lactobacillus plantarum* strains from spontaneous cassava fermentation and other microorganisms. International Journal of Food Microbiology, 53(2): 169–184.

Mlingi, N.L.V., Bainbridge, Z.A., Poulter, N.H. and Rosling, H. 1995. Critical stages in cyanogen removal during cassava processing in southern Tanzania. Food Chemistry, 53(1): 29–33.

Mohd Aripin, A., Mohd Kassim, A.S., Daud, Z. and Mohd Hatta, M.Z. 2013. Cassava peels for alternative fibre in pulp and paper industry: Chemical properties and morphology characterization. International Journal of Integrated Engineering, 5: 30–33.

Montagnac, J.A., Davis, C.R. and Tanumihardjo, S.A. 2009. Processing techniques to reduce toxicity and antinutrients of cassava for use as a staple food. Comprehensive Reviews in Food Science and Food Safety, 8(1): 17–27.

Moorthy, S.N. and Mathew, G. 1998. Cassava fermentation and associated changes in physicochemical and functional properties. Critical Reviews in Food Science and Nutrition, 38(2): 73–121.

Murugan, K., Sekar, K. and Al-Sohaibani, S. 2014. Detoxification of cyanides in cassava flour by linamarase of *Bacillus subtilis* KM05 isolated from cassava peel. African Journal of Biotechnology, 11(28): 7232–7237.

Muzanila, Y.C., Brennan, J.G. and King, R.D. 2000. Residual cyanogens, chemical composition and aflatoxins in cassava flour from Tanzanian villages. Food Chemistry, 70(1): 45–49.

Ngaba, P.R. and Lee, J.S. 1979. Fermentation of Cassava (*Manihot esculenta* Crantz). Journal of Food Science, 44(5): 1570–1571.

Nuraida, L. 2015. A review: Health promoting lactic acid bacteria in traditional Indonesian fermented foods. Food Science and Human Wellness, 4(2): 47–55.

Nwokoro, O. and Anya, F.O. 2011. Linamarase enzyme from *Lactobacillus delbrueckii* NRRL B-763: Purification and some properties of a β-glucosidase. Journal of the Mexican Chemical Society, 55(4): 246–250.

Obilie, E.M., Tano-Debrah, K. and Amoa-Awua, W.K. 2003. Microbial modification of the texture of grated cassava during fermentation into akyeke. International Journal of Food Microbiology, 89(2): 275–280.

Obilie, E.M., Tano-Debrah, K. and Amoa-Awua, W.K. 2004. Souring and break down of cyanogenic glucosides during the processing of cassava into akyeke. International Journal of Food Microbiology, 93(1): 115–121.

Oboh, G. and Akindahunsi, A.A. 2003. Biochemical changes in Cassava products (flour and gari) subjected to *Saccharomyces cerevisae* solid media fermentation. Food Chemistry, 82(4): 599–602.

Oduro, I., Ellis, W.O., Dziedzoave, N.T. and Nimako-Yeboah, K. 2000. Quality of gari from selected processing zones in Ghana. Food Control, 11(4): 297–303.

Ogbonnaya, N. 2016. Linamarase production by some microbial isolates and a comparison of the rate of degradation of cassava cyanide by microbial and cassava linamarases. Hemijska Industrija, 21-21.

Oguntoyinbo, F.A. and Dodd, C.E. 2010. Bacterial dynamics during the spontaneous fermentation of cassava dough in gari production. Food Control, 21(3): 306–312.

Okafor, N., Umeh, C. and Ibenegbu, C. 1998. Amelioration of gari, a cassava-based fermented food by the inoculation of microorganisms secreting amylase, lysine and linamarase into cassava mash. World Journal of Microbiology and Biotechnology, 14(6): 835–838.

Okpokiri, A.O., Ijioma, B.C., Alozie, S.O. and Ejiofor, M.A.N. 1985. Production of improved cassava fufu. Nigerian Food Journal, 2 (2,3): 145–148.

Padonou, S.W., Nielsen, D.S., Hounhouigan, J.D., Thorsen, L., Nago, M.C. and Jakobsen, M. 2009. The microbiota of Lafun, an African traditional cassava food product. International Journal of Food Microbiology, 133(1): 22–30.

Pandey, A. 2003. Solid-state fermentation. Biochemical Engineering Journal, 13(2): 81–84.

Poonpairoj, P. and Chitradon, L. 2011. Renewable utilization of cassava coat solid waste using fungal enzyme technology. Kasetsart Journal: Natural Sciences, 45(2): 260–267.

Purwadaria, T. 2014. Solid substrate fermentation of cassava peel for poultry feed ingredient. WARTAZOA. Indonesian Bulletin of Animal and Veterinary Sciences, 23(1): 15–22.

Rattanachomsri, U., Tanapongpipat, S., Eurwilaichitr, L. and Champreda, V. 2009. Simultaneous non-thermal saccharification of cassava pulp by multi-enzyme activity and ethanol fermentation by *Candida tropicalis*. Journal of Bioscience and Bioengineering, 107(5): 488–493.

Ray, R.C. and Sivakumar, P.S. 2009. Traditional and novel fermented foods and beverages from tropical root and tuber crops: Review. International Journal of Food Science and Technology, 44(6): 1073–1087.

Ray, R.C. and Swain, M.R. 2011. Bio-ethanol, bioplastics and other fermented industrial products from cassava starch and flour. Cassava: Farming, uses and economic impacts. (Ed. Colleen M. Pace), Nova Science Publishers Inc., Hauppauge, New York, USA, pp. 1–32.

Ray, R.C., Mohapatra, S., Panda, S. and Kar, S. 2007. Solid substrate fermentation of cassava fibrous residue for production of α-amylase, lactic acid and ethanol. Journal of Environmental Biology, 29(1): 111–115.

Sayre, R., Beeching, J.R., Cahoon, E.B., Egesi, C., Fauquet, C., Fellman, J. and Maziya-Dixon, B. 2011. The bio-cassava plus program: Biofortification of cassava for sub-Saharan Africa. Annual Review of Plant Biology, 62: 251–272.

Setter, T.L. and Fregene, M.A. 2007. Recent advances in molecular breeding of cassava for improved drought stress tolerance. pp. 701–711. *In*: Advances in Molecular Breeding Toward Drought and Salt Tolerant Crops. Springer, Netherlands.

Shittu, T.A., Dixon, A., Awonorin, S.O., Sanni, L.O. and Maziya-Dixon, B. 2008. Bread from composite cassava-wheat flour. II: Effect of cassava genotype and nitrogen fertilizer on bread quality. Food Research International, 41(6): 569–578.

Singhania, R.R., Patel, A.K., Soccol, C.R. and Pandey, A. 2009. Recent advances in solid-state fermentation. Biochemical Engineering Journal, 44(1): 13–18.

Sornyotha, S., Kyu, K.L. and Ratanakhanokchai, K. 2010. An efficient treatment for detoxification process of cassava starch by plant cell wall-degrading enzymes. Journal of Bioscience and Bioengineering, 109(1): 9–14.

Tijani, I.D.R., Jamal, P., Alam, M.Z. and Mirghani, M.E.S. 2012. Optimization of cassava peel medium to an enriched animal feed by the white rot fungi *Panus tigrinus* M609RQY. International Food Research Journal, 19(2): 427–432.

Tritscher, A., Miyagishima, K., Nishida, C. and Branca, F. 2013. Ensuring food safety and nutrition security to protect consumer health: 50 years of the Codex Alimentarius Commission. Bulletin of the World Health Organization, 91(7): 468–468.

Udoro, E.O., Kehinde, A.T., Olasunkanmi, S.G. and Charles, T.A. 2014. Studies on the physicochemical, functional and sensory properties of gari processed from dried cassava chips. Journal of Food Processing and Technology, 5: 293. doi:10.4172/2157-7110.1000293.

Umeh, S.O. and Odibo, F.J.C. 2014. Isolation of starter cultures to be used for cassava tuber retting to produce fufu. Journal of Global Biosciences ISSN, 3(2): 520–528.

Yao, A.A., Dortu, C., Egounlety, M., Pinto, C., Edward, V.A., Huch, M. and Thonart, P. 2009. Production of freeze-dried lactic acid bacteria starter culture for cassava fermentation into gari. African Journal of Biotechnology, 8(19): 4996–5004.

19

Enzymes in Dairy Processing

Ashok Kumar Mohanty[1],* and *Pradip Behare*[2]

1. Introduction

Milk and milk products contain a large number of enzymes, either natural or produced by certain microorganisms. The most common enzymes present in dairy products are proteinases, lipase, lactase, catalase, xanthine oxidase, lactoperoxidase, etc. These enzymes play a very important role not only in preserving natural values of milk, produce desirable products when used judiciously, but also cause spoilage in dairy products (Table 1). Among these, some enzymes have been traditionally used to manufacture various dairy products. The well-known dairy enzyme used in cheese manufacture is rennet, a collective name for commercial preparations containing acid proteases. Various other proteinases from plant, animal and microbial sources have been tried to replace conventional rennet/chymosin to fulfill the demands of the cheese industry. Recombinant chymosin has also been produced which is similar to calf rennet and better than other rennet substitutes. In addition to the milk-clotting enzymes which are used for production of various cheeses, the dairy sector uses enzymes such as lipases, lactase, lysozyme and lactoperoxidase for improving flavor, lactose hydrolysis, accelerated cheese ripening and control of microbial spoilage. One of the most promising applications of enzymes is in the cleaning of dairy equipment, removal of milk stones and biofilm formation which give alternative option to conventionally available detergents and other cleaning agents. In this chapter, the role of enzymes in cheese, fermented milk products, removal of off-flavors in milk products and as bio-detergents for cleaning dairy equipments is discussed.

[1] Principal Scientist, Animal Biotechnology Centre, ICAR-National Dairy Research Institute, Karnal-132001, Haryana, India.
[2] Scientist, Dairy Microbiology Division, ICAR-National Dairy Research Institute, Karnal-132001, Haryana, India.
 E-mail: pradip_behare@yahoo.com
* Corresponding author: ashokmohanty1@gmail.com

Table 1. Examples of Enzyme used in dairy processing.

Sr. No.	Name of Enzymes	Application
1	Acid proteinases	Milk coagulation
2	Neutral proteinases and peptidases	Accelerated cheese ripening, de-bittering, enzyme-modified cheese, production of hypoallergenic milk-based foods
3	Lipases	Accelerated cheese ripening, enzyme-modified cheese, flavor modified cheese, structurally-modified milk fat products
4	β-galactosidase	Reduced lactose whey products
5	Lactoperoxidase	Cold sterilization of milk, milk replacers for calves
6	Lysozyme	Nitrate replacer for washed curd cheeses and cheeses with eyes

2. Enzymes for Cheese Industry

Rennet and rennet substitutes are most important in the cheese industry.

2.1 Rennet and its substitutes

Cheese can be defined as a product or substance formed by coagulation of the milk of certain mammals by rennet or similar enzymes in the presence of lactic acid produced by added or adventitious microorganisms from which a portion of the moisture has been removed by cutting, warming and/or pressing that has been shaped in mould and then ripened by holding for some time at suitable temperature and humidity. Several types of cheese are manufactured all over the world with rennet being an essential component during its manufacture. The term 'rennet' has been used to describe the saline extract from the abomasums (4th stomach, veal or true stomach) of the young (< 30 days old) milk-fed calves. The term, however, may be used for a substitute preparation as well, provided a name modifier is used (e.g., pepsin rennet, bovine rennet, microbial rennet, fungal rennet, etc.). When the term 'rennet' is used without qualification, it refers to calf rennet. Calf rennet or standard rennet contains chymosin (rennin) as the principal enzyme. It is conventionally used as a milk-clotting agent for the manufacture of quality cheeses with good flavor and texture. Owing to an increase in demand for cheese production across the world in the last three decades, coupled with reduced supply of calf rennet, a search has begun for rennet substitutes. Various rennet substitutes including vegetable, plant, animal and microbial sources have been tried and some of those are available commercially (Table 2). Among these, microbial rennet has received wide popularity.

Rennin acts on the milk protein in two stages, by enzymatic and by non-enzymatic action, resulting in coagulation of milk. In the enzymatic phase, the resultant milk becomes a gel due to the influence of calcium ions and the temperature used in the process. Many microorganisms are known to produce rennet-like proteinases which can substitute the calf rennet. Microorganisms, like *Rhizomucor pusillus*, *Rhizomucor miehei*, *Cryphonectria parasitica* and *Aspergillus oryzae*, are used extensively for rennet production in cheese manufacture. One major drawback of microbial rennet use in cheese manufacture is the development of off-flavor and bitter taste in the non-ripened as well as in ripened cheeses.

Table 2. Major commercial milk-clotting enzymes.

Name	Source	Active-Principal
Rennet	Fourth stomach of calves, kids or lambs	Chymosin
Bovine rennet	Fourth stomach of bovine (cows, buffaloes)	Pepsin/Chymosin
Pepsin	Pig stomach	Pepsin
Fungal rennet *Rhizomucor* species *Cryphonectriaparasitica*	Fermentation liquor of *Rhizomucor miehei, Rhizomucor pusillus* Fermentation liquor of *Cryphonectria parasitica*	Protease Protease
Recombinant calf chymosin	Fermentation derived chymosin from *Escherichia coli*; *Kluyveromyces lactis*; *Aspergillus niger*	Chymosin

Source: (Upadhyay, 2003)

2.2 Recombinant rennet

The rennets from microbial sources are more proteolytic in nature in comparison to rennet from animal sources, resulting in production of some bitter peptides during the process of cheese ripening. Hence, attempts have been made to clone the gene for calf chymosin and to express it in selected bacteria, yeasts and moulds (Table 3). FDA has also given GRAS status to recombinant chymosin and approved its use in cheese manufacture. Many different laboratories have cloned the gene for calf prochymosin in *Escherichia coli,* and analyzed the structure of the gene as well as the properties of the recombinant chymosin (Mohanty et al., 1999). The expressed proenzyme in *E. coli* is present mainly as insoluble inclusion bodies, comprising reduced prochymosin as well as molecules that are interlinked by disulphide bridges. After disintegration of the cells, inclusion bodies are harvested by centrifugation. Individual laboratories have reported some differences in the procedure for renaturation of prochymosin from the inclusion bodies, but all have followed the same general scheme. The enzymatic properties of recombinant *E. coli* chymosin are indistinguishable from those of the native calf chymosin.

The gene for prochymosin has also been cloned in *Saccharomyces cerevisiae*; the levels of expression have been reported to be 0.5 to 2.0 per cent of total yeast protein. In yeast, about 20 per cent of the prochymosin can be released in soluble form which can be activated directly; the remaining 80 per cent is still associated with the cell debris. The prosequence is essential for correct folding of the polypeptide chain in prochymosin. Therefore, the recombinant chymosin produced by using only chymosin gene sequence (open reading frame) cannot clot the milk. The zymogen for the aspartic

Table 3. Some examples of expression of chymosin in bacteria, yeasts and moulds.

Species	Forms of Chymosin, Plasmid and Promoter Used	Expression host
Calf	Pro, p501, Tryptophan	*E. coli*
Bovine	Pro, p SM316, Neutral protease	*Bacillus subtilis*
Goat	Pro, pQSec1, pGK	*Kluyveromyces lactis*
Buffalo	Prepro, pro, chymosin, pGAPZáA, pGAP	*Pichia pastoris*
Camel	Pro, pGAMpR, glaA	*Asperigillus niger*

Source: Kumar et al. (2010)

proteases from *R. pusillus*, also called mucor rennin, has likewise been cloned and expressed in yeast. Studies on its conversion to active form showed that secretion of *R. pusillus* protease from recombinant yeast was dependent on glycosylation of the enzyme.

Compared to yeast, the filamentous fungi generally secrete larger quantities of proteins into the culture medium. Furthermore, filamentous fungi secrete the heterologous proteins with correct folding of the polypeptide chain and process correct pairing of sulphydryl groups. The gene for *R. miehei* protease has been expressed in *A. oryzae*. Prochymosin has been expressed in *Kluyveromyces lactis, A. nidulans, A. niger* and *Trichoderma reesei*. In most of the cases, the reported yields of the model systems were about 10–40 mg of enzyme per liter of culture medium. However, 3.3 g of enzyme per liter was also achieved. Yeast, *Kluyveromyces lactis,* has been used as an efficient host for the secretion of recombinant chymosin, which has led to a large-scale production process for chymosin. If produced on an industrial scale, the yields will perhaps be of a large magnitude. Most of the companies produce recombinant rennet of cattle calf origin in different microbial hosts. However in India, the major source of milk is buffalo, showing a different composition from that of the cow. The rennet from buffalo source has inherent compatibility for clotting buffalo milk (Mohanty et al., 2003). National Dairy Research Institute, Karnal, India has taken a lead in cloning the gene of buffalo chymosin. The buffalo chymosin cDNA has been cloned in *E. coli* and the sequence of buffalo chymosin indicates that it is highly homologous to the cattle chymosin. Since most of the rennet (> 90 per cent) added to cheese milk is lost in the whey, immobilization would considerably extend its catalytic life but efficiency as a milk coagulant has been questioned. So, there is a fairly general support for the view that immobilized enzymes cannot coagulate milk properly owing to inaccessibility of the Phe–Met peptide bond of k-casein, and that the apparent coagulating activity of immobilized rennets is due to leaching of the enzyme from the support.

Different types of conventional cheeses have been successfully made by using recombinant rennet on an experimental or pilot scale. No major differences have been detected between cheeses made with recombinant chymosins or natural enzymes, regarding cheese yield, texture, smell, taste or ripening. The recombinant chymosins are identical with calf rennet according to a report on biochemical and genetic evidences.

2.3 Enzymes in accelerated cheese ripening

Cheese ripening is a complex process mediated by biochemical and biophysical changes during which a bland curd is developed into a mature cheese with characteristic flavor, texture and aroma. The desirable attributes are produced by the partial and gradual break down of carbohydrates, lipids and proteins during ripening, mediated by several agents, viz. (i) residual coagulants, (ii) starter bacteria and their enzymes, (iii) nonstarter bacteria and their enzymes, (iv) indigenous milk enzymes, especially proteinases and (v) secondary microflora with their enzymes.

The proteinases used in cheese processing include (i) plasmin, (ii) rennet and (iii) proteinases (cell wall and/or intracellular) of the starter and nonstarter bacteria. Approximately 6 per cent of the rennet added to cheese milk remains in the curd after manufacture and contributes significantly to proteolysis during ripening. Combinations of individual neutral proteinases and microbial peptidases intensify cheese flavor and when used in combination with microbial rennets, reduce the intensity of bitterness caused by the latter. Acid proteases in isolation from microbial sources cause intense bitterness. Various animal or microbial lipases impart the characteristic cheese flavor, low bitterness and strong rancidity, while lipases in combination with proteinases and/or peptidases bestow good cheese flavor with low levels of bitterness. In a more balanced approach to the acceleration of cheese ripening using mixtures of proteinases and peptidases, attenuated starter cells or Cell-Free Extracts (CFE) are being preferred.

Cheese ripening is essentially an enzymatic process which can be accelerated by augmenting the activity of key enzymes. This has the advantage of initiating more specific action for flavor development compared to use of elevated temperatures that can result in accelerating undesirable nonspecific reactions, and consequently off-flavor development. Enzymes may be added to develop specific flavors in cheeses, for example, lipase addition for the development of Parmesan or Blue-type cheese flavors. The pathways leading to the formation of flavor compounds are largely unknown and therefore, the use of exogenous enzymes to accelerate ripening is mostly an empirical process. Different microbial enzymes used to accelerate cheese ripening are presented in Table 4.

Table 4. Microbial enzymes used to accelerate cheese ripening.

Enzyme	Source
Microbial lipases	*Aspergillus niger, A. oryzae*
Lactase	*L. lactis, Kluyveromyces* sp., *E. coli*
Microbial serine proteinases	*A. niger*
Neutral proteinases	*B. subtilis, A. oryzae*

Source: Neelakantan et al. (1998)

3. Enzymes for Fermentation of Milk

Lactase, proteinases and peptidases are important for fermentation of milk.

3.1 *Lactase (β-galactosidase)*

Lactases are classified as hydrolases, which catalyze the terminal residue of β-lactose galactopiranosil (Galb1-4Glc) and produce glucose and galactose (Carminatti, 2001). Milk and milk products are the exclusive source of lactose in human diet. Before absorption, lactose is hydrolyzed by the intestinal *β*-galactosidase (lactase) into glucose and galactose. These monosaccharides are absorbed and used as energy sources in the body. However, persons having impaired lactose metabolism cannot utilize lactose in milk, resulting in lactose maldigestion. Lactase is used to hydrolyze lactose as a digestive aid and to improve the solubility and sweetness of various dairy products. Lactose hydrolysis helps lactose-intolerant people to drink milk and eat various dairy products (Qureshi et al., 2015). Lactose, the sugar found in milk and whey, and its corresponding hydrolase (lactase or β-galactosidase) have been extensively researched during the past decade. The enzyme immobilization technique has created new and interesting possibilities for utilization of this sugar.

Among the dairy products, fermented milk made by use of lactic acid bacteria (*Lactococcus, Lactobacillus, Streptococcus* species, etc.) contains a good amount of lactase enzyme. Yoghurt bacteria produce functional lactase, the enzyme which hydrolyses 20–30 per cent of lactose to its absorbable monosaccharides. This expression may also contribute to better tolerance of lactose in yoghurt than of milk by persons with lactose maldigestion. Manufacturers of ice-cream, yoghurt and frozen desserts use lactase to improve scoop and creaminess, sweetness and digestibility and to reduce sandiness due to crystallization of lactose in concentrated preparations. Cheese manufactured from hydrolyzed milk ripens more quickly than the cheese manufactured from normal milk. The cheese-manufacturing industry produces large quantities of whey as a byproduct in which lactose present 70–75 per cent of the whey solids. The hydrolysis of lactose by lactase converts whey into more useful food ingredients. Lactases have also been used in the processing of dairy wastes and as a digestive aid taken by humans in tablet form when consuming dairy products.

Lactase can be obtained from various sources like plants, animal organs, bacteria, yeasts (intracellular enzyme), or moulds. Some of these sources are used for commercial enzyme preparations. Lactase preparations from *A. niger, A. oryzae* and *Kluyveromyces lactis* are considered safe. The most investigated *E. coli* lactase is not used in food processing because of its cost and safety issues. Properties of lactase (temperature and pH optima) from different microorganisms determine their applications. Fungal lactases, which work in the acidic range of pH 2.5–4.5, are used for acid whey hydrolysis. Yeast and bacterial lactase in neutral pH 6–7.5 are suitable for milk and sweet whey hydrolysis. Another important property determining the application of the enzyme is inhibition of enzyme activity by the product galactose. Enzyme, more susceptible to inhibition by galactose, is operative only in a dilute solution of whey (at low lactose level) in an immobilized column.

3.2 Proteinases and peptidases

During the manufacture of fermented milk, lactic acid bacteria possess various proteinases and peptidases that act upon the protein component present in milk. The proteolytic system of lactic acid bacteria is not only essential for their growth but also contributes significantly to flavor development in fermented milk products. The proteinases of lactic acid bacteria include extracellular proteinases, endopeptidases, aminopeptidases, tripeptidases and proline-specific peptidases, which are all serine proteases. Aminopeptidases are important for the development of flavor in fermented milk products, since they are capable of releasing single amino acid residues from oligopeptides formed by extracellular proteinase activity. The functional properties of milk proteins may be improved by limited proteolysis through the enzymatic modification of milk proteins. An acid-soluble casein, free of flavor and suitable for incorporation into beverages and other acid foods, has been prepared by limited proteolysis. The antigenicity of casein is destroyed by proteolysis and the hydrolysate is suitable for use in milk-protein-based foods for infants allergic to cow's milk.

4. Enzymes for Removal of Off-Flavour and Improving Shelf-Life of Milk

The other enzymes having limited applications in dairy processing include glucose oxidase, catalase, Sulfhydryl oxidase, lactoperoxidase and lysozymes.

4.1 Glucose oxidase

Glucose oxidase is an enzyme that catalyzes the oxidation of beta-D-glucose with the formation of D-gluconolactone. The enzyme contains the prosthetic group, flavin adenine dinucleotide (FAD), which enables the protein to catalyze oxidation-reduction reactions. It is often used together with catalase in selected foods for preservation (Law, 2002). The enzyme is mainly used in the food industry for the removal of harmful oxygen. Packaging materials and storage conditions are vital to maintain the quality of products containing probiotic microorganisms since the microbial group's metabolism is essentially anaerobic or microaerophilic (Mattila Sandholm et al., 2002). Oxygen level during storage should be consequently minimal to avoid toxicity, the organism's death and the consequent loss of the product's functionality. Glucose oxidase may be a biotechnological asset to increase stability of probiotic bacteria in yoghurt without chemical additives. It may thus be a biotechnological alternative.

4.2 Catalase

Catalase decomposes hydrogen peroxide. The ability of catalase to break down hydrogen peroxide into water and oxygen helps to prevent the breakdown of hydrogen peroxide into free radicals. Free radicals can go on to cause lipid oxidation, thus creating many off-flavors commonly found in milk and dairy ingredients. In this sense, catalase may have a positive effect on flavor (Campbell and Drake, 2013). However, catalase as a potential source of off-flavor in milk and dairy products has yet to be evaluated in detail. Catalase was among the first enzymes demonstrated in milk. Catalase can be an indigenous, exogenous, or endogenous enzyme, as it is found indigenously in milk and can also be produced during cheese manufacture by coryneform bacteria and yeasts; however, it may also be added exogenously in some dairy processes. In raw milk, high levels of catalase are associated with mastitic infection and high levels of somatic cell count in milk (Fox and Kelly, 2006). Similar to many other enzymes, catalase activity varies in milk due to feed, stage of lactation and mastitis.

4.3 Sulfhydryl oxidase

Sulfhydryl oxidase catalyzes the oxidation of free sulfhydryls, forming a disulfide bond and H_2O_2. This enzyme has been isolated from variety of sources such as bovine milk, bovine kidney, *Asperigillus niger* and other microbial species. Thermally induced generation of volatile sulphydryl groups is thought to be responsible for the cooked off-flavor in ultra high temperature (UHT) processed milk. Use of sulphydryl oxidase under aseptic conditions can eliminate the cooked flavor by catalyzing the oxidation of thiols responsible for this defect.

4.4 Lipases

Lipids that are present in dairy products can be enzymatically degraded by lipases either via oxidation or hydrolysis. Lipases can generally be categorized as enzymes that catalyze the hydrolysis of lipids, which are the major lipid component of milk (Deeth, 2006). Lipoprotein lipase accounts for most of the native lipolytic activity in bovine milk and is normally associated with the casein micelle. Lipases may also come from bacterial sources, such as *Pseudomonas* during cold storage of milk. Lipases from these bacteria are notably different from Lipoprotein lipase. Lipolysis in milk can alter both flavor and functionality of dairy products (Deeth, 2006). It makes an important contribution to Swiss cheese flavors, mainly due to the lipolytic enzymes of the starter cultures. The characteristic peppery flavour of Blue cheese is due to the short-chain fatty acids and methyl ketones. Most of the lipolysis in Blue cheese is catalysed by *Penicillium roqueforti* lipase, with a lesser contribution by indigenous milk lipase.

4.5 Phospholipases

Phospholipids, despite constituting approximately 0.5 per cent of the total lipid in bovine milk, have a critical role in stabilizing milk fat globules against coalescence. They also have different technological roles in dairy products, e.g., by coating powder particles, they provide higher foam volumes in aerated products and act as co-emulsifiers. Phospholipases can be used to modify phospholipid functionality in dairy processing by improving fat stability or increasing product yield. Partial hydrolysis of phospholipids increases cheese yield. Mozzarella cheese manufactured from milk hydrolyzed with fungal phospholipase A1 before renneting, reduced fat loss in whey and cooking water as well as increased cheese yield resulted due to improved fat and moisture retention. Phospholipase can also be used in the dairy industry to improve the foaming properties of whey protein.

4.6 Lactoperoxidase

A member of the oxidoreductase family, lactoperoxidase plays an important role in protecting newborn infants against pathogenic microorganisms (Seifu et al., 2005). Lactoperoxidase is a normal component of mammalian milk and has been found in all mammalian milks. It occurs naturally in raw milk, colostrum and saliva; it is thought to be part of the protective system for suckling animals against enteric infections. Lactoperoxidase is bactericidal to gram-negative bacteria, and bacteriostatic to gram-positive. It is a peroxidase that uses hydrogen peroxide to oxidise the thiocyanate ion to hypothiocyanate, the active bactericidal molecule. LP system irreversibly inhibits the membrane-energizing D-lactate dehydrogenase in gram-negative bacteria, leading to cell death. In gram-positive bacteria, the membrane ATP-ase is reversibly inhibited and may form the basis of bacteriostasis, The natural inhibitory mechanism in raw milk is due to the presence of low levels of lactoperoxidase, which can be activated by the external addition of traces of H_2O_2 and thiocyanate. It has been reported that the potential of LP-system and its activation enhance the keeping quality of milk.

4.7 Lysozyme

Lysozyme is a hydrolase widely distributed in Nature. It is bactericidal to many Gram-positive species because it breaks down their cell walls. The enzyme is a mucopeptide W-acetyl murarnoyihydrolase, available commercially from hen's egg white or *Micrococcus lysodiekticus*. The food-grade preparations are from egg albumin. Lysozyme is sold by major dairy enzyme suppliers as an alternative control agent for 'late blowing', the textural defect of slits and irregular holes caused by the butyric fermentation in Gouda, Emmental and other important hard and semi-hard cheese varieties. Traditionally the defect, caused by *Clostridium tyrobutyricum* in raw milk has been controlled by the addition of potassium nitrate to cheesemilk. However, this practice will be phased out because it is associated with the production of carcinogens, and lysozyme has become the preferred control agent. *Clostridium tyrobutyricum* is a spore-former and as such, cannot be killed by pasteurization, hence the lysozyme treatment to the milk can become an alternative method. Lysozyme kills vegetative cells and also inhibits outgrowth of spores in cheese. It is stable for long periods in the cheese matrix and because it binds to the cheese curd, little of the enzyme is lost on whey separation. Although lysozyme also inhibits the lactic acid bacteria used as starters in cheesemaking, they are less sensitive than Clostridia. Lysozyme also inhibits the growth of *Listeria monocytogenes* in yoghurt and fresh cheese with high acidity (< pH 5.0), but the effect is not consistent enough to rely for commercial fermented milk products and in any case, high acidity is usually sufficient in itself to inhibit these pathogens. Cow's milk can be provided with protective factors through the addition of lysozyme, making it suitable as infant milk. Lysozyme acts as a preservative by reducing bacterial counts in milk.

4.8 Transglutaminase

The only cross-linking enzyme that is currently available for catalysing covalent bond formation between protein molecules on a commercial scale is transglutaminase (Dickinson, 1997). With the recent availability of commercial microbially-derived transglutaminase preparations, there has been considerable interest in their application in the gelation of caseins and whey proteins. When applying transglutaminase in dairy products, it is possible to increase gel strength, surface viscosity, water-holding capacity, stability, rennetability and mechanical properties and decrease permeability (Özrenk, 2006). Transglutaminase is effective in reducing syneresis in acid milk gels and has been investigated as a method of improving the texture and shelf-life of yoghurt (Motoki and Seguro, 1998). Lorenzen and Schlimme (1998) suggested that potential field of application of transglutaminase in the dairy industry includes stabilizing products such as yoghurt, whipping cream, fresh cheese and novel products (e.g., spreads, low-calorie foods) and preparation of cross-linked caseinates as functional ingredients.

5. Enzymes for Cleaning of Dairy Equipments

In the dairy industry, a large quantity of milk is processed routinely using various types of equipments and in order to produce safe and wholesome products. The maintenance of hygienic conditions of equipment and environment within a dairy plant is of utmost importance and this can be achieved by the use of appropriate cleaning agents with or without sanitizers. Generally different types of detergents are employed for cleaning or removal of soil, dirt or foreign matter from the surface of the equipments. Improper cleaning of pipelines and equipments leads to accumulation of milk constituents, which become very hard over time and such milk stones, once formed, cannot be removed easily by the use of ordinary detergents and need to combine with sanitization process. However, in some cases, sanitization is not routinely possible and may be employed twice-thrice in a month. Adoption of membrane processes in dairy industry has created a new set of problems difficult to manage by the conventional cleaning process. The membranes employed are easily susceptible to fouling and clogging. These problems are not overcome by conventional detergents. Moreover, synthetic detergents are non-biodegradable, corrosive and toxic and their prolonged use in dairy plants leads to formation of slumps, which result in unhygienic conditions. On the other hand, enzyme-based detergent (also called bio-detergents or green chemicals) are fast emerging as better alternatives to ordinary detergents (Behare et al., 2010). Most bio-detergent preparations contain one or two types of enzymes that depend on the nature of the soil to be removed. Incorporation of enzymes into detergent formulations enhances cleaning by break down of large and hard pieces to remove fragments. There are three basic types of enzymes used in detergents viz., proteases, lipases and amylases.

5.1 Removal of milk stones

Formation of milk stones in dairy equipment during high heat treatment of milk, especially during pasteurization and UHT sterilization of milk, is one of the major problems in the industry encounter. The milk stones mainly consist of calcium phosphate, precipitated and denatured milk proteins and insoluble salts from hard water and washing solutions. Denatured whey proteins are very hard and difficult to remove even by using strong alkali solution practiced under normal 'Cleaning in Place' (CIP) procedure. In this case, application of protease-based detergent formulation, alacalse (Novo Nordisk, Denmark), optimase (Solvay, Germany), savinase (Novo Nordisk, Denmark), etc. commercially available (Table 5) may seen a better alternative to an ordinary detergent.

5.2 Cleaning of membranes

The fractionation processes, like microfiltration, ultrafiltration, nanofiltration and reverse osmosis provide a unique opportunity for accomplishing both fractionation and concentration of milk components in liquid system without phase change, while retaining desirable physical and chemical characteristics. Due to continuous and prolonged use of these membranes, a major problem encountered is fouling which

Table 5. Commercially available protease, lipase and amylase-based detergents.

Trade Name	Source organism	Optimum pH	Optimum temp. (°C)	Manufacturer
Protease-based detergents				
Alcalase	*Bacillus licheniformis*	8–9	60	Novo Nordisk, Denmark
Optimase	*Bacillus licheniformis*	9–10	60–65	Solvay, Germany
Savinase	*Bacillus lentus*	9–11	55	Novo Nordisk, Denmark
Purafect	*Alcalophilic* species	10	40–65	Genencor, USA
Protosol	*Alcalophilic* species	10	50	Advanced Biochemicals, Thane, India
Lipase-based detergents				
Lipomax	*Pseudomonas alcaligens*	8.5	60	The GistBrocades, The Netherlands
Lipolase	*Humicolalanu ginasa*	10.5–11	40	Novo Nordisk, Denmark
Lumafast	*Pseudomonas glumai*	9–9.5	60	Genencor, USA
Amylase-based detergent				
Maxamyl	*Bacillus subtilis*	6–8.5	100	The GistBrocades, The Netherlands
Solvay amylase	Thermostable *Bacillus*	5–8	75–90	Solvay, Germany
Amylase Mt	*Bacillus subtilis* *Bacillus licheniformis*	5–8	-	Novo Nordisk, Denmark
BAN	*Bacillus amyloliquefaciens*	6–7	70	Novo Nordisk, Denmark
Purafect Oxam	*Bacillus subtilis* *Bacillus licheniformis*	5–8	-	Genencor, USA

Source: Behare et al. (2010)

consequently affects its shelf-life. The key phenomenon in membrane fouling by dairy solutions are protein adsorption, protein or particle deposition (including casein or cheese fines, microbial and protein aggregates) and deposition of fats and minerals. Chemical cleaning involves strong acid and alkaline treatment with HNO_3 or phosphoric acid and NaOH followed by flushing with water and sanitization. However, strong acids and alkalis can damage the membranes at higher temperature and pH conditions thereby shortening the life of the membrane. In this case, most common protease enzyme may be the cleaning component to supplement an alkaline detergent cleaner. Membrane cleaning with lipases alone and in combination with protease effectively reduce the amount of lipids and proteins. Further, enzymes are less aggressive to the membrane and as they are highly substrate and reaction specific, they are believed to lengthen the membrane lifespan. Rinsing volumes are also reduced, leading to lower waste water volumes. Moreover, neutralization of effluent is not required.

5.3 Removal of bacterial biofilms

Biofilms consist primarily of viable and non-viable microorganisms embedded in polyanionic extracellular polymeric substances anchored to a surface. Extracellular polymeric substances may contain polysaccharides, proteins, phospholipids and other polymeric substances hydrated to 85 to 95 per cent of water. Dairy industry biofilms may also have high food residues and mineral contents that originate from the product and process water. Biofilm formation can cause mechanical blockage in the fluid-handling system, reduction of heat transfer and corrosion of metal surfaces. Bacteria within biofilms are more resistant to disinfectants and even CIP procedure cannot prevent the accumulation of microorganisms, which may assist the survival of spoilage and other food-borne pathogens in the food processing environment. Bio-detergents and bio-cleaners have proved effective in cleaning the extracellular polymers which form the biofilm matrix and thus help in removal of bacterial biofilm. The specific enzymes that are required vary according to the type of microflora making up the biofilm. A blend of enzyme mixture consisting of proteases, amylases and glucanse is very effective in cleaning a simulated industrial biofilm.

6. Conclusion

Proteases, lipases, lactases and other natural enzymes are present in milk and milk products. Naturally-occurring proteases contribute to the flavor characteristics of cheese and fermented dairy products made from raw milk. Other uses of enzymes in the dairy industry involve hydrolysis of lactose by lactases and milk coagulation by adding rennet. Calf rennet is most widely used in the dairy sector, while active enzyme present is an acid protease, commonly designated as rennin or chymosin. Microbial rennet has also been used in many countries as an alternative coagulant in the production of a number of cheeses. With the advancement in biotechnology, calf chymosin has been produced by recombinant microorganisms, like *E. coli* and yeasts. There are other particular applications of enzyme technology in the production of highly specialized milk products. Adequate blends of microbial lipases are used to produce low-carbon-chain-length fatty acids to confer a strong cheese flavor. Neutral fungal proteinases are most suited to the generation of strong savory flavor without bitterness in some mould-ripened cheeses. Enzyme-based detergents are gaining popularity in cleaning of dairy equipments and prove cost-effective alternatives to conventional cleaning procedures.

References

Behare, P.V., Nagpal, R., Kumar, M., Mohania, D., Rana, R., Mishra, V., Arora, S., Pawshe, R. and Singh, R. 2010. Bio-detergents: Green chemicals for dairy industry. Dairy Planner, 6: 31–35.

Carminatti, C.A. 2001. Ensaios de hidrólise enzimática da lactose em reator a membrana utilizando beta-galactosidase *Kluyveromyces lactis*. Florianópolis: Universidade Federal de Santa Catarina.

Campbell, R.E. and Drake, M.A. 2013. Invited review: The effect of native and nonnative enzymes on the flavor of dried dairy ingredients. Journal of Dairy Science, 96: 4773–4783.

Deeth, H.C. 2006. Lipoprotein lipase and lipolysis in milk. International Dairy Journal, 16: 555–562.

Dickinson, E. 1997. Enzymic crosslinking as a tool for food colloid rheology control and interfacial stabilization. Trends in Food Science and Technology, 10: 333–339.

Fox, P.F. and Kelly, A.L. 2006. Indigenous enzymes in milk: Overview and historical aspects—Part 1. International Dairy Journal, 16: 500–516.

Kumar, A., Grover, S., Sharma, J. and Batish, V.K. 2010. Chymosin and other milk coagulants: Sources and biotechnological interventions. Critical Reviews in Biotechnology, 30(4): 243–258.

Law, B. 2002. Enzymes in the manufacture of dairy products. pp. 91–108. In: R.J. Whitehurst and B.A. Law (eds.). Enzymes in Food Technology. Sheffield Academic Press Ltd. Mansion House, 19 Kingfield Road Sheffield S1 1 9AS, UK.

Lorenzen, P.C. and Schlimme, E. 1998. Properties and potential fields of application of transglutaminase preparations in dairying. IDF Bulletin No. 332. Brussels: International Dairy Federation.

Mattila-Sandholm, T., Crittenden, R., Mogensen, G., Fondén, R. and Saarela, M. 2002. Technological challenges for future probiotic foods. International Dairy Journal, 12: 173–182.

Mohanty, A.K., Mukhopadhyay, U.K., Grover, S., Batish, V.K. 1999. Bovine chymosin: Production by rDNA technology and application in cheese manufacture. Biotechnology Advances, 17(2-3): 205–17.

Mohanty, A.K., Mukhopadhyay, U.K., Kaushik, J.K., Grover, S., Batish, V.K. 2003. Isolation, purification and characterization of chymosin from riverine buffalo (Bubalos bubalis). Journal of Dairy Research, 70(1): 37–43.

Motoki, M. and Seguro, K. 1998. Transglutaminase and its use in food processing. Trends in Food Science & Technology, 8: 204–10.

Neelakantan, S., Mohanty, A.K. and Kaushik, J.K. 1998. Production and use of microbial enzymes for dairy processing. Current Science, 77(1): 143–148.

Özrenk, E. 2006. The use of transglutaminase in dairy products. International Journal of Dairy Technology, 59: 1–7.

Qureshi, M.A., Khare, A.K., Pervez, A. and Uprit, S. 2015. Enzymes used in dairy industries. International Journal of Applied Research 2015, 1(10): 523–527.

Seifu, E., Buys, E.M. and Donkin, E.F. 2005. Significance of the lactoperoxidase system in the dairy industry and its potential applications: A review. Trends in Food Science & Technology, 16: 137–154.

Upadhyay, K.G. 2003. Essentials of cheese making. The Alumni Association S.M.C. College of Dairy Science, GAU, Anand.

PART 4

ADVANCEMENT IN MICROBIAL ENZYME TECHNOLOGY

20

Recombinant Enzymes in the Meat Industry and the Regulations of Recombinant Enzymes in Food Processing

Kelly Dong,[1,#] *Yapa A. Himeshi Samarasinghe,*[1,#]
Wenjing Hua,[1] *Leah Kocherry*[1] *and Jianping Xu*[1,*]

1. Introduction

The meat industry is among the largest agriculture sectors in many countries, especially in Japan, Australia, New Zealand and the developed countries of Europe and the Americas. For example, in the US, the retail equivalent value of meat was worth \$85 billion in 2012. Globally, about one-third of the agricultural land (~14 billion hectares) is used to grow animal feed (~one billion tons). While the demand for meat is increasing rapidly in developing countries, such as China and Africa, in many developed countries of Western Europe and North America, the demand for meat has been largely saturated. However, the demand for high quality meat still has a significant potential in both the developed and developing countries. In the production of high-quality meat, enzymes often play a significant role in increasing the efficiency and reducing the costs.

The meat industry employs enzymes in four specific areas: meat tenderization, cross-linking meats, flavor development and nutrition improvement. Many of these processes operate at a specific pH, temperature and/or moisture conditions. A number

[1] Department of Biology, McMaster University, Hamilton, Ontario, L8S 4K1, Canada.
  E-mails: dongky@mcmaster.ca; samaraya@mcmaster.ca; huaw2@mcmaster.ca; kocheric@mcmaster.ca
* Corresponding author: jpxu@mcmaster.ca
# These two authors contributed equally to the work.

of microorganisms produce the required enzymes at fairly optimal temperatures or pH conditions in which these processes take place. However, genetic engineering is often needed in order to optimize the enzyme activities. In addition, using another host to produce the desired proteins cuts down on production time and eases the purification process in conditions where the enzyme is difficult to obtain due to resource limitations. For instance, proteases used in meat tenderization, naturally require proteases that are found in fruits, such as pineapples and kiwis. However, extracting enzymes from these fruits is a waste of the food resources and time sensitive–it requires the fruit to be in production season in order to harvest the enzymes. In contrast, bacteria require only 12–24 hours to produce enzymes as compared to the months required to produce proteases in fruits. What is more, bacterial production of the enzymes can be made at any point in the year. The purification of the enzyme is also more efficient and less onerous when using bacteria by having a fusion tagged protein that is engineered to bind to a specified matrix (Amid and Arshad, 2015).

Below is given the description of the structure of meat, followed by different types of meat processing and the enzymes used in these processes, with the focus on recombinant enzymes. The chapter is concluded with a summary of the regulations that govern the use of recombinant enzymes in food processing.

2. Muscle to Meat

Muscle in animals is fibrous due to its role to contract and expand for tensile strength. As a result, even with extensive tenderization, the innate fibrous texture of the meat remains. Soon after the animal is slaughtered, biochemical changes are initiated in the muscle. After death, the fibers continue to metabolize and the ATP level is maintained through degradation of polysaccharide glycogen and anaerobic glycolysis. This causes a drop in pH from 7.2 to 5.5 and a drop in temperature, which perpetuates a gradual decrease in ATP (Kemp and Parr, 2008). The main contractile protein complexes, which are usually separated in live animals, irreversibly bind together, creating a macroscopic change called *rigor mortis*.

The muscle proteins are then degraded and tenderized with proteases, which shear the two main components that encompass the fibrous texture of the meat—myofibrils and connective tissue. Myofibrils are the basic units of muscle fibers and are made of myosin and actin filaments. The connective tissue consists primarily of collagen. Changes in these components greatly affect both the texture and the water-holding capabilities of meat (Lantto et al., 2010).

Once the animals are slaughtered, a series of enzymatic actions take place within the meat. Two types of enzymes are specifically active in the process: calpains and cathepsins. Calpains are members of calcium-dependent, non-lysosomal cysteine proteases (proteolytic enzymes); while in contrast, cathepsins are proteases found primarily in lysosomes and are activated by low pH in lysosomes. Both types of enzymes are expressed ubiquitously in mammals and many other organisms (Sentandreu et al., 2002). Another group of enzymes are caspases, which are cysteine-aspartic proteases or cysteine-dependent aspartate-directed proteases. They play essential roles in apoptosis (programmed cell death), necrosis and inflammation (Ouali et al., 2006). Kemp and Parr (2008) investigated the use of human recombinant caspase 3 (rC3) expressed in *Escherichia coli* as a method of weakening myofibrillar proteins at post mortem in porcine muscle. The recombinant protein was generated from a cDNA insert of the full-length human caspase 3 into a vector and contained a poly-histidine tag at the C-terminus of the insert. The *E. coli* was activated using 0.2 mM isopropyl-beta-D-thiogalactopyranoside and purified using a liquid chromatography system. The recombinant caspase was found capable of cleaving many proteins, including alpha-actin, troponin T, myosin, myofibrillar proteins, desmin and troponin I (Kemp and Parr, 2008).

3. Meat Tenderization

Both texture and flavor are essential indicators of meat quality. The two main components that contribute to tenderness are myofibril integrity and connective tissue robustness. Proteolytic enzymes are extensively used for meat tenderness (Koohmaraie and Geesink, 2006). Bromelain, which is a mixture of proteolytic enzymes from pineapples, is used to tenderize tough meat due to its innate ability to hydrolyze fibrous proteins and connective tissue (Amid and Arshad, 2015). The three methods to introduce proteolytic enzymes into meat include dipping the meat in a solution of proteolytic enzymes, pumping enzyme solution into major blood vessels and rehydration of freeze-dried meat. These three methods are shown to be less than ideal since they can result in uneven tenderization, with overtenderization of the surface meat. The most effective method of introducing the protease solution is to inject the enzyme preparation into the animal pre-slaughter, which helps to create even tenderization (Lin et al., 2009).

The effect of tenderization is commonly evaluated through Warner-Bratzler shear and sensory evaluation, while the mode of action of the enzyme is either myofibrillar or collagen proteins. Proteases derived from *Aspergillus oryzae* are among the most active proteolytic enzymes for meat (Benito et al., 2006). However, these enzymes do not usually penetrate the meat. A combined treatment of protease enzymes from *Bacillus subtilis* and *A. oryzae* shows to have better penetration and a greater ameliorating effect on the tenderness of the meat than either when used alone. Both species preferentially degrade myofibrillar proteins over collagen proteins and give ameliorated sensory results (Sullivan and Calkins, 2010).

Optimization is needed for meat tenderization in order to reduce the toughness from myofibrillar proteins but without over-tenderizing the meat. Bromelain and papain, the two enzymes widely used for tenderization, have broad substrate specificities. As a result, peptides associated with bitter tastes can be produced, leading to an off flavor. A microbial elastase YaB from *Bacillus* sp. has been investigated and modified to optimize its activity for meat tenderization. Subtilisin YaB normally acts on tyrosine and phenylalaine, but has been engineered by replacing amino acids Gly124 and Gly159 so that it acts only on alanine—an amino acid that is prominent in connective tissue proteins (Yeh et al., 2002). The engineered YaB had an activity superior to natural proteases such as bromelain and papain, as well as a preference to elastin and collagen rather than myofibrillar proteins (Takagi et al., 1992). Furthermore, an elastase from *Bacillus* sp. EL31410 not only has specificities for collagen and elastin in the connective tissue, but is also active at the lower pH and at both low temperatures where the meat is typically stored and at high temperatures when the meat is cooked (Bruins et al., 2001; Qihe et al., 2006; Marques et al., 2010). With the continued discovery and exploitation of cold-adapted microorganisms, proteases with optimal activity at low temperatures will continue to be discovered and applied to meat tenderization (Yeh, 2009).

Other proteases evaluated for meat tenderization include the elastase from *Pseudomonas aeruginosa,* an opportunistic human pathogen. This elastase can hydrolyze insoluble elastin much more efficiently than any other proteases. However, the pathogenicity in *P. aeruginosa* makes enzymes in its native host unfit for food application. As a result, the elastase gene *lasB* has been expressed in *Escherichia coli* and *Pseudomonas putidas.* However, the enzyme was found to be difficult to purify from these hosts, making it unrealistic for commercial use. In contrast, the methylotrophic yeast, *Pichia pastoris*, was found to be a good host for expressing elastase. Lin et al. (2009) found the modified and recombinant elastase extracted from *P. aeruginosa* and expressed in *P. pastoris* had an extremely high activity, about 26-fold of that from *P. aeruginosa* and is very resistant to heat.

4. Dry-cured Meats

Dry-cured meats are uncooked meat products that are made after eight to 24 months of ripening, during which an uncontrolled microbial population proliferates on the surface of the meat. The use of proteases can shorten the ripening time. *Penicillium aurantiogriseum* has a high proteolytic activity on dry-cured meats. *Penicillium chrysogenum* has also been isolated from dry-cured meats and its enzymes investigated. The resultant enzyme, EPg222, is very active against myofibrillar proteins, which are hydrolyzed in the drying stage. In comparison, the collagen, the dominant protein in connective tissue, remained relatively unaltered (Benito et al., 2002). The gene coding for Epg222 has been cloned and expressed in *P. pastoris* and the recombinant protease has shown the same properties as those in the original host *P. chrysogenum* (Benito et al., 2006).

A long ripening process for cured meat is also required for the production of aromatic compounds from amino acids and fatty acids. However, using proteases and lipases alone will not create the final aroma or flavor, since lipid oxidation and amino acid catabolism require different enzymes and occur at different rates (Gerday et al., 2000). Both types of enzymes are required for aroma production. Application of purified proteinases (PrA and PrB) and aminopeptidases (arginyl aminopeptidase and prolyl aminopeptidase) from *Debaryomyces hansenii* CECT 12487 can produce the desired sensory quality. These enzymes catalyze the hydrolysis of sarcoplasmic proteins to produce ammonia, increase pH and accelerate the proteolytic pathway (Bolumar et al., 2006).

Fermented sausages are produced primarily by using bacteria, such as *Lactobacillus curvatus, Lactobacillus sakei, Pediococcusacidilactici* and *Enterococcus faecalis* (Lantto et al., 2010). Various recombinant enzymes from these species have been utilized to increase the fermentation processes. The catalase gene of katA of *Lactobacillus sakei* SR911 was cloned and expressed in *Lactobacillus plantarum* TISTR850, a normally catalase-deficient strain. The recombinant strain showed a catalase activity about three times higher than that in the natural strain and significantly lowered the lipid oxidation level in the recombinant strain. Lactic acid bacteria are usually needed to produce the acid that lowers the pH of the meat to produce the flavor and texture. Under certain environmental conditions, such as in the presence of abundant oxygen, the meat can quickly produce hydrogen peroxide and turn rancid. The produced hydrogen peroxide can alternatively react with iron during lipid oxidation and spoil the meat, especially during prolonged storage. Therefore, a high catalase activity in the recombinant strain can lower lipid oxidation and decrease the spoilage of meat (Noopakdee et al., 2004).

5. Cross-linked Meats

Protein cross-linking enzymes, such as transglutaminases, have been used to improve the texture, flavor, and shelf-life of meat products. Transglutaminase has the inherent capability to adhere to the surface of meat and through an acyl-transfer reaction, cross-link amines to the glutamines in proteins. Industrial production of transglutaminase is mainly from the bacterium *Streptoverticillium mobaraense*. Since transglutaminase exhibits activity even in chilled conditions, it can bind raw meat during frozen meat storage. Therefore, transglutaminase is commonly used in sausages, ham and fish storage. Cross-linked meat is an area of interest as it utilizes the muscles that are of poorer quality, such as the trimmings, and combines them to form whole meats that are more appealing to the consumer market. However, due to the nature of the processing to create cross-linked meats, the products are often frozen and/or can lose color, which can make them more difficult to market (Nielsen et al., 1995).

In addition to transglutaminase, other oxidative enzymes, such as tyrosinases and laccases, have been researched as putative candidates to cross-linking meat. However, none of these enzymes are currently used in the meat processing industry (Jus et al., 2012). The recombinant transglutaminase F XIIIa, produced through fermentation by *Saccharomyces cerevisiae*, can restructure raw minced meat. During a treatment at 37°C for 90 minutes, this enzyme increased cohesion, elasticity and hardness of minced meat. The effect of cohesion was augmented by salt and phosphate, which exaggerated the results. However, the addition of F XIIIa caused color deterioration that was deemed undesirable. In contrast, a treatment at 10°C for 23 hours did not cause major changes to the meat (Kuraishi et al., 1997). Future research attempting to reduce the color deterioration by F XIIIa would be of significant consumer importance.

6. Fresh and Frozen Meat Storage

Fresh meat is subject to spoilage from a variety of different species of bacteria, such as those from the genus *Pseudomonas, Acinetobacter, Lactobacillus* and a number of yeasts and moulds. The environment in which the meat is stored influences the production of particular bacterial flora. For instance, in vacuum-packaged meat, the growth of *Lactobacillus curvatus* and *Lactobacillus sake* are favored, which causes the glucose has to be metabolized into lactic acid. Similarly, the amino acids—leucine and valine—are metabolized into isovaleric and isobutyric acids. These alterations create a cheesy odor that disappears after the package is opened. Psychrotrophic *Clostridium laramie* also causes spoilage that affects the proteolysis. As a result, the meat produces a hydrogen sulfide odor, loses its texture, accumulates liquid and causes the red meat to become green due to the reductive action of myoglobin by hydrogen sulfide (Ray and Bhunia, 2007).

Recombinant antifreeze proteins have been used to prevent ice crystallization during freezing without affecting the actual freezing process itself. By reducing ice crystal formation, there is less drip, less cellular damage and minimal loss of nutrients. As a result, the texture is better preserved. Use of the recombinant antifreeze protein rAFP in *Lactococcus lactis* has greatly extended meat storage economically during all seasons (Yeh et al., 2009).

7. Fish Meat

In contrast to other meats, fish meat softens much more rapidly than mammalian or poultry meat. Thus, maintaining firmness is a key objective during the treatment and storage of fish meat. The natural tenderization and softening processes of fish meat are independent of rigor mortis and matrix metallo proteineases and matrix serine proteinases have been implicated to be involved in the auxiliary tenderization (Kubota et al., 2001).

To maintain and/or increase fish-meat firmness, the cross-linking enzyme transglutaminase has been used for the production of various fish meat. However, natural transglutaminases from fish have shown limited success when expressed in microorganisms. At present, a transglutaminase extracted from *S. mobaraense* is commercially used to stimulate gel formation and operates independently of calcium. While there are advantages of extracellular transglutaminases, the recombinant enzyme expressed by *Streptomyces lividans* and *E. coli* demonstrated a lower productivity than in the original strain of *S. mobaraense*. The recombinant enzyme was purified through cation-exchange chromatography and gel filtration, which is much easier than obtaining the enzyme from the original strain. In addition, the recombinant transglutaminase had a lower thermal stability and cross-linking activity, making it a less than ideal candidate for commercial production. If these characteristics can be improved upon, the expression of the enzyme in a different host would be ideal for producing a high quantity and high purity of the protein (Yokoyama et al., 2003).

One speciality fish dish is surimi, a fish mince procured by chopping, detendoning and leaching of the fish skeletal muscle. Surimi can form a firm structure through cross-linking of unfolded actomyosin in the presence of salt. Surimi's unique features are gel-forming, water-binding, and oil-binding abilities which make it a valuable base component for a number of food products (Marx et al., 2008). In several types of fish, the hydrolysis of endogenous proteinases hinders the gel-forming process in surimi. Recombinant chicken cystatin, a protease inhibitor, can inhibit gel softening. The recombinant cystatin has been successfully produced by *E. coli*, but whether it's safe for human consumption still remains unknown. The same enzyme has been successfully expressed in *Pichia pastoris* and further modified to improve the freezing stability, thermal stability and pH stability through a glycosylation modification (Tzeng and Jiang, 2004). The recombinant cystatin can be purified through centrifugation and column chromatography (Chen et al., 2001). Similarly, although the characteristics of cystatin expressed in yeast *P. pastoris* are similar to the wild type, the safety of consumption by humans warrants further investigation.

8. Animal Feed

Alfalfa, a major component of animal feed, is used to give poultry its distinct yellow color, which originates from xanthophylls in alfalfa. However, alfalfa is not as desirable for providing high fiber content in meat, which in turn affects the quality and nutritional value of the meat. Many experiments have been conducted to improve animal feed, with variable successes. For instance, Ponte et al. attempted to use a mixture of cellulase from *Cellvibrio mixtus* and the recombinant xylanse GH11-CBM6 from *Clostridium thermocellum* to improve the fiber content of the feed. The final results indicated that not only did the recombinant enzymes not contribute to the nutrition of the poultry, the enzymes were individually detrimental, causing negative impact on the final weight gain and feed intake. Neither did they increase the efficiency of the feed being converted into nutritional supplements for the animals (Ponte et al., 2004).

9. Meat Safety

The determination of safety of meat-processing enzyme preparations is based on scientific evidence and is assessed under the conditions in which it is used directly or indirectly. The manufacturer must provide accurate information with regard to its composition, method of manufacture, usage, dietary exposure, toxicology and other characteristics. The enzyme source must also be non-pathogenic and non-toxigenic. The enzyme is then submitted to the regulatory agencies, such as FDA in the US, for review and approval before it can be released into the market. In the approval process, the FDA often consults the Food Safety and Inspection Service (FSIS), which evaluates the suitability of the enzyme preparation so that the minimum levels of the enzyme are used to produce the proposed effect in meat (Gaynor, 2006).

Currently, transglutaminase is allowed in certain meat if the purpose of the enzyme application is binding and reforming meat cuts. The product must include labelling to indicate that it has been reformed. In addition, protease preparations from *Aspergillus oryzae, Aspergillus niger, Bacillus subtilis*, and *Bacillus subtilis* var. *amyloliquefaciens* are approved for use as meat tenderizers (Nazina et al., 2001; Ha et al., 2013).

10. General Regulations on Recombinant Proteins Used in Food Processing

The regulations and policies regarding enzyme preparations used in the processing of meat and other foods and drinks vary widely between countries around the world. The increasing use of recombinant enzymes and Genetically Modified Microorganisms (GMM) has contributed to this heterogeneity as different regions hold varying views on pre-market safety assessment of recombinant enzymes from thorough evaluations to no evaluations at all. An important distinction is made between food additives—enzymes present in the final food product as an ingredient, usually for color, taste, texture—and the processing aids or the enzymes used during the food manufacturing process to catalyze reactions but are either removed or inactivated in the final product. However, approval for commercial use may depend on the specific enzyme categorization system used in the country in question. For instance, in the United States, Canada and Japan, all food-processing enzymes are regulated as food additives whereas in New Zealand and Australia, they are considered as processing aids (Agarwal and Sahu, 2014).

Processing aids are used at very low levels during food production and they are removed from the final product or inactivated during baking/cooking. A major aspect of their safety assessment is the level of consumer exposure to residual enzymes in the food product. This is calculated as the amount of Total Organic Solids (TOS) present in the concentrated enzyme before formulation: TOS includes the enzyme and other organic materials that originated from the source microbial cultures (Olempska-Beer et al., 2006). The safety of the final enzyme preparation is then evaluated based on whether it meets the standards and specifications established for enzyme preparations by the Food Chemicals Codex (FCC). FCC is a compilation of internationally recognized standards for the purity and identity of food ingredients. It is revised and published in the United States every two years by the US Pharmacopeial Convention and includes a section on food-processing enzymes. FCC is recognized by regulatory agencies, manufacturers, vendors and consumers worldwide. In the following sections, regulatory frameworks on enzyme preparations around the world are briefly outlined.

10.1 European union (EU)

In the EU, food enzymes are currently regulated by Regulation (EC) No. 1332/2008. EU recognizes enzymes used during food preparation as processing aids that serve no technological purpose in the final product and are exempted from the requirement that they be listed as ingredients on the product label (EUR-LEX, 2008). EU requires the producer of an enzyme to submit an application with all necessary data such as its source and intended use to the European Commission where safety assessments are conducted by the European Food Safety Authority (EFSA). Once an enzyme has been declared safe by EFSA for its intended use, it is included in the Public Register of all food enzymes following the procedure in Regulation EC 1331/2008. However, EFSA requires that data submitted by the producer in support of the application includes the necessary information to verify that there is a technological use for the enzyme and that the intended use does not mislead the consumer. In the case that residual enzymes are present in the final product, EFSA still grants approval if it does not react with the food.

Article 12 of Regulation (EC) No. 1829/2003 requires products containing ingredients derived from genetically-modified organisms to be labelled as GMO products. However, this excludes recombinant processing aids as they are only used during the production process and are not present in the final product. Therefore, food articles manufactured by using recombinant processing aids are not given GMO labelling. Their website is: (http://ec.europa.eu/food/food/fAEF/enzymes/ guidance_docs_en.htm).

10.2 North America

United States

In the US, food-processing enzymes are regulated as secondary food additives under the Federal Food, Drug and Cosmetic Act (Gaynor, 2006). The regulatory status of a substance is established through the petition process. For pre-market approval of an enzyme preparation, the producer must file a food additive petition proposing its safety approval. All technological and chemical data needed by the FDA for a thorough safety assessment are also to be submitted. FDA reviews this information and determines if the safety of the enzyme has been successfully proven by the producer and then grants approval for its use in food processing. To obtain GRAS (Generally Regarded As Safe) status for a source, microbial strain or an enzyme, for a particular use, the producer must submit a GRAS affirmation petition along with scientific evidence for its safety and basis for concluding that this information is commonly acknowledged by qualified experts. Until 1997, FDA reviewed the data submitted by the producer for safety affirmation. This has now been replaced by a voluntary notification program where an organization outside of the government can evaluate the safety of an enzyme preparation and notify the government through a GRAS notification. In both pre-market and GRAS status submissions, approval is granted only for the intended use of the enzyme.

Recommended sections by FDA to be included in the notices/petitions are the identity of enzyme/microbial host, the composition of enzyme preparation, the manufacturing process, purity, the intended technical use and the estimated dietary exposure to the enzyme. Even though pre-market approval of recombinant enzymes follows the same procedure, additional information must be submitted by the producer. The Genetically Modified Microbe (GMM) and the introduced DNA must be thoroughly characterized, including the gene of interest, the selectable markers, and the regulatory regions. FDA does not impose any restrictions on the source of the enzymes: they can be derived from known organisms, unknown organisms isolated from the environmental samples or a pool of genes from various sources. It could also be synthesized or modified by site-directed mutagenesis and other molecular engineering techniques. The host microorganism can also be modified for better heterologous expression of the enzymes. All these approaches and steps in the production of GMM and the recombinant enzymes should be thoroughly described (Gaynor, 2006).

Canada

In Canada, food-processing enzymes are regulated as food additives under the Food and Drugs Regulation Act (Government of Canada, 2014). Health Canada is responsible for the pre-market safety approval of enzyme preparations. The safety of the source organism is the main consideration in safety evaluation. It requires the source microorganisms to be thoroughly characterized and that the organisms do not produce any toxins, pathogenic substances or antibiotics. Toxicity tests conducted on the enzyme preparations and the commercial manufacturing process of the enzyme are also evaluated before safety approval is granted. In assessing recombinant enzymes and their source GMMs, Health Canada expands on the above procedures by requesting additional information, specifically regarding the safety of the newly introduced DNA in the host. Techniques used to transfer DNA from the source to the host microorganism are also evaluated. The genome of the host microorganism is also required to be fully characterized before safety evaluation is completed and pre-market approval for the recombinant enzyme's intended use is granted. The list of approved food enzymes to date can be found at Health Canada's website: http://www.hc-sc.gc.ca/fn-an/securit/addit/list/5-enzymes-eng.php.

11. Conclusion

In this chapter, we reviewed the major characteristics of meat and the enzymes used in different types of meat processing. We described how recombinant enzymes are used in the processing of meat and how such food-processing enzymes in general are regulated in Europe and North America. With increasing demands for high quality meat, we expect that there will be a great demand for recombinant meat-processing enzymes in the years to come.

Acknowledgement

Research in our lab on microbial diversity and their roles in natural environments and human services is supported by the Natural Sciences and Engineering Research Council of Canada.

References

Agarwal, S. 2014. Safety and Regulatory Aspects of Food Enzymes : An Industrial Perspective, 1: 253–267.
Amid, A., Ismail, N.A. and Arshad, Z.I.M. 2015. Case study: Recombinant bromelain selection. pp. 143–158. *In*: A. Amid (ed.). Recombinant Enzymes—From Basic Science to Commercialization. Springer, Switzerland.
Benito, M.J., Connerton, I.F. and Córdoba, J.J. 2006. Genetic characterization and expression of the novel fungal protease, EPg222 active in dry-cured meat products. Applied Microbiology and Biotechnology, 73: 356–365.
Benito, M.J., Rodríguez, M., Núñez, F., Asensio, M.A., Bermúdez, M.E. and Córdoba, J.J. 2002. Purification and characterization of an extracellular protease from *Penicillium chrysogenum* Pg222 active against meat proteins. Applied and Environmental Microbiology, 68: 3532–3536.
Bolumar, T., Sanz, Y., Flores, M., Aristoy, M.C., Toldrá, F. and Flores, J. 2006. Sensory improvement of dry-fermented sausages by the addition of cell-free extracts from *Debaryomyceshansenii* and *Lactobacillus sakei*. Meat Science, 72: 457–466.
Bruins, M.E., Janssen, A.E. and Boom, R.M. 2001. Thermozymes and their applications. Applied Biochemistry and Biotechnology, 90: 155–186.
Chen, G.H., Tang, S.J., Chen, C.S. and Jiang, S.T. 2001. High-level production of recombinant chicken cystatin by *Pichia pastoris* and its application in mackerel surimi. Journal of Agricultural and Food Chemistry, 49: 641–646.
Gaynor, P. 2006. How U.S. FDA's GRAS notification program works. Food Safety Magazine, 11: 16–19.

Gerday, C., Aittaleb, M., Bentahir, M., Chessa, J.P., Claverie, P., Collins, T. and Feller, G. 2000. Cold-adapted enzymes: From fundamentals to biotechnology. Trends in Biotechnology, 18: 103–107.

Government of Canada. 2014. Food and Drugs Act. Justice Law Website.

Ha, M., Bekhit, A.E.D., Carne, A. and Hopkins, D.L. 2013. Comparison of the proteolytic activities of new commercially available bacterial and fungal proteases toward meat proteins. Journal of Food Science, 78: 170–177.

Jus, S., Stachel, I., Fairhead, M., Meyer, M., Thoeny-Meyer, L. and Guebitz, G.M. 2012. Enzymatic cross-linking of gelatine with laccase and tyrosinase. Biocatalysis and Biotransformation, 30: 86–95.

Kemp, C.M. and Parr, T. 2008. The effect of recombinant caspase 3 on myofibrillar proteins in porcine skeletal muscle. The Animal Consortium, 8: 1254–1264.

Koohmaraie, M. and Geesink, G.H. 2006. Contribution of postmortem muscle biochemistry to the delivery of consistent meat quality with particular focus on the calpain system. Meat Science, 74: 34–43.

Kubota, M., Kinoshita, M., Kubota, S., Yamashita, M., Toyohara, H. and Sakaguchi, M. 2001. Possible implication of metalloproteinases in post-mortem tenderization of fish muscle. Fisheries Science, 67: 965–968.

Kuraishi, C., Sakamoto, J., Yamazaki, K., Susa, Y., Kuhara, C. and Soeda, T. 1997. Production of restructured meat using microbial transglutaminase without salt or cooking. Journal of Food Science, 62: 488–490.

Lantto, R., Kruus, K., Puolanne, E., Honkapää, K., Roininen, K. and Buchert, J. 2010. Enzymes in meat processing. pp. 264–291. *In*: R. Whitehurst and M. Oort (eds.). Enzymes in Food Technology. Wiley-Blackwell, Iowa.

Lin, X., Xu, W., Huang, K., Mei, X., Liang, Z., Li, Z. and Luo, Y. 2009. Cloning, expression and characterization of recombinant elastase from *Pseudomonas aeruginosa* in *Picha pastoris*. Protein Expression and Purification, 63: 69–74.

Marques, A.C., Maróstica, M.R. and Pastore, G.M. 2010. Some Nutritional, Technological and Environmental Advances in the Use of Enzymes in Meat Products. Enzyme Research.

Marx, C.K., Hertel, T.C. and Pietzsch, M. 2008. Purification and activation of a recombinant histidine-tagged pro-transglutaminase after soluble expression in *Escherichia coli* and partial characterization of the active enzyme. Enzyme and Microbial Technology, 42: 568–575.

Nazina, T.N., Tourova, T.P., Poltaraus, A.B., Novikova, E.V., Grigoryan, A.A., Ivanova, A.E., Lysenko, A.M., Petrunyaka, V.V., Osipov, G.A., Belyaev, S.S. and Ivanov, M.V. 2001. Taxonomic study of aerobic thermophilic bacilli: Descriptions of *Geobacillussubterraneus* gen. nov., sp. nov. and *Geobacillussuzenensis* sp. nov. from petroleum reservoirs and transfer of *Bacillus stearothermophilus, Bacillus thermocatenulatus, Bacillus thermoleovorans, Bacillus kaustophilus, Bacillus thermoglucosidasius* and *Bacillus thermodenitriWcans* to *Geobacillus* as the new combinations *G. stearothermophilus, G. thermocatenulatus, G. thermoleovorans, G. kaustophilus, G. thermoglucosidasius and G. thermodenitriWcans*. Int. J. Syst. Evolutionary Microbiology, 51: 433–446.

Nielsen, G.S., Petersen, B.R. and Møller, A.J. 1995. Impact of salt, phosphate and temperature on the effect of a transglutaminase (F XIIIa) on the texture of restructured meat. Meat Science, 41: 293–299.

Noonpakdee, W., Sitthimonchai, S., Panyim, S. and Lertsiri, S. 2004. Expression of the catalase gene katA in starter culture *Lactobacillus plantarum* TISTR850 tolerates oxidative stress and reduces lipid oxidation in fermented meat product. International Journal of Food Microbiology, 95: 127–135.

Olempska-Beer, Z.S., Merker, R.I., Ditto, M.D. and DiNovi, M.J. 2006. Food-processing enzymes from recombinant microorganisms—A review. Regul. Toxicol. Pharmacol., 45: 144–158.

Ouali, A., Herrera-Mendez, C.H., Coulis, G., Becila, S., Boudjellal, A., Aubry, L. and Sentandreu, M.A. 2006. Revisiting the conversion of muscle into meat and the underlying mechanisms. Meat Science, 74: 44–58.

Ponte, P.I.P., Ferreira, L.M.A., Soares, M.A.C., Aguiar, M.A.N.M., Lemos, J.P.C., Mendes, I. and Fontes, C.M.G.A. 2004. Use of cellulases and xylanases to supplement diets containing alfalfa for broiler chicks: Effects on bird performance and skin color. The Journal of Applied Poultry Research, 13: 412–420.

Qihe, C., Guoqing, H., Yingchun, J. and Hui, N. 2006. Effects of elastase from a *Bacillus* strain on the tenderization of beef meat. *Food Chemistry,* 98: 624–629.

Ray, B. and Bhunia, A. 2007. Fundamental Food Microbiology. CRC Press.

Sentandreu, M.A., Coulis, G. and Ouali, A. 2002. Role of muscle endopeptidases and their inhibitors in meat tenderness. Trends in Food Science & Technology, 13: 400–421.

Sullivan, G.A. and Calkins, C.R. 2010. Application of exogenous enzymes to beef muscle of high and low-connective tissue. Meat Science, 85: 730–734.

Takagi, H., Kondou, M., Hisatsuka, T., Nakamori, S., Tsai, Y.C. and Yamasaki, M. 1992. Effects of an alkaline elastase from an alcalophilic *Bacillus* strain on tenderization of beef meat. Journal of Agricultural and Food Chemistry, 40: 2364–2368.

The European Parliament and the Council of the European Union. 2008. Regulation (EC) No. 1332/2008 of the European Parliament and of the Council. EUR-Lex Website [accessed 2015 Sep. 07]. http:// eur-lex.europa.eu/legal.

Tzeng, S.S. and Jiang, S.T. 2004. Glycosylation modification improved the characteristics of recombinant chicken cystatin and its application on mackerel surimi. Journal of Agricultural and Food Chemistry, 52: 3612–3616.

Yeh, C.M., Kao, B.Y. and Peng, H.J. 2009. Production of a recombinant Type 1 antifreeze protein analogue by *L. lactis* and its applications on frozen meat and frozen dough. Journal of Agricultural and Food Chemistry, 57: 6216–6223.

Yeh, C.M., Yang, M.C. and Tsai, Y.C. 2002. Application potency of engineered G159 mutants on P1 substrate pocket of subtilisin YaB as improved meat tenderizers. Journal of Agricultural and Food Chemistry, 50: 6199–6204.

Yokoyama, K., Ohtsuka, T., Kuraishi, C., Ono, K., Kita, Y., Arakawa, T. and Ejima, D. 2003. Gelation of food protein induced by recombinant microbial transglutaminase. Journal of Food Science, 68: 48–51.

21

Recombinant Enzymes Used in Fruit and Vegetable Juice Industry

*Yapa A. Himeshi Samarasinghe, Wenjing Hua, Kelly Dong,
Leah Kocherry and Jianping Xu**

1. Introduction

Fruit and vegetable juices are among the most consumed liquids in the world with orange, apple, tomato and grape juices being some of the most popular (Singh et al. 2015). Although specific steps may differ, all juices are generally prepared by mechanically pressing or macerating fresh fruits and vegetables, in the absence of heat or solvents, to extract their naturally-occurring liquids (Gil-Izquierdo et al., 2002). In commercial juice production, enzymes capable of degrading plant cell-wall components are applied to facilitate extraction and increase the juice yield (Puri et al., 2012). Recombinant gene technology is now the most popular choice for genetically modifying these enzymes to optimize the juice manufacturing process. Manufacturers often concentrate the juice by removing about two-third of the water content, which significantly increases its shelf-life (SSP PVT LTD, 2015). Water and lost flavors are re-added to the concentrated juices before they are released into the market or are sold as bulk concentrated juice preparations where consumers add water before consumption.

Due to public perception of juices as a healthy, easily-consumed alternative to eating fresh fruits and vegetables, health-conscious people, mainly in the West, have embraced juices as substitutes for sugar-sweetened beverages, such as Coca-Cola

Department of Biology, McMaster University, Hamilton, Ontario, L8S 4K1, Canada.
  E-mails: samaraya@mcmaster.ca; huaw2@mcmaster.ca; dongky@mcmaster.ca; kocheric@mcmaster.ca
* Corresponding author: jpxu@mcmaster.ca

and Pepsi-Cola. The availability of many juices being marketed as containing no added sugars by manufacturers has further fuelled this trend. However, consumers are becoming increasingly aware of the high content of naturally-occurring sugars in fruit juices which could lead to tooth decay, development of dental cavities and obesity (Dennison et al., 1997; Griep et al., 2013; Imamura et al., 2015; Wojcicki and Heyman, 2012). For example, Health Canada is currently reviewing dietary evidence for fruit juices in the face of concern raised by many dieticians and physicians who specialize in general obesity and paediatric obesity, which may result in juices being removed from Canada's Food Guide (Vogel, 2015). However, well-documented health benefits of fruit juices when taken in moderation keep the consumers interested in it as a rich source of vitamins, minerals and antioxidants (Basu and Penugonda, 2009; Bub et al., 2003; Duthie et al., 2006; Van Duyn and Pivonka, 2000). Starting in the 1990s, juice cleanses, in which a limited diet consisting entirely of fruit and vegetable juices is consumed for a certain time period with the intention of physical detoxification, have become popular in the West (Newman, 2010).

As a recent systematic assessment revealed, fruit juice consumption is strongly correlated with the average income of countries where the highest consumption rates were reported, as in Australia, high-income North America and Central Latin America (Singh et al., 2015). However, the juice industry has been steadily growing in the last few years due to the improving economic status of the developing world coupled with the increasing health-consciousness of the developed world. The global juice industry revenue is expected to grow at an annual rate of 3.0 per cent in the next five years to reach \$315 billion by 2020 (IBIS, 2015). Genetic engineering is increasingly applied to improve the manufacturing process as producers strive to meet the growing global demand for nutritious juices. Below are described the major enzymes used in the fruit juice industry and how genetic engineering has helped to improve the efficiency of juice production.

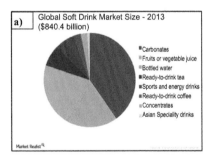
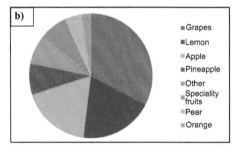

Figure 1. Statistics relevant to fruit and vegetable juice industry in 2013. (a) The proportions of major non-alcoholic beverages consumed on a global scale. Fruit and vegetable juices are the second most-consumed non-alcoholic beverages in the world, second only to the consumption of carbonated soft drinks (Marketrealist, 2015). (b) The relative proportions of different types of fruit juices based on their market share.

2. Pectinases

Pectinases, also known as pectinolytic enzymes, are the most abundantly used enzymes in juice preparation. They catalyze the degradation of pectin molecules found in the primary cell wall and middle lamella of plant cells. Pectins are complex glycosidic macromolecules mainly consisting of galacturonic acid residues that add strength to cell walls by supporting the cohesion of other cell wall components (Willats et al., 2001). Pectinases are traditionally classified into subtypes based on their substrate preferences: polymethylgalacturonases (PMG) catalyze the hydrolysis of α-1,4-glycosidic bonds in the pectin backbone whereas polygalacturonase (PGase) hydrolyzes the same bond in polygalacturonic acid. Other members of the pectinase family include deesterifying enzymes, such as pectin methyl esterase (PME), pectin lyases (PL) that randomly cleave glycosidic bonds in pectin, and pectatelyases (PGL) that preferentially cleave glycosidic linkages in polygalacturonic acid. Pectinase preparations used during juice production generally comprise a mixture of PGases, PGLs and pectin esterases.

2.1 Use in juice extraction

During juice extraction, the pectinase mixture is applied to the fruit pulp to degrade insoluble pectins in the cell walls that hinder free flow of the juice. As a result, the viscosity of the juice decreases, which in return reduces pressing time and increases juice yield (Heerd et al., 2012). Application of pectinase mixtures has shown to result in over 90 per cent juice yields (Lien and Man, 2010) from mashed puree when compared to conventional extraction methods, such as mechanical crushing and heating the release juices through plasmolysis of plant cells. Pectinolytic enzymes are naturally produced by a wide range of plants, microbes, insects and nematodes during infection of plants, decomposition of dead plant material, etc. However, microbial sources are exclusively used in commercial production due to low cost and ease of genetic manipulation that allow for optimization of enzyme production and functionality. The filamentous fungus *Aspergillus niger*, which possesses GRAS (Generally Regarded As Safe) status, issued from U.S. Federal and Drugs Administration (FDA), is the most abundantly used to produce industrial pectinase preparations through submerged fermentation (SmF). Traditionally, the fungal strains are cultured under different combinations of physical conditions and substrates to obtain pectinase mixtures of varying compositions. However, the activity of some pectinases in the mixture could lead to the presence of undesirable substances in the food product, e.g., the activity of pectin esterases during wine production leads to the formation of methanol which is toxic to humans. Furthermore, even under optimum conditions, wild-type strains do not overproduce proteins to an extent sufficient for industrial purposes (Rojas et al., 2011). With the advent of recombinant technology, it has become feasible to genetically manipulate the microbial strains to produce pectinolytic enzymes catering to specific needs.

2.2 Heterologous expression of pectinases

Heterologous expression of eukaryotic pectinases in a prokaryotic host is a promising avenue to produce a single enzyme in large quantities. The recombinant enzymes from different sources can be mixed in desired proportions to create pectinase preparations devoid of any unwanted substances and fitted to requirements of various applications. *Escherichia coli* is the most widely used prokaryotic host for heterologous expression of both bacterial and fungal pectinases due to its high yield of recombinant proteins and a thoroughly studied annotated genome (Ferrer-Miralles et al., 2009; Gummadi and Panda, 2003). Even though *E. coli* is unable to perform post-transcriptional and post-translational modifications of proteins, the recombinant pectinases produced by it are generally functional (Wang et al., 2011). Indeed, PGases, PMEs, PLs and other pectinolytic enzymes have been successfully expressed and purified in recombinant *E. coli* strains (Chen et al., 2014; Damak et al., 2013; Kumar et al., 2014; Massa et al., 2007; Parisot et al., 2003; Wang et al., 2011). However, over-expression of recombinant proteins in *E. coli* tends to promote the formation of insoluble protein aggregates called Inclusion Bodies (IB) that hinder the recovery of active proteins. The partially denatured proteins in IBs must be re-solubilized and refolded *in vitro* to recover the bioactive form, a process which can be too costly for commercial purposes. Attempts at optimizing the recovery process has revealed that mild solubilisation conditions (Singh and Panda, 2005), culture incubation at room temperature rather than at 37°C (Damak et al., 2013), and the use of smart polymers and affinity precipitation (Gautam et al., 2012) could enhance the recovery of active pectinases from IBs.

The use of eukaryotic hosts for heterologous expression of eukaryotic proteins presents an attractive prospect as they are capable of performing post-translational modifications and produce high levels of recombinant proteins (Porro et al., 2011). *Saccharomyces cerevisiae*, the GRAS-status-holding yeast and one of the best-studied organisms, has been successfully used as the host of recombinant PGases, PLs and other pectinolytic enzymes of fungal, bacterial and yeast origins. The majority of *S. cerevisiae* strains do not possess natural pectinolytic activity and thus are ideal for the overexpression and purification of a single recombinant pectinase (Blanco et al., 1999). An endo-PGase from *A. niger* RH5344 was cloned and expressed in *S. cerevisiae* and the resulting recombinant protein exhibited superior thermostability than the wild-type PGase (Lang and Looman, 1995). The yield of the heterologous protein was also significantly higher due to plasmid stability and increased plasmid copy number. Similarly, an acidic PGase (PG1) from *Aspergillus kawachii* (IFO 4308) (Rojas et al., 2011), a pectatelyase (PelE) and a PGase from *Erwinia chrysanthemi* (Laing and Pretorius, 1993) and a PME from *A. aculeatus* (Christgau et al., 1996) have been successfully expressed in *S. cerevisiae.* Several oenological *S. cerevisiae* strains expressing recombinant pectinolytic enzymes have also been developed as alternatives for fungal pectinases that are currently used for industrial wine filtration and clarification. A *S. cerevisiae* strain engineered to express a PGase (PG1) demonstrated faster wine filtration than traditionally-used strains without any change in wine composition (Vilanova et al., 2000), a strain engineered to express a PGase of yeast origin improved the yield of wine extraction compared to industrial strains (Fernandez-Gonzalez et al., 2005) and recombinant *S. cerevisiae* strains producing pectinolytic enzymes significantly improved the yield of banana wine fermentation compared to commercial enzyme preparations (Byaruagaba-Bazirake and Rensburg, 2013). *Pichia pastoris* is another yeast widely used as a heterologous host where remarkably high levels of constitutive expression of recombinant pectinases have been achieved. A PGase (EPG1-2) from *Kluyveromyces marxianus* CECT1043 was cloned into *P. pastoris* and the recombinant strain produced 200-fold more enzyme than the wild-strain (Sieiro et al., 2009). The eukaryotic host's glycosylation of the protein usually differs from that of the native protein: in some instances, the change in glycosylation patterns have imparted favorable characteristics to the recombinant protein (Sieiro et al., 2012). When a PGase (PG7FN) from the thermophilic fungus *Thielavia arenaria* XZ7 was expressed in *P. pastoris*, the recombinant PG7FN exhibited increased functional temperature (60°C), good pH stability and increased catalytic efficiency (Tu et al., 2014b).

2.3 Current trends in recombinant pectinase production

Currently, many research efforts are focused on the development of thermostable pectinases, bi-functional pectinolytic enzymes and optimization of yeast and hybrid strains as heterologous hosts to increase the efficiency of enzyme production and to tailor pectinase functionality to commercial production processes.

2.3.1 Thermostable pectinases

The extraction and clarification of most juices from fruits, such as apples, oranges and other citrus fruits are performed at temperatures ranging from 45°C–90°C and acidic pH conditions (Kashyap et al., 2001). Thus, efforts to optimize the juice production process has focused on engineering microbial strains to overexpress acidic pectinases with optimum functionality in the above-mentioned temperature range. For example, a thermostable pectatelyase (Pel1) from *Penicillium occitanis* was heterologously expressed in *E. coli* BL21 with a high yield and was discovered to retain its thermo-activity in high NaCl and imidazole concentrations (Damak et al., 2013). The gene *ACM61449* from thermophilic *Caldicellulosiruptor bescii* was expressed in *E. coli* DE3 which produced a highly active pectinolytic protein (CbPelA) with an optimum temperature of 72°C and a pH of 5.2 (Chen et al., 2014). When a gene encoding a thermostable pectinase, isolated from a soil metagenomic sample, was expressed in *E. coli* M15, the resulting recombinant protein was optimally active at 70°C and was active over a broad range of pH—a highly useful characteristic in industrial applications (Singh et al., 2012).

2.3.2 Yeasts as hosts

Within the last two decades, interest in pectinases produced by unicellular fungi (yeasts) has been increasing as potential commercial pectinases (Alimardani-Theuil et al., 2011). Out of the 678 yeast species identified (Barnet et al., 2000), only a few, including some strains of *S. cerevisiae* that was initially thought to be devoid of pectinolytic activity, produce pectinolytic enzymes. However, a range of yeast pectinases has now been discovered including PGases, PLs, and Pectin esterases. Some of these enzymes can act over a broad range of temperatures (0–60°C), a property not seen in pectinases of multicellular-fungal origin (Alimardani-Theuil et al., 2011). Psychrophilic yeasts isolated from frozen soil in Iceland and Japan produced an acidic PGase and a PL that were optimally active at 5°C–14°C (Birgisson et al., 2003; Nakagawa et al., 2002). These yeasts included species belonging to *Cryptococcus* and *Cystofilobasidium* genera, such as *Cystofilobasidium capitatum* and *Cryptococcus cylindricus*. Development of recombinant strains engineered to overexpress cold-adapted pectinases can enable juice production to be carried out at low temperatures that minimize the risk of microbial contamination and energy costs.

2.3.3 Development of hybrid strains

Protoplast fusion has been successfully performed to fuse pectinase-producing microbial cells to develop hybrid strains that incorporate the desirable pectinolytic activities of the parental strains. Protoplast fusion was performed between pectinolytic fungi *Aspergillus* sp. CH-Y-1043 (A13) *ade⁻* and *Aspergillus flavipes* ATCC-16795 (F7) *lys⁻* (Solis et al., 1997). Four of the resulting hybrid strains showed enhanced production of PGase and PL where the maximum production was 160 per cent that of the wild-type parental strains. The fusion between complementary mutant strains of *Penicillium griseoroseum* and *P. expansum* resulted in recombinant hybrids that had increased production of PGase (3-fold) and PL (1.2-fold) (Varavallo et al., 2007). The HZ hybrid resulting from the fusion between mutant pectinolytic *Aspergillus flavipes* and *Aspergillus niveus* CH-Y-1043 strains showed an increase of 450 per cent and 1300 per cent in PL production compared to its two parents respectively: the hydrolysis of orange peel by the recombinant PL resulted in a yield of 92 per cent of available substrate (Solís et al., 2009).

2.3.4 Bi-functional pectinolytic enzymes

Within the last decade, the discovery of natural bifunctional pectinolytic enzymes that contain two catalytic domains with different enzymatic activity has brought the prospect of a major advance in juice/wine extraction processes into the near horizon. A gene containing both PL and PME catalytic regions was isolated from the genome of the alkaliphilic gram-positive bacterium *Bacillus* KSM-P358 and was heterologously expressed in *Bacillus subtilis* (Kobayashi et al., 2003). The recombinant protein expressed optimum activity for the two catalytic regions between pH of 8.5–10 and a temperature of 40°C–45°C. Interestingly, when the gene sequences of the catalytic domains were separately expressed in *E. coli,* the functionality and catalytic properties were identical to that of the intact enzyme, suggesting that the bi-functional enzyme could function as a substitute for the two separate catalytic functions (Kobayashi et al., 2003). A similarly bi-functional enzyme with PL and PME activity was isolated from the marine bacterium *Pseudoalteromonas haloplanktis* ANT/505 that grows at low temperatures from 0–29°C (Truong and Schweder, 2006). The cold-adapted protein showed a high PME activity in a broad range of temperatures from 5°C–30°C at pH 7.5. Both these enzymes are alkaline as is generally seen with bacterial pectinases

(Favela-Torres et al., 2006) and have limited applications in the juice industry unless engineered to be acid-stable which would incur extra costs.

Recently, S6A, a multimodular pectinase consisting of an N-terminal PME catalytic domain and a C-terminal PGase domain, was identified in the fungus *Penicillium oxalicum* (Tu et al., 2014a). When cloned into *P. pastoris,* the recombinant S6A showed both PGase and PME activity with the PME specific activity (271.1 U/ mg) being higher than that of most fungal PMEs tested for citrus pectin degradation. The optimum temperature and pH were 50°C and 5.0 respectively while retaining excellent stability at pH 3.5–6 and at 40°C. Majority of the metal ions tested (including Na^+, K^+, Ca^{2+}, Li^+, Co^{2+}, Cr^{3+}) enhanced PME activity without any inhibitory effects on PGase functionality (Tu et al., 2014a). If the production and functionality of these bifunctional pectinases could be optimized with further genetic engineering, their utility in commercial juice and wine production would be invaluable as they are cost-effective, eco-friendly and highly efficient.

3. Amylases

Amylases are a group of enzymes that catalyze the hydrolysis of glycosidic linkages in starch to break them down into smaller units (El-fallal et al., 2012). Since starch is a common component of many biomaterials in human consumption, amylases are widely used for starch degradation in industries, such as food, textile, detergent, paper and many more. Amylases are used in juice industry for the clarification of hazy, cloudy juices to produce clear juices (Ceci and Lozano, 1998; Kahle et al., 2005). Along with pectin, starch is a major contributor to the cloudiness and turbidity of juice extracts which hinder filtration, cause gelling when concentrated and also cause post-concentration haze (Carrín et al., 2004). Therefore, both de-pectinization and de-starching are essential in industrial juice production. Amylases are categorized into two groups based on the nature of the glycosidic bonds they hydrolyze. Endoamylases cleave α-1,4-glycosidic bonds in the inner part of the amylose (starch polymer) chain: α-amylase, used in juice clarification, belongs to this category (El-fallal et al., 2012). Exoamylases, such as β-amylase used in detergent and textile industries, cleave the external α-1,4-glycosidic bonds in the starch polymers.

3.1 Natural microbial sources of amylases

Amylases are present in a vast majority of living organisms but microbial sources are exclusively used for industrial amylase production due to cost effectiveness and ease of genetic manipulation. The majority of microbial strains in industrial use are derived from a small number of bacterial and fungal species that have been extensively studied and naturally produce high levels of the enzyme. For example, the bacterial sources are mainly from *Bacillus* spp., such as *B. subtilis*, *B. amyloliquefaciens* and *B. licheniformis* (El-Banna et al., 2007) which can produce high levels of thermostable α-amylases (Sivaramakrishnan et al., 2006). Strains derived from *A. niger, A. oryzae* (Aunstrup, 1979) and *Penicillium expansum* (Doyle et al., 1989) comprise the major fungal counterparts.

When producing clear juices, amylases are added to the juice extract following de-pectinization with the incubation temperature usually kept at around 50°C (Dey et al., 2014). The untreated juice extracts are acidic with the pH of apple juice typically at 3.5 and pH of lemon juice at 2.2–2.8 (Grassin et al., 2009). Even though α-amylases from *Aspergillus* spp. dominate industrial applications, their low thermal and acid stability pose a major problem to improving the juice production process. Therefore, most research efforts have attempted to develop amylases with high thermostability and functionality at low pH conditions. Furthermore, most wild-type α-amylases identified to date contain a calcium (Ca^{2+})-binding site (Boel et al., 1990): the enzyme only reaches highest functionality when bound to Ca^{2+} (El-fallal et al., 2012). Since this requires the addition of Ca^{2+} salts to the juice extracts which need to be removed afterwards, Ca^{2+}-independent α-amylases are highly desirable.

3.2 Heterologous expression in prokaryotic and eukaryotic hosts

Heterologous expression of α-amylases in various prokaryotic and eukaryotic hosts to drive overexpression of the recombinant enzyme has been successfully achieved. The α-amylase gene from *B. amyloliquefaciens* was cloned into *B. subtilis* where the recombinant enzyme production was 2500-fold higher than wild-type *B. subtilis* and five-fold higher than the donor *B. amyloliquefaciens* (Palva, 1982). The functionality of the recombinant protein was unchanged and virtually the entire produced enzyme was secreted into the culture medium for easy purification. More recently, the α-amylase encoding gene amy1 from a strain of *B. licheniformis* isolated from a starch farm, was cloned and expressed in *B. subtilis* WB800 where a 1.48-fold increase in productivity was observed in the recombinant amy1 protein compared to the donor strain (Chen et al., 2015). One drawback of using *B. subtilis* as a heterologous host is that it secretes a variety of proteins, including many proteases, into the culture medium that is capable of degrading the recombinant α-amylase protein (Olempska-Beer et al., 2006). Various molecular techniques, such as deletion of protease genes and replacement of wild-type protease genes with mutants have been utilized to develop protease-deficient *B. subtilis* strains (Kodama et al., 2012; Stephenson and Harwood, 1998; Yang et al., 1984). In a novel approach, the number of translocons that transport proteins across the cell membrane was increased to determine its effect on the amount of proteins secreted into the culture medium. An artificial operon containing secYEG gene, the heterotrimeric protein complex required for translocation of proteins, was fused into an inducible promoter in *B. subtilis* (Mulder et al., 2013). In the recombinants, the increase in the level of secYEG proteins was accompanied by a significant increase in the α-amylase secretion. This strategy can be used in commercial production of recombinant α-amylases to avoid jamming the cell membranes due to a shortage of translocons, a phenomenon that often accompanies overexpression of proteins.

Several gram-negative bacterial species, including *E. coli* and *Pseudomonas fluorescens* Biovar I, have also been successfully utilized as hosts for heterologous α-amylase expression (Hmidet et al., 2008; Jorgensen et al., 1997; Landry et al., 2003; Olempska-Beer et al., 2006). However, due to the formation of intracellular IB aggregates, extra steps are required to purify the recombinant proteins from these hosts. *Aspergillus* spp., *Pichia pastoris* and *S. cerevisiae* strains are the common eukaryotic hosts used in commercial α-amylase production. In most cases, wild-type strains of *Aspergillus* spp. have been genetically modified to improve the yield and functional stability of the recombinant proteins. *A. oryzae* mutants with double deletions of CreA and CreB genes that are involved in carbon catabolite repression showed a ten-fold increase in α-amylase secretion (Ichinose et al., 2014). Strains of *A. oryzae* have also been engineered to contain additional copies of the α-amylase gene (Spohr et al., 1998). While *P. pastoris* is not a natural producer of starch-degrading amylases, it is commonly used for heterologous expression due to its efficient protein production and secretory systems. When cloned into *P. pastoris*, α-amylase from *A. niger* was functionally stable at the industrial temperature of 50°C and the pH range of 3–6 (Zeng et al., 2011). An acid and heat-stable α-amylase from *Rhizopus oryzae* was cloned into P. pastoris and successfully expressed under the induction of methanol (Li et al., 2011). Gene amplification has been attempted in a bid to increase the production of α-amylases in the industrial strains. The α-amylase gene from *B. licheniformis* (BLA) was subjected to homolog-mediated chromosomal amplification and expressed in the homologous host *B. licheniformis* B0204 (Niu et al., 2009). Recombinants with two to five copies of the gene expressed significantly more α-amylase compared to wild-type B0204. In heterologous expression, the amylase gene is generally fused to a molecular tag, such as poly-histidine (His) tag for easy purification of the recombinant protein through affinity chromatography. For instance, His-tag fusion at the C-terminus of the α-amylase from *Bacillus subtilis* CN7 improved its turnover rate by 59 per cent (Wang et al., 2014).

3.3 Recombinant amylases with unique characteristics

α-Amylases with desirable characteristics isolated from unconventional sources, such as extremophiles, are heterologously expressed to make their production economically feasible. Heat and acid-stable, Ca^{2+}-independent α-amylases are highly desirable in the juice industry due to their stability over a wide range of conditions, longer shelf-life and their potential to lower production costs. Thermostable α-amylase genes identified in thermophilic microbes are often cloned into industrial bacterial strains, such as *B. subtilis* and *E. coli*, for their overexpression. A highly heat-tolerant α-amylase gene isolated from hyperthermophilic archaeon *Pyrococcus furiosus* was cloned into *B. subtilis* and *E. coli* where the recombinant protein showed optimum activity at pH 4.5 and a temperature of almost 100°C (Jorgensen et al., 1997). Similarly, other thermostable amylases isolated from various microbial sources have been successfully overexpressed in heterologous hosts (Emtenani et al., 2015; Grzybowska et al., 2004; Haki and Rakshit, 2003; Mehta and Satyanarayana, 2013). A Ca^{2+}-independent α-amylase from acidophilic *B. acidicola* was cloned and expressed in *E. coli* BL21 where its production was increased 15-fold and demonstrated high thermostability (30°C–100°C) and acid stability (pH 3–7) (Sharma and Satyanarayana, 2012). Several other Ca^{2+}-independent α-amylases from thermophilic bacteria have been identified and characterized (Atsbha et al., 2015; Malhotra et al., 2000; Sajedi et al., 2005; Singh et al., 2015). Attempts to produce Ca^{2+}-independent α-amylases have not been limited to enzymes identified in natural sources. Site-directed mutagenesis was employed to alter the Ca^{2+}-binding site of the α-amylase from *B. licheniformis* MTCC 6598 to eliminate its dependence on Ca²⁺ for optimum functionality (Priyadharshini and Gunasekaran, 2007). When the mutant genes were cloned and expressed in *E. coli*, mutant amylase N104D showed significantly improved specific activity at pH 5 and 70°C compared to the wild-type enzyme. Efforts to optimize α-amylases for use in juice industry are continuing.

3.4 Current trends in recombinant amylase production

Due to the use of amylases in a wide range of commercial processes, vigorous research efforts continue to optimize the production and functionality of amylase enzymes. The latest advances, described briefly below, include the development of cold-tolerant amylases, use of halophilic microbes as hosts for heterologous expression and utilization of metagenomic approaches to discover novel amylase genes.

3.4.1 Cold-active amylases

Interest in cold-active α-amylases to be used in juice production along with cold-active pectinases (discussed above) has been growing due to their potential applicability in preventing microbial contamination and of decreasing energy usage. Amylases produced by psychrophilic microbes contain polypeptides with higher flexibility which makes accommodation of substrates easier at low temperatures (Kuddus et al., 2011). α-Amylase from Antarctic psychrophile *Alteromonas haloplanctis* was the first cold-active α-amylase to be successfully crystallized and its 3D-structure resolved (Aghajari et al., 1996). It was successfully cloned into the mesophilic *E. coli* and when expressed, the recombinant protein retained the psychrophilic functionality of the wild-type enzyme (Feller et al., 1998). A cold-adapted α-amylase was heterologously expressed in filamentous fungi for the first time when the amylase gene from the psychrotolerant fungus *Geomyces pannorum* was engineered into *A. oryzae* (Mao et al., 2015). The recombinant α-amylase was optimally active at 40°C but retained over 20 per cent of its maximal activity in the 0–20°C range. Production of cold-adapted enzymes often requires highly specific physical parameters since their natural hosts have adapted to survive and reproduce under extreme environmental conditions. Recreating these conditions for commercial enzyme production is not economically feasible. Therefore, the cold-adapted α-amylase production in a mesophilic background has to be optimized before being utilized in industrial juice extraction and other commercial applications.

3.4.2 Halophiles as heterologous hosts

The moderately halophilic microbes, especially those belonging to the genus Halomonas, have emerged as efficient cell factories to be used as heterologous hosts for recombinant amylases due to their elementary nutritional requirements and the ability to grow in extreme salt conditions (Frillingos et al., 2000). In addition to producing their own extracellular α-amylases (amyH), *H. meridian* and *H. elongate* were able to secrete the thermostable α-amylase from *B. licheniformis* (Coronado et al., 2000). The α-amylase from hyper-thermophilic archaeon, *P. woesei* was heterologously expressed in *H. elongata* where the recombinant protein was comparable in theromostability and functionality to the wild-type protein (Frillingos et al., 2000).

3.4.3 Use of metagenomics in search of novel amylases

The ever-increasing demand for biocatalysts with industrial value has prompted the exploration of metagenomic data in search of novel genes. Soil metagenomics, especially from extreme environments, hold the maximum potential as soil ecosystems are among the highest in microbial diversity and most of the resident microbes are unculturable (Lee and Lee, 2013). Since the gene sequences of many known fungal and bacterial α-amylases are now readily available, searching the metagenomic libraries for homologous genes has become a highly efficient way of identifying novel α-amylases. For example, an amylase gene (*pAMY*) identified in a soil metagenomic library derived from the north-western Himalayas was cloned into *E. coli* cells. It was discovered to be an efficient cold-active α-amylase that retained 90 per cent of its activity at low temperatures (Sharma et al., 2010). A thermostable and Ca^{2+}-dependent α-amylase was isolated from a soil metagenomic library of Western Ghats (a mountain range) in India and expressed in *E. coli* (Vidya et al., 2011). The recombinant protein was optimally active at 60°C and pH of 5.0 which demonstrated its potential to be used industrially in juice extraction.

4. Cellulases and hemicellulases

Cellulases and hemicellulases are two other component enzymes of the juice extraction cocktail, included to hydrolyze cellulose and hemicellulose molecules respectively, that give structural integrity to plant cells. The genetic modifications attempted to improve commercial production of these enzymes are described below.

4.1 Cellulases

Cellulases are a group of hydrolytic enzymes that break down cellulose which is the main component of the plant cell wall and the main contributor to its structural integrity. Cellulases are produced by a vast range of microorganisms, mainly fungi, bacteria and protozoans, that grow on cellulosic materials (Kuhad et al., 2011). Cellulases are often added to the macerating enzyme mixtures along with pectinases and amylases during commercial fruit juice production to increase the juice yield from mashed puree. Addition of cellulases generally results in a 5–10 per cent increase in the juice yield (Haitang et al., 2010; Karmakar and Ray, 2011; Kuhad et al., 2011).

4.1.1 Microbial sources of recombinant cellulases

The main sources of commercial cellulases are filamentous fungi, namely *Aspergillus* species (*A. niger, A. oryzae*) and *Trichoderma* species including *T. reesei* and *T. longibrachiatum* (Kuhad et al., 2011; Olempska-Beer et al., 2006). These species naturally secrete high amounts of cellulases into the extracellular environment, which along with their efficient secretion system and higher specific growth, makes them ideal hosts for heterologous expression of recombinant cellulases. Protease-deficient *A. niger* D15 strain was used as the host to express several recombinant cellulases and was shown to yield recombinant proteins of high homogeneity (Rose and Zyl, 2008). *T. reesei* strains were engineered to express cbh1 gene, an important subunit of cellulase, under a strong promoter upon which the recombinant cellulase production became significantly higher with enhanced enzymatic activity (Fang and Xia, 2013). Recombinant cellulases have also been successfully expressed in *E. coli* (Sahasrabudhe and Ranjekar, 1990), B. subtilis (Liu et al., 2012) and the yeast *P. pastoris* (Chen et al., 2007).

Optimizing of commercial juice extraction process requires cellulases to be as equally thermostable and acid-stable as the other components in the enzyme cocktail used during juice production. A thermostable cellulase from the thermophilic *Thermomonospora fusca* was expressed in *E. coli* and the recombinant cellulase E3 was discovered to be active at temperatures greater than 60°C and showed greater stability to proteolysis (Wilson et al., 1997). Similarly, recombinant thermostable cellulases have been expressed with high productivity in *B. subtilis* (Yang et al., 2010), the thermophilic fungus *Talaromycesemer sonii* (Murray et al., 2003) and the thermotolerant yeast *Kluyveromyces marxianus* NBRC1777 (Hong et al., 2007).

4.2 Hemicellulases

Hemicelluloses comprise a group of structural heteropolymers including xylans, glucuronoxylan, arabinoxylan and xyloglucan that are present in the plant cell-wall along with cellulose. They are different from cellulose in several aspects: (i) they are easily hydrolyzed by weak acids and bases whereas cellulose is highly resistant to hydrolysis; and (ii) hemicelluloses are branched polymers whereas cellulose is not. Hemicellulases, capable of hydrolyzing hemicelluloses, are often included in the enzyme cocktails used in juice production to facilitate greater juice extraction from mashed fruit puree.

5. Xylanases

The enzyme xylanase degrades β-1,4-xylan, the most prevalent component of the cell wall next to cellulose, into xylose, thus facilitating the breaking down of the plant cell-wall matrix. One of its many commercial uses is in extraction and clarification of juices and wine (Bajaj and Manhas, 2012; Kuhad et al., 2011). Commercial xylanases are predominantly produced by cultures of *Aspergillus* spp., *Trichoderma* spp., and some bacterial species, such as *Bacillus* spp. and *Streptomyces* spp. Use of recombinant DNA technology to enhance production and desired characteristics of xylanases is becoming increasingly common and the number of recombinant xylanases available in the market for commercial purposes is steadily increasing. Xylanases are heterologously expressed in both eukaryotic and prokaryotic hosts, such as *E. coli* (Zhang et al., 2010), *P. pastoris* (Damaso et al., 2003; Ruanglek et al., 2007) and *T. reesei* (de Faria et al., 2002) to obtain higher production rates. High extracellular expression of a recombinant xylanase with attractive biochemical properties from *Streptomyces* sp. S38 was achieved through codon optimization in *P. pastoris* (Fu et al., 2011). In a bid to increase extracellular secretion, xylanase from *A. niger* was expressed in P. pastoris under several secretion signal sequences from a variety of sources (Karaoglan et al., 2014). The highest xylanase secretion was achieved with a single-copy PIR which can be used in commercial xylanase production.

Thermostable xylanases have been isolated from several thermophilic microbes and cloned into heterologous hosts for over-expression. Highly efficient production of thermostable xylanase from Thermomyces lanuginosus was achieved by expressing the gene under the AOX1 promoter in *P. pastoris* (Damaso et al., 2003). The recombinant protein retained the native protein's optimum temperature of 75°C. Similarly, thermostable xylanases from a deep-sea thermophilic *GeoBacillus* spp. MT-1 (Wu et al., 2006), *Rhodothermus marinus* (Karlsson et al., 1999), *Nonomuraea flexuosa* (Leskinen et al., 2005) and *Thermoascus aurantiacus* (Zhang et al., 2011) have been successfully expressed in heterologous hosts and they hold great potential to be utilized in industrial applications.

6. Naringinase

Juices extracted from citrus fruit family, a genus to which many popular fruits such as oranges, grapefruits, lemons, limes and tangerines belong, contain a characteristic bitter taste caused by the presence of a molecule called naringin. Citrus juices are often treated with naringinase, an enzyme capable of degrading naringin during production to eliminate the excessive bitterness in the juice (Ni et al., 2014). Nariginase preparations immobilized on food contact—approved films or columns are commonly used in commercial sector to reduce the activation energy and Michaelis constant for the naringin hydrolysis reaction (Soares and Hotchkiss, 1998; Lei et al., 2011; Ni et al., 2012). Traditionally, filamentous fungi have been used as the main microbial source of commercial naringinase (Puri, 2012). These include *A. niger* (Bram et al., 1966; Manzanares et al., 1997; Thammawat et al., 2008), *A. kawachii* (Koseki et al., 2008), *A. terrus* (Gallego et al., 2001), *A. nidulans* (Orejas et al., 1999) and *Penicillium* sp. (Young et al., 1989). However, to the best of our knowledge, no recombinant naringinases have been used to date in commercial juice production.

7. Conclusions and Perspectives

In this chapter, we reviewed the major groups of enzymes, especially recombinant enzymes, that are involved in the production of fruit and vegetable juices. We described the characteristics of these enzymes, including their roles in enhancing juice preparation. The juices discussed cover a broad spectrum of items, from apples to oranges, grapes, citrus fruits, various combinations of these fruits and a diversity of vegetables. Recombinant enzymes are increasingly developed for commercial juice production. An emerging trend is the continued development of recombinant enzymes with high specificity in certain fruits and the juice production conditions. With the discovery of such enzymes from organisms living in diverse ecological niches through genomic and metagenomic tools, such a theme will become increasingly evident in the years to come.

Acknowledgement

Research in our lab on microbial diversity and their roles in natural environments and human services is supported by the Natural Sciences and Engineering Research Council of Canada.

References

Aghajari, N., Feller, G., Gerday, C. and Haser, R. 1996. Crystallization and preliminary X-ray diffraction studies of alpha-amylase from the antarctic psychrophile *Alteromonas haloplanctis* A23. Protein Science: Publication of the Protein Society, 5: 2128–9. doi: 10.1002/pro.5560050921.

Alimardani-Theuil, P., Gainvors-Claisse, A. and Duchiron, F. 2011. Yeasts: An attractive source of pectinases—From gene expression to potential applications: A review. Process Biochemistry, 46: 1525–1537. doi: 10.1016/j.procbio.2011.05.010.

Atsbha, T., Haki, G., Abera, S. and Gezmu, T. 2015. Thermo-stable, calcium independent alpha amylase from two *Bacillus* species in Afar, Ethiopia. International Research Journal of Pure and Applied Chemistry, 6: 9–18. doi: 10.9734/IRJPAC/2015/14970.

Aunstrup, K. 1979. Production, isolation and economics of extracellular enzymes, in Applied Biochemistry and Bioengineering. (ed.). L. Wingard (Elsevier).

Bajaj, B.K. and Manhas, K. 2012. Production and characterization of xylanase from *Bacillus licheniformis* P11(C) with potential for fruit juice and bakery industry. Biocatalysis and Agricultural Biotechnology, 1: 330–337. doi: 10.1016/j.bcab.2012.07.003.

Barnet, J.A., Payne, R.W. and Yarrow, D. 2000. Yeasts: Characteristics and Identification. Cambridge University Press.

Basu, A. and Penugonda, K. 2009. Pomegranate juice: A heart-healthy fruit juice. Nutrition Reviews, 67: 49–56. doi: 10.1111/j.1753-4887.2008.00133.x.

Birgisson, H., Delgado, O., Arroyo, L.G., Hatti-Kaul, R. and Mattiasson, B. 2003. Cold-adapted yeasts as producers of cold-active polygalacturonases. Extremophiles, 7: 185–193. doi: 10.1007/s00792-002-0310-7.

Blanco, P., Sieiro, C. and Villa, T.G. 1999. Production of pectic enzymes in yeasts. FEMS Microbiology Letters, 175: 1–9. doi: 10.1016/S0378-1097(99)00090-7.

Boel, E., Brady, L., Brzozowski, A.M., Derewenda, Z., Dodson, G.G., Jensen, V.J. et al. 1990. Calcium binding in alpha-amylases: An X-ray diffraction study at 2.1-A resolution of two enzymes from *Aspergillus*. Biochemistry, 29: 6244–6249.

Bram, B., Solomons, G.L. and Unit, F. 1966. Production of the enzyme naringinase by *Aspergillus niger*. Applied Microbiology, 14: 477. Available at: http://www.ncbi.nlm.nih.gov/pubmed/16349665.

Bub, A., Watzl, B., Blockhaus, M., Briviba, K., Liegibel, U., Müller, H. et al. 2003. Fruit juice consumption modulates antioxidative status, immune status and DNA damage. Journal of Nutritional Biochemistry, 14: 90–98. doi: 10.1016/S0955-2863(02)00255-3.

Byaruagaba-Bazirake, G. and Rensburg, P. 2013. Characterisation of banana wine fermented with recombinant wine yeast strains. American Journal of Food and Nutrition, 3: 105–116. doi: 10.5251/ajfn.2013.3.3.105.116.

Carrín, M.E., Ceci, L.N. and Lozano, J.E. 2004. Characterization of starch in apple juice and its degradation by amylases. Food Chemistry, 87: 173–178. doi: 10.1016/j.foodchem.2003.10.032.

Ceci, L. and Lozano, J. 1998. Determination of enzymatic activities of commercial pectinases for the clarification of apple juice. Food Chemistry, 61: 237–241. doi: 10.1016/S0308-8146(97)00088-5.

Chen, J., Chen, X., Dai, J., Xie, G., Yan, L., Lu, L. et al. 2015. Cloning, enhanced expression and characterization of an α-amylase gene from a wild strain in *B. subtilis* WB800. International Journal of Biological Macromolecules, 80: 6–13. doi: 10.1016/j.ijbiomac.2015.06.018.

Chen, X., Cao, Y., Ding, Y., Lu, W. and Li, D. 2007. Cloning, functional expression and characterization of *Aspergillus sulphureus* beta-mannanase in *Pichia pastoris*. Journal of Biotechnology, 128: 452–461. doi: 10.1016/j.jbiotec.2006.11.003.

Chen, Y.b., Sun, D., Zhou, Y., Liu, L., Han, W., Zheng, B. et al. 2014. Cloning, expression and characterization of a novel thermophilic polygalacturonase from *Caldicellulosiruptor bescii* DSM 6725. International Journal of Molecular Sciences, 15: 5717–5729. doi: 10.3390/ijms15045717.

Christgau, S., Kofod, L.V., Halkier, T., Andersen, L.N., Hockauf, M., Dorreich, K. et al. 1996. Pectin methyl esterase from *Aspergillus aculeatus*: Expression cloning in yeast and characterization of the recombinant enzyme. Biochemical Journal, 319: 705–712. Available at: <Go to ISI>:// A1996VT07700007.

Coronado, M.J., Vargas, C., Mellado, E., Tegos, G., Drainas, C., Nieto, J.J. et al. 2000. The α-amylase gene amyH of the moderate halophile *Halomonas meridiana*: Cloning and molecular characterization. Microbiology, 146: 861–868. doi: 10.1099/00221287-146-4-861.

Damak, N., Abdeljalil, S., Koubaa, A., Trigui, S., Ayadi, M., Trigui-Lahiani, H. et al. 2013. Cloning and heterologous expression of a thermostable pectate lyase from *Penicillium occitanis* in *Escherichia coli*. International Journal of Biological Macromolecules, 62: 549–556. doi: 10.1016/j. ijbiomac.2013.10.013.

Damaso, M.C.T., Almeida, M.S., Kurtenbach, E., Martins, O.B., Pereira, N., Andrade, C.M.M.C. et al. 2003. Optimized expression of a thermostable xylanase from thermomyces lanuginosus in *Pichia pastoris*. Applied and Environmental Microbiology, 69: 6064–6072. doi: 10.1128/AEM.69.10.6064-6072.2003.

de Faria, F.P., Te'O, V.S.J., Bergquist, P.L., Azevedo, M.O. and Nevalainen, K.M.H. 2002. Expression and processing of a major xylanase (XYN2) from the thermophilic fungus *Humicola grisea* var. *thermoidea* in *Trichoderma reesei*. Letters in Applied Microbiology, 34: 119–23. doi: 10.1046/j.1472-765x.2002.01057.x.

Dennison, B.A., Rockwell, H.L. and Baker, S.L. 1997. Excess fruit juice consumption by preschool-aged children is associated with short stature and obesity. Pediatrics, 99: 15–22.

Dey, T.B., Adak, S., Bhattacharya, P. and Banerjee, R. 2014. Purification of polygalacturonase from *Aspergillus awamori* Nakazawa MTCC 6652 and its application in apple juice clarification. LWT—Food Science and Technology, 59: 591–595. doi: 10.1016/j.lwt.2014.04.064.

Doyle, E.M., Kelly, C.T. and Fogarty, W.M. 1989. The high maltose-producing α-amylase of *Penicillium expansum*. Applied Microbiology and Biotechnology, 30: 492–496.

Duthie, S.J., Jenkinson, A.M., Crozier, A., Mullen, W., Pirie, L., Kyle, J. et al. 2006. The effects of cranberry juice consumption on antioxidant status and biomarkers relating to heart disease and cancer in healthy human volunteers. European Journal of Nutrition, 45: 113–122. doi: 10.1007/s00394-005-0572-9.

El-Banna, T.E., Abd-Aziz, A.A., Abou-Dobara, M.I. and Ibrahim, R.I. 2007. Production and immobilization of alpha-amylase from *Bacillus subtilis*. Pakistan Journal of Biological Sciences, PJBS, 10: 2039–2047.

El-fallal, A., Dobara, M.A., El-sayed, A. and Omar, N. 2012. Starch and microbial α-amylases: From concepts to biotechnological applications. Carbohydrates—Comprensive Studies on Glycobiology and Glycotechnology, 459–489. doi: 10.5772/51571.

Emtenani, S., Asoodeh, A. and Emtenani, S. 2015. Gene cloning and characterization of a thermostable organic-tolerant α-amylase from *Bacillus subtilis* DR8806. International Journal of Biological Macromolecules, 72: 290–8. doi: 10.1016/j.ijbiomac.2014.08.023.

Fang, H. and Xia, L. 2013. High activity cellulase production by recombinant *Trichoderma reesei* ZU-02 with the enhanced cellobiohydrolase production. Bioresource Technology, 144: 693–697. doi: 10.1016/j.biortech.2013.06.120.

Favela-Torres, E., Volke-Sepúlveda, T. and Viniegra-González, G. 2006. Production of hydrolytic depolymerising pectinases. Food Technology and Biotechnology, 44: 221–227. doi: ISSN 1330-9862.

Feller, G., Bussy, O.L.E., Gerday, C. and Le Bussy, O. 1998. Expression of psychrophilic genes in mesophilic hosts: Assessment of the folding state of a recombinant alpha-amylase. Applied and Environmental Microbiology, 64: 1163–5. doi: .

Fernandez-Gonzalez, M., Ubeda, J.F., Cordero-Otero, R.R., Thanvanthri Gururajan, V. and Briones, A.I. 2005. Engineering of an oenological *Saccharomyces cerevisiae* strain with pectinolytic activity and its effect on wine. Int. J. Food Microbiol., 102: 173–183. doi: 10.1016/j.ijfoodmicro.2004.12.012.

Ferrer-Miralles, N., Domingo-Espín, J., Corchero, J.L., Vázquez, E. and Villaverde, A. 2009. Microbial factories for recombinant pharmaceuticals. Microbial Cell Factories, 8: 17. doi: 10.1186/1475-2859-8-17.

Frillingos, S., Linden, A., Niehaus, F., Vargas, C., Nieto, J.J., Ventosa, A. et al. 2000. Cloning and expression of alpha-amylase from the hyperthermophilic archaeon *Pyrococcus woesei* in the moderately halophilic bacterium *Halomonas elongata*. Journal of Applied Microbiology, 88: 495–503. doi: jam988 [pii].

Fu, X.-Y., Zhao, W., Xiong, A.-S., Tian, Y.-S. and Peng, R.-H. 2011. High expression of recombinant *Streptomyces* sp. S38 xylanase in *Pichia pastoris* by codon optimization and analysis of its biochemical properties. Molecular Biology Reports, 38: 4991–7. doi: 10.1007/s11033-010-0644-7.

Gallego, M.V., Pinaga, F., Ramon, D. and Valles, S. 2001. Purification and characterization of an alpha-L-rhamnosidase from *Aspergillus terreus* of interest in winemaking. Journal of Food Science, 66: 204–209. Available at: <Go to ISI>://000168880200004.

Gautam, S., Dubey, P., Varadarajan, R. and Gupta, M.N. 2012. Role of smart polymers in protein purification and refolding. Bioengineered, 3: 286–8. doi: 10.4161/bioe.21372.

Gil-Izquierdo, A., Gil, M.I. and Ferreres, F. 2002. Effect of processing techniques at industrial scale on orange juice antioxidant and beneficial health compounds. Journal of Agricultural and Food Chemistry, 50: 5107–5114. doi: 10.1021/jf020162+.

Grassin, C., Fauquembergue, P. and Flickinger, M. 2009. Enzymes, Fruit Juice Processing. Available at: http://dx.doi.org/10.1002/9780470054581.eib292.

Griep, L.M.O., Stamler, J., Chan, Q., Van Horn, L., Steffen, L.M., Miura, K. et al. 2013. Association of raw fruit and fruit juice consumption with blood pressure: The INTERMAP study. American Journal of Clinical Nutrition, 97: 1083–1091. doi: 10.3945/ajcn.112.046300.

Grzybowska, B., Szweda, P. and Synowiecki, J. 2004. Cloning of the thermostable alpha-amylase gene from *Pyrococcus woesei* in *Escherichia coli*: Isolation and some properties of the enzyme. Molecular Biotechnology, 26: 101–110. doi: 10.1385/MB:26:2:101.

Gummadi, S.N. and Panda, T. 2003. Purification and biochemical properties of microbial pectinases—A review. Process Biochemistry, 38: 987–996. doi: 10.1016/S0032-9592(02)00203-0.

Haitang, W., Yan, L., Ying, Z., Xiaojie, Z., Dongmei, C. and Lidan, W. 2010. Effects of pectase and cellulase on the juice extraction rate of eureka lemon [J]. Academic Periodical of Farm Products Processing, 3: 19.

Haki, G.D. and Rakshit, S.K. 2003. Developments in industrially important thermostable enzymes: A review. Bioresource Technology, 89: 17–34. doi: 10.1016/S0960-8524(03)00033-6.

Heerd, D., Yegin, S., Tari, C. and Fernandez-Lahore, M. 2012. Pectinase enzyme-complex production by *Aspergillus* spp. in solid-state fermentation: A comparative study. Food and Bioproducts Processing, 90: 102–110. doi: 10.1016/j.fbp.2011.08.003.

Hmidet, N., Bayoudh, A., Berrin, J.G., Kanoun, S., Juge, N. and Nasri, M. 2008. Purification and biochemical characterization of a novel alpha-amylase from *Bacillus licheniformis* NH1. Cloning, nucleotide sequence and expression of amyN gene in *Escherichia coli*. Process Biochemistry, 43: 499–510. doi: 10.1016/j.procbio.2008.01.017.

Hong, J., Wang, Y., Kumagai, H. and Tamaki, H. 2007. Construction of thermotolerant yeast expressing thermostable cellulase genes. Journal of Biotechnology, 130: 114–123. doi: 10.1016/j.jbiotec.2007.03.008.

IBIS. 2015. Global Fruit and Vegetable Processing. IBIS World. Available at: http://clients1.ibisworld.com/reports/gl/industry/default.aspx?entid=357 [Accessed July 7, 2015].

Ichinose, S., Tanaka, M., Shintani, T. and Gomi, K. 2014. Improved α-amylase production by *Aspergillus oryzae* after a double deletion of genes involved in carbon catabolite repression. Apppl. Microb. Biotech., 98: 335–343. doi: 10.1007/s00253-013-5353-4.

Imamura, F., O'Connor, L., Ye, Z., Mursu, J., Hayashino, Y., Bhupathiraju, S.N. et al. 2015. Consumption of sugar sweetened beverages, artificially sweetened beverages and fruit juice and incidence of type 2 diabetes: Systematic review, meta-analysis and estimation of population attributable fraction. BMJ12. doi: 10.1136/bmj.h3576.

Jorgensen, S., Vorgias, C.E. and Antranikian, G. 1997. Cloning, sequencing, characterization, and expression of an extracellular alpha-amylase from the hyperthermophilic archaeon *Pyrococcus furiosus* in

Escherichia coli and *Bacillus subtilis*. J. Biol. Chem., 272: 16335–16342. Available at: http://www. ncbi.nlm.nih.gov/pubmed/9195939.

Kahle, K., Kraus, M. and Richling, E. 2005. Polyphenol profiles of apple juices. Molecular Nutrition and Food Research, 49: 797–806. doi: 10.1002/mnfr.200500064.

Karaoglan, M., Yildiz, H. and Inan, M. 2014. Screening of signal sequences for extracellular production of *Aspergillus niger* xylanase in *Pichia pastoris*. Biochemical Engineering Journal, 92: 16–21. doi: 10.1016/j.bej.2014.07.005.

Karlsson, E.N., Holst, O. and Tocaj, A. 1999. Efficient production of truncated thermostable xylanases from *Rhodothermus marinus* in *Escherichia coli* fed-batch cultures. Journal of Bioscience and Bioengineering, 87: 598–606. doi: S1389-1723(99)80121-2 [pii].

Karmakar, M. and Ray, R.r. 2011. Current trends in research and application of microbial cellulases. Journal of Microbiology, 6: 41–53. doi: 10.3923/jm.2011.41.53.

Kashyap, D.R., Vohra, P.K., Chopra, S. and Tewari, R. 2001. Applications of pectinases in the commercial sector: A review. Bioresource Technology, 77: 215–227. doi: 10.1016/S0960-8524(00)00118-8.

Kobayashi, T., Sawada, K., Sumitomo, N., Hatada, Y., Hagihara, H. and Ito, S. 2003. Bifunctional pectinolytic enzyme with separate pectate lyase and pectin methylesterase domains from an alkaliphilic *Bacillus*. World Journal of Microbiology & Biotechnology, 19: 269–277. Available at: <Go to ISI>:// BIOSIS:PREV200300343081.

Kodama, T., Sekiguchi, J., Ara, K., Ozaki, K., Manabe, K., Liu, S. et al. 2012. Approaches for improving protein production in multiple protease-deficient *Bacillus subtilis* host strains. INTECH Open Access Publisher.

Koseki, T., Mese, Y., Nishibori, N., Masaki, K., Fujii, T., Handa, T. et al. 2008. Characterization of an alpha-L-rhamnosidase from *Aspergillus kawachii* and its gene. Applied Microbiology and Biotechnology, 80: 1007–13. doi: 10.1007/s00253-008-1599-7.

Kuddus, M., Roohi, Arif, J.M. and Ramteke, P.W. 2011. An overview of cold-active microbial A-amylase: Adaptation strategies and biotechnological potentials. Biotechnology, 10: 246–258. doi: 10.3923/ biotech.2011.246.258.

Kuhad, R.C., Gupta, R. and Singh, A. 2011. Microbial cellulases and their industrial applications. Enzyme Research, 2011: 280696. doi: 10.4061/2011/280696.

Kumar, S., Jain, K.K., Singh, A., Panda, A.K. and Kuhad, R.C. 2014. Characterization of recombinant pectate lyase refolded from inclusion bodies generated in *E. coli* BL21(DE3). Protein Expression and Purification, 97: 61–71. doi: 10.1016/j.pep.2014.12.003.

Laing, E. and Pretorius, I.S. 1993. Co-expression of an *Erwinia chrysanthemi* pectate lyase-encoding gene (pelE) and an *Erwinia carotovora* polygalacturonase-encoding gene (peh1) in *Saccharomyces cerevisiae*. Applied Microbiology and Biotechnology, 39: 181–188. Available at: <Go to ISI>:// BIOSIS:PREV199396039191.

Landry, T.D., Chew, L., Davis, J.W., Frawley, N., Foley, H.H., Stelman, S.J. et al. 2003. Safety evaluation of an alpha-amylase enzyme preparation derived from the archaeal order *Thermococcales* as expressed in *Pseudomonas fluorescens* biovar I. Regulatory Toxicology and Pharmacology, RTP 37: 149–168.

Lang, C. and Looman, A.C. 1995. Efficient expression and secretion of *Aspergillus niger* RH5344 polygalacturonase in *Saccharomyces cerevisiae*. Applied Microbiology and Biotechnology, 44: 147–56. doi: 10.1007/BF00164494.

Lee, M.H. and Lee, S.-W. 2013. Bioprospecting potential of the soil metagenome: Novel enzymes and bioactivities. Genomics & Informatics, 11: 114–20. doi: 10.5808/GI.2013.11.3.114.

Lei, S., Xu, Y., Fan, G., Xiao, M. and Pan, S. 2011. Immobilization of naringinase on mesoporous molecular sieve MCM-41 and its application to debittering of white grapefruit. Applied Surface Science, 257: 4096–4099. doi: http://dx.doi.org/10.1016/j.apsusc.2010.12.003.

Leskinen, S., Mäntylä, A., Fagerström, R., Vehmaanperä, J., Lantto, R., Paloheimo, M. et al. 2005. Thermostable xylanases, Xyn10A and Xyn11A, from the actinomycete *Nonomuraea flexuosa*: Isolation of the genes and characterization of recombinant Xyn11A polypeptides produced in *Trichoderma reesei*. Applied Microbiology and Biotechnology, 67: 495–505. doi: 10.1007/s00253-004-1797-x.

Li, S., Zuo, Z., Niu, D., Singh, S., Permaul, K., Prior, B.A. et al. 2011. Gene cloning, heterologous expression and characterization of a high maltose-producing α-amylase of *Rhizopus oryzae*. Applied Biochemistry and Biotechnology, 164: 581–92. doi: 10.1007/s12010-011-9159-5.

Lien, N.L.P. and Man, L.V.V. 2010. Application of commercial enzymes for jicama pulp treatment in juice production. Science and Technology Development, 13: 64–76.

Liu, J.M., Xin, X.J., Li, C.X., Xu, J.H. and Bao, J. 2012. Cloning of thermostable cellulase genes of *Clostridium thermocellum* and their secretive expression in *Bacillus subtilis*. Applied Biochemistry and Biotechnology, 166: 652–662. doi: 10.1007/s12010-011-9456-z.

Malhotra, R., Noorwez, S.M. and Satyanarayana, T. 2000. Production and partial characterization of thermostable and calcium-independent alpha-amylase of an extreme thermophile *Bacillus thermooleovorans* NP54. Letters in Applied Microbiology, 31: 378–84. doi: 10.1046/j.1472-765x.2000.00830.x.

Manzanares, P., de Graaff, L.H. and Visser, J. 1997. Purification and characterization of an alpha-L-rhamnosidase from *Aspergillus niger*. Fems Microbiology Letters, 157: 279–283. Available at: <Go to ISI>://000072333700010.

Mao, Y., Yin, Y., Zhang, L., Alias, S.A., Gao, B. and Wei, D. 2015. Development of a novel *Aspergillus* uracil deficient expression system and its application in expressing a cold-adapted α-amylase gene from Antarctic fungi *Geomyces pannorum*. Process Biochemistry, 50: 1581–1590. doi: 10.1016/j.procbio.2015.06.016.

Massa, C., Degrassi, G., Devescovi, G., Venturi, V. and Lamba, D. 2007. Isolation, heterologous expression and characterization of an endo-polygalacturonase produced by the phytopathogen *Burkholderia cepacia*. Protein Expression and Purification, 54: 300–8. doi: 10.1016/j.pep.2007.03.019.

Mehta, D. and Satyanarayana, T. 2013. Biochemical and molecular characterization of recombinant acidic and thermostable raw-starch hydrolysing α-amylase from an extreme thermophile Geo*Bacillus thermoleovorans*. Journal of Molecular Catalysis B: Enzymatic, 85-86: 229–238. doi: 10.1016/j.molcatb.2012.08.017.

Mulder, K.C.L., Bandola, J. and Schumann, W. 2013. Construction of an artificial secYEG operon allowing high-level secretion of α-amylase. Protein Expression and Purification, 89: 92–96. doi: 10.1016/j.pep.2013.02.008.

Murray, P.G., Collins, C.M., Grassick, A. and Tuohy, M.G. 2003. Molecular cloning, transcriptional and expression analysis of the first cellulase gene (cbh2), encoding cellobiohydrolase II from the moderately thermophilic fungus *Talaromyces emersonii* and structure prediction of the gene product. Biochemical and Biophysical Research Communications, 301: 280–286.

Nakagawa, T., Yamada, K., Miyaji, T. and Tomizuka, N. 2002. Cold-active pectinolytic activity of psychrophilic-basidiomycetous yeast *Cystofilobasidium capitatum* strain PPY-1. Journal of Bioscience and Bioengineering, 94: 175–177. doi: 10.1016/S1389-1723(02)80140-2.

Newman, J. 2010. The juice cleanse: A strange and green journey. The New York Times. Available at: http://www.nancykalish.com/files/The_Juice_Cleanse_-_NY_Times.pdf.

Ni, H.B.C.D., Yang, Y.F.C., Chen, F.B., Ji, H.F.C., Yang, H.D., Ling, W.D. et al. 2014. Pectinase and naringinase help to improve juice production and quality from pummelo (Citrus grandis) fruit. Food Science and Biotechnology, 23: 739–746. doi: 10.1007/s10068-014-0100-x.

Ni, H., Chen, F., Cai, H., Xiao, A., You, Q. and Lu, Y. 2012. Characterization and preparation of *Aspergillus niger* naringinase for debittering citrus juice. Journal of Food Science, 77: C1–7. doi: 10.1111/j.1750-3841.2011.02471.x.

Niu, D., Shi, G. and Wang, Z. 2009. Genetic improvement of alpha-amylase producing *Bacillus licheniformis* by homolog-mediated alpha-amylase gene amplification. Sheng wu gong cheng xue bao. Chinese Journal of Biotechnology, 25: 375–80. Available at: http://www.ncbi.nlm.nih.gov/pubmed/19621577.

Olempska-Beer, Z.S., Merker, R.I., Ditto, M. D. and DiNovi, M.J. 2006. Food-processing enzymes from recombinant microorganisms: A review. Regulatory Toxicology and Pharmacology, 45: 144–158. doi: 10.1016/j.yrtph.2006.05.001.

Orejas, M., Ibanez, E. and Ramon, D. 1999. The filamentous fungus *Aspergillus nidulans* produces an alpha-L-rhamnosidase of potential oenological interest. Letters in Applied Microbiology, 28: 383–388. Available at: <Go to ISI>://000080391200011.

Palva, I. 1982. Molecular cloning of α-amylase gene from *Bacillus amyloliquefaciens* and its expression in *B. subtilis*. Gene, 19: 81–87. doi: 10.1016/0378-1119(82)90191-3.

Parisot, J., Langlois, V., Sakanyan, V. and Rabiller, C. 2003. Cloning, expression and characterization of a thermostable exopolygalacturonase from *Thermotoga maritima*. Carbohydrate Research, 338: 1333–1337 ST—Cloning expression and characters. doi: 10.1016/S0008-6215(03)00165-4.

Porro, D., Gasser, B., Fossati, T., Maurer, M., Branduardi, P., Sauer, M. et al. 2011. Production of recombinant proteins and metabolites in yeasts: When are these systems better than bacterial production systems? Applied Microbiology and Biotechnology, 89: 939–48. doi: 10.1007/s00253-010-3019-z.

Priyadharshini, R. and Gunasekaran, P. 2007. Site-directed mutagenesis of the calcium-binding site of alpha-amylase of *Bacillus licheniformis*. Biotechnology Letters, 29: 1493–9. doi: 10.1007/s10529-007-9428-0.

Puri, M. 2012. Updates on naringinase: Structural and biotechnological aspects. Applied Microbiology and Biotechnology, 93: 49–60. doi: 10.1007/s00253-011-3679-3.

Puri, M., Sharma, D. and Barrow, C.J. 2012. Enzyme-assisted extraction of bioactives from plants. Trends in Biotechnology, 30: 37–44. doi: 10.1016/j.tibtech.2011.06.014.

Rojas, N.L., Ortiz, G.E., Chesini, M., Baruque, D.J. and Cavalitto, S.F. 2011. Optimization of the production of polygalacturonase from *Aspergillus kawachii* cloned in *Saccharomyces cerevisiae* in batch and fed-batch cultures. Food Technology and Biotechnology, 49: 316–321. doi: 10.1007/s10295-010-0929-9.

Rose, S.H. and Zyl, W.H. Van. 2008. Exploitation of *Aspergillus niger* for the heterologous production of cellulases and hemicellulases. The Open Biotechnology Journal, 167–175. doi: 10.2174/1874070700802010167.

Ruanglek, V., Sriprang, R., Ratanaphan, N., Tirawongsaroj, P., Chantasigh, D., Tanapongpipat, S. et al. 2007. Cloning, expression, characterization and high cell-density production of recombinant endo-1,4-β-xylanase from *Aspergillus niger* in *Pichia pastoris*. Enzyme and Microbial Technology, 41: 19–25. doi: 10.1016/j.enzmictec.2006.11.019.

Sahasrabudhe, N.A. and Ranjekar, P.K. 1990. Cloning of the cellulase gene from *Penicillium funiculosum* and its expression in *Escherichia coli*. FEMS Microbiology Letters, 54: 291–293.

Sajedi, R.H., Naderi-Manesh, H., Khajeh, K., Ahmadvand, R., Ranjbar, B., Asoodeh, A. et al. 2005. A Ca-independent α-amylase that is active and stable at low pH from the *Bacillus* sp. KR-8104. Enzyme and Microbial Technology, 36: 666–671. doi: 10.1016/j.enzmictec.2004.11.003.

Sharma, A. and Satyanarayana, T. 2012. Production of acid-stable and high-maltose-forming α-amylase of *Bacillus acidicola* by solid-state fermentation and immobilized cells and its applicability in baking. Applied Biochemistry and Biotechnology, 168: 1025–34. doi: 10.1007/s12010-012-9838-x.

Sharma, S., Khan, F.G. and Qazi, G.N. 2010. Molecular cloning and characterization of amylase from soil metagenomic library derived from Northwestern Himalayas. Applied Microbiology and Biotechnology, 86: 1821–1828. doi: 10.1007/s00253-009-2404-y.

Sieiro, C., García-Fraga, B., López-Seijas, J., Silva, A.F. Da and Villa, T.G. 2012. Microbial pectic enzymes in the food and wine industry. Food Industrial Processes—Methods and Equipment, 201–218. doi: 10.5772/2491.

Sieiro, C., Sestelo, A.B.F. and Villa, T.G. 2009. Cloning, characterization and functional analysis of the EPG1-2 gene: A new allele coding for an endopolygalacturonase in *Kluyveromyces marxianus*. Journal of Agricultural and Food Chemistry, 57: 8921–6. doi: 10.1021/jf900352q.

Singh, G.M., Micha, R., Khatibzadeh, S., Shi, P., Lim, S., Andrews, K.G. et al. 2015. Global, Regional and National Consumption of Sugar-sweetened Beverages, Fruit Juices, and Milk: A Systematic Assessment of Beverage Intake in 187 Countries. PLoS ONE, 10: e0124845. doi: 10.1371/journal.pone.0124845.

Singh, R., Dhawan, S., Singh, K. and Kaur, J. 2012. Cloning, expression and characterization of a metagenome derived thermoactive/thermostable pectinase. Molecular Biology Reports, 39: 8353–8361. doi: 10.1007/s11033-012-1685-x.

Singh, R., Sharma, D. and Shrivastav, V. 2015. Production and characterization of a thermostable and Ca++ independent amylase enzyme from soil bacteria. British Microbiology Research Journal, 8: 329–342. doi: 10.9734/BMRJ/2015/13993.

Singh, S.M. and Panda, A.K. 2005. Solubilization and refolding of bacterial inclusion body proteins. Journal of Bioscience and Bioengineering, 99: 303–10. doi: 10.1263/jbb.99.303.

Sivaramakrishnan, S., Gangadharan, D., Madhavan, K., Soccol, C.R. and Pandey, A. 2006. α-amylases from microbial sources—An overview on recent developments. Production, 44: 173–184.

Soares, N.F.F. and Hotchkiss, J.H. 1998. Naringinase immobilization in packaging films for reducing naringin concentration in grapefruit juice. Journal of Food Science, 63: 61–65. doi: 10.1111/j.1365-2621.1998.tb15676.x.

Solis, S., Flores, M.E. and Huitron, C. 1997. Improvement of pectinase production by interspecific hybrids of *Aspergillus* strains. Letters in Applied Microbiology, 24: 77–81. Available at: <Go to ISI>://A1997WH48900001.

Solís, S., Loeza, J., Segura, G., Tello, J., Reyes, N., Lappe, P. et al. 2009. Hydrolysis of orange peel by a pectin lyase-overproducing hybrid obtained by protoplast fusion between mutant pectinolytic *Aspergillus flavipes* and *Aspergillus niveus* CH-Y-1043. Enzyme and Microbial Technology, 44: 123–128. doi: 10.1016/j.enzmictec.2008.11.003.

Spohr, A., Carlsen, M., Nielsen, J. and Villadsen, J. 1998. α-Amylase production in recombinant *Aspergillus oryzae* during fed-batch and continuous cultivations. Journal of Fermentation and Bioengineering, 86: 49–56. doi: 10.1016/S0922-338X(98)80033-0.

SSP PVT LTD. 2015. Fruit Juice, Pulp & Concentration Plants. Sspindia.com. Available at: http://www.sspindia.com/fruit-juice-processing.html [Accessed October 8, 2015].

Stephenson, K. and Harwood, C.R. 1998. Influence of a cell-wall-associated protease on production of alpha-amylase by *Bacillus subtilis*. Applied and Environmental Microbiology, 64: 2875–2881.

Thammawat, K., Pongtanya, P., Juntharasri, V. and Wongvithoonyaporn, P. 2008. Isolation, preliminary enzyme characterization and optimization of culture parameters for production of naringinase isolated from *Aspergillus niger* BCC 25166. Kasetsart Journal—Natural Science, 42: 61–72. Available at: http://www.scopus.com/inward/record.url?eid=2-s2.0-38849091276&partnerID=40&md5=8245d8 7ee264bd0698ab279d819ddcdd.

Truong, L. Van and Schweder, T. 2006. Characterization of the Pectinolytic Enzymes of the Marine Psychrophilic Bacterium *Pseudoalteromonas haloplanktis* Strain ANT/505.

Tu, T., Bai, Y., Luo, H., Ma, R., Wang, Y., Shi, P. et al. 2014a. A novel bifunctional pectinase from *Penicillium oxalicum* SX6 with separate pectin methylesterase and polygalacturonase catalytic domains. Applied Microbiology and Biotechnology, 98: 5019–28. doi: 10.1007/s00253-014-5533-x.

Tu, T., Meng, K., Huang, H., Luo, H., Bai, Y., Ma, R. et al. 2014b. Molecular characterization of a thermophilic endo-polygalacturonase from *Thielavia arenaria* XZ7 with high catalytic efficiency and application potential in the food and feed industries. Journal of Agricultural and Food Chemistry, 62: 12686–12694. doi: 10.1021/jf504239h.

Van Duyn, M.A. and Pivonka, E. 2000. Overview of the health benefits of fruit and vegetable consumption for the dietetics professional: Selected literature. Journal of the American Dietetic Association, 100: 1511–1521. doi: 10.1016/S0002-8223(00)00420-X.

Varavallo, M.A., De Queiroz, M.V., Lana, T.G., De Brito, A.T.R., Gonçalves, D.B. and De Araújo, E.F. 2007. Isolation of recombinant strains with enhanced pectinase production by protoplast fusion between *Penicillium expansum* and *Penicillium griseoroseum*. Brazilian Journal of Microbiology, 38: 52–57. doi: 10.1590/S1517-83822007000100011.

Vidya, J., Swaroop, S., Singh, S.K., Alex, D., Sukumaran, R.K. and Pandey, A. 2011. Isolation and characterization of a novel α-amylase from a metagenomic library of Western Ghats of Kerala, India. Biologia, 66: 939–944. doi: 10.2478/s11756-011-0126-y.

Vilanova, M., Blanco, P., Cortés, S., Castro, M., Villa, T.G. and Sieiro, C. 2000. Use of a PGU1 recombinant *Saccharomyces cerevisiae* strain in oenological fermentations. Journal of Applied Microbiology, 89: 876–883. doi: 10.1046/j.1365-2672.2000.01197.x.

Vogel, L. 2015. Food Guide Under Fire at Obesity Summit. Canadian Medical Association Journal, E256.

Wang, C., Wang, Q., Liao, S., He, B. and Huang, R. 2014. Thermal stability and activity improvements of a Ca-independent α-amylase from *Bacillus subtilis* CN7 by C-terminal truncation and hexahistidine-tag fusion. Biotechnology and Applied Biochemistry, 61: 93–100. doi: 10.1002/bab.1150.

Wang, H., Fu, L. and Zhang, X. 2011. Comparison of expression, purification and characterization of a new pectate lyase from *Phytophthora capsici* using two different methods. BMC Biotechnology, 11: 32. doi: 10.1186/1472-6750-11-32.

Willats, W.G.T., Mccartney, L., Mackie, W. and Knox, J.P. 2001. Pectin: Cell biology and prospects for functional analysis. Plant Molecular Biology, 47: 9–27. doi: 10.1023/A:1010662911148.

Wilson, D.B., Walker, L.P. and Zhang, S. 1997. Thermostable Cellulase from a Thermomonospora Gene. Available at: https://www.google.com/patents/US5677151.

Wojcicki, J.M. and Heyman, M.B. 2012. Reducing childhood obesity by eliminating 100% fruit juice. American Journal of Public Health, 102: 1630–1633. doi: 10.2105/AJPH.2012.300719.

Wu, S., Liu, B. and Zhang, X. 2006. Characterization of a recombinant thermostable xylanase from deep-sea thermophilic Geo*Bacillus* sp. MT-1 in East Pacific. Applied Microbiology and Biotechnology, 72: 1210–1216. doi: 10.1007/s00253-006-0416-4.

Yang, D., Weng, H., Wang, M., Xu, W., Li, Y. and Yang, H. 2010. Cloning and expression of a novel thermostable cellulase from newly isolated *Bacillus subtilis* strain I15. Molecular Biology Reports, 37: 1923–1929. doi: 10.1007/s11033-009-9635-y.

Yang, M.Y., Ferrari, E. and Henner, D.J. 1984. Cloning of the neutral protease gene of *Bacillus subtilis* and the use of the cloned gene to create an *in vitro*-derived deletion mutation. Journal of Bacteriology, 160: 15–21.

Yang, D., Weng, H., Wang, M., Xu, W., Li, Y. and Yang, H. 2010. Cloning and expression of a novel thermostable cellulase from newly isolated *Bacillus subtilis* strain I15. Molecular Biology Reports, 37: 1923–1929. doi: 10.1007/s11033-009-9635-y.

Young, N.M., Johnston, R.A.Z. and Richards, J.C. 1989. Purification of the α-L-rhamnosidase of *Penicillium decumbens* and characterisation of two glycopeptide components. Carbohydrate Research, 191: 53–62.

Zeng, Q., Wei, C., Jin, J., Wu, C. and Huang, B. 2011. Cloning of the gene encoding acid-stable alpha-amylase from *Aspergillus niger* and its expression in *Pichia pastoris*. African Journal of Food Science, 5: 668–675.

Zhang, J., Siika-Aho, M., Puranen, T., Tang, M., Tenkanen, M. and Viikari, L. 2011. Thermostable recombinant xylanases from *Nonomuraea flexuosa* and *Thermoascus aurantiacus* show distinct properties in the hydrolysis of xylans and pretreated wheat straw. Biotechnology for Biofuels, 4: 12. doi: 10.1186/1754-6834-4-12.

Zhang, M., Jiang, Z., Yang, S., Hua, C. and Li, L. 2010. Cloning and expression of a *Paecilomyces thermophila* xylanase gene in *E. coli* and characterization of the recombinant xylanase. Bioresource Technology, 101: 688–695. doi: 10.1016/j.biortech.2009.08.055.

22

Marine Microbial Enzymes in Food Application

Arunachalam Chinnathambi and *Chandrasekaran Muthusamy**

1. Introduction

The food and beverage industries manufacture a wide range of food and drinks to cater to the needs of the growing world population. The products manufactured include dairy and milk, grain and cereal, fruits and vegetables, beer and alcoholic beverages, meat and poultry, seafood, packaged or convenience foods and packaged drinks (Chandrasekaran et al., 2015). Application of enzymes in the food processing industry (Figs. 1 and 2) results in superior products with improved yields in addition to reducing carbon footprint, energy consumption and environmental pollution besides accounting for low greenhouse gas emissions and low raw material wastage. Major targets for modern enzyme technology continue to be preservation of foods and food components, efficient use of raw materials, improvement of food quality by way of texture and taste, manufacture of dietetic foods, elimination of anti-nutritive substances from certain nutritional raw materials, utilization of raw materials for preparation of animal feed and optimization of the process to reduce processing costs. Thus, several traditional chemical processes have been partially or completely replaced by enzymatic methods wherein enzymes are used largely in the food processing industry (Chandrasekaran and Rajeevkumar, 2002).

Department of Botany and Microbiology, College of Science, King Saud University, P.B. 2455, Riyadh-11451, Saudi Arabia.
E-mail: dr.arunmicro@gmail.com
* Corresponding author: chansek10@gmail.com

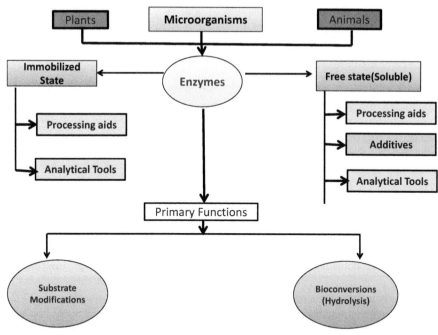

Figure 1. Source of enzymes and types of applications in food industries.

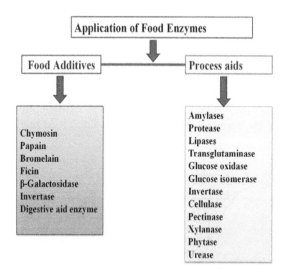

Figure 2. Application of enzymes in food and beverage industries.

2. Marine Environment as Source of Food Grade Enzymes

The marine environment ranges from nutrient-rich regions to nutritionally sparse locations where only a few organisms can survive. Complexity of the marine environment involving high salinity, high pressure, low temperature and special lighting conditions make the enzymes generated by marine microorganisms significantly different from homologous enzymes generated by terrestrial microorganisms (David and Michel, 1999). Moreover, during the course of evolution, marine microorganisms have excellently adapted themselves to diverse environmental parameters, such as high salt concentration, extreme temperatures, acidic and alkaline pH, extreme barometric pressure and low nutrient availability. They have developed unique metabolic capabilities to ensure survival in diverse habitats and produce a range of metabolites atypical to terrestrial microbial products (Faulkner, 2000). Marine microorganism take active part in the mineralization of complex organic matter through degradative pathways of their metabolism in the marine environment and contribute to secondary production in sea and are considered important ecological components in the marine environment due to their performance in biogeochemical processes (Chandrasekaran, 1997; Chandrasekaran and Rajeevkumar, 2002; Sowell et al., 2008).

Microbes live in various habitats of marine environment that include neuston, plankton, nekton, seston and epibiotic, endobiotic, pelagic and benthic environments (Fig. 3). These habitats harbor a diverse range of microbes including archaebacteria, cyanobacteria, eubacteria, actinomycetes, yeasts, filamentous fungi, microalgae, algae and protozoa (Fig. 4). Almost all of these groups are potential sources of useful enzymes that remain unexplored (Chandrasekaran and Rajeevkumar, 2002). Moreover, when compared with the terrestrial environment, the marine environment harbours marine microorganisms with unique genetic structures and life habitats (Stach et al., 2003). Marine microorganisms are attracting more and more attention as a resource for new enzymes because the microbial enzymes are relatively more stable than the corresponding enzymes derived from plants and animals (Kin, 2006). Further, marine microbial enzymes are of special interest due to their better stability, activity and tolerance to extreme conditions that most of the other proteins cannot withstand. These properties are often helpful in industrial processes (Burg, 2003). Bacteria and fungi from marine environments secrete different enzymes based on their habitat and their ecological functions. So far archaea, extremophiles and symbiotic microorganisms are known as sources of most of the interesting marine enzymes with

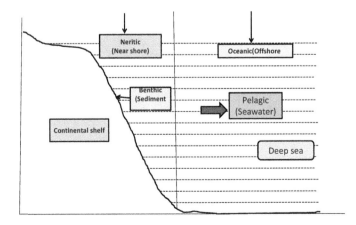

Figure 3. Marine Environments.

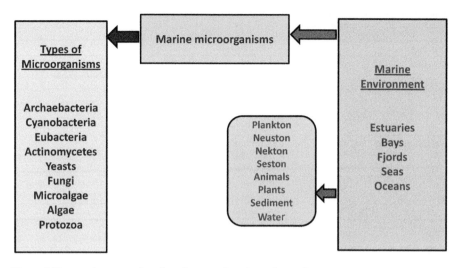

Figure 4. Types and sources of marine microorganisms in marine environments.

distinct structure, novel chemical properties and biocatalytic activity. But they are yet to be fully harnessed for food applications on a commercial scale.

3. Food Grade Marine Microbial Enzymes

Marine microbial enzymes are of special interest due to their habitat-related distinct features. Interestingly the enzymes reported from marine environments belong to one or more of the major classes of enzymes, viz: Oxidoreductase, Transferases, Hydrolases, Lyases, Isomerases and Ligases. Enzymes such as protease, amylase, alginate lyases, chitinase, cellulase, ligninase, pectinase, xylanase, lipases, proteases, glutaminase, asparaginase, arylsulphatase, phosphatase, beta lactamase and nucleases (DNAses, RNAses, and restriction enzymes), etc. by marine bacteria and fungi are known (Chandrasekaran and Rajeevkumar, 2002). Some of the food-grade enzymes isolated from marine microorganisms summarized in Table 1 are briefly discussed in the following sections.

3.1 Amido hydrolases-L-glutaminase

L-glutaminase (EC 3.5.1.2) is used in the food industry for the improvement of flavor in the food. *Pseudomonas fluorescens, Vibrio cholerae, Vibrio costicola* (Renu and Chandrasekaran, 1992) and marine fungi *Beauveria* sp. produced L-glutaminase extra-cellularly in copious amounts (Keerthi et al., 1999). *Pseudomonas flurorescens* produced salt tolerant L-glutaminase (Renu and Chandrasekaran, 1992). Marine *Micrococcus luteus* K-3 constitutively produced 2 salt-tolerant glutaminases (Moriguchi et al., 1994), which may be used to increase glutamic acid in foods at neutral or alkaline pH (Chandrasekaran and Rajeevkumar, 2002).

Table 1. Food grade Enzymes obtained from marine microorganisms (Adapted from Chandrasekaran and Rajeevkumar 2002; Fernandes, 2014).

Enzymes	Organisms
Proteases	*Bacillus licheniformis* (Halophilic), *Haloacterium halobium, Haloacterium salinarium* *Marinobacter* sp. *Vibrio gazogenes* *Vibrio* sp. 5709 *Metschnikowia reukaufii* W6b (yeast)
Alkaline protease	*Aureobasidium pullulans* *Bacillus mojavensis* A21 *Psychrobacter*
Subtilisin (EC 3.4.21.62)-like serine protease	*Vibrio* sp. PA-44 (psychrophilic) *Streptomyces* sp.
Neutral protease BBP-127	*Bacillus* sp. JT0127
Neutral proteinases	*Vibrio harveyi* (luminescent) *Vibrio splendidus* (luminescent)
Thiol protease	*Alteromonas haloplanktis* strain S5B *Pyrococcus* sp. KOD1
Collagenase	*Achromobacter iophagus* *Clostridium histolyticum* *Vibrio alginolyticus*
Cold resistant lipases	*Moraxella* (from the Antarctic seawater)
Lipase (EC 3.1.1.3)	*Bacillus smithii* *Vibrio* sp. VB-5 *Candida quercitrusa* JHSb, *Candida rugosa* wl8, *Candida intermedia* YA01a, *Candida parapsilosis* 3eA2, *Lodderomyces elongisporus* YF12c, *Pichia guilliermondii* N12c, *Rhodotorula mucilaginosa* L10-2, *Yarrowia lipolytica* N9a, *Aureobasidium pullulans* HN2.3 *Aspergillus awamori, A. sydowii* *Penicillium oxalicum*
Agarose (EC 3.2.1.81)	*Aeromonas* sp. *Alteromonas* sp. *Alteromonas agarlyticus* GJ1B *Bacillus cereus* *Cytophaga* sp. *Microbulbifer maritimus* *Pseudomonas* sp. *Pseudomonas stutzeri* *Pseudoalteromonas* sp. BL-3 *Pseudoalteromonas* sp. *Thalassomonas* sp. JAMB-A33 *Vibrio* sp.

Table 1. contd....

Table 1. contd....

Enzymes	Organisms
Alginate-lyase (EC 4.2.2.3)	*Alteromonas* sp., *Bacillus* sp. *Photobacterium* sp.
κ-Carrageenanase (EC 3.2.1.83)	A *Cytophaga*-referred to as strain *Dsij*, isolated from the marine red alga *Delesseria sanguinea*
α–Amylase (EC 3.2.1.1)	*Alteromonas rubra* *Desulfurococcus mucosus* *Fervidobacterium pennavorans* *Halobacterium halobium* and *Halob. Sodomense* *Halomonas* sp. strain AAD21 *Pseudomonas*-like strain MS300 *Pyrococcus woesei, Pyrococcus furiosus* *Thermococcus celer, Thermotoga maritime;* *Vibrio gazogenes* (psychrotrophic), *Streptomyces* sp. *Mucor* sp.
Glucoamylase (EC 3.2.1.3)	*Aspergillus* sp. JAN-25 *Aureobasidiumpullulans* N13d
α-Glucosidase (EC 3.2.1.2)	*Pyrococcus furiosus* *Pyrococcus wesei*
β-glucosidase (3.2.1.21)	*Aeromonas* sp. HC11e-3 *Martelella mediterranea* *Pyrococcus abyssi* *Pyrococcus furiosus* *Thermotoga maritima*
Pullulanases (EC 3.2.1.41)-Debranching Enzymes	*Fervidobacterium pullulanolyticum* (thermophilic) *Fervidobacterium pennavorans* *Desulfurococcus mucosus* *Pyrococcus furiosus* (Pf) DSM 3638 *Pyrococcus woesei,* *Thermococcus hydrothermalis* *Thermococcus litoralis* (Tl) DSM 5473 *Thermococcus celer* *Aureobasidiumpullulans*
Cyclomaltodextrin-glucanotransferase (Cgase EC 2.4.1.19)	*Bacillus subtilis* sp. (from a deep-sea mud)
α-Galactosidase (EC 3.2.1.22)	*Alteromonas* spp. (isolated from sponges and alga) *Photobacterium damsela* JT0160
β –galactosidase EC 3.2.1.23	*Pyrococcus furiosus* *Sulfolonbus solataricus* MT4
β –glucanases, (EC 3.2.1.75 & EC 3.2.1.39)	*Alteromonas* sp. *Bacillus circulans* *Cytophaga*-NK-5

Table 1. contd....

Table 1. contd....

Enzymes	Organisms
Chitinases EC 3.2.1.14	*Acinetobacter* sp. *Aeromonas hydrophila, Aeomonas hydophila* H-2330 *Alteromonas* sp., *Arthrobacter* sp. *Bacillus* sp., *Chromobacterium* sp., *Clostridium* sp. *Enterobacter* sp., *Flavobacterium* sp., *Klebsiella* sp. *Moraxella* sp., *Pseudomonas* sp., *Serratia* sp. *Vibrio* sp., *V. alginolyticus, Vibrio alginolyticus* H-8 *Vibrio harveyii, Vibrio fluvialis, Vibrio ischeri* *V. parahaemolyticus, V. mimicus, Vibrio vulnificus* *Listonella anguillarum* *Myxobacter* sp., *Sporocytophaga* sp. *Streptomyces* sp., *Streptomyces plicatus* *Aspergillus* sp., *Beauveria bassiana* *Penicillium* sp., *Rhizopus* sp.
Cellulases (EC 3.2.1.4)	A symbiotic bacterium found in the gland of *Deshayes* of the marine shipworm, *Aspergillus terreus*
exo-1,4-β-cellobiohydrolase (EC3.2.1.91)	*Thermotoga* sp.
Inulinases (EC 3.2.1.7)	*Bacillus cereus* MU-31 *Marinimicrobium* sp. LS-A18 *Nocardiopsis* sp. DN-K15 *Pichia guilliermondii* OUC1 *Yarrowia lipolytica* OUC2
Phytases	*Kodamea ohmeri* BG3
Tannase EC 3.1.1.20	*Aspergillus awamori* BTMFW032 *Phormidium valderianum* BDU140441
Xylanase EC 3.2.1.8)	*Thermotoga* sp. *Thermotoga maritima,* *T. neapolitana* and *T. thermarum*
β-xylosidase EC 3.2.1.37	*Pyrococcus furiosus* *Sulfolonbus solataricus* MT4
β -fucosidase EC 3.2.1.38	*Sulfolonbus solataricus* MT4
L-glutaminase (EC 3.5.1.2)	*Micococcus luteus* K-3 *Pseudomonas fluorescens,* *Vibrio cholerae, Vibio costicola,* *Beauveria* sp.
Aminopeptidases	*A. flavus*
Superoxide-dismutase (SOD, EC 1.15.1.1)	*Debaryomyces hansenii* (Halotolerant yeast) *Synechococcus* sp. NKBG 042902
Thermostable enzymes	*Thermotoga* sp., *Thermus* sp., *Thermococcus, Pyrococcus,* *Bacillus* sp., and *Sulfolobus* family. *Pyrococcus abyssi-*, a deep-sea hyperthermophilic archaebacterium, (thermostable beta-glucosidase (I) (EC 3.2.1.21) & thermostable esterase.

Table 1. contd....

Table 1. contd....

Enzymes	Organisms
Thermostable enzymes	*Pyrococcus* sp. KOD1 (thermostable thiol protease). *Moraxella* sp. TA144 and TA317 (lipase, EC 3.1.1.3), *Alteromonas haloplanctis* A23 (alpha-amylase, EC 3.2.1.1) *Psychrobacter immobilis* A5 (beta-lactamase, EC 3.5.2.6); *Bacillus* sp. TA39 (protease), *Rhodothermus marinus*(endo-1,4-beta-D-xylanase) *Thermotoga neopolitanai* (cellulase and xylanase) *Pyrococcus furiosus* (β-xylosidase, β-galactosidase, and β-glucosidase); *Sulfolonbus solataricus* MT4 (β-xylosidase, β-galactosidase, and β–fucosidase)
Cold adapted enzymes	*Pseudoalteromonas* sp. SM9913 from deep-sea (protease MCP-01 with collagenolytic activity) *Myroides profundi* D25 from deep sea (collagenase), A *Pseudoalteromonas* sp. CF6-2 from deep-sea sediment *Alteromonas* sp. (psychrostable metallo protease (almelysin); *Alteromonas haloplanctis* (α amylase); *Vibrio* sp (isocitrate dehydrogenase and lipase) from., *Moraxella* sp. (beta lactamase, and triose phosphate isomerase); *Bacillus* sp. (subtilisin); *Pseudomonas* and *Psychrobacter* species (cold-active lipases)
Alkalophilic enzymes	*Alteromonas macleodii* KMM162 (thermo stable alkaline phosphatase (EC 3.1.3.1) and endo-1,3-beta -D-glucanase (EC 3.2.1.39); *Alteromonas* sp. KMM 518 (alkaline phosphatase EC 3.1.3.1); *Vibrio* sp. NUF-BPP1(alkaline metallo endopeptidase) *Streptomyces* sp. (alpha-amylase)
Halophilic and halo tolerant enzymes	*Halobacterium halobium* and *Halob. Sodomense* (amylases), *Halob. salinarium* and *Halob.halobium* (proteases), several halobacteria (lipases). produced by *Micrococcus varians* subsp. *halophilus* (Nuclease H) Several halotolerant bacteria(salt resistant enzymes, such as the produced by a *Bacillus* sp. (amylase)

3.2 Lipase (EC 3.1.1.3)/esterases (EC 3.1.1.X)

Lipases and esterases (EC 3.1.1.X) are carboxyl hydrolases which cleave to form ester bonds. Esterases and lipases are widely used in the food industry for improving the quality of bread through changes in flour lipids; flavor enhancement of butter, cheese, and margarine; in production of crackers, pasta, degumming of vegetable oils; in increasing the titer of polyunsaturated fatty acids in vegetable oils and improving the digestibility of natural lipids (Visser et al., 2012; Ferreira-Dias et al., 2013). Lipases are important in the food processing industry due to their distinct catalytic properties in reactions, such as esterification, hydrolysis and trans-esterification (Shahidi and Wanasundara, 1998).

Lipases from marine sources have been used for hydrolysis of mono- and di-acylglycerols in the production of food emulsifiers. Marine *Vibrio* sp. VB-5 produces a lipase (EC 3.1.1.3) that hydrolyzes n-3 polyunsaturated fatty acid (PUFA)-containing fish oil. Saturated and monoenoic fatty acids were liberated easily from fish oil by lipase. Marine fungus *Aspergillus awamori* BTMFW032 (Soorej et al., 2011) from sea-water and *A. sydowii* were reported to produce extracellular lipase (Bindiya and Ramana, 2012). Marine *Bacillus smithii* BTMS 11 isolated from marine sediments was also noted for lipase production (Lailaja and Chandrasekaran, 2013).

3.3 Protease

Proteases are used in cheese manufacturing, baking, meat tendering and soy protein hydrolysis. Microbial proteases are used in brewing, meat processing and as a digestive aid, etc. Proteases hydrolyze peptide bonds in proteins and polypeptides (Ward et al., 2009). They are used as relatively crude preparations in a wide array of processes in the food industry, over a wide range of operational conditions (viz. temperature, pH, osmolarity) and may or may not require a high specificity in enzyme action (Sawant and Nagendran, 2014).

Several proteases are being used as digestive enzyme. A protease from marine yeast *Metschnikowia reukaufii* W6b was reported to have potential application in cheese, food and fermentation industries (Li et al., 2010). It showed high skimmed milk coagulability in skimmed milk clotting test. Proteases isolated from marine sources may be used in some of these processes. Several marine bacteria were observed to produce proteases. Halophilic *Bacillus licheniformis, V. gazogenes* and *Vibrio* sp. 5709 from marine sediments produced protease (Manachini and Fortina, 1994). Subtilisin (EC 3.4.21.62)-like serine protease from the psychrophilic marine *Vibrio* sp. PA-44, neutral protease BBP-127 from *Bacillus* sp. JT0127, metal neutral proteinases from marine luminescent *Vibrio harveyi* and *Vibrio splendidus*, and a novel thiol protease hydrolyzing benzoyl-arginyl-4-methylcoumaryl-7-amide from *Alteromonas haloplanktis* strain S5B were obtained (Chandrasekaran and Rajeevkumar, 2002).

An alkaline protease obtained from the marine *Aureobasidium pullulans* HN2-3 was cloned and expressed in *Yarrowia lipolytica* and used for the production of bioactive peptides used as nutraceuticals (Ni et al., 2009). Aminopeptidases isolated from a marine *Aspergillus flavus* are used in the food industry for debittering and enhancing the functional properties of protein-based products and for cheese flavoring (Sriram et al., 2012). Psychrophilic leucyl amino peptidases, with chymotrypsin and trypsin activities have also been identified within the spectrum of extra-cellular enzymes from heterotrophic microorganisms in the waters of an Arctic fjord (Steen and Arnosti, 2013). A relatively thermostable protease has also been isolated from *Marinobacter* sp. in the Indian Ocean (Fulzele et al., 2011).

3.4 Starch hydrolyzing enzymes

Starch hydrolysates and other derivatives are widely used in the food and beverage industry. The enzymes involved in the conversion of starch to low molecular-weight compounds, such as glucose, maltose and oligosaccharides are α-amylase, β-amylase, glucoamylase, debranching enzymes (pullulanase) and α-glucosidase. Amylases and glucosidases are used in glucose or maltose syrup production from starch. Glucose isomerase catalyzes the conversion of glucose to fructose (Guzman-Maldonado and Paredes-Lopez, 1995).

3.4.1 Amylases

Amylases act on starch and starch derivatives. Starch consists of a mixture of amylose and amylopectin, where α-D-glucopyranosyl units are bound together, either by essentially 1, 4 links or by 1, 4 links and 1, 6 links, respectively. Amylopectin is the major form on a 3 to 1 ratio to amylose in most starches (Bertoft, 2004). Amylases partially hydrolyze starch to dextrins during the liquefaction stage. Dextrins are then further hydrolyzed either to high and very high maltose syrups or to high dextrose syrups.

3.4.1.1 α-Amylase (EC 3.2.1.1)

Several α-amylases were reported from a marine *Bacillus* sp. that can degrade raw starch from corn, cassava, sago, potato, rice and maize to produce simple sugars, such as maltose, maltotriose, maltotetraose, maltopentaose, maltohexaose and glucose (Puspasari et al., 2011). α-Amylase has been obtained from thermophilic archaea *Pyrococcus woesei, Pyrococcus furiosus, Thermococcus celer, Fervidobacterium pennavorans, Desulfurococcus mucosus* and *Thermotoga maritime*; psychrotrophic *Vibrio* isolated from a deep-sea mud, *Vibrio gazogenes, Alteromonas rubra* and *Mucor* sp. *Pseudomonas*-like strain MS300 from the deep-sea, produces two major and two minor maltotetraohydrolases (G4-amylase) (Chandrasekaran and Rajeevkumar, 2002). *Bacterium* sp. isolated from a marine salt farm was also observed to produce α-amylase (Pancha et al., 2010). Calcium-independent α-amylases with alkaline and halophilic nature was produced by a salt tolerant *Streptomyces* strain isolated from marine sediments (Chakraborty et al., 2012). *Halomonas* sp. strain AAD21 produced α-amylase with halo- and thermal stability (Uzyol et al., 2012). Fuelzyme®, an α-amylase produced and commercialized by Verenium Corporation, has been used in a blend for starch processing, where addition of another amylase allowed for obtaining dextrose adequate for sweetener production (Nedwin et al., 2011).

3.4.1.2 Glucoamylases (EC 3.2.1.3)

Glucoamylase, also known as 'amyloglucosidase' or 'gamma amylase', has the potential for use in commercial applications for the production of high dextrose syrups. Production of glucoamylases by marine microorganisms was reported with the marine yeast *Aureobasidium pullulans* N13d (Li et al., 2007). Marine endophytic fungus *Aspergillus* sp. JAN-25 produced thermostable glucoamylase, which displays maximal activity between 50 and 60°C (El-Gendy, 2012).

3.4.1.3 Pullulanases (EC 3.2.1.41)

Pullulanases are mainly used in the starch industry similar to α-amylases and α-glucosidase. They are commonly termed debranching enzymes since they act over α-1, 6 glycosidic bonds in amylopectin and related polysaccharides (Hii et al., 2012).

Pullulanase type I is a typical bacterial enzyme that is specific for α-1, 6 linkages in branched oligosaccharides. It is unable to attack α-1, 4-linkages in α-glucan. Pullulanases type I with a temperature optimum of 90°C has been reported only in *Fervidobacterium pullulanolyticum*, while Pullulanase type II (amylo pullulanase), which is capable of hydrolyzing the α-1, 4-linkages and branching points (α-1, 6-linkages) in polysaccharides and limit dextrins, is mainly found in anaerobic bacteria and is widely distributed among thermophilic bacteria and archaea. Marine thermophilic *Thermococcus litoralis* (Tl) DSM 5473 (optimal growth temp. 90°C) and *Pyrococcus furiosus* (Pf) DSM 3638 (98°C) produce pullulanase type II. Each of these enzymes was able to hydrolyze, in addition to the alpha-1, 6 linkages in pullulan, alpha-1, 4 linkages in amylose and soluble starch. The enzymes appear to represent highly thermostable amylopullulanase versions of those which have been isolated from less thermophilic organisms (Chandrasekaran and Rajeevkumar, 2002). *Pyrococcus woesei, Thermococcus celer, Fervidobacterium pennavorans,* and *Desulfurococcus mucosus* have also been reported to produce pullulanase type II. *Thermococcus hydrothermalis*, an archaeon isolated from a deep-sea hydro thermal vent, produces a hyperthermophilic amylopullulunase (Gantelet and Duchiron, 1998) *Thermococcus* sp. HJ21, hyperthermophilic archaeon, was reported to produce a pullulanase (Xu et al., 2009). *Aureobasidiumpullulans*, yeast-like fungus isolated from sea mud produces pullulanase (Wu and Chen, 2014).

3.4.2 Glucosidases

Glucosidases can be used to enhance the aroma of some wines by freeing glycosidically-bound volatile terpenes and flavor precursors.

3.4.2.1 α-Glucosidase (EC 3.2.1.2)

α-Glucosidase is mainly used in the starch industry along with α-amylase. An extremely thermostable α-glucosidase (EC 3.2.1.2) was isolated from the hyperthermophilic marine archaebacterium, *Pyrococcus furiosus* and *Pyrococcus wesei* (Chandrasekaran and Rajeevkumar, 2002).

3.4.2.2 β-Glucosidases (EC 3.2.1.21)

β-*Glucosidases* hydrolyze the terminal, non-reducing β-D-glucosyl residues of alkyl and aryl β-glycosides as well as disaccharides and short-chain oligosaccharides, with release of β-D-glucose residues (Bhatia et al., 2002). β-Glucosidases are widely used for the processing of fruit juices, wine and beer, namely aiming to improve organoleptic properties and release aroma and flavor compounds (Mojsov, 2013; Trincone, 2013; Wang et al., 2013). β-glucosidases were screened from marine *Thermotoga maritima* (Goyal et al., 2001). Apart from *Thermotogae* species, production of β-glucosidases from *Martelella mediterranea* (Mao et al., 2010) and from *Aeromonas* sp. HC11e-3 (Huang et al., 2012) after cloning and expression in *E. coli* was reported. The enzyme proved surprisingly stable in an alkaline environment and exhibited 50 per cent of activity at 4°C, a feature that is useful in industrial processes where low temperatures are required to maintain the flavor of the product (Kumar et al., 2011). The glucosidase

(BglA) from *Aeromonas* sp. HC11e-3 was cloned and expressed in *E. coli* BL21 (Huang et al., 2012).

3.4.3 Cyclomaltodextrin-glucanotransferase (Cgase EC 2.4.1.19)

Cyclomaltodextrin-glucanotransferase has potential application in cyclodextrin production. Cgase was isolated from alkalophilic *Bacillus subtilis* sp. from a deep-sea mud (Chandrasekaran and Rajeevkumar, 2002).

3.4.4 Agarase (EC 3.2.1.81)

Agarases are used in hydrolysis of agar, a colloid extracted from seaweeds to be commonly used as an additive in food processing, since it can act as an emulsifying, gelling and stabilizing agent. Agar is a mixture of agarose and agaropectin. Agarose can be hydrolyzed to oligosaccharides, which hold a particular interest in the production of functional foods (Ramnani et al., 2012). Hydrolysis can be promoted by either α-agarases (EC 3.2.1.158), which cleave α-1,3 linkages to produce agaro oligosaccharides 3,6-anhydro-L-galactose (Anr) residues at their reducing ends; or by β-agarases (EC 3.2.1.81), which cleave β-1,4 linkage to produce neo-agaro oligosaccharides with D-galactose residues at their reducing ends (Fu and Kim, 2010; Chi et al., 2012). Agarases may display either endo- or exo-activity, or both activities (Chi et al., 2012). Most of the agarases identified to date are β-agarases (Chi et al., 2012; Han et al., 2013), while α-agarase was isolated only from *Alteromonas agarlyticus* GJ1B (Potin et al., 1993), *Thalassomonas* sp. JAMB-A33 (Ohta et al., 2005) and *Pseudoalteromonas* sp. BL-3 (Lee et al., 2005) all from marine sources. β-Agarases have been isolated from several marine sources, most notably *Alteromonas* sp., *Bacillus cereus*, *Cytophaga* sp., *Pseudoalteromonas* sp., *Pseudomonas* sp. and *Vibrio* sp. (Hu et al., 2009; Fu and Kim, 2010; Chi et al., 2012; Vijayaraghavan and Rajendran, 2012). The gram-negative *Microbulbifer maritimus* bacterium was also identified as producer of an extracellular agarase of 75.2 kDa (Vijayaraghavan and Rajendran, 2012). β-agarase was isolated from *Alteromonas* GNUM-1 (Jonghee and Hon, 2012). *Pseudomonas stutzeri*, *Aeromonas* sp. and *Vibrio* sp., isolated from sea produced agarase and agarase-0107, an endo-type β-agarase that hydrolyzed the β-1, 4-linkage of agarose to yield neoagarotetraose and neoagarobiose at a pH of around 8, have been isolated from *Vibrio* sp. JT0107 (Chandrasekaran and Rajeevkumar, 2002).

3.4.5 α-Galactosidase (EC 3.2.1.22)

α-Galactosidase is used in the sugar industry for enhancing the yield of sucrose by hydrolysis of raffinose. A-galactosidase was obtained from marine *Alteromonas* spp., which was isolated from sponges and alga *Polysiphonia* sp. and from bacteria associated with mussel (*Crenomytilus grayanus*) and scallop (*Patinopecten jessoensis*). A novel beta-galactoside-alpha-2, 6-sialyltransferase (EC 2.4.99.1) is produced by *Photobacterium damsela* JT0160 (FERM BP-4900), ATCC 33539 or ATCC 35083 (Chandrasekaran and Rajeevkumar, 2002).

3.4.6 Cellulases

A symbiotic bacterium found in the gland of *Deshayes* of the marine shipworm and *Aspergillus terreus* isolated from the seawater produced cellulase. Thermophilic *Thermotoga* sp. produced thermostable exo-1, 4-β-cellobiohydrolase (105°C) (Chandrasekaran and Rajeevkumar, 2002).

3.4.7 Glucanases

β-Glucanases are used in the preparation of alcohol, animal and chicken feed, beer, breadmaking and preservation, preparation of wine and for yeast hydrolysis. *Alteromonas* sp. produced Laminarin-induced endo-1, 3-β-D-glucanase (I) (EC 3.2.1.39), whereas marine *Cytophaga*-NK-5 produce endo-1, 6-β-D-glucanase (EC 3.2.1.75). Marine *Bacillus circulans* produced an enzyme that degraded glucans consisting of α-1-3 and or α-106 linkages (Chandrasekaran and Rajeevkumar, 2002).

3.4.8 Inulinases

Inulinases are enzymes with hydrolytic activity over inulin, a polyfructan present in several plants, where fructosyl units are bound together by β (2, 1) linkages. Total or partial hydrolysis of inulin leads to the production of fructose or of fructooligosaccharides, respectively. These are widely used in the food industry, either as sweetener or as prebiotic, respectively (Li et al., 2007; Lima et al., 2011). Inulinases are classified either as endo-inulinases (EC 3.2.1.7) or exo inulinases (EC 3.2.1.80) whether they cleave the inulin chain into smaller oligosaccharides or hydrolyze the terminal fructose from the inulin chain (Basso et al., 2010). Extracellular inulinase produced by marine *Pichia guilliermondii* OUC1 could be used for the production of ultra-high fructose syrups, whereas inulinase of marine *Yarrowia lipolytica* OUC2 could be used for the production of inulooligosaccharides (Guo, 2009; Lima et al., 2011). *Marinimicrobium* sp. LS-A18 produced extracellular inulinase which exhibited an alkali-tolerant nature (Li et al., 2012). Alkali tolerance (pH optimum of 8.0, and activity within pH 5.0–11.0) and thermostable nature (optimum 60°C) was displayed by an extracellular inulinase from *Nocardiopsis* sp. DN-K15 (Lu et al., 2014). Both enzymes displayed exo-inulinase activity, indicating a potential use for the production of ultra-high fructose syrups. *Bacillus cereus* MU-31 produced an extracellular exo-inulinase (Meenakshi et al., 2013).

3.4.9 Phytase

Phytases act upon phytate stored in plants. Their action results in the break down of phytic acid which is indigestible to non-ruminants and therefore, make available phosphorus as well as calcium and other nutrients stored as phytate. Hence exogenous supplementation of phytase is an efficient way to improve human and animal diets and digestion in the alimentary tract (Afinah et al., 2010; Mittal et al., 2013). The marine yeast *Kodamea ohmeri* BG3 was shown to display significant extracellular phytase activity (Li et al., 2008).

3.4.10 Tannase (EC 3.1.1.20)

Tannases hydrolyze the ester and depside linkages of tannic acid, yielding glucose and gallic acid. It is mostly used for preparing gallic acid, which is used in the preparation of the food anti-oxidant propyl gallate; producing coffee-flavored soft drinks and instant tea; for clarification of fruit juice and beer; and, for food detannification (Aguilar and Gutierrez-Sanchez, 2001; Molkabadi et al., 2013). Several tannases produced from marine sources have been identified and these include *Aspergillus awamori* BTMFW032 (Beena et al., 2010) and *Phormidium valderianum* BDU140441 (Palanisami et al., 2012).

3.4.11 Xylanases

Xylanases have found application in the food and feed industries, such as in clarification of fruit juices, in reducing viscosity leading to improved filterability of the juice liquefaction of fruits and vegetables, maceration of fruits and vegetables and improved extraction of fermentable sugars from barley facilitating production of beer, etc. (Harris and Ramalingam, 2010; Sharma and Kumar, 2013). Thermostable xylanase B from *Thermotoga maritima* is known in the production of frozen partially baked bread (Jiang et al., 2008). The addition of the xylanase to wheat flour dough resulted in softer bread, with higher volume. Crumb firmness decreased during storage, concomitant with anti-stalling effect. The enzyme also decreased the changes in total shrinking during aging. *Thermotoga* sp., *Thermotoga maritima*, *T. neapolitana* and *T. thermarum* produce endo-1, 4-β–xylanase (Chandrasekaran and Rajeevkumar, 2002).

3.4.12 Chitinases (EC 3.2.1.14)

Chitosan is approved for use as a food additive or dietary supplement in countries, such as Japan, England, USA, Italy, Portugal and Finland. In 1992, chitosan was accepted as a functional food ingredient by Japan's health department. The main producers of chitosan are Japan and the United states with smaller operations in India, Italy and Poland (Chen and Subirade, 2005). Chitinase have been isolated from several marine bacteria, including species of *Alteromonas, Pseudomonas, Moraxella, Acinetobacter, Vibrio* and *Vibrio ischeri*. A solid-state fermentation (SSF) was developed for extra-cellular chitinase production, using chitinous solid waste of the shellfish-processing industry as a solid substrate employing *Beauveria* sp. isolated from marine sediment (Suresh and Chandrasekaran, 1998). *Vibrio harveyii*, when exposed to different types of chitin, excreted several chitin-degrading proteins into the culture medium for efficient utilization of different forms of chitin and chitin by-products. *Vibrio alginolyticus* H-8 produces chitinase C1 and C3 along with chitobiose-deacetylase. *Aeromonas hydophila* H-2330 secreted chitinase only in the presence of chitin as a C-source, suggesting that the chitinase is an inducible enzyme. N-acetylglucosamine-deacetylase (AGD, EC 3.5.1.33) from *Alteromonas* sp. is useful as a drug and as a biocatalyst in the production of D-glucosamine, which is useful in the production of glucosamino-oligosaccharides with antimicrobial and anti-tumor activities. N-acetylglucosamine-6-

phosphate-deacetylase (EC 3.5.1.25) is produced by *Vibrio* sp. P2K-5 (Chandrasekaran and Rajeevkumar, 2002).

3.4.13 Superoxide-dismutase (SOD, EC 1.15.1.1)

Superoxide-dismutase reduces oxygen-free radicals and significantly inhibits lipoxidation of milk. Hence this enzyme has potential application in the dairy industry. Very few bacteria have been reported to produce SOD. Halo-tolerant yeast *Debaryomyces hansenii* is an alternative source of Cu/Zn superoxide-dismutase (SOD), which compares well with other superoxide-dismutase sources, e.g., brewer's yeast, cattle liver and human erythrocyte. *D. hansenii* is considered an ideal alternative for economic production of superoxide-dismutase. Marine *Synechococcus* sp. NKBG 042902 is also reported to produce superoxide-dismutase (Chandrasekaran and Rajeevkumar, 2002).

4. Extremozymes (Enzymes from Extremophiles)

Marine ecosystem includes hot and cold streams, acidic and alkaline water flow, bright surface, dark deep sea with high barometric pressure, hyper-saline hydrothermal vents, or cold seeps enriched with different minerals/gases and deep sea volcanoes. These environments harbor diverse flora and fauna in the water as well as on the sea-floor besides extremely diverse microbial populations, including several extremophiles (Marx et al., 2007). Extremophiles are the primary source of extremozymes that are active in extreme conditions of life. α-Amylase, pullulanase and α-glucosidase from archaea are active in the same pH and high temperature range, hence they could be used in a one-step process for the industrial bioconversion of starch.

4.1 Thermostable enzymes

Extremozymes are considered most significant from the industrial point of view since many industrial processes are facilitated by higher solubility, lower viscosity, better mixing, faster reaction rate and decreased risk of microbial contamination at high temperatures. For instance, thermostable enzymes obtained from thermophilic marine microorganisms are capable of catalyzing these reactions, like a thermostable α-amylase (Fuelzyme1) isolated from a deep sea microorganism by Verenium1 Corporation (Verenium-Fuelzym1, 2012) was found to be active over a broad range of temperatures and increased fuel ethanol yields due to improved starch hydrolysis at much lower concentrations.

Thermophilic microbes in the deep-sea thermal vents and hyperthermophilic bacteria that grow at 80–100°C are a potential source of highly specific and extremely thermostable enzymes. Thermostable polysaccharolytic enzymes are frequently used in food processing industries for conversion of starch into oligosaccharides, cyclodextrins, and maltose or glucose, while acid-stable enzymes can potentially improve starch liquefaction (Synowiecki et al., 2006). Extremozyme in commercial use has increased the efficiency with which compounds called cyclodextrins are produced from cornstarch. Cyclodextrins help to stabilize volatile substances (such as flavorings in foods), to improve the uptake of medicines by the body, to reduce bitterness and mask unpleasant odors in foods and medicines.

Thermostable marine microbial enzymes are reported mostly from thermophilic members of *Thermotoga, Thermus, Thermococcus, Pyrococcus, Bacillus* and *Sulfolobus* family. *Pyrococcus abyssi*, a deep-sea hyperthermophilic archaebacterium, produces thermostable β-glucosidase (I) (EC 3.2.1.21) under anaerobic conditions, which will be useful in the cellulose industry and in sugar derivative production. This species also produces thermostable esterase. Another hyperthermophilic *Pyrococcus* sp. KOD1 produces a thermostable thiol protease. *Moraxella* sp. TA144 and TA317 (lipase, EC 3.1.1.3), *Alteromonas haloplanctis* A23 (alpha-amylase, EC 3.2.1.1), *Psychrobacter immobilis* A5 (β-lactamase, EC 3.5.2.6), *Bacillus* sp. TA39 (protease), *Rhodothermus marinus* (endo-1, 4-β-D-xylanase) and hyperthermophile *Thermotoga neopolitanai* (cellulase and xylanase) were reported to produce thermostable enzymes (Chandrasekaran and Rajeevkumar, 2002). *Pyrococcus furiosus* was observed to produce β-xylosidase, β-galactosidase and β-glucosidase simultaneously. Similarly *Sulfolonbus solataricus* MT4 also produced β-xylosidase, β-galactosidase and β-fucosidase (Chandrasekaran and Rajeevkumar, 2002).

4.2 Cold adapted enzymes

The oceans, which maintain an average temperature of 1° to 3°C (34 to 38°F) make up over half the Earth's surface. The vast land areas of the Arctic and Antarctic are permanently frozen or are unfrozen for only a few weeks in summer. Microbial communities populate Antarctic ice-ocean water that remains frozen for much of the year. These communities include photosynthetic eukarya, notably algae and diatoms, as well as a variety of bacteria. These cold-loving organisms have attracted manufacturers who need enzymes that work at refrigerator temperatures, such as food processors (whose products often require cold temperatures to avoid spoilage). Cold-adapted enzymes are abundantly reported from deep-sea, cold streams, cold seeps and polar seas. Cold-adapted enzymes catalyze reactions at low temperatures resulting in the reduction of energy consumption, besides stabilizing thermolabile reactants/products, minimizing evaporation and diminishing the likelihood of microbial contamination. In addition, specific activities of cold-adapted enzymes are very high in comparison to their mesophilic counterparts.

Cold-adapted protease MCP-01, an abundant extracellular serine protease produced by the deep-sea psychrophilic *Pseudoalteromonas* sp. SM9913, was active down to 0°C, while being deactivated at 40°C, due to autolysis (Chen et al., 2003, 2007a,b). The later feature avoids the risk of over-tenderization during/after cooking. This collagenase proved effective in tenderization of beef meat at 4°C, since it reduced the beef meat shear force by up to 23 per cent, where it proved as effective as bromelain, a well-disseminated meat tenderizer (Zhao et al., 2012a). Another collagenase from a deep sea source, *Myroides profundi* D25, was used to further clarify the mechanism of collagen degradation (Ran et al., 2014). A pseudoalterin secreted by *Pseudoalteromonas* sp. CF6-2 from deep-sea sediment was reported to be used in elastin degradation (Zhao et al., 2012b).

A novel psychrostable metallo protease (almelysin) was produced from *Alteromonas* sp. Extra-cellular hydrolase, e.g., protease, chitinase, glucanase, esterase, lipase, phospholipase and DNA-degrading enzymes have been obtained from marine bacteria isolated from different sites in permanently cold Arctic and Antarctic habitats. Some of the cold-adapted enzymes from bacteria include α-amylase from *Alteromonas haloplanctis*, isocitrate dehydrogenase and lipase from *Vibrio* sp., β lactamase triose phosphate isomerase from *Moraxella* sp. and subtilisin from *Bacillus* sp. Potential application of psychrophilic xylanases from Antarctic microorganisms in the baking industry was reported and the enzymes are already patented (Collins et al., 2006). Several cold-active lipases were reported from Antarctic deep-sea sediments metagenome and microbial isolates, including *Pseudomonas* and *Psychrobacter* species (Zhang et al., 2007; Zhang and Zeng, 2008).

4.3 Alkalophilic enzymes

Several reactions of food processing warrant performances at extremely acidic or alkaline conditions. Enzymes that are active at highly alkaline pH are termed alkaline enzymes and such enzymes are produced mainly by alkalophiles, which have made a great impact in industrial applications. An important application is the industrial production of cyclodextrin with alkaline cyclomaltodextrin-glucanotransferase (EC 2.4.1.19). This enzyme reduced the production cost and paved the way for use in large quantities in foodstuffs. Alkaline phosphatase is a useful alkaliphilic enzyme frequently reported from marine microorganisms (Plisova et al., 2005; Sebastian and Ammerman, 2009). Some of the alkaline enzymes known from marine environments include thermo stable alkaline phosphatase (EC 3.1.3.1) with a wide pH range and endo-1, 3-beta-D-glucanase (EC 3.2.1.39) from *Alteromonas macleodii* KMM162; alkaline phosphatase (EC 3.1.3.1) from the marine *Alteromonas* sp. KMM 518; heat-labile alkaline phosphatase from a Gram-negative Antarctic marine bacterium; alkaline metallo endopeptidase from a marine *Vibrio* sp. NUF-BPP1 and alkaline protease from marine shipworm bacterium. A marine *Streptomyces* was reported to produce a novel alpha-amylase that is stable in the pH range of 9–11 for 48 hour besides retaining about 50 per cent of its activity at 85°C (Chakraborty et al., 2009).

4.4 Halophilic and halo tolerant enzymes

Exoenzymes produced by some halobacteria have optimal activity at high salinities and the potential for application in harsh industrial processes where the concentrated salt solutions would otherwise inhibit many enzymatic conversions. Several extracellular enzymes from marine microorganisms have optimal activity in the presence of high salt concentrations. These enzymes have applications in industrial processes, where reaction mixture contains high salt concentrations that inhibit most common industrial enzymes. Halophilic proteolytic enzymes are useful in peptide synthesis and also have a potential application in fish- and meat-processing industries. Further, halo-tolerant enzymes are uniquely adapted to function in low water availability and eventually many of them are reported to be organic solvent tolerant (Marhuenda-Egea and Bonete,

2002). Halobacterial exoenzymes include the amylases produced by *Halobacterium halobium* and *H. sodomense*, the proteases from *H. salinarium* and *H. halobium*, and lipases from several halobacteria. Nuclease H produced by *Micrococcus varians* subsp. *halophilus* is used in the industrial production of flavoring agents 5'-guanylic acid and 5'-inosinic acid in a bioreactor system with a column of flocculated cells. Several halotolerant bacteria produce salt-resistant enzymes, such as the amylase produced by a *Bacillus* sp. which is stable at 60°C and 5M NaCl and could be used in the treatment of effluents containing starchy or cellulosic residues (Chandrasekaran and Rajeevkumar, 2002).

5. Microbial Enzyme Production Methods

The enzyme industry largely survives on bulk production of enzymes at reasonable cost which in turn depends on development of feasible large-scale production of enzymes with suitable bioprocesses and economical downstream processing of enzymes. Unfortunately, such ventures are grossly neglected in the field of marine microbial technology. Commercial production of enzymes is primarily done by employing submerged fermentation (SmF) and by SSF. SmF is the usual bioprocess employed in enzyme industries for commercial-scale production. Nevertheless, extracellular enzymes (proteases, amylase, cellulases, tannases and lipases, etc.) are produced in large scale easily by both bacteria and fungi under SSF, whereas intracellular enzymes that have a wide range of applications are usually produced under SmF and subjected to intensive downstream processing.

However, literature on marine microorganisms, especially on fermentation production of enzymes, is rather scanty and limited to a few reports. These include L-gutamianse production by *Vibrio costicola* under SSF, using polystyrene as inert support (Nagendra Prabhu and Chandrasekaran, 1996, 1997), L-glutaminase (Sabu et al., 2000) and chitinase (Suresh and Chandrasekaran, 1998, 1999) production by *Beauvaeria* sp. under SSF, continuous production of extracellular L-glutaminase by Ca-alginate-immobilized marine *Beauveria bassiana*-BTMF-10 in packed bed reactor (Sabu et al., 2000), continuous production of L-glutaminase by an immobilized marine *Pseudomonas* sp. BTMS-51 (Rajeevkumar and Chandrasekaran, 2003) and extracellular β-glucosidase by a marine *Aspergillus sydowii* BTMFS 55 under SSF (Madhu et al., 2009), acidophilic tannase from marine *Aspergillus awamori* BTMFW032 (Beena et al., 2010), extracellular alkaline serine protease from marine *Engyodontium album* BTMF S10 (Sreeja et al., 2006), production of lipase from marine *Aspergillus awamori* BTMFW032 (Soorej et al., 2011), detergent-compatible alkaline lipase by marine *Bacillus smithii* BTMS 11 (Lailaja and Chandrasekaran, 2013) and glucoamylase by marine endophytic *Aspergillus* sp. JAN-25 under optimized SSF conditions on agro residues (El-Gendy, 2012).

Although, as indicated in earlier pages of this chapter, several marine bacteria, yeasts, cyanobacteria were observed to produce enzymes of economic importance studies pertaining to fermentation production of the same were not attempted. In fact from the literature available on enzyme production by marine microorganisms it was understood that these marine microorganisms require sea-water as the solvent in enzyme production besides complex media with a vast range of nutrients that are ideally found in sea-water for their growth and enzyme production. This requirement is not practically feasible for transfer of technology to commercial-scale operation. Moreover, downstream processing of enzymes from salt-rich cultivation media poses serious hurdles for purification of enzymes and the volume of enzyme released by these organisms is also a critical factor for choice of sea-water-based cultivation medium for enzyme production. Ideal bioreactors/fermenters for cultivation of marine microorganisms and large-scale enzyme production are yet another major hurdle in the transfer of technology to industries.

Nevertheless, there is no dearth of interest in exploring the marine biodiversity to harness marine microorganisms with the potential for novel enzymes in varied applications, including food and beverage processing. A detailed account of fermentation technologies on enzyme production is out of context in this chapter. Of course, more details on bioprocess/fermentation technologies for enzyme production by microorganisms are available in literature and the reader may refer to such research reviews.

6. Current Trend in Resourcing Marine Microbial Enzymes

Search for marine microbial enzymes, novel microbial products and novel genes have drawn utmost attention of scientific investigators around the world in the new millennium. In this context, development in the field of molecular biology, recombinant DNA technology and bioinformatics play a key role in the identification of microorganisms, gene-encoding novel biocatalysts and products, prediction of structure and function of unknown proteins and in primer design and mutant construction for protein engineering. Recent advances in genomics, metagenomics, proteomics, efficient expression systems and emerging recombinant DNA techniques have facilitated the discovery of new microbial enzymes from Nature (through genome and metagenome) or by creating (or evolving) enzymes with improved catalytic properties. These advanced biological techniques are also used in current exploration and harnessing of marine biodiversity for novel biocalaysts and bioproducts of industrial applications.

Despite the fact that more and more novel enzymes are yet to be discovered in marine microorganisms from diverse marine environments around the world, the prospects of isolating the genes coding for the novel enzymes with potential applications have already attracted the attention of genetic engineers and biotechnologists for rapid production and exploitation. Cloning and over-expression of marine-derived enzymes is a common practice now since gene cloning has paved the way for easy isolation of the concerned genes, their characterization and cloning and expression of the same in hosts, such as *Escherichia coli, Pichia pastoris* or any other suitable vectors successfully. It would not be an exaggeration to state that in the coming years, all the novel enzymes from marine microorganisms could be cloned and expressed or amplified in suitable hosts (Chandrasekaran and Rajeevkumar, 2002).

7. Conclusion and Future Prospects

Marine microorganisms require special culture conditions at times, for instance, a high hydrostatic pressure in the case of deep-sea bacteria and suitable fermentation media for the desired product or biomass. They may require entirely different kinds of complex nutrients in the production medium identical or closer to the type of complex substances that exist in marine environments, unlike traditional sources, such as soya bean meal, corn steep liquor or molasses, or a chemically-defined medium with known inducers. Further, ideal mode of operation of the bioprocesses, batch or fed batch, or continuous fermentation or immobilized conditions for enhanced yield is an inevitable prerequisite for enhanced enzyme yield. Hence, there is an urgent need, not only to harness the positive qualities of marine microorganisms for enzyme production, but also to exploit them for industrial use, employing SSF. Interestingly marine microorganisms, by virtue of their basic characteristics to adsorb on to solid particles, such as sestons and triptons, offer ideal candidates for use in SSF (Chandrasekaran, 1996).

References

Afinah, S., Yazid, A.M., Anis Shobirin, M.H. and Shuhaimi, M. 2010. Phytase: Application in food industry. International Food Research Journal, 17: 13–21.

Aguilar, C.N. and Gutierrez-Sanchez, G. 2001. Review: Sources, properties, applications and potential uses of tannin acyl hydrolase. Food Science and Technology International, 7: 373–382.

Basso, A., Spizzo, P., Ferrario, V., Knapic, L., Savko, N. and Braiuca, P. 2010. Endo- and exo-inulinases: Enzyme-substrate interaction and rational immobilization. Biotechnology Progress, 26: 397–405.

Beena, P.S., Soorej, M.B., Elyas, K.K., Sarita, G.B. and Chandrasekaran, M. 2010. Acidophilic Tannase from Marine *Aspergillus awamori* BTMFW032. Journal of Microbiology and Biotechnology, 20: 1403–1414.

Bertoft, E. 2004. Analysing starch structure. pp. 57–96. *In*: A.C. Eliasson (ed.). Starch in Food-Structure, Function and Applications. Woodhead Publishing Limited, Cambridge.

Bhatia, Y., Mishra, S. and Bisaria, V.S. 2002. Microbial beta-glucosidases: Cloning, properties and applications. Critical Reviews in Biotechnology, 22: 375–407.

Bindiya, P. and Ramana, T. 2012. Optimization of lipase production from an indigenously isolated marine *Aspergillus sydowii* of Bay of Bengal. Journal of Biochemical Technology, 3: S203–S211.

Burg, B.V.D. 2003. Extremophiles as a source of novel enzymes. Current Opinion in Microbiology, 6: 213–218.

Chakraborty, S., Khopade, A., Kokare, C., Mahadik, K. and Chopade, B. 2009. Isolation and characterization of novel α-amylase from marine *Streptomyces* sp. D1. Journal of Molecular Catalysis B: Enzymatic, 58: 17–23.

Chakraborty, S., Raut, G., Khopade, A., Mahadik, K. and Kokare, C. 2012. Study on calcium ion independent α-amylase from halo alkaliphilic marine *Streptomyces* strain A3. Indian Journal of Biotechnology, 11: 427–437.

Chandrasekaran, M. and Rajeevkumar, S. 2002. Marine microbial enzymes. *In*: Encyclopedia of Life Support Systems (EOLSS-UNESCO publication). Firsthard copy, CD ROM and online published in 2002.

Chandrasekaran, M. 1996. Harnessing of marine microorganisms through solid state fermentation. Journal of Scientific and Industrial Research, 155(5&6): 468–471.

Chandrasekaran Muthusamy, Soorej M. Basheer, Sreeja Chellappan, Jissa G. Krishna and Beena, P.S. 2015. Enzymes in food and beverage production: An overview. pp. 117–138. *In*: Muthusamy Chandrasekaran (ed.). Enzymes in Food and Beverage Processing. CRC Press, Taylor & Francis Group, Florida Boca Raton, Florida, U.S.A.

Chandraskaran, M. 1997. Industrial enzymes from marine microorganisms: Indian scenario. Journal of Marine Biotechnology, 5: 86–89.

Chen, X.L., Zhang, Y.Z., Gao, P.J. and Luan, X.W. 2003. Two different proteases produced by a deep-sea psychrotrophic strain *Pseudoaltermonas* sp. SM9913. Marine Biotechnology, 143: 989–993.

Chen, L. and Subirade, M. 2005. Chitosan/β-lactoglobulin core–shell nanoparticles as nutraceutical carriers. Biomaterials, 26(30): 6041–6053.

Chen, X.L., Xie, B.B., Lu, J.T., He, H.L. and Zhang, Y.Z. 2007a. A novel type of subtilase from the psychrotolerant bacterium *Pseudoalteromonas* sp. SM9913: Catalytic and structural properties of deseasin MCP-01. Microbiology, 153: 2116–2125.

Chen, X.L., Zhang, Y.Z., Lu, J.T., Xie, B.B., Sun, C.Y. and Guo, B. 2007b. Autolysis of a novel multi-domain subtilase-cold-adapted deseasin MCP-01is pH-dependent and the surface loops in its catalytic domain, the linker and the P-proprotein domain are susceptible to proteolytic attack. Biochemical and Biophysical Research Communications, 358: 704–709.

Chi, W.J., Chang, Y.K. and Hong, S.K. 2012. Agar degradation by microorganisms and agar-degrading enzymes. Applied Microbiology and Biotechnology, 94: 917–930.

Collins, T., Hoyoux, A., Dutron, A. et al. 2006. Use of glycoside hydrolase family 8 xylanases in baking. Journal of Cereal Science, 43: 79–84.

David, W. and Michel, J. 1999. Extremozymes. Current Opinion in Chemical Biology, 3: 39–46.

El-Gendy, M.M.A. 2012. Production of glucoamylase by marine endophytic *Aspergillus* sp. JAN-25 under optimized solid-state fermentation conditions on agroresidues. Australian Journal of Basic and Applied Sciences, 6: 41–54.

Faulkner, D.J. 2000. Highlights of marine natural products chemistry (1972–1999). Natural Product Reports, 17: 1–6.

Fernandes, P. 2014. Marine enzymes and food industry: Insight on existing and potential interactions. Frontiers in Marine Science; Marine Biotechnology, October 2014, Article 46, Vol. 1: 1–18. doi: 10.3389/fmars.2014.00046.

Ferreira-Dias, S., Sandoval, G., Plou, F. and Valero, F. 2013. The potential use of lipases in the production of fatty acid derivatives for the food and nutraceutical industries. Electronic Journal of Biotechnology, 16: 1–38.

Fu, X.T. and Kim, S.M. 2010. Agarase: Review of major sources, categories, purification method, enzyme characteristics and applications. Marine Drugs, 26: 200–218.

Fulzele, R., DeSa, E., Yadav, A., Shouche, Y. and Bhadekar, R. 2011. Characterization of novel extracellular protease produced by marine bacterial isolate from the Indian Ocean. Brazilian Journal of Microbiology, 42: 1364–1373.

Gantelet, H. and Duchiron, F. 1998. Purification and properties of a thermoactive and thermostable pullulanase from *Thermococcus hydrothermalis*, a hyperthermophilic archaeon isolated from a deep-sea hydrothermal vent. Applied Microbiology and Biotechnology, 49: 770–777.

Goyal, K., Selvakumar, P. and Hayashi, K. 2001. Characterization of a thermostable β-glucosidase (BglB) from *Thermotoga maritima* showing transglycosylation activity. Journal of Molecular Catalysis B: Enzymatic, 15: 45–53.

Guo, B., Chen, X.L., Sun, C.Y., Zhou, B.C. and Zhang, Y.Z. 2009. Gene cloning, expression and characterization of a new cold-active and salt-tolerant endo-beta-1,4-xylanase from marine *Glaciecola mesophila* KMM 241. Applied Microbiology and Biotechnology, 84: 1107–1115.

Guzman-Maldonado, H. and Paredes-Lopez, O. 1995. Amylolytic enzymes and products derived from starch: A review. Critical Reviews in Food Science and Nutrition, 35: 373–403.

Han, W.J., Gu, J.Y., Liu, H.H., Li, F.C., Wu, Z.H. and Li, Y.Z. 2013. An extra peptide within the catalytic module of a β-agarase affects the agarose degradation pattern. Journal of Biological Chemistry, 288: 9519–9531.

Harris, A.D. and Ramalingam, C. 2010. Xylanases and its application in food industry: A review. Journal of Experimental Sciences, 1: 1–11.

Hii, S.L., Tan, J.S., Ling, T.C. and Ariff, A.B. 2012. Pullulanase: Role in starch hydrolysis and potential industrial applications. Enzyme Research, Article ID: 921362, pp. 1–14.

Hu, Z., Lin, B.-K., Xu, Y., Zhong, M.Q. and Liu, G.-M. 2009. Production and purification of agarose from a marine agarolytic bacterium *Agarivorans* sp. HZ105. Journal of Applied Microbiology, 106: 181–190.

Huang, X., Zhao, Y., Dai, Y., Wu, G., Shao, Z., Zeng, Q. et al. 2012. Cloning and biochemical characterization of α-glucosidase from a marine bacterium *Aeromonas* sp. HC11e-3. World Journal of Microbiology and Biotechnology, 28: 3337–3344.

Jiang, Z., Lebail, A. and Wu, A. 2008. Effect of the thermostable xylanase B (XynB) from *Thermotoga maritima* on the quality of frozen partially baked bread. Journal of Cereal Science, 47: 172–179.

Jonghee, K. and Hon, S.-K. 2012. Isolation and characterization of an agarase-producing bacterial strain, *Alteromonas* sp. GNUM-1 from the West Sea, Korea. Journal of Microbiology and Biotechnology, 22: 1621–1628.

Keerthi, T.R., Suresh, P.V., Sabu, A., Rajeevkumar, S. and Chandrasekaran, M. 1999. Extracellular production of L-glutaminase by alkalophilic *Beauveria bassiana* BTMF S10 isolated from marine sediments. World Journal of Microbiology and Biotechnology, 15: 751–752.

Kin, S.L. 2006. Discovery of novel metabolites from marine actinomycetes. Current Opinion in Microbiology, 9: 245–251.

Kumar, P.S., Ghosh, M., Pulicherla, K.K. and Rao, K.R.S.S. 2011. Cold active enzymes from the marine psychrophiles: Biotechnological perspective. Adv. Biotech., 10: 16–20.

Lailaja and Chandrasekaran, M. 2013. Detergent compatible alkaline lipase produced by marine *Bacillus smithii* BTMS 11. World Journal of Microbiology and Biotechnology, 29(8): 1349–1360.

Lee, Y.H., Jun, S.E. and Shin, H.D. 2005. Low molecular weight agarose-specific alpha-agarase from agarolytic marine microorganism *Pseudoalteromonas* sp. BL-3 which hydrolyzes alpha-1,3-glycoside bond of agaroragarose to produce agarobiose and agarotetraose. Patent KR2005079035–A.

Li, A.-X., Guo, L.-Z. and Lu, W.-D. 2012. Alkaline inulinase production by a newly isolated bacterium *Marinimicrobium* sp. LS–A18 and inulin hydrolysis by the enzyme. World Journal of Microbiology and Biotechnology, 28: 81–89.

Li, H., Chi, Z., Duan, X., Wang, L., Sheng, J. and Wu, L. 2007. Glucoamylase production by the marine yeast *Aureobasidium pullulans* N13d and hydrolysis of potato starch granules by the enzyme. Process Biochemistry, 42: 462–465.

Li, X., Chi, Z., Liu, Z., Yan, K. and Li, H. 2008. Phytase production by a marine yeast *Kodameaohmeri* BG3. Applied Biochemistry and Biotechnology, 149: 183–193.

Li, J., Peng, Y., Wang, X. and Chi, Z. 2010. Optimum production and characterization of an acid protease from marine yeast *Metschnikowia reukaufii* W6. Journal of Ocean University of China, 9: 359–364.

Lima, D.M., Fernandes, P., Nascimento, D.S., Ribeiro, R.C.L.F. and deAssis, S.A.A. 2011. Fructose syrup: A biotechnology asset. Food Technology and Biotechnology, 49: 424–434.

Lu, W.-D., Li, A.-X. and Guo, Q.-L. 2014. Production of novel alkali tolerant and thermostable inulinase from marine actinomycete *Nocardiopsis* sp. DN-K15 and inulin hydrolysis by the enzyme. Annals of Microbiology, 64: 441–449.

Madhu, K.M., Beena, P.S. and Chandrasekaran, M. 2009. Extracellular β-glucosidase production by a marine *Aspergillus sydowii* BTMFS 55 under solid state fermentation using statistical experimental design. Biotechnology and Bioprocess Engineering, 14: 457–466.

Manachini, P.L. and Fortina, M.G. 1994. Proteinase production by halophilic isolates from marine sediments. Folia Microbiologica, 39(5): 378–80.

Mao, X., Hong, Y., Shao, Z., Zhao, Y. and Liu, Z. 2010. A novel cold-active and alkali-stable β-glucosidase gene isolated from the marine bacterium *Martelella mediterranea*. Applied Biochemistry and Biotechnology, 162: 2136–2148.

Marhuenda-Egea, F.C. and Bonete, M.J. 2002. Extreme halophilic enzymes in organic solvents. Current Opinion in Biotechnology, 13: 385–389.

Marx, J.C., Collins, T., D'Amico, S., Feller, G. and Gerday, C. 2007. Cold adapted enzymes from marine Antarctic microorganisms. Marine Biotechnology, 9: 293–304.

Meenakshi, S., Umayaparvathi, S., Manivasagan, P., Arumugam, M. and Balasubramanian, T. 2013. Purification and characterization of inulinase from marine bacterium *Bacillus cereus* MU-31. Indian Journal of Geo-Marine Sciences, 42: 510–515.

Mittal, A., Gupta, V., Singh, G., Yadav, A. and Aggarwal, N.K. 2013. Phytase: A boom in food industry. Octa Journal of Biosciences, 1: 158–169.

Mojsov, K. 2013. Use of enzymes in wine making: A review. International Journal of Marketing and Technology, 3: 112–127.

Molkabadi, E.Z., Hamidi-Esfahani, Z., Sahari, M.A. and HoseinAzizi, M. 2013. A new native source of tannase producer, *Penicillium* sp. EZ-ZH190: Characterization of the enzyme. Iranian Journal of Biotechnology, 11: 244–250.

Moriguchi, M., Sakai, K., Tateyama, R., Furuta, Y. and Wakayama, M. 1994. Isolation and characterization of salt-tolerant glutaminases from marine *Micrococcus luteus* K-3. Journal of Fermentation and Bioengineering, 77(6): 621–25.

Nagendra Prabhu, G. and Chandrasekaran, M. 1996. L-Glutaminase production by marine *Vibrio costicola* under solid state fermentation using different substrates. Journal of Marine Biotechnology, 4: 176–179.

Nagendra Prabhu, G. and Chandrasekaran, M. 1997. Impact of process parameters on L-glutaminase production by marine *Vibrio costicola* under solid state fermentation using polystyrene as inert support. Process Biochemistry, 32(4): 285–289.

Nedwin, G., Sharma, V. and Shetty, J. 2011. α-Amylase blend for starch processing and use there of Patent application. WO 2011017093 A1.

Ni, X., Yue, L., Chi, Z., Li, J., Wang, X. and Madzak, C. 2009. Alkaline protease gene cloning from the marine yeast *Aureobasidiumpullulans* HN2-3 and the protease surface display on *Yarrowialipolytica* for bioactive peptide production. Marine Biotechnology, 11: 81–89.

Ohta, Y., Hatada, Y., Miyazaki, M., Nogi, Y., Ito, S. and Horikoshi, K. 2005. Purification and characterization of a novel α-agarase from a *Thalassomonas* sp. Current Microbiology, 50: 212–216.

Palanisami, S., Kannan, K. and Lakshmanan, U. 2012. Tannase activity from the marine cyanobacterium *Phormidium valderianum* BDU140441. Journal of Applied Phycology, 24: 1093–1098.

Pancha, I., Jain, D., Shrivastav, A., Mishra, S.K., Shethia, B., Mishra, S. et al. 2010. A thermoactive α-amylase from a *Bacillus* sp. isolated from CSMCRI salt farm. International Journal of Biological Macromolecules, 47: 288–291.

Plisova, E.Y., Balabanova, L.A. and Ivanova, E.P. 2005. A highly active alkaline phosphatase from the marine bacterium Cobetia. Marine Biotechnology, 7: 173–178.

Potin, P., Richard, C., Rochas, C. and Kloareg, B. 1993. Purification and characterization of the α-agarase from *Alteromonas agarlyticus* (Cataldi) comb. nov., strain GJ1B. European Journal of Biochemistry, 214: 599–607.

Puspasari, F., Nurachman, Z., Noer, A.S., Radjasa, O.K., van der Maarel, M.J.E.C. and Natalia, D. 2011. Characteristics of raw starch degrading a-amylase from *Bacillus aquimaris* MKSC 6.2 associated with soft coral *Sinularia* sp. Starch, 63: 462–467.

Rajeevkumar, S. and Chandrasekaran, M. 2003. Continuous production of L-glutaminase production by an immobilized marine *Pseudomonas* sp. BTMS-51 in a packed bed reactor. Process Biochemistry, 38: 1431–1436.

Ramnani, P., Chitarrari, R., Tuohy, K., Grant, J., Hotchkiss, S., Philp, K. et al. 2012. *In vitro* fermentation and prebiotic potential of novel low molecular weight polysaccharides derived from agar and alginate seaweeds. Anaerobe, 18: 1–6.

Ran, L.Y., Su, H.N., Zhou, M.Y., Wang, L., Chen, X.L. and Xie, B.B. 2014. Characterization of a novel subtilisin-like protease myroicolsin from deep sea bacterium *Myroides profundi* D25 and molecular insight into its collagenolytic mechanism. Journal of Biological Chemistry, 289: 6041–6053.

Renu, S. and Chandrasekaran, M. 1992. Extracellular L-glutaminase production by marine bacteria. Biotechnology Letters, 14(6): 471–474.

Sabu, A., Keerthi, T.R., Rajeevkumar, S. and Chandrasekaran, M. 2000. L-glutaminase production by marine *Beauveria* sp. under solid state fermentation. Process Biochemistry, 35(7): 705–710.

Sawant, R. and Nagendran, S. 2014. Protease: An enzyme with multiple industrial applications. World Journal of Pharmaceutical Sciences, 3: 568–579.

Sebastian, M. and Ammerman, J.W. 2009. The alkaline phosphatase PhoX is more widely distributed in marine bacteria than the classical PhoA. ISME Journal, 3: 563–572.

Shahidi, F. and Wanasundara, U.N. 1998. Omega-3 fatty acid concentrates: Nutritional aspects and production technologies. Trends in Food Science and Technology, 9: 230–240.

Sharma, M. and Kumar, A. 2013. Xylanases: an overview. British Biotechnology Journal, 3: 1–28.

Soorej, M. Basheer, Chellappan, S., Beena, P.S., Sukumaran, R.K., Elyas, K.K. and Chandrasekaran, M. 2011. Lipase from marine *Aspergillus awamori* BTMFW032: Production, partial purification and application in oil effluent treatment. New Biotechnology, 28: 627–638.

Sowell, S.M., Norbeck, A.D., Lipton, M.S., Nicora, C.D., Callister, S.J. and Smith, R.D. 2008. Proteomic analysis of stationary phase in the marine bacterium Candidatus Pelagibacter ubique. Applied and Environmental Microbiology, 74: 4091–4100.

Sreeja Chellappan, Jasmin C., Soorej, M. Basheer, Elyas, K.K., Sarita, G. Bhat and Chandrasekaran, M. 2006. Production, purification and partial characterization of a novel protease from Marine *Engyodontium album* BTMFS10 under solid state fermentation. Process Biochemistry, 41(4): 956–961.

Sriram, N., Priyadharshini, M. and Sivasakthi, S. 2012. Production and characterization of amino peptidase from marine *Aspergillus flavus*. International Journal of Microbiology Research, 3: 221–226.

Stach, J.E.M., Maldonado, L.A., Ward, A.C., Good fellow, M. and Bull, A.T. 2003. New primers for the class Actinobacteria: Application to marine and terrestrial environments. Environmental Microbiology, 5: 828–841.

Steen, A.D. and Arnosti, C. 2013. Extracellular peptidase and carbohydrate hydrolase activities in an Arctic fjord (Smeerenburgfjord, Svalbard). Aquatic Microbial Ecology, 69: 93–99.

Suresh, P.V. and Chandrasekaran, M. 1998. Utilization of prawn waste for chitinase production by marine *Beauveria bassiana* under solid state fermentation. World Journal of Microbiology and Biotechnology, 14(5): 655–660.

Suresh, P.V. and Chandrasekaran, M. 1999. Impact of process parameters on chitinase production by an alkalophilic marine *Beauveria bassiana* under solid state fermentation. Process Biochemistry, 34: 257–267.

Synowiecki, J., Grzybowska, B. and Zdzieblo, A. 2006. Sources, properties and suitability of new thermostable enzymes in food processing. Critical Reviews in Food Science and Nutrition, 46: 197–205.

Trincone, A. 2013. Angling for uniqueness in enzymatic preparation of glycosides. Biomolecules, 3: 334–350.

Uzyol, K.S., Sarıyar-Akbulut, B., Denizci, A.A. and Kazan, D. 2012. Thermostable α-amylase from moderately halophilic *Halomonas* sp. AAD21. Turkish Journal of Biology, 36: 327–338.

Verenium-Fuelzym. 2012. Fuelzyme enzyme is a next generation alpha amylase for starch liquefaction. Available online: http://www.verenium.com/ products_fuelzyme.html (accessed on December 22, 2015).

Vijayaraghavan, R. and Rajendran, S. 2012. Identification of a novel agarolytic γ-proteobacterium *Microbulbifer maritimus* and characterization of its agarase. Journal of Basic Microbiology, 52: 705–712.

Visser, J., Hinz, S., Verij, J., Visser, J., Joosten, V., Koetsier, M. et al. 2012. Novel fungal esterases. WIPO Patent Application WO/2012/078741A2.

Wang, Y., Zhang, C., Li, J. and Xu, Y. 2013. Different influences of β-glucosidases on volatile compounds and anthocyanins of Cabernet Gernischt and possible reason. Food Chemistry, 140: 245–254.

Ward, O.P., Rao, M.B. and Kulkarni, A. 2009. Proteases. pp. 495–511. *In*: M. Schaechter (ed.). Encyclopedia of Microbiology, Vol. 1, Elsevier, Academic Press, Amsterdam.

Wu, S.-J. and Chen, J. 2014. Preparation of malto triose from fermentation broth by hydrolysis of pullulan using pullulanase. Carbohydrate Polymers, 107: 94–97.

Xu, J.-L., Lu, M.-S., Wang, S.-J., Li, H.Z., Sun, Y.-Y. and Fang, Y.-W. 2009. Production and characterization of pullulanase from hyperthermophilic archaeon *Thermococcus* sp. HJ21. Food Science and Biotechnology, 2: 243–249.

Zhang, J. and Zeng, R. 2008. Molecular cloning and expression of a cold-adapted lipase gene from an Antarctic deep sea psychrotrophic bacterium *Pseudomonas* sp. 7323. Marine Biotechnology, 10: 612–621.

Zhang, J., Lin, S. and Zeng, R. 2007. Cloning, expression and characterization of a cold-adapted lipase gene from an Antarctic deep-sea psychrotrophic bacterium, *Psychrobacter* sp. 7195. Journal of Microbiology and Biotechnology, 17: 604–610.

Zhao, G.Y., Zhou, M.Y., Zhao, H.L., Chen, X.L., Xie, B.B. and Zhang, X.Y. 2012a. Tenderization effect of cold-adapted collagenolytic protease MCP- 01 on beef meat at low temperature and its mechanism. Food Chemistry, 134: 1738–1744.

Zhao, H.L., Chen, X.L., Xie, B.B., Zhou, M.Y., Gao, X. and Zhang, X.Y. 2012b. Elastolytic mechanism of a novel M23 metalloprotease pseudoalterin from deep-sea *Pseudoalteromonas* sp. CF6-2: Cleaving not only glycyl bonds in the hydrophobic regions but also peptide bonds in the hydrophilic regions involved in cross-linking. Journal of Biological Chemistry, 287: 39710–39720.

23

Extremophiles as Potential Resource for Food Processing Enzymes

Archana Sharma[1] and *T. Satyanarayana*[2,] *

1. Introduction

Microorganisms which grow and survive in extreme environments are known as extremophiles. The term extremophile was used for the first time by MacElroy in the year 1974 (MacElroy, 1974). According to Madigan and Marrs (1997) and Rothschild and Manicinelli (2001), extreme environments include those with high (55 to 121°C) and low (–2 to 20°C) temperatures, alkaline (pH > 8) and acidic pH (pH < 4). Certain extremophiles can tolerate extreme conditions, such as high pressure, high levels of radiation or toxic compounds. The limits of temperature, pressure, pH, salinity and water activity at which life can thrive have not yet been precisely defined. Extreme environment is a relative term; an environment which is extreme for the growth of some organisms may be essential for others. Most extremophiles which have been identified to date belong to the domain Archaea. Many extremophiles have, however, been identified and characterized as belonging to bacterial, archaeal and eukaryotic kingdoms.

Extremozymes are enzymes obtained from extremophilic microbes which function in extreme conditions that were previously thought to be too harsh for proteins. Such conditions are commonly encountered in industrial processes; therefore, an increasing demand for extremozymes (Elleuche and Antranikian, 2013). Extremophilic microbes

[1] Department of Biophysics and University of Delhi, South Campus, Benito Juarez Road, New Delhi-110021, India.
[2] Department of Microbiology, University of Delhi, South Campus, Benito Juarez Road, New Delhi-110021, India.
* Corresponding author: tsnarayana@gmail.com

are superior to their mesophilic counterparts in this regard, being a vast source of naturally-tailored enzymes for applications under extreme conditions. Only a few extremozymes are commercially available—an example being DNA-polymerase from *Thermus aquaticus* and related organisms. Every category of extremophiles has unique characteristics that can be exploited to provide enzymes with a wide range of application possibilities (Adams et al., 1995; Yano and Poulos, 2003). Developments in molecular biology and protein engineering have led to the evolution of many novel enzymes and their applications in industrial biotechnology in turn is contributing to the growth of this multi-billion-dollar commercial sector.

The use of enzymes in food-processing industries has many advantages. Enzymes are highly specific in their action and enzyme-catalyzed processes have fewer side reactions and by-products, resulting in high quality products with less pollution. A number of companies are coming up for manufacturing enzymes. Denmark is leading with Novozymes (45 per cent) and Danisco (17 per cent) capturing the world's enzyme market. Other companies are Genencor (USA), DSM (The Netherlands) and BASF (Germany) (Ogawa and Shimizu, 2002; Berka and Cherry, 2006; Binod et al., 2008). The rate of development in emerging markets suggests that companies from India and China can also join the race in the near future (Carrez and Soetaert, 2005; Chandel et al., 2007).

The food industry uses more than 55 enzyme products in food processing (Oort, 2010). This number will go up if more extremophiles are explored for enzymes that can be used in food processing. The main industrial processes where microbial enzymes are employed are discussed in this review.

2. Starch Industry

The first enzyme discovered in 1833 by the French chemist Anselme Payen was diastase, a starch-degrading enzyme. Starch is one of the most abundant sources of energy. This complex polysaccharide is a main additive in food and a source of various sugar syrups which are widely used in fermentation and the pharmaceutical and confectionary industries. It is a heterogeneous polysaccharide composed of glucose units linked by α-1, 4 and α-1, 6-glycosidic bonds forming the insoluble linear polymer amylose and the soluble branched amylopectin. Synergistic action of enzymes is needed for the complete hydrolysis of starch to glucose. α-Amylases, β-amylases, glucoamylases, α-glucosidases and pullulanases are enzymes involved in the hydrolysis of starch to sugar. These enzymes account for 25 per cent of the global enzyme market and are widespread among microorganisms, including extremophiles. Starch saccharification is a multistep process (Fig. 1). The starch is first gelatinized at 105°C for 5 min. in a jet cooker followed by liquefaction (95°C for one hour at pH 6.0) of raw starch granules in the presence of bacterial α-amylase. The next is the saccharification step (60°C for three hours at pH 4.5) which is carried out in the presence of glucoamylase in combination with pullulanase. The α-amylase presently used in starch processing (from *Bacillus amyloliquefaciens* and *B. licheniformis*) is optimally active at 95°C and pH 6.8 and is stabilized by Ca^{2+}. Therefore, the industrial processes using these enzymes cannot be carried out at low pH, i.e., 3.2–4.5, which

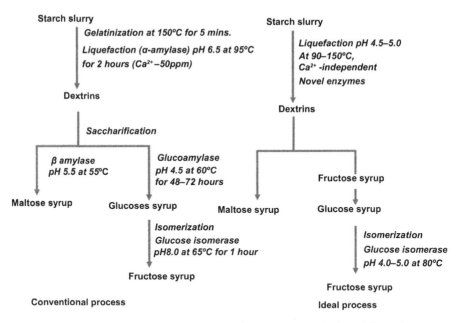

Figure 1. Conventional and ideal starch saccharification processes (Adapted from Sharma and Satyanarayana, 2013).

is the pH of native starch. In order to be well-suited to the optimal pH of the enzyme required for liquefaction, the pH of the starch slurry is increased to 5.8–6.2, and further, Ca^{2+} is supplemented to increase the activity and/or stability of the enzyme. The second step, saccharification using glucoamylase, requires readjustment of pH to 4.2–4.5 and salt removal, both of which are time consuming and increase the cost of the process (Sharma et al., 2012). Therefore, enzymes ideal for this process need several properties in combination, such as activity and stability at pH 3.0–4.0, 90–110°C and Ca^{2+}-independence.

α-Amylase is an endo-acting enzyme that randomly cleaves α-1, 4-linkages of the starch and related substrates, forming branched and linear α-anomeric oligo- and poly-saccharides of different chain lengths. A number of α-amylases have been identified from different extremophiles. α-Amylase from *Alicyclobacillus acidocaldarius* is one of the first examples of heat and acid-stable enzymes with an optimum of pH 3.0 and 75°C (Matzke et al., 1997; Bertoldo et al., 2004). There are very few reports of thermo-acid-stable α-amylases (Bai et al., 2012; Sharma and Satyanarayana, 2010; Liu and Xu, 2008). Thermo-acid-stable α-amylases from *A. acidocaldarius* and *Bacillus acidicola* are useful in the starch industry (Sharma and Satyanarayana, 2012; Bai et al., 2012). The α-amylase from a hyperthermophilic archaeon *Pyrococcus furiosus* is optimally active at 100°C (Koch et al., 1990). Another thermostable α-amylase has been reported from *Methanococcus jannaschii* with an optimum temperature at 120°C (Kim et al., 2001). *Pyrococcus woesei* α-amylase is the first crystallized enzyme from hyperthermophilic archaea with optimum growth above 100°C (Linden et al., 2003; Frillingos et al., 2000). Biophysical analysis of amylases from thermophilic microorganisms shows a highly compact structure and an increased number of salt bridges (Linden and Wilmanns, 2004). Some recent reports describe Ca^{2+}-independent α-amylases which do not require Ca^{2+} for their activity or stability (Sharma and Satyanarayana, 2010; Rao and Satyanarayana, 2007). As evident from the above reports, α-amylases from extremophiles showing thermostability, acid stability or Ca^{2+}-independence have been discovered, paving the way for engineering an industrial enzyme with all the characteristics in combination.

Glucoamylases are exo-acting enzymes which successively release single β-D-glucose units from non-reducing ends by cleaving α-1, 4-glycosidic bonds from starch and other related polysaccharides. This glucose can be used in a number of food and beverage industries to produce ethanol, amino acids, organic acids and others (Polakovic and Bryjak, 2004). It is also converted into crystalline dextrose and high glucose syrups or isomerized to fructose in the presence of glucose isomerase (Crabb and Shetty, 1999). Most of the known thermophilic glucoamylases are from fungal sources with very few reports from bacteria and archaea. The first gene-encoding glucoamylase was identified from *Aspergillus niger* (Svensson et al., 1983). The most thermostable glucoamylase from bacteria was identified in *Thermoanaerobacter tengcongensis* which is optimally active at pH 5.0 and 75°C on maltooligosaccharides with only four monosaccharide units (Zheng et al., 2010). The first thermoacidophilic archaeal glucoamylases are known from *Picrophilus torridus, P. oshimae* and *Thermoplasma acidophilum*. Archaeal glucoamylases are reported to be optimally active at pH 2.0 and 90°C (Serour and Antranikian, 2002), while glucoamylases produced by fungi, yeast and bacteria are optimally active at 70°C and in the pH range of 3.5–6.0. A neutral glucoamylase optimally active at pH 7.0 and 60°C from the thermophilic mould *Thermomucor indicae-seudaticae* has been shown to be useful in starch saccharification (Kumar and Satyanarayana, 2003).

β-Amylase, also an exo-acting enzyme, attacks α-1, 4-glucosidic linkage of the starch and produces dimeric sugar β-maltose. β-Amylase, most active on *p*NP-α-maltopyranoside, was reported from hyperthermophilic archeaon *P. furiosus* (Comfort et al., 2008). A thermo-acid-stable β-amylase is known from *Clostridium thermosulfurogenes*, which is optimally active at pH 5.5 and 75°C (Shen et al., 1988).

Pullulanases are endo-acting enzymes capable of hydrolyzing α-1, 6-glycosidic linkages in starch, pullulan, amylopectin and related oligosaccharides. Amylopullunases are bifunctional enzymes that cleave both α-1, 4 and α-1, 6 linkages in starch, amylose and other oligosaccharides, and α-1, 6 linkages in pullulan. This enzyme finds application in the starch industry for making various sugar syrups, such as maltose and maltooligosaccharides syrups used in the food industry (Nisha and Satyanarayana, 2013). A thermostable pullulanase with a half-life of 45 min. at 100°C has been reported from a hyperthermophilic archaeon *Thermococcus kodakarensis* KOD1 (Han et al., 2013). A thermostable and Ca^{2+}-independent amylopullulanase from a thermophile *Geobacillus thermoleovorans* with a $T_{1/2}$ of 7.8 hours at 90°C is known to hydrolyze starch, pullulan and malto-oligosaccharides (Nisha and Satyanarayana, 2014).

The characteristics of thermoacid-stable starch hydrolyzing enzymes from extremophiles are summarized in Table 1.

Table 1. Starch hydrolyzing enzymes from bacteria and archaea.

Enzymes	Organism	$T_{(Opt)}$ °C	$pH_{(Opt)}$	References
α-Amylase				
	Alicyclobacillus acidocaldarius	75	3.0	Schwermann et al., 1994
	Bacillus acidicola	60	4.0	Sharma and Satyanarayana, 2010
	Bacillus sp. Ferdowsicous	70	4.5	Asoodeh et al., 2010
	Bacillus sp. YX1	-	5.0	Liu and Xu, 2008
	Dictyoglomus thermophilum	90	5.5	Fukusumi et al., 1988
	Geobacillus thermodenitrificans	80	5.5	Ezeji and Bahl, 2006
	Geobacillus sp. LH8	80	5.0–7.0	Mollania et al., 2010
	Lipomyces starkeyi	70	4.0	Kelly et al., 1985
	Pyrococcus furiosus	100	5.5–6.0	Koch et al., 1990
	Pyrococcus Woesei	100	5.5	Koch et al., 1991
	Thermococcus profundus	.5.0–6.0	80	Chung et al., 1995
	Thermococcus sp.	95	5.0	Wang et al., 2008
	Thermotoga maritima MSB8	85–90	7.0	Liebel et al., 1997
Glucoamylase				
	Thermomucor indicae-seudaticae	60	7.0	Kumar and Satyanarayana, 2003
	Thermoplasma acidophilum	90	6.5	Serour and Antranikian, 2002
	Picrophilus oshimae	90	2.0	Serour and Antranikian, 2002
	Picrophilus torridus	90	2.0	Serour and Antranikian, 2002
β-Amylase				
	Bacillus stearothermophilus	6.9	57	Srivastava, 1987
Pullulanase				
	Fervidobacterium pennavorans Ven5	80	6	Bertoldo et al., 1999
	Thermotoga maritima	90	6.0	Bibel et al., 1998
	Thermus caldophilus GK24	75	5.5	Kim et al., 1996
	Desulfurococcus mucosus	85	5.0	Duffner et al., 2000
	Pyrococcus woesei	100	6.0	Rudiger et al., 1995
	Pyrococcus furiosus	100	6.0	Brown and Kelly, 1993
	Thermococcus celer	90	5.5	Canganella et al., 1994
	Thermococcus litoralis	98	5.5	Brown and Kelly, 1993
	Thermoanaerobacter ethanolicus	90	5.5	Lin and Liu, 2002

3. Baking Industry

Bread is one of the most common and traditional foods around the world. Drastic improvements have been made with the usage of enzymes in baking, in terms of flavor, texture, dough flexibility, machinability, stability, loaf volume, crumb structure and shelf-life. Among the enzymes used in food applications, nearly one-third of the total enzyme market is shared by the baking industry. Baking enzymes are used as flour additives which work as dough conditioners. Baking industry mainly uses five types of enzymes, such as amylase, protease, xylanase, oxidase and lipase. All these enzymes together improve the organoleptic properties of bread. Amylases used in the baking industry need different properties than those used by the starch industry. Acid-stable and maltogenic amylases with intermediate thermostability are required in baking. The maltogenic nature of the enzyme shows antistaling effect on bread, while intermediate thermostability leads to inactivation of enzyme at the end of baking, preventing residual enzyme activity and product deterioration. Acid-stability of α-amylase is important in baking as the pH of the dough is acidic. An acidic α-amylase of *Bacillus acidicola*, optimally active at pH 4.5, has been shown to be useful in baking (Sharma and Satyanarayana, 2012).

Xylanases were introduced in the baking industry in 1970 and are most often used in combination with amylases, lipases and many oxidoreductases to improve the rheological properties of dough and organoleptic properties of bread (Collins et al., 2006). These enzymes have also been used to improve the quality of biscuits, cakes and other baked products (Poutanen, 1997). The important function of xylanases in baking is the break down of hemicellulose present in wheat flour and redistribution of water, leaving the dough soft and easy to knead. Its supplementation in dough aids in the absorption of water, resistance to fermentation and increase in the volume of bread. The use of xylanases removes insoluble arabinoxylans that interfere with the formation of the gluten network. It forms high molecular weight solubilized arabinoxylans that result in increased viscosity. Therefore, stable, flexible and easily manageable dough is formed, resulting in improved oven spring, increased loaf volume, soft crumb, better texture and crumb structure (Rouau et al., 1994; Courtin and Delcour, 2001; Sorensen et al., 2004; Heldt-Hansen, 2006). In addition, the incorporation of xylanases during dough processing is expected to enhance the concentration of arabinoxylo-oligosaccharides in bread, which have useful effects on human health (Bhat, 2000). The GH family eight psychrophilic xylanase from *Pseudoalteromonas haloplanktis* showed a positive effect on the loaf volume (Collins et al., 2002a, 2003; Van Petegem et al., 2003). Shah et al. (2006) reported the use of acidstable xylanases (optimum pH 5.3) from acidophilic fungus *Aspergillus foetidus* as bread improver for making whole wheat bread. Very few thermoacid-stable xylanases are known from extremophiles, such as *Pyrococcus furiosus* (102°C, pH 6.0) (Bauer et al., 1999), *Sulfolobus solfataricus* (105°C, pH 5.3) (Nucci et al., 1993) and *Thermotoga neapolitana* (102°C, pH 5.5) (Zverlov et al., 1996). A thermo-alkali-stable xylanase of polyextremophilic *Bacillus halodurans* has recently been shown to be useful in whole wheat bread (Kumar and Satyanarayana, 2014).

Protease is used in baking to decrease mixing time, to reduce dough consistency for dough uniformity, to regulate gluten strength, improve bread texture and flavor. In bread production, a fungal acid, protease, is used to modify mixtures containing high gluten content. The addition of proteases in the blend makes it soft and easy to pull and knead (Di Cagno et al., 2003). These enzymes have great impact on dough rheology and the quality of bread, possibly due to its effects on the gluten network or on gliadin (Salleh et al., 2006). They act on the proteins of wheat flour, reducing gluten elasticity. Thus, shrinkage of dough or paste after moulding and sheeting is reduced (Cauvain and Young, 2006; Kara et al., 2005). Proteases are also used in the manufacture of pastries, biscuits and cookies. Hydrolysis of glutenin proteins present in the flour is responsible for the elasticity of dough and improves the spread ratio of cookies (Kara et al., 2005).

The action of amylopullulanase on starch and other related polysaccharides produces maltotriose and maltooligosaccharide syrups that are used in baking to prevent retrogradation of starch and as an antistaling agent (Nisha and Satyanarayana, 2013).

4. Fruit Juice Industry

The global fruit and vegetable processing industries have seen consistent demand over the last five years with an annual growth rate of 1.3 per cent. Increase in demand for processed foods, especially in developing countries, is expected to increase the industry revenue by 0.4 per cent from \$271.3 billion to \$273.1 billion by 2020 (www. ibisworld.com).

In fruit-juice processing, the important steps are extraction, clarification and stabilization. Traditionally pectinases have been used before the extraction step to facilitate pressing and thus maximize juice yield. Pressing is generally followed by concentration step or a 'hot break' step at high temperatures, say, up to 90°C. The enzyme treatment thus requires the addition of enzymes from extremophiles, since commercial enzymes are not thermostable (Grassin and Coutel, 2010).

Pectin is present in all fruits that gel after crushing. It holds the juice within the mash, decreases pressability, reduces juice yield and hinders clarification and evaporation for concentration. The main objective of enzymatic treatments is to reduce the pectin viscosity in order to speed up processing. Pectin is not just one substance, but a group of polysaccharides with a large molecular structure. Pectin is composed of long chains of galacturonic acid residues. On each residue is a carboxyl (–COOH) group, sometimes esterified by the addition of methyl groups (–COOCH$_3$). Pectin, in which half or more of the acid residues, are methoxylated being called high methoxyl pectin; and where there are fewer are known as low methoxyl pectin (Madden, 2000). In Nature, nearly 80 per cent of carboxylic groups of galacturonic acid are esterified with methanol.

On the basis of the mode of action, three major types of pectinases are known. Pectinesterase (PE), also known as pectinmethylhydrolase, catalyses deesterification of the methoxyl group present in pectin to pectic acid. The second group is of depolymerizing enzymes which are divided into two subclasses: hydrolyzing enzymes, such as polymethylgalacturonase (PMG) and polygalacturonase (PG), and cleaving enzymes, such as polymethylgalacturonate (PMGL) and polygalacturonate lyase (PGL). PMG catalyses the hydrolysis of α-1, 4-glycosidic bonds in pectin, especially of highly esterified forms. PG catalyses hydrolysis of α-1, 4-glycosidic linkages in polygalacturonic (pectic) acid. PMGL catalyses break down of pectin by trans-eliminative cleavage and PGL cleaves the α-1, 4-glycosidic linkage in pectic acid by trans-elimination. A third group, known as protopectinase, solubilizes protopectin (a parent pectic substance that produces pectin and pectinic acid on restricted hydrolysis) to highly polymerized soluble pectin.

The juice of red berries, such as black currant, cherries, grapes and others, has strong acidity and high phenolic content. Following a 'hot break' process at 90°C for two min. pectinase treatment at 50°C is carried out for one to two hours. Use of thermo-acid-stable pectinases could expedite this process and potentially prevent time-dependent oxidation. High acidity of lemon juice (pH 2.0) inhibits most of the fungal pectinases and makes enzymatic clarification of lemon juice very difficult.

With a number of enzymes, such as pectinases, xylanases, α-amylases, cellulases and others, the yields of fruit juices have improved due to reduction in viscosity and turbidity, along with enhanced recovery of aroma, essential oils, vitamins and mineral salts. Among all the enzymes used in fruit juice processing, the most important are pectinases which aid in obtaining clear juice in high yields (Dupaigne, 1974; Viquez et al., 1981; Girard and Fukumoto, 1999; Lee et al., 2006; Sandri et al., 2012). Their commercial application was first observed in 1930 in the preparation of wines and fruit juices (Oslen, 2000). Pectic enzymes from fungi, such as *A. niger*, *Penicillium notatum* or *Botrytis cinerea* are mainly used in wine making. Acidic pectinases used in fruit juice processing industry and winemaking mainly come from fungal sources, especially from *Aspergillus* which is optimally active at the natural pH of fruits that ranges between 2.5–5.5. The other sources of acidstable pectinases are listed in Table 2. Acidic pectinases are used commercially in the production of clear juices of apple, pear, grapes, tangerine, plum, peach, apricot juice and banana (Girard and Fukumoto, 1999; Mondor et al., 2000; Vaillant et al., 1999; Lee et al., 2006). Many researchers have reported that depectinization using pectinase can successfully produce clear banana pulp (Viquez et al., 1981; Koffi et al., 1991; Yusof and Ibrahim, 1994; Brasil et al., 1995; Alvarez et al., 1998; Ceci and Lozano, 1998; Vaillant et al., 1999; Kaur et al., 2004; Lee et al., 2006).

In pear araban, hemicellulose and cellulose contents are high as compared to apple; therefore, pectinases with high arabanase and arabinosidase activities are needed. PL, PME, PG and arabanases pectinases improve the extraction of juice.

There are three categories of juices produced by industries commercially: (1) sparkling, clear juices (apple, pear and grapes), (2) juices with clouds including citrus, prune, tomato juices and nectars, and (3) unicellular products.

Starch is a potential contributor of haziness to fruit juices. Green apples contain up to 2 g L^{-1} starch. When the juice is heated to 75–80°C, starch gets gelatinized and reforms into amorphous aggregates on cooling. Therefore, it gives a cloudy appearance.

Table 2. Characteristics of microbial pectinases.

Organism	Type of Pectinases	pH[(Opt)]	Temp[(Opt)]	References
Aspergillus niger CH4	Endo-pectinase,	4.5 ± 6.0	Below 50	Acuna-Arguelles et al., 1995
	Exo-pectinase	3.5 ± 5.0		
Penicillium frequentans	Endopolygalacturonase (Endo-PG)	4.5 ± 4.7	50	Borin et al., 1996
Sclerotium rolfsii	Endo-PG	3.5	55	Channe and Shewal, 1995
Rhizoctonia solani	Endo-PG	4.8	50	Marcus et al., 1986
Mucor pusillus	PG	5.0	40	Al-Obaidi et al., 1987
Clostridium thermosaccharolyticum	Polygalacturonate hydrolase	5.5 ± 7.0	30 ± 40	Rijssel et al., 1993

Only few thermo-acid-stable amylases are known (Table 1). There is a need to explore new amylases from extremophiles that can be used in fruit-juice processing. The known thermo-acid-stable amylases may also be used for this application.

Pectic enzymes are generally added in apple juice preparations to facilitate pressing or juice extraction and as an aid in separating flocculent precipitate by using different methods, such as sedimentation, filtration and centrifugation. Application of pectinases improved the clarification of apple juice with 35 per cent viscosity drop (Girard and Fukumoto, 1999; Mondor et al., 2000).

Aspergillus species also produce another type of pectinase known as rhamnogalacturonase, which is used in the maceration of apple juice (Schols et al., 1990). Maceration is the transformation of organized tissues into a suspension of whole cells and the product formed is used as the base material for pulpy juices and nectars, baby foods, components for dairy products, such as pudding and yoghurt. Clarification of the fruit juice is mainly affected by pH, temperature, time and enzyme concentration (Neubeck, 1975; Baumann, 1981; Lanzarini and Pifferi, 1989). Enzymes, active at acidic pH and temperature (40–60°C), produce clear juice as compared to those active at alkaline pH. Therefore, thermo-acid-stable enzymes are needed in fruit-juice processing.

5. Animal Feeds

In the present farming systems, the use of enzymes in animal nutrition plays an important role (Choct, 2006). The potential significance of exogenous enzymes to improve nutrient utilization and performance in poultry is already known. The poultry industry is the biggest user of feed enzymes. The integrated nature of the poultry sector has helped in the rapid uptake of new technologies and exogenous enzymes in the feed industry to improve nutrient digestibility and its efficient utilization. In countries like United Kingdom, Australia, New Zealand and Canada, more than 90 per cent of broiler diets contain feed enzymes and nearly 70 per cent of wheat and barley-based poultry feeds are supplemented with glycanases (xylanases and β-glucanases). Feed enzymes increase the digestibility of nutrients and degradation of undesirable components in feed, which are harmful or of little or no value (Ravindran and Son, 2011). Diets based on cereals, such as barley, rye and wheat contain high amounts of non-starch polysaccharides (NSPs), which might reduce the intestinal methane production when supplemented with NSP enzymes. Animal feed supplementation with enzymes, such as xylanases, amylases, cellulases, pectinases, phytases and proteases causes reduction in unwanted residues, such as phosphorus, nitrogen, copper and zinc in the excreta which plays an important role in reducing environmental contamination (Li et al., 2012). Xylanases are added in animal feed to hydrolyze arabinoxylans present in the feed. Arabinoxylans are found in the cell walls of grains and show anti-nutrient effect in poultry. Application of cellulases in feed processing improves the feed digestibility and animal performance (Lewis et al., 1996; Kung et al., 1997). Bedford et al. (2003) reported the use of *Trichoderma* cellulases as feed additive in improving the feed conversion ratio and/or increasing the digestibility of a cereal-based feed.

Microbial phytases are being widely used in poultry and pig diets in response to increasing phosphorus (P) pollution from effluents due to animal operations. The ban on the use of meat and bone meal (major sources of P) in the European Union also raised the use of microbial phytases. As a result, the demand for microbial phytase over glycanases as the primary feed enzyme has increased worldwide. Phytases act on phytic acid (myo-inositol1, 2, 3, 4, 5, 6-hexakis dihydrogen phosphate), a main stored form of P in plant-derived ingredients (Ravindran and Son, 2011).

According to Frost and Sullivan (2007), the global market for feed enzymes was estimated around $344 million in 2007 and was expected to reach $727 million in 2015. Currently, feed enzymes available commercially by catalytic types are: 3-phytase, 6-phytase, subtilisin, α-galactosidase, glucanase, xylanase, α-amylase and polygalacturonase (Selle and Ravindran, 2007). Microbial phytases are mainly added to animal (swine and poultry) feeds and human foodstuffs to improve mineral bioavailability and food processing. Xylanases and phytases are generally reported from fungi and yeast. Recently, thermo-acid-stable phytase from the yeast *Pichia anomala* has been reported to efficiently reduce the phytic acid content of various broiler feeds (Joshi and Satyanarayana, 2015). There are very few reports of xylanases and phytases from acidophilic bacteria and archaea. Some of the reports of acidstable xylanases include production of enzyme from *S. solfataricus* that displayed activity on carboxymethyl cellulose with optimum activity at pH 3.5 and 95°C. Another report of xylanases is from *Acidobacterium capsulatum* that exhibits optimum activity and stability in the acidic range (Inagaki et al., 1998).

Available exogenous enzymes not significantly stable when processing temperatures increase beyond 70°C. Therefore, enzymes are included as liquids through post-pelleting application systems to avoid thermostability problems at high pelleting temperatures. Application of liquid enzymes accurately after pelleting may be a complex and costly process. Therefore, thermostable enzymes are needed which will simplify the pre-pelleting application of dry product and promote the use of the enzyme in pelleted diets (Ravindran and Son, 2011). There are reports of thermo-acid-stable phytases from *Pichia anomala* (Vohra and Satyanarayana, 2002) and *Sporotrichum thermophile* (Singh and Satyanarayana, 2009). There are no reports of stable phytases from acidophilic bacteria and archaea. There is a need to explore acidic phytases from extremophilic bacteria and archaea for use in animal feeds, as they are expected to have better thermo-acid-stability, higher substrate specificity and catalytic efficiency than the fungal phytases (Rodriguez et al., 1999; Kim et al., 2003). The application of different enzymes in various industrial sectors is listed in Table 3.

6. Conclusions

Although the addition of enzymes in foods and feeds is well established, there is still a need for more effective combinations of enzymes which will reduce time and cost of the processes. New/improved thermo-acid-stable biocatalysts which are unaffected by metal ions and less susceptible to inhibitory agents and harsh environmental conditions are required. Developments in molecular biology, enzyme engineering, computational tools and high-throughput methodologies will permit development of efficient and useful biocatalysts from extremophiles, which retain activity under harsh conditions that are employed in the processing of certain foods and feeds.

Table 3. Applications of enzymes in different industrial processes.

Industry	Enzymes	Advantages
Starch processing	α-Amylase	cleaves α-1, 4-linkages of the starch and related substrates, forming branched and linear α-anomeric oligo and polysaccharides of different chain lengths.
	Glucoamylase	releases single β-D-glucose from non reducing ends by cleaving α-1, 4-glycosidic bonds from starch and other related polysaccharides
	β-Amylase	attacks α-1, 4-glucosidic linkage of the starch molecule producing the dimeric sugar β-maltose.
	Pullulanase	hydrolyses α-1, 6-glycosidic linkages in starch, pullulan, amylopectin, and related oligosaccharide
	Amylopullulanses	cleaves both α-1, 4 and α-1, 6 linkages in starch, amylose and other oligosaccharides, and α-1,6 linkages in pullulan
Baking		
	α-Amylase	degrades starch in flour, control the volume and crumb structure of bread,
	Xylanases	improves dough stability and handling
	Oxidoreductases	increases gluten strength
	Proteases	reduces the protein in flour
	Lipases	improves stability of gas cells in the dough
	Amylases/glucoamylases	breaks down starch to glucose, clarifies cloudy juice
Juice		
	Pectinases	degrades pectin the structural polysaccharide present in the cell wall
	Cellulases	Release of the antioxidants from fruit and vegetable pomace, improvement of yields in starch and protein extraction, improved maceration, pressing, clarification of fruit juices, improved aroma of wines
	Laccase	increases the susceptibility of browning during storage
	Naringinase and limoninase	act on compounds that cause bitterness in citrus juices
Feed		
	Xylanases	degrades fiber in viscous diets
	Phytases	degrades phytic acid to release phosphorous, calcium and magnesium
	Proteases	degrades protein into peptides and amino acids to overcome antinutritional factors
	α-Amylases	digests starch
	Cellulase	production of energy-rich animal feed, improved nutritional quality of animal feed, improved feed digestion and absorption

References

Acuna-Arguelles, M.E., Gutierrez-Rajas, M., Viniegra-Gonzalez, G. and Favela-Toress, E. 1995. Production and properties of three pectinolytic activities produced by *A. niger* in submerged and solid state fermentation. Applied Microbiology and Biotechnology, 43: 808–814.

Adams, M.W.W., Perler, F.B. and Kelly, R.M. 1995. Extremozymes: Expanding the limits of biocatalysis. Biotechnology, 13: 662–668.

Al-Obaidi, Z.S., Aziz, G.M. and Al-Bakir, A.Y. 1987. Screening of fungal strains for polygalacturonase production. Journal of Agriculture and Water Resources Research, 6: 125–182.

Alvarez, S., Alvarez, R., Riera, F.A. and Coca, J. 1998. Influence of depectinization on apple juice ultra filtration. Colloids and Surfaces A: Physicochemical and Engineering Aspects, 138: 377–382.

Asoodeh, A., Chamani, J. and Lagzian, M. 2010. A novel thermostable, acidophilic α-amylase from a new thermophilic '*Bacillus* sp. Ferdowsicous' isolated from Ferdows hot mineral spring in Iran: Purification and biochemical characterization. International Journal of Biological Macromolecules, 46: 289–297.

Bai, Y., Huang, H., Meng, K., Shi, P., Yang, P., Luo, H., Luo, C., Feng, Y., Zhang, W. and Yao, B. 2012. Identification of an acidic α-amylase from *Alicyclobacillus* sp. A4 and assessment of its application in the starch industry. Food Chemistry, 131: 1473–1478.

Bauer, M., Driskil, L., Callen, W., Snead, M., Mathur, E. and Kelly, R. 1999. An endoglucanase EglA, from the hyperthermophilic archaeon *Pyrococcus furiosus* hydrolyzes a-1, 4 bonds in mixed linkage (1-3), (1-4)-β-D-glucans and cellulose. Journal of Bacteriology, 181: 284–290.

Baumann, J.W. 1981. Application of enzymes in fruit juice technology. pp. 129–147. *In*: G.G. Birch, N. Blakebrough and K.J. Parker (eds.). Enzymes and Food Processing. Applied Science Publishers Ltd., London.

Bedford, M.R., Koepf, E., Lanahan, M., Tuan, J. and Street, P.F.S. 2003. Relative efficacy of a new, thermotolerant phytase in wheat-based diets for broilers. Poultry Science 82(Suppl. 1): 36 (Abstract).

Berka, R.M. and Cherry, J.R. 2006. Enzyme biotechnology. pp. 477–498. *In*: C. Ratledge and B. Kristiansen (eds.). Basic Biotechnology. Cambridge University Press, Cambridge, UK.

Bertoldo, C., Dock, C. and Antranikian, G. 2004. Thermoacidophilic microorganisms and their novel biocatalysts. Engineering in Life Sciences, 4: 521–531.

Bertoldo, C., Duffner, F., Jorgensen, P.L. and Antranikian, G. 1999. Pullulanase type I from *Fervidobacterium pennavorans* Ven5: Cloning, sequencing and expression of the gene and biochemical characterization of the recombinant enzyme. Applied & Environmental Microbiology, 65: 2084–91.

Bhat, M.K. 2000. Cellulases and related enzymes in biotechnology. Biotechnology Advances, 18: 355–383.

Bibel, M., Brettl, C., Gosslar, U., Kriegshauser, G. and Liebl, W. 1998. Isolation and analysis of genes for amylolytic enzymes of the hyperthermophilic bacterium Thermotoga maritima. FEMS Microbiology Letters, 158: 9–15.

Binod, P., Singhania, R.R., Soccol, C.R. and Pandey, A. 2008. Industrial enzymes. pp. 291–320. *In*: A. Pandey, C. Larroche, C.R. Soccol and C.-G. Dussap (eds.). Advances in Fermentation Technology, Asiatech Publishers, New Delhi, India.

Borin, M.D.F., Said, S. and Fonseca, M.J.V. 1996. Purification and biochemical characterization of extracellular endopolygalacturonase from *Penicillium frequentans*. Journal of Agriculture and Food Chemistry, 44: 1616–1620.

Brasil, I.M., Maia, G.A. and Figuiredo, R.W. 1995. Physical-chemical changes during extraction and clarification of guava juice. Food Chemistry, 54: 383–386.

Brown, S.H. and Kelly, R.M. 1993. Characterization of amylolytic enzymes having both α-1,4 and α-1,6 hydrolytic activity from the thermophilic archaea *Pyrococcus furiosus* and *Thermococcus litoralis*. Applied & Environmental Microbiology, 59: 2614–2621.

Canganella, F., Andrade, C. and Antranikian, G. 1994. Characterisation of amylolytic and pullulytic enzymes from thermophilic archaea and from a new *Fervidobacterium* species. Applied Microbiology and Biotechnology, 42: 239–245.

Carrez, D and Soetaert, W. 2005. Looking ahead in Europe: White biotech by 2025. Industrial Biotechnology, 1: 95–101.

Cauvain, S. and Young, L. 2006. Ingredients and their influences. pp. 72–98. *In*: S. Cauvain and L. Young (eds.). Baked Products: Science, Technology and Practice. Oxford: Blackwell Publishing.

Ceci, L. and Lozano, J. 1998. Determination of enzymatic activities of commercial pectinases for the clarification of apple juice. Food Chemistry, 61: 237–241.

Chandel, A.K., Rudravaram, R., Rao, L.V., Ravindra, P. and Narasu, M.L. 2007. Industrial enzymes in bioindustrial sector development: An Indian perspective. Journal of Commercial Biotechnology, 13: 283–291.

Channe, P.S. and Shewal, J.G. 1995. Pectinase production by *Sclerotium rolfsii*: Effect of culture conditions. Folia Microbiologica, 40: 111–117.

Choct, M. 2006. Enzymes for the feed industry: Past, present and future. World's Poultry Science Journal, 62: 5–16.

Chung, Y.C., Kobayashi, T., Kanai, H., Akiba, T. and Kudo, T. 1995. Purification and properties of extracellular amylase from the hyperthermophilic archaeon *Thermococcus profundus* DT5432. Applied & Environmental Microbiology, 61: 1502–1506.

Collins, T., Hoyoux, A., Dutron, A., Georis, J., Genot, B., Dauvrin, T., Arnaut, F., Gerday, C. and Feller, G. 2006. Use of glycoside hydrolase family 8 xylanases in baking. Journal of Cereal Science, 43: 79–84.

Comfort, D.A., Chou, C.J., Conners, S.B., VanFossen, A.L. and Kelly, R.M. 2008. Functional-genomics-based identification and characterization of open reading frames encoding α-glucoside-processing enzymes in the hyperthermophilic archaeon *Pyrococcus furiosus*. Applied Environmental Microbiology, 74: 1281–1283.

Courtin, C.M. and Delcour, J.A. 2001. Relative activity of endoxylanases towards water-extractable and water-unextractable arabinoxylan. Journal of Cereal Science, 33: 301–312.

Crabb, W.D. and Shetty, J.K. 1999. Commodity-scale production of sugars from starches. Current Opinion in Microbiology, 2: 252–256.

Di Cagno, R., De Angelis, M., Corsettic, A., Lavermicocca, P, Arnault, P., Tossut, P., Gallo, G. and Gobbetti, M. 2003. Interactions between sourdough lactic acid bacteria and exogenous enzymes: effects on the microbial kinetics of acidification and dough textural properties. Food Microbiology, 20: 67–75.

Duffner, F., Bertoldo, C., Andersen, J.T., Wagner, K and Antranikian, G. 2000. A new thermoactive pullulanase from Desulfurococcus mucosus: Cloning, sequencing, purification and characterization of the recombinant enzyme after expression in Bacillus subtilis. Journal of Bacteriology, 182: 6331–6338.

Dupaign, P. 1974. The aroma of bananas. Fruits, 30: 783–789.

Elleuche, S. and Antranikian, G. 2013. Starch-hydrolyzing enzymes from thermophiles. pp. 509–533. *In*: T. Satyanarayana, J. Littlechild and Y. Kawarabayasi (eds.). Thermophilic Microbes in Environmental and Industrial Biotechnology: Biotechnology of Thermophiles. Springer.

Ezeji, T.C. and Bahl, H. 2006. Purification, characterization, and synergistic action of phytate-resistant α-amylase and α-glucosidase from *Geobacillus thermodenitrificans* strain HRO10. Journal of Biotechnology, 125: 27–38.

Frillingos, S., Linden, A., Niehaus, F., Vargas, C., Nieto, J.J., Ventosa, A. et al. 2000. Cloning and expression of α-amylase from the hyperthermophilic archaeon *Pyrococcus woesei* in the moderately halophilic bacterium *Halomonas elongata*. Journal of Applied Microbiology, 88: 495–503.

Frost and Sullivan. 2007. Feed enzymes: The global scenario. https://www.frost.com/sublib/display-market-insight.do?id=115387658.

Fukusumi, S., Kamizono, A., Horinouchi, S. and Beppu, T. 1988. Cloning and nucleotide sequence of a heat stable amylase gene from an anaerobic thermophile, *Dictyoglomus thermophilum*. European Journal of Biochemistry, 174: 15–23.

Girard, B. and Fukumoto, L.R. 1999. Apple juice clarification using microfiltration and ultrafiltration polymeric membranes. Lebensmittel-Wissenschaft und-Technologie. Food Science and Technology, 32: 290–298.

Grassin, C. and Coutel, Y. 2010. Enzymes in fruit and vegetable processing and juice extraction. pp. 236–261. *In:* Whitehurst, R.J. and M. Van Oort (eds.). Enzymes in Food Technology. Wiley Blackwell.

Han, T., Zeng, F., Li, Z., Liu, L., Wei, M., Guan, Q., Liang, X., Peng, Z., Liu, M., Qin, J., Zhang, S. and Jia, B. 2013. Biochemical characterization of a recombinant pullulanase from *Thermococcus kodakarensis* KOD1. Letter in Applied Microbiology, 57: 336–43.

Heldt-Hansen, H.P. 2006. Macromolecular Interactions in Enzyme Applications for Food Products. pp. 363–388. *In*: Gaonkar A.G. and A. McPherson (eds.). Ingredients Interactions: Effects on Food Quality, second ed. Boca Raton: Taylor & Francis.

Inagaki, K., Nakahira, K., Mukai, K., Tamura, T. and Tanaka, H. 1998. Gene cloning and characterization of an acidic xylanase from *Acidobacterium capsulatum*. Bioscience, Biotechnology and Biochemistry, 62: 1061–1067.

Joshi, S. and Satyanarayana, T. 2015. Bioprocess for efficient production of recombinant *Pichia anomala* phytase and its applicability in dephytinizing chick feeds and whole wheat flat Indian breads. Journal of Industrial Microbiology and Biotechnology, 42: 1389–1400.

Kara, M., Sivri, D. and Koksel, H. 2005. Effects of high protease-activity flours and commercial proteases on cookie quality. Food Research International, 38: 479–486.

Kaur, G., Kumar, S. and Satyanarayana, T. 2004. Production, characterization and application of a thermostable polygalacturonase of thermophilic mould *Sporotrichum thermophile* Apinis. Bioresource Technology, 94: 239–243.

Kelly, C.T., Moriarty, M.E. and Fogarty, W.M. 1985. Thermostable extracellular α-amylase and α-glucosidase of *Lipomyces starkeyi*. Applied Microbiology and Biotechnology, 22: 352–358.

Kim, C.-Ho., Nashirua, O. and Ko, J.H. 1996. Purification and biochemical characterization of pullulanase type I from *Thermus caldophilus* GK-24. FEMS Microbiology Letters, 138: 147–152.

Kim, H.W., Kim, Y.O., Lee, J.H., Kim, K.K. and Kim, Y.J. 2003. Isolation and characterization of a phytase with improved properties from *Citrobacter braakii*. Biotechnology Letters, 25: 1231–1234.

Kim, J.W., Flowers, L.O., Whiteley, M. and Peeples, T.M. 2001. Biochemical confirmation and characterization of the family-57 like alpha amylase of *Methanococcus jannaschii*. Folia Microbiologica, 46: 467–473.

Koch, R., Spreinat, A., Lemke, K. and Antranikan, G. 1991. Purification and properties of a hyperthermoactive α-amylase from the archaeobacterium *Pyrococcus woesei*. Archives of Microbiology, 155: 572–578.

Koch, R., Zablowski, P., Spreinat, A. and Antranikian, G. 1990. Extremely thermophilic amylolytic enzyme from the archaebacterium *Pyrococcus furiosus*. FEMS Microbiology Letters, 71: 21–26.

Koffi, E.K., Sims, C.A. and Bates, R.P. 1991. Viscosity reduction and prevention of browning in the preparation of clarified banana juice. Journal of Food Quality, 14: 209–218.

Kumar and Satyanarayana, T. 2003. Purification and kinetics of a raw starch-hydrolyzing, thermostable and neutral glucoamylase of a thermophilic mould *Thermomucor indicae-seudaticae*. Biotechnology Progress, 19: 936–944.

Kumar, V. and Satyanarayana, T. 2014. Production of thermo-alkali-stable xylanase by a novel polyextremophilic *Bacillus halodurans* TSEV1 in cane molasses medium and its applicability in making 3 whole-wheat bread. Bioprocess Biosystems Engineering, 37: 1043–1053.

Kung, L. Jr., Kreck, E.M., Tung, R.S., Hession, A.O., Sheperd, A.C., Cohen, M.A., Swain, H.E. and Leedle, J.A.Z. 1997. Effects of a live yeast culture and enzymes on *in vitro* ruminal fermentation and milk production of dairy cows. Journal of Dairy Science, 80: 2045–2051.

Lanzarini, G. and Pifferi, P.G. 1989. Enzymes in the fruit juice industry. pp. 189–222. *In*: C. Cantarelli and G. Lanzarini (eds.). Biotechnology Applications in Beverage Production. Elsevier Science, London, United Kingdom.

Lee, W.C., Yusof, S., Hamid, N.S.A. and Baharin, B.S. 2006. Optimizing conditions for enzymatic clarification of banana juice using response surface methodology (RSM). Journal of Food Engineering, 73: 55–63.

Lewis, G.E., Hunt, C.W., Sanchez, W.K., Treacher, R., Pritchard, G.T. and Feng, P. 1996. Effect of direct-fed fibrolytic enzymes on the digestive characteristics of a forage-based diet fed to beef steers. Journal of Animal Science, 74: 3020–3028.

Li, S., Yang, X., Yang, S., Zhu, M. and Wang, X. 2012. Technology prospecting on enzymes: application, marketing and engineering. Computational and Structural Biotechnology Journal, 2: 1–11.

Liebl, W., Stemplinger, I. and Ruile, P. 1997. Properties and gene structure of the *Thermotoga maritima* alpha-amylase AmyA, a putative lipoprotein of a hyperthermophilic bacterium. Journal of Bacteriology, 179: 941–8.

Lin, F.P. and Liu, K.L. 2002. Cloning, expression and characterization of thermostable region of amylopullulanase gene from *Thermoanaerobacter ethanolicus* 39E. Applied Biochemistry and Biotechnology, 97: 33–44.

Linden, A. and Wilmanns, M. 2004. Adaptation of class-13 alpha-amylases to diverse living conditions. Chembiochem, 5: 231–239.

Linden, A., Mayans, O., Meyer-Claucke, W., Antranikian, G. and Wilmanns, M. 2003. Differential regulation of a hyperthermophilic α-amylase with a novel (Ca, Zn) two metal center by zinc. Journal of Biological Chemistry, 278: 9875–9884.

Liu, X.D. and Xu, Y. 2008. A novel raw starch digesting α-amylase from a newly isolated *Bacillus* sp. YX-1: Purification and characterization. Bioresource Technology, 99: 4315–4320.

MacElroy, R.D. 1974. Some comments on the evolution of extremophiles. Biosystems, 6: 74–75.

Madden, D. 2000. In a jam and out of juice. National Centre for Biotechnology Education. ISBN: 0704913739.

Madigan, M.T. and Marrs, B.L. 1997. Extremophiles. Scientific American, 276: 66–71.

Marcus, L., Barash, I., Sneh, B., Koltin, Y. and Finker, A. 1986. Purification and characterization of pectolytic enzymes produced by virulent and hypovirulent isolates of *Rhizoctonia solani* KUHN. Physiological and Molecular Plant Pathology, 29: 325–336.

Matzke, J., Schwermann, B. and Baker, E.P. 1997. Acidstable and acidophilic proteins: The example of the alpha amylase from *Alicyclobacillus acidocaldarius*. Comparative Biochemistry and Physiology—Part A: Molecular and Integrative Physiology, 118: 411–419.

Mollania, N., Khajeh, K., Hosseinkhani, S. and Dabirmanesh, B. 2010. Purification and characterization of a thermostable phytate resistant α-amylase from *Geobacillus* sp. LH8. International Journal of Biological Macromolecules, 46: 27–36.

Mondor, M., Girard, B. and Moresoli, C. 2000. Modeling flux behavior for membrane filtration of apple juice. Food Research International, 33: 539–548.

Neubeck, C.E. 1975. Fruits, fruit products, and wines. pp. 397–442. *In*: R. Gerald (ed.). Enzymes in Food Processing. Academic Press San Francisco, London.

Nisha, M. and Satyanarayana, T. 2013. Recombinant bacterial amylopullulanases developments and perspectives. Bioengineered, 4: 388–400.

Nisha, M. and Satyanarayana, T. 2014. Characterization and multiple applications of a highly thermostable and Ca²⁺-independent amylopullulanase of the extreme thermophile *Geobacillus thermoleovorans*. Applied Biochemistry and Biotechnology, 174: 2594–615.

Nucci, R., Moracci, M., Vaccaro, C., Vespa, N. and Rossi, M. 1993. Exo-glucosidase activity and substrate specificity of the β-glycosidase from *Sulfolobus solfataricus*. Biotechnology and Applied Biochemistry, 17: 239–250.

Ogawa, J. and Shimizu, S. 2002. Industrial microbial enzymes: Their discovery by screening and use in large-scale production of useful chemicals in Japan. Current Opinion in Biotechnology, 13: 367–375.

Ogawa, J. and Shimizu, S. 2002. Industrial microbial enzymes: their discovery by screening and use in large-scale production of useful chemicals in Japan. Current Opinion in Biotechnology, 13: 367–75.

Oort, M.V. 2010. Enzymes in food technology-introduction. pp. 1–17. *In*: R.J. Whitehurst and M.V. Oort (eds.). Enzymes in Food Technology. Wiley Blackwell Publishers, Singapore.

Oslen, H.S. 2000. Enzymes at Work—A Concise Guide to Industrial Enzymes and Their Use. Novozymes A/S Bagsvaerd, Denmark.

Polakovic, M. and Bryjak, J. 2004. Modelling of potato starch saccharification by an *Aspergillus niger* glucoamylase. Biochemical Engineering Journal, 18: 57–64.

Poutanen, K. 1997. Enzymes: An important tool in the improvement of the quality of cereal foods. Trends in Food Science and Technology, 8: 300–306.

Rao, J.L.U.M. and Satyanarayana, T. 2007. Improving production of hyperthermostable and high maltose-forming-amylase by an extreme thermophile *Geobacillus thermoleovorans* using response surface methodology and its applications. Bioresource Technology, 98: 345–52.

Ravindran, V. and Son, Jang-Ho. 2011. Feed enzyme technology: Present status and future developments. Recent Patents in Food, Nutrition and Agriculture, 3: 102–109.

Rijssel, M.W., Gerwig, J.G.J. and Hausen, T.A. 1993. Isolation and characterization of an extracellular glycosylated protein complex from *Clostridium thermosaccharolyticum* with pectin methylesterase and polygalacturonate hydrolase activity. Applied & Environmental Microbiology, 59: 828–836.

Rodriguez, E., Han, Y. and Lei, X.G. 1999. Cloning, sequencing and expression of an *Escherichia coli* acid phosphatase, phytase gene *(app*A2*)* isolated from pig colon. Biochemical & Biophysical Research Communications, 257: 117–123.

Rothschild, L.J. and Manicinelli, R.L. 2001. Life in extreme environments. Nature, 409: 1092–1101.

Rouau, X., El-Hayek, M.L. and Moreau, D. 1994. Effect of an enzyme preparation containing pentosanases on the breadmaking quality of flours in relation to changes in pentosan properties. Journal of Cereal Science, 19: 259–272.

Rudiger, A., Jorgensen, P.L. and Antranikian, G. 1995. Isolation and characterization of a heat-stable pullulanase from the hyperthermophilic archaeon *Pyrococcus woesei* after cloning and expression of its gene in *Escherichia coli*. Applied & Environmental Microbiology, 61: 567–575.

Salleh, A.B., Razak, C.N.A., Rahman, R.N.Z.R.A. and Basri, M. 2006. Protease: Introduction. pp. 23–39. *In*: A.B. Salleh, R.N.Z.R.A. Rahman and M. Basri (eds.). New Lipases and Proteases, Nova Science Publishers, New York.

Sandri, I.G., Fontana, R.C., Barfknecht, D.M. and Da Silveira, M.M. 2012. Clarification of fruit juices by fungal pectinases. Food Science and Technology, 44: 2217–2222.

Schols, H., Geraeds, C., Searle-Van-Leeuwen, M., Komelink, F. and Voragen, A. 1990. Rhamnogalacturonase: A novel enzyme that degrades the hairy region of pectins. Carbohydrate Research, 206: 105–115.

Schwermann, B., Pfau, K., Liliensiek, B., Schleyer, M., Fischer, T. and Bakker, E.P. 1994. Purification, properties and structural aspects of the thermoacidophilic a-amylase from *Alicyclobacillus acidocaldarius* ATCC 27009. Insight into acidostability of proteins. European Journal of Biochemistry, 226: 981–991.

Selle, P.H. and Ravindran, V. 2007. Microbial phytase in poultry nutrition. Animal Feed Science and Technology, 135: 1–41.

Serour, E. and Antranikian, G. 2002. Novel thermoactive glucamylases from the thermoacidophilic Archaea *Thermoplasma acidophilum*, *Picrophilus torridus* and *Picrophilus oshimae*. Antonie van Leewenhock, 81: 73–83.

Shah, A.R., Shah, R.K. and Madamwar, D. 2006. Improvement of the quality of whole wheat bread by supplementation of xylanase from *Aspergillus foetidus*. Bioresource Technology, 97: 2047–2053.

Sharma, A. and Satyanarayana, T. 2010. High maltose-forming, Ca^{2+}-independent and acid stable α-amylase from a novel acidophilic bacterium *Bacillus acidicola* TSAS1. Biotechnology Letters, 32: 1503–1507.

Sharma, A. and Satyanarayana, T. 2012b. Production of acid-stable and high-maltose-forming α-amylase of *Bacillus acidicola* by solid-state fermentation and immobilized cells and its applicability in baking. Applied Biochemistry and Biotechnology, 168: 1025–1034.

Sharma, A. and Satyanarayana, T. 2013. Microbial acid stable amylases: characteristics, genetic engineering and applications. Process Biochemistry, 48: 201–211.

Sharma, A., Kawarabayasi, Y. and Satyanarayana, T. 2012a. Acidophilic bacteria and archaea: Acid-stable biocatalysts and their potential applications. Extremophiles, 16: 1–19.

Shen, G., Saha, B., Lee, Y., Bhatnagar, L. and Zeikus, J. 1988. Purification and characterization of a novel thermostable β-amylase from *Clostridium thermosulfurogenes*. Biochemical Journal, 254: 835–840.

Singh, B. and Satyanarayana, T. 2009. Characterization of HAP-phytase from a thermophilic mould *Sporotrichum thermophile*. Bioresource Technology, 100: 2046–2051.

Sorensen, J.F., Kragh, K.M., Sibbesen, O., Delcour, J., Goesaert, H., Svensson, B., Tahir, T.A., Brufau, J., Perez-Vendrell, A.M., Bellincampi, D., D'Ovidio, R., Camardella, L., Giovane, A., Bonnin, E. and Juge, N. 2004. Potential role of glycosidase inhibitors in industrial biotechnological applications. Biochimica et Biophysica Acta-Proteins and Proteomics, 1696: 275–287.

Srivastava, R.A.K. 1987. Purification and chemical characterization of thermostable amylase produced by *Bacillus stearothermophilus*. Enzyme & Microbial Technology, 9: 749–754.

Svensson, B., Larsen, K., Svendsen, I. and Boel, E. 1983. Amino acid sequence of tryptic fragments of glucoamylase G1 from *Aspergillus niger*. Carlsberg Research Communication, 48: 529–544.

Takebe, I., Otsuki, Y. and Aoki, S. 1968. Isolation of tobacco mesophyll cells in intact and active state. Plant and Cell Physiology, 9: 115.

Vaillant, F., Millan, P.O., Brien, G., Dornier, M., Decloux, M. and Reynes, M. 1999. Cross flow microfiltration of passion fruit juice after partial enzymatic liquefaction. Journal of Food Engineering, 42: 215–224.

Van Petegem, F., Collins, T., Meuwis, M.A., Gerday, C., Feller, G. and Van, Beeumen. J. 2003. The structure of a cold-adapted family 8 xylanase at 1.3 A resolution. Structural adaptations to cold and investigation of the active site. Journal of Biological Chemistry, 28: 7531–7539.

Viquez, F., Laetreto, C. and Cooke, R.D. 1981. A study of the production of clarified banana juice using pectinolytic enzymes. International Journal of Food Science and Technology, 16: 115–125.

Vohra, A. and Satyanarayana, T. 2002. Purification and characterization of a thermostable and acid-stable phytase from *Pichia anomala*. World Journal of Microbiology & Biotechnology, 18: 687–691.

Wang, S.J., Lu, Z.X., Lu, M.S., Qin, S., Liu, H.F., Deng, X.Y., Lin, Q. and Chen, J.N. 2008. Identification of archaeon producing hyperthermophilic α-amylase and characterization of the α-amylase. Applied Microbiology& Biotechnology, 80: 605–614.

Yano, J.K. and Poulos, T.L. 2003. New understandings of thermostable and piezostable enzymes. Current Opinion in Biotechnology, 14: 360–365.

Yusof, S. and Ibrahim, N. 1994. Quality of sour sop juice after pectinase enzyme treatment. Food Chemistry, 51: 83–88.

Zheng, Y., Xue, Y., Zhang, Y., Zhou, C., Schwaneberg, U. and Ma, Y. 2010. Cloning, expression and characterization of a thermostable glucoamylase from *Thermoanaerobacter tengcongensis* MB4. Applied Microbiology and Biotechnology, 87: 225–33.

Zverlov, V., Piotukh, K., Dakhova, O., Velikodvorskaya, G. and Borriss R. 1996. The multidomain xylanase A of the hyperthermophilic bacterium *Thermotoga neapolitana* is extremely thermo-resistant. Applied Microbiology and Biotechnology, 45: 245–247.

24

Production of Microbial Enzymes by Solid-state Fermentation for Food Applications

Quentin Carboué,* Marie-Stéphane Tranier,[a]
Isabelle Perraud-Gaime[b] and Sevastianos Roussos[c]

1. Introduction

Many studies have been published in recent years supporting the application of Solid-State Fermentation (SSF) in valorization of agricultural byproducts and in production of fine chemicals and enzymes. Solid-state processes are, therefore, of special economic interest for countries with an abundance of biomass and agro-industrial residues, as these can be used as cheap raw materials (Gombert et al., 1999; De la Cruz Quiroz et al., 2015). In addition to the feasibility of the process, SSF has several advantages concerning molecules production, especially enzymes, over submerged fermentation (SmF) (Panda et al., 2016). Indeed, numerous studies based on multiple hydrolases of industrial interest show the quantitative and qualitative advantages of SSF-produced enzymes over those produced by SmF (Acuña-Argüelles et al., 1995; Barrios-González, 2012). Indeed, hydrolases – EC 3 – represent the main group of enzymes used in the

Equipe d'Eco technologies et Bioremédiation, Aix Marseille Université; IMBE-UMR; CNRS-7263/IRD-237, Case 421, Campus Etoile, Faculté St Jérôme; 13397 Marseille cedex 20; France.
[a] E-mail: marie.tranier@gmail.com
[b] E-mail: isabelle.gaime-perraud@imbe.fr
[c] E-mail: sevastianos.roussos@imbe.fr
* Corresponding author: Quentin.carboue@imbe.fr

industry; they function without requiring the addition of cofactors and are massively secreted by filamentous fungi under certain conditions, as they feed on complex polymeric substrate. Concerning the differences between SSF and SmF, they originate due to the fact that the type of culture depends on complex physiological interactions between the microorganisms and the medium occurring during fermentation (Pandey, 2003). These differences occur at multiple levels—from the physical aspect of the microorganisms to its genetic regulation (Marzluf, 1997).

In this chapter, we briefly focus on SSF strategies for production of vital enzymes required for food applications.

2. World Demand for Enzymes

Enzyme technology is an interdisciplinary field recognized by the Organization for Economic Cooperation and Development (OECD) as an important component of sustainable industrial development. Its applications range from straightforward industrial processes to pharmaceutical discovery and development. Thus industrial enzymes represent the heart of biotechnology (Thomas et al., 2013). The world market for industrial enzymes was estimated at US\$ 625–700 million for 1989–1990. More recently it was estimated at US\$ 3,3 billion in 2010 and was expected to reach US\$ 4,4 billion by 2015 (Jaramillo et al., 2015). Amongst the various industrial sectors, 75 per cent of the industrial enzymes are hydrolases (Bhat, 2000). Proteases constitute one of the important groups accounting for about 60 per cent of the total enzyme utilization (Sawant and Nagendran, 2014). Cellulases account for approximately 20 per cent (Juwaied et al., 2011) while lipases account for one per cent (Mala and Takeuchi, 2008).

During the early stages, the commercial use of enzymes in the food industry was limited to a small number of applications, such as production of fermented food products upon the action of endogenous proteases under appropriate conditions. Today, enzymatic methods constitute an important and essential part of the processes used by the modern food industries to produce a large and diversified range of products for human consumption (Shahidi and Kamil, 2001). In 2010, the global enzyme market was dominated by the food and beverage industry, which has benefitted with the expansion of the middle class in rapidly developing economies. Indeed, consumer demand requires higher levels of quality in food in terms of natural flavor, taste, digestibility and nutritional value not only in the US and Europe, but also in the developing countries where consumption shifts away from staple sources of calories towards more demanding requirements. This trend triggered the need for development of enzymes applications in food processing. In the world enzyme market, food industry comes before household care, animal food and bioenergy, and is expected to continue its growth (Li et al., 2012; Miguel et al., 2013).

3. Solid State Fermentation (SSF)

Throughout history, enzyme technology is notably linked to development in the fermentation field. As a consequence, just like enzyme technology, SSF is a very ancient process primarily used to meet human needs. Typical examples of it are fermentation of rice by *Aspergillus oryzae* to initiate the koji process and *Penicillium roquefortii* for cheese production. In China, SSF has been used extensively to produce brewed food (Rodrìguez Coute and Sanromán, 2006); nowadays, SSF is widely used in the entire fermentation industry, particularly for the production of enzymes for which it holds tremendous potential (Pandey et al., 1999). SSF is a process involving microbial growth on the surface and inside a porous matrix in the absence or near-absence of free liquid in the inter-particle volume (Lonsane et al., 1985; Hesseltine, 1987; Raghavarao et al., 2003). The low moisture content means that fermentation can only be carried out by a limited number of microorganisms, mainly yeasts and fungi, although some bacteria have also been used (Rodrìguez Coute and Sanromán, 2006). In addition, concerning the inoculation, spores are usually preferred over vegetative or mycelial cells in the SSF system due to the ease in mixing the inoculum with autoclaved moist solids (Roussos et al., 1991). Even though the free liquid runoff is very limited, water is one the most important factors to consider (Oriol et al., 1988). It is indeed essential for microbial growth to occur and its role in biological systems are numerous. It allows, for example, stability and function of the biological structures organized at the molecular and cellular levels. In addition, solutes as well as dissolved-gas transfers take place in the aqueous film surrounding the microorganisms. So, even in SSFs, the microorganisms are in a liquid medium (Graminha et al., 2008). Indeed, solute diffusion in the substrate must occur in the liquid phase and gaseous diffusion in the substrate can occur in the liquid as well as in the gaseous phase (Gervais and Molin, 2003). In the near-absence of free running water, its presence depends only on the water-retention abilities of the medium (Manpreet et al., 2005). The osmotic gradient due to heterogeneous distribution of solutes and adsorption forces can be recognized as key factors in SSF. More specifically, the low level of water activity (a_w) of the solid substrate has a significant effect on the physiological activity of microorganisms and enzyme production (Antier et al., 1993).

The medium is also an important factor to be taken into account following two considerations—as substrate, it has to efficiently cater to microbial nutritional needs and as a support of the culture, it has to possess favorable physical properties (Pandey et al., 2000). Smaller substrate particles provide larger surface area for microbial attack and, thus, are a desirable factor. However, too small substrate particles may result in substrate accumulation, which may interfere with gaseous exchanges and hence in microbial respiration. Therefore, the result is poor growth. In contrast, larger particles provide better medium aeration efficiency due to increased inter-particle space, but provide limited surface for microbial attack. This necessitates a compromised particle size for a particular process (Pandey et al., 1999; Guan et al., 2014). Following the same logic, if the moisture content is too high, the void spaces in the solid are filled with water, resulting in oxygen limitation. On the other extreme, if the moisture content is too low, microorganism growth will be hindered (Vitcosque et al., 2012).

There is a distinction between the two types of medium used in SSF:

- It can be synthetic and work as a support only. Therefore it must be complemented with nutritive substances or
- Be of natural origin, working both as support and substrate and in this case, it's not an obligation for it to be complemented with nutritive supplements

Generally, SSF uses natural products – typically starch or ligno-cellulose-based agricultural products, which may have a disadvantage. The carbon source constitutes part of their structure. During the growth of the microorganisms, the solid medium is degraded and as a result, the geometric and physical characteristics of the medium change. Consequently, heat and mass transfer can be reduced (Ooijkaas et al., 2000). Mass transfer includes nutrient consumption and gaseous exchanges. Most of the processes being aerobic, the limitation of oxygen after its uptake during the exponential phase may limit the growth at the industrial scale if the gaseous phase inside the inter-particle void is not regenerated (Rajagopalan and Modak, 1995).

Concerning the heat transfer, a large amount of metabolic heat is generated during SSF and its rate is directly proportional to the level of metabolic activity in the system (Liu, 2013). Heat transfer in SSF reactors is not as efficient as in SmF, mainly due to the solidness of the substrate—the solid matrix has a low heat conductivity—and the lack of free water available during fermentation, since the thermal conductivity of air is very poor compared to water (Jou and Lo, 2011; Robinson and Nigam, 2003). As a consequence, heat removal is one of the major constraints in SSF, especially in large-scale processes (Figueroa-Montero et al., 2011). Numerous strategies to overcome this issue have been adopted, from which a wide diversity of bioreactor designs have emerged: static or stirred, aerated or not (Roussos et al., 1993; Durand, 2003). Finally, using a complex medium, SSF processes rarely yield to purified compounds; instead molecule extraction yields crude complexes. In the case of industrial utilization, its particularity not a drawback: indeed, industrial utilization of purified enzymes is not economically justified. In this case, the use of a crude enzyme is most adequate, bearing in mind the obviously lower commercial value for industrial use. The effective dosage of crude enzyme costs about one per cent of the cost of pure enzymes (Leite et al., 2008).

Some common enzymes used in the food industry and produced under SSF are given in Table 1.

4. Comparison Between SSF and SmF

In order to fully understand the strong points of this process, it is useful to compare SSF with its liquid equivalent. The advantages are based on two main aspects—first, an economical aspect using agricultural byproducts, which are naturally rich in carbohydrates and other nutrients and produced in excess all over the world, as the culture medium in SSF is cheaper than using the synthetic liquid substrate (Rodrìguez Coute and Sanromán, 2006; Balkan and Ertan, 2007). Some of the substrates include sugarcane bagasse, wheat bran, rice bran, maize bran, gram bran, wheat straw, rice straw, rice husk, soy hull, corncobs, banana waste, tea waste, cassava waste, apple pomace, etc. to name only a few. Wheat bran, however, is the most commonly used substrate (Sangeetha et al., 2004). Moreover, the absence of free water reduces liquid volume used for the process and the downstream effluent treatment costs (Ghosh et al., 2013). Economic analysis of the production of lipase by *Penicillium restrictum* in SSF and SmF cultures showed that the investment capital necessary for SmF was 78 per cent higher than that required for SSF. Consequently, the price of the SSF product was

Table 1. Some example microbial enzymes commonly used in food processing produced by SSF.

Enzyme	Food Industries	Microorganism	By-products	References
Amylase	Sugar industry	*Penicillium chrysogenum*	Corncob leaf, rye straw, ye straw, wheat bran	Balkan and Ertan (2007)
B-galactosidase	Milk industry	*Aspergillus oryzae*	Wheat bran and rice husk	Nizamuddin et al. (2008)
Cellulase	Wine and brewery industry, fruit and vegetables juice	*Trichoderma reesei* *Trichoderma viride*	Wheat bran Sugar cane bagass	Singhania et al. (2007) Juwaied et al. (2011) Reddy et al. (2003)
		Pleurotus sp.	Banana wastes (leaf biomass and pseudostems) Grape pomace	
		Aspergillus awamori *Aspergillus niger*	Kitchen waste residues (corn cobs, carrot peelings, composite, grass, leaves, orange peelings, pineapple peelings, potato peelings, rice husk, sugarcane baggage, saw dust, wheat bran, wheat straw)	Botella et al. (2005) Bansal et al. (2012)
Chitinase	Food preservatives	*Beauveria bassiana* *Metarhizium anisopliae* *Aspergillus terreus*	Prawn waste Silkworm chrysalis Parrot fish scales waste	Suresh and Chandrasekaran (1998) Barbosa Rustiguel et al. (2012) Ghanem et al. (2013)
Fructosyl transferase	Prebiotic	*Aspergillus oryzae* *Bacillus subtilis*	Rice bran, wheat bran Starch	Sangeetha et al. (2004) (Esawy et al. (2013)
Lipases	Food additives	*Penicillium restrictum*	Babassu oil cake	Gombert et al. (1999)
Pectinase	Wine and brewery industry, fruit and vegetables juice, oil	*Aspergillus niger*	Coffee pulp Wheat bran and soy bran	Antier et al. (1993) Castilho et al. (2000)
Protease	Fruit and vegetables juice, coffee extraction	*Aspergillus oryzae* *Pleurotus ostreatus* and *Trametes versicolor* *Aspergillus oryzae*	De-oiled seedcakes from *Jatropha curcas* Tomato pomace Canola Cake	Thanapimmetha et al. (2012) Iandolo et al. (2011) Freitas et al. (2013)
Rennet	Cheese industry	*Mucor miehei*	Complemented wheat bran	Thakur et al. (1990)
Xylanase	Fruit and vegetables juice, coffee extraction	*Aspergillus niger* *Scytalidium thermophilum*	De-oiled seedcakes from *Jatropha curcas*	Ncube et al. (2012) Joshi and Khare (2011)

47 per cent lower. The investment return from the SSF process can reach 68 per cent within five years (Graminha et al., 2008). This economical advantage also induces an important concept: valorization. Indeed, the agricultural industry produces important amounts of byproducts whose main part is unused and raises environmental issues (Arumugam et al., 2014). SSF allows the production of value-added molecules from unused resources, thus decreasing the pollution volume and its toxicity, as is the case of *Jatropha curcas* byproducts. The seeds of this plant contain high amounts of oil containing triglycerides which, after certain treatments, including transesterification, lead to the formation of biodiesel (Mazumdar et al., 2013). The byproduct is a de-oiled seedcake which is highly toxic due to the presence of phorbol esters and other anti-nutrients. If, left to decay, its accumulation would lead to environmental problems. However, its composition – 60 per cent protein, 0,6 per cent fat, 9 per cent ash, 4 per cent fiber and 26 per cent carbohydrates – makes its utilization with the right strain suitable for numerous SSF processes (Joshi and Khare, 2011; Kumar and Kanwar, 2012; Ncube et al., 2012; Thanapimmetha et al., 2012). The second main advantage of SSF concerns qualitative and quantitative aspects of production of valuable molecules. Indeed, it has been shown that certain molecules produced under SSF often have different and more interesting properties than the ones obtained with SmF. Some of them includes, for example, more thermostable or show higher pH stability (Mateos-Diaz et al., 2006; Mienda et al., 2011; Barrios-González, 2012; Saqib et al., 2012). In their study, Acuña-Argüelles et al. (1995) produced pectinases with both SSF and Smf. According to the electrophoretic patterns of proteins, the culture method appeared to induce differences in the mobility of pectinases, besides variations between the substrate affinities (K_m) for some pectinases. They explained that this difference is related to the various levels of glycosylation – the attachment of the sugar molecule oligosaccharide, known as glycan, to an amide nitrogen atom of a protein – induced by the cultural method, which changed pectinolytic activities and also thermostability. Later studies show that glycosylation could improve the protein thermostability (Zhu et al., 2014). The yields are higher as well. But it is true for a majority of molecules in enzymes that a greater quantity of fungal exoenzymes accumulated when the growth occurred under SSF than under SmF (Hernandez et al., 1992; Kar et al., 2013). This may be explained by the fact that the same observation can be made concerning the fungal biomass, which is remarkably higher, during the same incubation time, when produced by SSF. The morphological difference between biofilm and pellet growth indicates difference in physical structure and also implies dissimilar mass transfer patterns. Indeed, by comparing the area to volume ratio, it appears that SSF cultures have a much larger air to liquid interphase than conventional SmF pellets and are thus more likely exposed to passive gas exchanges (Viniegra-González et al., 2003). Furthermore, some studies have highlighted the variation of proteome, indicating a differential gene expression ruled by physical interactions between the fungus and its support (Gamarra et al., 2010). It has also been shown that SSF reduces the level of catabolite repression or end-product feedback repression compared to the SmF system; yet the exact reason is not fully understood. This is cost-effective over liquid fermentation, which needs to use lower substrate concentrations in the fermentation media or use cost-intensive fermentation operation strategies in order to overcome catabolite repression (Nandakumar et al., 1999). For example, in a homogeneous liquid

medium, *Aspergillus niger*'s endo- and exopectinases' synthesis is repressed if the medium contains glucose (3 per cent). Conversely, in SSF endo- and exopectinases' synthesis is not affected for a medium much richer in glucose (upto 10 per cent). Pectinases' activities even increase when pectin is added to the medium (Solís-Pereira et al., 1993).

5. Bioreactors

As said before, in bioreactor models, the two most important environmental variables are temperature and water activity of the bed and both of them are intimately tied to the metabolic activity of the microorganism. Generally, literature describes many kinds of SSF bioreactors, depending on whether air is forcefully blown into the bed and/or the substrate is agitated: some of them include tray, packed bed, rotating drum, gas–solid fluidized bed, stirred aerated bed, and rocking drum bioreactors (Jou and Lo, 2011). Furthermore, as an endothermic process, water evaporation is really an efficient method to overcome heat generation. Thus aeration with water-saturated air may be used to avoid dryness in the cultural medium and regulate the heat generated during the growth (Saucedo-Castañeda, 1994; Utpat et al., 2014). With regard to mixing three strategies are available: the substrate bed may either be left static or mixed only very infrequently or it might be mixed reasonably frequently with static periods in between, or alternatively it might be mixed continuously (Mitchell et al., 2000). Each of them has been successfully used to produce enzymes.

The first categories of bioreactors are referred as static bioreactors as they are mixed only very infrequently – once or twice a day. The tray bioreactor, also known as koji bioreactor, is the most widely used bioreactor for SSF. It also has the simplest design: it consists of substrate placed on a tray placed in a room where atmosphere is controlled (Rosales et al., 2007; Brijwani et al., 2010). As there is no forced aeration in the medium, mass and heat transfers occur by natural diffusion and convection. Therefore, the substrate thickness on the tray is the major limitation parameter (Figueroa-Montero et al., 2011).

Rodrìguez Coute et al. (2003) have shown that on barley bran as medium, the tray configuration led to the highest laccase activity when compared to packed-bed bioreactor. However, for a better monitoring, other bioreactors may be used. In addition, Raghavarao et al. (1993) determined the critical height of substrate on a perforated bottom tray as 4.78 cm; over this height, the oxygen concentration may fall to zero during fermentation. Such a value suggests that an important surface must be used in order to achieve industrial-scale production when using this technique. The packed-bed has advantages over the tray because the forced aeration allows better control of environmental conditions in the bed, due to the ability to manipulate the temperature and flow rate of the process air (Mitchell and von Meien, 2000). Glass columns, also known as 'Raimbault column', are the typical packed-bed bioreactor at laboratory scale (Rodríguez-León et al., 2013). They have a perforated base plate on the bottom which allows the air flow and can be oriented vertically or horizontally. Castro et al. (2015), in order to overcome the scale-up limitations of conventional tray-type bioreactors, used a cylindrical fixed-bed bioreactor with forced aeration – with a working volume of 1,8 1 – for the production of a pool of industrially relevant enzymes by *Aspergillus awamori*, using babassu cake as the raw material. They reported that despite significant temperature gradients, it was possible to obtain good titers with production of the six enzyme groups evaluated: exo-amylases, endo-amylases, proteases, xylanases and cellulases.

Another category consists of bioreactors in which the bed is continuously mixed or mixed intermittently with a frequency of minutes to hours and air is circulated around the bed, but not blown forcefully through it. Two bioreactors that have this mode of operation, using different mechanisms to achieve the agitation, are 'rotating drum bioreactors' and 'stirred drum bioreactors'. In rotating drum bioreactors, the drum is partially filled with a bed of substrate and the whole drum rotates around its central axis to mix the bed. In stirred drums, the bioreactor body remains stationary with paddles or scrapers mounted on a shaft running along the central axis of the bioreactor rotating within the drum (Mitchell et al., 2006). Such bioreactors increase mass and heat transfers by improving homogeneity of the bed, using agitation of the medium. This kind of a process is relevant only if the fermentation is carried with a microorganism resistant to damages from the shear forces caused by agitation and able to grow on a substrate with lower porosity after compaction of the medium (Fujian et al., 2002). Typically, fungi with non-septate hyphae – like zygomycetes – are less resistant to agitation (Durand, 2003). This issue may be overcome if the mixing periods are relatively short, as in the case of intermittently mixed bioreactors, since many fungal processes can tolerate infrequent mixing (von Meien et al., 2004).

The last category includes bioreactors with both aeration and agitation inside. The agitated bioreactor is generally preferred on an industrial scale because it maximizes the mass and heat transfers. A forcefully aerated and intermittently mixed rotating-drum bioreactor, named *FERMSOSTAT*, utilizing the following agitation parameters—5 minutes of agitation at 0.5 rotation per minute every day—showed promising results in the production of cellulases (Lee et al., 2011). This category also includes the gas-solid fluidized bed, in which solid substrate is placed on a porous plate or metal net and sterilized air is blown in under the plate. When the airflow rate is high enough, solid substrate particles get suspended in the gas phase, allowing a very good rate of heat and mass transfer (Cen and Xia, 1999). Amylases and proteases produced from *Eupenicillium javanicum* under fluidized culture were reported to have activities two times higher than that obtained in a stationary culture (Tanaka et al., 1986). However fluidization requires the use of fine particles and the minimum air velocity for fluidization is high in high-power requirements. Therefore, large, coarse and sometimes sticky particles, which usually appear in SSF are difficult to fluidize. A spouted bed requires a lower gas-flow rate and may provide good mass and heat transfer for solid materials that are too coarse or dense for stable fluidization (Wang and Yang, 2007). Silva and Yang (1998) have shown that amylase activities from *A. oryzae* grown on rice were equivalent when a packed bed bioreactor and gas-solid spouted-bed bioreactor with intermittent spouting of air were used and these activities were higher than those obtained in a tray-type reactor. However, the fermented rice obtained from the gas-solid spouted-bed bioreactor was homogeneously fermented and did not show rice aggregates knitted together with fungal mycelia, as in the packed bed bioreactor.

Continuous SSF also exists. In this case, the solid phase within the bioreactor can be assimilated to a flow wherein there is no heat and gaseous gradient. However, there are specific challenges to operate continuous SSF bioreactors and this process is currently scarcely used in the industry. Consequently, batch production is more common (Ramos Sánchez et al., 2015).

6. Techno-economic Feasibility of Enzyme Production in SSF and Limitations

Despite all these advantages – the biggest one being enzymes produced in higher levels in SSF – SmF remains, in the vast majority of cases, the preferable method to produce molecules on an industrial scale (Mitchell et al., 2006). Indeed, SSF is still suffering from lack of industrial-scale bioreactors (Farias et al., 2015; Ramos Sánchez et al., 2015). SSF up-scaling, necessary for use on an industrial scale, raises severe engineering problems due to the build up of temperature, pH, oxygen, substrate and moisture gradients (Hölker et al., 2004). On the contrary, in SmF, the media is made up essentially of water. In this environment, the temperature and pH regulation are trivial and pose no problem during the scaling-up of the process. The only one major difficulty encountered is the transfer of oxygen to the microorganisms; this can easily be overcome on an industrial scale with an adapted shape of the bioreactor and the presence of an agitation/aeration system, while in the SSF, the system is multiphasic and thus more complex. So, above some critical quantity of substrate, the metabolic heat removal becomes difficult to solve and restricts the design strategies that are available. The solid medium becomes compacted or creates air channeling, leading to a system with inefficient heat and mass transfers (Durand, 2003).

Various possibilities may be explored to overcome the problems encountered during the up-scaling of the process, like increasing the dimensions of the bioreactor. In this case, the most adapted process is forced aeration combined with agitation. Indeed, aeration efficiently removes metabolic heat with water evaporation and the agitation in association with sprayed water maintains the water activity homogeneous everywhere in the substrate. However, as dimensions increase, maintaining asepsis for efficiency of the process – sterilization being a process that requires important handling – may be difficult. Another possibility is to reduce the maintenance operations as it is the case for the *FMS-unique* bioreactor, a single-use packed-bed bioreactor which can process 5 kg of dry substrate to autoclave directly inside the bioreactor. In this case, the relatively low quantity of fermented substrate enables aeration to efficiently counterbalance the metabolic heat generation. The small amount of product can also be offset by the number of easily operating bioreactors working in parallel (Roussos et al., 2014).

By the way, examples of industrial production of enzymes using SSF and which are economically competitive do exist. Japan is one of the major actors in this domain. Indeed, it is a very large producer of soya sauce and SSF is an integral part of the sauce production process. Therefore, this particular industry has enabled the country to make technological advances in the field of SSF bioreactors, consequently driving the other industrial SSF processes, such as enzyme production (Suryanarayan, 2003).

7. Conclusion

SSF is successfully employed in enzyme production. The increased production of many induced enzymes using solid-state fermentation is related to a significant resistance of organisms cultivated in a solid medium to catabolic repression (Stoykov et al., 2015). Furthermore, the enzymes produced possess interesting properties, such as thermostability, which can be particularly valuable in various industries (Zamost et al., 1991; Mateos-Diaz et al., 2006). Indeed there is evolution of higher fungi on solid substrates, with a wide majority of these known fungi being terrestrial. As a result, the cultivation of such fungi in an aqueous medium doesn't optimize their metabolism and yield (Hölker et al., 2004). Many studies can be found in literature relating to successful enzyme productions using SSF. However, SmF remains the most widely used process to produce enzyme on larger scales, but as the engineering improvements progress, SSF, driven by environment-friendly and economic considerations, should create a growing enthusiasm in industries to become the next common way to produce enzymes on an industrial level in the future.

References

Acuña-Argüelles, M.E., Gutiérrez-Rojas, M., Viniegra-González, G. and Favela-Torres, E. 1995. Production and properties of three pectinolytic activities produced by *Aspergillus niger* in submerged and solid-state fermentation. Applied Microbiology and Biotechnology, 43(5): 808–814.

Antier, P., Minjares, A., Roussos, S., Raimbault, M. and Viniegra-Gonzalez, G. 1993. Pectinase hyperproducing mutants of *Aspergillus niger* C28B25 for solid-state fermentation of coffee pulp. Enzyme and Microbial Technology, 15(3): 254–260.

Arumugam, G.K., Selvaraj, V., Gopal, D. and Ramalingam, K. 2014. Solid-State Fermentation of Agricultural Residues for the Production of Antibiotics. pp. 139–162. *In*: S.K. Brar, G.S. Dhillon and C.R. Soccol (eds.). Biotransformation of Waste Biomass into High Value Biochemicals. Springer Berlin-Heidelberg, Berlin, Germany.

Balkan, B. and Ertan, F. 2007. Production of α-amylase from *Penicillium chrysogenum* under solid state fermentation by using some agricultural by-products. Food Technology and Biotechnology, 45(4): 439–442.

Bansal, N., Tewari, R., Soni, R. and Soni, S.K. 2012. Production of cellulases from *Aspergillus niger* NS-2 in solid state fermentation on agricultural and kitchen waste residues. Waste Management, 32(7): 1341–1346.

Barbosa Rustiguel, C., Atílio Jorge, J. and Henrique Souza Guimarães, L. 2012. Optimization of the chitinase production by different *Metarhizium anisopliae* strains under solid-state fermentation with silkworm chrysalis as substrate using CCRD. Advances in Microbiology, 2(3): 268–276.

Barrios-González, J. 2012. Solid-state fermentation: Physiology of solid medium, its molecular basis and applications. Process Biochemistry, 47: 175–185.

Bhat, M.K. 2000. Cellulases and related enzymes in biotechnology. Biotechnology Advances, 18(5): 355–383.

Botella, C., De Ory, I., Webb, C., Cantero, D. and Blandino, A. 2005. Hydrolytic enzyme production by *Aspergillus awamori* on grape pomace. Biochemical Engineering Journal, 26(2): 100–106.

Brijwani, K., Oberoi, H.S. and Vadlani, P.V. 2010. Production of a cellulolytic enzyme system in mixed-culture solid-state fermentation of soybean hulls supplemented with wheat bran. Process Biochemistry, 45(1): 120–128.

Castilho, L.R., Medronho, R.A. and Alves, T.L. 2000. Production and extraction of pectinases obtained by solid state fermentation of agro-industrial residues with *Aspergillus niger*. Bioresource Technology, 71(1): 45–50.

Castro, A.M., Castilho, L.R. and Freire, D.M.G. 2015. Performance of a fixed-bed solid-state fermentation bioreactor with forced aeration for the production of hydrolases by *Aspergillus awamori*. Biochemical Engineering Journal, 93: 303–308.

Cen, P. and Xia, L. 1999. Production of cellulase by solid-state fermentation. pp. 70–91. *In*: T. Scheper and G.T. Tsao (eds.). Recent Progress in Bioconversion of Lignocellulosics. Springer Berlin-Heidelberg, Berlin, Germany.

De la Cruz Quiroz, R., Roussos, S., Hernández, D., Rodríguez, R., Castillo, F. and Aguilar, C.N. 2015. Challenges and opportunities of the bio-pesticides production by solid-state fermentation: Filamentous fungi as a model. Critical Reviews in Biotechnology, 35: 326–333.

Durand, A. 2003. Bioreactor designs for solid state fermentation. Biochemical Engineering Journal, 13: 113–125.

Esawy, M.A., Abdel-Fattah, A.M., Ali, M.M., Helmy, W.A., Salama, B.M., Taie, H.A.A., Hashem, A.M. and Awad, G.E.A. 2013. Levan sucrase optimization using solid state fermentation and levan biological activities studies. Carbohydrate Polymers, 96: 332–341.

Farias, C.M., de Souza, O.C., Souza, M.A., Cruz, R., Magalhães, O.M.C., de Medeiros, E.V., Moreira, K.A. and de Souza-Motta, M. 2015. High-level lipase production by *Aspergillus candidus* URM 5611 under solid state fermentation (SSF) using waste from *Siagruscoronata* (Martius) Becari. African Journal of Biotechnology, 14: 820–828.

Figueroa-Montero, A., Esparza-Isunza, T., Saucedo-Castañeda, G., Huerta-Ochoa, S., Gutiérrez Rojas, M. and Favela-Torres, E. 2011. Improvement of heat removal in solid-state fermentation tray bioreactors by forced air convection. Journal of Chemical Technology and Biotechnology, 86(10): 1321–1331.

Freitas, A.C., Castro, R.J.S., Fontenele, M.A., Egito, A.S., Farinas, C.S. and Pinto, G.A.S. 2013. Canola cake as a potential substrate for proteolytic enzymes production by a selected strain of *Aspergillus oryzae*: Selection of process conditions and product characterization. ISRN Microbiology, 2013: 1–8.

Fujian, X., Hongzhang, C. and Zuohu, L. 2002. Effect of periodically dynamic changes of air on cellulase production in solid-state fermentation. Enzyme and Microbial Technology, 30(1): 45–48.

Gamarra, N.N., Villena, G.K. and Gutiérrez-Correa, M. 2010. Cellulase production by *Aspergillus niger* in biofilm, solid-state and submerged fermentations. Applied Microbiology and Biotechnology, 87: 545–551.

Gervais, P. and Molin, P. 2003. The role of water in solid-state fermentation. Biochemical Engineering Journal, 13: 85–101.

Ghanem, K.M., Al-Garni, S.M. and Al-Makishah, N.H. 2013. Statistical optimization of cultural conditions for chitinase production from fish scales waste by *Aspergillus terreus*. African Journal of Biotechnology, 9(32): 5135–5146.

Ghosh, S., Murthy, S., Govindasamy, S. and Chandrasekaran, M. 2013. Optimization of Lasparaginase production by *Serratiamarcescens* (NCIM 2919) under solid state fermentation using coconut oil cake. Sustainable Chemical Processes, 1: 1–9.

Gombert, A.K., Pinto, A.L., Castilho, L.R. and Freire, D.M. 1999. Lipase production by *Penicillium restrictum* in solid-state fermentation using babassu oil cake as substrate. Process Biochemistry, 35: 85–90.

Graminha, E.B.N., Gonçalves, A.Z.L., Pirota, R.D.P.B., Balsalobre, M.A.A., Da Silva, R. and Gomes, E. 2008. Enzyme production by solid-state fermentation: Application to animal nutrition. Animal Feed Science and Technology, 144: 1–22.

Guan, J., Yang, G., Yin, H., Jia, F. and Wang, J. 2014. Particle size for improvement of peptide production in mixed-culture solid-state fermentation of soybean meal and the corresponding kinetics. American Journal of Agriculture and Forestry, 2(1): 1–6.

Guimarães, L.H.S., Peixoto-Nogueira, S.C., Michelin, M., Rizzatti, A.C.S., Sandrim, V.C., Zanoelo, F.F., Aquino, A.C.M.M., Altino, J.B., de Lourdes, M. and Polizeli, M.D.L. 2006. Screening of filamentous fungi for production of enzymes of biotechnological interest. Brazilian Journal of Microbiology, 37(4): 474–480.

Hernandez, M.R.T., Raimbault, M., Roussos, S. and Lonsane, B.K. 1992. Potential of solid state fermentation for production of ergot alkaloids. Letters in Applied Microbiology, 15: 156–159.

Hesseltine, C.W. 1987. Solid state fermentation—An overview. International Biodeterioration, 23: 79–89.

Hölker, U., Höfer, M. and Lenz, J. 2004. Biotechnological advantages of laboratory-scale solid state fermentation with fungi. Applied Microbiology and Biotechnology, 64(2): 175–186.

Iandolo, D., Piscitelli, A., Sannia, G. and Faraco, V. 2011. Enzyme production by solid substrate fermentation of *Pleurotusostreatus* and *Trametes versicolor* on tomato pomace. Applied Biochemistry and Biotechnology, 163(1): 40–51.

Jaramillo, P.M.D., Gomes, H.A.R., Monclaro, A.V., Silva, C.O.G. and Filho, E.X.F. 2015. Lignocellulose-degrading enzymes: An overview of the global market. pp. 73–85. *In*: V.K. Gupta, R.L. Mach and S. Sreenivasaprasad (eds.). Fungal Biomolecules: Sources, Applications and Recent Developments. John Wiley & Sons, Ltd., Chichester, UK.

Joshi, C. and Khare, S.K. 2011. Utilization of deoiled *Jatrophacurcas* seed cake for production of xylanase from thermophilic *Scytalidium thermophilum*. Bioresource Technology, 102(2): 1722–1726.

Jou, R.Y. and Lo, C.T. 2011. Heat and mass transfer measurements for tray-fermented fungal products. International Journal of Thermophysics, 32: 523–536.

Juwaied, A.A., Al-Amiery, A.A.H., Abdumuniem, Z. and Anaam, U. 2011. Optimization of cellulase production by *Aspergillus niger* and *Tricoderma viride* using sugarcane waste. Journal of Yeast and Fungal Research, 2(2): 19–23.

Kar, S., Sona Gauri, S., Das, A., Jana, A., Maity, C., Mandal, A., Das Mohapatra, P.K., Pati, B.R. and Mondal, K.C. 2013. Process optimization of xylanase production using cheap solid substrate by *Trichoderma reesei* SAF3 and study on the alteration of behavioral properties of enzymes obtained from SSF and SmF. Bioprocess and Biosystems Engineering, 36(1): 57–68.

Kumar, A. and Kanwar, S.S. 2012. Lipase production in solid-state fermentation (SSF): Recent developments and biotechnological applications. Dynamic Biochemistry, Process Biotechnology and Molecular Biology, 6(1): 13–27.

Lee, C.K., Darah, I. and Ibrahim, C.O. 2011. Production and optimization of cellulase enzyme using Aspergillus niger USM AI 1 and comparison with Trichoderma reesei via solid state fermentation system. Biotechnology Research International, Volume 2011, Article ID 658493, 6 pages.

Lee, C.K., Darah, I. and Ibrahim, C.O. 2010. Production and optimization of cellulase enzyme using *Aspergillus niger* USM AI 1 and comparison with *Trichoderma reesei* via solid state fermentation system. Biotechnology Research International, 2011: 1–6.

Leite, R.S.R., Alves-Prado, H.F., Cabral, H., Pagnocca, F.C., Gomes, E. and Da-Silva, R. 2008. Production and characteristics comparison of crude β-glucosidases produced by microorganisms *Thermoascus aurantiacus* and *Aureobasidium pullulans* in agricultural wastes. Enzyme and Microbial Technology, 43: 391–395.

Li, S., Yang, X., Yang, S., Zhu, M. and Wang, X. 2012. Technology prospecting on enzymes: Application, marketing and engineering. Computational and Structural Biotechnology Journal, 2(3): 1–11.

Liu, L. 2013. Process engineering of solid-state fermentation. pp. 55–94. *In*: J. Chen and Y. Zhu (eds.). Solid State Fermentation for Foods and Beverages. CRC Press, Boca Raton, Florida, USA.

Lonsane, B.K., Ghildyal, N.P., Budiatman, S. and Ramakrishna, S.V. 1985. Engineering aspects of solid state fermentation. Enzyme and Microbial Technology, 7: 258–265.

Mala, J.G.S. and Takeuchi, S. 2008. Understanding structural features of microbial lipases—An overview. Analytical Chemistry Insights, 3: 9–19.

Manpreet, S., Sawraj, S., Sachin, D., Pankaj, S. and Banerjee, U.C. 2005. Influence of process parameters on the production of metabolites in solid-state fermentation. Malalaysian Journal of Microbiology, 1(2): 1–9.

Marzluf, G.A. 1997. Genetic regulation of nitrogen metabolism in the fungi. Microbiology and Molecular Biology Reviews, 61: 17–32.

Mateos-Diaz, J.C., Rodriguez, J.A., Roussos, S., Cordova, J., Abousalham, A., Carriere, F. and Baratti, J. 2006. Lipase from the thermotolerant fungus *Rhizopus homothallicus* is more thermostable when produced using solid state fermentation than liquid fermentation procedures. Enzyme and Microbial Technology, 39(5): 1042–1050.

Mazumdar, P., Dasari, S.R., Borugadda, V.B., Srivasatava, G., Sahoo, L. and Goud, V.V. 2013. Biodiesel production from high free fatty acids content *Jatropha curcas* L. oil using dual step process. Biomass Conversion and Biorefinery, 3(4): 361–369.

Miguel, A.S.M., Souza, T., Costa Figueiredo, E.V. da, Paulo Lobo, B.W. and Maria, G. 2013. Enzymes in bakery: Current and future trends. pp. 287–321. *In*: I. Muzzalupo (ed.). Food Industry. InTech.

Mienda, B.S., Idi, A. and Umar, A. 2011. Microbiological features of solid state fermentation and its applications—An overview. Research in Biotechnology, 2(6): 21–26.

Mitchell, D.A. and von Meien, O.F. 2000. Mathematical modeling as a tool to investigate the design and operation of the zymotis packed-bed bioreactor for solid-state fermentation. Biotechnology and Bioengineering, 68: 127–135.

Mitchell, D.A., Krieger, N., Stuart, D.M. and Pandey, A. 2000. New developments in solid-state fermentation. Process Biochemistry, 35: 1211–1225.

Mitchell, D.A., Krieger, N. and Berovič, M. 2006. Introduction to solid-state fermentation bioreactors. pp. 33–43. *In*: D.A. Mitchell, N. Krieger and Berovič (eds.). Solid-State-Fermentation Bioreactors. Springer Berlin-Heidelberg, Berlin, Germany.

Nandakumar, M.P., Thakur, M.S., Raghavarao, K.S.M.S. and Ghildyal, N.P. 1999. Studies on catabolite repression in solid state fermentation for biosynthesis of fungal amylases. Letters in Applied Microbiology, 29(6): 380–384.

Ncube, T., Howard, R.L., Abotsi, E.K., van Rensburg, E.L.J. and Ncube, I. 2012. *Jatropha curcas* seed cake as substrate for production of xylanase and cellulase by *Aspergillus niger* FGSCA733 in solid-state fermentation. Industrial Crops and Products, 37(1): 118–123.

Nigam, P. and Singh, D. 1994. Solid-state (substrate) fermentation systems and their applications in biotechnology. Journal of Basic Microbiology, 34(6): 405–423.

Nizamuddin, S., Sridevi, A. and Narasimha, G. 2008. Production of β-galactosidase by *Aspergillus oryzae* in solid-state fermentation. African Journal of Biotechnology, 7(8): 1096–1100.

Ooijkaas, L.P., Weber, F.J., Buitelaar, R.M., Tramper, J. and Rinzema, A. 2000. Defined media and inert supports: Their potential as solid-state fermentation production systems. Trends in Biotechnology, 18(8): 356–360.

Oriol, E., Raimbault, M., Roussos, S. and Viniegra-Gonzales, G. 1988. Water and water activity in the solid state fermentation of cassava starch by *Aspergillus niger*. Applied Microbiology and Biotechnology, 27: 498–503.

Otero, J.M., Panagiotou, G. and Olsson, L. 2007. Fueling industrial biotechnology growth with bioethanol. pp. 1–40. *In*: L. Olsson (ed.). Biofuels. Springer Berlin-Heidelberg, Berlin, Germany.

Panda, S.K., Mishra, S.S., Kayitesi, E and Ray, R.C. 2016. Microbial-processing of fruit and vegetable wastes for production of vital enzymes and organic acids: Biotechnology and scopes. Environmental Research, 146: 161–172.

Pandey, A. 2003. Solid-state fermentation. Biochemical Engineering Journal, 13: 81–84.

Pandey, A., Selvakumar, P., Soccol, C.R. and Poonam, N. 1999. Solid state fermentation for the production of industrial enzymes. Current Science, 77(1): 149–162.

Pandey, A., Soccol, C.R. and Mitchell, D. 2000. New developments in solid state fermentation. I. Bioprocesses and products. Process Biochemistry, 35(10): 1153–1169.

Rajagopalan, S. and Modak, J.M. 1995. Modeling of heat and mass transfer for solid state fermentation process in tray bioreactor. Bioprocess Engineering, 13: 161–169.

Raghavarao, K.S.M.S., Gowthaman, M.K., Ghildyal, N.P. and Karanth, N.G. 1993. A mathematical model for solid state fermentation in tray bioreactors. Bioprocess Engineering, 8: 255–262.

Raghavarao, K.S.M.S., Ranganathana, T.V. and Karanth, N.G. 2003. Some engineering aspects of solid-state fermentation. Biochemical Engineering Journal, 13(2-3): 127–135.

Ramos Sánchez, L.B., Cujilema-Quitio, M.C., Julian-Ricardo, M.C., Cordova, J. and Fickers, P. 2015. Fungal lipase production by solid-state fermentation. Journal of Bioprocessing and Biotechniques, 5(2): 1–9.

Reddy, G.V., Babu, P.R., Komaraiah, P., Roy, K.R.R.M. and Kothari, I.L. 2003. Utilization of banana waste for the production of lignolytic and cellulolytic enzymes by solid substrate fermentation using two *Pleurotus* species (*P. ostreatus* and *P. sajor-caju*). Process Biochemistry, 38(10): 1457–1462.

Robinson, T. and Nigam, P. 2003. Bioreactor design for protein enrichment of agricultural residues by solid state fermentation. Biochemical Engineering Journal, 13(2): 197–203.

Rodriguez Coute, S.R., Moldes, D., Liébanas, A. and Sanromán, A. 2003. Investigation of several bioreactor configurations for laccase production by *Trametes versicolor* operating in solid-state conditions. Biochemical Engineering Journal, 15: 21–26.

Rodriguez Coute, S.R. and Sanromán, M.A. 2006. Application of solid-state fermentation to food industry—A review. Journal of Food Engineering, 76(3): 291–302.

Rodríguez-León, J.A., Rodríguez-Fernández, D.E. and Soccol, C.R. 2013. Laboratory and industrial bioreactors for solid-state fermentation. pp. 181–199. *In*: C.R. Soccol, A. Pandey and C. Larroche (eds.). Fermentation Processes Engineering in the Food Industry, Contemporary Food Engineering. CRC Press, Boca Raton, Florida, USA.

Rosales, E., Rodrìguez Coute, S. and Sanromán, M.A. 2007. Increased laccase production by *Trameteshirsuta* grown on ground orange peelings. Enzyme and Microbial Technology, 40: 1286–1290.

Roussos, S., Labrousse, Y., Tranier, M.-S. and Lakhtar, H. 2014. Dispositif de fermentation en milieu solide et produits obtenus. Brevet français # WO 2014118757 A1.

Roussos, S., Olmos, A., Raimbault, M., Saucedo-Castañeda, G. and Lonsane, B.K. 1991. Strategies for large-scale inoculum development for solid state fermentation system: Conidiospores of *Trichoderma harzianum*. Biotechnology Techniques, 5: 415–420.

Roussos, S., Raimbault, M., Prebois, J.-P. and Lonsane, B.K. 1993. Zymotis, a large-scale solid state fermenter design and evaluation. Applied Biochemistry and Biotechnology, 42: 37–52.

Sangeetha, P.T., Ramesh, M.N. and Prapulla, S.G. 2004. Production of fructosyltransferase by *Aspergillus oryzae* CFR 202 in solid-state fermentation using agricultural byproducts. Applied Microbiology and Biotechnology, 65(5): 530–537.

Saqib, A.A.N., Farooq, A., Iqbal, M., Hassan, J.U., Hayat, U. and Baig, S. 2012. A thermostable crude endoglucanase produced by *Aspergillus fumigatus* in a novel solid state fermentation process using isolated free water. Enzyme Research, 2012: 1–6.

Saucedo-Castañeda, G., Trejo-Hernandez, M.d.R., Lonsane, B.K., Navarro, J.M., Roussos, S., Dufour, D. and Raimbault, M. 1994. Online automated monitoring and control-systems for CO_2 and O_2 in aerobic and anaerobic solid-state fermentations. Process Biochemistry, 29: 13–24.

Sawant, R. and Nagendran, S. 2014. Protease: An enzyme with multiple industrial applications. World Journal of Pharmaceutical Sciences, 3: 568–579.

Shahidi, F. and Kamil, Y.J. 2001. Enzymes from fish and aquatic invertebrates and their application in the food industry. Trends in Food Science & Technology, 12(12): 435–464.

Silva, E.M. and Yang, S.T. 1998. Production of amylases from rice by solid-state fermentation in agas-solid spouted-bed bioreactor. Biotechnology Progress, 14: 580–587.

Singhania, R.R., Sukumaran, R.K. and Pandey, A. 2007. Improved cellulase production by *Trichoderma reesei* RUT C30 under SSF through process optimization. Applied Biochemistry and Biotechnology, 142(1): 60–70.

Solís-Pereira, S., Favela-Torres, E., Viniegra-González, G. and Gutiérrez-Rojas, M. 1993. Effects of different carbon sources on the synthesis of pectinase by *Aspergillus niger* in submerged and solid state fermentations. Applied Microbiology and Biotechnology, 39(1): 36–41.

Stoykov, Y.M., Pavlov, A.I. and Krastanov, A.I. 2015. Chitinase biotechnology: Production, purification, and application. Engineering in Life Sciences, 15(1): 30–38.

Suresh, P.V. and Chandrasekaran, M. 1998. Utilization of prawn waste for chitinase production by the marine fungus *Beauveria bassiana* by solid state fermentation. World Journal of Microbiology and Biotechnology, 14(5): 655–660.

Suryanarayan, S. 2003. Current industrial practice in solid state fermentations for secondary metabolite production: The Biocon India experience. Biochemical Engineering Journal, 13(2): 189–195.

Tanaka, M., Kawaide, A. and Matsuno, R. 1986. Cultivation of microorganisms in an air-solid fluidized bed fermentor with agitators. Biotechnology Bioengineering, 28: 1294–1301.

Thanapimmetha, A., Luadsongkram, A., Titapiwatanakun, B. and Srinophakun, P. 2012. Value added waste of *Jatropha curcas* residue: Optimization of protease production in solid state fermentation by Taguchi DOE methodology. Industrial Crops and Products, 37(1): 1–5.

Thakur, M.S., Karanth, N.G. and Nand, K. 1990. Production of fungal rennet by *Mucor miehei* using solid state fermentation. Applied Microbiology and Biotechnology, 32: 409–413.

Thomas, L., Larroche, C. and Pandey, A. 2013. Current developments in solid-state fermentation. Biochemical Engineering Journal, 81: 146–161.

Utpat, S.S., Kinnige, P.T. and Dhamole, P.B. 2014. Effect of periodic water addition on citric acid production in solid state fermentation. Journal of the Institution of Engineers (India): Series E, 94(2): 67–72.

Viniegra-González, G., Favela-Torres, E., Aguilar, C.N., Rómero-Gomez, S. de J., Díaz Godínez, G. and Augur, C. 2003. Advantages of fungal enzyme production in solid state over liquid fermentation systems. Biochemical Engineering Journal, 13: 157–167.

Vitcosque, G.L., Fonseca, R.F., Rodríguez-Zúñiga, U.F., Bertucci Neto, V., Couri, S. and Farinas, C.S. 2012. Production of biomass-degrading multienzyme complexes under solid-state fermentation of soybean meal using a bioreactor. Enzyme Rresearch, 2012.

Von Meien, O.F., Luz Jr., L.F., Mitchell, D.A., Pérez-Correa, J.R., Agosin, E., Fernández-Fernández, M. and Arcas, J.A. 2004. Control strategies for intermittently mixed, forcefully aerated solid-state fermentation bioreactors based on the analysis of a distributed parameter model. Chemical Engineering Science, 59(21): 4493–4504.

Wang, L. and Yang, S.T. 2007. Solid state fermentation and its applications. pp. 465–489. *In*: S.T. Yang (ed.). Bioprocessing for Value-added Products from Renewable Resources—New Technologies and Applications. Elsevier, Amsterdam, Netherlands.

Zamost, B.L., Nielsen, H.K. and Starnes, R.L. 1991. Thermostable enzymes for industrial applications. Journal of Industrial Microbiology, 8: 71–81.

Zhu, J., Liu, H., Zhang, J., Wang, P., Liu, S., Liu, G. and Wu, L. 2014. Effects of Asn-33 glycosylation on the thermostability of *Thermomyces lanuginosus* lipase. Journal of Applied Microbiology, 117: 151–159.

25

Scaling-up and Modelling Applications of Solid-state Fermentation and Demonstration in Microbial Enzyme Production Related to Food Industries

An Overview

Steve C.Z. Desobgo,[1,4,]* *Swati S. Mishra,*[2] *Sunil K. Behera*[3] and *Sandeep K. Panda*[4]

1. Introduction

Solid-state fermentation (SSF) is regarded as one of the most acceptable and convenient technologies for production of microbial enzymes. SSF is understood as the cultivation of microorganisms on solid, moist substrates in the absence of free aqueous phase, at average water activity significantly below 1 (Pandey, 2003). SSF is advantageous over submerged fermentation for the production of enzymes in numerous aspects, especially, fungal enzyme production is preferred in SSF as fungi show higher germination rates, spore formation and hyphae penetration in moist solid substrates (Holker and Lenz, 2005). Keeping in view the market size of the enzymes (£ 2100 million in 2010) and compound annual growth rate (CAGR) of 6 per cent forecast over the next five years, researchers and industrialists are showing tremendous interest in the techno-economic

For affiliaton see at the end of the chapter.
For Nomenclature see at the end of the chapter.

feasibility of production of different microbial enzymes (Saxena, 2015). Microbial enzymes are very commonly applied in modern times for the production of different food products. The most important microbial enzymes used in food industries are α-amylase, gluco-amylase, lipase, protease, invertase and pectinase, etc. Several studies in the last decade have developed technologies for overproduction of these enzymes. Researchers have also focused on economical production of the enzymes from cheaper substrates to reduce the cost of the final product (Panda and Ray, 2015). Genetically-engineered strains have been proved to be useful for enhanced production of the enzymes (Panda et al., 2016).

Although several review articles have covered the history and microbiological production of different enzymes related to food processing on the laboratory scale, hardly any article covers the exact scenario in the scaling-up process. The current article covers the developments in modelling, scaling-up equipments and production of commercially important enzymes related to food industries in different types of bioreactors with selected examples.

2. Scale up and Enzymes

The production of bulk chemicals and enzymes has secured an advantage with SSF (Soccol et al., 1994). Several researchers have demonstrated the bulk production of enzymes using SSF. Pectolytic enzymes were produced in large scale by solid-state fermentation using *Aspergillus carbanerius* strain on wheat bran medium. Fermentation for 21 hours with the tray temperature of 30°C gave maximum production of enzymes. Further, steaming of wheat bran at 15 Pa for 45 min. enhanced the enzyme production (Ghildyal et al., 1981). Roussos et al. (1993) described a novel, large-scale solid-state fermenter designated as *Zymotis*, which aids the upscaling studies at 4–12 kg substrate dry matter (SDM) or 15–55 kg moist solid medium capacity, depending on the initial moisture content of the medium. Sugarcane bagasse and wheat bran [80:20 (w/w)] were used in fermentation for production of cellulase. The fermentation was conducted for 64 hours at 28°C and the medium was aerated at the rate of 300:1 humidified air/h/ compartment during the first 12 hours, and further the aeration rate was doubled for rest of the fermentation time. The study was carried out to compare the production of cellulolytic enzymes by *Trichoderma harzianum* in *Zymotis* and a medium-sized laboratory column fermenter. It was observed that the enzyme titres were marginally higher in *Zymotis* at all the substrates when compared with parallel fermentation in the laboratory-scale column fermenter. The improved performance of the *Zymotis* is attributed to better control of cultural parameters in it.

3. Scale-up Approaches and Models

In the natural environment, SSFs occur everywhere—in cultivated grounds, the compost or the ensilage, but the term 'fermentation' is not naturally associated with what occurs in these mediums. This term is used with more elaborate processes developed in laboratory and implemented in bioreactors.

The scale-up of a bioprocess is a critical stage which guarantees the economic viability of the bio-product concerned. It consists in increasing the size of the bioreactor in order to reach sufficient productivity, and this, while maintaining as much as possible the outputs obtained at the time of tests in laboratory or on a pilot scale. For most biotechnological products (like enzymes production), a viable process is reached only for volumes of bioreactor of a few hundreds of thousands of liters (Lonsane et al., 1992). Technically, the definition of the scale-up understands several aspects (Trilli, 1986; Glenn and Rogers, 1989; Kossen and Oosterhuis, 1985; Oosterhuis et al., 1985). It would not only be about the increase in the size of the bioreactor, but should take into account the costs, the outputs and the simplicity of the device (Oosterhuis et al., 1985). In the case of microbial fermentation for the production of enzymes, a broad multidisciplinary field is shared (Banks, 1984) and the scale up then becomes an important link for the bioprocess transfer from the laboratory scale to the industrial scale for commercial needs.

The scale up is carried out in several stages (Banks, 1984; Trilli, 1986): from Erlenmeyer flask to the laboratory bioreactor, from laboratory bioreactor to the pilot bioreactor and from pilot bioreactor to the industrial bioreactor. It is impossible to preserve all the parameters identical or proportional to the scaling. Each transfer to a greater scale is complex because various parameters, such as the scale of sterilization, aeration and agitation are modified when one increases volume, the diameter and the height of the bioreactor (Lonsane et al., 1992; Mitchell et al., 2006). Similar physicochemical conditions must be maintained in the environment of each cell in spite of the increase in the volume of culture. With each stage of the scale up, various parameters are analyzed and modified because the physicochemical and enzymatic reactions occurring inside the bioreactor vary according to the volume of the bioreactor used. Moreover, during the scaling, it is essential to take into account the investment costs of equipment (bioreactor, techniques of extraction and purification) and of operation (culture medium and energy); to carry out an automation of equipment if possible; to reduce the production of waste; to obtain products corresponding to desired quality (Mitchell et al., 2006).

The representation of the working system in terms of mathematical expression is what constitutes the modelling in SSF. The application of mathematical modelling techniques to describe the biological and transport phenomena within the system made significant improvements in understanding how to design, operate and scale-up SSF bioreactors. Various features of the conceptually divided microscale and macroscale phenomena that occur within SSF bioreactors are described by equations of the mathematical models of bioreactor (Mitchell et al., 2000). The prediction models for the scale up is taking into account the growth kinetic, rotating-drum, traditional and *Zymotis* packed-bed, intermittently-mixed forcefully-aerated and well-mixed bioreactors. The aspects globally developed are growth kinetic, energy and water balance. Some assumptions are made to handle the models.

3.1 Growth kinetic models

The growth kinetic models are one of the first important models concerning the scale up and especially on modelling and controlling. It can be used to handle the conditions in the reactor. These conditions take into consideration the parameters involved during biomass growth. Table 1 presents a non-exhaustive number of models used for growth modelling. In a study conducted by Rodriguez-Fernandez et al. (2011), the behavior of

Table 1. Describing equations for microbial kinetics.

Model	Equations	
Arrhenius	$\mu_{\max}(T) = k_g^0 \exp\left(\dfrac{-E_g}{RT}\right) - k_g \exp\left(\dfrac{-E_d}{RT}\right)$	[5]
Bovill et al.	$\dfrac{dX}{dt} = \mu X \left(1 - \dfrac{X}{X_m}\right)\left(\dfrac{Q}{1+Q}\right)$ (Bovill et al., 2000)	[6]
Esener	$\mu_{\max}(T) = \dfrac{A \exp\left(-\dfrac{\Delta H_1}{RT}\right)}{1 + k \exp\left(-\dfrac{\Delta H_2}{RT}\right)}$ (Saucedo-Castaneda et al., 1990)	[7]
Fanaei and Vaziri	$\mu = \left(\dfrac{\mu_s + \left(T_{\max} - T_{opt}\right)}{T_{\max} - T_{opt}}\right)\left(\dfrac{\mu_{opt}\left(T_{\max} - T\right)}{\mu_s + \left(T_{\max} - T\right)}\right)$ (Fanaei and Vaziri, 2009)	[8]
Okazaki et al.	$X = \dfrac{X_m}{1 + \left[\left(\dfrac{X_m}{X_0}\right) - 1\right] e^{-\mu t}}$ (Okazaki et al., 1980)	[9]
Saucedo-Castaneda	$\mu = \dfrac{2.964 \times 10^{11} \exp\left(-\dfrac{70225}{RT}\right)}{1 + 1.3 \times \exp\left(-\dfrac{283356}{RT}\right)}$ $X_m = -127.08 + 7.95(T - 273) - 0.016(T - 273)^2$ $+ 4.03 \times 10^{-3}(T - 273)^3 + 4.73 \times 10^{-5}(T - 273)^4$ (Saucedo-Castaneda et al., 1990)	[10] [11]
Ratkowsky et al.	$\mu_{\max}(T) = c_0\left(T - T_{\min}\right)\left[1 - \exp\left(c_1\left(T - T_{\max}\right)\right)\right]^2$ (Ratkowsky et al., 1983; Weber et al., 2002)	[12]
Exponential	$\dfrac{dX}{dt} = \mu X$ (Mitchell et al., 2004; Sosa et al., 2012)	[13]
Linear	$\dfrac{dX}{dt} = K$ (Mitchell et al., 2004; Sosa et al., 2012)	[14]
Logistic	$\dfrac{dX}{dt} = \mu X \left(1 - \dfrac{X}{X_m}\right)$ (Mitchell et al., 2004; Sosa et al., 2012)	[15]

Table 1. contd....

Table 1. contd....

Model	Equations	
Polynomials	$\mu_{max}(T) = -s_0 + s_1T - s_2T^2 + s_3T^3 - s_4T^4 + s_5T^5$	[16]
	$\mu_{max}(T) = -b_0 + b_1T - b_2T^2$	[17]
		[18]
	$X_m(T) = -e_0 + e_1T - e_2T^2 + e_3T^3 - e_4T^4$	
	(Mitchell et al., 2004; Sosa et al., 2012)	
Two phase	$\dfrac{dX}{dt} = \mu X,\ t \leq t_a$	[19]
		[20]
	$\dfrac{dX}{dt} = \left[\mu L e^{-k(t-t_a)}\right]X,\ t \geq t_a$ (Mitchell et al., 2004; Sosa et al., 2012)	

kinetic parameters in production of pectinase and xylanase by solid-state fermentation was observed. *Aspergillus niger* F3 was applied for production of pectinase and xylanase in a 2 kg bioreactor and citrus peel was used as a substrate. It was revealed that pectinase production was highest at 72 hours whereas the xylanase production increased after 72 hours, which is because of the reduction of pectin in the medium and forcing the microorganism to use xylan as the carbon source. The best air flow intensity noted for the microorganism growth and optimum production of pectinase (265 U/g) and xylanase (65 U/g) was 1 V kg M (volumetric air flow per kilogram of medium). This is because of the sufficient amount of O_2 incorporated into the medium. The following equation (1) was applied to obtain the values for metabolic O_2 balance:

$$\frac{dO_2}{dt} = \frac{1}{Y_{x/o}}\frac{dx}{dt} + mX \qquad [1]$$

where dX/dt is the biomass production rate; X, biomass synthesised during the time interval; Yx/o, yield based on O_2 consumption for biomass synthesis; dO_2, differential O_2 consumed during the differential time interval; m, maintenance coefficient; dt, differential time interval.

Along with the growth of microorganisms and consumption of substrates, researchers also studied the modelling of product formation, such as enzymes and bio-ethanol. One of the general modelling used for the formation of products in biological processes is to assume the growth and non-growth-associated components. Therefore, the general equation (2) for a product (*P*) is:

$$\frac{dp}{dt} = Y_{PX}\frac{dX}{dt} + m_PX \qquad [2]$$

where Y_{PX} is the stoichiometric coefficient and m_P is the maintenance coefficient.

Mass balances are also used to develop the product formation models. Hashemi et al. (2011) proposed equation (3) to model different phases of bacterial growth curve and the production of α-amylase in the SSF process using wheat bran as substrate and *Bacillus* sp. as inoculum.

$$\frac{dW}{dt} = \frac{dS}{dt} + \frac{dB}{dt} + \frac{dP}{dt} \qquad [3]$$

In the study it was assumed that the changes in total dry fermenting medium weight (W) correspond to substrate consumption rate (dS/dt), biomass growth rate (dB/dt) and product formation rate (dP/dt).

Experimental data generated from a series of batch fermentations were applied to validate the proposed models. Prediction of the production of α-amylase during the fermentation process was attempted by using equation (4) which makes a correlation between P (enzyme production) and t (time) based on the variation in the fermented medium weight. According to the equation, α and β coefficients are estimated experimentally to predict the product kinetics derived from $\frac{dW}{dt}$ data:

$$P = (\alpha - \frac{\gamma}{\delta}) \int_0^t B dt + \frac{\beta}{\delta} \int_0^t \frac{dW}{dt} dt \qquad [4]$$

3.2 Rotating-drum bioreactors models for scale-up and applications

The rotating-drum bioreactors comprise a horizontally rotating drum, that may or may not have a paddle mixer and rotates slowly for proper mixing of fermentation substrate. For scaling-up purposes, many assumptions need to be made concerning the rotating-drum bioreactors (Fig. 1). The most important ones can be summarized as follows: the bioreactor is cylindrical (with a length L and diameter D) and partially filled; since the solid materials are degraded during fermentation, it will be considered that only the density of the bed is affected; the dry gas remains constant in the headspace; the gas flow rates remain the same between the inlet and the outlet of the bioreactor; the solid particles and gas phase are in equilibrium (moisture and thermal) and the diffusion from the axe is negligible (Mitchell et al., 2002).

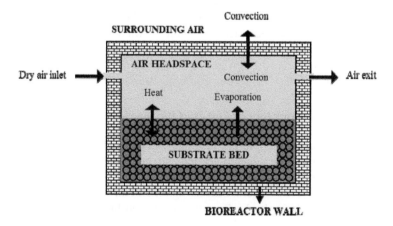

Figure 1. Rotating-drum bioreactor.

The models used for the rotating drum bioreactors scale-up take into consideration mass and heat transfer balances. The mass transfer (for water and air) occurring over the bed and headspace must be considered.

3.2.1 Microbial growth

Initially, at $t = 0$, $X = X_0$, and the microbial growth kinetic which is a logistic equation can be expressed as follows (Mitchell et al., 2002; Wang et al., 2010):

$$\frac{dX}{dt} = \mu X \left(1 - \frac{X}{X_m}\right)$$ [21]

where: X, is the biomass concentration, μ, the specific growth rate; X_m, maximal concentration of biomass and X_0, is the initial biomass concentration.

3.2.2 Applications

The theoretical aspects of the rotating-drum bioreactors were applied for different enzymes production. Several successful stories have been reported for upscaling of enzyme production using rotating-drum bioreactors. In the early 19th century manufacture of amylase by *Aspergillus oryzae* on wheat bran, in rotating-drum bioreactors was scaled up to industrial level (Takamine, 1914). It reported the use of fuzzy logic control for amylase and protease enzymes production by *A. oryzae*. That fuzzy logic control system as a scale-up aspect exhibited the highest enzyme activity (Sukumprasertsri et al., 2013). While using the same microorganism, the scale up on a pilot-scale was realized and concerned the oxygen uptake and agitation. Achievement in terms of O_2 utilization was better at 9 rpm than at 2 rpm. This was apparently as a result of the effect of the rotational speed on the performance of the stirring inside the bed (Stuart, 1996). A comparative study was attempted by Alam et al. (2009) for the production of cellulase enzyme in both Erlenmeyer flask (500 mL) and a horizontal rotating-drum bioreactor (50 L) by using empty fruit bunches as substrate. It was observed that the highest cellulase activity on the fourth day of fermentation in the 500 mL flask was 8.2 filter paper activity (FPA)/gram dry solids (gds) of empty fruit bunches, while the same was 10.1 FPA/gds for rotating-drum bioreactor on the second day of fermentation.

3.3 Traditional and Zymotis packed-bed bioreactors models

The scale up purpose for this bioreactor assumes that the traditional packed-bed bioreactor is a cylindrical unit which is aerated with humid air from the bottom (Fig. 2). The humid packed-bed (substrate) is then inoculated and introduced in the unit and that time is considered as zero with no agitation (static bed) (Mitchell et al., 1999; Mitchell et al., 2003). In the *Zymotis* packed-bed bioreactor (Fig. 3) which was modelled by Mitchell and Von-Mein (2000), the heat transfer is realized in two directions (the horizontal direction is normal to air flow direction and the direction

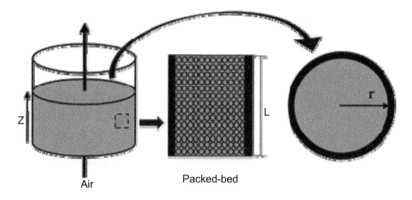

Figure 2. Traditional packed-bed bioreactor.

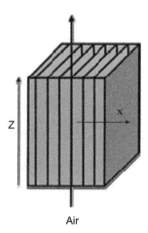

Figure 3. *Zymotis* packed-bed bioreactor.

which is co-linear to air flow exhibiting the removal of evaporative and convective heat). The front-to-back gradients are considered negligible (Mitchell et al., 2003).

3.3.1 Microbial growth

The logistic equation is used to express the kinetic growth of the biomass for both packed-bed and is as follows (Mitchell et al., 1999):

$$\frac{dX}{dt} = \mu X \left(1 - \frac{X}{X_m} \right)$$

[22]

The packed-bed bioreactor presents a limitation on the height of the bed which can reduce significantly the performance. That bed height limit is not constant and varies with the microbe growth rate and superficial air velocity (Mitchell et al., 1999).

3.3.2 Applications

Successful upscaling applications for enzyme production were carried out using packed-bed bioreactor. The production of pectinase was scaled up from 12 g to 30 kg of dry substrate; the best outcome was achieved with a 40-cm high bed having 27 kg of wheat bran and 3 kg of sugarcane bagasse (Pitol et al., 2016). The manufacture of lipase from *Penicillium simplicissimum* was performed also in packed-bed bench-scale bioreactors and the scale up regarding the yield of the enzyme was profitably completed by adjusting the temperature and air circulation rate in the bioreactor, utilizing bagasse and sugarcane molasses (6.25 per cent). The highest lipase activity reached 26.4 U/g at 27°C and a flow rate of 0.8 L/min (Cavalcanti et al., 2005). Also, a model has been established with N-tanks in the series approach as a scale up for a packed-bed bioreactor, using solid-state fermentation. The model investigated the production of protease enzyme applying *A. niger* and was confirmed across experimental research accessible in literature (Sahir et al., 2007). Similarly, scale up was realized in a laboratory packed-bed bioreactor to obtain α-amylase from *Bacillus* sp. KR-8104 using SSF by prospecting the control of temperature. Wheat bran (WB) was used as substrate. The coexisting impacts of aeration rate, initial humidity of substrate and temperature incubation on α-amylase manufacture were checked out. The optimum circumstances in which the maximum α-amylase production was attained were 37°C, 72 per cent (w/w) initial substrate humidity and 0.15 L/min aeration. The typical enzyme activity reached under best conditions was 473.8 U/g dry substrate (Derakhti et al., 2012). In another study, α-amylase production by *Bacillus amyloliquefaciens* was executed in 300 mL and 3 L packed-bed bioreactors working volume. The observations pointed out high rates of aeration demand to boost enzyme yield. That yield exhibited a linear relationship with air flow rate. The highest activity of 41.4 U/(mL·h) was attained in 14 hours of fermentation in 300-mL and a similar activity of 40 U/(mL·h) was reached after 12 hours in 3 L (Gangadharan et al., 2011).

3.4 Intermittently-mixed forcefully-aerated bioreactor

The bioreactor is advantageous over the packed-bed bioreactors as it is relatively simple to add water uniformly to the substrate (Mitchell et al., 2006). In this type of bioreactor (Fig. 4), two types of behaviors are to be considered—the static and mixing systems (Durand and Chereau, 1988; Xue et al., 1992; Chamielec et al., 1994; Agosin et al., 1997; Bandelier et al., 1997). The packed-bed system is assumed to be the one observed during the static phase. Other assumptions are: the heat transfer from the wall is negligible so that the heat and mass transfer are in one direction (axial); the solid and air are not in thermal and moisture equilibrium. No degradation of microorganism growth is included in the model (Mitchell et al., 2003; Von-Meien and Mitchell, 2002; Mitchell et al., 2006).

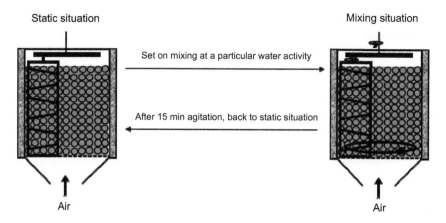

Figure 4. Intermittently-mixed forcefully-aerated bioreactor.

The calculation of changes in the solid and liquid phase takes into consideration the driving forces, when the mass and heat transfer balanced equation is presented. Since the temperature and the water activity affect the microbial growth, the models are used to calculate the water activity of the solid, using its water content and temperature.

3.4.1 Microbial growth

The microbial growth is written as (Mitchell et al., 2003; Von-Meien and Mitchell, 2002; Mitchell et al., 2006):

$$\frac{\partial X}{\partial t} = \mu X \left(1 - \frac{X}{X_m} \right)$$
[23]

The consumption of solid during microbial fermentation can be expressed as follows (Von-Meien and Mitchell, 2002; Mitchell et al., 2006):

$$\frac{\partial M_s}{\partial t} = \left[1 - \frac{1}{Y_s} \right] \frac{\partial (XM_s)}{\partial t}$$
[24]

3.4.2 Applications

An intermittent spouting with air-bed bioreactor was developed to produce enzymes from rice substrate using *A. oryzae*. The enzyme production was realized in a large scale. The scaling up was to overcome problems like mass transfer, heat transfer and solid handling. High levels of α-amylase, β-amylase and glucoamylase were obtained. It was noticed that an increase of spouting frequency decreased enzyme production (Silva and Yang, 1998). The production of β-mannanase from palm kernel cake was scaled up in a laboratory glass-column bioreactor (consisting of a jacketed vessel with a height of 50 cm, an inner diameter of 16 cm and loaded with 100 g of palm kernel cake) in terms of maximizing the yield. The optimal conditions from the model stated that for the highest level of β-mannanase, the incubation temperature must be 32°C, initial humidity 59 per cent and aeration 0.5 L/min. This allowed β-mannanase activity of 2231.26 U/g (Abdeshahian et al., 2010).

3.5 Well-mixed bioreactor

In a well-mixed bioreactor, no concentration gradients are observed either in gas or in liquid phase. The assumptions made for the well-mixed bioreactor are: it is cylindrical, with water-jacketed on its sides (Fig. 5); the microbial growth is considered as a logistic equation, with specific growth rate constant affected by the water activity and the temperature of the particles; the gas and air phases are considered as different subsystems (Mitchell and Krieger, 2006).

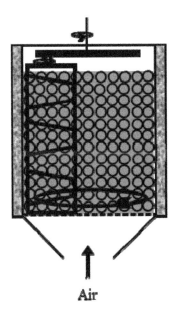

Figure 5. Well-mixed bioreactor.

The microbial growth, solid consumption, water and energy balances are treated here and the equations are written (Mitchell and Krieger, 2006; Von-Meien and Mitchell, 2002; dos-Santos et al., 2004; Marques et al., 2006):

3.5.1 Microbial growth

The logistic equation used for this bioreactor is:

$$\frac{dX}{dt} = \mu X \left(1 - \frac{X}{X_m}\right) \qquad [25]$$

The specific growth rate (fractional) can be influenced by water activity and is obtained by using the following equation:

$$\mu_{aw} = 1.0538 \exp\left(-131.6a_w^3 + 94.996a_w^2 + 214.219a_w - 177.668\right) \qquad [26]$$

where: μ_{aw} is fraction of water activity dependence specific growth rate; a_w, water activity.

3.5.2 Applications

Like the other type of bioreactors, the well-mixed bioreactor was used and scaled up for the production of food enzymes. It was a low-cost process developed for apple pomace utilisation in order to obtain pectinolytic enzyme. That enzymes from *A. niger* as well as pectinesterase and polygalacturonase were studied. The scale-up process was from the laboratory tray bioreactor to a 15 L horizontal stirred bioreactor. Process specifications, such as inoculation, mixing, aeration, temperature and humidity on pectolytic enzymes production, were investigated. The highest quantities of 15 g.kg^{-1} of substrate of polygalacturonase, 200 mg.kg^{-1} pectinesterase at activity up to 900 AJDA U ml-1 of enzyme mixture was attained on an average (Berovic and Ostroversnik, 1997). Also, investigations were conducted on the manufacture of protease by slightly halophilic *Bacillus* sp. on agro-industrial waste. The optimal pH (7.0), temperature (30°C) and agitation rate (50 rpm) were found to be the best conditions for protease production during the scale-up fermentation (Prasad et al., 2014). Multiple isolates of *A. niger* were also screened from polluted oil soil samples and isolated for lipase production, using tributyrin medium. The lipase production was successfully scaled up on a 5 L pilot scale stirred fermenter, preserving expansion parameters at pH (7), temperature (28°C), stirring (120 rpm), airflow rate (30 L/hr), O$_2$ concentration (50 per cent) and pressure (0.05 MPa) (Rai et al., 2014). A model for a well-mixed bioreactor was also executed to investigate the problems that could occur on a large scale. It was suggested that, using high flow rate, up to 85 per cent of the enzyme obtained from the microorganism could be denatured by the end of fermentation, exhibiting the problems of large-scale well-mixed bioreactor (dos-Santos et al., 2004).

4. Limitations

Scale up, product purification and biomass estimation are major challenges that need to be established before SSF can become industrially viable. The reason for scale up is difficult because of transportation limitations within a large reactor. Other major problems encountered are control over different parameters, such as pH, temperature, aeration, oxygen transfer and moisture. Also the used microorganisms are limited as they grow in reduced levels of humidity. Lack of experimental data hampers parameter identification and thus a broader use of mathematical modelling is essential.

5. Conclusion and Future Perspectives

The current article critically discusses the upscaling, modelling, microbiology and biochemical aspects of solid-state fermentation for obtaining enzymes related to the food industry. In addition microbial enzymes are used for treatment in the food industry. Although much work has been done on solid-fermentation bioreactors, the important aspect is to explore the production of enzymes in these industries. Work on different kinetics in bioreactors needs to be investigated. Also, optimization of enzyme production can be one of the primary axes as the ability to obtain high yields within short intervals during fermentation in solid media has been recognized. There is another important research on enzymatic degradation of the substrates. Therefore, optimization of the physical parameters in bioreactors and appropriate kinetics are crucial. The continuous cultivation of microorganisms (using genetically modified strains) and high density culture of cells in order to produce food enzymes needs to be investigated as well as the scale-up bioreactor. More research is needed on physical techniques and computational approaches to allow better model validation and ensure progress in rational bioengineering of SSF.

Nomenclature

A	Area of heat transfer coefficient bioreactor wall	m^2
a_w	Water activity	
b_0, b_1, b_2	Constants	
M_s	Dry mass of substrate bed	kg
m_W	Water production maintenance coefficient	
$S_0, S_1, S_2, S_3, S_4, S_5,$	Constants	
t	Time	s
t_c	Contact time between substrate particles and wall	s
T_a	Air phase temperature	^{o}C
T_{max}	Maximal temperature for growth	^{o}C
T_{min}	Minimal temperature for growth	^{o}C
T_{opt}	Optimal growth temperature	^{o}C
X	Biomass concentration	$kg\ biomass/kg\ substrate$
X_0	Initial biomass concentration	$kg\ biomass/kg\ substrate$
X_m	Maximal concentration of biomass	
Y_X	Heat yield metabolic coefficient	$J/kg\ biomass$
μ	Specific growth rate	s^{-1}

contd....

contd....

μ_{aw}	Fraction of water activity dependence specific growth rate	s^{-1}
μ_{opt}	Optimal specific growth rate	s^{-1}
μ_T	Fraction of temperature dependence specific growth rate	s^{-1}
ρ_a	Air density	kg/m^3

References

Abdeshahian, P., Samat, N., Hamid, A.A. and Yusoff, W.M. 2010. Utilization of palm kernel cake for production of beta-mannanase by *Aspergillus niger* FTCC 5003 in solid substrate fermentation using an aerated column bioreactor. Journal of Industrial Microbiology and Biotechnology, 37(1): 103–109.

Agosin, E., Perez-Correa, R., Fernandez, M., Solar, I. and Chiang, L. 1997. An Aseptic Pilot Bioreactor for Solid Substrate Cultivation Processes. Dordrecht: Kluwer Academic Publishers, pp. 233.

Alam, M.Z., Mamun, A.A., Qudsieh, I.Y., Muyibi, S.A., Salleh, H.M. and Omar, N.M. 2009. Solid state bioconversion of oil palm empty fruit bunches for cellulase enzyme production using a rotary drum bioreactor. Biochemical Engineering Journal, 46(1): 61–64.

Bandelier, S., Renaud, R. and Durand, A. 1997. Production of gibberellic acid by fed-batch solid state fermentation in an aseptic pilot-scale reactor. Process Biochemistry, 32: 141–145.

Banks, G.T. 1984. Scale up of fermentation processes. pp. 170–266. *In*: A. Wiseman (ed.). Topics in Enzyme and Fermentation Biotechnology. Chichester: Ellis Horwood Limited.

Berovic, M. and Ostroversnik, H. 1997. Production of *Aspergillus niger* pectolytic enzymes by solid state bioprocessing of apple pomace. Journal of Biotechnology, 53(1): 47–53.

Bovill, R., Rew, J., Cook, N., D'Agostino, N., Wilkinson, N. and Baranyi, J. 2000. Predictions of growth for *Listeria* monocytogenes and *Salmonella* during fluctuation temperature. International Journal of Food Microbiology, 59(3): 157–165.

Cavalcanti, E.d'A.C., Gutarral, M.L.E., Freirel, D.M.G., dos Reis Castilho II, L. and Sant'Anna Júnior II, G.L. 2005. Lipase production by solid-state fermentation in fixed-bed bioreactors. Brazilian Archives of Biology and Technology, 48: 79–84.

Chamielec, Y., Renaud, R., Maratray, J., Almanza, S., Diezand, M. and Durand, A. 1994. Pilot-scale reactor for aseptic solid-state cultivation. Biotechnology Techniques, 8: 245–248.

Derakhti, S., Shojaosadati, S.A., Hashemi, M. and Khajeh, K. 2012. Process parameters study of α-amylase production in a packed-bed bioreactor under solid-state fermentation with possibility of temperature monitoring. Preparative Biochemistry and Biotechnology, 42(3): 203–216.

dos-Santos, M.M., da-Rosa, A.S., Dal'Boit, S., Mitchell, D.A. and Krieger, N. 2004. Thermal denaturation: Is solid-state fermentation really a good technology for the production of enzymes? Bioresource Technology, 93: 261–268.

Durand, A. and Chereau, D. 1988. A new pilot reactor for solid-state fermentation: Application to the protein enrichment of sugar beet pulp. Biotechnology and Bioengineering, 31: 476–486.

Fanaei, M.A. and Vaziri, B.M. 2009. Modelling of temperature gradients in packed-bed solid-state bioreactors. Chemical Engineering and Processing, 48(1): 446–451.

Gangadharan, D., Nampoothiri, M. and Pandey, A. 2011. Alpha-amylase produced by *B. amyloliquefaciens.* Food Technology and Biotechnology, 49(3): 336–340.

Ghildyal, N.P., Ramakrishna, S.V., Nirmala Devi, P., Lonsane, B.K. and Asthana, H.N. 1981. Large-scale production of pectolytic enzyme by solid state fermentation. Journal of Food Science and Technology, 18(6): 248–251.

Glenn, D.R. and Rogers, P.L. 1989. Industrialization of indigenous food processes: Technological aspects. *In*: K.H. Steinkraus (ed.). Industrialization of Indigenous Fermented Foods. New York: Marcel Dekker, Inc.

Holker, U. and Lenz, J. 2005. Solid-state fermentation: Are there any biotechnological advantages? Current Opinion in Microbiology, 8: 301–306.

Hashemi, M., Mousavi, S.M., Razavi, S.H. and Shojaosadati, S.A. 2011. Mathematical modelling of biomass and α-amylase production kinetics by *Bacillus* sp. in solid-state fermentation based on solid dry weight variation. Biochemical Engineering Journal, 53: 159–164.

Kossen, N.W.F. and Oosterhuis, N.M.G. 1985. Modelling and scaling up of bioreactors. pp. 572–605. *In*: H.-J. Rehm and G. Reed (eds.). Biotechnology. Weinheim: VCH.

Lonsane, B.K., Saucedo-Castaneda, G., Raimbault, M., Roussos, S., Viniegra-González, G., Ghildyal, N.P., Ramakrishna, M. and Krishnaiah, M.M. 1992. Scale-up strategies for solid-state fermentation systems. Process Biochemistry, 27: 259–270.

Marques, C.B., Barga, M.C., Balmant, W., Luz-Jr, L.F., Krieger, N. and Mitchell, D.A. 2006. A model of the effect of the microbial biomass on the isotherm of the fermenting solids in solid-state fermentation. Food Technology and Biotechnology, 44(4): 457–463.

Mitchell, D.A., Berovic, M. and Krieger, N. 2006. Solid-state Bioreactors. Germany: Springer-Verlag Berlin Heidelberg.

Mitchell, D.A. and Krieger, N. 2006. A model of a well-mixed SSF bioreactor. pp. 295–314. *In*: D.A. Mitchell, N. Krieger and M. Berovic (eds.). Solid-State Fermentation Bioreactors: Fundamentals of Design and Operation. Germany: Springer-Verlag Berlin, Heidelberg.

Mitchell, D.A., Pandey, A., Sangsurasak, P. and Krieger, N. 1999. Scale-up strategies for packed-bed bioreactors for solid-state fermentation. Process Biochemistry, 35: 167–178.

Mitchell, D.A., Krieger, N., Stuart, D.M. and Pandey, A. 2000. New developments in solid-state fermentation. II. Rational approaches to the design, operation and scale-up of bioreactors. Process Biochemistry, 35: 1211–1225.

Mitchell, D.A., Tongta, A., Stuart, D.M. and Krieger, N. 2002. The potential for establishment of axial temperature profiles during solid-state fermentation in rotating drum bioreactors. Biotechnology and Bioengineering, 80(1): 114–122.

Mitchell, D.A. and Von-Meien, O.F. 2000. Mathematical modelling as a tool to investigate the design and operation of the *Zymotis* packed-bed bioreactor for solid-state fermentation. Biotechnology and Bioengineering, 68(2): 127–135.

Mitchell, D.A., Von-Meien, O.F. and Krieger, N. 2003. Recent developments in modeling of solid-state fermentation: Heat and mass transfer in bioreactors. Biochemical Engineering Journal, 13: 137–147.

Mitchell, D.A., Von-Meien, O.F., Krieger, N. and Dalsenter, F.D.H. 2004. A review of recent developments in modelling of microbial growth kinetics and intraparticle phenomena in solid-state fermentation. Biochemical Engineering Journal, 17: 15–26.

Mitchell, D.A., Von-Meien, O.F., Luiz, F.L., Luz, Jr. and Krieger, N. 2006. A model of an intermittently-mixed forcefully-aerated bioreactor. pp. 349–362. *In*: D.A. Mitchell, N. Krieger and M. Berovic (eds.). Solid-state Fermentation Bioreactors: Fundamentals of Design and Operation. Germany: Springer-Verlag Berlin Heidelberg.

Okazaki, N., Sugama, S. and Tanaka, T. 1980. Mathematical model for surface culture of Koji mold: Growth of Koji mold on the surface of steamed rice grains (IX). Journal of Fermentation Technology, 58(5): 471–476.

Oosterhuis, N.M.G., Kossen, N.W.F., Oliver, A.P.C. and Schenk, E.S. 1985. Scale-down and optimisation studies of the gluconic acid fermentation by *Gluconobacter oxydans*. Biotechnolgy and Bioengineering, 27: 711–720.

Panda, S.K. and Ray, R.C. 2015. Microbial processing for valorization of horticultural wastes. pp. 203–221. *In*: L.B. Sukla, N. Pradhan, S. Panda and B.K. Mishra (eds.). Environmental Microbial Biotechnology. Springer International Publishing.

Panda, S.K., Mishra, S.S., Kayitesi, E. and Ray, R.C. 2016. Microbial-processing of fruit and vegetable wastes for production of vital enzymes and organic acids: Biotechnology and scopes. Environmental Research, 146: 161–172.

Pandey, A. 2003. Solid-state fermentation. Biochemical Engineering Journal, 13(2-3): 81–84.

Pitol, L.O., Biz, A., Mallmann, E., Krieger, N. and Mitchell, D.A. 2016. Production of pectinases by solid-state fermentation in a pilot-scale packed-bed bioreactor. Chemical Engineering Journal, 283: 1009–1018.

Prasad, R., Abraham, T.K. and Nair, A.J. 2014. Scale up of production in a bioreactor of a halotolerant protease from moderately halophilic *Bacillus* sp. isolated from soil. Brazilian Archives of Biology and Technology, 57(3): 448–455.

Rai, B., Shrestha, A., Sharma, S. and Joshi, J. 2014. Screening, optimization and process scale up for pilot scale production of lipase by *Aspergillus niger*. Biomedicine and Biotechnology, 2(3): 54–59.

Ratkowsky, D.A., Lowry, R.K., McMeekin, T.A., Stokes, A.N. and Chandler, R.E. 1983. Model for bacterial culture growth rate throughout the entire biokinetic temperature range. Journal of Bacteriology, 154(3): 1222–1226.

Rodríguez-Fernández, D.E., Rodríguez-León, J.A., De Carvalho, J.C., Sturm, W. and Soccol, C.R. 2011. The behavior of kinetic parameters in production of pectinase and xylanase by solid-state fermentation. Bioresource Technology, 102(22): 10657–10662.

Roussos, S., Raimbault, M., Prebois, J.P. and Lonsane, B.K. 1993. *Zymotis*, a large-scale solid-state fermentation: Design and evaluation. Applied Biochemistry and Biotechnology, 42: 37–52.

Sahir, A.H., Kumar, S. and Kumar, S. 2007. Modelling of a packed bed solid-state fermentation bioreactor using the *N*-tanks in series approach. Biochemical Engineering Journal, 35(1): 20–28.

Saucedo-Castaneda, G., Gutierrez-Rojas, M., Bacquet, G., Raimbault, M. and Viniegra-Gonzalez, G. 1990. Heat transfer simulation in solid-substrate fermentation. Biotechnolgy and Bioengineering, 35(8): 802–808.

Saxena, S. 2015. Microbial enzymes and their industrial applications. pp. 121–154. *In*: Applied Microbiology. Springer India.

Soccol, C.R., Iloki, I., Marin, B. and Raimbault, M. 1994. Comparative production of alpha-amylase, glucoamylase and protein enrichment of raw and cooked cassava by Rhizopus strains in submerged and solid-state fermentations. Journal of Food Science and Technology, 31: 320–323.

Silva, E.M. and Yang, S.T. 1998. Production of amylases from rice by solid-state fermentation in a gas-solid spouted-bed bioreactor. Biotechnology Progress, 14(4): 580–587.

Sosa, D., Boucourt, R. and Dustet, J.C. 2012. Use of mathematical modeling on the solid-state fermentation processes of fibrous substrates for animal feeding. Cuban Journal of Agricultural Science, 46(2): 119–126.

Stuart, D.M. 1996. Solid-state Fermentation in Rotating Drum Bioreactor. Ph.D thesis, The University of Queensland, Queensland.

Sukumprasertsri, Monton, Unrean, P., Pimsamarn, J., Kitsubun, P.S. and Tongta, A. 2013. Fuzzy logic control of rotating-drum bioreactor for improved production of amylase and protease enzymes by *Aspergillus oryzae* in solid-state fermentation. Journal of Microbiology and Biotechnology, 23(3): 335–342.

Takamine, J. 1914. Enzymes for *Aspergillus oryzae* and the application of its amyloclastic enzyme to the fermentation industry. Industrial and Engineering Chemistry, 6: 824–828.

Trilli, A. 1986. Scale up of fermentation. pp. 227–307. *In*: A.L. Demain and N.A. Solomon (eds.). Industrial Microbiology and Biotechnology. Washington, USA: American Society of Microbiology.

Von-Meien, O.F. and Mitchell, D.A. 2002. A two-phase model for water and heat transfer within an intermittently-mixed solid-state fermentation bioreactor with forced aeration. Biotechnology and Bioengineering, 79: 416–428.

Wang, E.-Q., Li, S.-Z., Tao, L., Geng, X. and Li, T.-C. 2010. Modeling of rotating-drum bioreactor for anaerobic solid-state fermentation. Applied Energy, 87: 2839–2845.

Weber, F.J., Oostra, J., Tramper, J. and Rinzema, A. 2002. Validation of a model for process development and scale-up of packed-bed solid-state bioreactors. Biotechnology and Bioengineering, 77(4): 381–393.

Xue, M., Liu, D., Zhang, H., Hongyan, Q. and Lei, Z. 1992. A pilot process of solid-state fermentation from sugar beet pulp for the production of microbial protein. Journal of Fermentation and Bioengineering, 73: 203–220.

[1] Department of Food Process and Quality Control, University Institute of Technology of the University of Ngaoundere, P.O. Box 455, Ngaoundere, Cameroon.
[2] Department of Biodiversity and Conservation of Natural Resources, Central University of Orissa, Koraput-764020.
E-mail: swatisakambarimishra@gmail.com
[3] Department of Metallurgy, Faculty of Engineering and the Built Environment, University of Johannesburg, P.O. Box 17911, Doornfontein Campus, 2028, Johannesburg, South Africa.
E-mail: skbehera2020@gmail.com
[4] Department of Biotechnology and Food Technology, Faculty of Science, University of Johannesburg, P.O. Box 17011, Doornfontein Campus, Johannesburg, South Africa.
E-mail: sandeeppanda2212@gmail.com
* Corresponding author: desobgo.zangue@gmail.com

Enzyme Encapsulation Technologies and their Applications in Food Processing

Steva Lević,[1,a,]* *Verica Đorđević,*[2,d] *Zorica Knežević-Jugović,*[3]
Ana Kalušević,[1,b] *Nikola Milašinović,*[4] *Branko Bugarski*[2,e] *and*
Viktor Nedović[1,c]

1. Introduction

Applications of enzymes in the food industry have been intensively studied and improved in recent years. Moreover, some food processes without enzymatic reactions are too long or even not possible. Enzyme market is dynamic, not only regarding the economical aspect that is expressed by an annual growth of 7–9 per cent, but also concerning new innovations and studies that have been done in recent years.

[1] Department of Food Technology and Biochemistry, Faculty of Agriculture, University of Belgrade, Nemanjina 6, 11 080 Belgrade-Zemun, Serbia.
[a] E-mail: slevic@agrif.bg.ac.rs
[b] E-mail: ana.kalusevic@agrif.bg.ac.rs
[c] E-mail: vnedovic@agrif.bg.ac.rs
[2] Department of Chemical Engineering, Faculty of Technology and Metallurgy, University of Belgrade, Karnegijeva 4, 11 120 Belgrade, Serbia.
[d] E-mail: vmanojlovic@tmf.bg.ac.rs
[e] E-mail: branko@tmf.bg.ac.rs
[3] Department of Biochemical Engineering and Biotechnology, Faculty of Technology and Metallurgy, University of Belgrade, Karnegijeva 4, 11 120 Belgrade, Serbia.
E-mail: zknez@tmf.bg.ac.rs
[4] Academy of Criminalistic and Police Studies, Department of Forensics, Cara Dušana 196, 11080 Belgrade, Serbia.
E-mail: nikola.milasinovic@kpa.edu.rs
* Corresponding author

The application of immobilized enzyme is strictly related to economic benefits of immobilization. Regardless of numerous patents and published articles, application of immobilized enzymes in food industry is relatively limited to several processes. The production of fructose syrup is one of these costly enzyme processes requiring application of glucose isomerase to convert glucose into fructose. The enzyme glucose isomerase should be reused in order to ensure economic sustainability of fructose syrup production. One of the solutions for enzymes usage in the several repeated cycles is to attach them to the surface or to embed them in the structure of suitable support material (DiCosimo et al., 2013).

The studies of enzyme immobilization were mainly focused on two major areas: (1) immobilized biocatalyst (i.e., enzyme) for application in the biotechnological processes (DiCosimo et al., 2013), and (2) immobilized enzyme as sensors (i.e., biosensors) for detection of chemical compounds (Wang, 2005).

In order to meet the specific demands of enzymatic processes, various methods have been developed as well as many carrier materials have been tested as support for the enzyme immobilization. Both, polymer materials (DiCosimo et al., 2013) and inorganic compounds are found to be suitable for enzyme immobilization (Subramanian et al., 1999; Costa et al., 2004; DiCosimo et al., 2013). More recently, the new class of so-called nanomaterials was introduced as support for enzyme immobilization.

Generally, enzymes can be immobilized using physical or chemical methods. In the food industry, four methods of enzyme immobilization are most widely applied: adsorption, covalent bonding, enzyme cross-linking and encapsulation (Costa et al., 2004) (Fig. 1).

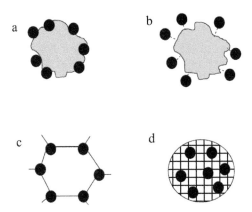

Figure 1. Methods of enzyme immobilization. Adsorption (a); covalent bonding; (b) enzyme cross-linking; (c) encapsulation (d).

Adsorption: Enzyme immobilization *via* adsorption is based on enzyme attachment to the solid carrier by weak interactions (van der Waals forces, ionic interactions, hydrophobic interactions, hydrogen bonding). The interaction between enzyme and carrier is defined by the structure of the enzyme and the surface properties of carrier material. The adsorption method is relatively simple and consisting of contact (mixing) of enzyme solution and carrier followed by removing excess enzyme by washing. The main disadvantage of this technique is that efficiency of enzyme loading is dependent on pH and ionic strength, which means that even small variations of these factors can lead to enzyme release (i.e., enzyme loss). For enzyme adsorption and immobilization various synthetic and natural materials can be used: porous glass, kaolinite, bentonite, ceramics, nylon, polystyrene, chitosan, dextran, gelatine, cellulose, etc.) (Costa et al., 2004).

Covalent bonding: This type of immobilization is based on enzyme covalent bonding to a carrier material (Ó'Fágáin, 2003). The covalent bond is formed between carrier functional groups located on its surface and groups present in the enzyme structure, such as amino, carboxyl, hydroxyl, and sulfhydryl groups. Bonding between enzyme and carrier material can be achieved by direct interaction or by spacer molecule. The spacer molecule provides better linking between enzyme and carrier that allows better enzyme orientation and consequently higher activity. However, the immobilization *via* covalent bonding could damage the enzyme molecule, leading to the loss of the enzyme activity. Numerous carrier materials are available for enzyme immobilization by covalent bonding and the selection of carrier depends on material characteristics and conditions of enzymatic reaction (Costa et al., 2004).

Cross-linked enzyme: Immobilization of enzyme molecules could be achieved by covalent bonding of the single enzyme molecules that leads to formation of particles (Ó'Fágáin, 2003). Moreover, these particles (i.e., aggregates) could be further stabilized by additional solid carrier (e.g., polymer structure) in order to improve mechanical properties of the immobilized enzyme. This approach is the base for many commercial enzyme immobilization processes (DiCosimo et al., 2013). Among many tested chemical compounds, glutaraldehyde is the most frequently used cross-linking agent for formation of cross-linked enzyme particles. The main advantage of enzyme cross-linking immobilization is that the particle formation process is relatively simple. On the other hand, cross-linking can cause the changes of enzyme structure which is usually followed by the loss of the enzyme activity. It should be pointed out that this procedure for enzyme immobilization is widely used for production of commercially available immobilized enzyme, especially for glucose isomerase used for production of high fructose syrup (Costa et al., 2004; DiCosimo et al., 2013).

Encapsulation: Immobilization of enzyme by encapsulation is based on the physical entrapment of enzyme into structure of carrier material (i.e., polymer structure) (Fig. 2). The mixture of enzyme and polymer material is submerged to cross-linking during which enzyme becomes entrapped inside the polymer matrix. Various shapes of polymeric forms with loaded enzyme can be produced by this method (e.g., particles, films or fibers). This technique has become popular since it provides a number of possibilities for application of various carrier materials for enzyme immobilization. Some restrictions of this procedure are related to possible diffusion limitation through carrier structure during enzyme application and loss of enzyme during immobilization/ exploitation caused by porosity of polymer structure (Costa et al., 2004). For production of encapsulate with desirable properties, carrier material could be made from a mixture of different components or can be formed of separated layers of different materials

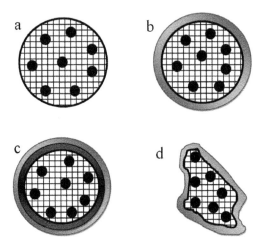

Figure 2. Main types of particles produced by encapsulation process: simple particle (a); single-layered particle (b); multi-layered particle (c); irregular-shaped particle (d).

around active compound. Modern encapsulation techniques provide productions of either spherical or irregular shaped particles (Fig. 2).

In addition, the particle size is also an important parameter and could be controlled by using an appropriate encapsulation technique. In order to produce particles with required morphological properties, numerous encapsulation techniques have been developed (Đorđević et al., 2014).

Another approach in the enzyme encapsulation is formation of protective layer(s) around the enzyme, using opposite charged polyelectrolytes. Protective capsule is created around the enzyme that is in the crystal form. The difference in the molecules charges provides layer formation in several steps by contacting opposite charged polymer solution with an already formed layer. The method is relatively simple and could be used for protection of enzyme crystals against negative environmental influences (Caruso et al., 2000).

The main techniques and carrier materials for enzyme immobilization as well as application of immobilized enzyme in food industry are reviewed below.

2. Enzyme Immobilization Using Inorganic Carrier Materials

Immobilization of enzyme onto inorganic carriers is a well-established technique, especially in the field of industrial applications. Numerous inorganic carrier materials have been tested for enzyme immobilization in the form of pure compounds or as composites of inorganic and organic materials. Immobilization of commercially available enzymes, such as lipase or glucose isomerase, is mainly based on inorganic or inorganic/polymer composites as carriers. This technology of enzyme immobilization is proved as reliable, especially under conditions of industrial bioreactors (DiCosimo et al., 2013). Even more, inorganic materials could be used for separation and purification of targeted enzyme (Dimitrijević et al., 2012). Some of the most important inorganic materials that have been used for enzyme immobilization, such as glass, zeolites, natural clay and materials with magnetic properties are reviewed below.

Glass has been used for enzyme immobilization due to its surface modification possibilities in order to improve enzyme binding. Commercially available aminopropyl glass beads, activated with glutaraldehyde, were used for peroxidase immobilization. Immobilized enzymes showed high activity in the repeated reaction cycles as well as high activity after storage for 70 days (Marchis et al., 2012). Porous glass beads functionalized by (3-aminopropyl) triethoxy silane and glutaraldehyde were reported as carriers for pectinesterase immobilization. Kinetic parameters (i.e., V_{max} and K_m) of immobilized pectinesterase were stable, while enzymes exhibited prolonged stability and high activity after storage for 30 days (Karakuş and Pekyardimci, 2006). Similar results were reported for enzyme immobilization on non-porous glass, using glutaraldehyde for enzyme binding. Chopda et al. (2014) immobilized glucose isomerase on non-porous glass surface modified by different activators, while glutaraldehyde was used as the cross-linking agent. This procedure provided optimal conditions for enzyme activity with relatively high immobilization yield (~80 per cent).

Cysteine was also found to be suitable as a binding agent for enzyme immobilization on to the glass surface. In this procedure, glass surface was modified by (3-mercaptopropyl) trimethoxysilane sol-gel and with cysteine-silver nanoparticles system. Further, enzyme (alkaline phosphatase) binding *via* amino and carboxyl groups of cysteine was achieved by carbodiimide and glutaraldehyde treatment. Immobilized alkaline phosphatase exhibited high stability (85 per cent) after eight repeated cycles of usage (Upadhyay and Verma, 2014). Also, the modification of glass beads could be done without using glutaraldehyde for enzyme binding. Treatment of glass beads with phthaloyl chloride and 3-aminopropyl-triethoxysilane showed promising results as agents for covalent binding of enzyme. Immobilization of α-amylase onto such modified glass beads provided high enzyme activity (81.4 per cent) after 25 reaction cycles (Kahraman et al., 2007).

Zeolites are another group of inorganic materials that have been applied in enzyme immobilization. Besides naturally occurring zeolites, the synthetic zeolites with specific surface properties can provide adequate support for enzyme immobilization (Knezevic et al., 1998). Also, zeolite can be designed for specific purposes in the form of extrudates, suitable for enzyme immobilization (i.e., adsorption) (Chang et al., 2006; Chang and Chu, 2007). However, zeolites have relatively small pore sizes that limit their application as enzyme carriers. The solution to this problem is modification of their structure and creation of mesoporous zeolites (Mitchell and Pérez-Ramírez, 2011; Kumari et al., 2015). Hydrothermal synthesis of zeolite with mesoporous structure showed great potential for preparation of enzyme carriers with required pore characteristics. Moreover, immobilized enzyme (i.e., protease) on such prepared zeolite exhibited greater efficiency as compared to mesoporous silica (Kumari et al., 2015). According to Calgaroto et al. (2011), the Si/Al ratio of zeolite carrier affects significantly the enzyme catalytic activity. This ratio is regulated by zeolite synthesis and consequently the enzyme immobilization is efficiently controlled by Si/Al ratio. Further, the efficiency of enzyme immobilization using zeolites as carrier can be improved by surface modifications of zeolites with functional polymers. The first step in this procedure is polymerization of polymer on the zeolite's surface that results in formation of polymer/zeolite composite, followed by enzyme covalent binding to such composite. It was reported that laccase immobilized on zeolite/polymer composite had better storage and thermal properties compared to free enzyme (Celikbicak et al., 2014).

Other inorganic carriers for enzyme immobilization, such as natural clay, showed promising results, especially in environmental applications (Kim et al., 2012). The immobilization onto clay could be performed either by adsorption or by covalent binding. Further improvement in enzyme immobilization efficiency, activity and stability is possible due to covalent enzyme binding onto modified clay. Modification of clay minerals could be performed by addition of cross-linking agents, such as glutaraldehyde, (3-aminopropyl) triethoxy silane, bentonite, sepiolite, etc. (An et al., 2015). In order to ensure better storage properties of immobilized enzymes, clay can be modified by formation of clay/polymer composite (Bayramoglu et al., 2013) or by addition of surfactants during immobilization procedures (Tzialla et al., 2009; Andjelković et al., 2015). Also, some studies reported the usage of chitosan/clay composites for immobilization of β-glucosidase (Chang and Juang, 2007), tyrosinase (Dinçer et al., 2012) as well as calcium alginate-clay composite for alpha-amylase, glucoamylase and cellulase immobilization (Rahim et al., 2013).

In recent years, application of magnetic particles as enzyme carriers has drawn increased attention because such supports provide simple recovery of immobilized enzymes from the solution with magnetic field and it usage in the repeated cycles (Franzreb et al., 2006). This is especially interesting since some important industrial enzymes, such as naringinase, can be successfully immobilized onto magnetic microparticles (Magario et al., 2008). Another advantage of enzyme immobilization onto magnetic particles is that repeated procedures are relatively simple, providing good catalytic properties of immobilized enzymes in the consecutive cycles of usage (Alftrén and Hobley, 2014). Another group of magnetic materials, so-called magnetic nanoparticles, have also been used for enzyme immobilization and results of these material application as enzyme carriers are reviewed below.

3. Enzyme Immobilization Using Synthetic Carrier Materials

Synthetic materials are widely used for enzyme immobilization, especially for immobilization of some industrially important enzymes, such as glucose isomerase (DiCosimo et al., 2013), peroxidase (Prodanović et al., 2012) or lipase (Bezbradica et al., 2009; Ognjanovic et al., 2009).

Commercially available polymeric materials, particularly those with functional surface properties (e.g., ion exchange resins), showed good results regarding enzyme activity and stability but also satisfactory mechanical properties required for bioreactor applications. The resin such as Purolite® A109 (polystyrene macroporous resin) could be used effectively for lipase immobilization in real bioreactor conditions since its structure and diameter of particles provide low pressure drops inside the reactor (Mihailović et al., 2014). The specific designed polymer carriers could be an alternative to conventional immobilization procedure, such as those that use glutaraldehyde for enzyme binding (Prlainović et al., 2011).

Some of the synthetic materials, such as nylon, have been used in commercially available forms for immobilization of enzymes, with adequate treatment for preventing enzyme leakage (Isgrove et al., 2001). Commercially available polymers, such as Eudragit® S (a copolymer of methacrylic acid and methyl methacrylate) has been already successfully used as the carrier for immobilization of trypsin and other serine proteases (Arasaratnam et al., 2000; Silva et al., 2006), lipases (Rodrigues et al., 2002), xylanase (Ai et al., 2005), α-amylase, β-glucosidase and alkaline phosphatase (Sardar et al., 1997). Eudragit® S 100 is suitable for enzyme immobilization since its solubility could be controlled by adjusting the solution pH value (Sardar et al., 1997). Thus, the reaction catalyzed by the enzymes immobilized on this carrier can be performed in a homogenous single phase, free from solid-liquid mass transfer problem, including pore diffusion limitations with respect to substrate or product. After the reaction, the immobilized enzymes on carrier can easily be made insoluble by changing the physical conditions, such as pH or temperature, and could be recovered by precipitation for reuse (Arasaratnam et al., 2000).

The covalent immobilization of enzymes onto epoxy-activated carriers has drawn considerable interest. Commercial epoxy-carriers, such as Eupergit® C, Sepabeads® or Immobeads-150® are very convenient for the multipoint covalent attachment of enzymes considering that the present epoxide groups could bind to amino, thiol and hydroxyl groups placed on enzyme molecules (Babich et al., 2012; Katchalski-Katzir and Kraemer, 2000; Knežević-Jugović et al., 2011). Kinetic and thermodynamic stabilization of the 3-D structure of the enzymes is achieved by formation of rigid enzyme-carrier linkages (Knežević et al., 2006; Tacelão et al., 2012).

Specific properties of polymer carrier material could be achieved by controlling the parameters of the polymerization process and/or polymer composition as well as the ingredients ratio. Design of synthetic carriers provides immobilized enzymatic systems with specific properties that respond to the external stimuli. For example, polymers based on N-isopropylacrylamide and itaconic acid could be used as enzyme carriers with wide pH range of enzyme stability and high enzyme loading efficiency (Milašinović et al., 2010a,b, 2012a,b). Further, synthetic polymers were easily shaped in the form of membrane with specific (functionalized) properties that allow high activity of immobilized enzymes and improved storage stability (Bayramoğlu et al., 2011).

A new process for synthesis of carriers, such as electrochemical synthesis of polyaniline, was found to be a promising procedure for production of stable enzyme support (Bezbradica et al., 2011). In addition, Gvozdenović et al. (2011) applied polyaniline for immobilization of enzymes on graphite electrode for electrochemical detection of glucose. Electrochemical polymerization of polyaniline on graphite was used in the first stage of electrode formation followed by glucose oxidase immobilization by glutaraldehyde. This procedure provides required electrode design as well as electrode device with good structural properties that allows preferable enzyme orientation and distribution. Polyaniline could also be used for modification of carrier material for improvement of enzyme immobilization efficiency. Bayramoğlu et al. (2009) investigated the possibilities for modification of polyacylonitrile fibers with polyaniline. Modified fibers were used for invertase immobilization in order to ensure better stability of immobilized enzymes. According to the authors, immobilized enzymes retained 83 per cent of their original activity as compared to 7 per cent of free enzyme over a period of two months.

4. Encapsulation of Enzymes in Hydrogels by Encapsulation Technique

Natural and synthetic polymer hydrogels have been frequently used for encapsulation of cells, enzymes and food compounds. The main advantages are their simple production, non-toxicity, good diffusion characteristics and the possibility to facilitate exploitation in bioreactor conditions. Hydrogels based on biodegradable and natural polymers are attractive carriers regarding ecological aspects of their application. Polysaccharides, such as alginate, pectin, guar gum or acacia gum, have been widely used for encapsulation of active compounds and biocatalysts. On the other hand, some hydrogels based on the synthetic polymers are found to be promising carriers for cells and enzyme encapsulation. Generally, the first phase in encapsulation of active compound within the structure of hydrogels is preparation of polymer/active compound solution (i.e., dispersion or emulsion). The formation of hydrogels could be performed via several mechanisms depending on the material's chemical properties (Lozinsky et al., 2003):

- Chemotropic gels-gels formed by intermolecular interactions (e.g., polyacrylamide gels)
- Ionotropic gels-gels formed by ion-exchange reactions (e.g., alginate-polylysine gels)
- Chelatotropic gels-gels formed by chelating reactions (e.g., calcium-alginate gels)
- Solvotropic gels-gels formed by changing solvent composition (e.g., cellulose acetate systems)
- Thermotropic gels-gels formed by heating polymer solution (e.g., hydrophobically modified hydroxyethyl cellulose gels)
- Psychrotropic gels-gels formed by chilling polymer solution (e.g., gelatine gels)
- Cryotropic gels-gels formed by freezing of polymer solution (e.g., poly(vinyl alcohol) cryogels)

Hydrogels could be processed into different shapes and sizes (e.g., spherical droplets or fibers), depending on the process demands. Hydrogel beads have been produced in the range of several nanometres to several millimetres. One of the commonly used hydrogel systems for encapsulation in biotechnology is calcium-alginate (Ca-alginate) beads. The ability of alginate (e.g., sodium-alginate) to form stable hydrogels by cross-linking with divalent cations (e.g., Ca^{2+}) has been used for encapsulation of cells (Nedović et al., 2001, 2002; Lalou et al., 2013), enzymes (Knezevic et al., 2002), food aromas (Levic et al., 2013; Lević et al., 2015), etc. Prüsse et al. (2008) showed that uniform Ca-alginate beads could be produced by several different techniques: vibration technique, coaxial air-flow technique, electrostatic extrusion and JetCutter® technology (liquid jet mechanical cutting) (Fig. 3).

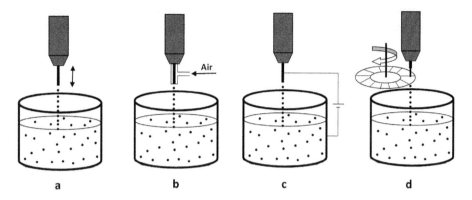

Figure 3. Schemes of encapsulation processes for production of Ca-alginate hydrogel beads: vibration technique (a), coaxial air-flow technique (b), electrostatic extrusion (c), JetCutter® technology (mechanical cutting of liquid jet) (d).

These techniques are based on application of physical forces (vibration, pressure, electrostatic or mechanical) for dispersion of liquid (e.g., sodium-alginate) into small droplets. When droplets drop into gelling solution (e.g., with Ca^{2+} cations), spherical hydrogel beads are formed. For example, spherical shaped Ca-alginate hydrogel beads could be produced even without applying any of the presented techniques. However, the listed techniques provide better uniformity and control of bead size and allow production of beads with diameter less than 1000 μm (Prüsse et al., 2008). Ca-alginate hydrogels beads were found to be promising carriers for encapsulation of enzymes, especially when high stability and repeated usage of biocatalyst is required (Milovanović et al., 2007). In order to achieve high activity of immobilized enzyme, alginate hydrogel beads could be coated with an additional layer of polymer material (e.g., chitosan). By this modification it is possible to reach selective permeability of hydrogels (Taqieddin and Amiji, 2004). One of the problems concerning application of Ca-alginate as carrier for enzyme encapsulation is the enzyme leakage during bead production and application of immobilized enzyme. In order to prevent leakage of enzyme from Ca-alginate beads, the increased concentration of both alginate and gelling solution (e.g., calcium-chloride) should be applied (Blandino et al., 2000). Blandino et al. (2001) showed that immobilization of glucose oxidase within Ca-alginate beads for oxidation of glucose to gluconic acid was optimized by control of alginate and $CaCl_2$ concentration as well as duration of the gelling process. However, the authors reported that immobilized enzyme showed significant lower maximum reaction rate (0.09 ± 0.01 mM/min) compared to free enzyme (1.03 ± 0.11 mM/min). This could be explained by limited diffusion of glucose through alginate matrix and consequently by reduced number of glucose molecules that could reach the active sites of the enzyme. On the other hand, Ca-alginate hydrogel matrix provides better environment for the enzyme regarding critical parameters, such as pH and temperature. Tanriseven and Doğan (2001) studied immobilization of invertase into Ca-alginate and showed that immobilized enzyme was more stable under broader pH and temperature range. Further, the alginate properties as enzyme carrier could be improved by formation of composites with inorganic compounds. Coradin and Livage (2003) confirmed this by immobilization of β-galactosidase into Ca-alginate/ mesoporous silica composite. Immobilized enzyme exhibited better stability during the ageing test as well as less leakage compared to enzyme immobilized into pure Ca-alginate hydrogel.

Synthetic materials, such as poly(vinyl alcohol), could be also processed in the form of hydrogel beads for enzyme immobilization. Rebroš et al. (2007) used poly(vinyl alcohol) in the commercially available form (LentiKats®) for immobilization of invertase. The immobilized enzyme was stable for 8 months and retained high activity even after 45 repeated batch reaction cycles. Akgöl et al. (2001) reported immobilization of invertase on magnetic poly(vinyl alcohol) microspheres. According to the authors, magnetic poly(vinyl alcohol) microspheres with immobilized invertase were mechanically stable and were effectively used in packed bed reactor for sucrose hydrolysis.

Among the techniques for production of hydrogel beads, the electrostatic extrusion has attracted great attention in recent years. This technique is based on application of opposite charged electrodes (i.e., needle and gelling solution) that create an electrostatic field sufficient for breaking of the liquid jet into small particles with uniform size distribution (Bugarski et al., 1994; Poncelet et al., 1999; Bugarski et al., 2006; Nedovic et al., 2006). Knezevic et al. (2002) used electrostatic extrusion for production of Ca-alginate with immobilized lipase. The enzyme was loaded inside Ca-alginate beads with diameter less than 1mm. Ca-alginate beads exhibited high enzyme immobilization efficiency (> 90 per cent) and were used in repeated cycles without significant decrease in enzymatic activity.

Chitosan, as another natural polymer, was also found to be a promising carrier for enzyme immobilization (Žuža et al., 2011). Chitosan has desirable properties, such as biocompatibility, non-toxicity, gel formation and hydrophilicity that allow high enzyme retention and activity (Krajewska, 2004). The main advantage of chitosan application is the relatively simple procedure of particles preparation with a broad range of particle size (i.e., macro-, micro- and nanosized particles). The formation of particles is based on different methods: precipitation, emulsion cross-linking and ionic gelation. By using these methods of immobilization, it is possible to achieve relatively low activity loss (less than 5 per cent) of immobilized enzyme during storage period (Biró et al., 2008).

5. Nanocarriers for Enzyme Immobilization/Encapsulation

The carriers for enzyme immobilization designed in the form of nanoparticles, nanofibers or mesoporous materials have been intensively studied in recent years. The main advantages of nanomaterials as enzyme carriers are the large surface area, optimal pores sizes, possibilities for surface fictionalization, better mass transfer, etc. (Kim et al., 2006). In spite of the benefits of using nanomaterial in biotechnology, some limitations must be considered. The main concerns regarding nanomaterials are small size of particles (e.g., SiO_2 nanoparticles) that could penetrate through the cell's membrane as well as insolubility of some nanoparticles that could accumulate in certain tissues and organs (Tran and Chaudhry, 2010). The application of nanomaterials in the food sector requires complex regulations (both national and international). Nevertheless, any new technology or material applied in the food sector must be safe and must ensure that the quality of final product is in agreement with the standards (Gergely et al., 2010).

Inorganic materials, such as porous silica carrier materials, have been widely used as supports for immobilization of enzymes in the field of biotechnology. The efficiency of silica as a carrier for enzyme immobilization is defined by pore size that must be large enough for enzyme entrapping. Also, the surface properties of silica carrier can be modified in order to improve stability of the immobilized enzyme (Fan et al., 2003).

Production of silica particles in broad sizes (50–2000 nm) by hydrolysis of tetraethylorthosilicate (TEOS) is a well-established method (Li et al., 2012). The selection of the precursor for silica carrier production is important since some of them provide more hydrophilic properties of resulted silica carrier. On the other hand, methyl trimethoxysilane (MTMOS) as precursor gives more hydrophobic properties to silica (B. Wang et al., 2000). Besides the used technique for silica production, these materials were found to be highly efficient as carriers for immobilization of pharmaceutical products and enzymes (Popat et al., 2011).

In the food industry silica has been used as an anticaking agent. Also, silica can be applied as a carrier for encapsulation of food active compounds. The encapsulation on silica is based on adsorption of active compounds on the silica surface (Zuidam and Heinrich, 2010). Moreover, commercially available silica showed better performance when compared to organic carrier in the conventional encapsulation processes (Silva et al., 2012). Hence, there is an increasing interest in application of silica-based carriers for immobilization of active compounds and as biocatalysts.

Silica mesoporous materials are suitable for enzyme immobilization by adsorption onto the carrier, using the immersion method. The procedure is relatively simple and includes two steps: preparation of carrier material and adsorption-immobilization of enzyme onto porous silica. Results showed that enzyme immobilized on mesoporous silica exhibits high enzymatic activity in the organic solvent as well as high thermal stability. The pore size was found to be a critical factor for maintaining optimal enzymatic activity (Takahashi et al., 2000). Additionally, nanoscale mesoporous silica particles were used for separation of biomolecules. The main advantages of these materials are high efficiency of enzyme adsorption as well as fast separation of targeted biomolecule (Sun et al., 2006).

Some of the earliest applications of sol-gel techniques for enzyme immobilization on silica-based carriers was construction of biosensors. Li et al. (1996) used tetramethoxysilane (TMOS) as the precursor for production of silica carrier loading horseradish peroxidase. The immobilized enzyme was used as biosensor for hydrogen-peroxide detection. This system stabilized immobilized enzyme for as many as 35 days (retained activity 60 per cent). David et al. (2006) used sol-gel technique for preparation of porous silica carriers and immobilization of invertase. They found that the carrier surface could be additionally modified by amine groups. Further, amine groups were activated with glutaraldehyde in order to attach the enzyme to the carrier surface. Zhao et al. (2013) used specially designed capsules with silica core (epoxy functionalized silica nanospheres) and protective layer around them for better protection and prevention of enzyme leakage. After immobilization of lipase on functionalized silica, the mesoporous silica shell was created by addition of TEOS under controlled conditions. In this way it is possible to create yolk-shell spheres with porous structures around encapsulated enzyme.

Another way for production of high porosity silica is by formation of silica aerogels. These gels can be further modified for improving the enzyme retention and stability by surface modifications (Gao et al., 2009).

In addition, silica-based enzymatic biosensors could be constructed by addition of natural polymers. Silica sol-gel/chitosan film was found to be an adequate material that provides stable conditions for immobilization of peroxidase as hydrogen-peroxide biosensor. Such a system showed high stability of immobilized enzyme that was stored over a period of 30 days (Miao and Tan, 2001). Li et al. (2008) modified chitosan/silica sol-gel hybrid membranes loading peroxidase by addition of gold nanoparticles (for better conductivity) and potassium-ferricyanide, for better electron transfer between enzyme and electrode. Moreover, chitosan was found to be an appropriate template for production of silica-based materials with high porosity. Sodium-silicate used in this procedure could be obtained from renewable resources, such as rice husk ash. Further synthesis includes sol-gel process followed by heat treatment. Obtained porous silica showed better surface properties and thermal stability as compared to materials produced without using chitosan as template (Witoon et al., 2008). Additional protection of enzyme adsorbed on porous silica is possible by formation of a polymer layer around the system silica/enzyme. Catalase immobilized on porous silica and additionally protected with polymers exhibited high stability against enzyme-degrading substances (Y. Wang and Caruso, 2005).

Another approach for enzyme immobilization onto small-scale carriers is based on application of nano-structured carbon, titanium and zinc. In recent years, in many fields of science and industry, these materials have attracted great attention. Carrier materials based on carbon nanotubes were found to be suitable for production of biosensor and biocatalysts (J. Wang, 2005; Cang-Rong and Pastorin, 2009; Feng and Ji, 2011; Miyake et al., 2011). Biosensors with glucose oxidase immobilized on carbon nanotubes have been constructed and successfully tested for glucose detection (S.G. Wang et al., 2003; Besteman et al., 2003). Unmodified, as well as functionalized multi-walled carbon nanotube showed high potential as a carrier for immobilization of various enzymes, such as β-glucosidase (Gómez et al., 2005) or lipase (Prlainović et al., 2013), providing high enzyme activity and loading. Even more, the carbon nanotubes provide a platform for production of enzyme biosensors with other materials, such as chitosan (M. Zhang et al., 2004), silica (Ivnitski et al., 2008) and poly(acrylonitrile-co-acrylic acid) (Z.G. Wang et al., 2006).

Other inorganic materials, such as $CaCO_3$, were tested and showed very good characteristics as carriers for enzyme immobilization (Ghamgui et al., 2007). Moreover, $CaCO_3$ exhibits excellent thermal properties, providing good protection of lipase at elevated temperatures. Also, high immobilization efficiency, possibilities for its use in repeated cycles and in different media (with or without organic solvent) give $CaCO_3$ an advantage over some other inorganic materials, such as silica gel or amberlite (Ghamgui et al., 2004).

As we pointed out above, the magnetic microparticles could be used for additional protection of the enzyme. Also, magnetic nanoparticles were found to be good materials for enzymes and protein immobilization (Koneracká et al., 1999; Xu et al., 2014). The simple recovery of magnetic particles (by magnetic field) is the major benefit of their application in biotechnology (Franzreb et al., 2006).

Mizuki et al. (2010) reported successful immobilization of α-amylase onto the surface of magnetic particles, without modification of the particle's surface prior to immobilization. Even more, the immobilized enzyme exhibited strong activity in the rotational magnetic field. These findings may be a promising step in the development of new enzyme reactor systems, such as micro reactors. According to Baskar et al. (2015), one of the main advantages of α-amylase immobilized onto magnetic nanoparticles is that this kind of carrier material provides simple recovery of immobilized enzyme from the reaction solution.

Additional modification of magnetic nanoparticles can lead to better retention and activity of the immobilized enzyme. Example of such modification is Cu-modified magnetic nanoparticles that showed better characteristics as carriers for lipase immobilization as compared to unmodified carrier (Woo et al., 2015). The properties of magnetic nanoparticles as carriers for enzyme immobilization could be further improved by formation of a composite based on natural polymers. Fe_3O_4/chitosan nanoparticles can be relatively easily produced by the co-precipitation method in which Fe^{2+} and Fe^{3+} ions are mixed with chitosan followed by precipitation with NaOH and addition of enzyme for formation of enzyme/carrier complex (Y.X. Wang et al., 2015). The same procedure can be applied for production of magnetic-chitin particles for loading α-amylase (Sureshkumar and Lee, 2011).

Besides, the nanocarriers for enzyme immobilization can be produced, using natural or synthetic polymer materials. These materials are relatively easily processed into nanoparticles or nanofibers using adequate techniques (Kim et al., 2006). Among natural polymer materials, chitosan, alginate, cellulose, gelatine, agarose, dextran and collagen have been intensively studied as nanocarriers for enzyme immobilization (Talbert and Goddard, 2012). Among these potential carriers, chitosan was found to be especially suitable for production of nanoparticle and nanofibers as support for enzyme immobilization (Tang et al., 2006; Zhao et al., 2011). Ionic gelation method is a relatively simple method for chitosan nanoparticles production, based on interaction between chitosan and tripolyphosphate and could be controlled by variation of process conditions, such as pH, concentration, method of mixing, etc. (Zhao et al., 2011). Also, chitosan nanoparticles can be prepared by using sodium sulfate for gelation, while immobilization of enzyme is performed after particles formation (Biró et al., 2008). Tang et al. (2006) tested chitosan nanoparticles prepared by the ionic gelation method for immobilization of neutral proteinase. Immobilized neutral proteinase showed better properties, especially resistance towards extreme process conditions compared to the free enzyme. Furthermore, chitosan could be processed into nanofibrous membrane for enzyme immobilization. Huang et al. (2007) used chitosan/poly(vinyl alcohol) membrane produced by an electrospinning process for lipase immobilization. The fiber diameter was in the range of 80–150 nm and was subsequently reinforced by chemical treatment. In this way the produced membranes were good carriers for preservation of enzyme activity (up to 30 days) and were used in repeated cycles.

Recently, the electrospinning process has been applied for production of various types of nanofibrous membranes for enzyme immobilization. Li et al. (2007) have described polyacrylonitrile membranes obtained by electrospinning process for lipase immobilization via covalent bonds. Membrane showed satisfactory mechanical properties while immobilized lipase retained high activity even after 10 repeated batches. It was shown that immobilized lipase on polyacrylonitrile nanofibrous membrane was a promising biocatalyst for soybean oil hydrolysis (Li and Wu, 2009). The properties of nanofibrous membrane produced by electrospinning depend on characteristics of polymer and process conditions. For example, poly(vinyl alcohol) can be relatively easily processed into nanofibrous membrane by electrospinning, without addition of other polymers and using relatively simple equipment (Lević et al., 2014). Y. Wang and Hsieh (2008) showed that electrospun poly(vinyl alcohol) is suitable for immobilization of lipase with high retained enzyme activity. However, chemical cross-linking of membrane with glutaraldehyde in ethanol caused reduced activity of immobilized lipase. Also, the activity of immobilized enzyme can be improved by addition of different additives into polymer solution prior to electrospinning (G.Z. Wang et al., 2006). Synthetic nanofibers are further modified by creating the biomimetic support on fiber surface in order to improve the enzyme retention. Gelatine and chitosan were found to be adequate as additional supports for enzyme immobilization on synthetic electrospun membrane (Ye et al., 2006).

6. Immobilization of Enzymes *via* Cross-Linked Enzyme Aggregates

Conventional techniques for enzyme immobilization are based on interaction between the enzyme and the carrier material. However, this approach involves application of a carrier that represents additional mass without catalytic properties and hence causes limited catalytic activity of immobilized system. The solution for this problem could be use of cross-linked enzyme as well as cross-linked enzyme aggregates, which are essentially immobilized enzymes without external solid carrier. Immobilization of enzymes via cross-linked enzyme aggregates is a relatively simple method since it can be performed in one single operation combined with enzyme isolation and purification (Sheldon, 2011). The main advantages of cross-linked enzymes are high stability and high activity of enzyme aggregates (Cao et al., 2003; Cao, 2005). For industrial applications, cross-linked enzyme aggregates could be a good alternative for conventional immobilization procedure because formation of aggregates is inexpensive (since the immobilized enzyme (i.e., aggregates) is formed without carrier material). Furthermore, cross-linking could be applied simultaneously for more different enzymes in order to provide immobilized biocatalysts capable of multiple enzymatic reactions (Sheldon, 2011).

Formation of stable cross-linked enzymes is a chemical process that includes enzyme treatment by chemical agents in order to change enzyme solubility and cause formation of aggregates that could be further processed (e.g., cross-linking and drying) (Tandjaoui et al., 2015). The chemicals used for enzyme aggregates formation are substances commonly applied for protein purification (without denaturation of protein structure) (Cao, 2005). Organic solvents, such as ethanol, acetone, isopropanol, or salt, such as ammonium-sulfate, have been used for enzyme precipitation during formation of enzyme aggregates (Kartal et al., 2011). It is pointed out that there is no unique procedure for enzyme aggregate formation since selection of adequate precipitation agent must be carried out regarding enzyme properties, cross-linking procedures, etc. (Kartal et al., 2011; Tandjaoui et al., 2015). The size of enzyme aggregates produced by the cross-linking enzyme technique is generally below 10 μm (Cao, 2005). After aggregation, usually the cross-linking of enzyme aggregates is carried out in order to prevent aggregate degradation and loss of immobilized enzyme (Kartal et al., 2011; Tandjaoui et al., 2015). Furthermore, for production of cross-linked enzyme aggregates with high enzyme activity and stability, it is necessary to establish optimal conditions such as type of precipitation agents and concentration of cross-linking agent (Perez et al., 2009). Kartal et al. (2011) showed that lipase aggregates obtained by precipitation with isopropanol exhibited higher activity. However, after cross-linking with gutaraldehyde the aggregates produced by ethanol precipitation showed higher activity. According to the authors, these results could be explained by the differences that occurred in the conformation of enzymes during aggregation and effects of cross-linking process on such a structure.

Tandjaoui et al. (2015) prepared cross-linked enzyme aggregates of *Brassica rapa* peroxidase by solvent precipitation and glutaraldehyde cross-linking. Enzyme aggregates were formed by solvent precipitation from crude *Brassica rapa* extracted juice without any purification. The immobilized enzyme (after cross-linking with glutaraldehyde and drying) was in the form of aggregates with a large surface area. The activity and formation of enzyme aggregates were affected by properties of precipitation solvent. Immobilized *B. rapa* peroxidase exhibited better storage properties and higher activity as compared to free enzyme.

Co-immobilization of two or more enzymes by formation of cross-linked enzyme aggregates could be a promising alternative for conventional enzymatic processes. Namely, by combining several different enzymes into structure of aggregates can provide transformation of substrate to required products in the single vessel under relatively unfavorable conditions, such as elevated temperature. One of the examples of co-immobilized cross-linked enzyme aggregates is a system composed of α-amylase, glucoamylase and pullulanase used for starch hydrolysis. Co-immobilized enzymes exhibited excellent performances, such as high conversion of substrate, enhanced thermal stability and high activity even after five repeated cycles (Talekar et al., 2013).

Further, formation of spherical enzyme aggregates could be achieved by the emulsification process via creation of water-in-oil emulsion followed by cross-linking with glutaraldehyde. Prior to emulsification, the enzyme was dissolved in the water phase and then dispersed in the oil by mechanical stirring and finally being chemically cross-linked. This procedure provided spherical aggregates with diameter in the range of 20–60 μm and with high activity and stability of immobilized enzyme (Chen et al., 2014).

Some limitations of enzyme cross-linked aggregates can restrict their application in real industrial processes. These problems, such as negative influence of cross-linking agents on enzyme activity or small particle size can be overcome by introduction of a protective substance or additional carrier materials.

Stability of cross-linked enzyme aggregates could be further improved by adsorption, for example, onto silica gel. Such an approach shows better thermal stability of immobilized enzyme as well as higher activity (Nguyen and Yang, 2014). Production of enzyme aggregates with high porosity and enhanced enzyme activity can be achieved by co-immobilization using organic support, such as polyethyleneimine and cross-linking by glutaraldehyde (Vaidya et al., 2012). Addition of some compounds, such as bovine serum albumin can also improve stability of cross-linked enzyme aggregates (Kim et al., 2013; Tükel et al., 2013; Agyei and He, 2015). Some cations were also found to be protective agents for cross-linked enzymes. Torabizadeh et al. (2014) discovered that addition of calcium and sodium ions enhanced the catalytic activity of cross-linked α-amylase aggregates.

An alternative for support of cross-linked enzymes could be formation of magnetic cross-linked enzyme aggregates. These aggregates can be synthesized in the presence of functionalized magnetic materials (e.g., amino-functionalized materials). The results showed higher stability of magnetic cross-linked xylanases aggregates (up to 90 per cent) as compared to free or cross-linked enzyme aggregates without support (Bhattacharya and Pletschke, 2014). Talekar et al. (2012) showed that amino functionalized magnetic nanoparticles increased the cross-linking efficiency and provided better activity of immobilized α-amylase. Another advantage of magnetic nanoparticles with cross-linked enzyme is simple recovery from liquid using the magnetic field (Talekar et al., 2012; L. Liu et al., 2015). Additional performance improvement of magnetic cross-linked enzyme aggregates could be achieved by exposing the aggregate to magnetic field during enzyme reaction. According to literature data, lipase magnetic cross-linked aggregates exposed to alternating magnetic field showed improved catalytic activity (Y. Liu et al., 2015). Also, magnetic particles provide numerous possibilities for modification and improvement of their characteristics. For example, W.W. Zhang et al. (2015) used surfactants for surface activation of magnetic particles and production of lipase magnetic cross-linked aggregates with improved activity during storage.

7. Enzyme Encapsulation by Spray Drying

Spray drying is the most frequently used technique for encapsulation of active compounds in the food industry. This is a simple method for production of large quantities of encapsulates by rapid evaporation of solvent from solution. Prior to drying, the solution is dispersed into small droplets in the drying chamber, usually heated by hot air. At this time, hot gas transfers heat to droplets, causing solvent evaporation. During solvent evaporation, dried particles are formed with dimensions in the range 10–400 μm. Spray drying combined with natural carrier material is suitable for encapsulation of bacterial cells, vitamins, enzymes, aromas, etc. Materials, such as maltodextrins, gum Arabica, alginate and different proteins and protein isolates have been used as carrier materials in the spray-drying process. Relative simplicity and well-established procedures are main advantages of spray drying regarding encapsulation of food ingredients. However, applying high temperature is harmful for thermal sensitive compounds and could cause irreversible changes in their structure (Zuidam and Shimoni, 2010).

Nevertheless, under appropriate drying conditions, enzyme activity can be retained even at higher temperatures. Yamamoto and Sano (1992) showed that β-galactosidase, glucose oxidase and alkaline phosphatase in solutions of different sugars remain active in the form of suspended droplets during drying. Smaller droplets, heat treatment, type of sugar and specific design of droplets were found to be critical for preservation of enzyme activity. Generally, lower temperature, smaller droplets as well as foamed structure of droplets provide better conditions for enzyme retention. According to the authors, under appropriate conditions, it is possible to preserve enzyme activity even at temperatures above 80°C. Alloue et al. (2007) demonstrated that spray-dried lipase from *Yarrowia lipolytica* was stable over a period of one year. Also, the addition of different additives (e.g., skim milk powder, gum Arabic, maltodextrin and calcium-chloride) was found to be critical in successful spray drying and production of fluent powder. The activating and stabilizing role of Ca^{2+} ions (from calcium-chloride) is especially important for thermostability and catalytic activity of lipase.

Furthermore, stability of spray-dried enzymes can be improved with addition of protein compounds in the feed solution. The activity of spray-dried alcohol dehydrogenase may be preserved by addition of bovine serum albumin and β-lactoglobulin as protein compounds in the feed solution of trehalose (Yoshii et al., 2008). Also, the negative influence of high temperature on enzyme activity during spray drying could be overcome by careful choice of process conditions as well as a combination of carrier materials. Cui et al. (2006) showed that spray-dried microbial transglutaminase exhibited high activity after storage for 60 days. Moreover, the spray-dried transglutaminase showed higher activity after storage as compared to freeze-dried enzyme. High activity (90 per cent) of spray-dried transglutaminase was achieved by combination of maltodextrin and reduced glutathione as carrier material and under low outlet temperature (~60°C). However, in some spray-drying processes, sugar-based carriers are insufficient for preservation of enzyme activity. Branchu et al. (1999) showed that β-galactosidase activity is damaged by spray-drying in the presence of sucrose. On the other hand, use of hydroxypropyl-β-cyclodextrin as a carrier improved enzyme stability. Application of different protecting substances, such as carboxymethylcellulose sodium salt, polyacrylic acid sodium salt or lactose combined with sodium-alginate could prevent enzyme inactivation during the spray-drying process (Coppi et al., 2002).

8. Applications of Immobilized Enzymes in Food Processes

Enzymes are widely used in the food sector for fruit juices and sweetener production, food aroma modification, modification of lipids, etc. However, the price of enzymes and conditions of some enzymatic processes require application of immobilized enzymes. Simple recovery of enzyme after enzymatic transformation is critical for economic sustainability of some industrial processes. The most important applications of immobilized enzyme in the food industry as well as some of the promising solutions for food enzyme immobilization are reviewed below.

8.1 Production of high fructose syrup by immobilized enzyme

The production of high fructose syrup is one of the most important applications of immobilized enzyme technology in the industrial sector (Olsen, 1995). Chemical conversion of glucose to fructose is a long known process. However, this conversion requires relatively high pH as an adequate environment for synthesis of various undesirable products. For this reason, enzymatic isomerization of glucose to fructose is the main process for production of high fructose syrup nowadays (Bhosale et al., 1996).

High fructose syrup is a sweetener widely used in the food industry (Deshpande and Rao, 2006) and it already takes over one-third of the sweetener market in the US (8 million tons annual production) (Olsen, 1995). In the industrial production of high fructose syrup, starch from corn is commonly used as a precursor (Fig. 4). In the first step, starch is hydrolyzed to glucose by enzymes (i.e., α-amylase and glucoamylase), followed by isomerization of glucose to fructose. The isomerization of glucose to fructose is an enzymatic reaction catalyzed by glucose isomerase enzyme. The main advantage of high fructose syrup over sucrose (both have similar fructose and glucose ratio = 50:50) is that fructose is ~75 per cent sweeter than sucrose. This property of fructose is the main reason for wide application of high fructose syrup in the food industry as a sweetener. Usually, high fructose syrup is produced with 42 per cent fructose while other grades are also available (55 per cent and 90 per cent fructose) (Hanover and White, 1993; Y. Zhang et al., 2004).

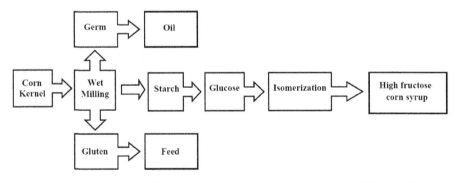

Figure 4. Simplified scheme for high fructose corn syrup production (Hanover and White, 1993).

However, the isomerization of glucose to fructose is an equilibrium-controlled reaction with the equilibrium constant depending on temperature. The fructose equilibrium concentration reaches 48 per cent at 45°C and 55 per cent at 85°C. The processing at elevated temperature, however, is not possible due to limited enzyme thermal stability. Hence, the ability to design an efficient immobilized glucose isomerse system being stable at elevated temperature for extended periods of time would be a major advance in this technology. The Clinton Corn Processing Company, IA, was the first company which commercially produced fructose syrups using glucose isomerase immobilized on a cellulose ion-exchange polymer in a packed-bed reactor in 1974 (Vasic-Racki, 2006). Since then, packed-bed reactor with immobilized glucose isomerase is in use in the industrial-scale production of high fructose syrup. This process depends on numerous factors, such as pH, temperature and quality of glucose (the quality must be maintained constantly) (Tomotani and Vitolo, 2007; DiCosimo et al., 2013). Other factors, such as Ca^{2+} and Mg^{2+} ion concentrations, the presence of impurities and dissolved oxygen are also important and could limit production and process efficiency. Additionally, the process must be carried out continually since products should be removed from the reactor in order to prevent formation of byproducts. It is important to emphasize that this production is considered economic only by using immobilized enzyme. The commercially available glucose isomerase, such as Sweetzyme® T (Novo Nordisk) is widely used for isomerization of glucose and production of high fructose syrup. According to literature data, this enzyme has been isolated from *Streptomyces murinus.* Further, the immobilization procedure consists of cells disruption, cross-linking with glutaraldehyde, flocculation, extrusion and drying to the final form (Olsen, 1995).

Besides commercially available immobilized glucose isomerase, the new immobilization procedures, reactor systems and carriers are constantly developed. For example, commercially available carriers provide good support to immobilization of glucose isomerase. Tükel and Alagöz (2008) applied glucose isomerase immobilized on Eupergit® C 250 L for fructose production. This procedure showed promising results regarding efficiency and reusability of immobilized enzyme.

Further improvement of glucose isomerization process is achieved by modification of the reactor system in order to produce a syrup with high fructose content. For example, the simulated moving bed reactor (SMBR) combines two basic processes: chemical/biochemical reaction and a chromatographic separation process. The whole system is designed as a single unit and both (i.e., chemical/biochemical reaction and separation) are integrated. The system is optimized for application of commercially available immobilized enzymes and for production of high fructose syrup with more than 90 per cent of fructose (Da Silva et al., 2006).

At this point, other alternatives for production of high fructose syrup from different sources that have been also investigated should be mentioned. One such attempt is based on using immobilized dead cells of *Kluyveromyces marxianus* immobilized in gelatine for hydrolysis of inulin from Jerusalem artichoke. Using this method of immobilization, inulinase is entrapped inside the gel structure (within cell biomass) of carrier and could be repeatedly used for up to 10 cycles (Bajpai and Margaritis, 1985). Wenling et al. (1999) used partially purified intracellular inulinase from *Kluyveromyces* sp. for continuous preparation of fructose syrup from Jerusalem artichoke. Prior to the enzymatic process, inulinase was immobilized on macroporous polystyrene beads. Enzyme was adsorbed onto polystyrene beads followed by cross-linking with glutaraldehyde. Immobilized enzyme was found to be stable in the packed continuous bed-column reactor for 32 days. Other carrier materials for inulinase immobilization have been reported in literature, such as Duolite™ A568 (weak base anion resin) and glutaraldehyde (Singh et al., 2007) or chitin (Nguyen et al., 2011). The system based on resin immobilized exoinulinase showed great potential for hydrolysis of inulin in the continuous reactor. Moreover, the final product of inulin hydrolysis had more than 95 per cent of fructose (Singh et al., 2008).

Also, it was noticed that inulinase from new sources could improve fructose production from inulin. The *Aspergillus niger* and *Candida guilliermondii* showed a high potential as sources of inulinase with high activity (Sirisansaneeyakul et al., 2007).

Padmaja and Moorthy (2009) reported using cassava and sweet potato roots for production of glucose and high fructose syrup by direct conversion techniques. Since conventional processes of starch hydrolysis from these plants are not economically sustainable, the direct conversion of cassava and sweet potato root slurry with new (improved) enzymes could be applied. After hydrolysis (saccharification) at room temperature using Stargen™ (for starch hydrolysis, Genencor International Inc., USA), glucose was converted to high fructose syrup by Sweetzyme® T (immobilized glucose isomerase from Novo Nordisk).

The other solutions for high fructose syrup production are based on sucrose hydrolysis by invertase. This process could be improved by using immobilized invertase in the continuous-membrane bioreactor. Invertase immobilized on the polystyrene anionic resin exhibits high stability and activity. Further, enzymatic-produced inverted sugar can be submitted to chromatographic separation for production of high fructose syrup with 70 per cent of fructose (Tomotani and Vitolo, 2007).

8.2 Debittering of fruit juices

Citrus juices are widely produced in many countries since there is a growing demand for this kind of food product. However, the problem of bitter substances that are usually present in citrus fruits affects the sensorial properties of final products (i.e., juices) and, consequently, the consumers. The two main compounds of citrus bitter taste have been identified as naringin (identified in the fruit membrane and albedo) and limonin (Puri et al., 1996a).

It should be emphasized that the problem of high bitterness is not limited to citrus products only. For example, olive bitterness that originates from phenolic compounds is also considered as a problem in industrial processing and a new procedure has been developed to overcome the disadvantages of conventional debittering (Habibi et al., 2016).

The primarily bitter taste of citrus juice comes from naringin (prime causer of bitter taste), while limonin is related to delayed bitterness that is created after fruit processing and juice production. During the years, the following processes for debittering of citrus juices have been developed (Puri et al., 1996a):

- Adsorption of bitter compounds or adsorption/ultrafiltration treatment—Adsorption of limonin by polyamides or application of other materials, such as nylon matrices, cellulose acetate and ion exchangers for corrections of citrus juices taste (i.e., bitterness reducing)
- Chemical treatments-correction of juice pH, treatment with CO_2 or ethylene treatment
- Application of polystyrene resins for debittering—efficient method for juice debittering (80–90 per cent of bitter compounds removal)
- Entrapment of naringin by cyclodextrin-formation of inclusive complex between cyclodextrin and naringin

- Debittering by immobilized microbial biomass-using microbial cells for degradation of substances that contribute to juice bitter taste
- Debittering by immobilized enzyme-application of immobilized naringinase isolated from different microorganisms

Although some of these processes provide high debittering efficiency and good quality final products, some limitations, such as removal of valuable compounds (Lee and Kim, 2003) as well as variation between batches or formation of additional waste, restrict their wider use for juice treatment (Puri et al., 1996a). The adsorption of bitter compounds onto porous resins is most frequently used in industrial processing of citrus juices (Stinco et al., 2013). In recent years, numerous studies to improve adsorption process for removal of bitter compounds using new adsorbents have been conducted (Liu and Gao, 2015; Bao et al., 2015). However, regarding antioxidant and biomolecule protection capacity of fresh grapefruit juice, enzymatic debittering using naringinase showed better results than adsorption by resins (Cavia-Saiz et al., 2011). The new approaches for naringinase purification are promising for improved enzyme characteristics and potential applications in industrial processes (Shanmugaprakash et al., 2015). Also, the new researches for introduction of methods for identification of potential sources of naringinase have been conducted (Caraveo et al., 2014).

One of the earliest techniques for immobilization of naringinase was based on cellulose triacetate as a carrier material. The tests in the column reactor showed great potential of such a system, which ensured better thermal stability of immobilized enzymes, satisfactory efficiency and good hydrodynamics of process (Tsen et al., 1989). Puri et al. (1996b) used naringinase immobilized in alginate beads for debittering kinnow juice. According to the authors, immobilized enzyme showed high efficiency (82 per cent naringin hydrolyzed for 3 hours), better thermal stability and good mechanical properties. More recently, Ribeiro et al. (2010) reported application of naringinase immobilized in Ca-alginate beads for removal of the bitter taste compound (i.e., naringin). The main difference compared to previous application of immobilized naringinase in Ca-alginate beads is that the authors applied high pressure at different temperatures for enzymatic hydrolysis of naringin. They discovered that optimal conditions for hydrolysis of naringin by immobilized naringinase were pressure of 205 MPa and 60°C for 30 min. Under optimal conditions, immobilized naringinase provided 81 per cent of naringin conversion. Busto et al. (2007) used poly(vinyl alcohol) cryogels for naringinase immobilization and hydrolysis of naringin. The carrier was produced in the form of beads by dropping polymer/enzyme mixture into liquid nitrogen. Immobilized enzyme exhibited satisfactory stability over six consecutive cycles. Nunes et al. (2014) immobilized naringinase in the poly(vinyl alcohol) lens-like carriers for application in bioreactor conditions. Results of naringin hydrolysis showed that immobilized enzyme could be used for 40 days in a continual process. Lei et al. (2011) reported immobilization of naringinase onto mesoporous molecular sieve (MCM-41) and applied immobilized enzyme for debittering white grapefruit. The immobilization was carried out by adsorption of naringinase onto the sieve in the solution that contained glutaraldehyde. The immobilized enzyme was used in six cycles, while thermal and storage stability of naringinase were also improved.

8.3 Enzymatic transformation of lipids

Interesterification of lipids is carried out in the food industry for modification of oils and fats physico-chemical properties. The enzymatic interesterification is performed by lipase under specific conditions. Although chemical interesterification is also used, the enzymatic process is more attractive since specific lipase could be used for targeted triglyceride modifications. Hence, the main interest in the immobilization of lipase for lipids interesterification is enzyme reuse in repeated cycles (Nakashima et al., 1988; DiCosimo et al., 2013).

The enzymatic interesterification of lipids by lipase is limited by water content. A high water content leads to hydrolysis of ester bonds of triglycerides, while low water content results in lipase-catalyzed esterification or transesterification of lipids. However, some water percentage is required for maintaining enzyme activity (Ison et al., 1988; DiCosimo et al., 2013). Usually, interesterification of lipids by lipase is controlled by adequate amounts of the organic solvent that controls enzyme activity as well as hydrodynamic parameters of enzyme reactor (Ison et al., 1988).

Although lipase isolated form microbes is widely used for interesterification of lipids, the characteristic of enzyme depends on sources (i.e., microorganism). Non-specific lipase from some *Candida* species can catalyze exchanges of groups at all three positions of triglycerides (i.e., on positions *sn*-1, *sn*-2 and *sn*-3). However, lipase from some *Mucor*, *Rhizopus* or *Asperillus* species is used for changes of acyl groups primarily on the *sn*-1,3 positions. Using *sn*-1,3 region-specific lipase can provide products that are chemically different when compared to products obtained by chemical interesterification (DiCosimo et al., 2013). Filamentous fungi from genus *Rhizopus* are good sources for *sn*-1,3 specific lipase. Nakashima et al. (1988) used cells of *Rhizopus chinensis* dried with acetone and immobilized on biomass particles for interestierification of olive oil.

However, from a practical point of view, immobilization of isolated lipase attracted more attention and has been more commercialized than immobilization of lipase-rich whole cells for interesterification of lipids. Numerous studies were conducted in the past decades and several immobilized lipases made commercially available. Some earliest attempts for lipase immobilization and interestierification of lipids were based on using diatomaceous earth as enzyme carrier (Wisdom et al., 1985; Ison et al., 1988; Mojović et al., 1993). The commercially available immobilized lipases (non-specific and *sn*-1,3 specific) have been immobilized on different inorganic materials, such as silica, diatomaceous earth or ceramic particles, while lipase immobilized on synthetic materials, like macroporous resin or polymer esters, is also available (DiCosimo et al., 2013).

Immobilized *sn*-1,3 region-specific lipase isolated from *Rhizopus arrhizus* could be used for production of modified lipids as a substitute for natural products, such as cocoa butter. Immobilized lipase was efficiently used for such modification and production of cocoa butter substitute by palm oil interesterification with stearic acid. Celite® 545 showed excellent properties as carrier for lipase immobilization, providing high activity of immobilized enzyme. Moreover, addition of a protective compound, such as soya lecithin, can increase process efficiency. As interesterification by lipase requires non-conventional media, such as water/non-polar solvent, it seems that soya lecithin additionally protects enzyme from solvent and also increases diffusion of substrate and reaction products (Mojović et al., 1993). Reactor design was also found to be important for efficiency of interesterification of lipids by immobilized lipase. Some reactor types, such as air-lift reactor, can provide better conditions for lipase activity with respect to reaching equilibrium state as compared to conventional reactors (Mojović et al., 1994). Production of lipids that can be used as substitutes for natural fats is also performed by lipase immobilized on macroporous resin. Such a catalyst showed 6.9 higher reaction rate when compared to free enzyme. Interesterification of tea-seed oil, methyl-palmitate and methyl-stearate with lipase immobilized on macroporous resin provided products that can be used as cocoa butter substitutes (X.H. Wang et al., 2006). Sol-gel technique has been used for immobilization of lipase for milkfat-soybean oil blends interesterification (Nunes et al., 2011).

In recent years, there is a growing interest in the food industry for replacement of organic solvents used in some processes with more economically sustainable solutions, such as processes with immobilized enzymes. Ilyasoglu (2013) reported interesterification of tripalmitin with extra virgin olive oil and flaxseed oil by Lipozyme® TL IM in the medium free from organic solvent. The purpose of such interesterification was to develop human fat milk analogue with α-linolenic acid. The main quality parameters of the final product were further improved by addition of antioxidants. Jenab et al. (2014) used supercritical carbon dioxide as an environment for interesterification between canola oil and fully-hydrogenated canola oil. Lipozyme® RM IM and TL IM showed high stability under high pressure batch-stirred reactor for up to four cycles. Moreover, Da Silva et al. (2012) introduced a continuous process of lard (70 per cent) and soybean oil (30 per cent) blend interesterification by Lipozyme® TL IM. The blend was submerged to interesterification under different flow rates and the quality of products was analyzed by DSC analysis. The higher flow rates showed better results regarding productivity of the process as compared to low flow rates.

Also, one of the efficient ways for immobilization of lipase is entrapping of enzyme inside the structure of Ca-alginate gel beads. Under optimized conditions, Ca-alginate beads loading mixed lipases can be used for converting waste cooking oil into biodiesel for 60 hours with high yield (over 90 per cent) at relatively low temperature of 35°C (Razack and Duraiarasan, 2015).

8.4 Other applications of immobilized enzyme in food processes

Pectinolytic enzymes are one of the most frequently used enzymes in the food industry. Some estimation suggests that up to 25 per cent enzymes used in the food industry are pectinolytic enzymes. These enzymes have been applied for fruit juice extraction and clarification, plant oil extraction, beverage production, tea and coffee treatment, etc. Pectinolytic enzymes hydrolyze the pectin substances in the structure of the plant material and are present in microorganisms and plants where they conduct many important physiological functions (Jayani et al., 2005).

Spagna et al. (1995) tested several organic and inorganic carriers for immobilization of pectinlyase from *Aspergillus niger*. The glutaraldehyde-activated bentonite was found to be a suitable carrier for pectinlyase immobilization since it provides high enzyme activity, while being relatively inexpensive and possessing good mechanical properties. Demir et al. (2001) revealed the high potential of commercial pectinase immobilized onto ion exchange resin particles for carrot puree treatment. Immobilized enzyme was used up to five cycles with only 6.5 per cent activity loss. Busto et al. (2006) used Ca-alginate beads for pectinlyase immobilization and fruit juice treatment. Both, free and immobilized enzymes exhibited similar pH and temperature stability. However, the immobilized enzyme was used in four cycles for juice treatment which is the main advantage of such a system as compared to the free enzyme.

One of the potential applications of immobilized enzyme in the food industry is enhancing of wine aroma. Since many volatile grape compounds are chemically bonded into the structure of non-volatile compounds, their release during wine production is considered necessary. The treatment of wine aroma precursors with enzyme extracted from *A. niger* showed increasing volatile compounds concentration in the final product (DiCosimo et al., 2013). Application of the immobilized enzyme could be the next step in the development of sustainable system for grape mash treatment during wine production. It was showed that immobilized fungal glucosidases could be used for extra release of volatile compounds from grape mash (Caldini et al., 1994; DiCosimo et al., 2013).

Additionally, pure food aroma compounds could be also produced by immobilized enzymes. Such systems provide production of high purified compounds compared to conventional chemical synthesis that requires additional separation and purification steps (Knežević-Jugović et al., 2008; Damnjanović et al., 2012).

9. Summary and Conclusion

Enzymes have a long tradition of being used in food processing, in processes such as beer and wine production, fruit juices clarification, cheese production and meat and sugar processing. For the last twenty years, immobilized/encapsulated enzyme technology has been widely investigated as a solution to a number of practical problems in the use of enzymes, especially, limited stability under conditions other than optimal. Immobilized/encapsulated enzymes provide many benefits compared to free enzymatic catalysts and these include increased product yield (due to improved activity, stability and selectivity), repeated use, possibility for continuous operation, simple handling and separation from the product. On the other hand, altered enzymatic activity and high operational costs are the main constrains in a wider use of immobilized/ encapsulated enzymes on an industrial scale. This chapter is a comprehensive review of the current state of immobilization/encapsulation technology for food enzymes and their application in food industrial processes. Different methods (classified as binding to a prefabricated support-carrier), entrapment in organic or inorganic matrices and cross-linking of enzyme molecules for the immobilization/encapsulation of enzymes were critically reviewed. Special focus was given to recent developments, such as the use of novel supports, e.g., mesoporous silicas, hydrogels, smart and nano-structured polymers, novel entrapment methods and cross-linked enzyme aggregates. Some industrial applications of immobilized/encapsulated enzymes, like production of high fructose corn syrup, clarifications and taste modifications of fruit juices and modification of food oils were also discussed.

References

Agyei, D. and He, L. 2015. Evaluation of cross-linked enzyme aggregates of *Lactobacillus* cell-envelope proteinases, for protein degradation. Food and Bioproducts Processing, 94: 59–69.

Ai, Z., Jiang, Z., Li, L., Deng, W., Kusakabe, I. and Li, H. 2005. Immobilization of *Streptomyces olivaceoviridis* E-86 xylanase on Eudragit S-100 for xylo-oligosaccharide production. Process Biochemistry, 40: 2707–2714.

Akgöl, S., Kaçar, Y., Denizli, A. and Arıca, M.Y. 2001. Hydrolysis of sucrose by invertase immobilized onto novel magnetic polyvinylalcohol microspheres. Food Chemistry, 74: 281–288.

Alftrén, J. and Hobley, J.T. 2014. Immobilization of cellulase mixtures on magnetic particles for hydrolysis of lignocellulose and ease of recycling. Biomass and Bioenergy, 65: 72–78.

Alloue, M.A.W., Destain, J., Amighi, K. and Thonart, P. 2007. Storage of Yarrowia lipolytica lipase after spray-drying in the presence of additives. Process Biochemistry, 42: 1357–1361.

An, N., Zhou, H.C., Zhuang, Y.X., Tong, S.D. and Yu, H.W. 2015. Immobilization of enzymes on clay minerals for biocatalysts and biosensors. Applied Clay Science, 114: 283–296.

Andjelković, U., Milutinović-Nikolić, A., Jović-Jovičić, N., Banković, P., Bajt, T., Mojović, Z., Vujčić, Z. and Jovanović, D. 2015. Efficient stabilization of *Saccharomyces cerevisiae* external invertase by immobilisation on modified beidellite nanoclays. Food Chemistry, 168: 262–269.

Arasaratnam, V., Galaev, I.Y. and Mattiasson, B. 2000. Reversibly soluble biocatalyst: Optimization of trypsin coupling to Eudragit S-100 and biocatalyst activity in soluble and precipitated forms. Enzyme and Microbial Technology, 27: 254–263.

Babich, L., Hartog, A.F., Van Der Horst, M.A. and Wever, R. 2012. Continuous-flow reactor-based enzymatic synthesis of phosphorylated compounds on a large scale. Chemistry—A European Journal, 18: 6604–6609.

Bajpai, P. and Margaritis, A. 1985. Immobilization of *Kluyveromyces marxianus* cells containing inulinase activity in open pore gelatin matrix: 2. Application for high fructose syrup production. Enzyme and Microbial Technology, 7: 459–461.

Bao, Y., Yuan, F., Zhao, X., Liu, Q. and Gao, Y. 2015. Equilibrium and kinetic studies on the adsorption debittering process of ponkan (*Citrus reticulata* Blanco) juice using macroporous resins. Food and Bioproducts Processing, 94: 199–207.

Baskar, G., Afrin Banu, N., Helan Leuca, G., Gayathri, V. and Jeyashree, N. 2015. Magnetic immobilization and characterization of α-amylase as nanobiocatalyst for hydrolysis of sweet potato starch. Biochemical Engineering Journal, 102: 18–23.

Bayramoğlu, G., Hazer, B., Altıntaş, B. and Arıca, M.Y. 2011. Covalent immobilization of lipase onto amine functionalized polypropylene membrane and its application in green apple flavor (ethyl valerate) synthesis. Process Biochemistry, 46: 372–378.

Bayramoğlu, G., Karakışla, M., Altıntaş, B., Metin, U.A., Saçak, M. and Arıca, M.Y. 2009. Polyaniline grafted polyacylonitrile conductive composite fibers for reversible immobilization of enzymes: Stability and catalytic properties of invertase. Process Biochemistry, 44: 880–885.

Bayramoglu, G., Senkal, F.B. and Arica, Y.M. 2013. Preparation of clay-poly(glycidyl methacrylate) composite support for immobilization of cellulose. Applied Clay Science, 85: 88–95.

Besteman, K., Lee, J., Wiertz, M.G.F., Heering, A.H. and Dekker, C. 2003. Enzyme-coated carbon nanotubes as single-molecule biosensors. Nano Letters, 3: 727–730.

Bezbradica, D., Jugović, B., Gvozdenović, M., Jakovetić, S. and Knežević-Jugović, Z. 2011. Electrochemically synthesized polyaniline as support for lipase immobilization. Journal of Molecular Catalysis B: Enzymatic, 70: 55–60.

Bezbradica, D., Mijin, D., Mihailović, M. and Knežević-Jugović, Z. 2009. Microwave-assisted immobilization of lipase from *Candida rugosa* on Eupergit® supports. Journal of Chemical Technology and Biotechnology, 84: 1642–1648.

Bhattacharya, A. and Pletschke, I.B. 2014. Magnetic cross-linked enzyme aggregates (CLEAs): A novel concept towards carrier free immobilization of lignocellulolytic enzymes. Enzyme and Microbial Technology, 61-62: 17–27.

Bhosale, H.S., Rao, B.M. and Deshpande, V.V. 1996. Molecular and industrial aspects of glucose isomerase. Microbiological Reviews, 60: 280–300.

Biró, E., Németh, S.A. Sisak, C., Feczkó, T. and Gyenis, J. 2008. Preparation of chitosan particles suitable for enzyme immobilization. Journal of Biochemical and Biophysical Methods, 70: 1240–1246.

Blandino, A., Macías, M. and Cantero, D. 2001. Immobilization of glucose oxidase within calcium alginate gel capsules. Process Biochemistry, 36: 601–606.

Blandino, A., Macías, M. and Cantero, D. 2000. Glucose oxidase release from calcium alginate gel capsules. Enzyme and Microbial Technology, 27: 319–324.

Branchu, S., Forbes, T.R., York, P., Petrén, S., Nyqvist, H. and Camber, O. 1999. Hydroxypropyl-β-cyclodextrin Inhibits spray-drying induced inactivation of β-Galactosidase. Journal of Pharmaceutical Sciences, 88: 905–911.

Bugarski, B., Li, Q., Goosen, F.A.M., Poncelet, D., Neufeld, J.R. and Vunjak, G. 1994. Electrostatic droplet generation: Mechanism of polymer droplet formation. AIChE Journal, 40: 1026–1031.

Bugarski, B., Obradovic, B., Nedovic, V. and Goosen, M.F.A. 2006. Electrostatic droplet generation techniquefor cell immobilization. pp. 869–886. *In*: J.P. Shu and A. Spasic (eds.). Finely Dispersed Systems. CRC Press, Boca Raton.

Busto, D.M., García-Tramontín, E.K., Ortega, N. and Perez-Mateos, M. 2006. Preparation and properties of an immobilized pectinlyase for the treatment of fruit juices. Bioresource Technology, 97: 1477–1483.

Busto, D.M., Meza, V., Ortega, N. and Perez-Mateos, M. 2007. Immobilization of naringinase from *Aspergillus niger* CECT 2088 in poly(vinyl alcohol) cryogels for the debittering of juices. Food Chemistry, 104: 1177–1182.

Caldini, C., Bonomi, F., Pifferi, G.P., Lanzarini, G. and Galante, M.Y. 1994. Kinetic and immobilization studies on fungal glycosidases for aroma enhancement in wine. Enzyme and Microbial Technology, 16: 286–291.

Calgaroto, C., Scherer, P.R., Calgaroto, S., Oliveira, V.J., De Oliveira, D. and Pergher, C.B.S. 2011. Immobilization of porcine pancreatic lipase in zeolite MCM 22 with different Si/Al ratios. Applied Catalysis A: General, 394: 101–104.

Cang-Rong, T.J. and Pastorin, G. 2009. The influence of carbon nanotubes on enzyme activity and structure: investigation of different immobilization procedures through enzyme kinetics and circular dichroism studies. Nanotechnology, 20: 255102.

Cao, L. 2005. Immobilised enzymes: Science or art? Current Opinion in Chemical Biology, 9: 217–226.

Cao, L., Van Langen, L. and Sheldon, A.R. 2003. Immobilised enzymes: Carrier-bound or carrier-free? Current Opinion in Biotechnology, 14: 387–394.

Caraveo, L., Medina, H., Rodríguez-Buenfil, I., Montalvo-Romero, C. and Evangelista-Martínez, Z. 2014. A simple plate-assay for screening extracellular naringinase produced by streptomycetes. Journal of Microbiological Methods, 102: 8–11.

Caruso, F., Trau, D., Möhwald, H. and Renneberg, R. 2000. Enzyme encapsulation in layer-by-layer engineered polymer multilayer capsules. Langmuir, 16: 1485–1488.

Cavia-Saiz, M., Muñiz, P., Ortega, N. and Busto, D.M. 2011. Effect of enzymatic debittering on antioxidant capacity and protective role against oxidative stress of grapefruit juice in comparison with adsorption on exchange resin. Food Chemistry, 125: 158–163.

Celikbicak, O., Bayramoglu, G., Yılmaz, M., Ersoy, G., Bicak, N., Salih, B. and Arica, Y.M. 2014. Immobilization of laccase on hairy polymer grafted zeolite particles: Degradation of a model dye and product analysis with MALDI–ToF-MS. Microporous and Mesoporous Materials, 199: 57–65.

Chang, K.Y. and Chu, L. 2007. A simple method for cell disruption by immobilization of lysozyme on the extrudate-shaped NaY zeolite. Biochemical Engineering Journal, 35: 37–47.

Chang, K.Y., Huang, Z.R., Lin, Y.S., Chiu, J.S. and Juan-Chin Tsai, C.J. 2006. Equilibrium study of immobilized lysozyme on the extrudate-shaped NaY zeolite. Biochemical Engineering Journal, 28: 1–9.

Chang, Y.M. and Juang, S.R. 2007. Use of chitosan-clay composite as immobilization support for improved activity and stability of β-glucosidase. Biochemical Engineering Journal, 35: 93–98.

Chen, Y., Xiao, P.C., Chen, Y.X., Yang, W.L., Qi, X., Zheng, F.J., Li, C.M. and Zhang, J. 2014. Preparation of cross-linked enzyme aggregates in water-in-oilemulsion: Application to trehalose synthase. Journal of Molecular Catalysis B: Enzymatic, 100: 84–90.

Chopda, R.V., Nagula, N.K., Bhand, V.D. and Pandit, B.A. 2014. Studying the effect of nature of glass surface on immobilization of glucose isomerase. Biocatalysis and Agricultural Biotechnology, 3: 86–89.

Coppi, G., Iannuccelli, V., Bernabei, M.T. and Cameroni, R. 2002. Alginate microparticles for enzyme peroral administration. International Journal of Pharmaceutics, 242: 263–266.

Coradin, T. and Livage, J. 2003. Mesoporous alginate/silica biocomposites for enzyme immobilization. C. R. Chimie, 6: 147–152.

Costa, S.A., Azevedo, H.S. and Reis, R.L. 2004. Enzyme immobilization in biodegradable polymers for biomedical applications. pp. 301–324. *In*: R.L. Reis and S.J. Román (eds.). Biodegradable Systems in Tissue Engineering and Regenerative Medicine. CRC Press, Boca Raton, Florida.

Cui, L., Zhang, D., Huang, L., Liu, H., Du, G. and Chen, J. 2006. Stabilization of a new microbial transglutaminase from *Streptomyces hygroscopicus WSH03-13* by spray drying. Process Biochemistry, 41: 1427–1431.

Damnjanović, J.J., Žuža, G.M., Savanović, K.J., Bezbradica, I.D., Mijin, Ž.D., Bošković-Vragolović, N. and Knežević-Jugović, D.Z. 2012. Covalently immobilized lipase catalyzing high-yielding optimized geranyl butyrate synthesis in a batch and fluidized bed reactor. Journal of Molecular Catalysis B: Enzymatic, 75: 50–59.

Da Silva, B.A.E., De Souza, U.A.A., De Souza, U.G.S. and Rodrigues, E.A. 2006. Analysis of the high-fructose syrup production using reactive SMB technology. Chemical Engineering Journal, 118: 167–181.

Da Silva, C.R., De Martini Soares, S.A.F., Hazzan, M., Capacla, R.I., Gonçalves, A.I.M. and Gioielli, A.L. 2012. Continuous enzymatic interesterification of lard and soybean oil blend: Effects of different flow rates on physical properties and acyl migration. Journal of Molecular Catalysis B: Enzymatic, 76: 23–28.

David, E.A., Wang, S.N., Yang, C.V. and Yang, J.A. 2006. Chemically surface modified gel (CSMG): An excellent enzyme-immobilization matrix for industrial processes. Journal of Biotechnology, 125: 395–407.

Demir, N., Acar, J., Sarıoğlu, K. and Mutlu, M. 2001. The use of commercial pectinase in fruit juice industry. Part 3: Immobilized pectinase for mash treatment. Journal of Food Engineering, 47: 275–280.

Deshpande, V. and Rao, M. 2006. Glucose isomerase. pp. 239–252. *In*: A. Pandey, C. Webb, C.R. Soccol and C. Larroche (eds.). Enzyme Technology. Springer, Asiatech Publishers, Inc. New Delhi.

DiCosimo, R., McAuliffe, J., Poulose, J.A. and Bohlmann, G. 2013. Industrial use of immobilized enzymes. Chemical Society Reviews, 42: 6437–6474.

Dimitrijević, A., Veličković, D., Bihelović, F., Bezbradica, D., Jankov, R. and Milosavić, N. 2012. One-step, inexpensive high yield strategy for *Candida antarctica* lipase A isolation using hydroxyapatite. Bioresource Technology, 107: 358–362.

Dinçer, A., Becerik, S. and Aydemir, T. 2012. Immobilization of tyrosinase on chitosan-clay composite beads. International Journal of Biological Macromolecules, 50: 815–820.

Đorđević, V., Balanč, B., Belščak-Cvitanović, A., Lević, S., Trifković, K., Kalušević, A., Kostić, I., Komes, D., Bugarski, B. and Nedović, V. 2014. Trends in encapsulation technologies for delivery of food bioactive compounds. Food Engineering Reviews, 1–39.

Fan, J., Lei, J., Wang, L., Yu, C., Tu, B. and Zhao, D. 2003. Rapid and high-capacity immobilization of enzymes based on mesoporous silicas with controlled morphologies. Chemical Communications, 17: 2140–2141.

Feng, W. and Ji, P. 2011. Enzymes immobilized on carbon nanotubes. Biotechnology Advances, 29: 889–895.

Franzreb, M., Siemann-Herzberg, M., Hobley, J.T. and Thomas, T.R.O. 2006. Protein purification using magnetic adsorbent particles. Applied Microbiology and Biotechnology, 70: 505–516.

Gao, S., Wang, Y., Wang, T., Luo, G. and Dai, Y. 2009. Immobilization of lipase on methyl-modified silica aerogels by physical adsorption. Bioresource Technology, 100: 996–999.

Gergely, A., Bowman, D. and Chaudhry, Q. 2010. Small ingredients in a big picture: Regulatory perspectives on nanotechnologies in foods and food contact materials. pp. 150–181. *In*: Q. Chaudhry, L. Castle and R. Watkins (eds.). Nanotechnologies in Food: The Royal Society of Chemistry. Cambridge.

Ghamgui, H., Karra-Chaâbouni, M. and Gargouri, Y. 2004. 1-Butyl oleate synthesis by immobilized lipase from *Rhizopus oryzae*: A comparative study between *n*-hexane and solvent-free system. Enzyme and Microbial Technology, 35: 355–363.

Ghamgui, H., Miled, N., Karra-Chaâbouni, M. and Gargouri, Y. 2007. Immobilization studies and biochemical properties of free and immobilized *Rhizopus oryzae* lipase onto CaCO$_3$: A comparative study. Biochemical Engineering Journal, 37: 34–41.

Gómez, J.M., Romero, M.D. and Fernández, T.M. 2005. Immobilization of β-Glucosidase on carbon nanotubes. Catalysis Letters, 101: 275–278.

Gvozdenović, M.M., Jugović, B.Z., Bezbradica, D.I., Antov, M.G., Knežević-Jugović, Z.D. and Grgur, B.N. 2011. Electrochemical determination of glucose using polyaniline electrode modified by glucose oxidase. Food Chemistry, 124: 396–400.

Habibi, M., Golmakani, T.M., Farahnaky, A., Mesbahi, G. and Majzoobi, M. 2016. NaOH-free debittering of table olives using power ultrasound. Food Chemistry, 192: 775–781.

Hanover, M.L. and White, S.J. 1993. Manufacturing, composition, and applications of fructose. The American Journal of Clinical Nutrition, 58: 724–732.

Huang, J.X., Ge, D. and Xu, K.Z. 2007. Preparation and characterization of stable chitosan nanofibrous membrane for lipase immobilization. European Polymer Journal, 43: 3710–3718.

Ilyasoglu, H. 2013. Production of human fat milk analogue containing α-linolenic acid by solvent-free enzymatic interesterification. LWT—Food Science and Technology, 54: 179–185.

Isgrove, F.H., Williams, R.J.H., Niven, G.W. and Andrews, A.T. 2001. Enzyme immobilization on nylon-optimization and the steps used to prevent enzyme leakage from the support. Enzyme and Microbial Technology, 28: 225–232.

Ison, A.P., Dunnill, P. and Lilly, M.D. 1988. Effect of solvent concentration on enzyme catalysed interesterification of fats. Enzyme and Microbial Technology, 10: 47–51.

Ivnitski, D., Artyushkova, K., Rincón, A.R., Atanassov, P., Luckarift, R.H. and Johnson, R.G. 2008. Entrapment of Enzymes and Carbon Nanotubes in Biologically Synthesized Silica: Glucose Oxidase-Catalyzed Direct Electron Transfer. Small, 4: 357–364.

Jayani, S.R., Saxena, S. and Gupta, R. 2005. Microbial pectinolytic enzymes: A review. Process Biochemistry, 40: 2931–2944.

Jenab, E., Temelli, F., Curtis, W.J. and Zha, Y.Y. 2014. Performance of two immobilized lipases for interesterification between canola oil and fully-hydrogenated canola oil under supercritical carbon dioxide. LWT—Food Science and Technology, 58: 263–271.

Kahraman, V.M., Bayramoğlu, G., Kayaman-Apohan, N. and Güngör, A. 2007. α-Amylase immobilization on functionalized glass beads by covalent attachment. Food Chemistry, 104: 1385–1392.

Karakuş, E. and Pekyardimci, Ş. 2006. Immobilization of apricot pectinesterase (*Prunus armeniaca* L.) on porous glass beads and its characterization. Journal of Molecular Catalysis B: Enzymatic, 56: 13–19.

Kartal, F., Janssen, H.A.M., Hollmann, F., Sheldon, A.R. and Kılınc, A. 2011. Improved esterification activity of *Candida rugosa* lipase in organic solvent by immobilization as cross-linked enzyme aggregates (CLEAs). Journal of Molecular Catalysis B: Enzymatic, 71: 85–89.

Katchalski-Katzir, E. and Kraemer, D.M. 2000. Eupergit® C, a carrier for immobilization of enzymes of industrial potential. Journal of Molecular Catalysis - B Enzymatic, 10: 157–176.

Kim, H.M., Park, S., Kim, H.Y., Won, K. and Lee, H.S. 2013. Immobilization of formate dehydrogenase from *Candida boidinii* through cross-linked enzyme aggregates. Journal of Molecular Catalysis B: Enzymatic, 97: 209–214.

Kim, J., Grate, W.J. and Wang, P. 2006. Nanostructures for enzyme stabilization. Chemical Engineering Science, 61: 1017–1026.

Kim, J.H., Suma, Y., Lee, H.S., Kim, A.J. and Kim, S.H. 2012. Immobilization of horseradish peroxidase onto clay minerals using soil organic matter for phenol removal. Journal of Molecular Catalysis B: Enzymatic, 83: 8–15.

Knezevic, Z., Bobic, S., Milutinovic, A., Obradovic, B., Mojovic, L.j. and Bugarski, B. 2002. Alginate-immobilized lipase by electrostatic extrusion for the purpose of palm oil hydrolysis in lecithin/isooctane system. Process Biochemistry, 38: 313–318.

Knezevic, Z., Mojovic, L.j. and Adnadjevic, B. 1998. Palm oil hydrolysis by lipase from *Candida cylindracea* immobilized on zeolite type Y. Enzyme and Microbial Technology, 22: 275–280.

Knezevic, Z., Milosavic, N., Bezbradica, D., Jakovljevic, Z. and Prodanovic, R. 2006. Immobilization of lipase from *Candida rugosa* on Eupergit® C supports by covalent attachment. Biochemical Engineering Journal, 30: 269–278.

Knežević-Jugović, D.Z., Damnjanović, J.J., Bezbradica, I.D. and Mijin, Ž.D. 2008. The immobilization of lipase on sepabeads: Coupling, characterization and application in geranyl butyrate synthesis in a low aqueous system. Chemical Industry & Chemical Engineering Quarterly, 14: 245–249.

Knežević-Jugović, Z.D., Bezbradica, D.I., Mijin, D.Ž. and Antov, M.G. 2011. The immobilization of enzyme on Eupergit® supports by covalent attachment. pp. 99–111. *In*: S.D. Minteer (ed.). Methods in Molecular Biology (Clifton, N.J.). Humana Press, Springer Science, NY.

Koneracká, M., Kopčansky, P., Antalík, M., Timko, M., Ramchand, C.N., Lobo, D., Mehta, R.V. and Upadhyay, R.V. 1999. Immobilization of proteins and enzymes to fine magnetic particles. Journal of Magnetism and Magnetic Materials, 201: 427–430.

Krajewska, B. 2004. Application of chitin- and chitosan-based materials for enzyme immobilizations: A review. Enzyme and Microbial Technology, 35: 126–139.

Kumari, A., Kaur, B., Srivastava, R. and Sangwan, S.R. 2015. Isolation and immobilization of alkaline protease on mesoporous silica and mesoporous ZSM-5 zeolite materials for improved catalytic properties. Biochemistry and Biophysics Reports, 2: 108–114.

Lalou, S., Mantzouridou, F., Paraskevopoulou, A., Bugarski, B., Levic, S. and Nedovic, V. 2013. Bioflavour production from orange peel hydrolysate using immobilized *Saccharomyces cerevisiae*. Applied Microbiology and Biotechnology, 97: 9397–9407.

Lee, S.H. and Kim, J.G. 2003. Effects of debittering on red grapefruit juice concentrate. Food Chemistry, 82: 177–180.

Lei, S., Xu, Y., Fan, G., Xiao, M. and Pan, S. 2011. Immobilization of naringinase on mesoporous molecular sieve MCM-41 and its application to debittering of white grapefruit. Applied Surface Science, 257: 4096–4099.

Levic, S., Djordjevic, V., Rajic, N., Milivojevic, M., Bugarski, B. and Nedovic, V. 2013. Entrapment of ethyl vanillin in calcium alginate and calcium alginate/poly(vinyl alcohol) beads. Chemical Papers, 67: 221–228.

Lević, S., Lijaković, I.P., Đorđević, V., Rac, V., Rakić, V., Knudsen, T.Š., Pavlović, V, Bugarski, B. and Nedović, V. 2015. Characterization of sodium alginate/D-limonene emulsions and respective calcium alginate/D-limonene beads produced by electrostatic extrusion. Food Hydrocolloids, 45: 111–123.

Lević, S., Obradović, N., Pavlović, V., Isailović, B., Kostić, I., Mitrić, M., Bugarski, B. and Nedović, V. 2014. Thermal, morphological and mechanical properties of ethyl vanillin immobilized in polyvinyl alcohol by electrospinning process. Journal of Thermal Analysis and Calorimetry, 118: 661–668.

Li, F.S., Chen, P.J. and Wu, T.W. 2007. Electrospun polyacrylonitrile nanofibrous membranes for lipase immobilization. Journal of Molecular Catalysis B: Enzymati, 47: 117–124.

Li, F.S. and Wu, T.W. 2009. Lipase-immobilized electrospun PAN nanofibrous membranes for soybean oil hydrolysis. Biochemical Engineering Journal, 45: 48–53.

Li, J., Tan, N.S. and Ge, H. 1996. Silica sol-gel immobilized amperometric biosensor for hydrogen peroxide. Analytica Chimica Acta, 335: 137–145.

Li, W., Yuan, R., Chai, Y., Zhou, L., Chen, S. and Li, N. 2008. Immobilization of horseradish peroxidase on chitosan/silica sol-gel hybrid membranes for the preparation of hydrogen peroxide biosensor. Journal of Biochemical and Biophysical Methods, 70: 830–837.

Li, Z., Barnes, C.J., Bosoy, A., Stoddart, F.J. and Zink, I.J. 2012. Mesoporous silica nanoparticles in biomedical applications. Chemical Society Reviews, 41: 2590–2605.

Liu, L., Cen, Y., Liu, F., Yu, J., Jiang, X. and Chen, X. 2015. Analysis of α-amylase inhibitor from corni fructus by coupling magnetic cross-linked enzyme aggregates of α-amylase with HPLC-MS. Journal of Chromatography B, 995-996: 64–69.

Liu, Q. and Gao, Y. 2015. Binary adsorption isotherm and kinetics on debittering process of ponkan (*Citrus reticulata* Blanco) juice with macroporous resins. LWT—Food Science and Technology, 63: 1245–1253.

Liu, Y., Guo, C. and Liu, Z.C. 2015. Enhancing the resolution of (*R,S*)-2-octanol catalyzed by magnetic crosslinked lipase aggregates using an alternating magnetic field. Chemical Engineering Journal, 280: 36–40.

Lozinsky, I.V., Galaev, Y.I., Plieva, M.F., Savina, N.I., Jungvid, H. and Mattiasson, B. 2003. Polymeric cryogels as promising materials of biotechnological interest. Trends in Biotechnology, 21: 445–451.

Magario, I., Ma, X., Neumann, A., Syldatk, C. and Hausmann, R. 2008. Non-porous magnetic micro-particles: Comparison to porous enzyme carriers for a diffusion rate-controlled enzymatic conversion. Journal of Biotechnology, 134: 72–78.

Marchis, T., Cerrato, G., Magnacca, G., Crocella, V. and Laurenti, E. 2012. Immobilization of soybean peroxidase on aminopropyl glass beads: Structural and kinetic studies. Biochemical Engineering Journal, 67: 28–34.

Miao, Y. and Tan, S.N. 2001. Amperometric hydrogen peroxide biosensor with silica sol-gel/chitosan film as immobilization matrix. Analytica Chimica Acta, 437: 87–93.

Mihailović, M., Stojanović, M., Banjanac, K., Carević, M., Prlainović, N., Milosavić, N. and Bezbradica, D. 2014. Immobilization of lipase on epoxy-activated Purolite® A109 and its post-immobilization stabilization. Process Biochemistry, 49: 637–646.

Milašinović, N., Kalagasidis Krušić, M., Knežević-Jugović, Z. and Filipović, J. 2010a. Hydrogels of N-isopropylacrylamide copolymers with controlled release of a model protein. International Journal of Pharmaceutics, 383: 53–61.

Milašinović, N., Knežević-Jugović, Z., Milosavljević, N., Filipović, J. and Kalagasidis Krušić, M. 2012a. Controlled release of lipase from *Candida rugosa* loaded into hydrogels of N-isopropylacrylamide and itaconic acid. International Journal of Pharmaceutics, 436: 332–340.

Milašinović, N., Milosavljević, N., Filipović, J., Knežević-Jugović, Z. and Kalagasidis Krušić, M. 2010b. Synthesis, characterization and application of poly(N-isopropylacrylamide-co-itaconic acid) hydrogels as supports for lipase immobilization. Reactive & Functional Polymers, 70: 807–814.

Milašinović, N., Milosavljević, N., Filipović, J., Knežević-Jugović, Z. and Kalagasidis Krušić, M. 2012b. Efficient immobilization of lipase from *Candida rugosa* by entrapment into poly(N-isopropylacrylamide-co-itaconic acid) hydrogels under mild conditions. Polymer Bulletin, 69: 347–361.

Milovanović, A., Božić, N. and Vujčić, Z. 2007. Cell wall invertase immobilization within calcium alginate beads. Food Chemistry, 104: 81–86.

Mitchell, S. and Pérez-Ramírez, J. 2011. Mesoporous zeolites as enzyme carriers: Synthesis, characterization, and application in biocatalysis. Catalysis Today, 168: 28–37.

Miyake, T., Yoshino, S., Yamada, T., Hata, K. and Nishizawa, M. 2011. Self-regulating enzyme-nanotube ensemble films and their application as flexible electrodes for biofuel cells. Journal of the American Chemical Society, 133: 5129–5134.

Mizuki, T., Watanabe, N., Nagaoka, Y., Fukushima, T., Morimoto, H., Usami, R. and Maekawa, T. 2010. Activity of an enzyme immobilized on superparamagnetic particles in a rotational magnetic field. Biochemical and Biophysical Research Communications, 393: 779–782.

Mojović, L.j., Šiler-Marinković, S., Kukić, G., Bugarski, B. and Vunjak-Novaković, G. 1994. *Rhizopus arrhizus* lipase-catalyzed interesterification of palm oil midfraction in a gas-lift reactor. Enzyme and Microbial Technology, 16: 159–162.

Mojović, L.j., Šiler-Marinković, S., Kukić, G. and Vunjak-Novaković, G. 1993. *Rhizopus arrhizus* lipase-catalyzed interesterification of the midfraction of palm oil to a cocoa butter equivalent fat. Enzyme and Microbial Technology, 15: 438–443.

Nakashima, T., Fukuda, H., Kyotani, S. and Morikawa, H. 1988. Culture conditions for intracellular lipase production by *Rhizopus chinensis* and its immobilization within biomass support particles. Journal of Fermentation Technology, 66: 441–448.

Nedović, V.A., Obradović, B., Leskošek-Čukalović, I., Trifunović, O., Pešić, R. and Bugarski, B. 2001. Electrostatic generation of alginate microbeads loaded with brewing yeast. Process Biochemistry, 37: 17–22.

Nedovic, V., Manojlovic, V., Pruesse, U., Bugarski, B., Djonlagic, J. and Vorlop, K.D. 2006. Optimization of the electrostatic droplet generation process for controlled microbead production—single nozzle system. Chemical Industry & Chemical Engineering Quarterly (CI&CEQ), 12: 53–57.

Nedović, V.A., Obradović, B., Poncelet, D., Goosen, M.F.A., Leskošek-Čukalović, I. and Bugarski, B. 2002. Cell immobilisation by electrostatic droplet generation. Landbauforschung Volkenrode SH 241: 11–17.

Nguyen, D.Q., Rezessy-Szabó, M.J., Czukor, B. and Hoschke, A. 2011. Continuous production of oligofructose syrup from Jerusalem artichoke juice by immobilized endo-inulinase. Process Biochemistry, 46: 298–303.

Nguyen, L.T. and Yang, L.K. 2014. Uniform cross-linked cellulase aggregates prepared in millifluidic reactors. Journal of Colloid and Interface Science, 428: 146–151.

Nunes, M.F.G., De Paula, V.A., De Castro, F.H. and Dos Santos, C.J. 2011. Compositional and textural properties of milkfat-soybean oil blends following enzymatic interesterification. Food Chemistry, 125: 133–138.

Nunes, P.A.M., Emilia Rosa, M., Fernandes, B.C.P. and Ribeiro, L.H.M. 2014. Operational stability of naringinase PVA lens-shaped microparticles in batch stirred reactors and mini packed bed reactors-one step closer to industry. Bioresource Technology, 164: 362–370.

Ó'Fágáin, C. 2003. Enzyme stabilization-recent experimental progress. Enzyme and Microbial Technology, 33: 137–149.

Ognjanovic, N., Bezbradica, D. and Knezevic-Jugovic, Z. 2009. Enzymatic conversion of sunflower oil to biodiesel in a solvent-free system: Process optimization and the immobilized system stability. Bioresource Technology, 100: 5146–5154.

Olsen, H.S. 1995. Enzymatic production of glucose syrups. pp. 26–64. *In*: M.W. Kearsley and S.Z. Dziedzic (eds.). Handbook of Starch Hydrolysis Products and their Derivatives, Springer, US.

Padmaja, G.J.R. and Moorthy, S.N. 2009. Comparative production of glucose and high fructose syrup from cassava and sweet potato roots by direct conversion techniques. Innovative Food Science and Emerging Technologies, 10: 616–620.

Perez, I.D., Van Rantwijk, F. and Sheldon, A.R. 2009. Cross-linked enzyme aggregates of chloroperoxidase: Synthesis, optimization and characterization. Advanced Synthesis & Catalysis, 351: 2133–2139.

Poncelet, D., Babak, V., Neufeld, R.J., Goosen, M. and Bugarski, B. *1999.* Theory of electrostatic dispersion of polymer solution in the production of microgel beds containing biocatalyst. Advances in Colloid Interface Science, 79: 213–228.

Popat, A., Hartono, B.S., Stahr, F., Liu, J., Qiao, Z.S. and Lu, M.Q.G. 2011. Mesoporous silica nanoparticles for bioadsorption, enzyme immobilisation, and delivery carriers. Nanoscale, 3: 2801–2818.

Prlainović, Ž.N., Bezbradica, I.D., Zorica, D., Knežević-Jugović, D.Z., Stevanović, I.S., Avramov Ivić, L.M., Uskoković, S.P. and Mijin, D. 2013. Adsorption of lipase from Candida rugosa on multi walled carbon nanotubes. Journal of Industrial and Engineering Chemistry, 19: 279–285.

Prlainović, Ž.N., Knežević-Jugović, D.Z., Mijin, Ž.D. and Bezbradica, I.D. 2011. Immobilization of lipase from *Candida rugosa* on Sepabeads®: The effect of lipase oxidation by periodates. Bioprocess and Biosystems Engineering, 34: 803–810.

Prodanović, O., Prokopijević, M., Spasojević, D., Stojanović, Ž., Radotić, K., Knežević-Jugović, Z.D. and Prodanović, R. 2012. Improved covalent immobilization of horseradish peroxidase on macroporous glycidyl methacrylate-based copolymers. Applied Biochemistry and Biotechnology, 168: 1288–1301.

Prüsse, U., Bilancetti, L., Bučko, M., Bugarski, B., Bukowski, J., Gemeiner, P., Lewinska, D., Manojlovic, V., Massart, B., Nastruzzi, C., Nedovic, V., Poncelet, D., Siebenhaar, S., Tobler, L. Tosi, A., Vikartovská, A. and Vorlop, K.D. 2008. Comparison of different technologies for alginate beads production. Chemical Papers, 62: 364–374.

Puri, M., Marwaha, S.S., Kothari, R.M. and Kennedy, J.F. 1996a. Biochemical basis of bitterness in citrus fruit juices and biotech approaches for debittering. Critical Reviews in Biotechnology, 16: 145–155.

Puri, M., Marwaha, S.S. and Kothari, M.R. 1996b. Studies on the applicability of alginate-entrapped naringinase for the debittering of kinnow juice. Enzyme and Microbial Technology, 18: 281–285.

Rahim, A.N.S., Sulaiman, A., Hamzah, F., Hamid, K.H.K. Rodhi, M.N.M., Musa, M. and Edama, A.N. 2013. Enzymes encapsulation within calcium alginate-clay beads: Characterization and application for cassava slurry saccharification. Procedia Engineering, 68: 411–417.

Razack, A.S. and Duraiarasan, S. 2015. Response surface methodology assisted biodiesel production from waste cooking oil using encapsulated mixed enzyme. Waste Management (In Press).

Rebroš, M., Rosenberg, M., Mlichová, Z. and Krištofíková, L. 2007. Hydrolysis of sucrose by invertase entrapped in polyvinyl alcohol hydrogel capsules. Food Chemistry, 102: 784–787.

Ribeiro, L.H.M., Afonso, C., Vila-Real, J.H., Alfaia, J.A. and Ferreira, L. 2010. Contribution of response surface methodology to the modeling of naringin hydrolysis by naringinase Ca-alginate beads under high pressure. LWT—Food Science and Technology, 43: 482–487.

Rodrigues, Á.R., Cabral, J.M.S. and Taipa, M.Â. 2002. Immobilization of *Chromobacterium viscosum* lipase on Eudragit S-100: Coupling, characterization and kinetic application in organic and biphasic media. Enzyme and Microbial Technology, 31: 133–141.

Sardar, M., Agarwal, R., Kumar, A. and Gupta, M.N. 1997. Noncovalent immobilization of enzymes on an enteric polymer Eudragit S-100. Enzyme and Microbial Technology, 20: 361–367.

Shanmugaprakash, M., Vinothkumarb, V., Ragupathy, J. and Amala Reddy, D. 2015. Biochemical characterization of three phase partitioned naringinase from *Aspergillus brasiliensis* MTCC 1344. International Journal of Biological Macromolecules, 80: 418–423.

Sheldon, A.R. 2011. Characteristic features and biotechnological applications of cross-linked enzyme aggregates (CLEAs). Applied Microbiology and Biotechnology, 92: 467–477.

Silva, C.J.S.M., Gübitz, G. and Cavaco-Paulo, A. 2006. Optimisation of a serine protease coupling to Eudragit S-100 by experiment design techniques. Journal of Chemical Technology and Biotechnology, 81: 8–16.

Silva, S.L., Da Silva, S.L., Brumano, L., Stringheta, C.P., De Oliveira Pinto, A.M., Dias, M.O.L., De Sá Martins Muller, C., Scio, E., Fabri, L.R., Castro, C.H. and Da Penha Henriques do Amaral, M. 2012. Preparation of dry extract of *Mikania glomerata* sprengel (Guaco) and determination of its coumarin levels by spectrophotometry and HPLC-UV. Molecules, 17: 10344–10354.

Singh, S.R., Dhaliwal, R. and Puri, M. 2007. Production of high fructose syrup from *Asparagus* inulin using immobilized exoinulinase from *Kluyveromyces marxianus* YS-1. Journal of Industrial Microbiology & Biotechnology, 34: 649–655.

Singh, S.R., Dhaliwal, R. and Puri, M. 2008. Development of a stable continuous flow immobilized enzyme reactor for the hydrolysis of inulin. Journal of Industrial Microbiology & Biotechnology, 35: 777–782.

Sirisansaneeyakul, S., Worawuthiyanan, N., Vanichsriratana, W., Srinophakun, P. and Chisti, Y. 2007. Production of fructose from inulin using mixed inulinases from *Aspergillus niger* and *Candida guilliermondii*. World Journal of Microbiology and Biotechnology, 23: 543–552.

Spagna, G., Pifferi, G.P. and Gilioli, E. 1995. Immobilization of a pectinlyase from *Aspergillus niger* for application in food technology. Enzyme and Microbial Technology, 17: 729–738.

Stinco, M.C., Fernández-Vázquez, R., Hernanz, D., Heredia, J.F., Meléndez-Martínez, J.A. and Vicario, M.I. 2013. Industrial orange juice debittering: Impact on bioactive compounds and nutritional value. Journal of Food Engineering, 116: 155–161.

Subramanian, A., Kennel, J.S., Oden, I.P., Jacobson, B.K., Woodward, J. and Doktycz, J.M. 1999. Comparison of techniques for enzyme immobilization on silicon supports. Enzyme and Microbial Technology, 24: 26–34.

Sun, J., Zhang, H., Tian, R., Ma, D., Bao, X., Su, S.D. and Zou, H. 2006. Ultrafast enzyme immobilization over large-pore nanoscale mesoporous silica particles. Chemical Communications, 12: 1322–1324.

Sureshkumar, M. and Lee, K.C. 2011. Polydopamine coated magnetic-chitin (MCT) particles as a new matrix for enzyme immobilization. Carbohydrate Polymer, 84: 775–780.

Takahashi, H., Li, B., Sasaki, T., Miyazaki, C., Kajino, T. and Inagaki, S. 2000. Catalytic activity in organic solvents and stability of iImmobilized enzymes depend on the pore size and surface characteristics of mesoporous silica. Chemistry of Materials, 12: 3301–3305.

Talbert, N.J. and Goddard, M.J. 2012. Enzymes on material surfaces. Colloids and Surfaces B: Biointerfaces, 93: 8–19.

Talekar, S., Ghodake, V., Ghotage, T., Rathod, P., Deshmukh, P., Nadar, S., Mulla, M. and Ladole, M. 2012. Novel magnetic cross-linked enzyme aggregates (magnetic CLEAs) of alpha amylase. Bioresource Technology, 123: 542–547.

Talekar, S., Pandharbale, A., Ladole, M., Nadar, S., Mulla, M., Japhalekar, K., Pattankude, K. and Arage, D. 2013. Carrier free co-immobilization of alpha amylase, glucoamylase and pullulanase as combined cross-linked enzyme aggregates (combi-CLEAs): A tri-enzyme biocatalyst with one pot starch hydrolytic activity. Bioresource Technology, 147: 269–275.

Tandjaoui, N., Tassist, A., Abouseoud, M., Couvert, A. and Amrane, A. 2015. Preparation and characterization of cross-linked enzyme aggregates (CLEAs) of *Brassica rapa* peroxidase. Biocatalysis and Agricultural Biotechnology, 4: 208–213.

Tang, X.Z., Qian, Q.J. and Shi, E.L. 2006. Characterizations of immobilized neutral proteinase on chitosan nano-particles. Process Biochemistry, 41: 1193–1197.

Tanriseven, A. and Doğan, Ş. 2001. Immobilization of invertase within calcium alginate gel capsules. Process Biochemistry, 36: 1081–1083.

Taqieddin, E. and Amiji. M. 2004. Enzyme immobilization in novel alginate-chitosan core-shell microcapsules. Biomaterials, 25: 1937–1945.

Tecelão, C., Guillén, M., Valero, F. and Ferreira-Dias, S. 2012. Immobilized heterologous *Rhizopus oryzae* lipase: A feasible biocatalyst for the production of human milk fat substitutes. Biochemical Engineering Journal, 67: 104–110.

Tomotani, J.E. and Vitolo, M. 2007. Production of high-fructose syrup using immobilized invertase in a membrane reactor. Journal of Food Engineering, 80: 662–667.

Torabizadeh, H., Tavakoli, M. and Safari, M. 2014. Immobilization of thermostable α-amylase from *Bacillus licheniformisby* cross-linked enzyme aggregates method using calcium and sodium ions as additives. Journal of Molecular Catalysis B: Enzymatic, 108: 13–20.

Tran, L. and Chaudhry, Q. 2010. Engineered nanoparticles and food: An assessment of exposure and hazard. pp. 120–133. *In:* Q. Chaudhry, L. Castle and R. Watkins (eds.). Nanotechnologies in Food. The Royal Society of Chemistry. Cambridge.

Tsen, Y.H., Tsai, Y.S. and Yu, K.G. 1989. Fiber entrapment of naringinase from *Penicillium* sp. and application to fruit juice debittering. Journal of Fermentation and Bioengineering, 67: 186–189.

Tükel, S.S. and Alagöz, D. 2008. Catalytic efficiency of immobilized glucose isomerase in isomerization of glucose to fructose. Food Chemistry, 111: 658–662.

Tükel, S.S., Hürrem, F., Yildirim, D. and Alptekin, Ö. 2013. Preparation of crosslinked enzyme aggregates (CLEA) of catalase and its characterization. Journal of Molecular Catalysis B: Enzymatic, 97: 252–257.

Tzialla, A.A., Kalogeris, E., Enotiadis, A., Taha, A.A., Gournis, D. and Stamatis, H. 2009. Effective immobilization of *Candida antarctica* lipase B in organic-modified clays: Application for the epoxidation of terpenes. Materials Science and Engineering B, 165: 173–177.

Upadhyay, B.S.L. and Verma, N. 2014. Dual immobilization of biomolecule on the glass surface using cysteine as a bifunctional linker. Process Biochemistry, 49: 1139–1143.

Vaidya, K.B., Kuwar, S.S., Golegaonkar, B.S. and Nene, N.S. 2012. Preparation of cross-linked enzyme aggregates of l-aminoacylase via co-aggregation with polyethyleneimine. Journal of Molecular Catalysis B: Enzymatic, 74: 184–191.

Vasic-Racki, D. 2006. History of industrial biotransformation-dreams and realities. pp. 1–36. *In*: A. Liese, K. Seelbach and C. Wandrey (eds.). Industrial Biotransformations, second, revised and extended edition. Wiley-VCH Verlag GmbH & Co. KgaA, Weinheim, Germany.

Wang, B., Zhang, J. and Dong, S. 2000. Silica sol-gel composite film as an encapsulation matrix for the construction of an amperometric tyrosinase-based biosensor. Biosensors & Bioelectronics, 15: 397–402.

Wang, G.Z., Wang, Q.J. and Xu, K.Z. 2006. Immobilization of lipase from *Candida rugosa* on electrospun polysulfone nanofibrous membranes by adsorption. Journal of Molecular Catalysis B: Enzymatic, 42: 45–51.

Wang, J. 2005. Carbon-nanotube-based electrochemical biosensors: A review. Electroanalysis, 17: 7–14.

Wang, S.G., Zhang, Q., Wang, R., Yoon, S.F., Ahn, J., Yang, D.J., Tian, J.Z., Li, J.Q. and Zhou, Q. 2003. Multi-walled carbon nanotubes for the immobilization of enzyme in glucose biosensors. Electrochemistry Communications, 5: 800–803.

Wang, X.H., Wu, H., Ho, T.C. and Weng, C.X. 2006. Cocoa butter equivalent from enzymatic interesterification of tea seed oil and fatty acid methyl esters. Food Chemistry, 97: 661–665.

Wang, Y. and Caruso, F. 2005. Mesoporous silica spheres as supports for enzyme immobilization and encapsulation. Chemistry of Materials, 17: 953–961.

Wang, Y. and Hsieh, L.Y. 2008. Immobilization of lipase enzyme in polyvinyl alcohol (PVA) nanofibrous membranes. Journal of Membrane Science, 309: 73–81.

Wang, Y.X., Jiang, P.X., Li, Y., Zeng, S. and Zhang, W.Y. 2015. Preparation Fe3O4@chitosan magnetic particles for covalent immobilization of lipase from *Thermomyces lanuginosus*. International Journal of Biological Macromolecules, 75: 44–50.

Wang, Z.G., Xu, K.Z., Wan, S.L., Wu, J., Innocent, C. and Seta, P. 2006. Nanofibrous membranes containing carbon nanotubes: Electrospun for redox enzyme immobilization. Macromolecular Rapid Communications, 27: 516–521.

Wenling, W., Huiying, L.W.W. and Shiyuan, W. 1999. Continuous preparation of fructose syrups from Jerusalem artichoke tuber using immobilized intracellular inulinase from *Kluyveromyces* sp. Y-85. Process Biochemistry, 34: 643–646.

Wisdom, R.A., Dunnill, P. and Lilly, M.D. 1985. Enzymic interesterification of fats: The effect of non-lipase material on immobilized enzyme activity. Enzyme and Microbial Technology, 7: 567–572.

Witoon, T., Chareonpanich, M. and Limtrakul, J. 2008. Synthesis of bimodal porous silica from rice husk ash via sol-gel process using chitosan as template. Materials Letters, 62: 1476–1479.

Woo, J.E., Kwon, S.H. and Lee, H.C. 2015. Preparation of nano-magnetite impregnated mesocellular foam composite with a Cu ligand for His-tagged enzyme immobilization. Chemical Engineering Journal, 274: 1–8.

Xu, J., Sun, J., Wang, J., Sheng, J., Wang, F. and Sun, M. 2014. Application of iron magnetic nanoparticles in protein immobilization. Molecules, 19: 11465–11486.

Yamamoto, S. and Sano, Y. 1992. Drying of enzymes: Enzyme retention during draying of a single droplet. Chemical Engineering Science, 47: 177–183.

Ye, P., Xu, K.Z., Wu, J., Innocent, C. and Seta, P. 2006. Nanofibrous poly(acrylonitrile-*co*-maleic acid) membranes functionalized with gelatin and chitosan for lipase immobilization. Biomaterials, 27: 4169–4176.

Yoshii, H., Buche, F., Takeuchi, N., Terrol, C., Ohgawara, M. and Furuta, T. 2008. Effects of protein on retention of ADH enzyme activity encapsulated in trehalose matrices by spray drying. Journal of Food Engineering, 87: 34–39.

Zhang, M., Smith, A. and Gorski, W. 2004. Carbon nanotube-chitosan system for electrochemical sensing based on dehydrogenase enzymes. Analytical Chemistry, 76: 5045–5050.

Zhang, W.W., Yang, L.X., Jia, Q.J., Wang, N., Hu, L.C. and Yu, Q.X. 2015. Surfactant-activated magnetic cross-linked enzyme aggregates (magnetic CLEAs) of *Thermomyces lanuginosus* lipase for biodiesel production. Journal of Molecular Catalysis B: Enzymatic, 115: 83–89.

Zhang, Y., Hidajat, K. and Ray, K.A. 2004. Optimal design and operation of SMB bioreactor: production of high fructose syrup by isomerization of glucose. Biochemical Engineering Journal, 21: 111–121.

Zhao, M.L., Shi, E.L., Zhang, L.Z., Chen, M.J., Shi, D.D., Yang, J. and Tang, X.Z. 2011. Preparation and application of chitosan nanoparticles and nanofibers. Brazilian Journal of Chemical Engineering, 28: 353–362.

Zhao, Y.Z., Liu, J., Hahn, M., Qiao, S., Middelberg, J.P.A. and He, L. 2013. Encapsulation of lipase in mesoporous silica yolk-shell spheres with enhanced enzyme stability. RSC Advances, 3: 22008–22013.

Zuidam, J.N. and Heinrich, E. 2010. Encapsulation of aroma. pp. 127–160. *In*: N.J. Zuidam and V.A Nedovic (eds.). Encapsulation Technologies for Food Active Ingredients and Food Processing. Springer, Dordrecht.

Zuidam, N.J. and Shimoni, E. 2010. Overview of microencapsulates for use in food products or processes and methods to make them. pp. 3–30. *In*: N.J. Zuidam and V.A Nedovic (eds.). Encapsulation Technologies for Food Active Ingredients and Food Processing. Springer, Dordrecht.

Žuža, M., Obradović, M.B. and Knežević-Jugović, D.Z. 2011. Hydrolysis of penicillin G by penicillin G acylase immobilized on chitosan microbeads in different reactor systems. Chemical Engineering & Technology, 34: 1706–1714.

Index

Printed and bound by CPI Group (UK) Ltd, Croydon, CR0 4YY

01/11/2024

01782622-0015